# Developments in Mathematics

Volume 64



**Aims and Scope**

The Developments in Mathematics (DEVM) book series is devoted to publishing well-written monographs within the broad spectrum of pure and applied mathematics. Ideally, each book should be self-contained and fairly comprehensive in treating a particular subject. Topics in the forefront of mathematical research that present new results and/or a unique and engaging approach with a potential relationship to other fields are most welcome. High-quality edited volumes conveying current state-of-the-art research will occasionally also be considered for publication. The DEVM series appeals to a variety of audiences including researchers, postdocs, and advanced graduate students.

More information about this series at http://www.springer.com/series/5834

Teresa W. Haynes • Stephen T. Hedetniemi
Michael A. Henning
Editors

# Topics in Domination in Graphs

*Editors*

Teresa W. Haynes
Department of Mathematics and Statistics
East Tennessee State University
Johnson City, TN, USA

Department of Mathematics and Applied Mathematics
University of Johannesburg
Johannesburg, South Africa

Stephen T. Hedetniemi
School of Computing
Clemson University
Clemson, SC, USA

Michael A. Henning
Department of Mathematics and Applied Mathematics
University of Johannesburg
Johannesburg, South Africa

ISSN 1389-2177 ISSN 2197-795X (electronic)
Developments in Mathematics
ISBN 978-3-030-51119-7 ISBN 978-3-030-51117-3 (eBook)
https://doi.org/10.1007/978-3-030-51117-3



This Springer imprint is published by the registered company Springer Nature Switzerland AG
The registered company address is: Gewerbestrasse 11, 6330 Cham, Switzerland

# Preface

While concepts related to domination in graphs can be traced back to the mid-1800s in connection with various chessboard problems, domination was first defined as a graph-theoretical concept in 1958. Domination in graphs experienced rapid growth since its introduction resulting in over 1200 papers published on domination in graphs by the late 1990s. Noting the need for a comprehensive survey of the literature on domination in graphs, in 1998 Haynes, Hedetniemi, and Slater published the first two books on domination, *Fundamentals of Domination in Graphs* and *Domination in Graphs: Advanced Topics*. We refer to these as Books I and II.

The explosive growth has continued and today more than 4000 papers have been published on domination in graphs, and the material in Books I and II is more than 20 years old. Thus, the authors think it is time for an update on the developments in domination theory since 1998. We also want to give a comprehensive treatment of the major topics in domination. This coverage of domination including both the fundamental major results and updates will be in the form of three books, which we shall call Books III, IV, and V.

Book III, *Domination in Graphs: Core Concepts*, is written by the authors and concentrates, as the title suggests, on the three main types of domination in graphs: domination, independent domination, and total domination. It contains major results on these basic domination numbers, including proofs of selected results that illustrate many of the proof techniques that are used in domination theory. For the companion books, Books IV and V, we invited leading researchers in domination to contribute chapters.

Book V has three parts, the first of which focuses on several domination-related concepts. The second part focuses on domination in (i) hypergraphs, (ii) chessboards, and (iii) digraphs and tournaments. The third part focuses on the development of algorithms and complexity of domination parameters.

The present volume, Book IV, concentrates on major domination parameters that were not covered in Book III. Although well over 70 types of dominating sets have been defined, Book IV focuses on the primary ones that have received the most attention in the literature. In particular, the chapters include such parameters as paired domination, connected domination, restrained domination, domination functions, Roman domination, and power domination.

The authors of Book IV provide a survey of known results with a sampling of proof techniques for each parameter. To avoid excessive repetition of definitions and notations, Chapter 1 provides a glossary of commonly used terms and Chapter  gives an overview of models of domination from which the parameters are defined.

This book is intended as a reference resource for researchers and is written to reach the following audience: First, the audience includes the established researchers in the field of domination who want an updated comprehensive coverage on domination. Second are the researchers in graph theory and graduate students who wish to become acquainted with topics in domination including major accomplishments in the field and proof techniques used. We anticipate that it could also be used in a seminar course on domination in graphs.

We wish to thank the authors who contributed chapters to this book as well as the reviewers of the chapters.

Johnson City, TN, USA — Teresa W. Haynes
Clemson, SC, USA — Stephen T. Hedetniemi
Johannesburg, South Africa — Michael A. Henning

# Contents

# Glossary of Common Terms

**Teresa W. Haynes, Stephen T. Hedetniemi, and Michael A. Henning**

## 1 Introduction

It is difficult to say when the study of domination in graphs began, but for the sake of this glossary let us say that it began in 1962 with the publication by Oystein Ore's book *Theory of Graphs* [15]. In *Chapter 13 Dominating Sets, Covering Sets and Independent Sets* of [15], we see for the first time the name *dominating set*, defined as follows: "A subset $D$ of $V$ is a *dominating set* for $G$ when every vertex not in $D$ is the endpoint of some edge from a vertex in $D$." Ore then defines the *domination number*, denoted $\delta(G)$, of a graph $G$, as "the smallest number of vertices in any minimal dominating set." So, at this point, and for the first time, domination has a "name" and a "number."

Of course, prior to this Claude Berge [3], in his book *Theory of Graphs and its Applications*, which was first published in France in 1958 by Dunod, Paris, had previously defined the same concept, but had, in *Chapter 4 The Fundamental*

T. W. Haynes (✉)
Department of Mathematics and Statistics, East Tennessee State University, Johnson City, TN 37614-0002, USA

Department of Mathematics and Applied Mathematics, University of Johannesburg, Auckland Park 2006, South Africa
e-mail: haynes@etsu.edu

S. T. Hedetniemi
Professor Emeritus: School of Computing, Clemson University, Clemson, SC 29634, USA
e-mail: hedet@clemson.edu

M. A. Henning
Department of Mathematics and Applied Mathematics, University of Johannesburg, Auckland Park 2006, South Africa
e-mail: mahenning@uj.ac.za



T. W. Haynes et al. (eds.), *Topics in Domination in Graphs*, Developments in Mathematics 64, https://doi.org/10.1007/978-3-030-51117-3_1

*Numbers of the Theory of Graphs* of [3], given it the name "the coefficient of external stability."

Before Berge, Dénes König, in his 1936 book *Theorie der Endlichen und Unendlichen Graphen* [13], had defined essentially the same concept, but in *VII Kapitel, Basisproblem für gerichtete Graphen*, König gave it the name "punktbasis," which we would today say is an independent dominating set.

And even before König, in the books by Dudeney in 1908 [8] and W. W. Rouse Ball in 1905 [2], one can find the concepts of domination, independent domination, and total domination discussed in connection with various chessboard problems. And it was Ball who, in turn, credited such people as W. Ahrens in 1910 [1], C. F. de Jaenisch in 1862 [7], Franz Nauck in 1850 [14], and Max Bezzel in 1848 [4] for their contributions to these types of chessboard problems involving dominating sets of chess pieces.

But it was Ore who gave the name *domination* and this name took root. Not long thereafter, Cockayne and Hedetniemi [6] gave the notation $\gamma(G)$ for the domination number of a graph, and this also took root and is the notation adopted here.

Since the subsequent chapters in this book will deal with domination parameters, there will be much overlap in the terminology and notation used. One purpose of this chapter is to present definitions common to many of the chapters in order to prevent terms being defined repeatedly and to avoid other redundancy. Also, since graph theory terminology and notation sometimes vary, in this glossary we clarify the terminology that will be adopted in subsequent chapters.

We proceed as follows. In Section 2.1, we present basic graph theory definitions. We discuss common types of graphs in Section 2.2. Some fundamental graph constructions are given in Section 2.3. In Section 3.1 and Section 3.2, we present parameters related to connectivity and distance in graphs, respectively. The covering, packing, independence, and matching numbers are defined in Section 3.3. Finally in Section 3.4, we define selected domination-type parameters that will occur frequently throughout the book.

For more details and terminology, the reader is referred to the two books *Fundamentals of Domination in Graphs* [10] and *Domination in Graphs, Advanced Topics* [11] written and edited by Haynes, Hedetniemi, and Slater, and the book *Total Domination in Graphs* by Henning and Yeo [12]. An annotated glossary, from which many of the definitions in this chapter are taken, was produced by Gera, Haynes, Hedetniemi, and Henning in 2018 [9].

## 2 Basic Terminology

In this section, we give basic definitions, common types of graphs, and fundamental graph constructions.

## 2.1 Basic Graph Theory Definitions

Before we proceed with our glossary of parameters, we need to define a few basic terms, which are used in the definitions in the following subsections. For $k \geq 1$ an integer, we use the standard notation $[k] = \{1, \ldots, k\}$ and $[k]_0 = [k] \cup \{0\} = \{0, 1, \ldots, k\}$.

A (finite, undirected) *graph* $G = (V, E)$ consists of a finite nonempty set of *vertices* $V = V(G)$ together with a set $E = E(G)$ of unordered pairs of distinct vertices called *edges*. Each edge $e = \{u, v\}$ in $E$ is denoted with any of $e$, $uv$, $vu$, and $\{u, v\}$. We say that a graph $G$ has *order* $n = |V|$ and *size* $m = |E|$.

Two vertices $u$ and $v$ in $G$ are *adjacent* if they are joined by an edge $e$, that is, $u$ and $v$ are adjacent if $e = uv \in E(G)$. In this case, we say that each of $u$ and $v$ is *incident* with the edge $e$. Further, we say that the edge $e$ *joins* the vertices $u$ and $v$. Two edges are *adjacent* if they share a common vertex. Two vertices in a graph $G$ are *independent* if they are not adjacent. A set of pairwise independent vertices in $G$ is an *independent set* of $G$. Similarly, two edges are *independent* if they are not adjacent.

A *neighbor* of a vertex $v$ in $G$ is a vertex $u$ that is adjacent to $v$. The *open neighborhood* of a vertex $v$ in $G$ is the set of neighbors of $v$, denoted $N_G(v)$. Thus, $N_G(v) = \{u \in V \mid uv \in E(G)\}$. The *closed neighborhood of* $v$ is the set $N_G[v] = \{v\} \cup N_G(v)$. For a set of vertices $S \subseteq V$, the *open neighborhood* of $S$ is the set $N_G(S) = \bigcup_{v \in S} N_G(v)$ and its *closed neighborhood* is the set $N_G[S] = N_G(S) \cup S$. If the graph $G$ is clear from the context, we omit it in the above expressions. For example, we write $N(v)$, $N[v]$, $N(S)$, and $N[S]$ rather than $N_G(v)$, $N_G[v]$, $N_G(S)$, and $N_G[S]$, respectively.

For a set of vertices $S \subseteq V$ and a vertex $v$ belonging to the set $S$, the *$S$-private neighborhood* of $v$ is defined by $\text{pn}[v, S] = \{w \in V \mid N_G[w] \cap S = \{v\}\}$, while its *open $S$-private neighborhood* is defined by $\text{pn}(v, S) = \{w \in V \mid N_G(w) \cap S = \{v\}\}$. As remarked in [12], the sets $\text{pn}[v, S] \setminus S$ and $\text{pn}(v, S) \setminus S$ are equivalent and we define the *$S$-external private neighborhood* of $v$ to be this set, abbreviated $\text{epn}[v, S]$ or $\text{epn}(v, S)$. The *$S$-internal private neighborhood* of $v$ is defined by $\text{ipn}[v, S] = \text{pn}[v, S] \cap S$ and its *open $S$-internal private neighborhood* is defined by $\text{ipn}(v, S) = \text{pn}(v, S) \cap S$. We define an *$S$-external private neighbor* of $v$ to be a vertex in $\text{epn}(v, S)$ and an *$S$-internal private neighbor* of $v$ to be a vertex in $\text{ipn}(v, S)$.

The *degree* $d_G(v)$ of a vertex $v$ is the number of neighbors $v$ has in $G$, that is, $d_G(v) = |N_G(v)|$. Again if the graph $G$ is clear from the context, we use $d(v)$ rather than $d_G(v)$. We remark that some books use $\deg(v)$ and $\deg v$ to denote the degree of $v$. We leave it to the authors to choose which of these notations to adopt in their chapters. For a subset of vertices $S \subseteq V$, the *degree of $v$ in $S$*, denoted $d_S(v)$, is the number of vertices in $S$ adjacent to the vertex $v$; that is, $d_S(v) = |N_G(v) \cap S|$. In particular, if $S = V$, then $d_S(v) = d_G(v)$. The *degree sequence* of a graph $G$ with vertex set $V = \{v_1, v_2, \ldots, v_n\}$ is the sequence $d_1, d_2, \ldots, d_n$, where $d_i = d(v_i)$

for $i \in [n]$. Often the degree sequence, $d_1, d_2, \ldots, d_n$, is written in non-increasing order, and so $d_1 \geq d_2 \geq \ldots \geq d_n$.

An *isolated vertex* is a vertex of degree 0 in $G$. A graph is *isolate-free* if it does not contain an isolated vertex. The minimum degree among the vertices of $G$ is denoted by $\delta(G)$, and the maximum degree by $\Delta(G)$. A *leaf* is a vertex of degree 1, while its neighbor is a *support vertex*. A *strong support vertex* is a (support) vertex with at least two leaf neighbors.

For subsets $X$ and $Y$ of vertices of $G$, we denote the set of edges that join a vertex of $X$ and a vertex of $Y$ in $G$ by $[X, Y]$.

Two graphs $G$ and $H$ are *isomorphic*, denoted $G \cong H$, if there exists a bijection $\phi\colon V(G) \to V(H)$ such that two vertices $u$ and $v$ are adjacent in $G$ if and only if the two vertices $\phi(u)$ and $\phi(v)$ are adjacent in $H$. A *parameter* of a graph $G$ is a numerical value (usually a non-negative integer) that can be associated with a graph such that whenever two graphs are isomorphic, they have the same associated numerical value.

By a *partition* of the vertex set $V$ of a graph $G$, we mean a family $\pi = \{V_1, V_2, \ldots, V_k\}$ of nonempty pairwise disjoint sets whose union equals $V$, that is, for all $1 \leq i < j \leq k$, $V_i \cap V_j = \emptyset$ and

$$\bigcup_{i=1}^{k} V_i = V.$$

For such a partition $\pi$, we will say that $\pi$ has *order* $k$.

A *walk* in a graph $G$ from a vertex $u$ to a vertex $v$ is a finite, alternating sequence of vertices and edges, starting with the vertex $u$ and ending with the vertex $v$, in which each edge of the sequence joins the vertex that precedes it in the sequence to the vertex that follows it in the sequence. A *trail* is a walk containing no repeated edges, and a *path* is a walk containing no repeated vertices. We will mainly be concerned with paths. A path joining two vertices $u$ and $v$ is called a $(u, v)$*-path* or a $u$-$v$ *path* or a $u, v$*-path* in the literature. The *length* of a walk equals the number of edges in the walk. A graph $G$ is *connected* if there is a path between every pair of vertices of $G$.

A *cycle* is a path in which the first and last vertices are the same and all other vertices are distinct. A *chord* of a cycle $C$ is an edge between two nonconsecutive vertices of $C$.

The *distance* $d(u, v)$ between two vertices $u$ and $v$, in a connected graph $G$, equals the minimum length of a $(u, v)$-path in $G$. A shortest, or minimum length, path between two vertices $u$ and $v$ is called a $(u, v)$*-geodesic*; a $v$*-geodesic* is any shortest path from $v$ to another vertex; a *geodesic* is any shortest path in a graph.

A graph $G' = (V', E')$ is a *subgraph* of a graph $G = (V, E)$ if $V' \subseteq V$ and $E' \subseteq E$. A subgraph $G'$ of a graph $G$ is called a *spanning subgraph* of $G$ if $V' = V$. If $G = (V, E)$ and $S \subseteq V$, then the *subgraph of* $G$ *induced by* $S$ is the graph $G[S]$, whose vertex set is $S$ and whose edges are all the edges in $E$ both of whose vertices are in $S$.

Let $F$ be an arbitrary graph. A graph $G$ is said to be *$F$-free* if $G$ does not contain $F$ as an induced subgraph.

If $G = (V, E)$ and $S \subseteq V$, the subgraph obtained from $G$ by deleting all vertices in $S$ and all edges incident with one or two vertices in $S$ is denoted by $G - S$; that is, $G - S = G[V \setminus S]$. If $S = \{v\}$, we simply denote $G - \{v\}$ by $G - v$. The *contraction* of an edge $e = xy$ in a graph $G$ is the graph obtained from $G$ by deleting the vertices $x$ and $y$ and adding a new vertex and edges joining this new vertex to all vertices that were adjacent to $x$ or $y$ in $G$.

A *component* of a graph is a maximal connected subgraph. An *odd (even) component* is a component of odd (even) order. Let $\text{oc}(G)$ equal the number of odd components of $G$. A vertex $v \in V$ is a *cut vertex* if the graph $G - v$ has more components than $G$. An edge $e = uv$ is a *bridge* if the graph $G - e$ obtained by deleting $e$ from $G$ has more components than $G$.

## 2.2 Common Types of Graphs

A graph of order $n = 1$ is called a *trivial graph*, while a graph with at least two vertices is called a *nontrivial graph*. A graph of size $m = 0$ is an *empty graph*, while a graph with at least one edge is a *nonempty graph*. Recall that a *connected graph* is a graph for which there is a path between every pair of its vertices.

A *$k$-regular graph* is a graph in which every vertex has degree $k$. A *regular graph* is a graph that is $k$-regular for some $k \geq 0$. A 3-regular graph is also called a *cubic graph*.

A graph of order $n$ that is a cycle is denoted by $C_n$ and a graph that is a path is denoted by $P_n$. Note that a cycle is a 2-regular graph.

A graph is *acyclic* if it does not contain a cycle. A *tree* is a connected acyclic graph. Equivalently, a tree is a connected graph having size one less than its order. Thus, if $T$ is a tree of order $n$ and size $m$, then $T$ is connected and $m = n - 1$. A *forest* is an acyclic graph. Thus, a forest is a disjoint union of trees. A *linear forest* is a forest in which every component is a path.

If $G$ is a vertex-disjoint union of $k$ copies of a graph $F$, we write $G = kF$.

A *complete graph* is a graph in which every two vertices are adjacent. A complete graph of order $n$ is denoted by $K_n$. A *triangle* is a subgraph isomorphic to $K_3$ or $C_3$, since $K_3 \cong C_3$.

A graph $G$ is *bipartite* if its vertex set can be partitioned into two independent sets $X$ and $Y$. The sets $X$ and $Y$ are called the *partite sets* of $G$. A *complete bipartite graph*, denoted $K_{r,s}$, is a bipartite graph with partite sets $X$ and $Y$, where $|X| = r$, $|Y| = s$, and every vertex in $X$ is adjacent to every vertex in $Y$. The graph $K_{r,s}$ has order $r + s$, size $rs$, $\delta(K_{r,s}) = \min\{r, s\}$, and $\Delta(K_{r,s}) = \max\{r, s\}$.

A *star* is a nontrivial tree with at most one vertex that is not a leaf. Thus, a star is a complete bipartite graph $K_{1,k}$ for some $k \geq 1$. A *claw* is an induced copy of the graph $K_{1,3}$. Thus, a *claw-free graph* is a $K_{1,3}$-free graph.

For $r, s \geq 1$, a *double star* $S(r, s)$ is a tree with exactly two (adjacent) vertices that are not leaves, one of which has $r$ leaf neighbors and the other $s$ leaf neighbors.

A *diamond* is an induced copy of the graph $K_4 - e$, which is obtained from a copy of the complete graph of order 4 by deleting an edge $e$.

A graph $G$ can be *embedded* on a surface $S$ if its vertices can be placed on $S$ and all of its edges can be drawn between the vertices on $S$ in such a way that no two edges intersect. A graph $G$ is *planar* if it can be embedded in the plane; a *plane graph* is a graph that has been embedded in the plane.

A *rooted tree* $T$ is a tree having a *distinguished vertex* labeled $r$, called the *root*. Let $T$ be a rooted tree with root $r$. For each vertex $v$, let $P(v)$ be the unique $(r, v)$-path in $T$. The *parent* of a vertex $v$ is its neighbor on $P(v)$, while the other neighbors of $v$ are called its *children*. The set of children of $v$ is denoted by $C(v)$. Note that the root $r$ is the only vertex of $T$ with no parent. A *descendant* of $v$ is any vertex $u \neq v$ such that the $P(u)$ contains $v$, while an *ancestor* of $v$ is a vertex $u \neq v$ that belongs $P(v)$ in $T$. In particular, every child of $v$ is a descendant of $v$, while the parent of $v$ is an ancestor of $v$. A *grandchild* of $v$ is a descendant of $v$ at distance 2 from $v$. We let $D(v)$ denote the set of descendants of $v$, and we define $D[v] = D(v) \cup \{v\}$. The *maximal subtree* at $v$, denoted $T_v$, is the subtree of $T$ induced by $D[v]$. The *depth* of a vertex $v$ in $T$ equals $d(r, v)$, and the *height* of $v$, denoted ht(v), is the maximum distance from $v$ to a descendant of $v$. Thus, $\text{ht(v)} = \max\{d(v, w) : w$ is a descendant of $v\}$.

For classes of graphs not defined here, we refer the reader to the definitive encyclopedia on graph classes, *Graph Classes: A Survey* [5] by Brandstädt, Le, and Spinrad.

## 2.3 Graph Constructions

Given a graph $G = (V, E)$, the *complement* of $G$ is the graph $\overline{G} = (V, \overline{E})$, where $uv \in \overline{E}$ if and only if $uv \notin E$. Thus the complement, $\overline{G}$, of $G$, is formed by taking the vertex set of $G$ and joining two vertices by an edge whenever they are not joined in $G$.

By a *graph product* $G \otimes H$ on graphs $G$ and $H$, we mean a graph whose vertex set is the Cartesian product of the vertex sets of $G$ and $H$ (that is, $V(G \otimes H) = V(G) \times V(H)$) and whose edge set is determined entirely by the adjacency relations of $G$ and $H$. Exactly how it is determined depends on what kind of graph product we are considering.

The *Cartesian product* $G \,\Box\, H$ of two graphs $G$ and $H$ is the graph with vertex set $V(G) \times V(H)$ where two vertices $(u_1, v_1)$ and $(u_2, v_2)$ are adjacent if and only if either $u_1 = u_2$ and $v_1v_2 \in E(H)$ or $v_1 = v_2$ and $u_1u_2 \in E(G)$.

The *direct product* (also known as the *cross product*, *tensor product*, *categorical product*, and *conjunction*) $G \times H$ of two graphs $G$ and $H$ is the graph with vertex set $V(G) \times V(H)$ where two vertices $(u_1, v_1)$ and $(u_2, v_2)$ are adjacent in $G \times H$ if and only if $u_1u_2 \in E(G)$ and $v_1v_2 \in E(H)$.

Given a graph $G = (V, E)$ and an edge $uv \in E$, the *subdivision* of edge $uv$ consists of (i) deleting the edge $uv$ from $E$, (ii) adding a new vertex $w$ to $V$, and (iii) adding the new edges $uw$ and $wv$ to $E$. In this case we say that the edge $uv$ has been *subdivided*. The *subdivision graph* $S(G)$ is the graph obtained from $G$ by subdividing every edge of $G$ exactly once.

Given a graph $G = (V, E)$, the *line graph* $L(G) = (E, E(L(G)))$ is the graph whose vertices correspond 1-to-1 with the edges in $E$, and two vertices are adjacent in $L(G)$ if and only if the corresponding edges in $G$ have a vertex in common, that is, if and only if the corresponding two edges are adjacent.

The *corona* $G \circ K_1$ of a graph $G$, also denoted cor($G$) in the literature, is the graph obtained from $G$ by adding for each vertex $v \in V$ a new vertex $v'$ and the edge $vv'$. The edge $vv'$ is called a *pendant edge*. The *k-corona* $G \circ P_k$ of $G$ is the graph of order $(k+1)|V(G)|$ obtained from $G$ by attaching a path of length $k$ to each vertex of $G$ so that the resulting paths are vertex-disjoint. In particular, the *2-corona* $G \circ P_2$ of $G$ is the graph of order $3|V(G)|$ obtained from $G$ by attaching a path of length 2 to each vertex of $G$ so that the resulting paths are vertex-disjoint. The *generalized corona* $G \circ H$ is the graph obtained by adding a copy of $H$ for each vertex $v$ of $G$ and joining $v$ to every vertex of $H$. Thus, a generalized corona $G \circ H$, where $H = K_1$, is the ordinary corona $G \circ K_1$. We note that whether $G \circ P_k$ is intended to denote a $k$-corona or a generalized corona will be clear from context or specifically stated by the author.

# 3 Graph Parameters

In this section, we present common graph parameters that may appear in this book.

## 3.1 Connectivity and Subgraph Numbers

In this subsection, we present parameters related to connectivity in graphs.

(a) *blocks* bl(G), number of blocks in $G$. A *block* of a graph $G$ is a maximal nonseparable subgraph of $G$, that is, a maximal subgraph having no cut vertices.
(b) *bridges* br(G), number of bridges in $G$.
(c) *circumference* cir(G), maximum length or order of a cycle in $G$.
(d) *clique number* $\omega(G)$, maximum order of a complete subgraph of $G$.
(e) *components* $c(G)$, number of maximal connected subgraphs of $G$.
(f) A *vertex cut* of a connected graph $G$ is a subset $S$ of the vertex set of $G$ with the property that $G - S$ is disconnected (has more than one component). A vertex cut $S$ is a *k-vertex cut* if $|S| = k$.

(g) *vertex connectivity* $\kappa(G)$, minimum cardinality of a vertex cut of $G$ if $G$ is not the complete graph and $\kappa(G) = n - 1$ if $G$ is a complete graph $K_n$ on $n \geq 2$ vertices. A graph $G$ is *k-vertex-connected* (or *k-connected*) if $\kappa(G) \geq k$ for some integer $k \geq 0$. Thus, $\kappa(G)$ is the smallest number of vertices whose deletion from $G$ produces a disconnected graph or the trivial graph $K_1$. A nontrivial graph has connectivity 0 if and only if it is disconnected.

(h) An *edge cut* of a nontrivial connected graph $G$ is a nonempty subset $F$ of the edge set of $G$ with the property that $G - F$ is disconnected (has more than one component). Thus, the deletion of an edge cut from the connected graph $G$ results in a disconnected graph. An edge cut $F$ is a *k-edge cut* if $|F| = k$.

(i) *edge connectivity* $\lambda(G)$, minimum cardinality of an edge cut of $G$ if $G$ is nontrivial, while $\lambda(K_1) = 0$. A graph $G$ is *k-edge-connected* if $\lambda(G) \geq k$ for some integer $k \geq 0$. Thus, $\lambda(G)$ is the smallest number of edges whose deletion from $G$ produces a disconnected graph or the trivial graph $K_1$. Hence, $\lambda(G) = 0$ if and only if $G$ is disconnected or trivial.

(j) *girth* of $G$, denoted *girth(G)* or $g(G)$ in the literature, the minimum length of a cycle in $G$.

## 3.2 Distance Numbers

This subsection contains the definitions of parameters which are defined in terms of the distances $d(u, v)$ between vertices $u$ and $v$ in a graph.

(a) *eccentricity* $\text{ecc}(v)$, of a vertex $v$ in a connected graph $G$, is the maximum of the distances from $v$ to the other vertices of $G$.

(b) *diameter* $\text{diam}(G)$, maximum distance among all pairs of vertices of $G$. Equivalently, the diameter of $G$ is the maximum length of a geodesic in $G$. Thus, the diameter of $G$ is the maximum eccentricity taken over all vertices of $G$. Two vertices $u$ and $v$ in $G$ for which $d(u, v) = \text{diam}(G)$ are called *antipodal* or *peripheral vertices* of $G$. A *diametral path* in $G$ is a geodesic whose length equals the diameter of $G$.

(c) The *periphery* of a graph $G$ is the subgraph of $G$ induced by its peripheral vertices.

(d) *radius* $\text{rad}(G)$, minimum eccentricity taken over all vertices of $G$.

(e) The *center* of a graph $G$, denoted $C(G)$, is the subgraph of $G$ induced by the vertices in $G$ whose eccentricity equals the radius of $G$. A vertex $v \in C(G)$ is called a *central vertex* of $G$.

## 3.3 Covering, Packing, Independence, and Matching Numbers

As previously defined, a set of pairwise independent vertices in $G$ is an *independent set* of $G$. An independent set $S$ is *maximal* if no superset of $S$ is independent.

A set of pairwise independent edges of $G$ is called a *matching* in $G$, while a matching of maximum cardinality is a *maximum matching*. Given a matching $M$, we denote by $V[M]$ the set of vertices in $G$ incident with an edge in $M$. A *perfect matching* is a matching in which every vertex is incident with an edge of the matching. Thus, if $G$ has a perfect matching $M$, then $G$ has even order $n = 2k$ for some $k \geq 1$ and $|M| = k$.

A vertex and an edge are said to *cover* each other in a graph $G$ if they are incident in $G$. A *vertex cover* in $G$ is a set of vertices that covers all the edges of $G$, while an *edge cover* in $G$ is a set of edges that covers all the vertices of $G$. Thus, a vertex cover in $G$ is a set of vertices that contains at least one vertex of every edge in $G$.

A subset $S$ of vertices in $G$ is a *packing* if the closed neighborhoods of vertices in $S$ are pairwise disjoint. Equivalently, $S$ is a packing in $G$ if the vertices in $S$ are pairwise at distance at least 3 apart in $G$. Thus, if $S$ is a packing in $G$, then $|N_G[v] \cap S| \leq 1$ for every vertex $v \in V(G)$. A packing is also called a *2-packing* in the literature. More generally, for $k \geq 2$, a *k-packing* in $G$ is a set of vertices in $G$ that are pairwise at distance at least $k + 1$ apart in $G$. Thus, if $S$ is a $k$-packing in $G$, then $d_G(u, v) > k$ for every two distinct vertices $u$ and $v$ that belong to $S$.

A subset $S$ of vertices in $G$ is an *open packing* if the open neighborhoods of vertices in $S$ are pairwise disjoint. Thus, if $S$ is an open packing in $G$, then $|N_G(v) \cap S| \leq 1$ for every vertex $v \in V(G)$.

All of the parameters in this subsection have to do with sets that are independent or cover other sets. These include some of the most basic of all parameters in graph theory.

(a) *vertex independence numbers* $i(G)$ and $\alpha(G)$, minimum and maximum cardinality of a maximal independent set in $G$. The lower vertex independence number, $i(G)$, is also called the *independent domination number* of $G$, while the upper vertex independence number, $\alpha(G)$, is also called the *independence number* of $G$. (While the notation $i(G)$ is fairly standard for the independent domination number, we remark that the independence number is also denoted by $\beta_0(G)$ in the literature.)
(b) *vertex covering numbers* $\beta(G)$ and $\beta^+(G)$, minimum and maximum cardinality of a minimal vertex cover in $G$. (We remark that the vertex covering number is also denoted by $\tau(G)$ or by $\alpha(G)$ in the literature.)
(c) *edge covering numbers* $\beta'(G)$ and $\beta'^+(G)$, minimum and maximum cardinality of a minimal edge cover in $G$.
(d) *k-packing number* $\rho_k(G)$, maximum cardinality of a $k$-packing in $G$ for $k \geq 2$. When $k = 2$, the $k$-packing number $\rho_k(G)$ is called the *packing number* of $G$, denoted by $\rho(G)$. Thus, $\rho(G)$ is the maximum cardinality of a packing in $G$.
(e) *open packing number* $\rho^o(G)$, maximum cardinality of an open packing in $G$.

(f) *matching numbers* $\alpha'^{-}(G)$ and $\alpha'(G)$, minimum and maximum cardinality of a maximal matching in $G$. The upper matching number, $\alpha'(G)$, is also called the *matching number* of $G$. Recall that a *perfect matching* is a matching in which every vertex is incident with an edge of the matching. Thus, if a graph $G$ of order $n$ has a perfect matching, then $\alpha'(G) = \frac{1}{2}n$. It should be noted that by a well-known theorem of Gallai, that if $G$ is a graph of order $n$ with no isolated vertices, then $\alpha(G) + \beta(G) = n = \alpha'(G) + \beta'(G)$. (The matching number is also denoted by $\beta_1(G)$ in the literature.)

## 3.4 Domination Numbers

A *dominating set* in a graph $G = (V, E)$ is a set $S$ of vertices of $G$ such that every vertex in $V \setminus S$ has a neighbor in $S$. Thus, if $S$ is a dominating set of $G$, then $N_G[S] = V$ and every vertex in $V \setminus S$ is therefore adjacent to at least one vertex in $S$. For subsets $X$ and $Y$ of vertices of $G$, if $Y \subseteq N_G[X]$, then the set $X$ *dominates* the set $Y$ in $G$. In particular, if $X$ dominates $V(G)$, then $X$ is a dominating set of $G$. If no proper subset of a dominating set $S$ is a dominating set of $G$, then $S$ is a *minimal dominating set* of $G$.

The many variations of dominating sets in a graph $G$ are based on (i) conditions which are placed on the subgraph $G[S]$ induced by a dominating set $S$, (ii) conditions which are placed on the vertices in $V \setminus S$, or (iii) conditions which are placed on the edges between vertices in $S$ and vertices in $V \setminus S$. We mention only the major domination numbers here.

A *total dominating set*, abbreviated TD-set, in a graph $G$ with no isolated vertices is a set $S$ of vertices of $G$ such that every vertex in $V$ is adjacent to at least one vertex in $S$. Thus, a subset $S \subseteq V$ is a TD-set in $G$ if $N_G(S) = V$. If no proper subset of $S$ is a TD-set of $G$, then $S$ is a *minimal TD-set* of $G$. Every graph without isolated vertices has a TD-set, since $S = V$ is such a set. If $X$ and $Y$ are subsets of vertices in $G$, then the set $X$ *totally dominates* the set $Y$ in $G$ if $Y \subseteq N_G(X)$. In particular, if $X$ totally dominates $V(G)$, then $X$ is a TD-set in $G$.

A *paired dominating set*, abbreviated PD-set, of $G$ is a set $S$ of vertices of $G$ such that every vertex is adjacent to some vertex in $S$ and the subgraph $G[S]$ induced by $S$ contains a perfect matching $M$. Two vertices joined by an edge of $M$ are said to be *paired* with respect to a perfect matching $M$ and are also called *partners* in $S$. A PD-set $S$ in a graph $G$ is *minimal* if no proper subset of $S$ is a PD-set of $G$.

A *connected dominating set*, abbreviated CD-set, in a graph $G$ is a dominating set $S$ of vertices of $G$ such that $G[S]$ is connected. A CD-set $S$ in a graph $G$ is *minimal* if no proper subset of $S$ is a CD-set of $G$.

(a) *domination numbers* $\gamma(G)$ and $\Gamma(G)$, minimum and maximum cardinalities of a minimal dominating set in $G$. The parameters $\gamma(G)$ and $\Gamma(G)$ are called the *domination number* and *upper domination number* of $G$, respectively. A dominating set of $G$ of cardinality $\gamma(G)$ is called a *$\gamma$-set* of $G$, while a minimal dominating set of cardinality $\Gamma(G)$ is called a $\Gamma$-set of $G$.

(b) *independent domination* $i(G)$, minimum cardinality of a dominating set in $G$ that is also independent. An independent dominating set of $G$ of cardinality $i(G)$ is called an $i$-set of $G$. We note that the maximum order of a minimal independent dominating set equals the vertex independence number $\alpha(G)$.

(c) *total domination numbers* $\gamma_t(G)$ and $\Gamma_t(G)$, minimum and maximum cardinalities of a minimal total dominating set of $G$. The parameters $\gamma_t(G)$ and $\Gamma_t(G)$ are called the *total domination number* and *upper total domination number* of $G$, respectively. A TD-set of $G$ of cardinality $\gamma_t(G)$ is called a *$\gamma_t$-set* of $G$, while a minimal TD-set of cardinality $\Gamma_t(G)$ is called a $\Gamma_t$-set of $G$.

(d) *paired domination numbers* $\gamma_{\rm pr}(G)$ and $\Gamma_{\rm pr}(G)$, minimum and maximum cardinalities of a minimal PD-set of $G$. The parameters $\gamma_{\rm pr}(G)$ and $\Gamma_{\rm pr}(G)$ are called the *paired domination number* and *upper paired domination number* of $G$, respectively. A PD-set of $G$ of cardinality $\gamma_{\rm pr}(G)$ is called a *$\gamma_{\rm pr}$-set* of $G$, while a minimal PD-set of cardinality $\Gamma_{\rm pr}(G)$ is called a $\Gamma_{\rm pr}$-set of $G$.

(e) *connected domination numbers* $\gamma_c(G)$ and $\Gamma_c(G)$, minimum and maximum cardinalities of a minimal CD-set of $G$. The parameters $\gamma_c(G)$ and $\Gamma_c(G)$ are called the *connected domination number* and *upper connected domination number* of $G$, respectively. A CD-set of $G$ of cardinality $\gamma_c(G)$ is called a *$\gamma_c$-set* of $G$, while a minimal CD-set of cardinality $\Gamma_c(G)$ is called a $\Gamma_c$-set of $G$.

## References


1. W. Ahrens, *Mathematische Unterhaltungen und Spiele*. Druck und Verlag von B. G. Teubner, Leipzig-Berlin, (1910), 311–312.
2. W. W. Rouse Ball, *Mathematical Recreations and Essays*, Fourth Edition, Macmillan, 1905.
3. C. Berge, *The Theory of Graphs and its Applications*. Methuen, London, 1962.
4. M. Bezzel, Schachfreund. *Berliner Schachzeitung* 3 (1848), 363.
5. A. Brandstädt, V. B. Le and J. P. Spinrad, *Graph Classes: A Survey*, SIAM Monographs on Discrete Mathematics and Applications, SIAM, Philadelphia, PA, 1999.
6. E.J. Cockayne and S.T. Hedetniemi, Towards a theory of domination in graphs. *Networks* 7 (1977), 247–261.
7. C. F. de Jaenisch, *Applications de l'Analyse Mathematique au Jeu des Echecs*. Petrograd, 1862.
8. H. E. Dudeney, *The Canterbury Puzzles and Other Curious Problems*. E. P. Dutton and Company, New York, 1908.
9. R. Gera, T. W. Haynes, S. T. Hedetniemi and M. A. Henning, An Annotated Glossary of Graph Theory Parameters with Conjectures in *Graph Theory, Favorite Conjectures and Open Problems, Volume 2*, R. Gera, T. W. Haynes and S. T. Hedetniemi, editors. Springer, 2018, pages 177–281.
10. T. W. Haynes, S. T. Hedetniemi and P. J. Slater, *Fundamentals of Domination in Graphs*, Marcel Dekker, New York, 1998.
11. T. W. Haynes, S. T. Hedetniemi and P. J. Slater, *Domination in Graphs, Advanced Topics*, Marcel Dekker, New York, 1998.
12. M. A. Henning and A. Yeo, *Total Domination in Graphs (Springer Monographs in Mathematics)*. Springer New York (2013). ISBN: 978-1-4614-6524-9 (Print) 978-1-4614-6525-6 (eBook).

13. D. König, *Theorie der Endlichen und Unendlichen Graphen*, Akademische Verlagsgesellschaft M. B. H., Leipzig, 1936, later Chelsea, New York, 1950.
14. F. Nauck, Briefwechsel mit allen für alle, *Illustrirte Zeitung* 15 (1850), 182.
15. O. Ore, *Theory of Graphs*, Amer. Math. Soc. Colloq. Publ., Vol 38, Amer. Math. Soc., Providence, RI, 1962.

# Models of Domination in Graphs

**Teresa W. Haynes, Stephen T. Hedetniemi, and Michael A. Henning**

**Keywords** Domination · Frameworks for domination

**AMS Subject Classification** 05C69

## 1 Introduction

As we have said before, a set $S \subseteq V$ is a *dominating set* of a graph $G = (V, E)$ if every vertex $v \in V$ is either an element of $S$ or is adjacent to an element of $S$. In this chapter, we will take a look at dominating sets from a variety of different perspectives. Each perspective suggests a variation in the domination theme and different types or aspects of dominating sets.

We will not attempt to be comprehensive here, only to provide a sufficient number of different models to reveal domination in a much broader view. Chapter 11 in the book *Fundamentals of Domination in Graphs* [5] presents ten logical structures or frameworks where the concept of domination naturally arises. The suggested frameworks range from integer programming to hypergraphs. We repeat a few of these frameworks in this chapter.

---

T. W. Haynes (✉)
Department of Mathematics and Statistics, East Tennessee State University, Johnson City, TN 37614-0002, USA

Department of Mathematics and Applied Mathematics, University of Johannesburg, Auckland Park 2006, South Africa
e-mail: haynes@etsu.edu

S. T. Hedetniemi
Professor Emeritus: School of Computing, Clemson University, Clemson, SC 29634, USA
e-mail: hedet@clemson.edu

M. A. Henning
Department of Mathematics and Applied Mathematics, University of Johannesburg, Auckland Park 2006, South Africa
e-mail: mahenning@uj.ac.za



T. W. Haynes et al. (eds.), *Topics in Domination in Graphs*, Developments in Mathematics 64, https://doi.org/10.1007/978-3-030-51117-3_2

In the most general sense, we are interested in sets of vertices in a graph having some property $\mathcal{P}$, called $\mathcal{P}$-sets. We are interested in finding $\mathcal{P}$-sets of minimum and maximum cardinalities, using a notational system of the form $a(G)$ for the minimum cardinality of a $\mathcal{P}$-set, and upper case $A(G)$ for the maximum cardinality of a $\mathcal{P}$-set. These parameters are sometimes referred to as lower and upper parameters.

But we will make a further distinction. As an example, recall that a set $S \subseteq V$ is *independent* if no two vertices in $S$ are adjacent. Since independence is an *hereditary property* (henceforth denoted H), meaning that every subset of an independent set is also independent, it would not make any sense to seek a minimum cardinality independent set, and so instead we seek a minimum cardinality maximal independent set (denoted *minimax*). Similarly, the property of being a dominating set is *superhereditary* (henceforth denoted SH), meaning that any superset of a dominating set is also a dominating set. Thus, it would not make any sense to seek a maximum cardinality dominating set, since the entire vertex set of any graph is a dominating set. So, instead, we seek the maximum cardinality of a minimal dominating set (denoted *maximin*). For concepts that are neither hereditary nor superhereditary, we generally seek a minimum cardinality $\mathcal{P}$-set, denoted *min*, and sometimes a maximum cardinality $\mathcal{P}$-set, denoted *max*.

To illustrate these examples, consider the double star $S(r, s)$ for $1 \le r \le s$ with two adjacent vertices $u$ and $v$, where $u$ is adjacent to $r$ leaves and $v$ is adjacent to $s$ leaves. Let $L(u)$ and $L(v)$ denote the set of leaves adjacent to $u$ and $v$, respectively. We note that $S(r, s)$ has exactly four minimal dominating sets, namely $S_1 = \{u, v\}$, $S_2 = L(u) \cup \{v\}$, $S_3 = L(v) \cup \{u\}$, and $S_4 = L(u) \cup L(v)$. It follows that $S_1$ is a minimum dominating set and $S_4$ is a maximin dominating set. Thus, the domination number $\gamma(S(r, s)) = |S_1| = 2$ and the upper domination number $\Gamma(S(r, s)) = |S_4| = r + s$. Further, the maximal independent sets of $S(r, s)$ are precisely the sets $S_2$, $S_3$, and $S_4$. Hence, $S_4$ is a maximum independent set and $S_2$ is a minimax independent set, and so the independent domination number $i(S(r, s)) = |S_2| = r + 1$ and the independence number $\alpha(S(r, s)) = |S_4| = r + s$.

In the remaining sections of this chapter, we will use the following notation, where recall for a vertex $v$ in $G$, the set $N_G(v)$ denotes the set of neighbors of $v$ in $G$, and the degree of $v$ in $G$ is denoted by $d_G(v) = |N_G(v)|$. Further recall that for a subset of vertices $S \subseteq V$, the degree of $v$ in $S$, denoted $d_S(v)$, is the number of vertices in $S$ adjacent to the vertex $v$; that is, $d_S(v) = |N_G(v) \cap S|$. In particular, if $S = V$, then $d_S(v) = d_G(v)$. If the graph $G$ is clear from context, we simply write $N(v)$ and $d(v)$ rather than $N_G(v)$ and $d_G(v)$, respectively. For $k \ge 1$ an integer, we use the standard notation $[k] = \{1, \dots, k\}$ and $[k]_0 = [k] \cup \{0\} = \{0, 1, \dots, k\}$. At a glance,

$\overline{S} = V \setminus S$, that is, $\overline{S}$ denotes the vertices in $V$ but not in $S$, called the *complement* of $S$ in $G$.

$d_S(v) = |N(v) \cap S|$.

$d_S[v] = |N[v] \cap S|$.

$d_{\overline{S}}(v) = |N(v) \cap \overline{S}|$.

$d_{\overline{S}}[v] = |N[v] \cap \overline{S}|$.

$G[S]$, the subgraph of $G$ *induced by* $S$.
$\delta(G) = \min\{d(v) \mid v \in V\}$.
$\Delta(G) = \max\{d(v) \mid v \in V\}$.

Also, to avoid excessive repetition, we will frequently list domination parameters in the following abbreviated format:

*parameter name*: concept definition; designation of being hereditary H, or superhereditary SH; if neither hereditary nor superhereditary no designation is given; notation for lower parameter and type of $\mathcal{P}$-set (min or minimax), notation for upper parameter and type $\mathcal{P}$-set (max or maximin).

For example, domination, where $\gamma(G)$ is the domination number and $\Gamma(G)$ is the upper domination number, is listed as:

*domination*: $N[S] = V$, that is, for every $v \in \overline{S}$, $d_S(v) \geq 1$; SH; $\gamma(G)$ (min), $\Gamma(G)$ (maximin).

## 2 Fundamental Domination Parameters

In this section, we present what are arguably the most basic of all parameters related to domination in graphs. From these basic parameters all others are derived in one way or another. We begin with a list of five fundamental domination parameters. Thereafter, we list seven related parameters. The designations $H$ hereditary and $SH$ superhereditary are given whenever they apply. Unless otherwise stated, $S$ always denotes a subset of $V$ and $F$ always denotes a subset of $E$.

### 2.1 Domination Parameters

(a) *domination*: $N[S] = V$, that is, for every $v \in \overline{S}$, $d_S(v) \geq 1$; SH; $\gamma(G)$ (min), $\Gamma(G)$ (maximin).
(b) *independent domination*: $N[S] = V$ and $S$ is independent; $i(G)$ (min), $\alpha(G)$ (max).
(c) *total domination*: $N(S) = V$, that is, for every $v \in V$, $d_S(v) \geq 1$; $\gamma_t(G)$ (min), SH; $\Gamma_t(G)$ (maximin);
(d) *paired domination*: $N[S] = V$ and $G[S]$ has a *perfect matching*, that is, an independent set of edges of cardinality $\frac{1}{2}|S|$; $\gamma_{pr}(G)$ (min), $\Gamma_{pr}(G)$ (maximin).
(e) *connected domination*: $N[S] = V$ and $G[S]$ is connected; $\gamma_c(G)$ (min), SH; $\Gamma_c(G)$ (maximin).

It is perhaps worth commenting why total domination and connected domination are both superhereditary properties (SH). Whereas a superset $S^*$ of a connected set $S$ might not be a connected set, if $S$ is also a dominating set, then every vertex in

$\overline{S}$ is adjacent to a vertex in $S$ implying that $S^*$ is also a connected dominating set. Similarly, total domination is superhereditary.

## 2.2 Related Parameters

(a) *vertex covering*: every edge $e \in E$ is incident to a vertex in $S$; SH; $\beta(G)$ (min), $\beta^+(G)$ (maximin). Note that for any graph $G$ of order $n = |V|$,

$$\alpha(G) + \beta(G) = n$$

and

$$i(G) + \beta^+(G) = n.$$

(b) *irredundance*: for every vertex $v \in S$, $N[v] \setminus N[S \setminus \{v\}] \neq \emptyset$; H; $ir(G)$ (minimax), $IR(G)$ (max).

(c) *enclaveless*: $S$ does not contain an *enclave*, that is, a vertex $v \in S$, such that $N[v] \subseteq S$; H; $\psi(G)$ (minimax), $\Psi(G)$ (max). Note that for every graph $G$ of order $n$,

$$\gamma(G) + \Psi(G) = n$$

and

$$\Gamma(G) + \psi(G) = n.$$

(d) *packing*: for every $u, v \in S$, $d(u, v) > 2$; H; $p_2(G)$ (minimax), $P_2(G)$ (max). The packing number $P_2(G)$ is also denoted $\rho(G)$ in the literature. Note that the packing number is a standard lower bound on the domination number for any graph $G$, that is, $P_2(G) \leq \gamma(G)$.

(e) *edge domination*: $F \subseteq E$ and every edge not in $F$ is adjacent to some edge in $F$; SH; $\gamma'(G)$ (min), $\Gamma'(G)$ (maximin).

(f) *matching*: $F \subseteq E$ and $F$ is an independent set of edges; H; $\alpha'^-(G)$ (minimax), $\alpha'(G)$ (max).

(g) *edge covering*: every vertex $v \in V$ is incident to an edge in $F \subseteq E$; SH; $\beta'(G)$ (min), $\beta'^+(G)$ (maximin). Note, it has been shown in [4] and [6], respectively, that for every graph $G$ of order $n$,

$$\alpha'(G) + \beta'(G) = n,$$

and

$$\alpha'^-(G) + \beta'^+(G) = n.$$

## 3 Conditions on the Dominating Set

Many domination parameters are formed by combining domination with another graph theoretical property $P$. In this section, we consider the parameters defined by imposing an additional constraint on the dominating set. In the next section, we will see that a condition may also be placed on the dominated set or on the method of dominating.

We list samples of types of dominating sets $S$ defined either by imposing a condition on the subgraph $G[S]$ induced by $S$ or requiring that every vertex in $S$ satisfy some added condition. Clearly, some of the basic types are defined within this framework. For example, if $G[S]$ has no edges, then the set $S$ is an independent dominating set, if $G[S]$ has no isolated vertices, then $S$ is a total dominating set, and if $G[S]$ is connected, then $S$ is a connected dominating set. Since all of the pairs of parameters in this section consist of the smaller as a minimum and the larger as a maximum of minimal, the designations (min) and (maximin) are omitted.

(a) *acyclic domination*: $N[S] = V$ and $G[S]$ is acyclic (contains no cycles); $\gamma_a(G)$, $\Gamma_a(G)$.
(b) *bipartite domination*: $N[S] = V$ and $G[S]$ is bipartite; $\gamma_{\text{bip}}(G)$, $\Gamma_{\text{bip}}(G)$.
(c) *clique domination*: $N[S] = V$ and $G[S]$ is a complete graph: $\gamma_{cl}(G)$, $\Gamma_{cl}(G)$.
(d) *private domination*: $N[S] = V$ and for every $u \in S$ there exists a vertex $v \in \overline{S}$ such that $N(v) \cap S = \{u\}$; $\gamma_{\text{pvt}}(G)$, $\Gamma_{\text{pvt}}(G)$. Note that a well-known theorem of Bollobás and Cockayne [1] shows that for every graph $G$ with no isolated vertices, $\gamma(G) = \gamma_{\text{pvt}}(G)$, that is, $G$ has a $\gamma$-set $S$ such that for each vertex $u \in S$, there is a vertex $v \in \overline{S}$ with $N(v) \cap S = \{u\}$.
(e) *semitotal domination*: $N[S] = V$ and for every vertex $u \in S$, there exists a vertex $v \in S$ with $d(u, v) \leq 2$; SH; $\gamma_{t2}(G)$, $\Gamma_{t2}(G)$.
(f) *weakly connected domination*: $N[S] = V$ and $G' = (V, E_S)$ is connected, where $E_S$ is the set of edges of $G$ incident to at least one vertex of $S$; SH; $\gamma_w(G)$, $\Gamma_w(G)$.
(g) *semipaired domination*: $N[S] = V$ and the vertices in $S$ can be partitioned into $|S|/2$ pairs $\{u, v\}$ such that $d(u, v) \leq 2$; $\gamma_{\text{pr2}}(G)$, $\Gamma_{\text{pr2}}(G)$.
(h) *convex domination*: $N[S] = V$ and for any two vertices $u, v \in S$, the vertices contained in all shortest paths between $u$ and $v$, called $u - v$ geodesics, belong to $S$; $\gamma_{\text{conv}}(G)$, $\Gamma_{\text{conv}}(G)$.
(i) *weakly convex domination*: $N[S] = V$ and for any two vertices $u, v \in S$, there exists at least one $u - v$ geodesic, all of whose vertices belong to $S$; $\gamma_{\text{wconv}}(G)$, $\Gamma_{\text{wconv}}(G)$.
(j) *cycle domination*: $N[S] = V$ and $G[S]$ has a Hamilton cycle; $\gamma_{\text{cy}}(G)$, $\Gamma_{\text{cy}}(G)$.
(k) *equivalence domination*: $N[S] = V$ and $G[S]$ is disjoint union of complete subgraphs; $\gamma_e(G)$, $\Gamma_e(G)$.
(l) *k-dependent domination*: $N[S] = V$ and $\Delta(G[S]) \leq k$; $\gamma_{[k]}(G)$, $\Gamma_{[k]}(G)$.

## 4 Conditions on $\overline{S} = V \setminus S$

The framework considered in this section encompasses dominating sets $S$ for which some condition is imposed on the vertices in the set $\overline{S}$ or the subgraph $G[\overline{S}]$ induced by $\overline{S}$. As before, we list a sampling of types of dominating sets in this framework. In many of them, we do not mention the upper parameters, which indicates that in general they have not been studied. As before, the designations $H$ hereditary and $SH$ superhereditary are given whenever they apply.

(a) *distance k-domination*: for every $v \in \overline{S}$, there exists a vertex $u \in S$ with $d(u, v) \leq k$; SH; $\gamma_{\leq k}(G)$, $\Gamma_{\leq k}(G)$.
(b) *k-step domination*: for every $v \in \overline{S}$, there exists a vertex $u \in S$ and a $(u, v)$-path of length equal to $k$; SH; $\gamma_{=k}(G)$.
(c) *k-domination*: for every vertex $v \in \overline{S}$, $d_S(v) \geq k$; SH; $\gamma_k(G)$.
(d) *restrained domination*: $N[S] = V$ and for every $v \in \overline{S}$, $d_{\overline{S}}(v) \geq 1$; $\gamma_r(G)$.
(e) *geodetic domination*: every vertex in $\overline{S}$ lies on a shortest path between two vertices in $S$; SH; $\gamma_g(G)$.
(f) *locating domination*: $N[S] = V$ and for every $v, w \in \overline{S}$, $N(v) \cap S \neq N(w) \cap S$; SH; $\gamma_L(G)$.
(g) *secondary domination*: every vertex $w \in \overline{S}$ is adjacent to at least one vertex $u \in S$ and is distance at most $k$ to a second vertex in $S$; SH; $\gamma_{(1,k)}(G)$. Note that for any nontrivial graph without isolated vertices, $\gamma(G) = \gamma_{(1,4)}(G)$ and $\gamma_2(G) = \gamma_{(1,1)}(G)$.
(h) *downhill domination*: for every vertex $v \in \overline{S}$, there exists a vertex $u \in S$ and a (downhill) path $u = v_1, v_2, \ldots v_k = v$ from $u$ to $v$, such that $d(v_i) \geq d(v_{i+1})$ for all $i \in [k-1]$; SH; $\gamma_{\text{down}}(G)$.
(i) *uphill domination*: for every vertex $v \in \overline{S}$, there exists a vertex $u \in S$ and an (uphill) path $u = v_1, v_2, \ldots v_k = v$ from $u$ to $v$, such that $d(v_i) \leq d(v_{i+1})$ for all $i \in [k-1]$; SH; $\gamma_{\text{up}}(G)$.
(j) *exponential domination*: for every vertex $v \in \overline{S}$, $\text{w}_s(v) \geq 1$, where

$$\text{w}_s(v) = \sum_{u \in S} \frac{1}{2^{\overline{d}(u,v)-1}},$$

and $\overline{d}(u, v)$ equals the length of a shortest $(u, v)$-path in $V \setminus (S \setminus \{u\})$ if such a path exists, and $\infty$ otherwise; SH; $\gamma_{\text{exp}}(G)$.
(k) *fair domination*: $N[S] = V$ and every two vertices $u, v \in \overline{S}$ have the same number of neighbors in $S$; fdom$(G)$.
(l) *H-forming domination*: every vertex $v \in \overline{S}$ is contained in a copy of a graph $H$ (not necessarily induced) with a subset of vertices in $S$; SH; $\gamma_H(G)$.
(m) *outer-connected domination*: $N[S] = V$ and $G[\overline{S}]$ is connected; $\gamma_{\overline{c}}(G)$.
(n) *b-disjunctive domination*: for every $v \in \overline{S}$, either $v$ is adjacent to a vertex $u \in S$ or there exist at least $b$ vertices in $S$ at distance 2 from $v$; SH; $\gamma_b^d(G)$.

(o) *secure domination*: $N[S] = V$ and for every vertex $u \in \overline{S}$, there is an adjacent vertex $v \in S$ such that the set $(S \setminus \{v\}) \cup \{u\}$ is a dominating set; SH; $\gamma_s(G)$.

## 5 Conditions on $V$

In this section, we consider a framework where the dominating set is defined by an added condition that is imposed on every vertex of $G$.

(a) *total domination*: $N(S) = V$, that is, for every vertex $v \in V$, $N(v) \cap S \neq \emptyset$; SH; $\gamma_t(G)$, $\Gamma_t(G)$.
(b) *odd domination*: $N[S] = V$, and for every $v \in V$, $|N[v] \cap S|$ is odd; $\gamma_{\text{odd}}(G)$. It is noteworthy that Sutner [7] was the first to observe that every graph $G$ has an odd dominating set.
(c) *even domination*: $N[S] = V$, and for every $v \in V$, $|N[v] \cap S|$ is even; $\gamma_{\text{even}}(G)$.
(d) *identifying code number*: $N[S] = V$, and for every $v \in V$, $N[v] \cap S$ is unique; SH; $\gamma_{\text{id}}(G)$.
(e) *total distance k-dominating*: for every vertex $v \in V$, there exists a vertex $u \in S$, $u \neq v$, such that $d(u, v) \leq k$; SH; $\gamma_t^k(G)$.
(f) *k-tuple domination*: for every $v \in V$, $|N[v] \cap S| \geq k$; SH; $\gamma_{\times k}(G)$.

## 6 Conditions on Vertex Degrees

As we will see in this section, many types of dominating sets can be defined in terms of how many neighbors a vertex must have in either $S$ or $\overline{S}$. These constraints are often perceived as requirements of access to the resources provided by members of a dominating set.

### *6.1 Degree Conditions on S and $\overline{S}$*

Degree conditions as a framework of domination was first suggested by Telle [8]. We present a slightly different form of his framework here. There are four possible values under consideration, namely, $d_S(v)$ and $d_{\overline{S}}(v)$ for $v \in S$, and $d_S(v)$ and $d_{\overline{S}}(v)$ for $v \in \overline{S}$. Table 1 illustrates how with using combinations of these four values, different domination parameters are defined. We only include a few of the many parameters which can be defined by various combinations of the four degree values. A blank entry in Table 1 implies that this condition is not relevant to the definition. Let D-set, TD-set, ID-set, and RD-set denote dominating set, total dominating set, independent dominating set and restrained dominating set, respectively.

**Table 1** Degree Conditions

| $S$ is | $v \in S, d_S(v)$ | $v \in S, d_{\overline{S}}(v)$ | $v \in \overline{S}, d_S(v)$ | $v \in \overline{S}, d_{\overline{S}}(v)$ |
|---|---|---|---|---|
| a D-set | | | $\geq 1$ | |
| an ID-set | $= 0$ | | $\geq 1$ | |
| a TD-set | $\geq 1$ | | $\geq 1$ | |
| a perfect dominating set | | | $= 1$ | |
| an RD-set | | | $\geq 1$ | $\geq 1$ |
| a $k$-dominating set | | | $\geq k$ | |
| a D-set and $\overline{S}$ is a D-set | | $\geq 1$ | $\geq 1$ | |
| a $[1, k]$-dominating set | | | $\geq 1$ and $\leq k$ | |
| an odd D-set | even | | odd | |
| an open odd D-set | odd | | odd | |
| an efficient D-set | =0 | | $= 1$ | |
| a 1-dependent D-set | $\leq 1$ | | $\geq 1$ | |

## 6.2 *Degree Conditions Per Vertex*

As in the previous section, the framework here is defined in terms of the minimum cardinality of a nonempty set $S$ satisfying the stated conditions based on degree. The difference is that the constraints now depend on comparative comparative values of degrees. Recall that the boundary of a set $S$ is $\partial(S) = N[S] \setminus S$.

(a) *alpha domination*: for every $v \in \overline{S}$, $d_S(v)/d(v) \geq \alpha$ where $0 < \alpha \leq 1$; SH; $\gamma_\alpha(G)$.
(b) *defensive alliance*: for every $v \in S$, $d_S[v] \geq d_{\overline{S}}(v)$; $a(G)$.
(c) *defensive k-alliance*: for every $v \in S$, $d_S(v) \geq d_{\overline{S}}(v) + k$; $a_k(G)$. Note that for $k = -1$, a defensive $k$-alliance is the standard defensive alliance, that is, $a_{-1}(G) = a(G)$.
(d) *global defensive alliance*: $N[S] = V$ and for every $v \in S$, $d_S[v] \geq d_{\overline{S}}(v)$; $\gamma_a(G)$.
(e) *offensive alliance*: for every $v \in \partial(S)$, $d_S(v) \geq d_{\overline{S}}[v]$; $a_o(G)$.
(f) *offensive k-alliance*: for every $v \in \partial(S)$, $d_S(v) \geq d_{\overline{S}}(v) + k$; $a_{ok}(G)$. Note that for $k = 1$, a $k$-offensive alliance is the normal offensive alliance.
(g) *global offensive alliance*: for every $v \in \overline{S}$, $d_S(v) \geq d_{\overline{S}}[v]$; $\gamma_{a_o}(G)$.
(h) *powerful alliance*: for every $u \in S$, $d_S[u] \geq d_{\overline{S}}(u)$ and for every $v \in \partial(S)$, $d_S(v) \geq d_{\overline{S}}[v]$; $a_p(G)$.
(i) *(static) monopoly*: for every vertex $v \in \overline{S}$, $d_S(v) \geq d_{\overline{S}}(v)$, that is, every vertex not in $S$ has at least $\lceil d(v)/2 \rceil$ neighbors in $S$, or equivalently, every vertex in $\overline{S}$ has at least as many neighbors in $S$ as it has in $\overline{S}$; SH; $m(G)$.
(j) *open, or total, monopoly*: for every vertex $v \in V$, $d_S(v) \geq d_{\overline{S}}(v)$, that is, every vertex in $V$ has at least as many neighbors in $S$ as it has in $\overline{S}$; SH; $m_t(G)$.
(k) *weak domination*: for every $v \in \overline{S}$, there exists a neighbor $u \in S$, $d(u) \leq d(v)$; SH; $\gamma_w(G)$.

(l) *strong domination*: for every $v \in \overline{S}$, there exists a neighbor $u \in S, d(u) \geq d(v)$; SH; $\gamma_s(G)$.
(m) *cost effective domination*: $N[S] = V$ and for every $v \in S, d_S(v) \leq d_{\overline{S}}(v)$; $\gamma_{ce}(G)$.
(n) *very cost effective domination*: $N[S] = V$ and for every $v \in S, d_S(v) < d_{\overline{S}}(v)$; $\gamma_{vce}(G)$.
(o) 1-*equitable domination*: $N[S] = V$ and for all $u, v \in S$, $|d_{\overline{S}}(u) - d_{\overline{S}}(v)| \leq 1$; $\gamma_{1eq}(G)$.
(p) 2-*equitable domination*: $N[S] = V$ and for all $u, v \in \overline{S}$, $|d_S(u) - d_S(v)| \leq 1$; $\gamma_{2eq}(G)$
(q) *equitable domination*: $N[S] = V$ and for all $u, v \in S$, $|d_{\overline{S}}(u) - d_{\overline{S}}(v)| \leq 1$, and for all $u, v \in \overline{S}$, $|d_S(u) - d_S(v)| \leq 1$; $\gamma_{eq}(G)$.
(r) *global distribution center*: $N[S] = V$ and for all $v \in \overline{S}$, there exists a vertex $u \in S$ such that $d_S[u] \geq d_{\overline{S}}[v]$; SH; gdc$(G)$.

## 7 Functions $f: V \to \mathbb{N}$

For every set $S \subseteq V$, there is a corresponding *characteristic function* $f_S: V \to \{0, 1\}$, such that $f(v) = 1$ if $v \in S$, and $f(v) = 0$ if $v \in \overline{S}$. This suggests a variety of options for the range $\mathbb{N}$ of a function $f: V \to \mathbb{N}$, in terms of domination. In this section, we present a sample of the functions that have been considered under this framework. The value of each of the following parameters equals the minimum weight of a function of the given type, where the *weight* w$(f)$ of such a function $f$ is the sum of all assigned values,

$$\mathrm{w}(f) = \sum_{v \in V} f(v).$$

### *7.1 Dominating Functions*

(a) *domination*: $f : V \to \{0, 1\}$, for every vertex $v \in V$, $f(N[v]) \geq 1$; $\gamma(G)$.
(b) *fractional domination*: $f : V \to [0, 1]$, for every vertex $v \in V$, $f(N[v]) \geq 1$; $\gamma_f(G)$.
(c) *signed domination*: $f : V \to \{-1, 1\}$, for every vertex $v \in V$, $f(N[v]) \geq 1$; $\gamma_s(G)$.
(d) *minus domination*: $f : V \to \{-1, 0, 1\}$, for every vertex $v \in V$, $f(N[v]) \geq 1$; $\gamma_m(G)$.
(e) $\{k\}$-*domination*: $f : V \to \{0, 1, \ldots, k\}$, for every vertex $v \in V$, $f(N[v]) \geq k$; $\gamma_{\{k\}}(G)$.

(f) *k-rainbow domination*: $f : V \to \mathcal{P}\{1, 2, \dots, k\}$, every vertex $v \in V$ is assigned a subset of $\{1, 2, \dots, k\}$ such that for every vertex $v \in V$ with $f(v) = \emptyset$, the union of the sets assigned to the closed neighborhood $N[v]$ equals $\{1, 2, \dots, k\}$; $\gamma_{rk}(G)$.

## 7.2 Roman Dominating Functions

The types of domination in this section are models of a military defense strategy instituted by Emperor Constantine, between 306 and 337 AD, in which the regions in the Roman Empire were defended by armies stationed at key locations. A region was secured by armies stationed there, and a region without an army was protected by sending mobile armies from neighboring regions. But Constantine decreed that a mobile field army could not be sent to defend a region, if doing so left its original region unsecured. This defense strategy gave rise to what is called *Roman domination*, given below. As in the previous section, the value of each of the following domination parameters equals the minimum weight of a function of the given type.

**Definition 1** Roman domination: $f : V \to \{0, 1, 2\}$, *for every vertex* $v$ *with* $f(v) = 0$, *there is a vertex* $u \in N(v)$ *with* $f(u) = 2$; $\gamma_R(G)$.

It is easy to see, for example, that for every graph $G$, $\gamma(G) \leq \gamma_R(G) \leq 2\gamma(G)$. From the initial definition of Roman domination as a framework, many varieties of domination can clearly be defined, and indeed, many have been defined. We only provide a sample here.

(a) *weak Roman domination*: $f : V \to \{0, 1, 2\}$, for every $v$ with $f(v) = 0$, there is a vertex $u \in N(v)$ with $f(u) > 0$ such that the function $f'$ with $f'(v) = 1$, $f'(u) = f(u) - 1$, and $f'(w) = f(w)$ otherwise, has no *undefended vertex*, meaning a vertex with $f'(N[w]) = 0$; $\gamma_r(G)$.
(b) *double Roman domination*: $f : V \to \{0, 1, 2, 3\}$, every vertex $w$ with $f(w) = 0$ either has a neighbor $u$ with $f(u) = 3$ or two neighbors $u, v$ with $f(u) = f(v) = 2$, and if $f(w) = 1$, then $w$ has at least one neighbor $u$ with $2 \leq f(u) \leq 3$; $\gamma_{dR}(G)$.
(c) *Roman {2}-domination, also called Italian domination*: $f : V \to \{0, 1, 2\}$, every vertex $v$ with $f(v) = 0$ has $f(N(v)) \geq 2$; $\gamma_{R2}(G)$ (also $\gamma_I(G)$).
(d) *Roman k-domination*: $f : V \to \{0, 1, 2\}$, every vertex $v$ with $f(v) = 0$ is adjacent to at least $k$ vertices $u$ with $f(u) = 2$; $\gamma_{kR}(G)$.
(e) *independent Roman domination*: $f : V \to \{0, 1, 2\}$, every vertex $v$ with $f(v) = 0$ has at least one neighbor $u$ with $f(u) = 2$ and the set of vertices $w$ with $f(w) > 0$ is an independent set; $i_R(G)$.
(f) *signed Roman domination*: $f : V \to \{-1, 1, 2\}$, for every vertex $v \in V$, $f(N[v]) \geq 1$, and every vertex $v$ with $f(v) = -1$ has at least one neighbor $u$ with $f(u) = 2$; $\gamma_{sR}(G)$.

(g) *total Roman domination*: $f : V \to \{0, 1, 2\}$, every vertex $w$ with $f(w) = 0$ has at least one neighbor $u$ with $f(u) = 2$ and every vertex $u$ with $f(u) > 0$ has at least one neighbor $v$ with $f(v) > 0$; $\gamma_{tR}(G)$.

## 8 Stratified Domination

A graph $G$ together with a fixed partition of its vertex set $V$ into nonempty subsets is called a *stratified graph*. If the partition is $V = \{V_1, V_2\}$, then $G$ is a 2-stratified graph and the sets $V_1$ and $V_2$ are called the *strata* or sometimes the *color classes* of $G$. A framework for domination based on coloring the vertices of a graph was defined in [2] as follows. Let $F$ be a 2-stratified graph with one fixed blue vertex $v$ specified; $F$ is said to be *rooted* at the blue vertex $v$. An *$F$-coloring* of a graph $G$ is defined to be a red-blue coloring of the vertices of $G$ such that every blue vertex $v$ is a root of a copy of $F$ (not necessarily induced) in $G$. The *$F$-domination number* $\gamma_F(G)$ of $G$ is the minimum number of red vertices in an $F$-coloring of $G$.

We note that if $F$ is a 2-stratified $K_2$ rooted at a blue vertex that is adjacent to a red vertex, then the set of red vertices in an $F$-coloring of $G$ is a dominating set of $G$ and $\gamma_F(G) = \gamma(G)$.

This extends to other 2-stratified graphs $F$ and encapsulates many types of domination related parameters, including the domination, total domination, restrained, total restrained, and $k$-domination numbers. For example, let $F$ be a 2-stratified $P_3$ rooted at a blue vertex $v$. The five possible choices for the graph $F$ are shown in Figure 1.

Let $G$ be a connected graph of order at least 3. It is shown in [2] that if $F = F_1$, then the set of red vertices of an $F$-coloring of $G$ is a total dominating set and $\gamma_F(G) = \gamma_t(G)$, while if $F = F_2$, then the set of red vertices is a dominating set of $G$ and $\gamma_F(G) = \gamma(G)$. Furthermore, if $F = F_4$, then the set of red vertices of an $F$-coloring of $G$ is a restrained dominating set and $\gamma_F(G) = \gamma_r(G)$, and if $F = F_5$, then the set of red vertices of an $F$-coloring of $G$ is a 2-dominating set and $\gamma_F(G) = \gamma_2(G)$.

On the other hand, the parameter $\gamma_{F_3}(G)$ defined a new domination parameter that had not been studied prior to considering domination from this framework.

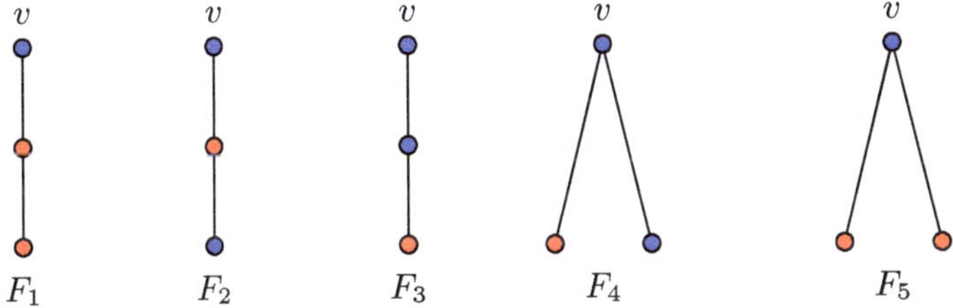


**Fig. 1** The five 2-stratified graphs $P_3$

Stratified domination encompasses many known domination parameters and suggests new avenues for study.

## 9 Domination Chain

The domination chain expresses relationships that exist among dominating sets, independent sets, and irredundant sets in graphs. Irredundance is the concept that describes the minimality of a dominating set. If a dominating set $S$ is *minimal*, then for every vertex $u \in S$ the set $S \setminus \{u\}$ is no longer a dominating set. This means that the vertex $u$ dominates some vertex, which could be itself, that no other vertex in $S$ dominates. Given a vertex set $S \subseteq V$ and a vertex $v \in S$, we make the following definitions.

(a) The vertex $v$ is a *self-private neighbor* if $v$ has no neighbors in $S$, that is, $N[v] \cap S = \{v\}$.
(b) The vertex $v$ has an $S$-*external private neighbor* if there exists a vertex $w \in \overline{S}$ such that $N(w) \cap S = \{v\}$.
(c) The vertex $v$ has an $S$-*internal private neighbor* if there exists a vertex $w \in S$ such that $N(w) \cap S = \{v\}$.

A nonempty set $S$ is *irredundant* if and only if every vertex $v \in S$ either is a self-private neighbor or has an $S$-external private neighbor. The *irredundance numbers*, $ir(G)$ and $IR(G)$, are the minimum and maximum cardinalities, respectively, of a maximal irredundant set.

The following two properties of a minimal and maximal dominating set yield the domination chain:

**Observation 1** *The following hold in a graph $G$.*

(a) *Every minimal dominating set in $G$ is a maximal irredundant set of $G$.*
(b) *Every maximal independent set in $G$ is a minimal dominating set of $G$.*

**Theorem 2 (The Domination Chain)** *For every graph $G$,*

$$ir(G) \leq \gamma(G) \leq i(G) \leq \alpha(G) \leq \Gamma(G) \leq IR(G).$$

Since its introduction by Cockayne, Hedetniemi, and Miller [3] in 1978, the domination chain of Theorem 2 has become one of the major focal points in the study of domination in graphs, inspiring several hundred papers. As a framework, it is possible to obtain inequality chains similar to the domination chain starting from a suitable seed property. Thus, almost any property of subsets could be considered, for example, the seed property that $S$ is a vertex cover.

## 10 Conditions Relating to Perfection

The concept of being dominated exactly once by the vertices in a set $S$ is generally referred to as perfect or efficient domination. In this section, we list a few parameters related to this model of domination.

Given a set $S \subseteq V$, a vertex $v \in V$ is *perfect* (with respect to $S$) if $|N[v] \cap S| = 1$, and is *almost perfect* if it is either perfect or is adjacent to a perfect vertex. A vertex $v \in V$ is *open perfect* (with respect to $S$) if $|N(v) \cap S| = 1$, and is *almost open perfect* if it is either open perfect or adjacent to an open perfect vertex. A set $S$ is a *perfect neighborhood set* if every vertex $v \in V$ is either perfect or almost perfect, with respect to $S$. A set $S$ is called *internally perfect* if every vertex $v \in S$ is perfect, with respect to $S$, that is, if $S$ is an independent set. And $S$ is called *externally perfect* if every vertex $w \in \overline{S}$ is perfect, that is, every vertex $w$ is adjacent to exactly one vertex in $S$. Externally perfect sets are also called *perfect dominating sets.* A set that is both internally and externally perfect is an *efficient dominating set.* A set $S$ is called *nearly perfect* if for every vertex $v \in \overline{S}$, $|N(v) \cap S| \leq 1$, that is, every vertex in $\overline{S}$ is dominated at most once by the vertices in $S$, or every vertex in $\overline{S}$ has at most one neighbor in $S$.

(a) *perfect domination*: for every vertex $v \in \overline{S}$, $d_S(v) = 1$; $\gamma_p(G)$.
(b) *efficient domination*: for every $v \in V$, $d_S[v] = 1$; $\gamma(G)$. Note, it can be shown that all efficient dominating sets have the same cardinality, namely, $\gamma(G)$. Efficient dominating sets are also called *perfect codes.* This means that if $S = \{v_1, v_2, \ldots, v_k\}$ is an efficient dominating set, then $\pi = \{N[v_1], N[v_2], \ldots, N[v_k]\}$ is a partition of $V$.
(c) *efficiency*: $\varepsilon(S) = |\{v \in \overline{S} : d_S(v) = 1\}|$; $\varepsilon(G) = max_{S \subseteq V}\{\varepsilon(S)\}$.
(d) *total efficiency*: $\varepsilon_t(S) = |\{v \in V : d_S(v) = 1\}|$; $\varepsilon_t(G) = max_{S \subseteq V}\{\varepsilon_t(S)\}$.

## 11 Criticality Parameters

For any parameter, such as $\alpha(G)$ or $\gamma(G)$, it is natural to consider how the value changes when a small change is made in the graph $G$, for example, by the deletion of a vertex or edge, the addition of an edge, the subdivision of an edge, the identification of two non-adjacent vertices (an *elementary homomorphism*), or the identification of two adjacent vertices (an *elementary contraction*). Most of the study along these lines involves families of graphs whose domination number changes whenever the given modification is made arbitrarily in the graph. For example, *domination edge critical* graphs $G$ have the property that the domination number decreases whenever any arbitrary edge is added, that is, $\gamma(G + e) < \gamma(G)$ for any $e \in E(\overline{G})$.

On the other hand, parameters that in some sense measure the degree of criticality have also been studied. Here we describe selected *criticality* parameters of this type that have been studied for domination. We note that this perspective deviates from our other frameworks in that it does not encompass dominating sets, but instead considers effects of a graph modification on the domination number.

(a) *reinforcement number*, $r(G)$: minimum number of edges that must be added to $G$ in order to decrease the domination number.
(b) *bondage number*, $b(G)$: minimum number of edges that must be deleted from $G$ in order to increase the domination number.
(c) *domination sensitivity*, $\gamma_{\pm}(G)$: minimum number of vertices that must be deleted to either increase or decrease the domination number.
(d) *domination subdivision number*, $sd_{\gamma}(G)$: minimum number of edges that must be subdivided in order to increase the domination number.
(e) *total domination subdivision number*, $sd_{\gamma_t}(G)$: minimum number of edges that must be subdivided in order to increase the total domination number.
(f) *paired domination subdivision number*, $sd_{\rm pr}(G)$: minimum number of edges that must be subdivided in order to increase the paired domination number.
(g) *forcing domination number*, $F_{\gamma}(G)$. A subset $T$ of a minimum dominating set $S$ is a *forcing subset* for $S$ if $S$ is the unique minimum dominating set containing $T$. The *forcing domination number* $F_{\gamma}(S)$ of a minimum dominating set $S$ is the minimum cardinality among the forcing subsets of $S$, and the *forcing domination number* $F_{\gamma}(G)$ of $G$ is the minimum forcing domination number among the minimum dominating sets $S$ of $G$. It follows from the definition that $F_{\gamma}(G) \leq \gamma(G)$.

## 12 Partitions

For any property $\mathcal{P}$ of interest, it is natural to consider partitions of the vertex set $V = \{V_1, V_2, \ldots, V_k\}$ such that every set $V_i$, where $i \in [k]$, is a $\mathcal{P}$-set; these are generally referred to as $\mathcal{P}$-colorings. The most often studied partitions of this type are called *proper colorings*, in which each set $V_i$ is an independent set.

In this section, we describe a variety of $\mathcal{P}$-colorings which have been studied, in which the property $\mathcal{P}$ is related to domination. As in the previous section, this perspective deviates from our other frameworks in that it does not encompass dominating sets, but instead considers parameters based on graph partitions involving dominating sets.

(a) *domatic number*, $d(G)$: maximum order of a vertex partition into dominating sets.
(b) *idomatic number*, $id(G)$: maximum order of a vertex partition into independent dominating sets, or the maximum number of vertex disjoint independent dominating sets.
(c) *capacitated domination*, $\gamma_{cap_k}(G)$: minimum order of a vertex partition into sets $V_i$ such that $G[V_i]$ has a spanning star of order at most $k + 1$.

(d) *iterated independence numbers*, $i^*(G)$ and $\alpha^*(G)$: minimum and maximum orders of partitions resulting from repeated removals of maximal independent sets.
(e) *iterated domination numbers*, $\gamma^*(G)$, $\Gamma^*(G)$: minimum and maximum orders of partitions resulting from repeated removals of minimal dominating sets.
(f) *iterated irredundance numbers*, $ir^*(G)$, $IR^*(G)$: minimum and maximum orders of partitions resulting from repeated removals of maximal irredundant sets.
(g) *dominator coloring number*, $\chi_d(G)$: minimum order of a vertex partition, such that every vertex $v \in V$ dominates at least one set $V_i$.
(h) *gamma-gamma domination*, $\gamma\gamma(G)$, $\Gamma\Gamma(G)$: minimum and maximum of $|S_1|+|S_2|$ for two disjoint (minimal) dominating sets in $G$.
(i) *gamma-i domination*, $\gamma i(G)$: minimum of $|S_1|+|S_2|$ for two disjoint dominating sets in $G$, one of which is an independent dominating set.
(j) *defensive alliance partition number*, $\Psi_a(G)$: maximum order of a vertex partition into defensive alliances.

## 13 Summary

In the preceding sections we have seen a wide variety of contexts in which aspects of dominating sets in graphs can be expressed and studied.

If a condition can be imposed on the vertices only in $S$ or only in $\overline{S}$, it can also be imposed to hold on all vertices in $V$. In this way we move from domination to total domination. If a condition can be imposed on the closed neighborhoods $N[v]$ of vertices, it can then be relaxed to hold only for open neighborhoods $N(v)$. All parameters involving sets $S \subseteq V$ can also be studied from the point of view of subsets of edges $F \subseteq E$.

One can consider the minimum cardinality of a set $S$ having some property, and also consider the maximum cardinality of a minimal set having the same property. One can consider the maximum cardinality of a set $S$ having some property, and also consider the minimum cardinality of a maximal set having the same property. Consider any hereditary property $\mathcal{P}$ of a set of vertices $S$, such as being an independent set. You can ask: what condition must exist for a set $S$ to be a maximal $\mathcal{P}$-set? This condition is a property $\mathcal{P}'$ in its own right, and every maximal set having property $\mathcal{P}$ must then also have property $\mathcal{P}'$. In the same way you can consider any superhereditary property $\mathcal{Q}$ of a set of vertices, such as being a dominating set. You can then ask: what condition must exist for a set $S$ to be a minimal $\mathcal{Q}$-set? This will then give rise to another property $\mathcal{Q}'$, which can be studied in its own right.

Among all subsets $S$ having some property $\mathcal{P}$, one can impose an additional condition, often that the set $S$ also be independent, but that the induced subgraph $G[S]$ have some common graph property, like having no isolated vertices, or being a connected subgraph. You can, of course impose an added condition on the set $\overline{S}$. In this way, for example, we get restrained domination and outer-connected domination. We have seen many examples where a condition is imposed on either $N_S(u)$ or $N_S[u]$, and likewise on $N_{\overline{S}}(v)$ or $N_{\overline{S}}[v]$, for vertices in either $S$ or $\overline{S}$.

For every set $S \subseteq V$, there is a corresponding characteristic function $f_S : V \to \{0, 1\}$, such that $f(v) = 1$ if $v \in S$, and $f(v) = 0$ if $v \in \overline{S}$. This suggests a variety of options for the range of a function $f$, such as the closed unit interval, $f : V \to [0, 1]$, from which we get fractional domination, or $f : V \to \{0, 1, 2\}$, from which we get Roman domination, or $f : V \to \{-1, 1\}$ from which we get signed domination.

It is natural to consider partitions $\pi = \{V_1, V_2, \ldots, V_k\}$ such that every set $V_i$ where $i \in [k]$ has some property $\mathcal{P}$, the most studied is that every set $V_i$ is an independent set. Such partitions are sometimes called *$\mathcal{P}$-colorings* of graphs, and one seeks either the minimum order of such a partition or the maximum order, usually depending on whether the property $\mathcal{P}$ is hereditary (minimum order, e.g. chromatic number) or superhereditary (maximum order, e.g. domatic number).

Real-world applications of dominating sets often suggest new and interesting models of domination. This was the case with Roman domination, in which a vertex $v$ with $f(v) = 2$ represents a location at which two armies are stationed, one of which can be used to defend a neighboring location by traveling along a single edge. This one application alone has suggested numerous other models for defending the vertices of a graph with different types of dominating sets.

In computer networks, a dominating set is viewed as a set of vertices, or nodes, each of which supplies, "in one hop" a needed resource to all neighboring vertices. But if one of these vertices becomes inoperative, or faulty, it might be helpful to have some sort of backup arrangement. One such arrangement could be to have a neighbor of the faulty node, also in the dominating set, so that a service could be provided in at most two hops while the fault can be fixed; this corresponds to a total dominating set, and is closely related to the model of a $(1, k)$-dominating set, in which every node either has one-hop service or secondary service at most $k$-hops away. Another arrangement might be to have a neighboring node to the faulty node serve temporarily as a backup in such a way that the resulting set of nodes is another dominating set. This leads to the model of secure dominating sets.

What models of domination have not we discussed? At the outset of this chapter, we said that space limitations would not permit us to be comprehensive in reviewing the many different models of domination that are being considered in the current literature. Some compensation for the limitations of this chapter, however, are provided by chapters in this volume and other books on domination. We list some of them here along with selected sources of information. Of course, there are many other application driven frameworks of domination, ranging from social networks to mathematical chemistry, that are beyond the scope of these sources.

(a) *Domination in hypergraphs*

Chapter 11 Domination in Hypergraphs, by M. A. Henning and A. Yeo, in *Structures of Domination in Graphs*, Springer, 2020.

(b) *Domination in linear and integer programming*

Chapter 1 LP-Duality, Complementarity, and Generality of Graphical Subset Parameters, by P. J. Slater, in *Domination in Graphs, Advanced Topics*, Marcel Dekker, 1998.

(c) *Domination in directed graphs and tournaments*

Chapter 15 Topics on Domination in Directed Graphs, by J. Ghoshal, R. C. Laskar, and D. Pillone, in *Domination in Graphs, Advanced Topics*, Marcel Dekker, 1998.

Chapter 13 Domination in Digraphs and Tournaments, by T. W. Haynes, S. T. Hedetniemi, and M. A. Henning, in *Structures of Domination in Graphs*, Springer, 2020.

(d) *Domination in chessboards*

Chapter 6 Combinatorial Problems on Chessboards: II, by S. M. Hedetniemi, S. T. Hedetniemi, and R. Reynolds, in *Domination in Graphs, Advanced Topics*, Marcel Dekker, 1998.

J.J. Watkins, *Across the Board: The Mathematics of Chessboard Problems*, Princeton University Press, 2004.

Chapter 12 Domination in Chessboards, by J. T. Hedetniemi and S. T. Hedetniemi, in *Structures of Domination in Graphs*, Springer, 2020.

(e) *Algorithms and complexity of domination in graphs*

Chapter 12 Domination Complexity and Algorithms, in *Fundamentals of Domination in Graphs*, Marcel Dekker, 1998.

Chapter 8 Algorithms, by D. Kratsch, in *Domination in Graphs, Advanced Topics*, Marcel Dekker, 1998.

Chapter 9 Complexity Results, by S. T. Hedetniemi, A. A. McRae, and D. A. Parks, in *Domination in Graphs, Advanced Topics*, Marcel Dekker, 1998.

Chapter 14 Algorithms and Complexity - Signed and Minus Domination, by S. T. Hedetniemi, A. A. McRae, and R. Mohan, in *Structures of Domination in Graphs*, Springer, 2020.

Chapter 15 Algorithms and Complexity - Power Domination, by S. T. Hedetniemi, A. A. McRae, and R. Mohan, in *Structures of Domination in Graphs*, Springer, 2020.

Chapter 16 Self-Stabilizing Domination Algorithms, by S. T. Hedetniemi, in *Structures of Domination in Graphs*, Springer, 2020.

(f) *Domination games on graphs*

Chapter 8 An Introduction to Game Domination in Graphs, by M. A. Henning, in *Structures of Domination in Graphs*, Springer, 2020.

(g) *Domination and eigenvalues in graph theory*

Chapter 9 Domination and Spectral Graph Theory, by C. Hoppen, D. Jacobs, and V. Trevisan, in *Structures of Domination in Graphs*, Springer, 2020.

## References

1. B. Bollobás and E. J. Cockayne, Graph theoretic parameters concerning domination, independence and irredundance. *J. Graph Theory* 3 (1979), 241–250.
2. G. Chartrand, T.W. Haynes, M.A. Henning, and P. Zhang, Stratification and domination in graphs. *Discrete Math.* 272 (2003), 171–185.
3. E. J. Cockayne, S. T. Hedetniemi, and D. J. Miller, Properties of hereditary hypergraphs and middle graphs. *Canad. Math. Bull.* 21 (1978), 461–468.
4. T. Gallai, Über extreme Punkt- und Kantenmengen, *Ann. Univ. Sci. Budapest. Eötvös Sect. Math.* 2 (1959), 133–138.
5. T.W. Haynes, S.T. Hedetniemi, and P.J. Slater, *Fundamentals of Domination in Graphs*, Marcel Dekker, Inc. New York, 1998.
6. S. T. Hedetniemi, A max-min relationship between matchings and domination in graphs. *Congr. Numer.* 40 (1983), 23–34.
7. K. Sutner, Linear cellular automata and the Garden-of-Eden. *Math. Intelligencer* 11 (1989), 49–53.
8. J. A. Telle, *Vertex Partitioning Problems: Characterization, Complexity and Algorithms on Partial k-trees*. Ph.D. Thesis. University of Oregon, 1994.

# Paired Domination in Graphs

**Wyatt J. Desormeaux, Teresa W. Haynes, and Michael A. Henning**

**AMS Subject Classification** 05C69

## 1 Introduction

Applications of dominating sets include security models where each vertex in the dominating set represents the location of a guard or an officer capable of protecting every vertex it dominates. Paired domination in graphs was introduced by Haynes and Slater [41, 42] as a model for assigning backups to police officers. To ensure the safety of each officer, it is common practice that officers are dispatched in pairs, that is, they are assigned partners so each can back up the other. This practice is modeled by paired domination, where each officer's location must be adjacent to his/her partner's location and they are "designated as backups" for each other.

Formally, a *paired dominating set*, abbreviated PD-set, of a graph $G = (V, E)$ is a set $S \subseteq V$ such that every vertex of $V$ is adjacent to some vertex in $S$ and the induced subgraph $G[S]$ contains a perfect matching $M$ (not necessarily induced). Two vertices joined by an edge of $M$ are said to be *paired* and are also called *partners* in $S$. The *paired domination number* of $G$, denoted by $\gamma_{\rm pr}(G)$, is

Research of the second and third authors supported in part by the University of Johannesburg.

W. J. Desormeaux (✉) · M. A. Henning
Department of Mathematics and Applied Mathematics, University of Johannesburg, Auckland Park, 2006, South Africa
e-mail: mahenning@uj.ac.za

T. W. Haynes
Department of Mathematics and Statistics, East Tennessee State University, Johnson City, TN, 37614-0002, USA
Department of Mathematics and Applied Mathematics, University of Johannesburg, Auckland Park, 2006, South Africa
e-mail: haynes@etsu.edu


T. W. Haynes et al. (eds.), *Topics in Domination in Graphs*, Developments in Mathematics 64, https://doi.org/10.1007/978-3-030-51117-3_3

the minimum cardinality of a PD-set of $G$, and a PD-set of $G$ having cardinality $\gamma_{\mathrm{pr}}(G)$ is called a $\gamma_{\mathrm{pr}}$-*set* of $G$. Every graph without isolated vertices has a paired dominating set, and hence a paired domination number.

A PD-set $S$ in a graph $G$ is *minimal* if no proper subset of $S$ is a PD-set of $G$. The *upper paired domination number* of $G$, denoted by $\Gamma_{\mathrm{pr}}(G)$, is the maximum cardinality of a minimal PD-set of $G$. A minimal PD-set of $G$ of cardinality $\Gamma_{\mathrm{pr}}(G)$ is called a $\Gamma_{\mathrm{pr}}$-*set* of $G$.

Recall that a *dominating set* of a graph $G = (V, E)$ is a set $S \subseteq V$ such that every vertex in $V \setminus S$ is adjacent to some vertex in $S$. In addition, if every vertex in $V$ is adjacent to some vertex in $S$, then $S$ is a *total dominating set*, abbreviated TD-set, of $G$. The *domination number*, denoted by $\gamma(G)$, is the minimum cardinality of a dominating set, while the *total domination number*, denoted by $\gamma_t(G)$, is the minimum cardinality of a TD-set of $G$. A dominating set of $G$ having cardinality $\gamma(G)$ is called a $\gamma$-*set* of $G$. Two vertices are neighbors if they are adjacent.

Total domination and paired domination are defined only for graphs with no isolated vertices. By definition every PD-set is a TD-set, and every TD-set is a dominating set. Hence, we have the following observation relating these three parameters.

**Observation 1 ([42])** *If $G$ is a graph with no isolated vertex, then*

$$\gamma(G) \leq \gamma_t(G) \leq \gamma_{\mathrm{pr}}(G).$$

Necessarily, the paired domination number of a graph is an even integer. We also note that every support vertex (vertex adjacent to an end vertex) of $G$ is contained in every PD-set of $G$. For example, for $k \geq 2$ the subdivided star $G = S(K_{1,k})$, which is obtained from a star $K_{1,k}$ by subdividing every edge exactly once, has domination number $\gamma(G) = k$, total domination number $\gamma_t(G) = k + 1$, and paired domination number $\gamma_{\mathrm{pr}}(G) = 2k$. On the other hand, the double star $G = S(r, s)$, which is a tree with exactly two (adjacent) non-leaf vertices one of which has $r \geq 1$ leaf neighbors and the other $s \geq 1$ leaf neighbors, has $\gamma(G) = \gamma_t(G) = \gamma_{\mathrm{pr}}(G) = 2$.

Bollobás and Cockayne [3] showed that every graph without isolated vertices has a $\gamma$-set $S$ in which every vertex $v \in S$ dominates a vertex $v' \in V \setminus S$ whose only neighbor in $S$ is $v$. Such a vertex $v'$ is called an *S-external private neighbor* of $v$ and is not necessarily unique. For each vertex $v \in S$, select one such vertex external private neighbor $v'$. Then the set $\cup_{v \in S}\{v, v'\}$ is a PD-set of $G$, where the vertex $v$ is paired with the vertex $v'$. Thus, we have the following result first observed by Haynes and Slater [42].

**Theorem 2 ([42])** *If $G$ is a graph with no isolated vertices, then $\gamma_{\mathrm{pr}}(G) \leq 2\gamma(G)$.*

Also given in [42] is the following interesting necessary condition for equality in the bound of Theorem 2.

**Theorem 3 ([42])** *If $\gamma_{\mathrm{pr}}(G) = 2\gamma(G)$, then every $\gamma$-set of $G$ is an independent set.*

Additional relationships between the paired domination number and other domination parameters will be presented in Section 9. The remainder of this chapter is organized as follows. Bounds on the paired domination are presented in Section 2. Section 3 covers graphs having maximum size with a given paired domination number, and Section 4 contains Nordhaus–Gaddum type results for paired domination. We discuss paired domination in Cartesian products in Section 5 and paired domination in other graph families including trees, claw-free graphs, and planar graphs in Section 6. For additional graph families, we refer the reader to [5, 6, 15, 67, 69, 71]. Criticality concepts for paired domination are presented in Section 7, and the upper paired domination number is covered in Section 8. Perfect concepts involving paired domination are presented in Section 10, while complexity issues are in Section 11. Conjectures and open questions are stated throughout the text where appropriate.

We note that this chapter gives an overview of paired domination, while focusing mainly on bounds. For variations and topics involving paired domination not covered here, we refer the reader to the survey [22]. Also, see [38] for a paired domination version of the domination game. Any terminology not defined herein can be found in the glossary of chapter "Glossary of Common Terms".

# 2 Bounds on the Paired Domination Number

As we shall see in Section 11, the decision problem associated with paired domination is NP-complete, so it is of interest to determine bounds on the paired domination number of a graph. In this section, we present major known bounds.

## 2.1 *Bounds Involving Order and Degree*

Since the paired domination number is not defined for graphs with isolated vertices, we consider only graphs with minimum degree $\delta(G) \geq 1$. Haynes and Slater [42] established the following lower bound on the paired domination number of a graph in terms of its order and maximum degree.

**Theorem 4 ([42])** *If $G$ is a graph of order $n$ with no isolated vertices and maximum degree $\Delta(G) = \Delta$, then $\gamma_{pr}(G) \geq \frac{n}{\Delta}$, and this bound is sharp.*

Sharpness of the bound of Theorem 4 can be seen with the graph $G = mK_2$. This graph also shows that the paired domination number may equal the order of the graph. However, this is the unique such graph with this property as the following result shows.

**Theorem 5 ([42])** *If $G$ is a graph of order $n$ with no isolated vertices, then $\gamma_{pr}(G) = n$ if and only if $G = mK_2$.*

The trivial upper bound of Theorem 5 can be improved slightly if we forbid $K_2$-components. Haynes and Slater [42] obtained the following upper bound on the paired domination number of a connected graph of order at least 3.

**Theorem 6 ([42])** *If $G$ is a connected graph of order $n \geq 3$, then $\gamma_{\rm pr}(G) \leq n - 1$ with equality if and only if $G$ is the cycle $C_3$, the cycle $C_5$, or a subdivided star $S(K_{1,k})$ for $k \geq 1$.*

If $G$ is a connected graph of order $n \geq 3$ and $G$ is none of the graphs listed in Theorem 6, then it follows that $\gamma_{\rm pr}(G) \leq n - 2$. The connected graphs $G$ of order $n \geq 4$ for which $\gamma_{\rm pr}(G) = n - 2$ are characterized by Ulatowski in [78].

Chellali and Haynes [9] obtained the following upper bound on the paired domination number of a graph in terms of its order and minimum degree.

**Theorem 7 ([9])** *If $G$ is a graph of order $n$ with no isolated vertices, then $\gamma_{\rm pr}(G) \leq n - \delta(G) + 1$.*

### 2.1.1 Minimum Degree at Least 2

For graphs of order $n \geq 6$ having minimum degree at least 2, the upper bound in Theorem 6 on the paired domination number of a graph can be improved from one less than its order to two-thirds its order. This result was first presented in [42], but the proof contained an error. Huang and Shan [56] provided a corrected proof, a slightly modified version of which we present here.

**Theorem 8 ([42, 56])** *If $G$ is a connected graph of order $n \geq 6$ with minimum degree $\delta(G) \geq 2$, then $\gamma_{\rm pr}(G) \leq \frac{2}{3}n$.*

**Proof.** Let $G$ be a connected graph with $\delta(G) \geq 2$. For a subset $S \subseteq V$, let $\lambda(S)$ be the number of edges in the subgraph induced by $V \setminus S$. Among all $\gamma_{\rm pr}$-sets of $G$, let $S$ be chosen so that $\lambda(S)$ is minimized, and let $M$ be a matching of $G[S]$. Label the vertices of $S$ as $u_i, v_i$ for $1 \leq i \leq \gamma_{\rm pr}(G)/2$ such that $\{u_i, v_i\} \in M$. We now define a weak partition of the set $S$ (where some sets in the partition may be empty) given by $S = (A, B, C, D)$ as follows:

- $A = \{u_i, v_i \in S \mid {\rm epn}(u_i, S) \neq \emptyset \text{ and } {\rm epn}(v_i, S) \neq \emptyset\}$.
- $B = \{u_i, v_i \in S \mid {\rm epn}(u_i, S) \neq \emptyset \text{ and } {\rm epn}(v_i, S) = \emptyset\}$.
- $C = \{u_i, v_i \in S \mid {\rm epn}(u_i, S) = \emptyset \text{ and } {\rm epn}(v_i, S) \neq \emptyset\}$.
- $D = \{u_i, v_i \in S \mid {\rm epn}(u_i, S) = \emptyset \text{ and } {\rm epn}(v_i, S) = \emptyset\}$.

Let $A'$, $B'$, and $C'$ be the $S$-external private neighbors of the vertices in sets $A$, $B$, and $C$, respectively. By definition, the $S$-private neighbors are in $V \setminus S$, and so

$$n - |S| = |V \setminus S| \geq |A'| + |B'| + |C'| \geq |A| + \frac{1}{2}|B| + \frac{1}{2}|C|. \qquad (1)$$

If $D = \emptyset$, then $|S| = |A| + |B| + |C|$. By Inequality (1), we have $n - |S| \geq |A| + \frac{1}{2}|B| + \frac{1}{2}|C| \geq \frac{1}{2}|S|$, and so $\gamma_{\rm pr}(G) = |S| \leq \frac{2}{3}n$. We may, therefore, assume that $D \neq \emptyset$, for otherwise the desired result holds. Relabeling the vertices of $S$ if necessary, we may assume that $D = \{u_i, v_i \mid i \in [\ell]\}$ for some $\ell \geq 1$. We proceed further with the following claim. □

**Claim 1** *There exists a subset* $X = \{x_1, \ldots, x_\ell\} \subseteq V \setminus S$ *such that* $x_i$ *is adjacent to* $u_i$ *or* $v_i$ *for each* $i \in [\ell]$.

**Proof.** We consider the pair $u_i$,$v_i$, where $i \in [\ell]$. Recall that $u_i$ and $v_i$ are adjacent and $\{u_i, v_i\} \subseteq D$, and so $\text{epn}(u_i, S) = \text{epn}(v_i, S) = \emptyset$. Furthermore, for each pair $u_i$,$v_i$, there exists some vertex $x_i \in V \setminus S$ such that $x_i$ is adjacent to $u_i$ or $v_i$. Otherwise, since $\delta(G) \geq 2$, $S \setminus \{u_i, v_i\}$ is a PD-set of $G$ with cardinality less than $\gamma_{\text{pr}}(G)$, a contradiction.

We now use induction to prove Claim 1. When $\ell = 1$, the result holds. This establishes the base case. Assume that there exists a subset $X = \{x_1, \ldots, x_k\} \subseteq V \setminus S$ such that $x_i$ is adjacent to $v_i$ or $u_i$ for some $k$, where $1 \leq k < \ell$. Without loss of generality, we assume that $x_i$ is adjacent to $v_i$ for $i \in [k]$.

We now consider the pair $u_{k+1}$ and $v_{k+1}$. As observed earlier, there exists a vertex $x_{k+1} \in V \setminus S$ such that $x_{k+1}$ is adjacent to $u_{k+1}$ or $v_{k+1}$. Renaming $u_{k+1}$ and $v_{k+1}$, if necessary, we may assume that $x_{k+1}$ is adjacent to $v_{k+1}$. If $x_{k+1} \notin X$, then the results hold for $k+1$. Hence, we may assume that $x_{k+1}$ cannot be chosen to belong to $V \setminus (S \cup X)$. Thus, $N(v_{k+1}) \subseteq S \cup X$ and $N(u_{k+1}) \subseteq S \cup X$. In particular, $x_{k+1} \in X$, and so $x_{k+1} = x_j$ for some $j \in [k]$. By our earlier assumptions, $x_j$ is adjacent to $v_j$. If there is a vertex $x'_j \in V \setminus (S \cup X)$ that is adjacent to $u_j$ or $v_j$, then we can replace $x_j$ in $X$ with the vertex $x'_j$ and add the vertex $x_{k+1}$ (where recall that $x_{k+1} = x_j$) to the resulting set $X$, and the result holds for $k+1$. Hence, we may assume that $N(u_j) \subseteq S \cup X$ and $N(v_j) \subseteq S \cup X$.

Suppose that $u_{k+1}$ has a neighbor in $X$, say $x_i$. If $x_i = x_j$, then $(S \setminus \{u_j, u_{k+1}, v_{k+1}\}) \cup \{x_j\}$ (with $v_j$ and $x_j$ paired) is a PD-set of $G$ with order less than $\gamma_{\text{pr}}(G)$, contrary to the minimality of $S$. If $x_i \neq x_j$, then $(S \setminus \{u_i, u_j, u_{k+1}, v_{k+1}\}) \cup \{x_i, x_j\}$ (with $v_j$ and $x_j$ paired, $v_i$ and $x_i$ paired) is a PD-set of $G$ with order less than $\gamma_{\text{pr}}(G)$, again contradicting the minimality of $S$. Hence, $u_{k+1}$ has no neighbor in $X$, implying by our earlier observations that all neighbors of $u_{k+1}$ belong to the set $S$.

Recall that $\delta(G) \geq 2$. Let $w$ be an arbitrary neighbor of $u_{k+1}$ different from $v_{k+1}$. As observed earlier, $w \in S$. If $w \neq u_j$, then $(S \setminus \{u_j, u_{k+1}, v_{k+1}\}) \cup \{x_j\}$ (with $v_j$ and $x_j$ paired) is a PD-set of $G$ of cardinality less than $\gamma_{\text{pr}}(G)$, a contradiction. Hence, $w = u_j$. Since $w$ is an arbitrary neighbor of $u_{k+1}$ different from $v_{k+1}$, this implies that $d(u_{k+1}) = 2$ and $N(u_{k+1}) = \{u_j, v_{k+1}\}$. A similar argument holds for $u_j$; that is, $d(u_j) = 2$ and $N(u_j) = \{v_j, u_{k+1}\}$.

We show next that $d(v_j) = 2$. Suppose, to the contrary, that $d(v_j) \geq 3$. Let $y \in N(v_j) \setminus \{u_j, x_j\}$. Since $u_j$ has a neighbor in $S$ different from $v_j$, by our earlier assumptions the vertex $v_j$ has no neighbor in $S$ except for its partner $u_j$. In particular, $y \notin S$. By our earlier observations, $N(v_j) \subseteq S \cup X$, implying that $y \in X$. Thus, $y = x_i$ for some $i \in [k] \setminus \{j\}$. Recall that $x_i$ is adjacent to $v_i$. Arguments analogous to those for the pairs $\{u_{k+1}, v_{k+1}\}$ and $\{u_j, v_j\}$ show that by considering the pairs $\{u_j, v_j\}$ and $\{u_i, v_i\}$, we deduce that $u_i$ is adjacent to $u_j$, which contradicts the fact that $d(u_j) = 2$. Hence, $d(v_j) = 2$, and so $N(v_j) = \{u_j, x_j\}$. Similarly, $d(v_{k+1}) = 2$ and $N(v_{k+1}) = \{x_j, u_{k+1}\}$.

Finally, we show that $d(x_j) = 2$. If this is not the case, then there exists a vertex $y \in N(x_j) \setminus \{v_j, v_{k+1}\}$. If $y \in S$, then $S \setminus \{v_j, v_{k+1}\}$ (with $u_j$ and $u_{k+1}$ paired) is a PD-set of $G$ with cardinality less than $\gamma_{\rm pr}(G)$, contradicting the minimality of the $\gamma_{\rm pr}$-set $S$. If $y \notin S$, then $S' = S \setminus \{v_{k+1}\} \cup \{x_j\}$ (with $u_j$ and $u_{k+1}$ paired and $v_j$ and $x_j$ paired) is a $\gamma_{\rm pr}$-set of $G$ having $|E(G[V \setminus S'])| < |E(G[V \setminus S])|$, contradicting our choice of $S$. Hence, $d(x_j) = 2$. Since $G$ is connected, the graph $G$ is determined and $G \cong C_5$, contradicting the supposition that $G$ has order $n \geq 6$. This completes the proof of Claim 1. □

By Claim 1, there exists a subset $X = \{x_1, \ldots, x_\ell\} \subseteq V \setminus S$ such that $x_i$ is adjacent to $u_i$ or $v_i$ for each $i \in [\ell]$. We note that $|X| = \ell = \frac{1}{2}|D|$. Thus,

$$\begin{aligned} n - |S| &= |V \setminus S| \\ &\geq |A'| + |B'| + |C'| + |X| \\ &\geq |A| + \tfrac{1}{2}|B| + \tfrac{1}{2}|C| + \tfrac{1}{2}|D| \\ &\geq \tfrac{1}{2}(|A| + |B| + |C| + |D|) \\ &= \tfrac{1}{2}|S|, \end{aligned}$$

implying that $\gamma_{\rm pr}(G) = |S| \leq \frac{2}{3}n$. This completes the proof of Theorem 8. □

It was remarked in [42] that the cycle $C_6$ attains the bound of Theorem 8, but no families of graphs were given to illustrate sharpness. However, they showed that the bound of Theorem 8 is asymptotically sharp for an infinite family of graphs as follows. If $G_k$ is any graph obtained from $k \geq 2$ vertex disjoint copies of $K_3$ by adding a new vertex and joining it to one vertex from each copy of $K_3$ (as illustrated in Figure 1), then the graph $G_k$ has order $n = 3k + 1$ and $\gamma_{\rm pr}(G_k) = 2k = \frac{2}{3}(n-1)$. For $k$ sufficiently large, $\gamma_{\rm pr}(G_k)$ can be made arbitrarily close to $\frac{2}{3}n$.

The graphs achieving equality in Theorem 8 were subsequently characterized by Henning [44]. This characterization showed that no infinite family attains the bound of Theorem 8. In fact, the only graphs attaining the bound are the ten graphs $F_1, F_2, \ldots, F_{10}$ shown in Figure 2. Let $\mathcal{F} = \{F_1, F_2, \ldots, F_{10}\}$.

**Theorem 9 ([44])** *If $G$ is a connected graph of order $n \geq 6$ with $\delta(G) \geq 2$, then $\gamma_{\rm pr}(G) \leq \frac{2}{3}n$, with equality if and only if $G \in \mathcal{F}$.*

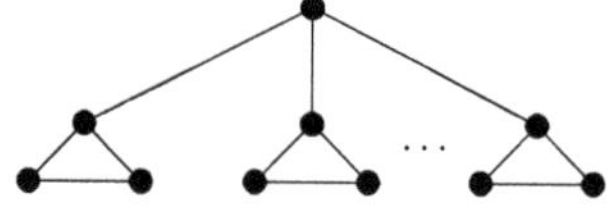

**Fig. 1** $\gamma_{\rm pr}(G)$ approaching $\frac{2}{3}n$ for large $n$

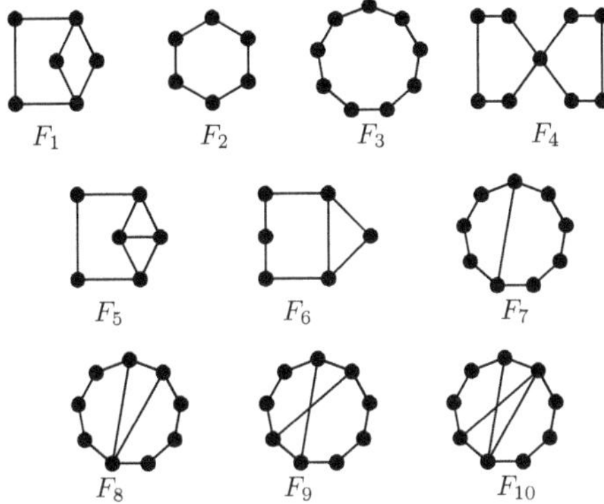


**Fig. 2** The ten graphs in $\mathcal{F}$

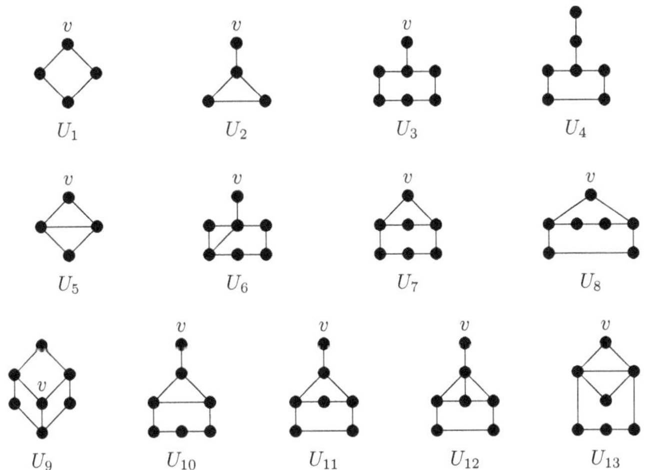


**Fig. 3** The thirteen units $U_i$, where $i \in [13]$

We remark that the maximum order of the graphs in $\mathcal{F}$ is 9. If the order is restricted to at least 10, then the upper bound in Theorem 9 can only be improved slightly as shown in [44]. To state this result, let $U_1, U_2, \ldots, U_{13}$ be the thirteen graphs shown in Figure 3. We define a *unit* to be a graph that is isomorphic to the graph $U_i$ for some $i \in [13]$. The vertex named $v$ in each unit in Figure 3 is called the *link vertex* of the unit. A unit is called a *type-i unit* for $i \in [13]$ if it isomorphic to the graph $U_i$.

Let $H$ be any graph obtained from the disjoint union of $t \geq 2$ units by identifying the $t$ link vertices, one from each unit, into one new vertex. Let $\mathcal{H}$ denote the family of all such graphs $H$. A graph $H \in \mathcal{H}$ with seven units and with identified vertex $v$ is shown in Figure 4.

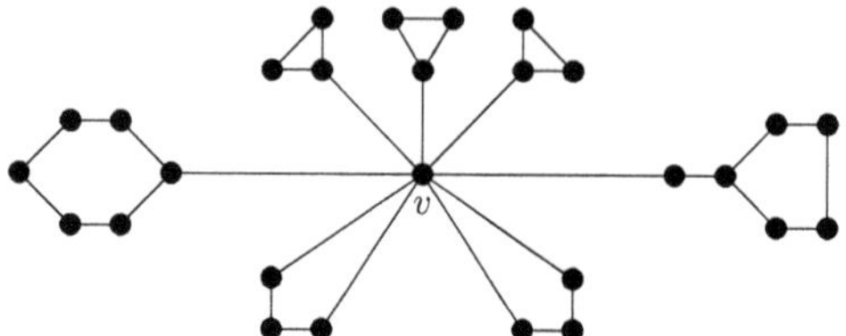


**Fig. 4** A graph $H$ in the family $\mathcal{H}$

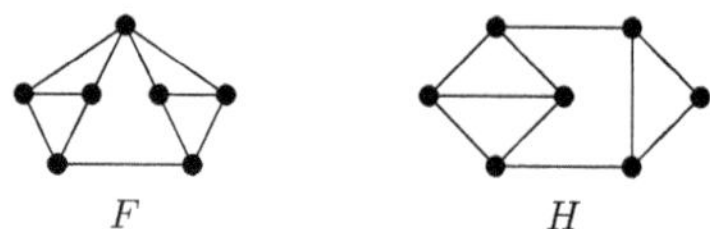


**Fig. 5** Graphs $F$ and $H$

We are now in a position to state the main result in [44].

**Theorem 10 ([44])** *If $G$ is a connected graph of order $n \geq 10$ with $\delta(G) \geq 2$, then*

$$\gamma_{\mathrm{pr}}(G) \leq \frac{2}{3}(n-1).$$

*Furthermore, for $n \geq 14$, $\gamma_{\mathrm{pr}}(G) = \frac{2}{3}(n-1)$ if and only if $G \in \mathcal{H}$.*

### 2.1.2 Minimum Degree at Least 3

Surprisingly, it remains an open problem to determine a tight upper bound on the paired domination number of a connected graph with minimum degree at least 3 in terms of the order of the graph. Chen, Sun, and Xing [14] posed the following conjecture.

**Conjecture 1 ([14])** *If $G$ is a connected graph of order $n \geq 11$ with $\delta(G) \geq 3$, then $\gamma_{\mathrm{pr}}(G) \leq \frac{4}{7}n$.*

A slightly stronger conjecture than Conjecture 1 is posed by Goddard and Henning [37]. Recall that the Petersen graph, which we shall denote by $G_{10}$, is the graph shown in Figure 7.

**Conjecture 2 ([37])** *If $G \neq G_{10}$ is a connected graph of order $n$ with $\delta(G) \geq 3$, then $\gamma_{\mathrm{pr}}(G) \leq \frac{4}{7}n$.*

If Conjecture 2 is true, then the bound is achieved, for example, by the graph $F$ shown in Figure 5.

We remark that although there is no known infinite family of graphs that achieve the upper bound of Conjecture 2, there is an infinite family of connected graphs of order $n$ with $\delta(G) \geq 3$ and with $\gamma_{\mathrm{pr}}(G)$ approaching $4n/7$ for large $n$. For example,

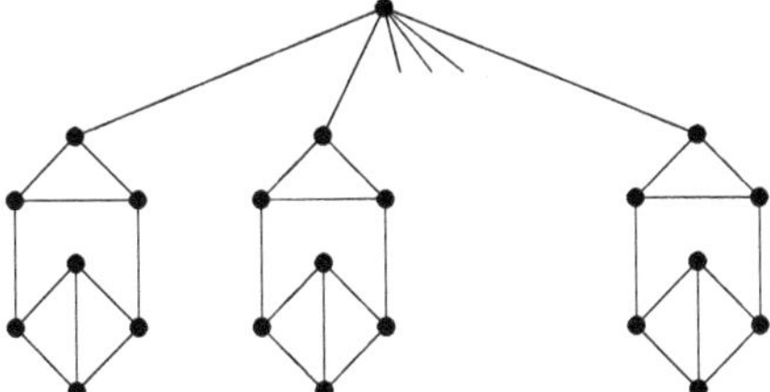

**Fig. 6** A graph $G$ in the family $\mathcal{G}$

let $\mathcal{G}$ be the family of all graphs $G$ that can be obtained from $k \geq 3$ disjoint copies of the graph $H$ shown in Figure 5 by adding a new vertex and joining it to the vertex of degree 2 in each of the $k$ copies of $H$. A graph $G$ in the family $\mathcal{G}$ is illustrated in Figure 6. Such a graph $G$ has order $n = 7k + 1$ and $\gamma_{\rm pr}(G) = 4k = 4(n-1)/7$.

Clark, Shekhtman, Suen, and Fisher [17] established the following upper bound on the paired domination number of a graph in terms of its minimum degree and order.

**Theorem 11 ([17])** *If $G$ is a connected graph of order $n$ with minimum degree $\delta$, then*

$$\gamma(G) \leq \left(1 - \prod_{k=1}^{\delta+1} \frac{k\delta}{k\delta + 1}\right) n.$$

Since

$$f(\delta) = \prod_{k=1}^{\delta+1} \frac{k\delta}{k\delta + 1}$$

is an increasing function and $f(9) \geq \frac{5}{7}$, Theorem 11 implies that Conjecture 2 holds for graphs having minimum degree $\delta \geq 9$, as noted by Lu, Wang, and Wang in [64].

**Theorem 12 ([64])** *If $G$ is a connected graph of order $n$ with $\delta(G) \geq 9$, then $\gamma_{\rm pr}(G) \leq \frac{4}{7}n$.*

Lu, Wang, and Wang [64] also showed that Conjecture 2 holds for $r$-regular graphs with $r \geq 4$.

**Theorem 13 ([64])** *If $G$ is a connected $r$-regular graph of order $n$ with $r \geq 4$, then $\gamma_{\rm pr}(G) \leq \frac{4}{7}n$.*

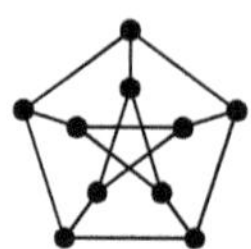

**Fig. 7** The Petersen graph $G_{10}$

### 2.1.3 Cubic Graphs

Chen, Sun, and Xing [14] established the following upper bound on the paired domination number of a cubic graph.

**Theorem 14 ([14])** *If $G$ is a cubic graph of order $n$, then $\gamma_{\rm pr}(G) \le \frac{3}{5}n$.*

Goddard and Henning [37] characterized the cubic graphs that achieve equality in the bound of Theorem 14.

**Theorem 15 ([37])** *If $G$ is a connected cubic graph of order $n$, then $\gamma_{\rm pr}(G) \le \frac{3}{5}n$ with equality if and only if $G$ is the Petersen graph $G_{10}$ (see Figure 7).*

We remark that Conjecture 2 has yet to be settled even in the special case of the class of cubic graphs.

**Conjecture 3 ([37])** *If $G \neq G_{10}$ is a connected cubic graph of order $n$, then $\gamma_{\rm pr}(G) \le \frac{4}{7}n$.*

The following conjecture and open problem were posed by Desormeaux and Henning in [22].

**Conjecture 4** *If $G$ is a bipartite cubic graph of order $n$, then $\gamma_{\rm pr}(G) \le \frac{1}{2}n$.*

**Problem 1** *Determine a tight upper bound on the paired domination number of a cubic graph that contains no 5-cycle. In particular, is it true that if $G$ is a cubic graph of order $n$ that contains no 5-cycle, then $\gamma_{\rm pr}(G) \le \frac{1}{2}n$?*

## 2.2 *Bounds in Terms of Radius and Diameter*

Lower bounds on the total domination number in a connected graph in terms of its radius or diameter are established by DeLaViña, Liu, Pepper, Waller, and West [20]. Since the paired domination number is at least the total domination number, the bounds stated below are an immediate consequence of results in [20].

**Theorem 16 ([20])** *If $G$ is a connected graph of order at least 2, then the following hold.*

(a) $\gamma_{\rm pr}(G) \ge \text{rad}(G)$.
(b) $\gamma_{\rm pr}(G) \ge \frac{1}{2}(\text{diam}(G) + 1)$.

We note that if $G$ is a path $P_n$ of order $n$, where $n$ is congruent to 0 modulo 4, then $\mathrm{rad}(G) = \gamma_{\mathrm{pr}}(G) = \frac{n}{2}$ and $\mathrm{diam}(G) = n - 1$, implying that the lower bounds in Theorem 16 are sharp.

## 2.3 Bounds in Terms of Girth

Chen, Shiu, and Chan [13] gave upper bounds on the paired domination number of a graph in terms of its order, minimum degree, maximum degree, and girth.

**Theorem 17 ([13])** *If $G$ is a connected graph of order $n$ with $\delta(G) \geq 2$ and girth $g(G) \geq 6$, then $\gamma_{\mathrm{pr}}(G) \leq \frac{1}{3}(2n - (\delta(G) - 1)(\delta(G) - 2))$.*

As observed in [13], the requirement that $g(G) \geq 6$ is necessary. For example, if $G = C_5$, then $\delta(G) = 2$, $g(G) = 5$, and $\gamma_{\mathrm{pr}}(G) = 4 > \frac{1}{3}(2n - (\delta(G) - 1)(\delta(G) - 2))$. If we impose a minimum degree at least 3 requirement, then we have the following upper bound on the paired domination number in terms of the maximum degree and order.

**Theorem 18 ([13])** *If $G$ is a connected graph of order $n$ with maximum degree $\Delta(G)$, minimum degree $\delta(G) \geq 3$, and girth $g(G) \geq 6$, then $\gamma_{\mathrm{pr}}(G) \leq \frac{2}{3}(n + 1 - \Delta(G))$.*

As before, the girth requirement of at least 6 is necessary. For example, as observed in [13], if $G$ is the Petersen graph, then $\Delta(G) = 3$, $g(G) = 5$, and $\gamma_{\mathrm{pr}}(G) = 6 > \frac{2}{3}(n + 1 - \Delta(G))$.

The following result provides an upper bound on the paired domination number of a graph with minimum degree at least 3 in terms of its order and girth.

**Theorem 19 ([13])** *If $G$ is a connected graph of order $n$ with $\delta(G) \geq 3$, then $\gamma_{\mathrm{pr}}(G) \leq \frac{2n}{3} - \frac{g(G)}{6} + \frac{5}{6}$.*

## 2.4 Bounds in Terms of Size

Henning [45] characterized the connected graphs with minimum degree at least 2 and size at least 18 that have maximum possible paired domination number. Before we state this result, we define a family of graphs.

A *type-1 unit* is a graph that is isomorphic to the graph shown in Figure 8(a) and a *type-2 unit* is a graph that is isomorphic to the graph shown in Figure 8(b). We define a *unit* to be a type-1 unit or a type-2 unit. The vertex named $v$ in each unit in Figure 8 is called the *link vertex* of the unit.

For $n_1 + n_2 \geq 2$, let $F = F(n_1, n_2)$ be the graph obtained from the disjoint union of $n_1$ units of type-1 and $n_2$ units of type-2 by identifying the $n_1 + n_2$ link vertices, one from each unit, into one new vertex which we call the *identified vertex* of $G$. Let $\mathcal{F}$ denote the family of all such graphs $F$. The graph $F(2, 2)$ with four units, two of type-1 and two of type-2, and with identified vertex $v$ is shown in Figure 9.

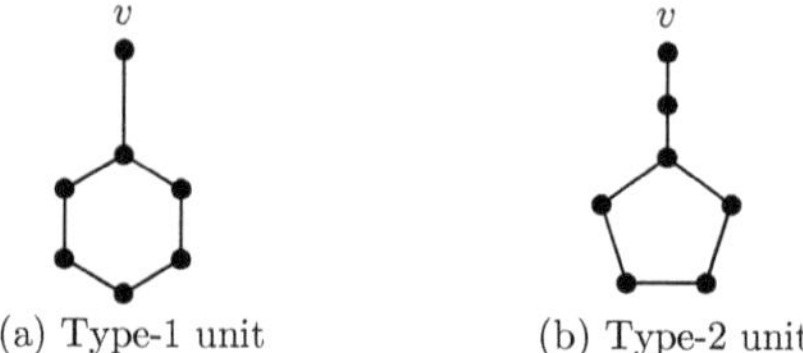


**Fig. 8** A type-1 unit and a type-2 unit

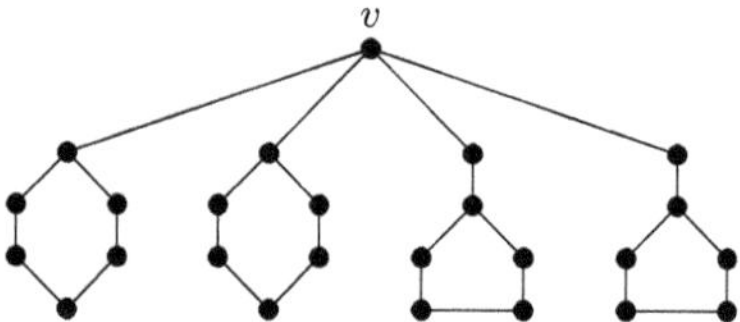


**Fig. 9** A graph $F$ in the family $\mathcal{F}$

**Theorem 20 ([45])** *Let $G$ be a connected graph of size $m \geq 18$ with $\delta(G) \geq 2$. Then, $\gamma_{\rm pr}(G) \leq \frac{4}{7}m$ with equality if and only if $G$ is the cycle $C_{21}$ or if $G \in \mathcal{F}$.*

# 3 Graphs with Maximum Size and Given Paired Domination Number

A classical result of Vizing [80] relates the size and the ordinary domination number of a graph of given order. Henning, McCoy, and Southey [47] determined a Vizing-like relation between the size and the paired domination number of a connected graph by determining the maximum possible number of edges in a graph with given order and given paired domination number. Furthermore, they characterized the infinite family of graphs that achieve this maximum possible size. Since complete graphs have the maximum number of edges of any graph having paired domination number equal to 2, it is only of interest to restrict our attention to graphs with paired domination number at least 4.

The following families of graphs were defined in [47]. For two sets $X$ and $Y$ of vertices, we say that $E[X, Y]$ is full if every vertex in $X$ is adjacent to every vertex in $Y$. Let $C_5(n_1, n_2, n_3, n_4, n_5)$ denote the graph that can be obtained from a 5-cycle $x_1x_2x_3x_4x_5$ by replacing each vertex $x_i$ for $i \in [5]$ with a nonempty clique $X_i$, where $|X_i| = n_i \geq 1$, and adding all edges between $X_i$ and $X_{i+1}$, where addition is taken modulo 5. For $n \geq 5$, let $\mathcal{F}_n = \{C_5(n_1, n_2, n_3, n_4, n_5) \mid n_1 = n_2 = n_3 = 1 \text{ and } n = n_4 + n_5 + 3\}$. A graph in the family $\mathcal{F}_n$ is illustrated in Figure 10, where in this diagram both $A$ and $B$ represent cliques, $E[A, B]$ is full, the vertex $x_1$ dominates $A \cup \{x_2\}$, the vertex $x_3$ dominates $B \cup \{x_2\}$, and $n = |A| + |B| + 3$.

For $n \geq 7$, let $\mathcal{G}_n$ be the family of graphs constructed as follows: Take a complete graph on $n-4$ vertices with vertex set $S$ and partition the set $S$ into three (nonempty)

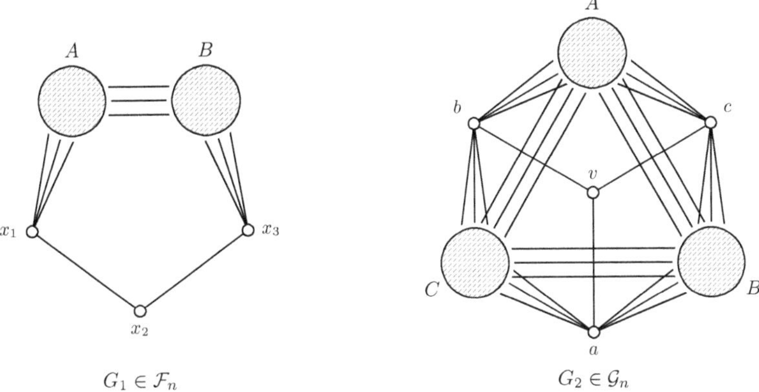


**Fig. 10** Graphs in the family $\mathcal{F}_n$ and $\mathcal{G}_n$

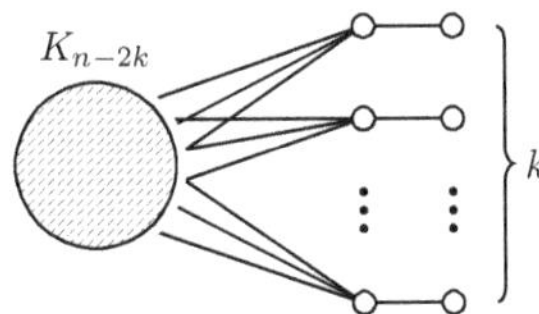


**Fig. 11** The graph $R_{n,k}$

sets $A$, $B$, and $C$. Add three new vertices $a$, $b$, and $c$, and join $a$ to every vertex in $S \setminus A$, join $b$ to every vertex in $S \setminus B$, and join $c$ to every vertex in $S \setminus C$. Finally, add a new vertex $v$ and join $v$ to $a$, $b$, and $c$. A graph in the family $\mathcal{G}_n$ is illustrated in Figure 10. In this diagram, $A$, $B$, and $C$ represent cliques and $E[A, B]$, $E[A, C]$, and $E[B, C]$ are full. The vertex $a$ dominates $B \cup C \cup \{v\}$, the vertex $b$ dominates $A \cup C \cup \{v\}$, and the vertex $c$ dominates $A \cup B \cup \{v\}$.

If $k \geq 3$ and $n = 2k$, let $R_{n,k}$ be the graph $kP_2$ consisting of $k$ disjoint copies of $P_2$. For $k \geq 3$ and $n \geq 2k+1$, let $R_{n,k}$ be the graph obtained from the disjoint union of $k$ copies of $P_2$ and a copy of the complete graph on $n - 2k$ vertices by joining one vertex from each copy of $P_2$ to every vertex in the complete graph. For $k \geq 3$, we note that $\gamma_{\text{pr}}(R_{n,k}) = 2k$ and that $R_{n,k}$ has size $\binom{n}{2} - k(n-2)$. The graph $R_{n,k}$ is illustrated in Figure 11.

We are now in a position to state the results in [47].

**Theorem 21 ([47, 51])** *Let $G$ be a graph of order $n \geq 4$ and size $m$ satisfying $\gamma_{\text{pr}}(G) = 4$. Then $m \leq \binom{n}{2} - 2(n-2) + 1$, with equality if and only if $G \in \mathcal{F}_n \cup \mathcal{G}_n$ or if $G$ is the complement of the Petersen graph.*

**Theorem 22 ([47])** *For $k \geq 3$, let $G$ be a graph of order $n \geq 2k$ and size $m$ satisfying $\gamma_{\text{pr}}(G) = 2k$. Then $m \leq \binom{n}{2} - k(n-2)$, with equality if and only if $G = R_{n,k}$ or $G = C_5 \cup P_2$.*

## 4 Nordhaus–Gaddum Type Results

The following Nordhaus–Gaddum result was determined in the introductory paper on paired domination.

**Theorem 23 ([42])** *If $G$ is a graph of order $n$ and neither $G$ nor $\overline{G}$ has isolated vertices, then*

$$\gamma_{\rm pr}(G) + \gamma_{\rm pr}(\overline{G}) \le \begin{cases} n+3 & \text{if } n \text{ is odd} \\ n+2 & \text{otherwise.} \end{cases}$$

As observed in [42], the cycle $C_5$ and the graph consisting of $m \ge 2$ disjoint copies of $K_2$ achieve the bounds of Theorem 23. Further, the following is noted in [42].

**Observation 24 ([42])** *For any graph $G$ with no isolated vertices, the following hold.*

(a) *If* $\operatorname{diam}(G) \ge 3$, *then* $\gamma_{\rm pr}(\overline{G}) = 2$.
(b) $\gamma_{\rm pr}(G) \ge 4$ *if and only if* $\operatorname{diam}(\overline{G}) = 2$.

Favaron, Karami, and Sheikholeslami [34] showed that if both $\gamma_{\rm pr}(G) > 4$ and $\gamma_{\rm pr}(\overline{G}) > 4$, then the bound of Theorem 23 can be improved.

**Theorem 25 ([34])** *For any graph $G$ of order $n \ge 6$ with $\gamma_{\rm pr}(G) > 4$ and $\gamma_{\rm pr}(\overline{G}) > 4$,*

$$\gamma_{\rm pr}(G) + \gamma_{\rm pr}(\overline{G}) \le 3 + \min\{\delta(G), \delta(\overline{G})\}.$$

**Proof.** Let $G$ be a graph of order $n \ge 6$ with $\gamma_{\rm pr}(G) > 4$ and $\gamma_{\rm pr}(\overline{G}) > 4$. Since the paired domination number is always even, $\gamma_{\rm pr}(G) \ge 6$ and $\gamma_{\rm pr}(\overline{G}) \ge 6$. Let $x$ be a vertex of degree $\delta(G)$, and let $X = V \setminus N[x]$. If $N(x) \cap N(y) = \emptyset$ for some $y \in X$, then in $\overline{G}$, the vertex $x$ dominates $X$ and the vertex $y$ dominates $N[x]$. Thus, $\{x, y\}$ is a PD-set of $\overline{G}$, and so $\gamma_{\rm pr}(\overline{G}) = 2$, a contradiction. Therefore, $N(x) \cap N(y) \ne \emptyset$ for all $y \in X$. Hence, $N(x)$ dominates $X$. Since $\gamma_{\rm pr}(G) \ge 6$, no vertex of $N(x)$ dominates $X$, for otherwise, $\gamma_{\rm pr}(G) = 2$. As an immediate consequence of $\gamma_{\rm pr}(G) \ge 6$, we have the following claim.

**Claim 2** *If $A$ is a subset of $N(x)$ that dominates the set $X$ and $B$ is a maximum subset of $A$ not dominating $X$, then $|A| \ge 3$ and $|B| \ge 2$.*

Let $S$ be a maximum subset of $N(x)$ that does not dominate all vertices in $X$ and let $T = N(x) \setminus S$. By the maximality of $S$, every vertex in the set $T$ dominates $X \setminus N(S)$, and by Claim 2, we have $|S| \ge 2$.

**Claim 3** *The set $T$ dominates $X$.*

**Proof.** Suppose, to the contrary, that $y \in X$ and $y$ is not dominated by $T$. By the definition of $S$, there exists a vertex $z \in X$ not dominated by $S$. Since $y \in N(S)$ and $z \in X \setminus N(S)$, we note that $y \neq z$. Now $\{x, z, y, w\}$ is a PD-set of $\overline{G}$, where $w$ is an arbitrary vertex in $T$ (where $x$ is paired with $z$ and $y$ is paired with $w$), and so $\gamma_{\rm pr}(\overline{G}) \leq 4$, a contradiction. □

**Claim 4** *The set $T$ has no isolated vertex.*

**Proof.** Suppose, to the contrary, that $y$ is an isolated vertex in $T$. As observed earlier, every vertex in $T$ dominates $X \setminus N(S)$. In particular, the vertex $y$ dominates $X \setminus N(S)$. By Claim 2, the vertex $y$ does not dominate $X$. Thus, there is a vertex $z \in N(S)$ that is not adjacent to $y$. Let $w$ be a vertex in $X \setminus N(S)$, and so $wy \in E(G)$ and $w \neq z$. Now $\{x, w, y, z\}$ is a PD-set of $\overline{G}$ (where $x$ is paired with $w$ and $y$ is paired with $z$), and so $\gamma_{\rm pr}(\overline{G}) \leq 4$, a contradiction. □

Let $M = \{x_1y_1, \ldots, x_sy_s\}$ be a maximum matching of $G[T]$, and let $U = T \setminus V(M)$. By Claim 4, we note that $s \geq 1$. By the maximality of $M$, the set $U$ is an independent set of $G$.

**Claim 5** *If $V(M)$ does not dominate $X$ and $u \in U$, then $V(M) \cup \{u\}$ dominates $X$.*

**Proof.** Suppose that $V(M)$ does not dominate $X$. By Claim 3, this implies that $U \neq \emptyset$. Let $u \in U$ and suppose, to the contrary, that $V(M) \cup \{u\}$ does not dominate $X$. Let $w_1 \in X$ be a vertex not dominated by $V(M) \cup \{u\}$, The vertex $w_1$ dominates $V(M)$ in $\overline{G}$ and the vertex $u$ dominates $U$ in $\overline{G}$ since $U$ is independent in $G$. If $w_1 \in X \setminus N(S)$, then let $w_2$ be an arbitrary vertex in $X$ different from $w_1$, while if $w_1 \in N(S)$, then let $w_2$ be a vertex in $X \setminus N(S)$. It follows that $\{x, w_2, u, w_1\}$ is a PD-set of $\overline{G}$ (where $x$ is paired with $w_2$ and $u$ is paired with $w_1$), and so $\gamma_{\rm pr}(\overline{G}) \leq 4$, a contradiction. □

Let $M'$ be a maximal submatching of $M$ such that $V(M')$ does not dominate $X$, and let $W$ be a maximum subset of $T$ containing $V(M')$ and not dominating $X$. By Claims 2 and 3 and the definition of $S$, we note that $M' \neq \emptyset$, $W \neq T$, and $|S| \geq |W|$. However, it is possible that $M' = M$, in which case $W = V(M)$ by Claim 5, or that $V(M') \subseteq W \neq V(M)$.

**Claim 6** *If $V(M') \cup \{y\}$ dominates $X$ for some $y \in N(x) \setminus V(M')$, then $\gamma_{\rm pr}(G) \leq |V(M')| + 2$ and $|V(M')| \geq 4$.*

**Proof.** If $V(M') \cup \{y\}$ dominates $X$, then $V(M') \cup \{x, y\}$ is a PD-set of $G$ (with $x$ and $y$ paired), implying that $\gamma_{\rm pr}(G) \leq |V(M')| + 2$. Since $\gamma_{\rm pr}(G) \geq 6$, this in turn implies that $|V(M')| \geq 4$. □

**Claim 7** $\gamma_{\rm pr}(G) \leq |V(M')| + 4$.

**Proof.** Suppose that $M' = M$. By Claim 5 if $y \in U$, then $V(M') \cup \{y\}$ dominates $X$, implying by Claim 6 that $\gamma_{\rm pr}(G) \leq |V(M')| + 2$. If $M' \neq M$, then let $x_sy_s \in M \setminus M'$. By the maximality of $M'$, the set $V(M') \cup \{x_s, y_s\}$ dominates $X$ and $V(M') \cup \{x_s, y_s, x, x'\}$, where $x'$ is an arbitrary vertex of $S$, is a PD-set of $G$ (where the vertices of $M'$ are paired as in the matching while $x$ and $x'$ are paired), implying that $\gamma_{\rm pr}(G) \le |V(M')| + 4$. □

Let $T_0 = T$ and let $S_0 = W$. Let $T_1 = T \setminus W$. If $T_1$ dominates $X$, let $S_1$ be a maximal subset of $T_1$ not dominating $X$ and let $T_2 = T_1 \setminus S_1$. If $T_2$ dominates $X$, let $S_2$ be a maximal subset of $T_2$ not dominating $X$ and let $T_3 = T_2 \setminus S_2$. We continue the process until a subset $T_k$ of $T$ not dominating $X$ is obtained. Thus, we construct a finite chain $T = T_0 \supset T_1 \supset \cdots \supset T_k$, where $k \ge 1$ such that

(a) the set $T_i$ dominates $X$ for $i \in \{0\} \cup [k-1]$.
(b) the set $S_i = T_i \setminus T_{i+1}$ does not dominate $X$, but $S_i \cup \{y\}$ dominates $X$ for each $y \in T_{i+1}$ for $i \in \{0\} \cup [k-1]$.
(c) the set $T_k$ does not dominate $X$.

We note that if $k \ge 2$, then $|S_i| \ge 2$ for $i \in [k-1]$ by Claim 2. For all $i \in \{0\} \cup [k-1]$, let $x_i$ be a vertex of $X$ not dominated by $S_i$, and let $y_i \in S_i$. Let $x_k$ be a vertex of $X$ not dominated by $T_k$ and $y_k \in T_k$. By (b), the vertex $x_i$ is adjacent to every vertex in $T_{i+1}$. Hence, all the vertices $x_i$ are distinct. Similarly, all the vertices $y_i$ are distinct since $S_0, S_1, \ldots, S_{k-1}, T_k$ are disjoint. We now let $y \in X \setminus N(S)$. We note that in the graph $\overline{G}$, the vertex $y$ dominates $S$ and the vertex $x$ dominates $X$ in $\overline{G}$. Further, we note that $x_i y_i$ is an edge in $\overline{G}$, and that the vertex $x_i$ dominates the set $S_i$ in $\overline{G}$ for $i \in \{0\} \cup [k]$. Thus, the set $\{x_0, y_0, x_1, y_1, \ldots, x_k, y_k, x, y\}$ is a PD-set of $\overline{G}$ (with $x$ and $y$ paired and with $x_i$ and $y_i$ paired for $i \in \{0\} \cup [k]$). Hence,

$$\gamma_{\rm pr}(\overline{G}) \le 2k + 4. \tag{2}$$

By the choice of $x$ and our earlier observations, we have

$$\begin{aligned} \delta(G) &= |N(x)| \\ &= |S| + |T| \\ &= |S| + \sum_{i=0}^{k-1} |S_i| + |T_k| \\ &\ge |S| + |W| + 2(k-1) + 1 \\ &\ge 2|W| + 2k - 1 \end{aligned}$$

or, equivalently, $2k \le \delta(G) - 2|W| + 1$. Thus, by Inequality (2), we have

$$\gamma_{\rm pr}(\overline{G}) \le \delta(G) - 2|W| + 5. \tag{3}$$

If $|W| \ge |V(M')| + 2$, then by Inequality (3), we have $\gamma_{\rm pr}(\overline{G}) \le \delta(G) - 2|V(M')| + 1$, implying by Claim 7 that $\gamma_{\rm pr}(G) + \gamma_{\rm pr}(\overline{G}) \le (|V(M')| + 4) + (\delta(G) - 2|V(M')| + 1) = \delta(G) - |V(M')| + 5 \le \delta(G) + 3$.

If $|W| = |V(M')| + 1$ and $V(M') \cup \{y\}$ dominates $X$ for some $y \in T \setminus V(M')$, then by Claim 6, $\gamma_{\rm pr}(G) \le |V(M')| + 2$ and $|V(M')| \ge 4$. Thus, in this case, $\gamma_{\rm pr}(G) \le |W| + 1$ and $|W| \ge 5$. Hence by Inequality (3), $\gamma_{\rm pr}(G) + \gamma_{\rm pr}(\overline{G}) \le (|W| + 1) + (\delta(G) - 2|W| + 5) = \delta(G) - |W| + 6 \le \delta(G) + 1$.

If $W = V(M')$, then by Claim 6, $\gamma_{pr}(G) \leq |V(M')| + 2$ and $|V(M')| \geq 4$. Thus, in this case, $\gamma_{pr}(G) \leq |W| + 2$ and $|W| \geq 4$. Hence, by Inequality (3), $\gamma_{pr}(G) + \gamma_{pr}(\overline{G}) \leq (|W|+2) + (\delta(G) - 2|W| + 5) = \delta(G) - |W| + 7 \leq \delta(G) + 3$.

Hence, we may assume that $|W| = |V(M')| + 1$ and $V(M') \cup \{y\}$ does not dominate $X$ for any $y \in T \setminus V(M')$, for otherwise the desired result follows. In this case by Claim 5, we note that $M' \neq M$. Let $w_1 w_2 \in M \setminus M'$. By the definition of $M'$, the set $V(M') \cup \{w_1, w_2\}$ dominates $X$. In this case, we may assume that $W$ is chosen to contain the vertex $w_1$; that is, $W = V(M') \cup \{w_1\}$.

If $V(M') \cup \{w_1, w_2\}$ dominates $N(x)$, then $\gamma_{pr}(G) \leq |V(M')| + 2 = |W| + 1$. Since $\gamma_{pr}(G) \geq 6$, this implies that $|W| \geq 5$. By Inequality (3), $\gamma_{pr}(G) + \gamma_{pr}(\overline{G}) \leq (|W|+1)+(\delta(G)-2|W|+5) = \delta(G)-|W|+6 \leq \delta(G)+1$. Hence, we may assume that there exists a vertex $z \in N(x)$ that is not dominated by $V(M') \cup \{w_1, w_2\} \subseteq V(M)$. We note that

$$z \in S \cup (\bigcup_{i=1}^{k-1} S_i) \cup T_k.$$

We now consider the vertices $y$, $x_i$, and $y_i$ as defined above (in the paragraph preceding Inequality (2)) with the supplementary property that if $z \in S_i$ (respectively, $z \in T_k$), then $y_i = z$ (respectively, $y_k = z$). If $z \notin S$, then $\{x_1, y_1, \ldots, x_k, y_k, x, y\}$ is a PD-set of $\overline{G}$ (with $x$ and $y$ paired, and with $x_i$ and $y_i$ paired for $i \in [k]$), and if $z \in S$, then $\{x_2, y_2, \ldots, x_k, y_k, y, z, x_1, x\}$ is a PD-set of $\overline{G}$ (with matching $\{x_i y_i, yz, x_1 x \mid 2 \leq i \leq k\}$). In both cases, Inequality (2) can be improved to

$$\gamma_{pr}(\overline{G}) \leq 2k + 2 \tag{4}$$

and Inequality (3) can be improved to

$$\gamma_{pr}(G) \leq \delta(G) - 2|W| + 3 = \delta(G) - 2|V(M')| + 1. \tag{5}$$

By Claim 7 and Inequality (5), we have

$$\begin{aligned} \gamma_{pr}(G) + \gamma_{pr}(\overline{G}) &\leq (|V(M')| + 4) + (\delta(G) - 2|V(M')| + 1) \\ &= \delta(G) - |V(M')| + 5 \\ &\leq \delta(G) + 3. \end{aligned}$$

By the symmetry between $G$ and $\overline{G}$, analogous arguments show that $\gamma_{pr}(G) + \gamma_{pr}(\overline{G}) \leq \delta(\overline{G}) + 3$. Thus, $\gamma_{pr}(G) + \gamma_{pr}(\overline{G}) \leq \min\{\delta(G), \delta(\overline{G})\} + 3$. This completes the proof of Theorem 25. □

As a consequence of Theorem 25, we have the following result.

**Corollary 26** *If $G$ is a graph of order $n \geq 6$ with $\gamma_{pr}(G) > 4$ and $\gamma_{pr}(\overline{G}) > 4$, then*

$$\gamma_{pr}(G) + \gamma_{pr}(\overline{G}) \leq \frac{1}{2}(n+5) \leq \frac{2}{3}(n-1).$$

**Proof.** Since the paired domination is an even integer, we note that $\gamma_{pr}(G) \geq 6$ and $\gamma_{pr}(\overline{G}) \geq 6$. Since $\delta(\overline{G}) \leq \Delta(\overline{G}) = n - 1 - \delta(G)$, we note that

$$\min\{\delta(G), \delta(\overline{G})\} \leq \frac{1}{2}(\delta(G) + \delta(\overline{G})) \leq \frac{1}{2}(n-1).$$

Thus, by Theorem 25, we have $12 \leq \gamma_{pr}(G) + \gamma_{pr}(\overline{G}) \leq \frac{1}{2}(n-1) + 3 = \frac{1}{2}(n+5)$. Moreover, $\frac{1}{2}(n+5) \geq 12$ implies that $n \geq 19$, and therefore that $\frac{1}{2}(n+5) \leq \frac{2}{3}(n-1)$. $\square$

For graphs of sufficiently large order, the bounds of Theorem 23 can be improved.

**Theorem 27 ([34])** *If $G$ is a graph of order $n \geq 6$ such that $\delta(G) \geq 2$ and $\delta(\overline{G}) \geq 2$, then*

$$\gamma_{pr}(G) + \gamma_{pr}(\overline{G}) \leq \frac{2}{3}n + 4,$$

*and the graphs $F_1$ and $F_5$ shown in Figure 2 are the only extremal graphs. Moreover, if $n \geq 14$, then*

$$\gamma_{pr}(G) + \gamma_{pr}(\overline{G}) \leq \begin{cases} \frac{2n+8}{3} & \text{if } n \in \{14, 17, 20\} \\ \frac{2}{3}n + 2 & \text{otherwise.} \end{cases}$$

*For $n \geq 25$, equality $\gamma_{pr}(G)+\gamma_{pr}(\overline{G}) = \frac{2}{3}n+2$ occurs if and only if each component of $G$ or $\overline{G}$ belongs to $\mathcal{F}\backslash\{F_1, F_5\}$, where $\mathcal{F}$ is the family of graphs shown in Figure 2.*

## 5 Paired Domination in Cartesian Products

The *Cartesian product* $G \,\square\, H$ of graphs $G$ and $H$ is the graph whose vertex set is $V(G) \times V(H)$, and where two vertices $(g_1, h_1)$ and $(g_2, h_2)$ are adjacent in $G \,\square\, H$ if either $g_1 = g_2$ and $h_1h_2$ is an edge in $H$, or $h_1 = h_2$ and $g_1g_2$ is an edge in $G$. The most famous open problem involving domination in graphs is the conjecture of Vizing [81] posed in 1968, which states the domination number of the Cartesian product of any two graphs is at least as large as the product of their domination numbers.

**Vizing's Conjecture** *For any graphs $G$ and $H$, $\gamma(G)\gamma(H) \leq \gamma(G \,\square\, H)$.*

Vizing's conjecture has yet to be settled, though it has been shown to be true for certain classes of graphs. A survey of what is known about Vizing's conjecture can be found in [4]. The best general upper bound to date on the product of the domination numbers of two graphs in terms of their Cartesian product is due to Clark and Suen [18].

**Theorem 28 ([18])** *For any graphs G and H,* $\gamma(G)\gamma(H) \leq 2\gamma(G \square H)$.

Here we discuss the analogous problem for the paired domination number. As an immediate consequence of Theorems 2 and 28, we have the following result.

**Corollary 29** *For any graphs G and H without isolated vertices,* $\gamma_{\rm pr}(G)\gamma_{\rm pr}(H) \leq 8\gamma_{\rm pr}(G \square H)$.

Hou and Jiang [55] improved the bound of Corollary 29 as follows.

**Theorem 30 ([55])** *For any graphs G and H without isolated vertices,* $\gamma_{\rm pr}(G)\gamma_{\rm pr}(H) \leq 7\gamma_{\rm pr}(G \square H)$.

Choudhary, Margulies, and Hicks [16] considered the product of more than two graphs, and their result for this generalized product improved the bound of Theorem 30 for the Cartesian product of two graphs.

**Theorem 31 ([16])** *For any graphs G and H without isolated vertices,* $\gamma_{\rm pr}(G)\gamma_{\rm pr}(H) \leq 6\gamma_{\rm pr}(G \square H)$.

Brešar, Henning, and Rall [6] improved the upper bound of Theorem 30 when one of the graphs is a tree. To present their upper bound, we first give a definition and a couple of relevant results.

A 3*-packing* in $G$ is a set of vertices of $G$ that are pairwise at a distance of greater than 3 apart in $G$. The maximum cardinality of a 3-packing in $G$ is called the 3*-packing number* $\rho_3(G)$ of $G$. The following relationship between the paired domination number and the 3-packing number was first observed in [6].

**Observation 32 ([6])** *If G is a graph without isolated vertices, then* $\gamma_{\rm pr}(G) \geq 2\rho_3(G)$.

Brešar et al. [6] showed that the class of trees achieves equality in the lower bound of Observation 32.

**Theorem 33 ([6])** *For every nontrivial tree T,* $\gamma_{\rm pr}(T) = 2\rho_3(T)$.

**Theorem 34 ([6])** *For any graphs G and H without isolated vertices,*

$$\gamma_{\rm pr}(G \square H) \geq \max\{\gamma_{\rm pr}(G)\rho_3(H), \gamma_{\rm pr}(H)\rho_3(G)\}.$$

As a consequence of Theorem 34, we have the following result.

**Theorem 35 ([6])** *For any graphs G and H without isolated vertices, if* $2\rho_3(G) = \gamma_{\rm pr}(G)$ *or if* $2\rho_3(H) = \gamma_{\rm pr}(H)$, *then* $\gamma_{\rm pr}(G)\gamma_{\rm pr}(H) \leq 2\gamma_{\rm pr}(G \square H)$ *and this bound is sharp.*

As an immediate consequence of Theorems 33 and 35, we have the following result.

**Corollary 36 ([6])** *If $T$ is a nontrivial tree and $H$ is any graph without isolated vertices, then $\gamma_{\rm pr}(T)\gamma_{\rm pr}(H) \le 2\gamma_{\rm pr}(T \,\Box\, H)$ and this bound is sharp.*

As shown in [6], sharpness in Corollary 36 is achieved when $T$ is a tree satisfying $\gamma_{\rm pr}(T) = 2\gamma(T)$ and $H$ consists of the union of disjoint copies of $K_2$.

As remarked in [6], it is *not true* in general that for isolate-free graphs $G$ and $H$, $\gamma_{\rm pr}(G \,\Box\, H) \ge 2\rho(G)\rho(H)$. For example, if $G = H = P_4$, then $\gamma_{\rm pr}(G) = \gamma_{\rm pr}(H) = 2$ and $\gamma_{\rm pr}(G \,\Box\, H) = 6$, and so $\gamma_{\rm pr}(G \,\Box\, H) < 2\rho(G)\rho(H)$. However, as a consequence of Observation 32 and Theorem 34, we have the following lower bound on $\gamma_{\rm pr}(G \,\Box\, H)$ in terms of the 3-packing number of both graphs.

**Theorem 37 ([6])** *For any graphs $G$ and $H$ without isolated vertices,*

$$\gamma_{\rm pr}(G \,\Box\, H) \ge 2\rho_3(G)\rho_3(H).$$

# 6 Special Classes of Graphs

In this section, we investigate the paired domination number for various classes of graphs, including trees, claw-free graphs, and net-free graphs. Among the standard classes of graphs, we have $\gamma_{\rm pr}(K_n) = \gamma_{\rm pr}(K_{r,s}) = \gamma_{\rm pr}(W_n) = 2$ and $\gamma_{\rm pr}(P_n) = \gamma_{\rm pr}(C_n) = 2\lceil \frac{n}{4} \rceil$. For the corona $G \circ K_1$, $\gamma_{\rm pr}(G \circ K_1) = |V(G)|$ if $G$ has even order, while $\gamma_{\rm pr}(G \circ K_1) = |V(G)| + 1$ if $G$ has odd order.

## *6.1 Trees*

In this section, we investigate paired domination in trees.

### 6.1.1 Bounds

Chellali and Haynes [8] presented the following bounds on the paired domination number of a tree.

**Theorem 38 ([8])** *If $T$ is a tree of order $n \ge 3$ with $s$ support vertices, then the following hold.*

(a) $\gamma_t(T) \le \gamma_{\rm pr}(T) \le \gamma_t(T) + s - 1$.
(b) $\gamma_{\rm pr}(T) \le \frac{1}{2}(n + 2s - 1)$, *and this bound is sharp.*

Sharpness of the bound of Theorem 38(b) is achieved, for example, with those trees obtained from a star $K_{1,k}$ by subdividing each edge exactly five times. In this case, $n = 6k + 1$, $s = k$, and $\gamma_{pr}(T) = 4k = (n + 2s - 1)/2$. As remarked in [8], in some cases the upper bound in Theorem 38(b) is better than the upper bound in Theorem 2. For example, let $H_k$ be the graph obtained from $P_2 \cup kP_{14}$ by adding a new vertex $v$ and adding $k + 1$ edges joining $v$ to a leaf from each of the paths. In this case, $\gamma(H_k) = 5k + 1$, $s = k + 1$, and $\gamma_{pr}(H_k) = 8k + 2 = (14k + 3 + 2k + 2 - 1)/2 = (n + 2s - 1)/2$. For sufficiently large $k$, the difference between $2\gamma(T)$ and $(n + 2s - 1)/2$ can, therefore, be made arbitrarily large.

Raczek [73] obtained a lower bound on the paired domination number of a tree in terms of its order and number of leaves and characterized trees for which this bound is obtained. If $T_1$ and $T_2$ are vertex disjoint trees, define $T_1 \oplus T_2$ to be the operation of adding an edge joining a leaf of $T_1$ to a leaf of $T_2$. Let $\mathcal{R}$ denote the family of trees that contains every double star, and if $T_1 \in \mathcal{R}$ and $T_2 \in \mathcal{R}$, then $T_1 \oplus T_2 \in \mathcal{R}$.

**Theorem 39 ([73])** *If $T$ is a tree of order $n \geq 2$ with $n_1$ leaves, then $\gamma_{pr}(T) \geq \frac{1}{2}(n + 2 - n_1)$ with equality if and only if $T \in \mathcal{R}$.*

The *annihilation number* $a(G)$ is the largest integer $k$ such that the sum of the first $k$ terms of the non-decreasing degree sequence of $G$ is at most the number of edges in $G$. Dehgardi, Sheikholeslami, and Khodkar [21] gave an upper bound on the paired domination number of trees in terms of their annihilation numbers.

**Theorem 40 ([21])** *If $T$ is a tree of order $n \geq 2$, then $\gamma_{pr}(T) \leq \frac{1}{3}(4a(T) + 2)$ with equality if and only if $T$ is the path $P_2$ or the subdivided star $S(K_{1,k})$, where $k \geq 3$ is odd.*

### 6.1.2 Equal Domination and Paired Domination Numbers

Recall that by Observation 1 and Theorem 2, we have that if $G$ is a graph with no isolated vertices, then $\gamma(G) \leq \gamma_{pr}(G) \leq 2\gamma(G)$. Qiao, Kang, Cardel, and Du [72] gave a characterization of trees $T$ for which $\gamma(T) = \gamma_{pr}(T)$.

Subsequently, Haynes, Henning, and Slater [40] gave a simpler characterization using labelings. In order to state the characterization, a *$\rho$-$\gamma_{pr}$-labeling* of a tree $T = (V, E)$ is defined in [40] as a weak partition $S = \{S_A, S_B, S_C, S_D\}$ of $V$ such that (i) $S_A \cup S_D$ is a $\gamma_{pr}(T)$-set, (ii) $S_C \cup S_D$ is a $\rho(T)$-set, and (iii) $|S_A| = |S_C|$. We refer to the pair $(T, S)$ as a *$\rho$-$\gamma_{pr}$-tree*. The *label* or *status* of a vertex $v$ is the letter $x \in \{A, B, C, D\}$ such that $v \in S_x$. By a *labeled* $P_4$, we mean a $P_4$ with the two leaves of status $C$ and the two support vertices of status $A$.

Let $\mathcal{F}_{\text{tree}}$ be the family of labeled trees that (i) contains a labeled $P_4$ and (ii) is closed under the four operations $\mathcal{F}_j$, $1 \leq j \leq 4$, listed below, which extend the tree $T$ by attaching a tree to the vertex $v \in V(T)$.

- **Operation $\mathcal{F}_1$.** Attach a vertex of status $B$ to vertex $v$, where $v$ has status $A$.

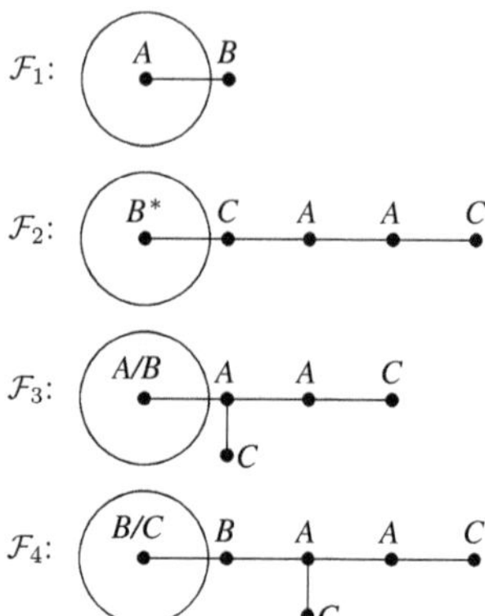


**Fig. 12** The four operations $\mathcal{F}_1$, $\mathcal{F}_2$, $\mathcal{F}_3$, and $\mathcal{F}_4$

- **Operation** $\mathcal{F}_2$. Add a labeled $P_4$ and join a leaf of the $P_4$ to vertex $v$, where $v$ has status $B$ and no neighbor of status $C$.
- **Operation** $\mathcal{F}_3$. Add a labeled $P_4$ and join a support vertex of the $P_4$ to vertex $v$, where $v$ has status $A$ or $B$.
- **Operation** $\mathcal{F}_4$. Add a labeled $P_4$ and a vertex, say $y$, of status $B$ and join $y$ to a support vertex of the $P_4$ and to vertex $v$, where $v$ has status $B$ or $C$.

The four operations $\mathcal{F}_1$, $\mathcal{F}_2$, $\mathcal{F}_3$, and $\mathcal{F}_4$ are illustrated in Figure 12, where by status $B^*$ we mean a status $B$ vertex that has no neighbor of status $C$.

We are now in a position to present the characterization using labelings in [40] of trees $T$ for which $\gamma(T) = \gamma_{pr}(T)$.

**Lemma 41 ([40])** *For a tree $T$, $\gamma_{pr}(T) = \gamma_t(T) = \gamma(T)$ if and only if $T$ has a $\rho$-$\gamma_{pr}$-labeling.*

**Theorem 42 ([40])** *A labeled tree $(T, S)$ is a $\rho$-$\gamma_{pr}$-tree if and only $(T, S) \in \mathcal{F}_{\mathrm{tree}}$.*

As a consequence of Lemma 41 and Theorem 42, we have the following result.

**Corollary 43 ([40])** *For any tree $T$, $\gamma_{pr}(T) = \gamma(T)$ if and only if $(T, S) \in \mathcal{F}_{\mathrm{tree}}$ for some labeling $S$ of $V(T)$.*

### 6.1.3 Paired Domination Number Twice the Domination Number

As observed earlier, if $G$ is a graph with no isolated vertices, then $\gamma_{pr}(G) \le 2\gamma(G)$. Haynes and Henning [39] characterized trees that achieve equality in this bound. For this purpose, they define an *almost paired dominating set* of $G$ relative to a vertex $v$ as a set $S \subseteq V(G)$ such that either (i) $v \notin S$ and $S$ is a PD-set of $G - v$ or (ii) $v \in S$, $S$ dominates $V(G)$, and $G[S]$ or $G[S \setminus \{v\}]$ contains a perfect matching. The almost paired domination number of $G$ relative to $v$, denoted $\gamma_{pr}(G; v)$, is the minimum cardinality of an almost paired dominating set of $G$ relative to $v$.

Let $\mathcal{T}_{\rm tree}$ be the family of trees obtained from $P_2$ by a finite sequence of operations $\mathcal{T}_1$, $\mathcal{T}_2$, and $\mathcal{T}_3$ listed below.

- **Operation** $\mathcal{T}_1$. Attach a vertex to a vertex $v$ of $T$, where $v$ is in every $\gamma_{\rm pr}$-set of $T$.
- **Operation** $\mathcal{T}_2$. Attach a leaf of a path $P_2$ to a vertex $v$ of $T$, where $\gamma_{\rm pr}(T; v) = \gamma_{\rm pr}(T)$.
- **Operation** $\mathcal{T}_3$. Attach a leaf of a path $P_3$ to a vertex $v$ of $T$, where $\gamma_{\rm pr}(T - v) = \gamma_{\rm pr}(T)$.

We now give the characterization of trees $T$ for which $\gamma_{\rm pr}(T) = 2\gamma(T)$.

**Theorem 44 ([39])** *For every nontrivial tree* $T$, $\gamma_{\rm pr}(T) = 2\gamma(T)$ *if and only if* $T \in \mathcal{T}_{\rm tree}$.

It is of interest to note that trees with paired domination number equal to twice their domination number were also independently characterized by Hou in [53] and by Henning and Vestergaard using a different method in [52].

### 6.1.4 Paired Domination Number Equal to Other Parameters

Recall that by Observation 1, for every isolate-free graph $G$, $\gamma_t(G) \le \gamma_{\rm pr}(G)$. Henning [43] gave a constructive characterization of the trees $T$ for which $\gamma_t(T) = \gamma_{\rm pr}(T)$. Using a different approach, Shan, Kang, and Henning [76] also gave a constructive characterization of these trees.

### 6.1.5 Vertices in Every or in No $\gamma_{\rm pr}$-Set

Note that the support vertices of every tree $T$ must be contained in every $\gamma_{\rm pr}$-set of $T$. For many trees, however, there exist vertices that are contained in no $\gamma_{\rm pr}$-set or vertices that belong to every $\gamma_{\rm pr}$-set. For instance in a double star the two non-leaves form the unique $\gamma_{\rm pr}$-set. For a graph $G$, let $\mathcal{A}_{\rm pr}(G)$ and $\mathcal{N}_{\rm pr}(G)$ be the sets defined as follows:

- $\mathcal{A}_{\rm pr}(G) = \{v \in V(G) \mid v \text{ is in all } \gamma_{\rm pr}\text{-sets}\}$, and
- $\mathcal{N}_{\rm pr}(G) = \{v \in V(G) \mid v \text{ is in no } \gamma_{\rm pr}\text{-set}\}$.

Using a technique of tree pruning, Henning and Plummer [49] characterized the sets $\mathcal{A}_{\rm pr}(T)$ and $\mathcal{N}_{\rm pr}(T)$. Such a characterization is useful, for example, in constructive characterizations of those trees with equal total domination and paired domination numbers and of those trees for which the paired domination number is twice the matching number.

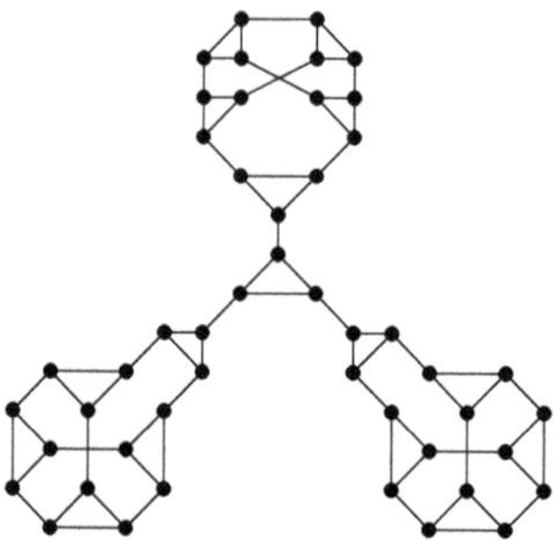

**Fig. 13** The unique connected cubic $(K_{1,3}, K_4 - e, C_4)$-free graph $G$ with $\gamma_{\text{pr}}(G) = \frac{3}{8}n$

## 6.2 Claw-Free Cubic Graphs

A graph is *claw-free* if it contains no induced $K_{1,3}$. In this section, we investigate paired domination in claw-free cubic graphs. The upper bound on the paired domination number of a cubic graph in Theorem 15 can be improved if we restrict our attention to cubic graphs with certain forbidden subgraphs. Recall that a *diamond* is the graph $K_4 - e$, where $e$ denotes an arbitrary edge of the complete graph.

**Theorem 45 ([32])** *If $G$ is a connected $(K_{1,3}, K_4-e, C_4)$-free cubic graph of order $n \geq 6$, then there exists a PD-set of $G$ of cardinality at most $3n/8$ that contains at least one vertex from each triangle of $G$. Furthermore, $\gamma_{\text{pr}}(G) = \frac{3}{8}n$ if and only if $G$ is the graph shown in Figure 13.*

**Theorem 46 ([32])** *If $G$ is a connected claw-free cubic graph of order $n \geq 6$ that contains $k \geq 0$ diamonds, then there exists a PD-set of $G$ of cardinality at most $\frac{2}{5}(n + 2k)$ that contains at least one vertex from each triangle of $G$. Furthermore, $\gamma_{\text{pr}}(G) = \frac{2}{5}(n + 2k)$ if and only if $G \in \{G_0, G_1, G_2, G_3\}$, where $G_0$, $G_1$, $G_2$, and $G_3$ are the four graphs shown in Figure 14.*

**Theorem 47 ([32])** *If $G$ is a connected claw-free and diamond-free cubic graph of order $n \geq 6$, then there exists a PD-set of $G$ of cardinality at most $2n/5$ that contains at least one vertex from each triangle of $G$. Furthermore, $\gamma_{\text{pr}}(G) = \frac{2}{5}n$ if and only if $G = G_0$, where $G_0$ can be seen in Figure 14.*

**Theorem 48 ([32])** *If $G$ is a connected claw-free cubic graph of order $n$, then $\gamma_{\text{pr}}(G) \leq \frac{1}{2}n$ with equality if and only if $G \in \{K_4, H_1, H_2, H_3, G_3\}$, where $H_1, H_2$, and $H_3$ are the graphs seen in Figure 15 and $G_3$ is the graph shown in Figure 14.*

The upper bound on the paired domination number of a claw-free cubic graph presented in Theorem 48 can be improved if we add the restriction that the graph is 2-connected.

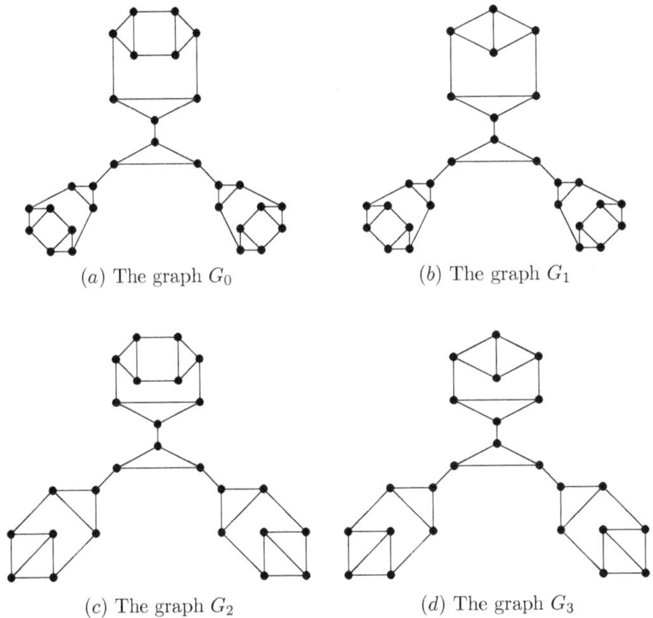

(a) The graph $G_0$ (b) The graph $G_1$

(c) The graph $G_2$ (d) The graph $G_3$

**Fig. 14** The four connected cubic claw-free graph $G_k$, $0 \le k \le 3$, with $k$ copies of $K_4 - e$ and with $\gamma_{\rm pr}(G_k) = \frac{2}{5}(n + 2k)$

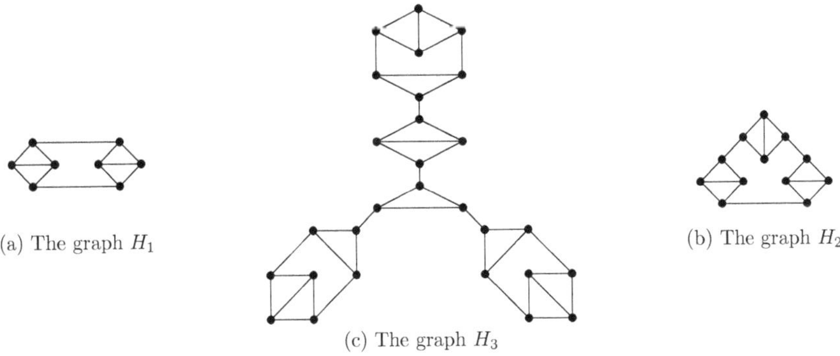

(a) The graph $H_1$ (b) The graph $H_2$

(c) The graph $H_3$

**Fig. 15** Three connected cubic claw-free graphs $H_1$, $H_2$, and $H_3$

**Theorem 49 ([32])** *If G is a 2-connected claw-free cubic graph of order $n \ge 6$, that contains $k \ge 0$ diamonds, then $\gamma_{\rm pr}(G) \le \frac{1}{3}(n + 2k)$.*

As an immediate consequence of Theorem 49, we have the following result.

**Theorem 50 ([32])** *If G is a 2-connected claw-free and diamond-free cubic graph of order $n \ge 6$, then $\gamma_{\rm pr}(G) \le \frac{1}{3}n$.*

In [31, 33], it is noted that the bound of Theorem 45 can be improved for graphs of sufficiently large order.

**Theorem 51 ([31])** *If $G$ is a cubic graph of order $n \geq 48$ such that $G$ is $(C_4, K_{1,3}, K_4 - e)$-free, then $\gamma_{\mathrm{pr}}(G) \leq (10n + 6)/27$.*

## 6.3 Claw-Free Graphs

Inspired by the work on cubic graphs, Huang, Kang, and Shan [58] studied paired domination in claw-free graphs with minimum degree at least 3. They obtained a slightly better bound on the paired domination number for claw-free graphs than that for cubic graphs.

**Theorem 52 ([58])** *If $G$ is a connected claw-free graph of order $n$ with $\delta(G) \geq 3$, then $\gamma_{\mathrm{pr}}(G) \leq \frac{1}{5}(3n - 1)$, and this bound is sharp.*

The graph $F$ in Figure 5 has order $n = 7$ and $\gamma_{\mathrm{pr}}(F) = 4 = \frac{1}{5}(3n - 1)$, attaining the bound of Theorem 52.

In 2019 Lu, Wang, Wang, and Wu [63] improved the result of Theorem 52.

**Theorem 53 ([63])** *If $G$ is a connected claw-free graph of order $n$ with $\delta(G) \geq 3$, then $\gamma_{\mathrm{pr}}(G) \leq \frac{4}{7}n$.*

We note that if $n \geq 8$, then the bound of Theorem 53 is an improvement of the result of Theorem 52. Increasing the minimum degree by only 1, in 2018 Lu, Mao, and Wang [62] improved the $\frac{4}{7}$-bound in Theorem 53 to a $\frac{1}{2}$-bound.

**Theorem 54 ([62])** *If $G$ is a connected claw-free graph of order $n$ with $\delta(G) \geq 4$, then $\gamma_{\mathrm{pr}}(G) \leq \frac{1}{2}n$, and this bound is sharp.*

## 6.4 Generalized Claw-Free Graphs

A graph is *generalized claw-free* if it contains no induced $K_{1,a+2}$, where $a \geq 1$. In this section, we investigate paired domination in generalized claw-free graphs. Dorbec, Gravier, and Henning [25] obtained the following upper bound.

**Theorem 55 ([25])** *For $a \geq 0$ an integer, if $G$ is a connected $K_{1,a+2}$-free graph of order $n \geq 2$, then $\gamma_{\mathrm{pr}}(G) \leq \frac{2(an+1)}{2a+1}$ and this bound is sharp.*

That the bound of Theorem 55 is sharp may be seen by considering the graphs $G_a$, $a \geq 0$, constructed in [25] as follows. Let $p \geq 2$ be an arbitrary integer. If $a = 0$, let $G_a = K_p$. For $a \geq 1$, form $G_a$ from the disjoint union of $p$ copies of the subdivided star $S(K_{1,a})$ and one copy of the star $K_{1,1} = P_2$, by taking the $p$ central vertices of the subdivided stars and one vertex of the star $K_{1,1}$ and forming a clique

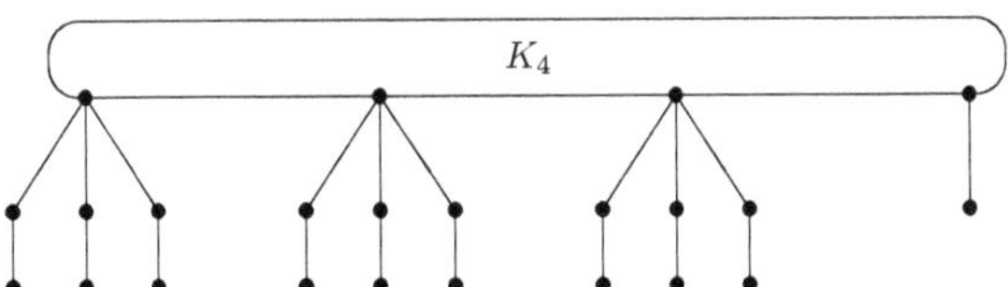


**Fig. 16** The graph $G_3$ when $p = 3$

on these $p+1$ vertices. In this case, $G_a$ is a connected $K_{1,a+2}$-free graph of order $n = (2a+1)p+2$. Since every PD-set in a graph contains all its support vertices and since the set of $ap+1$ support vertices in $G_a$ dominate $V(G_a)$ and form an independent set in $G_a$, the graph $G_a$ satisfies $\gamma_{\rm pr}(G_a) = 2(ap+1) = \frac{2(an+1)}{2a+1}$. By taking $p = 3$, for example, the graph $G_3$ is illustrated in Figure 16.

Dorbec and Gravier [23] continued the study on a topic closely related to generalized claw-free graphs. In contrast to looking at graphs that had no induced $K_{1,a+2}$ for $a \ge 1$, they studied bounds on the paired domination number of graphs that contained no induced subdivided star $S(K_{1,a+2})$ for $a \ge 1$. Dorbec and Gravier [23] established the following upper bound on the paired domination number of such graphs.

**Theorem 56 ([23])** *For an integer $a \ge 1$, if $G$ is a connected graph of order $n \ge 3$ that contains no induced $S(K_{1,a+2})$, then*

$$\gamma_{\rm pr}(G) \le \frac{2(an+1)}{2a+1},$$

*and this bound is sharp.*

That the bound of Theorem 56 is sharp may once again be seen by considering the graphs $G_a$, $a \ge 1$, constructed earlier since we note that $G_a$ contains no induced $S(K_{1,a+2})$ and $\gamma_{\rm pr}(G_a) = 2(ap+1) = \frac{2(an+1)}{2a+1}$. Dorbec and Gravier [24] also determined an upper bound on the paired domination number of graphs with no induced $P_5$ as follows.

**Theorem 57 ([24])** *Let $G$ be a connected graph of order $n \ge 2$. If $G \ne C_5$ and $G$ contains no induced $P_5$, then $\gamma_{\rm pr}(G) \le \frac{1}{2}n + 1$ and this bound is sharp.*

As described in [24], sharpness of the bound of Theorem 57 can be seen by the corona $G = K_k \circ K_1$, where $k$ is an odd integer. Then $G$ has $n = 2k$ vertices, no induced $P_5$, and any PD-set of $G$ must contain all $k$ vertices of the clique and at least one leaf. In particular, $\gamma_{\rm pr}(G) \ge k+1$. However, the vertices of the clique and one leaf form a PD-set of $G$, and so $\gamma_{\rm pr}(G) \le k+1$. Consequently, $\gamma_{\rm pr}(G) = k+1 = \frac{1}{2}n + 1$, and therefore the bound of Theorem 57 is sharp.

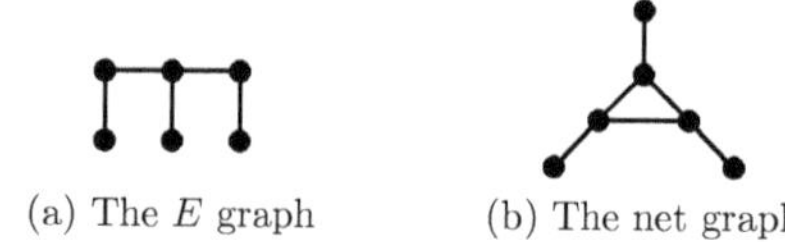


**Fig. 17** The $E$ graph and the net graph

## 6.5 E-Free and Net-Free Graphs

The $E$ *graph* is the corona $P_3 \circ K_1$ and the *net graph* is the corona $K_3 \circ K_1$. Both graphs are illustrated in Figure 17.

A graph is $E$-free if it does not contain the $E$ graph as an induced subgraph, while a graph is net-free if it does not contain the net graph as an induced subgraph. Schaudt [74] studied paired domination of graphs that are both $E$-free and net-free.

**Theorem 58 ([74])** *If $G$ is a connected $(E, net)$-free graph of order $n \geq 2$, then $\gamma_{\rm pr}(G) \leq 2\lceil \frac{n}{4} \rceil$ and this bound is sharp.*

As remarked in [74], nontrivial paths $P_n$ attain the bound of Theorem 58.

## 6.6 Planar Graphs with Diameter 2

MacGillivray and Seyffarth [65] proved that planar graphs with diameter 2 have bounded domination numbers. In particular, this implies that the domination number of such a graph can be determined in polynomial time. On the other hand, they observed that in general graphs with diameter 2 have unbounded domination number, and hence the paired domination number is also unbounded. Specifically, MacGillivray and Seyffarth [65] established the following result.

**Theorem 59 ([65])** *If $G$ is a planar graph with* $\text{diam}(G) = 2$, *then $\gamma(G) \leq 3$.*

The bound of Theorem 59 is sharp as may be seen by considering the graph $G_9$ of Figure 18 constructed by MacGillivray and Seyffarth [65]. The graph $G_9$ of Figure 18 is in fact the unique planar graph of diameter 2 with domination number 3 as shown by Goddard and Henning in [36].

**Theorem 60 ([36])** *If $G$ is a planar graph with* $\text{diam}(G) = 2$, *then $\gamma(G) \leq 2$ or $G = G_9$, where $G_9$ is the graph of Figure 18.*

As an immediate consequence of Theorem 60, we have the following result.

**Theorem 61 ([36])** *If $G$ is a planar graph with* $\text{diam}(G) = 2$, *then $\gamma_{\rm pr}(G) \leq 4$.*

A characterization of planar graphs with diameter 2 and paired domination number 4 seems to be difficult to obtain since there are infinitely many such graphs. If we restrict our attention to planar graphs with certain structural properties, then

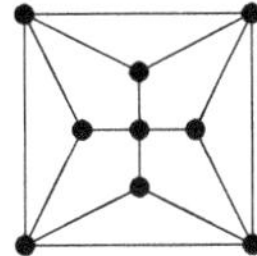

**Fig. 18** A planar graph $G_9$ of diameter 2 with domination number 3

a characterization of such planar graphs is possible. Henning and McCoy [46] say that a graph $G$ satisfies the *domination-cycle property* if there is some $\gamma$-set not contained in any induced 5-cycle of $G$. Further, they define a family which they call $\mathcal{F}_{\text{planar}}$ consisting of thirty four graphs of small orders (at most 11) and characterize the planar graphs with diameter 2 and paired domination number 4 that satisfy the domination-cycle property as follows.

**Theorem 62 ([46])** *If $G$ is a planar graph of diameter 2 that satisfies the domination-cycle property, then $\gamma_{\text{pr}}(G) = 4$ if and only if $G \in \mathcal{F}_{\text{planar}}$.*

# 7 Criticality Concepts for Paired Domination

In this section, we discuss criticality concepts for paired domination.

## 7.1 Paired Domination Edge-Critical Graphs

Edwards, Gibson, Henning, and Mynhardt [30] introduced the concept of paired domination edge-critical graphs. They define a graph $G$ as *paired domination edge-critical*, denoted $\gamma_{\text{pr}}EC$, if for every $e \in E(\overline{G}) \neq \emptyset$, $\gamma_{\text{pr}}(G + e) < \gamma_{\text{pr}}(G)$. Further, if $G$ is a $\gamma_{\text{pr}}EC$ graph and $\gamma_{\text{pr}}(G) = k$, we say that $G$ is $k_{\text{pr}}EC$. As observed in [30], if $G$ is a $\gamma_{\text{pr}}EC$ graph, then $\gamma_{\text{pr}}(G+e) = \gamma_{\text{pr}}(G)-2$ for every $e \in E(\overline{G})$. The family of $\gamma_{\text{pr}}EC$ trees consist precisely of subdivided stars with at least three leaves.

**Theorem 63 ([30])** *A tree $T$ is $\gamma_{\text{pr}}EC$ if and only if $T = S(K_{1,t})$, where $t \geq 3$.*

The maximum diameter of a $k_{\text{pr}}EC$ graph is at least $k - 2$. To see this, let $G_\ell$, for $\ell \geq 1$, be the graph constructed as follows. For each $i \in [\ell]$, let $H_i \cong P_4$ be the path $x_i u_i v_i y_i$. We now construct $G_\ell$ recursively as follows. Let $G_0 = K_2$ with $V(G_0) = \{u_0, v_0\}$ and once $G_{i-1}$ has been constructed, let $G_i$ be the graph with $V(G_i) = V(G_{i-1}) \cup V(H_i)$ and $E(G_i) = E(G_{i-1}) \cup E(H_i) \cup \{u_{i-1}x_i, u_{i-1}y_i, v_{i-1}x_i, v_{i-1}y_i\}$. We are now in a position to state the result in [30].

**Proposition 64 ([30])** *For every $\ell \geq 1$, $G_\ell$ is a $(2\ell + 2)_{\text{pr}}$-EC graph with diameter $2\ell$.*

If $G$ is connected and $\gamma_{\text{pr}}(G) = k$, then as shown in [30], we have $\text{diam}(G) \leq 2k - 1$. This bound can be improved for $k_{\text{pr}}EC$ graphs.

**Theorem 65 ([30])** *For $k \geq 6$, if $G$ is a connected $k_{\text{pr}}EC$ graph, then* $\text{diam}(G) \leq 2k - 6$.

As remarked in [30], it is not yet known if the bound of Theorem 65 is the best possible if $G$ is $6_{\text{pr}}EC$. It remains an open problem to find a $6_{\text{pr}}EC$ graph with diameter 5 or 6 or to improve the bound of Theorem 65. For $k \geq 10$, the bound of Theorem 65 can be improved.

**Theorem 66 ([30])** *For $k \geq 10$, if $G$ is a connected $k_{pr}EC$ graph, then* $\text{diam}(G) \leq \frac{3k}{2} + 3$.

Let $d_k$ denote the maximum value of the diameter for any $k_{\text{pr}}EC$ graph. It remains an open problem to determine sharp upper bounds on $d_k$. We remark that there exist $k_{\text{pr}}EC$ graphs of diameter 2 for all $k \geq 4$ even.

**Proposition 67 ([30])** *For every even $k \geq 4$, there exists a $k_{\text{pr}}EC$ graph of diameter* 2.

The existence of graphs in Proposition 67 is evident by taking the Cartesian product of the complete graph $K_k$ with itself. The authors in [30] ask the question: What is the spectrum of diameters for $k_{\text{pr}}EC$ graphs? In particular, is it true that there exists a $k_{\text{pr}}EC$ graph of diameter $\ell$ for every $2 \leq \ell \leq d_k$?

## 7.2 *Paired Domination Vertex Critical Graphs*

In 2006 in her master's thesis, Edwards [29] studied paired domination vertex critical graphs. A graph $G$ is *paired domination vertex critical*, abbreviated *$\gamma_{\text{pr}}$-vertex critical*, if for every $v \in V(G) \setminus S_G$, $\gamma_{\text{pr}}(G - v) < \gamma_{\text{pr}}(G)$, where $S_G$ denotes the set of support vertices of $G$. The restriction of the removal of vertices to those that are not support vertices is necessary to avoid creating isolated vertices. If $G$ is a $\gamma_{\text{pr}}$-vertex critical graph and $\gamma_{\text{pr}}(G) = k$, then $G$ is called *$k$-$\gamma_{\text{pr}}$-vertex critical.* As observed by Henning and Mynhardt [48], if $G$ is a $\gamma_{\text{pr}}$-vertex critical graph, then $\gamma_{\text{pr}}(G - v) = \gamma_{\text{pr}}(G) - 2$ for every $v \in V(G) \setminus S_G$. Furthermore, a $\gamma_{\text{pr}}$-set $G - v$ contains no neighbor of $v$.

Bounds on the diameter of $\gamma_{\text{pr}}$-vertex critical graphs are studied in [48]. The connected graphs with minimum degree 1 that are paired domination vertex critical are characterized and sharp bounds on their maximum diameter obtained. Recall that $\alpha'(G)$ denotes the edge independence number of $G$.

**Theorem 68 ([48])** *Let $G$ be a connected graph of order at least* 3 *with at least one leaf. Then $G$ is a $\gamma_{\text{pr}}$-vertex critical graph if and only if $G$ is the corona $H \circ K_1$ for some connected graph $H$ for which $\alpha'(H) = \alpha'(H - v)$ for every $v \in V(H)$. Further, if $G$ is $\gamma_{\text{pr}}$-vertex critical graph, then* $\text{diam}(G) \leq \frac{1}{2}(\gamma_{\text{pr}}(G) + 2)$, *and this bound is sharp.*

Tightness of the diameter bound in Theorem 68 can be seen with the graph $H = C_{k-1}$. As observed in [48], there are infinite families of connected graphs $H$ satisfying $\alpha'(H) = \alpha'(H - v)$ for every $v \in V(H)$. For example, let $H$ be any Hamiltonian graph of odd order. As shown in [48], if $H$ is a connected graph satisfying $\alpha'(H) = \alpha'(H - v)$ for every $v \in V(H)$, then $H$ is a 2-edge-connected graph. In particular, this implies the following result.

**Theorem 69 ([48])** *No tree is $\gamma_{\rm pr}$-vertex critical.*

The following lower bound on the maximum diameter of a $k$-$\gamma_{\rm pr}$-vertex critical graph is established in [48].

**Theorem 70 ([48])** *For every even integer $k \geq 4$, there exists a connected $k$-$\gamma_{\rm pr}$-vertex critical graph of diameter $\frac{3}{2}(k-2)$.*

As an immediate consequence of Theorem 70, we have the following result.

**Corollary 71** *The maximum diameter of a connected $k$-$\gamma_{\rm pr}$-vertex critical graph is at least $\frac{3}{2}(k-2)$.*

For small $k$, we have the following result.

**Theorem 72 ([48])** *For $k \leq 8$, the diameter of a connected $k$-$\gamma_{\rm pr}$-vertex critical graph is at most $\frac{3}{2}(k-2)$.*

The following question is posed in [48].

**Question 1** ([48]) *If $G$ is a connected $\gamma_{\rm pr}$-vertex critical graph, then is it true that $\mathrm{diam}(G) \leq \frac{3}{2}(\gamma_{\rm pr}(G) - 2)$?*

Note that by Theorem 72, Question 1 is true for $\gamma_{\rm pr}(G) \leq 8$. By Corollary 71, if Question 1 is true, then this bound is sharp. Hou and Edwards [54] answered Question 1 in the affirmative. Their result, together with Theorem 70, establishes the following maximum diameter of a connected $k$-$\gamma_{\rm pr}$-vertex critical graph.

**Theorem 73 ([54])** *The maximum diameter of a connected $k$-$\gamma_{\rm pr}$-vertex critical graph is $\frac{3}{2}(k-2)$.*

Utilizing a construction technique due to Brigham, Chinn, and Dutton [7], Hou and Edwards [54] presented the following method for constructing $\gamma_{\rm pr}$-vertex critical graphs from two smaller $\gamma_{\rm pr}$-vertex critical graphs. Suppose that $F$ and $H$ are nonempty graphs. Let $u$ and $w$ be non-isolated vertices of $F$ and $H$, respectively. Then $(F \cdot H)(u, w : v)$ denotes the graph obtained from $F$ and $H$ by identifying $u$ and $w$ in a vertex labeled $v$. We call $(F \cdot H)(u, w : v)$ the *coalescence* of $F$ and $H$ via $u$ and $w$.

**Theorem 74 ([54])** *Let $F$ and $H$ be two graphs with no isolated vertex and let $G = (F \cdot H)(u, w : v)$. If $u \notin S(F)$ and $w \notin S(H)$, then $G$ is $\gamma_{\rm pr}$-vertex critical if and only if both $F$ and $H$ are $\gamma_{\rm pr}$-vertex critical. Furthermore, if $G$ is $\gamma_{\rm pr}$-vertex critical, then $\gamma_{\rm pr}(G) = \gamma_{\rm pr}(F) + \gamma_{\rm pr}(H) - 2$.*

A graph $G$ is *vertex diameter $k$-critical* if $\mathrm{diam}(G) = k$ and $\mathrm{diam}(G - v) > k$ for any $v \in V(G)$. Hou and Edwards [54] gave the following characterization of 4-$\gamma_{\mathrm{pr}}$-vertex critical graphs.

**Theorem 75 ([54])** *A connected graph $G$ is 4-$\gamma_{\mathrm{pr}}$-critical if and only if $\overline{G}$ is vertex diameter 2-critical or $G = K_3 \circ K_1$.*

Further results on 4-$\gamma_{\mathrm{pr}}$-vertex critical and 6-$\gamma_{\mathrm{pr}}$-vertex critical graphs can be found, for example, in Huang, Shan, and Kang [57], who study properties of $K_{1,5}$-free, 4-$\gamma_{\mathrm{pr}}$-vertex critical graphs, as well as $K_{1,4}$-free, 6-$\gamma_{\mathrm{pr}}$-vertex critical graphs.

# 8 The Upper Paired Domination Number

To determine the upper paired domination number of even relatively simple classes of graphs is more challenging than determining the paired domination number. The upper paired domination number of a path is given in the following result by Dorbec, Henning, and McCoy [28].

**Theorem 76 ([28])** *For $n \geq 2$ an integer,*

$$\Gamma_{\mathrm{pr}}(P_n) = 8\left\lfloor \frac{n+1}{11} \right\rfloor + 2\left\lfloor \frac{(n+1)(\bmod 11)}{3} \right\rfloor.$$

Recall that a PD-set $S$ in a graph $G$ is *minimal* if no proper subset in $S$ is a PD-set of $G$. The following elementary property of a minimal PD-set of a graph is often useful in determining results on the upper paired domination number.

**Proposition 77 ([28])** *If $S$ is a PD-set in a graph $G$ with no isolated vertex, then $S$ is a minimal PD-set of $G$ if and only if for every pair $\{u, v\} \subseteq S$, the set $S \setminus \{u, v\}$ is not a PD-set.*

The upper paired domination number of a graph is very sensitive to the removal of a vertex from the graph as the following result indicates.

**Proposition 78 ([28])** *For every even integer $k \geq 2$, there exists a graph $G$ such that $\Gamma_{\mathrm{pr}}(G - v) - \Gamma_{\mathrm{pr}}(G) = k$ for every vertex $v$ in $V(G)$.*

The graph used to show existence in Proposition 78 was formed by taking two disjoint copies of a complete graph $K_{k+5}$, where $k \geq 2$ is an even integer, and then adding the edges of a perfect matching between these two copies of $K_{k+5}$. Thus, $G$ is the Cartesian product $K_{k+5} \,\square\, K_2$ of a complete graph $K_{k+5}$ and a $K_2$, and $\Gamma_{\mathrm{pr}}(G) = 4$. However, $\Gamma_{\mathrm{pr}}(G - v) = k + 4$ for every vertex $v \in V(G)$.

## 8.1 Upper Total Domination Versus Upper Paired Domination

Dorbec et al. [28] investigated the relationship between the upper total domination and upper paired domination numbers of a graph. They observed that it is not always the case that every minimal PD-set of a graph is a minimal TD-set of the graph and is not always the case that every minimal TD-set of a graph is a minimal PD-set of the graph.

**Theorem 79 ([28])** *For every integer $k \geq 1$, there exist connected graphs $G$ and $H$ such that $\Gamma_{\rm pr}(G) - \Gamma_t(G) \geq k$ and $\Gamma_t(H) - \Gamma_{\rm pr}(H) \geq k$.*

The following result shows that the upper total domination is bounded below by some constant multiple of the upper paired domination number.

**Theorem 80 ([28])** *For every graph $G$ with no isolated vertices, $\Gamma_t(G) \geq \frac{1}{2}\Gamma_{\rm pr}(G) + 1$.*

The trees achieving equality in the bound of Theorem 80 are characterized in [28].

**Theorem 81 ([28])** *If $T$ is a tree on at least two vertices satisfying $\Gamma_t(T) = \frac{1}{2}\Gamma_{\rm pr}(T) + 1$, then $T$ is obtained from a star (possibly trivial) by attaching any number of pendant edges, but at least one, to every vertex of the star.*

When restricted to the class of trees, the upper total domination is bounded above by the upper paired domination number.

**Theorem 82 ([28])** *For every nontrivial tree $T$, $\Gamma_t(T) \leq \Gamma_{\rm pr}(T)$.*

## 8.2 Upper Paired Domination in Claw-Free Graphs

Even if the minimum degree is large, the upper paired domination number can be made arbitrarily close to the order of the graph.

**Theorem 83 ([27])** *If $G$ is a connected graph of order $n \geq 3$, then $\Gamma_{\rm pr}(G) \leq n-1$. Furthermore, if $G$ has minimum degree $\delta \geq 2$, then $\Gamma_{\rm pr}(G) \leq n - \delta + 1$, and this bound is sharp.*

**Corollary 84 ([27])** *If $G$ is a connected graph of order $n$, then $\Gamma_{\rm pr}(G) \leq n - {\rm O}(1)$ and this bound is best possible even for arbitrarily large, but fixed (with respect to $n$), minimum degree.*

That the bound of Theorem 83 is sharp, may be seen as follows [27]. For $x \geq 1$ and $\delta \geq 2$, let $G$ be obtained from a complete bipartite graph $K_{2x,\delta-1}$ with partite sets $X$ and $Y$, where $|X| = 2x$ by adding a perfect matching between the vertices of $X$ and forming a clique on the vertices of $Y$. We note that $G[X] = xK_2$ and $G[Y] = K_{\delta-1}$. Then, $G$ is a connected graph of order $n$ with minimum degree $\delta$. Since $X$ is

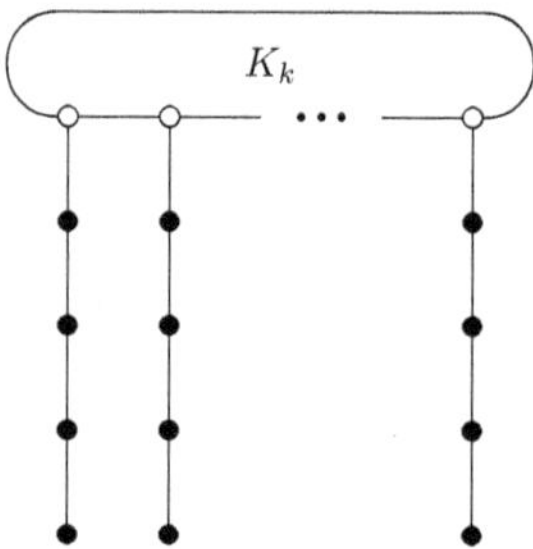


**Fig. 19** A graph $G \in \mathcal{G}_1$

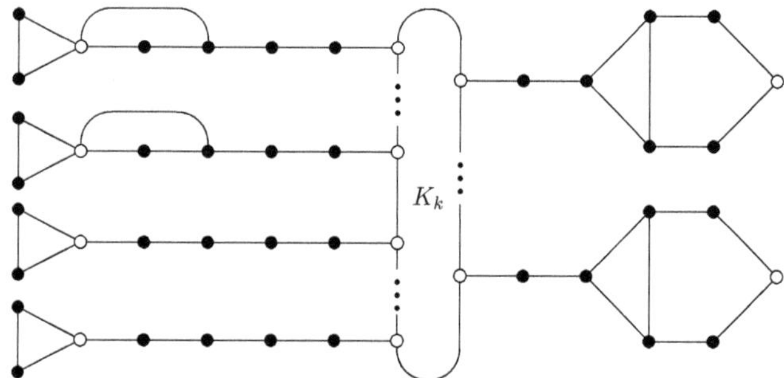


**Fig. 20** A graph $G \in \mathcal{G}_2$

a minimal PD-set of $G$, $\Gamma_{\text{pr}}(G) \geq n - \delta + 1$. Consequently, $\Gamma_{\text{pr}}(G) = n - \delta + 1$. Note that for $x$ arbitrarily large, we have $\Gamma_{\text{pr}}(G) = n - \text{O}(1)$.

Dorbec and Henning [27] showed that the upper bound of Theorem 83 can be improved if we restrict our attention to claw-free graphs. In order to state their result, they constructed four families of connected claw-free graphs as follows. Recall that the $k$-corona of a graph $G$ is the graph of order $(k+1)|V(G)|$ obtained from $G$ by attaching a path of length $k$ to each vertex of $G$ so that the resulting paths are vertex disjoint.

**The Family** $\mathcal{G}_1$ Let $\mathcal{G}_1$ be the family of connected claw-free graphs of order $n = 5k$ obtained from the 4-corona of a complete graph $K_k$ on $k \geq 1$ vertices. A graph $G$ in the family $\mathcal{G}_1$ is shown in Figure 19. The darkened vertices in $G$ form a minimal PD-set in $G$, and so $\Gamma_{\text{pr}}(G) \geq 4k = \frac{4}{5}n$.

**The Family** $\mathcal{G}_2$ Let $\mathcal{G}_2$ be the family of connected claw-free graphs of order $n = 8k$ obtained from a 7-corona of complete graph $K_k$ on $k \geq 2$ vertices as follows: For every attached path $vv_1v_2 \dots v_7$, where $v$ is a vertex of the complete graph $K_k$, add the edge $v_5v_7$ or add the edges $v_3v_5$ and $v_5v_7$ or add the edges $v_2v_7$ and $v_3v_7$. A graph $G$ in the family $\mathcal{G}_2$ is shown in Figure 20. The darkened vertices in $G$ form a minimal PD-set in $G$, and so $\Gamma_{\text{pr}}(G) \geq 6k = \frac{3}{4}n$.

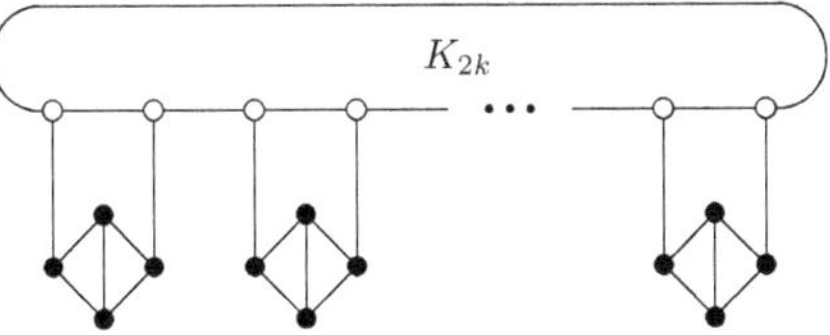


**Fig. 21** A graph $G \in \mathcal{F}_3$

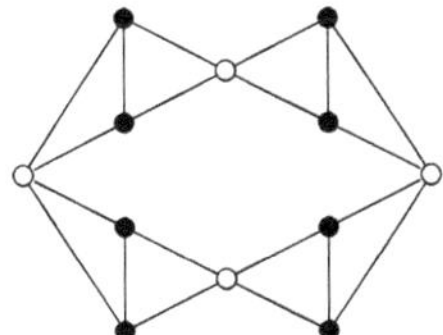

**Fig. 22** A graph $G \in \mathcal{G}_3$

**The Family** $\mathcal{F}_3$ Let $\mathcal{F}_3$ be the family of connected claw-free graphs obtained from the disjoint union of $k \geq 2$ copies of $K_4 - e$ by attaching a pendant edge to each vertex of degree 2 and then forming a clique $K_{2k}$ on the resulting $2k$ vertices of degree 1. A graph $F$ in the family $\mathcal{F}_3$ is shown in Figure 21. The darkened vertices in $F$ form a minimal PD-set in $F$, and so $\Gamma_{\rm pr}(F) \geq 4k = \frac{2}{3}|V(F)|$.

**The Family** $\mathcal{G}_3$ Let $\mathcal{G}_3$ be the family of connected claw-free graphs $G = (V, E)$ that admit a vertex partition $V = X \cup C$ such that (i) $G[X] = kK_2$, (ii) each vertex of $C$ is adjacent to vertices of $X$ from exactly two $K_2$s and possibly to other vertices of $C$, (iii) $|X| = 2|C|$, and (iv) $\delta(G) \geq 3$. A graph $G$ in the family $\mathcal{G}_3$ is shown in Figure 22, where the darkened vertices form the set $X$ and the white vertices the set $C$. The set $X$ is a minimal PD-set in $G$, and so $\Gamma_{\rm pr}(G) \geq 2k = \frac{2}{3}|V(G)|$.

We are now in a position to present the main result in [27] that shows that the upper bound of Theorem 83 can be improved if we restrict our attention to claw-free graphs.

**Theorem 85 ([27])** *If $G$ is a connected claw-free graph of order $n$ and minimum degree $\delta$, then*

$$\Gamma_{\rm pr}(G) \leq \begin{cases} \frac{4}{5}n & \textit{if } \delta = 1 \textit{ and } n \geq 3 \\ \frac{3}{4}n & \textit{if } \delta = 2 \textit{ and } n \geq 6 \\ \frac{2}{3}n & \textit{if } \delta \geq 3. \end{cases}$$

*Furthermore, if $n \geq 3$ and $\Gamma_{pr}(G) = \frac{4}{5}n$, then $G \in \mathcal{G}_1$. If $\delta = 2$, $n \geq 9$, and $\Gamma_{pr}(G) = \frac{3}{4}n$, then $G \in \mathcal{G}_2$.*

As remarked earlier, if $G_3 \in \mathcal{G}_3 \cup \mathcal{F}_3$ is a graph of order $n$, then $\delta(G_3) \geq 3$ and $\Gamma_{pr}(G_3) \geq \frac{2}{3}n$. However by Theorem 85, $\Gamma_{pr}(G_3) \leq \frac{2}{3}n$. Consequently, $\Gamma_{pr}(G_3) = \frac{2}{3}n$, implying that the bound of Theorem 85 for the case when $\delta \geq 3$ is sharp.

The claw-free graphs given to establish that the upper bounds of Theorem 85 are tight have small minimum degree $\delta \in \{1, 2, 3\}$. It remains an open problem to determine tight upper bounds on the upper paired domination in claw-free graphs for $\delta \geq 4$. The following conjecture is posed in [22].

**Conjecture 5 ([22])** *If $G$ is a connected claw-free graph of order $n$ and minimum degree $\delta \geq 4$, then*

$$\Gamma_{pr}(G) \leq \begin{cases} \left(\dfrac{\delta}{2\delta - 1}\right) n & \text{if } \delta \text{ is even} \\[2ex] \left(\dfrac{\delta + 1}{2\delta + 1}\right) n & \text{if } \delta \text{ is odd.} \end{cases}$$

If Conjecture 5 is true, then the bound is sharp. To see this, we construct a family $\mathcal{G}_\delta$ of graphs $G$ for even $\delta \geq 4$ satisfying $\Gamma_{pr}(G) = \delta n/(2\delta - 1)$, and a family $\mathcal{G}_\delta$ of graphs $G$ for odd $\delta \geq 5$ satisfying $\Gamma_{pr}(G) = (\delta + 1)n/(2\delta + 1)$.

Let $\delta \geq 4$ be an even integer. Let $F_\delta$ be the graph obtained from two disjoint copies of a complete graph $K_{\delta-1}$ by adding a perfect matching between them, and then adding a new vertex, called a *link vertex*, and joining it to every vertex in one of the two copies of $K_{\delta-1}$. For $k \geq 2$, let $\mathcal{G}_\delta$ be the graph obtained from the disjoint union of $k$ copies of $F_\delta$ by forming a clique $K_k$ on the $k$ link vertices, and then in the resulting graph forming a clique $K_{k(\delta-1)}$ on the $k(\delta - 1)$ vertices of degree $\delta - 1$. Then, $\mathcal{G}_\delta$ is a connected claw-free graph of order $n = k(2\delta - 1)$ with $\delta(\mathcal{G}_\delta) = \delta$. A graph $G$ in the family $\mathcal{G}_\delta$ is shown in Figure 23. The darkened vertices in $G$ form a minimal PD-set in $G$, and so $\Gamma_{pr}(G) \geq k\delta = \delta n/(2\delta - 1)$.

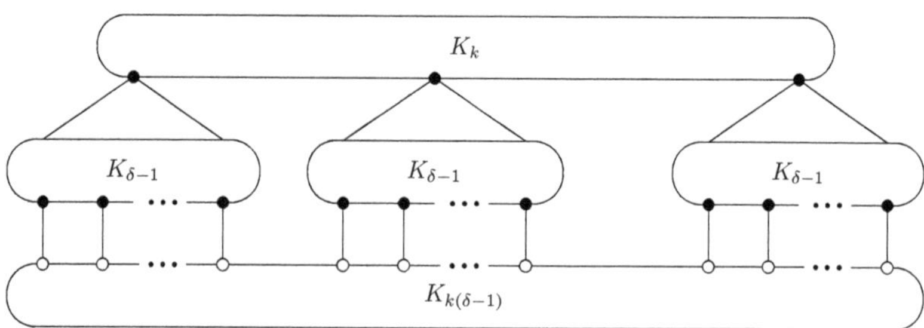


**Fig. 23** A graph $G \in \mathcal{G}_\delta$ for even $\delta \geq 4$

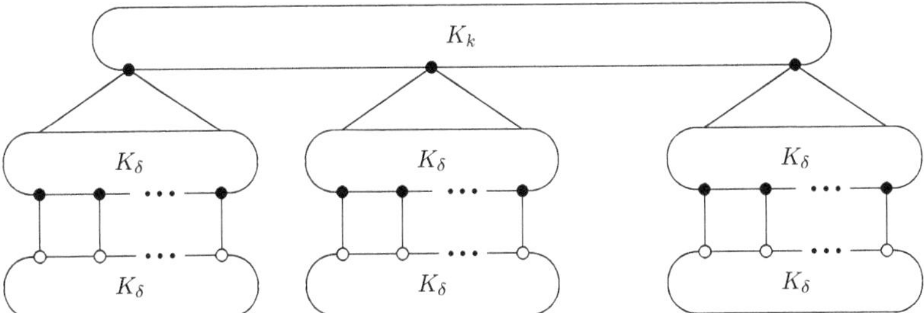


**Fig. 24** A graph $G \in \mathcal{G}_\delta$ for odd $\delta \geq 5$

Let $\delta \geq 5$ be an odd integer. Let $H_\delta$ be the graph obtained from two disjoint copies of a complete graph $K_\delta$ by adding a perfect matching between them, and then adding a new vertex and joining it to every vertex in one of the two copies of $K_\delta$. For $k \geq 1$, let $\mathcal{G}_\delta$ be the graph obtained from the disjoint union of $k$ copies of $H_\delta$ by forming a clique $K_k$ on the $k$ added vertices of degree $\delta$. Then, $\mathcal{G}_\delta$ is a connected claw-free graph of order $n = k(2\delta + 1)$ with $\delta(\mathcal{G}_\delta) = \delta$. A graph $G$ in the family $\mathcal{G}_\delta$ is shown in Figure 24. The darkened vertices in $G$ form a minimal PD-set in $G$, and so $\Gamma_{\rm pr}(G) \geq k(\delta + 1) = (\delta + 1)n/(2\delta + 1)$.

# 9 Relating Paired Domination to Other Parameters

In this section, we discuss how the paired domination number is related to other parameters, including the domination number, the total domination number, and the upper total domination number.

## 9.1 Total Domination Versus Paired Domination

By Observation 1, if $G$ is a graph with no isolated vertex, then $\gamma_t(G) \leq \gamma_{\rm pr}(G)$. However, the difference $\gamma_{\rm pr}(G) - \gamma_t(G)$ can be made arbitrarily large. For example, if $G$ is a subdivided star $S(K_{1,r-1})$ for $r \geq 3$, then $\gamma_t(G) = r$ and $\gamma_{\rm pr}(G) = 2(r-1)$, and so $\gamma_{\rm pr}(G) - \gamma_t(G) = r - 2 \geq 1$. Even for arbitrarily large minimum degree, the difference $\gamma_{\rm pr}(G) - \gamma_t(G)$ can be made arbitrarily large. Hence, it is more profitable to study ratios of the two parameters. Let $G$ be a graph with no isolated vertex. By Observation 1, $\gamma(G) \leq \gamma_t(G)$, while by Theorem 2, $\gamma_{\rm pr}(G) \leq 2\gamma(G)$, implying the following ratio of the paired domination and total domination numbers observed by Schaudt [75] and elsewhere.

**Observation 86 ([75])** *If $G$ is a graph with no isolated vertex, then* $\frac{\gamma_{\rm pr}(G)}{\gamma_t(G)} \leq 2$.

Schaudt [75] showed that the bound of Observation 86 can be improved by restricting our attention to graphs with certain forbidden subgraphs. In particular, Schaudt examines graphs with combinations of no induced $C_5$, star $K_{1,r}$, subdivided star $S(K_{1,r})$, corona $K_3 \circ K_1$, and corona $P_3 \circ K_1$.

**Theorem 87 ([75])** *If G be a $K_{1,r}$-free graph for some $r \geq 3$, then*

$$\frac{\gamma_{\rm pr}(G)}{\gamma_t(G)} \leq 2 - \frac{2}{r} \quad \textit{and} \quad \frac{\Gamma_{\rm pr}(G)}{\Gamma_t(G)} \leq 2 - \frac{2}{r}.$$

*The bound on $\gamma_{\rm pr}(G)/\gamma_t(G)$ is sharp for each $r \geq 3$.*

As observed earlier, if $G$ is a subdivided star $S(K_{1,r-1})$ for $r \geq 3$, then $\gamma_{\rm pr}(G)/\gamma_t(G) = 2 - \frac{2}{r}$, implying that the bound on $\gamma_{\rm pr}(G)/\gamma_t(G)$ in Theorem 87 is sharp. However, Schaudt [75] remarks that it is not known whether the bound on $\Gamma_{\rm pr}(G)/\Gamma_t(G)$ in Theorem 87 is sharp. As observed by Schaudt [75], if we restrict our attention to graphs which are $(C_5, S(K_{1,r}))$-free, then both bounds in Theorem 87 are sharp, as may be seen by considering the corona $G = K_{1,r-1} \circ K_1$ for which $\gamma_{\rm pr}(G) = \Gamma_{\rm pr}(G) = 2r - 2$ and $\gamma_t(G) = \Gamma_t(G) = r$.

Schaudt [75] established the following upper bound on the ratio of the paired domination number versus the total domination number, and the upper paired domination number versus the upper total domination number.

**Theorem 88 ([75])** *If G is a graph with no isolated vertex and maximum degree $\Delta$, then*

$$\gamma_{\rm pr}(G) \leq \left(\frac{2\Delta}{\Delta+1}\right)\gamma_t(G) \quad \textit{and} \quad \Gamma_{\rm pr}(G) \leq \left(\frac{2\Delta}{\Delta+1}\right)\Gamma_t(G).$$

As remarked by Schaudt [75], the upper bound on $\gamma_{\rm pr}(G)/\gamma_t(G)$ given in Theorem 88 is tight for all $\Delta \geq 2$, as may be seen by taking $G$ to be the subdivided star $S(K_{1,\Delta})$. Such a graph $G$ satisfies $\gamma_{\rm pr}(G) = 2\Delta$ and $\gamma_t(G) = \Delta + 1$. Cyman, Dettlaff, Henning, Lemanska, and Raczek [19] presented a slightly stronger result and a different proof of Schaudt's bound in order to characterize the regular graphs that achieve equality in this upper bound.

**Theorem 89 ([19])** *If G is a graph with no isolated vertex and maximum degree $\Delta$, then*

$$\gamma_{\rm pr}(G) \leq \left(\frac{2\Delta}{\Delta+1}\right)\gamma_t(G).$$

*Further, if $\gamma_{\rm pr}(G) = \left(\frac{2\Delta}{\Delta+1}\right)\gamma_t(G)$, then every minimum total dominating set in G induces a graph whose components are isomorphic to $K_{1,\Delta}$.*

We observe that for the extremal family of graphs provided by Schaudt [75], the difference between the maximum and minimum degrees is large. The connected, $k$-regular graphs that achieve equality in Theorem 88 were characterized in [19].

**Theorem 90 ([19])** *For $k \geq 2$ and $k \neq 57$, if $G$ is a connected, $k$-regular graph of girth at least 5, then*

$$\frac{\gamma_{\text{pr}}(G)}{\gamma_t(G)} \leq \frac{2k}{k+1}$$

*with equality if and only if*

(a) $k = 2$ *and* $G \cong C_5$, *or*
(b) $k = 3$ *and* $G$ *is the Petersen graph.*

The authors in [19] conjectured that the girth condition can be dropped in Theorem 90.

The following result of Schaudt [75] establishes a relationship between the paired domination number and the upper total domination number of a graph. A PD-set $S$ of a graph $G$ with the additional property that $G[S]$ is a union of disjoint copies of $K_2$ is called an *induced paired dominating set*. Induced paired dominating sets were introduced by Studer, Haynes, and Lawson in [77].

**Theorem 91 ([75])** *The following statements are equivalent in a graph $G$.*

(a) *Every induced subgraph $H$ of $G$ contains an induced paired dominating set.*
(b) $\max_{H \prec G} \left\{ \frac{\gamma_{\text{pr}}(H)}{\Gamma_t(H)} \right\} = 1.$
(c) *$G$ is $\{C_5, K_3 \circ K_1, P_3 \circ K_1\}$-free.*

The maximum value of the ratio $\gamma_{\text{pr}}(H)/\Gamma_t(H)$, taken over all induced subgraphs $H$ of a graph $G$, is determined in the following result.

**Theorem 92 ([75])** *If $G$ is a graph and $\lambda = \max\{2, \min\{r \mid G$ is $K_{1,r} \circ K_1$-free$\}\}$, then*

$$\max_{H \prec G} \frac{\gamma_{\text{pr}}(H)}{\Gamma_t(H)} = \begin{cases} 2 - \frac{2}{\lambda} & \text{if } G \text{ is } (C_5, K_3 \circ K_1)\text{-free} \\ \max\{\frac{4}{3}, 2 - \frac{2}{\lambda}\} & \text{otherwise.} \end{cases}$$

As a consequence of Theorem 92, we have the following two corollaries.

**Corollary 93 ([75])** *If $G$ is a $K_{1,r} \circ K_1$-free graph for some $r \geq 3$, then*

$$\frac{\gamma_{\text{pr}}(G)}{\Gamma_t(G)} \leq 2 - \frac{2}{r},$$

*and this bound is sharp for each $r \geq 3$.*

**Corollary 94 ([75])** *If $G$ is a connected graph with maximum degree $\Delta \geq 2$ that is not isomorphic to $C_5$, then*

$$\frac{\gamma_{\rm pr}(G)}{\Gamma_t(G)} \leq 2 - \frac{2}{\Delta},$$

*and this bound is sharp for each $\Delta \geq 2$.*

Sharpness in the bound in Corollary 93 is attained, for example, by the corona $K_{1,r-1} \circ K_1$, while sharpness in the bound in Corollary 94 is attained by the corona $K_{1,\Delta-1} \circ K_1$.

## 9.2 Double Domination Versus Paired Domination

In [9], Chellali and Haynes investigated relationships between the paired and double domination numbers of a graph. They obtained the following bound for graphs having minimum degree at least 2.

**Theorem 95 ([9])** *If $G$ is a graph of order $n$ with $\delta(G) \geq 2$, then*

$$\gamma_{\times 2}(G) \leq \frac{1}{2}(n + \gamma_{\rm pr}(G)).$$

A bound on the paired domination number of claw-free graphs is also given in [9].

**Theorem 96 ([9])** *If $G$ is a claw-free graph with no isolated vertices, then $\gamma_{\rm pr}(G) \leq \gamma_{\times 2}(G)$.*

The result of Theorem 96 was extended by Dorbec, Hartnell, and Henning [26].

**Theorem 97 ([26])** *For $r \geq 2$, if $G$ is a $K_{1,r}$-free graph with no isolated vertices, then*

$$\gamma_{\rm pr}(G) \leq \left(\frac{2r^2 - 6r + 6}{r(r-1)}\right) \gamma_{\times 2}(G),$$

*and this bound is asymptotically best possible.*

We note that when $r \in \{2, 3\}$, the upper bound in Theorem 97 simplifies to $\gamma_{\rm pr}(G) \leq \gamma_{\times 2}(G)$. When $r = 2$, every $K_{1,r}$-free graph $G$ with no isolated vertex is a disjoint union of complete graphs, each on at least two vertices, and $\gamma_{\rm pr}(G) = \gamma_{\times 2}(G)$ for such graphs $G$.

To show tightness of the bound in Theorem 97 when $r = 3$, the authors in [26] provide the following construction. For $r = 3$, let $G_k$ be the graph obtained from a complete graph $K_k$, where $k \geq 1$, as follows: for each vertex $v$ of the complete graph, add a 3-cycle and join $v$ to two vertices of this cycle. The resulting graph $G_k$ satisfies $\gamma_{\rm pr}(G_k) = 2k = \gamma_{\times 2}(G_k)$. Hence, the bound in Theorem 97 is tight for $r \in \{2, 3\}$.

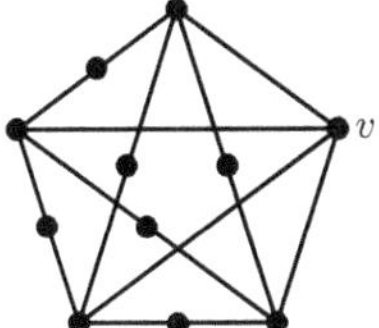


**Fig. 25** The graph $F_5$

To show that the bound in Theorem 97 is asymptotically best possible when $r \geq 4$, the following construction is given in [26]. For $r \geq 4$, let $F_r$ be the graph obtained from a complete graph $K_r$ as follows: select an arbitrary vertex $v$ and subdivide all edges not incident with $v$. The resulting graph $F_r$ is a $K_{1,r}$-free graph with no isolated vertex of order $r + \binom{r-1}{2}$. The graph $F_5$, for example, is illustrated in Figure 25.

**Proposition 98 ([26])** *For all $r \geq 4$, the graph $F_r$ is a $K_{1,r}$-free graph with no isolated vertex satisfying*

$$\gamma_{\mathrm{pr}}(F_r) = \left(\frac{2r^2 - 6r + 4}{r(r-1)}\right) \gamma_{\times 2}(F_r).$$

By Proposition 98, the upper bound of Theorem 97 is asymptotically best possible for all $r \geq 4$.

Chellali and Haynes made the following conjecture.

**Conjecture 6 ([9])** *For any nontrivial tree $T$, $\gamma_{\mathrm{pr}}(T) \leq \gamma_{\times 2}(T)$.*

In [2], Blidia, Chellali, and Haynes showed that not only is Conjecture 6 true, but they also provided both constructive and descriptive characterizations of trees having equal paired and double domination numbers. Their descriptive characterization is as follows.

**Theorem 99 ([2])** *For any nontrivial tree $T$, $\gamma_{\mathrm{pr}}(T) = \gamma_{\times 2}(T)$ if and only if $T = P_2$ or every support vertex of $T$ is adjacent to exactly one leaf, the support vertices of $T$ form an independent set, and $T$ has a unique $\gamma_{\times 2}$-set consisting of the support vertices and leaves of $T$.*

## 10 Perfect Graphs Involving Paired Domination

Given two graph parameters $\mu$ and $\psi$ related by a simple inequality $\mu(G) \leq \psi(G)$ for every graph $G$ having no isolated vertices, a graph is $(\mu, \psi)$-perfect if every induced subgraph $H$ with no isolated vertices satisfies $\mu(H) = \psi(H)$. Alvarado, Dantas, and Rautenbach [1] consider classes of $(\mu, \psi)$-perfect graphs, where $\mu$

and $\psi$ are domination parameters including $\gamma$, $\gamma_t$, and $\gamma_{pr}$. In this section, we discuss such results involving the paired domination number. If $\mu$ and $\psi$ are graph parameters related by $\mu(G) \leq \psi(G)$ for every graph $G$ having no isolated vertices, then let

$$\mathcal{G}(\mu, \psi) = \{G \text{ isolate-free graph} \mid \forall H \subseteq_{\text{ind}} G : \delta(H) \geq 1 \Rightarrow \mu(H) = \psi(H)\},$$

where $H \subseteq_{\text{ind}} G$ indicates that $H$ is an induced subgraph of $G$. Thus, $\mathcal{G}(\mu, \psi)$ is the family of $(\mu, \psi)$-perfect graphs. Since the parameters $\gamma_t$ and $\gamma_{\text{pr}}$ are defined only for graphs with no isolated vertices, it makes sense to restrict the induced subgraph $H$ of the graph $G$ in the definition of $\mathcal{G}(\mu, \psi)$ to satisfy $\delta(H) \geq 1$. Further, for a positive integer $k$, let

$$\begin{aligned}&\mathcal{G}_k(\mu, \psi)\\ &= \{G \text{ isolate-free graph} \mid \forall H \subseteq_{\text{ind}} G : (\delta(H) \geq 1 \text{ and } \gamma(H) \geq k) \Rightarrow \mu(H) = \psi(H)\}.\end{aligned}$$

Recall that the net graph $N$ is the corona $K_3 \circ K_1$, as illustrated in Figure 17.

The following results on $(\gamma_t, 2\gamma)$-perfect graphs, $(\gamma_{\text{pr}}, 2\gamma)$-perfect graphs, and $(\gamma, \gamma_{\text{pr}})$-perfect graphs were proven by Alvarado et al. [1], where $2K_2$ denotes the disjoint union of two copies of $K_2$.

**Theorem 100 ([1])** *A graph $G$ belongs to $\mathcal{G}(\gamma_t, 2\gamma)$ if and only if $G$ belongs to $\mathcal{G}(\gamma_{\text{pr}}, 2\gamma)$ if and only if $G$ is $\{C_4, P_4\}$-free.*

**Theorem 101 ([1])** *A graph $G$ belongs to $\mathcal{G}(\gamma_t, \gamma_{\text{pr}})$ if and only if $G$ is $\{C_5, P_5, N\}$-free.*

**Theorem 102 ([1])** *A graph $G$ belongs to $\mathcal{G}_2(\gamma, \gamma_{\text{pr}})$ if and only if $G$ is $\{C_5, 2K_2, N\}$-free.*

## 11 Complexity and Algorithmic Results

The well-known decision problem for the domination number is NP-complete (see [35]).

**Minimum Dominating Set (Min-DS)**

**Instance** A graph $G = (V, E)$ and a positive integer $k \leq |V|$.

**Question** Does $G$ have a dominating set of cardinality at most $k$?

The basic complexity question concerning the decision problem for the paired domination number is stated as follows.

**Table 1** Complexity results for computing the paired domination number for specific families

| Graph Family | Complexity | Authors, Citation |
|---|---|---|
| Trees | $O(n)$ | Qiao, Kang, Cardel, and Du [72] |
| Weighted Trees | $O(n)$ | Chen, Lu, and Zeng [12] |
| Inflated Trees | $O(n)$ | Kang, Sohn, and Cheng [59] |
| Strongly Chordal Graphs | $O(n+m)$ | Chen, Lu, and Zeng [11] |
| Permutation Graphs | $O(n)$ | Lappas, Nikolopoulous, and Palios [60] |
| Convex Bipartite Graphs | $O(n+m)$ | Panda and Pradhan [68] |
| Block Graphs | $O(n+m)$ | Chen, Lu, and Zeng [10] |
| Interval Graphs | $O(n+m)$ | Chen, Lu, and Zeng [10] |
| Circular-arc Graphs | $O(n+m)$ | Lin and Tu [61] |
| Strongly Orderable Graphs | $O(n+m)$ | Pradhan and Panda [70] |

**Minimum Paired Dominating Set (Min-PD-set)**

**Instance** A graph $G = (V, E)$ and a positive (even) integer $k \leq |V|$.

**Question** Does $G$ have a PD-set of cardinality at most $k$?

It was first shown in [42] that **Min-PD-set** is NP-complete. The proof of this complexity result shows that for any integer $k$, $\gamma(G) \leq k$ if and only if $\gamma_{\mathrm{pr}}(G) \leq 2k$, and then relies on the fact that **Min-DS** is known to be NP-complete.

Recall that a graph is *chordal* if every cycle of length greater than 3 has a chord (i.e., an edge joining two non-consecutive vertices in the cycle). A graph $G$ is a *split graph* if the vertex set of $G$ can be partitioned into a clique and an independent set. Chen, Lu, and Zeng [10] proved that **Min-PD-set** remains NP-complete when restricted to bipartite graphs, chordal graphs, or split graphs.

**Theorem 103 ([10])** *Min-PD-set is NP-complete for bipartite, chordal, and split graphs.*

On the other hand, polynomial time algorithms for calculating the paired domination number of several specific families of graphs have been developed. We summarize the best known results for selected families of graphs with order $n$ and size $m$ in Table 1. We note that some of these algorithms require a given ordering as input. The reader is referred to the referenced literature for definitions of these families of graphs.

We note that the polynomial time algorithm for **Min-PD-set** of strongly orderable graphs [70] shows that the class of strongly orderable graphs is a class for which **Min-PD-set** is solvable in polynomial time, whereas **Min-DS** is NP-hard [66]. For algorithms and complexity issues involving the upper paired domination number, see [27, 28, 50, 79].

## References


1. J. D. Alvarado, S. Dantas, and D. Rautenbach, Perfectly relating the domination, total domination, and paired domination numbers of a graph. *Discrete Math.* **338** (2015), 1424–1431.
2. M. Blidia, M. Chellali, and T. W. Haynes, Characterizations of trees with equal paired and double domination numbers. *Discrete Math.* **306** (2006), 1840–1845.
3. B. Bollobás and E. J. Cockayne, Graph theoretic parameters concerning domination, independence and irredundance. *J. Graph Theory* **3** (1979), 241–250.
4. B. Brešar, P. Dorbec, W. Goddard, B. Hartnell, M. A. Henning, S. Klavžar, and D. F. Rall, Vizing's conjecture: A survey and recent results. *J. Graph Theory* **69** (2012), 46–76.
5. B. Brešar, S. Klavžar, and D. F. Rall, Dominating direct products of graphs. *Discrete Math.* **307** (2000), 1636–1642.
6. B. Brešar, M. A. Henning, and D. F. Rall, Paired domination of Cartesian products of graphs. *Util. Math.* **73** (2007), 255–265.
7. R. C. Brigham, P. Z. Chinn, and R. D. Dutton: Vertex domination-critical graphs. *Networks* **18** (1988), 173–179.
8. M. Chellali and T. W. Haynes, Total and paired domination numbers of a tree. *AKCE Int. J. Graphs Comb.* **1** (2004), 69–75.
9. M. Chellali and T. W. Haynes, On paired and double domination in graphs. *Util. Math.* **67** (2005), 161–171.
10. L. Chen, C. Lu, and Z. Zeng, Labeling algorithms for paired domination problems in block and interval graphs. *J. Combin. Optim.* **19** (2010), 457–470.
11. L. Chen, C. Lu, and Z. Zeng, A linear time algorithm for paired domination problem in strongly chordal graphs. *Inform. Process. Lett.* **110** (2009), 20–23.
12. L. Chen, C. Lu, and Z. Zeng, Hardness results and approximation algorithms for (weighted) paired domination in graphs. *Theoret. Comput. Sci.* **410** (2009), no. 47–49, 5063–5071.
13. X. G. Chen, W. C. Shiu, and W. H. Chan, Upper bounds on the paired domination number. *Appl. Math. Lett.* **21** (2008), 1194–1198.
14. X. G. Chen, L. Sun, and H. M. Xing, Paired domination numbers of cubic graphs. (Chinese) *Acta Math. Sci. Ser. A Chin. Ed.* **27** (2007), 166–170.
15. T. C. E. Cheng, L. Kang, and C. T. Ng, Paired domination on interval and circular-arc graphs. *Discrete Appl. Math.* **155** (2007), 2077–2086.
16. K. Choudhary, S. Margulies, and I. V. Hicks, A note on total and paired domination of Cartesian product graphs. *Electron. J. Combin.* **20** (2013), no. 3, Paper 25, 9 pp.
17. W. E. Clark, B. Shekhtman, S. Suen, and D. C. Fisher, Upper bounds for the domination number of a graph. *Congr. Numer.* **132** (1998), 99–123.
18. W. E. Clark and S. Suen, An inequality related to Vizing's conjecture. *Electron. J. Combin.* **7** (2000), no.1, Note 4, 3pp.
19. J. Cyman, M. Dettlaff, M.A. Henning, M. Lemańska, and J. Raczek, Total domination versus paired-domination in regular graphs. *Discuss. Math. Graph Theory* **38** (2018), 573–586.
20. E. DeLaViña, Q. Liu, R. Pepper, B. Waller, and D. B. West, Some conjectures of Graffiti.pc on total domination. *Congr. Numer.* **185** (2007), 81–95.
21. N. Dehgardi, S.M. Sheikholeslami, and A. Khodkar, Bounding the paired-domination number of a tree in terms of its annihilation number. *Filomat* **28** (2014), 523–529.
22. W.J. Desormeaux and M.A. Henning, Paired domination in graphs: a survey and recent results. *Util. Math.* **94** (2014), 101–166.
23. P. Dorbec and S. Gravier, Paired domination in subdivided star-free graphs. *Graphs Combin.* **26** (2010), 43–49.
24. P. Dorbec and S. Gravier, Paired domination in $P_5$-free graphs. *Graphs Combin.* **24** (2008), 303–308.
25. P. Dorbec, S. Gravier, and M. A. Henning, Paired domination in generalized claw-free graphs. *J. Combin. Optim.* **14** (2007), 1–7.

26. P. Dorbec, B. Hartnell, and M.A. Henning, Paired versus double domination in $K_{1,r}$-free graphs. *J. Comb. Optim.* **27** (2014), 688–694.
27. P. Dorbec and M. A. Henning, Upper paired domination in claw-free graphs. *J. Combin. Optim.* **22** (2011), 235–251.
28. P. Dorbec, M. A. Henning, and J. McCoy, Upper total domination versus upper paired domination.*Quaest. Math.* **30** (2007), 1–12.
29. M. Edwards, *Criticality concepts for paired domination in graphs.* MS Thesis. University of Victoria, (2006).
30. M. Edwards, R. G. Gibson, M. A. Henning, and C. M. Mynhardt, Diameter of paired domination edge-critical graphs. *Australas. J. Combin.* **40** (2008), 279–291.
31. O. Favaron, Bounds on total and paired domination parameters in graphs and claw-free graphs. *Erster Aaachner Tag der Graphentheorie*, 59–73, Rheinisch-Westfalisch Tech. Hochsch. Lehrstufl II Math Aachen, 2004.
32. O. Favaron and M. A. Henning, Paired domination in claw-free cubic graphs. *Graphs Combin.* **20** (2004), 447–456.
33. O. Favaron and M. A. Henning, Bounds on total domination in claw-free cubic graphs. *Discrete Math.* **308** (2008), 3491–3507.
34. O. Favaron, H. Karami, and S. M. Sheikholeslami, Paired domination number of a graph and its complement. *Discrete Math.* **308** (2010), 6601–6605.
35. M. R. Garey and D. S. Johnson, *Computers and Intractability: A guide to the Theory of NP-Completeness*, Freeman, New York, 1979.
36. W. Goddard and M. A. Henning, Domination in planar graphs with small diameter. *J. Graph Theory* **40** (2002), 1–25.
37. W. Goddard and M. A. Henning, A characterization of cubic graphs with paired domination number three fifths their order. *Graphs Combin.* **25** (2009), 675–692.
38. T. W. Haynes and M. A. Henning, Paired-domination game played in graphs. *Commun. Comb. Optim.* **4** (2019), 79–94.
39. T. W. Haynes and M. A. Henning, Trees with large paired domination number. *Util. Math.* **71** (2006), 3–12.
40. T. W. Haynes, M. A. Henning, and P. J. Slater, Trees with equal domination and paired domination numbers. *Ars. Combin.* **76** (2005), 169–175.
41. T. W. Haynes and P. J. Slater, Paired domination and the paired domatic number. *Congr. Numer.* **109** (1995), 65–72.
42. T. W. Haynes and P. J. Slater, Paired domination in graphs. *Networks* **32** (1998), 199–206.
43. M. A. Henning, Trees with equal total domination and paired domination numbers. *Util. Math.* **69** (2006), 207–218.
44. M. A. Henning, Graphs with large paired domination number. *J. Combin. Optim.* **13** (2007), 61–78.
45. M. A. Henning, An upper bound on the paired domination number in terms of the number of edges in the graph. *Discrete Math.* **310** (2010), 2847–2857.
46. M. A. Henning and J. McCoy, Total domination in planar graphs of diameter two. *Discrete Math.* **309** (2009), 6181–6189.
47. M. A. Henning, J. McCoy and J. Southey, Graphs with maximum size and given paired domination number. *Discrete Appl. Math.* **170** (2014), 72–82.
48. M. A. Henning and C. M. Mynhardt, The diameter of paired domination vertex critical graphs. *Czechoslovak Math. J.* **58(133)** (2008), no. 4, 887–897.
49. M. A. Henning and M. D. Plummer, Vertices contained in all or in no minimum paired dominating set of a tree. *J. Combin. Optim.* **10** (2005), 283–294.
50. M. A. Henning and D. Pradhan, Algorithmic aspects of upper paired-domination in graphs. *Theoret. Comput. Sci.* **804** (2020), 98–114.
51. M. A. Henning and J. Southey, A characterization of graphs of minimum size having diameter two and no dominating vertex, manuscript (2012).
52. M. A. Henning and P. D. Vestergaard, Trees with paired domination number twice their domination number. *Util. Math.* **74** (2007), 187–197.

53. X. Hou, A characterization of $(2\gamma, \gamma_p)$-trees. *Discrete Math.* **308** (2008), 3420–3426.
54. X. Hou and M. Edwards, Paired domination vertex critical graphs. *Graphs Combin.* **24** (2008), 453–459.
55. X. M. Hou and F. Jiang, Paired domination of Cartesian products of graphs. *J. Math. Res. Exposition* **30** (2010), 181–185.
56. S. Huang and E. Shan, A note on the upper bound for the paired domination number of a graph with minimum degree at least two. *Networks* **57** (2011), 115–116.
57. S. Huang, E. Shan, and L. Kang, Perfect matchings in paired domination vertex critical graphs. *J. Combin. Optim.* **23** (2012), 507–518.
58. S. Huang, L. Kang, and E. Shan, Paired-domination in claw-free graphs. *Graphs Combin.* **29** (2013), 1777–1794.
59. L. Kang, M. Y. Sohn, and T. C. E. Cheng, Paired domination in inflated graphs. *Theoret. Comput. Sci.* **320** (2004), 485–494.
60. E. Lappas, S. D. Nikolopoulos, and L. Palios, An O($n$)-time algorithm for the paired domination problem on permutation graphs. *European J. Combin.* **34** (2013), 593–608.
61. C. Lin and H. Tu, A linear-time algorithm for paired-domination on circular-arc graphs. *Theoret. Comput. Sci.* **591** (2015), 99–105.
62. C. Lu, R. Mao, and B. Wang, Paired-domination in claw-free graphs with minimum degree at least four. *Ars Combin.* **139** (2018), 393–409.
63. C. Lu, B. Wang, K. Wang, and Y. Wu, Paired-domination in claw-free graphs with minimum degree at least three. *Discrete Appl. Math.* **257** (2019), 250–259.
64. C. Lu, C. Wang, and K. Wang, Upper bounds for the paired-domination numbers of graphs. *Graphs Combin.* **32** (2016), 1489–1494.
65. G. MacGillivray and K. Seyffarth, Domination numbers of planar graphs. *J. Graph Theory* **22** (1996), 213–229.
66. H. Müller and A. Brandstädt, The NP-completeness of Steiner tree and dominating set for chordal bipartite graphs, *Theor. Comput. Sci.* **53** (1987), 257–265.
67. C. M. Mynhardt and M. Schurch, Paired domination in prisms of graphs. *Discuss. Math. Graph Theory* **31** (2011), 5–23.
68. B.S. Panda and D. Pradhan, A linear time algorithm for computing a minimum paired-dominating set of a convex bipartite graph. *Discrete Appl. Math.* **161** (2013), 1776–1783.
69. P. Paulraja and S. Sampathkumar, A note on paired domination number of tensor product of graphs. *Bull. Inst. Combin. Appl.* **60** (2010), 79–85.
70. D. Pradhan and B. S. Panda, Computing a minimum paired-dominating set in strongly orderable graphs. *Discrete Appl. Math.* **253** (2019), 37–50.
71. K. E. Proffit, T. W. Haynes, and P.J. Slater, Paired domination in grid graphs. *Congr. Numer.* **150** (2001), 161–172.
72. H. Qiao, L. Kang, M. Cardel, and D. Z. Du, Paired domination of trees. Dedicated to Professor J.B. Rosen on his 80th birthday. *J. Global Optim.* **25** (2003), 43–54.
73. J. Raczek, Lower bound on the paired domination number of a tree. *Australas. J. Combin.* **34** (2006), 343–347.
74. O. Schaudt, Paired and induced paired domination in (E, net)-free graphs. *Discuss. Math. Graph Theory* **32** (2012), 473–485.
75. O. Schaudt, Total domination versus paired domination. *Discuss. Math. Graph Theory* **32** (2012), 435–447.
76. E. Shan, L. Kang, and M. A. Henning, A characterization of trees with equal total domination and paired domination numbers. *Australas. J. Combin.* **30** (2004), 31–39.
77. D. S. Studer, T. W. Haynes, and L. M. Lawson, Induced paired domination in graphs. *Ars Combin.* **57** (2000), 111–128.
78. W. Ulatowski, All graphs with paired-domination number two less than their order. *Opusc. Math.* **33** (2013), 763–783.
79. W. Ulatowski, The paired-domination and the upper paired-domination numbers of graphs. *Opusc. Math.* **35** (2015), 127–135.

80. V. G. Vizing, A bound on the external stability number of a graph. *Dokl. Akad. Nauk SSSR* **164** (1965), 729–731.
81. V. G. Vizing, Some unsolved problems in graph theory. *Uspehi Mat. Nauk* **23** (1968), (144), 117–134.

# Connected Domination

**Mustapha Chellali and Odile Favaron**

## 1 Introduction and Terminology

For a simple undirected graph $G = (V, E)$, we denote by $n(G) = |V|$ its *order*, by $m(G) = |E|$ its *size*, by $\delta(G)$ and $\Delta(G)$ its *minimum* and *maximum degrees*, and by $N(v)$ the neighborhood of the vertex $v$. More generally, we use the definitions and notation given in the glossary and we refer the reader to it. In particular, $\kappa$, $\kappa'$, $\alpha'$, $\beta$, diam, $g$, respectively, denote the *connectivity*, *edge connectivity*, *matching number*, *vertex covering number*, *diameter*, and *girth*. Concerning the domination parameters, $i(G)$ and $\alpha(G)$ denote the minimum and maximum cardinalities of a maximal independent set of $G$, while $\gamma(G)$ and $\Gamma(G)$ ($\gamma_t(G)$ and $\Gamma_t(G)$, $\gamma_{pr}(G)$ and $\Gamma_{pr}(G)$, respectively) denote the minimum and maximum cardinalities of a minimal dominating set (minimal total, or paired, dominating set) of $G$. Clearly $\gamma(G) \le \gamma_t(G)$. A subset $X$ of $V$ is irredundant if $N[x] - N[X - x] \neq \emptyset$ for every vertex $x$ of $X$. The minimum and maximum cardinalities of a maximal irredundant set of $G$ are denoted ir$(G)$ and IR$(G)$. We recall the well known inequalities chain ir$(G) \le \gamma(G) \le i(G) \le \alpha(G) \le \Gamma(G) \le IR(G)$, which are valid for all graphs.

A *cycle* (*path*) on $n$ vertices is denoted $C_n$ ($P_n$). A *star* is a complete bipartite graph $K_{1,k}$ and a *subdivided star* $S(K_{1,k})$ is obtained from a star by inserting a new vertex on each edge. A *double star* is a tree that has exactly two vertices that are not leaves. The *supports* of a tree are the neighbors of the leaves. The *corona* $G \circ H$ of two graphs $G$ and $H$ is obtained from one copy of $G$ and $n(G)$ copies of $H$ by

M. Chellali (✉)
LAMDA-RO Laboratory, Department of Mathematics, University of Blida, B.P. 270, Blida, Algeria
e-mail: m_chellali@yahoo.com

O. Favaron
University Paris-Saclay (honorary), LRI-CNRS, Orsay, France



T. W. Haynes et al. (eds.), *Topics in Domination in Graphs*, Developments in Mathematics 64, https://doi.org/10.1007/978-3-030-51117-3_4

joining by an edge the $i$th vertex of $G$ to every vertex of the $i$th copy of $H$ for every vertex $i$.

In 1979, on a suggestion of S.T. Hedetniemi, Sampathkumar and Walikar [133] introduced the concept of connected domination. A *connected dominating set* of a connected graph $G$ is a dominating set of $G$ whose induced subgraph is connected. The minimum (resp. maximum) cardinality of a minimal connected dominating set is the *connected dominating number* (resp. *upper connected dominating number*) and is denoted by $\gamma_c(G)$ (resp. $\Gamma_c(G)$). If $n > 1$ and $G$ has a universal vertex, i.e., if $\Delta = n - 1$, then $\gamma(G) = \gamma_c(G) = 1$ and $\gamma_t(G) = 2$. Otherwise, $2 \leq \gamma(G) \leq \gamma_t(G) \leq \gamma_c(G)$.

The *connected domatic number* $d_c(G)$ is the maximum order of a partition of $V(G)$ into connected dominating sets of $G$. Clearly, $d_c(G) \leq \delta$ if $G \neq K_n$, and $d_c(G)\gamma_c(G) \leq n$. Baogen et al. [20] observed that the last inequality implies $\gamma_c(G) + d_c(G) \leq \lfloor \frac{n}{2} \rfloor + 2$ as soon as $\gamma_c(G) \geq 2$ and $d_c(G) \geq 2$ (consider the product $(\gamma_c(G) - 2)(d_c(G) - 2)$). Otherwise, $\gamma_c(G) + d_c(G) \leq n + 1$ with equality if and only if $G = K_n$ [79].

In a Wireless Sensor Network (WSN) server nodes send data through multi-hops involving intermediate relay nodes. The use of a virtual backbone (VB) subset $D$ of sensors such that every sensor not in $D$ can directly communicate with a sensor in $D$ and two sensors in $D$ can communicate only through sensors in $D$, can reduce the energy depletion and the risk of transmission of redundant information. This leads to model a WSN by a graph $G$ and a VB by a connected dominating set of $G$. Finding a small VB is interesting to simplify the transmission and save energy. Unit Disk Graphs (Unit Ball Graphs), which are the intersection graphs of equal-sized circles in the plane (spheres in $R^3$), are particularly interesting models. A large literature is devoted to this domain, especially the research of efficient algorithms.

In their book, Du and Wan [58] give the following application of the connected domatic number in order to maximize the lifetime of a WSN. When a very large number of sensors are randomly deployed in target field, the addition of redundant sensors can create disjoint connecting dominating sets. By properly scheduling the activation and sleep times of sensors, the disjoint connected dominating sets can be organized to work in different times periods as VBs, thereby increasing the lifetime of the sensor networks by the factor of the number of connected dominating sets.

Hedetniemi and Laskar [79] proved that a subset $S$ of vertices of $G$ is a minimal connected dominating set if and only if $S$ is the set of the non-leaf vertices of a spanning tree of $G$ (the part "if" was previously observed by Sampathkumar and Walikar in [133]). Hence $\gamma_c(G) = n - \varepsilon_T(G)$, where $\varepsilon_T(G)$ denotes the maximum number of leaves of a spanning tree $T$ of $G$ (this relation is similar to $\gamma(G) = n - \varepsilon_F(G)$ where $\varepsilon_F(G)$ is the maximum number of leaves in a spanning forest $F$ of $G$). Many papers on connected domination take the point of view of the maximum number of leaves of a spanning tree. We express here all results in terms of $\gamma_c$ rather than in terms of $\varepsilon_T$.

The remainder of this chapter is structured as follows: Section 2 provides various upper and lower bounds on the connected domination parameters as well as their relationships with some graph parameters. Section 3 reviews some results on the connected domination number when the structure of the graph is modified by elementary operations on vertices or edges (addition, deletion, contraction). Section 4 presents seven of the most important variants of the connected domination number. Section 5 reviews the complexity and algorithmic aspects of the connected domination problem, and Section 6 presents some conjectures about the connected domination number.

# 2 Properties of the Connected Domination Parameters

The purpose of this section is to study the relationships between $\gamma_c$, $\Gamma_c$, $d_c$, and other graph parameters.

## *2.1 Order n, size m, and Maximum Degree Δ*

In general connected graphs, $\gamma_c(G) \leq n-2$ with equality if and only if $G$ is a path or a cycle. Smaller bounds are known under additional conditions on $G$. For instance, Desormeaux et al. [53] showed that if $G$ has diameter 2, then $\gamma_c(G) \leq \frac{1}{2}(1+3\sqrt{n\ln n})$, and $\gamma_c(G) \leq 3$ if moreover $G$ is planar, $\gamma_c(G) \leq 4$ if moreover $\overline{G}$ is planar. Kleitman and West [98] proved that if every edge of $G$ belongs to a triangle, then $\gamma_c(G) \leq \frac{2n-5}{3}$. A chain obtained from $p$ $K_4 - e$ with vertices of degree 2 denoted $x_i$ and $y_i$ by letting $y_i = x_{i+1}$ for $1 \leq i \leq p-1$ shows that this bound is tight. An example of a chain of $K_4 - e$ is illustrated in Figure 1.

Concerning the connected domatic number $d_c$, Hartnell and Rall [75] showed that if the connected graph $G$ is planar, then $d_c(G) \leq 4$ with equality if and only if $G$ is the clique $K_4$. Moreover, if $d_c(G) = 3$, then each class of a 3-domatic partition induces a path in $G$.

The bounds on $\gamma_c$ in terms of the size $m$ or the maximum degree $\Delta$ convey the intuitive idea that $\gamma_c$ and $m$ or $\Delta$ cannot be simultaneously large. The first ones were established by Sampathkumar and Walikar in [133] and completed by Hedetniemi and Laskar in [79].

**Theorem 1** *Let $G$ be a connected graph of order $n$, size $m$, and maximum degree $\Delta \geq 2$. Then*

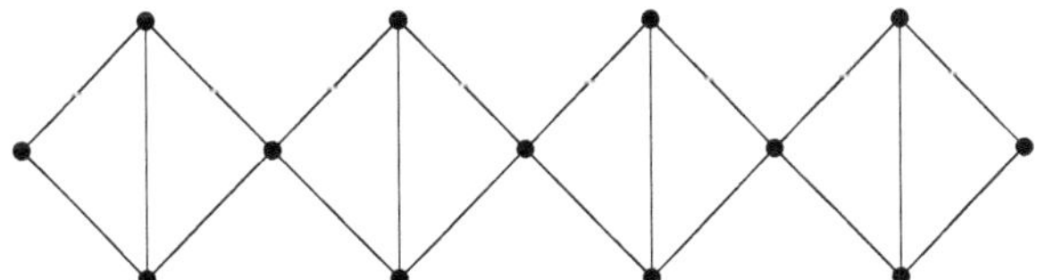

**Fig. 1** A graph $G$ with $\gamma_c(G) = (2n-5)/3$

$$\frac{n}{\Delta+1} \leq \gamma(G) \leq \gamma_c(G) = n - \varepsilon_T(G) \leq n - \Delta \leq n-2 \leq 2m-n.$$

Moreover, $\gamma_c(G) = \frac{n}{\Delta+1}$ if and only if $\Delta = n-1$, i.e., $G$ has a universal vertex. Also $\gamma_c(G) = 2m-n$ if and only if $G$ is a tree (since $m = n-1$) with $\Delta = 2$, i.e., a path. For a tree $T$, $\gamma_c(T) = n - \Delta$ if and only if $T$ is a subdivided star. When $\Delta \leq 3$, a better lower bound on $\gamma_c$ is given in Theorem 14.

We observe that the inequalities $2m \leq n\Delta \leq n(n-\gamma_c(G))$ imply $\gamma_c(G) \leq \frac{n^2-2m}{n}$ or, equivalently, $m \leq \frac{n(n-\gamma_c(G))}{2}$. Arumugam and Velammal [14] showed that $m = \lfloor \frac{n(n-\gamma_c(G))}{2} \rfloor$ if and only if $G$ is a clique or a clique minus a minimum edge cover, or a cycle. For these graphs, $\gamma_c(G) = 1, 2$ and $n-2$, respectively.

For graphs $G$ with $\gamma_c(G) \geq 3$, Sanchis improved the previous bound $B_1 = \frac{n(n-\gamma_c(G))}{2}$ on $m$.

**Theorem 2 ([131])** *Let $G$ be a connected graph of order $n$, size $m$, and $\gamma_c(G) \geq 3$. Then $m \leq B_2 = \binom{n-\gamma_c(G)+1}{2} + \gamma_c(G) - 1$, and the bound is best possible.*

When $\gamma_c(G) > 3$, $m = B_2$ if and only if $G$ is the union of a clique $K_{n-d}$ and a path $P_d$, where each vertex in the clique is adjacent to exactly one of the endpoints of the path, and each endpoint has at least one clique vertex adjacent to it. When $\gamma_c(G) = 3$, $m = B_2$ if and only if $G$ has the previous form or $G$ has a particular structure described in [131]. The reader can check that $2(B_2 - B_1) = (\gamma_c(G) - (n-2))(\gamma_c(G) - 1) \leq 0$ for every graph.

Another relationship between $\gamma_c$ and $m$ was given by Ding et al. [56]. They proved that if $m \geq n + \frac{t(t-1)}{2}$ and $n \neq t+2$, then $\gamma_c(G) < n - t$ and this is best possible.

In triangle-free graphs, Mukwembi lowered the bound of Theorem 2.

**Theorem 3 ([122])** *Let $G$ be a connected triangle-free graph of order $n$. Then $m(G) \leq \frac{(n-\gamma_c(G))^2}{4} + n - 1$ and thus $\gamma_c(G) \leq n - \frac{4m}{n} + 2$.*

The first bound is sharp as shown by joining the endvertices of a path $P_p$, respectively, to all vertices of different classes of a complete bipartite graph $K_{\frac{n-p}{2},\frac{n-p}{2}}$. Conjecture 2 of Written on the Wall II [52] states that $\gamma_c(G) \leq n - 2\frac{\Sigma_v \alpha(G[N(v)])}{n} + 2$ in every connected graph $G$, where $\alpha(G[N(v)])$ is the independence number of the subgraph induced by the neighbors of $v$. Since $\Sigma_v \alpha(G[N(v)]) = 2m$ in every triangle-free graph, Theorem 3 proves this conjecture in the particular case of triangle-free graphs.

Hartnell and Rall gave an upper bound, attained, for instance, by $K_{n/2,n/2}$, on the size of $G$ in terms of the connected domatic number.

**Theorem 4 ([75])** *Let $G$ be a connected graph of order $n$ and size $m$. Then $m \leq \frac{1+d_c(G)}{2}n - d_c(G)$.*

## 2.2 Minimum Degree $\delta$ and Degree Sequence

Let $l(n, k)$ be the maximum number of leaves of a spanning tree of a connected graph $G$ of order $n$ and minimum degree $\delta \geq k$, and let $\gamma_c(n, k) = n - l(n, k)$. In 1988, Linial conjectured that $l(n, k) \geq \frac{k-2}{k+1}n + c_k$, i.e., $\gamma_c(n, k) \leq \frac{3n}{k+1} - c_k$, where $c_k$ is a constant depending on $k$ (unpublished, cited by many authors). The family of graphs described below shows that Linial's bound would be sharp if correct.

**Example 1** A *necklace* $L_k(q)$ is a $k$-regular graph consisting of $q$ graphs $H_i$, where $H_i$ is a clique $K_{k+1}$ minus one edge $x_i y_i$, together with $q$ edges of the form $y_i x_{i+1}$ (mod $q$). For this graph, $n = q(k+1)$ and $\gamma_c(L_k(q)) = 3q - 2 = \frac{3n}{k+1} - 2$. For example, the graph $L_3(4)$ is illustrated in Figure 2.

Linial's Conjecture fails for large values of $n$ and $k$. Thomassé and Yeo [140] proved that for each value of $k$ there exists a $k$-uniform hypergraph $H$ with $n$ vertices and $n$ edges such that $\beta(H) \geq (1-\epsilon)\frac{\ln k}{k}n$, where $\beta(H)$ is the minimum cardinality of a vertex cover of $H$. Consider a $k$-regular graph $G$ with $V(G) = V(H)$ and such that $\{N(v); v \in V(G)\} = E(H)$. The vertex covers of $H$ are the total dominating sets of $G$ and thus $\gamma_c(G) \geq \gamma_t(G) \geq (1-\epsilon)\frac{\ln k}{k}n$ (cf [80]).

Kleitman and West [98] proved that if $k$ is sufficiently large, there is an algorithm that constructs a spanning tree with at least $(1 - b\frac{\ln k}{k})n$ leaves in any graph with minimum degree $k$, where $b$ is any constant exceeding 2.5. This shows that the previous lower bound on $\gamma_c(G)$ is asymptotically its good value. Caro et al. [33] lowered the constant $b$ and gave two proofs of an upper bound on $\gamma_c(G)$. The first one is probabilistic. The second one is a polynomial-time algorithm constructing a spanning tree with many leaves.

**Theorem 5 ([33])** *Let $G$ be a connected graph of order $n$ and minimum degree $\delta \geq k$. Then*

1. $\gamma_c(G) \leq n\frac{145+0.5\sqrt{\ln(k+1)}+\ln(k+1)}{k+1}$. *Hence* $\gamma_c(G) \leq (1 + o_k(1))\frac{\ln(k+1)}{k+1}n$.
2. *Given* $\epsilon > 0$ *and $k$ sufficiently large in terms of* $\epsilon$, $\gamma_c(G) \leq (1+\epsilon)\frac{\ln k}{k}n$.

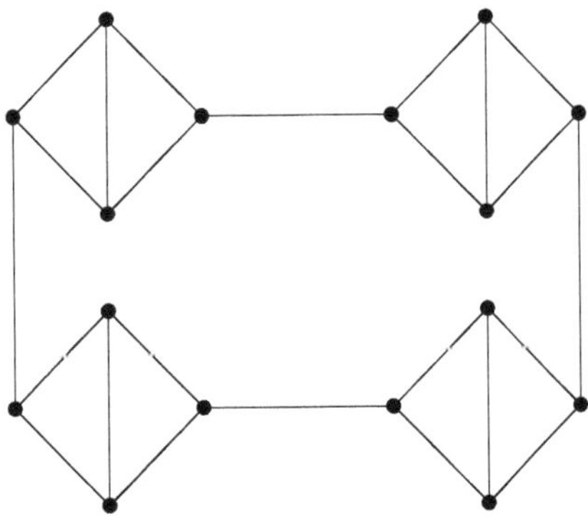

**Fig. 2** The necklace $L_3(4)$

For small values of $\delta$, Linial's type bounds have been established until $\delta = 5$. We give below the main known results.

**Theorem 6 ([98])** *Every connected graph $G$ of order $n$ and minimum degree $\delta \geq 3$ satisfies $\gamma_c(G) \leq \frac{3n}{4} - 2$.*

The result is sharp as shown by the necklace $L_3(q)$ described in Example 1. It was already proved for cubic graphs by Storer [137], Payan et al. [126], and Griggs et al. [69].

Griggs et al. [69], Bonsma [23], and Mafuta et al. [116] lowered the bound of Theorem 6 by forbidding some induced subgraphs. A claw is the graph $K_{1,3}$, a diamond is the graph $K_4 - e$, and a paw is the graph $K_{1,3} + e$ of order 4. A diamond of a graph $G$ is a cubic diamond if its four vertices have degree 3 in $G$.

**Theorem 7 ([69])** *Every connected diamond-free cubic graph $G$ of order $n$ satisfies $\gamma_c(G) \leq \frac{2n-4}{3}$.*

**Theorem 8 ([23])** *Let $G$ be a connected graph of order $n$ and minimum degree $\delta \geq 3$.*

*1. If $G$ contains $D \geq 0$ cubic diamonds, then $\gamma_c(G) \leq \frac{5n+D-12}{7}$.*
*2. If $G$ is triangle-free, then $\gamma_c(G) \leq \frac{2n-4}{3}$.*

**Theorem 9 ([116])** *Let $G$ be a connected graph of order $n$ and minimum degree $\delta \geq 3$.*

*1. If $G$ is paw-free, then $\gamma_c(G) \leq \frac{2n-4}{3}$.*
*2. If $G$ is paw-free and claw-free, then $\gamma_c(G) \leq \frac{n-1}{2}$.*

For $p \equiv 0 \pmod 3$, the necklace $G$ constructed from $p$ cycles $C_6 = x_1^i x_2^i \cdots x_6^i x_1^i$ by adding the edges $x_2^i x_5^i$, $x_3^i x_6^i$, $x_4^i x_1^{i+1}$ for $1 \leq i \leq p \pmod p$ satisfies $\gamma_c(G) = \frac{2n}{3} - 2 = \lfloor \frac{2n-4}{3} \rfloor$ and shows that the common bound of Theorems 7, 8-2, and 9-1 is tight. This example contains cycles $C_4$. By also forbidding $C_4$, Mafuta et al. [117] slightly lowered the bound of Theorem 8-2 by showing that if $G$ is $(C_3, C_4)$-free, then $\gamma_c(G) \leq \frac{2n-5}{3}$.

In a necklace $L_3(q)$ defined in Example 1 with $q = D + \ell$, replace each of the first $\ell$ diamonds $K_4 - e$ by a subgraph of vertices $x, a, b, z, c, d, y$ and edges $xa, xb, ab, za, zb, zc, zd, cd, yc, yd$. In this graph, $\delta = 3$, $n = 7\ell + 4D$, and $\gamma_c = 5\ell + 3D - 2 = \lfloor \frac{5n+D-12}{7} \rfloor$. This proves that the bound of Theorem 8-1 is tight.

Linial's bound was established by Kleitman and West for $\delta = 4$ and by Griggs and Wu for $\delta = 4$ and $\delta = 5$.

**Theorem 10 ([70, 98])** *Let $G$ be a connected graph of order $n$ and minimum degree $\delta \geq 4$. Then $\gamma_c(G) \leq \frac{3n-8}{5}$.*

The squares $C_6^2$ and $C_8^2$ of the cycles $C_6$ and $C_8$ satisfy $\gamma_c(G) = \lfloor \frac{3n-8}{5} \rfloor$. Griggs and Wu [70] showed that if other graphs satisfy this equality, they are necessarily

4-regular and each edge belongs to a triangle. This leads Mafuta et al. to forbid triangles.

**Theorem 11 ([117])** *Let $G$ be a connected triangle-free graph of order $n$ and minimum degree $\delta \geq 4$. Then $\gamma_c(G) \leq \frac{4n-12}{7}$. If moreover $G$ is $C_4$-free, then $\gamma_c(G) \leq \frac{4n}{7} - 3$.*

**Theorem 12 ([70])** *Let $G$ be a connected graph of order $n$ and minimum degree $\delta \geq 5$. Then $\gamma_c(G) \leq \frac{n}{2} - 2$.*

The 5-regular necklaces $L_5(q)$ of Example 1 show that the bound of Theorem 12 is tight. The authors of Theorem 11 gave an example of a 4-regular graph $G$ of order $n = 10$ with $\gamma_c(G) = 4 = \frac{4n-12}{7}$ and constructed a 5-regular triangle-free graph $G$ of order $n = 12$ with $\gamma_c(G) = \frac{n}{2} - 2$.

**Theorem 13 ([116])** *Let $G$ be a connected claw-free paw-free graph of order $n$ and minimum degree $\delta \geq 5$. Then $\gamma_c(G) \leq \frac{3n-6}{7}$.*

Stronger results can be obtained by considering the degree sequence of $G$ rather than its minimum degree. Let $n_i$ ($n_{\geq i}$) be the number of vertices of $G$ of degree $i$ (at least $i$). Payan et al. [126] established the following framework on $\gamma_c(G)$ for subcubic graphs $G$.

**Theorem 14 ([126])** *Let $G$ be a connected graph such that $n_{\geq 4} = 0$, i.e., $\Delta \leq 3$. Then $\frac{n}{2} - 1 \leq \gamma_c(G) \leq n - \frac{n_3}{4} - 2$.*

The lower bound is tight as shown by the graph constructed from a cycle $C_{2p}$ by adding the chords $x_i x_{2p-i+2}$, $2 \leq i \leq p$, and $x_1 x_{p+1}$. For cubic graphs, the upper bound is the same as in Theorem 6.

Let $G$ be a connected graph of order $n \geq 3$ with $n_2$ vertices of degree 2.

**Theorem 15** *1. Bankevich and Karpov [19] $\gamma_c \leq \frac{3n+n_2-6}{4}$.*
*2. Bankevich [18] If $G$ is triangle-free, then $\gamma_c \leq \frac{2n+n_2-4}{3}$.*

Paths show that the previous two bounds are sharp. Another limit example for Theorem 15.1 can be constructed from $p$ triangles $x_i y_i z_i$ and $p + 2$ vertices $y_0, x_{p+1}, t_i$, $1 \leq i \leq p$ by adding the edges $y_i x_{i+1}$ for $0 \leq i \leq p - 1$ and $z_i t_i$ for $1 \leq i \leq p$. Then $n_2 = 0$ and $\gamma_c(G) = 3p = \frac{3(n-2)}{4}$.

A tight upper bound on $\gamma_c$ in terms of the maximum length of a path of $G$ formed by vertices of degree 2 can also be found in [19]. Zhuang [154] proved that if $G$ is maximal outerplanar, then $\gamma_c(G) \leq \min\left\{\left\lfloor \frac{n+n_2}{2} \right\rfloor - 2, \left\lfloor \frac{2(n-n_2)}{3} \right\rfloor\right\}$.

In the following theorems due to Karpov, the graph $H$ is obtained from a cycle $C_6 = x_1 x_2 \ldots x_6 x_1$ and two vertices $y$ and $z$ by adding the edges $yx_2, yx_3, yx_5, yx_6, zx_1, zx_4, zx_5, zx_6, x_1x_3, x_2x_4$.

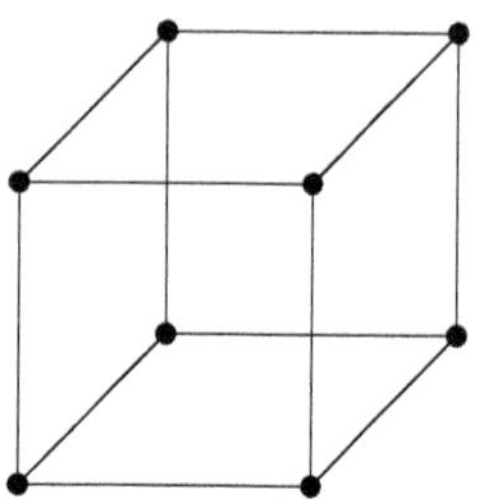

**Fig. 3** Graph $Q_3$

**Theorem 16** *Let G be a connected graph of order n.*

1. *Karpov [95]* $\gamma_c(G) \le n - \frac{n_1+n_3}{4} - \frac{n_{\ge 4}}{3} - \frac{3}{2}$.
2. *Karpov [94]* $\gamma_c(G) \le n - \frac{n_3+2n_{\ge 4}}{5} - c$*, where* $c = 8/5$ *if* $G = C_6^2$, $c = 9/5$ *if* $G = C_8^2$ *or* $G = H$, $c = 2$ *otherwise.*

The chain constructed from $k$ cycles $C_6 = x_i a_i y_i b_i z_i c_i x_i$ and $k+2$ vertices $z_0, x_{k+1}, t_i, 1 \le i \le k$, by adding the edges $z_i x_{i+1}, a_i b_i, b_i c_i, c_i a_i, z_i t_i, 1 \le i \le k$, is an example showing the tightness of Theorem 16-1. The cycle constructed from $k$ cliques $K_4$ of vertex sets $\{x_i, y_i, z_i, t_i\}$ and $2k$ vertices $a_i, b_i$ by adding the edges $a_i x_i$,$a_i y_i$, $b_i z_i$, $b_i t_i$, $b_i a_{i+1}$ (mod $k$) is an example showing the tightness of Theorem 16-2.

Before giving the main results of Theorem 17 due to Bonsma and Zickfeld [26], we need some new definitions of forbidden structures. A 2-*blossom* of endvertices $c_1$ and $c_2$ is formed by a cycle $C_6 = c_1 x y c_2 z t c_1$, a vertex $u$ and the edges $ux, uy, uz, ut$. A *diamond chain* (called 2-necklace in [26]) of endvertices $u_1$ and $v_k$ is formed by any number $k$ of diamonds consisting in a $C_4 = u_i a_i v_i b_i u_i$ plus the chord $a_i b_i$ by letting $v_i = u_{i+1}$ for $1 \le i \le k-1$. A 2-blossom $H$ or a diamond chain $H$ is said to be contained in a graph $G$ if $d_G(v) = d_H(v)$ for every vertex $v$ different from an endvertex and $d_G(v) = 3$ for each endvertex. Note that a cubic-diamond of $G$ is a diamond chain of $G$ containing one diamond. The graph $G_7$ of Theorem 17-1 is obtained from a 2-blossom by adding the two edges $c_1 c_2$ and $yz$. Let $Q_3$ be the 3-dimensional cube illustrated in Figure 3.

**Theorem 17 ([26])** *Let G be a connected graph of order* $n \ge 3$.

1. *If G contains no 2-blossom nor diamond chain, then* $\gamma_c(G) \le n - \frac{n_{\ge 3}}{3} - c$ *with* $c = 4/3$ *if* $G = Q_3$, $c = \frac{5}{3}$ *if G is cubic different from* $Q_3$ *or* $G = G_7$ *and* $c = 2$ *otherwise.*
2. *If G contains no cubic diamond and G is different from* $K_4$ *and from* $Q_3$*, then* $\gamma_c(G) \le n - \frac{2n_{\ge 3}}{7} - 2$.
3. *If G contains no diamond chain, then* $\gamma_c(G) \le n - \frac{4n_{\ge 3}}{13} - c$ *with* $c = 20/13$ *if G is cubic and* $c = 24/13$ *otherwise.*

Theorem 17-1 generalizes Theorem 7. Hence the limit examples for Theorem 7 still work for Theorem 17-1. Non-cubic graphs attaining the bound of Theorem 17-1 are also given in [26]. Theorem 17-2 generalizes Theorem 8-1 with $D = 0$.

Hence the tight limit example given after Theorem 8 still works for Theorem 17-2. For cubic graphs, the conditions of Theorems 17-2 and 17-3 are the same and the smallest bound of Theorem 17-3 is the good one. This explains why the graphs attaining the bound in Theorem 8-1 cannot be cubic. The graphs attaining the bound in Theorem 17-3 are described in [26] under the name of flower trees.

A stronger result taking into account the number of special 2-blossoms and diamond chains is also proved in [26].

Desormeaux et al. [53] considered the non-increasing degree sequence $d_1 = \Delta \geq d_2 \geq \cdots \geq d_n = \delta$ of $G$ and defined $\mathrm{ord}_c(G)$ as the smallest integer $k$ such that $\Sigma_{i=1}^{k} d_i \geq n + k - 2$.

**Theorem 18 ([53])** *For every connected graph $G$ of order $n \geq 3$ with $\Delta \leq n - 2$, $\gamma_c(G) \geq ord_c(G)$. If moreover $G$ is a tree $T$, then $\gamma_c(T) = ord_c(T)$.*

Concerning the connected domatic number, Zelinka [153] observed that $d_c(G) \leq \frac{n+n_{n-1}}{2}$.

## 2.3 Other Domination Parameters

Recall that if $\Delta(G) \leq n - 2$, then $\gamma(G) \leq \gamma_t(G) \leq \gamma_c(G)$, otherwise $\gamma(G) = \gamma_c(G) = 1$ and $\gamma_t(G) = 2$. Moreover, if some cliques dominate $G$, then $\gamma_{cli}(G)$ is the minimum cardinality of such a clique and $\gamma_c(G) \leq \gamma_{cli}(G)$.

The first results concern the equality $\gamma = \gamma_c$ are duc to Arumugam and Paulraj Joseph [13]. Let $G_1$ (resp. $G_2$) be the graph obtained from a cycle $x_1 x_2 \cdots x_8 x_1$ by adding the edges $x_1x_6, x_2x_5, x_3x_7, x_4x_8$ (resp. $x_1x_4, x_2x_5, x_3x_7, x_6x_8$).

**Theorem 19 ([13])** *Let $G$ be a connected graph of order $n \geq 3$.*

1. *If $G$ is cubic, then $\gamma_c(G) = \gamma(G)$ if and only if $G \in \{K_4, \overline{C_6}, K_{3,3}, G_1, G_2\}$.*
2. *If $G$ is a tree, then $\gamma_c(G) = \gamma(G)$ if and only if every non-leaf vertex of $G$ is a support.*

The authors of [13] also characterized unicyclic graphs satisfying $\gamma_c = \gamma$. In [41], Chen et al. generalized to block graphs and to cacti the characterization of trees and unicyclic graphs with this property. A block of a graph is a maximal induced subgraph without cutvertex. A block graph is a graph all blocks of which are complete. An endcutvertex $v$ of a block graph is such that at least one component of $G - v$ is a block. A cactus is a graph all blocks of which are cycles or $P_2$. Trees are particular block graphs and unicyclic graphs are particular cacti.

**Theorem 20 ([41])** *Let $G$ be a connected block graph of order $n \geq 3$. Then $\gamma_c(G) = \gamma(G)$ if and only if every cutvertex of $G$ is an endcutvertex.*

The description of connected unicyclic graphs or cacti such that $\gamma_c(G) = \gamma(G)$ is rather complicated. We refer the reader, respectively, to [13] and [41]. It is worth noting that Alvarado, Dantas, and Rautenbach [2] showed that it is NP-hard to decide that $\gamma_c(G) = \gamma(G)$ for a given graph $G$, which leads to the conclusion that these graphs do not have a simple structure.

Chen [38] characterized the trees and the unicyclic graphs for which $\gamma_c = \gamma_t$.

**Theorem 21 ([38])** *Let $T$ be a tree different from a star. Then $\gamma_c(T) = \gamma_t(T)$ if and only if for every internal vertex $v$ different from a support, at least one component of $T - v$ exactly contains one internal vertex of $T$.*

The cycles $C_4$, $C_5$ or the unicyclic graph obtained by attaching at least one leaf at two vertices at distance 2 on a $C_6$ are examples of graphs for which $\gamma_c = \gamma_t$. However, the complete description of all these graphs is long and can be found in [38].

Igor Zverovich [155] characterized the perfect $(\gamma - \gamma_c)$-graphs, i.e., the connected graphs $G$ such that $\gamma_c(H) = \gamma(H)$ for every induced subgraph $H$ of $G$.

**Theorem 22 ([155])** *For a connected graph G, the following properties are equivalent :*

1. *$G$ is $\{P_5, C_5\}$-free.*
2. *$\gamma_c(H) = \gamma(H)$ for every connected induced subgraph $H$ of $G$.*

From Theorem 22, every $\{P_5, C_5\}$-free graph $G$ has a minimum dominating set which is connected. Bacsó and Tuza, and independently Goddard and Henning, observed that the $\{P_5, C_5\}$-freeness implies that this connected dominating set is actually a clique. This shows the equivalence between the $\gamma - \gamma_c$ perfection and the $\gamma - \gamma_{\mathrm{cli}}$ perfection.

**Theorem 23 ([17, 68])** *For a connected graph G, the following properties are equivalent:*

1. *$G$ is $\{P_5, C_5\}$-free;*
2. *every connected induced subgraph of G has a dominating clique;*
3. *$\gamma(H) = \gamma_{cli}(H)$ for every connected induced subgraph $H$ of $G$.*

As noticed in [68], the previous condition is also equivalent to $\gamma(H) = \gamma_t(H)$ for every connected induced subgraph $H$ of $G$ such that $\gamma(H) \geq 2$. Because of the problem of the graphs $G$ with $\gamma_c(G) = 1$ and $\gamma_t(G) = 2$, it is not possible to ask for $\gamma_c(H) = \gamma_t(H)$ for every connected induced subgraph $H$ of $G$ in the definition of the $\gamma_t - \gamma_c$ perfection. So Schaudt studied the condition $\gamma_c(H) \leq \gamma_t(H)$. In Theorem 24, $G_1$ is obtained from a triangle $abc$ by attaching a path $auv$ at $a$; $G_2$ is obtained from two triangles $a_i b_i c_i$, $1 \leq i \leq 2$, by letting $a_1 = a_2$; $F_1$ (resp. $F_2$) is obtained from $G_1$ (resp. $G_2$) by attaching a pendant vertex at $b, c, v$ (resp. $b_1, c_1, b_2, c_2$). The *leaf graph* of a graph $G$ is obtained by attaching a pendant vertex at each non-cutvertex of $G$. Hence $F_i$ is the leaf graph of $G_i$ for $i = 1, 2$.

**Theorem 24 ([134])** *For a connected graph G the following are equivalent.*

1. *$\gamma_c(H) \leq \gamma_t(H)$ for every nontrivial connected subgraph $H$;*
2. *$G$ is $\{P_7, C_7, F_1, F_2\}$-free;*

3. *Any connected induced subgraph $H$ of $G$ has a connected dominating set $X$ which is $\{P_5, G_1, G_2\}$-free.*

Moreover, since $\gamma_t(H) \leq \min\{\Gamma_t(H), \gamma_{pr}(H), \Gamma_{pr}(H)\}$ for all graphs $H$ and since $\Gamma_t(H) = \gamma_{pr}(H) = \Gamma_{pr}(H) = 4 < \gamma_c(H) = 5$ for all $H \in \{P_7, C_7, F_1, F_2\}$, the previous conditions are also equivalent to $\gamma_c(H) \leq \Gamma_t(H)$ or $\gamma_c(H) \leq \gamma_{pr}(H)$ or $\gamma_c(H) \leq \Gamma_{pr}(H)$ for every nontrivial connected subgraph $H$ of $G$. The equivalence between the two last conditions of Theorem 24 is due to a structural result independently obtained by Bacsó [16] and Tuza [142]. Let $\mathcal{D}$ be a class of connected graphs closed under taking connected induced subgraphs. Dom($\mathcal{D}$) is the class of connected graphs $G$ such that each connected induced subgraph of $G$ contains a dominating subgraph belonging to $\mathcal{D}$. Bacsó and Tuza proved that the minimum induced subgraphs of Dom($\mathcal{D}$) are the cycle $C_{t+2}$ if $P_{t-1} \in \mathcal{D}$ but $P_t \notin \mathcal{D}$ and the leaf graphs of the minimal forbidden subgraphs of $\mathcal{D}$. Here $\mathcal{D}$ is the set of connected $\{P_5, G_1, G_2\}$-free graphs and Dom($\mathcal{D}$) is the set of connected $\{C_7, P_7, F_1, F_2\}$-free graphs.

The first upper bound on $\gamma_c$ in terms of $\gamma$ and other domination parameters was given by Duchet and Meyniel in [59].

**Theorem 25 ([59])** *In every connected graph $G$, $\gamma_c(G) \leq min\{3\gamma(G) - 2, 2\alpha(G) - 1\}$.*

The paths and cycles show that the bounds of Theorem 25 are asymptotically sharp.

Camby and Schaudt lowered the upper bound $3\gamma - 2$ on $\gamma_c$ of Theorem 25 in some hereditary classes of graphs.

**Theorem 26 ([32])** *Let $G$ be a connected graph.*

1. *$G$ is $\{P_6, C_6\}$-free if and only if $\gamma_c(H) \leq \gamma(H) + 1$ for every connected induced subgraph $H$ of $G$.*
2. *If $G$ is $\{P_8, C_8\}$-free, then $\gamma_c(G) \leq 2\gamma(G)$.*

The graph obtained from $K_{1,k}$ by subdividing each edge exactly once shows that the bound of Theorem 26-1 is tight. Let $F_k$ (resp. $G_k$) be the $\{P_7, C_7\}$-free (resp. $\{P_9, C_9\}$-free) graph obtained by attaching a path of length 2 (resp. 3) at each vertex of a clique $K_k$. Then $\gamma(F_k) = k$, $\gamma_c(F_k) = 2k$, $\gamma(G_k) = k + 1$, $\gamma_c(G_k) = 3k$. This shows that the bound 2 on $\gamma/\gamma_c$ in Theorem 26-2 is attained even in the class of $\{P_7, C_7\}$-free graphs and that the coefficient 3 of the bound on $\gamma/\gamma_c$ in Theorem 25 remains sharp in the class of $\{P_9, C_9\}$-free graphs.

In different articles due to Favaron and Kratsch, Wang, Bo and Liu, Sun, the bound $\gamma_c \leq 3\gamma - 2$ of Theorem 25 was improved by replacing $\gamma$ by $ir$ and bounds relating $\gamma_c$ to $\gamma_t$ or to $i$ were found.

**Theorem 27** *Let $G$ be a connected graph of order $n \geq 3$.*

1. *Favaron and Kratsch [64], Wang [147], and Bo and Liu [22] $\gamma_c(G) \leq 3ir(G)$–$2$.*
2. *Favaron and Kratsch [64] $\gamma_c(G) \leq 2\gamma_t(G) - 2$.*
3. *Sun [138] $\gamma_c(G) + i(G) \leq \lceil \frac{4n}{3} \rceil - 2$.*

The bounds of Theorem 27 are tight as shown by paths $P_{3k}$, $P_{4k}$, and $P_n$, respectively. Restricted to graphs with diameter two, Desormeaux et al. [53] proved the following.

**Theorem 28 ([53])** *Let $G$ be a diameter-2 graph of order $n$. Then $\gamma_c(G) \leq 3\gamma_t(G)/2 - 1$, implying $\gamma_c(G) \leq (1 + 3\sqrt{n \ln n})/2$.*

In Unit Disk Graphs, Wu et al. [150] proved that $\alpha(G) \leq 3.8\gamma_c(G) + 1.2$.

The following bound involves the cardinality $\gamma_k(G)$ of a minimum $k$-dominating set of $G$, i.e., a minimum dominating set $S$ of $G$ such that every vertex not in $S$ has at least $k$ neighbors in $S$. Chellali et al. [36] showed that every tree $T$ with maximum degree $\Delta(T) \geq k \geq 3$ satisfies $\gamma_c(T) \leq \gamma_p(T) - k + 1$ with equality if and only if $G$ is a subdivided star, or $\Delta(T) = 3$ and the vertices of degree 3 form an independent set.

The study of the upper connected domination parameter $\Gamma_c$ is more complicated since a minimal connected dominating set is not necessarily a minimal (resp. minimal total) dominating set. The ratios $\Gamma/\Gamma_c$ and $\Gamma_t/\Gamma_c$ are not bounded above. This can be seen from two adjacent vertices $x$ and $y$ by joining $x$ to every vertex of a path $P_k(x)$ and $y$ to every vertex of a path $P_k(y)$. Theorem 29 gives upper bounds on $\Gamma_c/\Gamma_t$ and on $\Gamma_c/\Gamma$, respectively, obtained by Favaron and Kratsch and by Ghoshal et al.

**Theorem 29** *Let $G$ be a connected graph.*

1. *Favaron and Kratsch [64]* $\Gamma_c(G) \leq 2\Gamma_t(G) - 2$.
2. *Ghoshal [67]* $\Gamma_c(G) \leq 2\Gamma(G) - 1$.

The first bound of Theorem 29 is tight as shown by attaching a pendant edge at each vertex $x_i$ with $i \equiv 2, 3 \pmod 4$ of a path $x_1 x_2 \ldots x_{4k}$. The second bound, which was conjectured in [64], is attained, for instance, by cycles $C_{4k+3}$.

The class $\mathcal{C}$ of connected graphs $G$ for which $\gamma_c(G) = \Gamma_c(G)$ is difficult to study. Arseneau et al. [12] studied the subclass $\mathcal{D}$ of connected graphs whose all spanning trees have the same number of leaves. Clearly $\mathcal{D} \subseteq \mathcal{C}$ but the inclusion is strict as shown, for instance, by the graph $H$ consisting of a $C_4 = xyztx$ plus a pendant edge $xu$. The graph $H$ admits spanning trees with two leaves, $\{y, u\}$ or $\{t, u\}$, and spanning trees with three leaves, $\{y, z, u\}$ or $\{t, z, u\}$. But only the sets $V \setminus \{y, z, u\}$ and $V \setminus \{t, z, u\}$ are minimal connected dominating sets and $\gamma_c(H) = \Gamma_c(H)$. Hence $H \in \mathcal{C} \setminus \mathcal{D}$.

**Theorem 30 ([12])** *A connected graph $G$ has the property that all spanning trees have the same number of leaves if and only if and only if both of the following conditions hold.*

1. *About each cycle in the graph $G$, the vertices are either all cutvertices, all non-cutvertices or alternating cut and non-cutvertices.*
2. *Every vertex of degree 3 or more is a cutvertex.*

## 2.4 Other Classical Graph Parameters

Since any path joining two diametral vertices has at least diam$(G) - 1$ internal vertices, it is clear that $\gamma_c(G) \geq \text{diam}(G) - 1$ for every graph.

Let $\kappa$ be the connectivity of a graph $G \neq K_n$ and let $A$ be a minimum cutset of $G$. Then every $\gamma_c(G)$-set contains at least one vertex of $A$. Thus $d_c(G) \leq \kappa(G)$ as noticed by Zelinka [153] and Sun [138]. Clearly $\gamma_c(G) \leq n - \Delta(G) \leq n - \kappa(G)$. The extremal graphs were characterized by Joseph and Arumugam [83] as follows.

**Theorem 31 ([83])** *Let $G$ be a connected graph of order $n$. Then $\gamma_c(G) \leq n-\kappa(G)$ with equality if and only if $G = C_n$, $K_n$, or $K_n - M$ with $n$ even and $M$ a perfect matching in $K_n$.*

Hedetniemi and Laskar [79] showed that every connected graph admits a maximum matching $M$ inducing a connected subgraph. Let $T$ be a spanning tree of $G$ admitting $V(M)$ as a connected dominating set and let $\mathcal{C}$ be a maximal $M$-alternating chain of $T$. As observed by Fajtlowicz (cf [52]), at least one endvertex $x$ of $\mathcal{C}$ belongs to $V(M)$ and has degree 1 in $T$ since $M$ is a maximum matching of the tree. Hence $V(M) - \{x\}$ is a connected dominating set of $G$. This implies the following bound on $\gamma_c$, attained, for instance, by odd cycles.

**Theorem 32 ([52])** *Let $G$ be a connected graph with maximum matching number $\alpha'(G)$. Then $\gamma_c(G) \leq 2\alpha'(G) - 1$.*

About the girth $g(G)$, the family $\mathcal{F}_k$ of graphs is defined by Desormeaux et al. [53] as follows. $\mathcal{F}_3$ is the family of graphs with $\Delta = n - 1$ and at least one triangle. $\mathcal{F}_4$ is the family of graphs obtained from a double star with central vertices $x$ and $y$ by adding at least one edge joining a leaf-neighbor of $x$ and a leaf-neighbor of $y$. For $k \geq 5$, $\mathcal{F}_k$ is the family of graphs constructed from a $k$-cycle $x_1x_2\dots x_kx_1$ by adding zero or more pendant edges incident to each $x_i$, $i \geq 3$; moreover, for $k = 5$ and 6, add zero or more edges joining $x_3$ and $x_k$ and subdivide each such added edge twice. Examples of graphs in the families $\mathcal{F}_k$ are illustrated in Figure 4.

**Theorem 33 ([53])** *Let $G$ be a connected graph with finite girth $g$.*

1. *$\gamma_c(G) \geq g - 2$ with equality if and only if $G \in \mathcal{F}_k$.*
2. *If $\Delta \leq n - 2$ and $g \geq 5$, then $\gamma_c(G) \geq \delta + 1$.*

The eccentricity ecc$(v)$ of a vertex $v$ of $G$ is the distance from $v$ to a vertex furthest away from $v$. The eccentricity of $G$ is $\text{ecc}(G) = \Sigma_{v \in V} \text{ecc}(v)$. Dankelmann and Mukwembi established the following upper bound on ecc$(G)$ in terms of the order and connected domination number.

**Theorem 34 ([47])** *Every connected graph $G$ satisfies* $\text{ecc}(G) \leq n(\gamma_c(G) + 1) - \frac{(\gamma_c(G))^2}{4} - \frac{\gamma_c(G)}{2}$.

The tree obtained by adding $n - k - 1$ pending edges at an endpoint of a path $P_{k+1}$ shows that this bound is sharp.

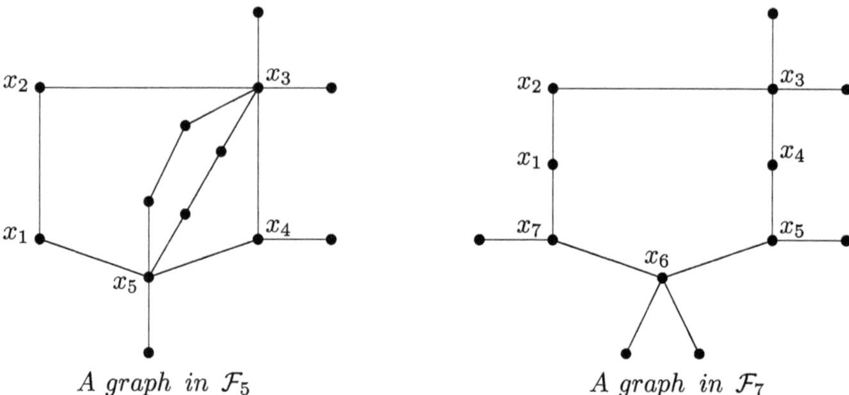


**Fig. 4** Graphs in the families $\mathcal{F}_k$

DeLaViña, Fajtlovicz, and Waller studied some properties of $\gamma_c$ in relation with Graffiti conjectures. The *bipartite number* $b(G)$ of $G$ is the maximum order of an induced bipartite subgraph of $G$ (clearly $b(G) \leq 2\alpha(G)$). The *local independence number* $\max_v \alpha(N(v))$ is the maximum value of the independence number of the neighborhood $N(v)$ taken on all the vertices $v \in V(G)$.

**Theorem 35** *Let G be a connected graph of order n.*

1. *DeLaViña et al. [51]* $\gamma_c(G) \leq 2\alpha - max_v \alpha(N(v)) + 1$.
2. *DeLaViña and Waller [50]* $\gamma_c(G) \leq b - \lceil \frac{max_v \alpha(N(v))}{2} \rceil$ *and* $\gamma_c(G) \leq n - 2\alpha + b - 1$.

The previous inequalities are tight, for instance, for odd cycles for the first two ones, for odd paths for the third one. The first one improves the bound $\gamma_c \leq 2\alpha - 1$ of Theorem 25. The two last ones constitute advances towards the solution of Conjectures 174 and 177 of graffiti.pc [52] (cf Conjectures 3 and 4 in Section 6).

## 2.5 Traceability and Hamiltonicity

A graph $G$ of order $n$ is *Hamiltonian* (resp. *traceable*) if it contains a cycle (resp. path) on $n$ vertices. The study of these problems in relation to the maximum number of leaves of a spanning tree was motivated by Graffiti.pc Conjectures 190 and 190a given in Section 6. The hypotheses of these conjectures are in terms of $\delta'(G)$, the second minimum smallest degree of $G$ in the non-decreasing degree sequence of $G$. However, the numerous papers written on the subject since 2013 replace $\delta'$ by $\delta$. Hence the results are similar to, but weaker than, Conjecture 190. We only cite the last ones, due to Mafuta et al., which give the strongest results.

**Theorem 36** *Let G be a connected graph of order n, minimum degree $\delta$ and connected dominating set $\gamma_c$.*

1. *Mafuta et al. [115] If $\delta \geq \frac{n-\gamma_c+1}{2}$, then G is traceable.*
2. *Mafuta [118] If $\delta \geq \frac{n-\gamma_c+2}{2}$, then G is Hamiltonian.*
3. *Mafuta [114] If G is 2-connected, $C_4$-free, $\delta \geq 4$ and $\delta \geq \frac{n-\gamma_c}{2}$, then G is Hamiltonian.*

The complete bipartite graphs $K_{p,p+2}$ and $K_{p,p+1}$, $p \geq 2$, respectively, show that the first two bounds of Theorem 36 are sharp. Note that for these graphs, $\delta' = \delta$, showing the sharpness of Graffiti.pc 190, if true.

## 2.6 Connected Domination in G and in Its Complement $\overline{G}$

For any parameter $\lambda$, we denote $\lambda(\overline{G})$ by $\overline{\lambda}$ and $\min\{\lambda, \overline{\lambda}\}$ by $\lambda^*$. We only consider nontrivial graphs. In $\overline{G}$, $\overline{\delta} = n - \Delta - 1$ and $\overline{\Delta} = n - \delta - 1$. Hence $\delta^* \leq \frac{\delta+\overline{\delta}}{2} \leq \frac{n-1}{2}$. When both $G$ and $\overline{G}$ are connected, then $n \geq 4$, $\delta^* \geq 1$, $\Delta \leq n-2$, and for $H \in \{G, \overline{G}\}$, $\gamma_c(H) \geq 2$ and $\text{diam}(H) \geq 2$. Moreover, $\gamma_c(\overline{H}) = 2$ if and only if $\text{diam}(H) \geq 3$. This explains the interest of considering graphs $G$ with $\text{diam}(G) = \text{diam}(\overline{G}) = 2$.

We first present some results on $\gamma_c$ and $\text{d}_c$ in $\overline{G}$ due to Desormeaux et al. [54], Sun [138], and Yu and Wang [152].

**Theorem 37 ([54])** *Let G be a graph of order n.*

1. *If $\gamma < \overline{\gamma}$, then $\gamma_c \leq \gamma + 1$, implying $\gamma_c \leq (1 + \sqrt{4n+1})/2$.*
2. *If both G and $\overline{G}$ are connected and $\gamma \leq \overline{\gamma} + 1$, then $\gamma_c \leq \gamma + 1$ or $\overline{\gamma}_c \leq \overline{\gamma} + 1$.*

**Theorem 38 ([138])** *If G and $\overline{G}$ are both connected, then $\overline{\gamma}_c \leq \kappa + 1$ where $\kappa$ is the connectivity of G.*

Equality is attained, for instance, with $G = P_n$.

The following result, sharp for $C_5$, was conjectured in [79].

**Theorem 39 ([152])** *If G and $\overline{G}$ are both connected, then $\gamma_c \leq 3\overline{d}_c$.*

A good survey on Nordhaus–Gaddum results on $\gamma_c$ until 2012 can be found in [10]. From $\gamma_c(G) \leq n - \Delta$, $\text{d}_c(G) \leq \delta$ and $\delta + \overline{\Delta} = n - 1$, it is clear that $\gamma_c + \overline{\gamma}_c \leq n - \Delta + \delta + 1 \leq n + 1$ and $\text{d}_c + \overline{\text{d}}_c \leq n - \Delta + \delta - 1 \leq n - 1$. These bounds, given in [79], were improved by Laskar and Peters for $\gamma_c$, Paulraj Joseph and Arumugam for $d_c$.

**Theorem 40 ([101])** *Let G and $\overline{G}$ be connected graphs of order $n \geq 4$. Then $\gamma_c + \overline{\gamma}_c = n + 1$ if $G = C_5$, $\gamma_c + \overline{\gamma}_c = n$ if $G = P_n$, $G = C_n$ for $n \geq 6$ or G is obtained by adding the two edges $x_1x_3$ and $x_2x_4$ to a cycle $C_6 = x_1x_2 \cdots x_6x_1$, $\gamma_c + \overline{\gamma}_c \leq n - 1$ otherwise.*

**Theorem 41 ([124])** *Let G and $\overline{G}$ be connected graphs of order n. Then* $d_c + \overline{d}_c \leq n - 2$.

All extremal graphs for Theorem 41 have order at most 8. Their list can be found in Theorem 3.25 of [10].

Karami et al. obtained smaller upper bounds on $\gamma_c + \overline{\gamma}_c$ by adding some conditions on $\delta^*$ or on $\gamma_c^*$, and an upper bound on the product $\gamma_c\overline{\gamma}_c$.

**Theorem 42 ([93])** *Let G be a graph of order n such that G and $\overline{G}$ are connected. Then*

1. $\gamma_c + \overline{\gamma}_c \leq \delta^* + 4 - (\gamma_c - 3)(\overline{\gamma}_c - 3)$.
2. $\gamma_c + \overline{\gamma}_c \leq 3n/4$ *when* $\delta^* \geq 3$ *and* $n \geq 14$.
3. $\gamma_c + \overline{\gamma}_c \leq \delta^* + 2$ *when* $\gamma^* \geq 4$.
4. *If* $n \geq 7$, *then* $\gamma_c\overline{\gamma}_c \leq 2n - 4$ *with equality if and only if G or $\overline{G}$ is a path or a cycle.*

Let $H$ be constructed from two complete graphs $H_1 = H_2 = K_r$ with $V(H_2) = \{v_1, \cdots, v_r\}$ and a star $K_{1,r}$ with leaves $x_i$ by adding an edge between $x_i$ and all vertices of the Cartesian product $H_1 \Box H_2$ with second coordinate $v_i$ for $1 \leq i \leq r$. For $H$, $\delta^* = r$, $\overline{\gamma}_c = 3$, and $\gamma_c = r + 1$. This family shows that the bound of Theorem 42-1 is attained for each value of $\delta^* \geq 2$. The cubic necklace $L_3(q)$ described in Example 1 for which $n = 4q$, $\gamma_c = 3q - 2$, and $\overline{\gamma}_c = 2$ shows that the bound of Theorem 42-2 is sharp. For the part 3, the authors showed that the equality can occur only if $\delta^* = 6$ and $\{\gamma_c, \overline{\gamma}_c\} = \{4, 4\}$ or $\{4, 5\}$, and constructed such a graph attaining the bound with about 15,000 vertices.

Desormeaux et al. showed that if both $G$ and $\overline{G}$ are connected with diameter 2, then $\gamma_c \leq 1 + \lceil \frac{\delta}{\overline{\gamma}_c - 2} \rceil$ and $\overline{\gamma}_c \leq \lceil \frac{\delta}{\gamma_c - 2} \rceil$, which lead to the following results.

**Theorem 43 ([53])** *Let G and $\overline{G}$ be connected graphs of order* $n \geq 2$.

1. $(\gamma_c - 2)(\overline{\gamma}_c - 2) < \delta^*$.
2. *If* $\gamma_c^* \geq 4$, *then* $\gamma_c\overline{\gamma}_c < 3(n - 1)/2$.

## 3 Connected Domination in Modified Graphs

Many studies have been done on graph parameters when the structure of the graph is slightly modified by the addition of edges/vertices, the deletion of edges/vertices and the contraction of edges or the identification of vertices. Thus, the notions of criticality and stability were introduced according to whether the parameter increases, decreases or remains unchanged.

For domination number, Walikar and Acharya [145] were the first to study those graphs for which the domination number changes upon the deletion of any edge, while Dutton and Brigham [61] were the first to study those graphs for which the domination number remains unchanged upon the deletion of any edge. Bauer et al.

[21] began the study of those graphs where the domination number increases on the removal of any vertex, while Brigham et al. [28] began the study of graphs for which the domination number decreases on the removal of any vertex. In 2006, Burton and Sumner [31] initiated the study of graphs for which the domination number decreases upon the identification of the vertices comprising any edge. The study of these types of problems was extended later to other domination parameters such as the total domination and connected domination numbers.

In this section we focus primarily on the effects of various edge and vertex operations on the connected domination number of a graph.

## 3.1 Edge Addition

We start by noting that adding an edge to a graph cannot increase its connected domination number. So, a graph $G$ is called *$\gamma_c$-critical* (*$\gamma_c$-stable*, respectively) if the addition of any edge to $G$ decreases (does not change, respectively) $\gamma_c(G)$. The study of $\gamma_c$-critical graphs was introduced in 2004 by Chen, Sun, and Ma [40], while the study of $\gamma_c$-stable graphs was initiated in 2015 by Desormeaux, Haynes, and van der Merwe [55]. We will also refer to $\gamma$-critical and $\gamma_t$-criticalgraphs that are defined similarly to $\gamma_c$-critical graphs.

### 3.1.1 $\gamma_c$-critical Graphs

In the sequel, a graph $G$ is said to be $k-\gamma_c$-critical if $\gamma_c(G)=k$ but $\gamma_c(G+e)<k$ for each edge $e$ belonging to the complement graph $\overline{G}$ of $G$. It was observed in [40] that adding an edge to a connected graph $G$ can decrease the connected domination number by at most 2. Moreover, if $\gamma_c(G+e)=\gamma_c(G)-2$ for some edge $e\in E(\overline{G})$, then every $\gamma_c(G+e)$-set contains the endvertices of $e$. Obviously, the only $1-\gamma_c$-critical graph is the complete graph $K_n$. Chen et al. [40] showed that no tree of order at least three is $\gamma_c$-critical, and established that a connected graph $G$ is $2-\gamma_c$-critical if and only if $\overline{G}$ is a forest with at least two components, each one is a nontrivial star. Ananchuen [3] was the first to observe that a connected graph is $3-\gamma_c$-critical if and only if it is $3-\gamma_t$-critical. In [89], Kaemawichanurat et al. showed that connected $4-\gamma_c$-critical graphs and $4-\gamma_t$-critical graphs are also the same, but for $k\geq 5$, they have shown that there are $k-\gamma_c$-vertex critical graphs which are not $k-\gamma_t$-vertex critical. So far, characterizing $k-\gamma_c$-critical graphs remains an open problem for $k\geq 3$. Therefore, to better understand the structure of $k-\gamma_c$-critical graphs, researchers focused on studying these graphs with respect to some properties such as number of cutvertices, matching, and hamiltonicity. Some results in this framework will be given later. We first give some basic properties of $k-\gamma_c$-critical graphs. For a set $S\subseteq V$, let $\omega(G-S)$ denote the number of components of a graph $G-S$.

**Theorem 44 ([40])** *Let G be a* $3-\gamma_c$*-critical graph. Then*

**(i)** *If S is a cutset of G, then* $\omega(G-S)\leq|S|+1$.
**(ii)** *If G has even order, then G contains a perfect matching.*
**(iii)** $diam(G)\leq 3$.

Bounds on the number of cutvertices were given by Ananchuen [3] and by Kaemawichanurat and Ananchuen [85] who, respectively, showed that it is at most one for $3-\gamma_c$-critical graph and at most two for $4-\gamma_c$-critical graphs. These two results were generalized by Kaemawichanurat and Ananchuen in [86] as follows.

**Theorem 45 ([95])** *For* $k\geq 3$, *let G be a* $k-\gamma_c$*-critical graph with* $\zeta(G)$ *cutvertices. Then* $\zeta(G)\leq k-2$.

It should be noted that unlike $3-\gamma_c$-critical graphs, Ananchuen and Plummer [8] showed that connected $3-\gamma$-critical graphs may contain more than one cutvertex. The first problem that can arise by looking at Theorem 45 is the characterization of $k-\gamma_c$-critical graphs with $k-2$ cutvertices. This problem was treated for $k=3$ by Ananchuen in [3], and for $k=4$ and $k\geq 5$ by Kaemawichanurat and Ananchuen in [85] and [86], respectively.

For $k-\gamma_c$-critical graphs with exactly one cutvertex, Ananchuen [3] established the following properties.

**Theorem 46 ([3])** *For* $k\geq 3$, *let G be a* $k-\gamma_c$*-critical graph with exactly one cutvertex* $x$. *Then*

**(i)** $x$ *belongs to every* $\gamma_c(G)$*-set.*
**(ii)** $G-x$ *contains exactly two components, say* $C_1$ *and* $C_2$.
**(iii)** *For each* $i\in\{1,2\}$, $\gamma_c(C_i)\leq k-1$ *and* $G[N(x)\cap V(C_i)]$ *is complete.*
**(iv)** *If* $C_i$ *is a non-singleton component of* $G-x$ *with* $\gamma_c(C_i)=k-1$, *then* $C_i$ *is* $(k-1)-\gamma_c$*-critical.*

**Theorem 47 ([3])** *For* $k\geq 3$, *let G be a* $k-\gamma_c$*-critical graph with exactly one cutvertex* $x$. *Suppose* $C_1$ *and* $C_2$ *are the components of* $G-x$. *Let* $A=G[V(C_1)\cup\{x\}]$ *and* $B=G[V(C_2)\cup\{x\}]$. *Then*

**(i)** $k-1\leq\gamma_c(A)+\gamma_c(B)\leq k$.
**(ii)** $\gamma_c(A)+\gamma_c(B)=k$ *if and only if exactly one of* $C_1$ *and* $C_2$ *is a singleton.*

Taylor et al. [139] and Kaemawichanurat et al. [86] independently established the maximum number of leaves of $k-\gamma_c$-critical graphs $G$ with $\gamma_c(G)\geq 3$.

**Theorem 48 ([86, 139])** *For* $k\geq 3$, *every* $k-\gamma_c$*-critical graph has at most one leaf.*

Item (iii) of Theorem 44 has also been generalized independently in [85] and [42] as follows.

**Proposition 49 ([42, 85])** *Let G be a* $k-\gamma_c$*-critical graph. Then* $diam(G)\leq k$.

The following example given in [42] shows that the upper bound of Proposition 49 is sharp. For $k \geq 3$, $G_{k-2}$ is the connected graph obtained from a path of order $k-1$ in which one leaf is labelled $x$, and a cycle $C_4$ whose vertices are labelled in order $a, b, c, d$ by adding the edges $xa$ and $xd$. Moreover, Chengye and Feilong [42] asked the question if every $k-\gamma_c$-critical graph with diameter $k$ has the graph $G_{k-2}$ as an induced subgraph. This question was answered by Taylor and van der Merwe [139] who provided for $k \geq 4$ a class of $k-\gamma_c$-critical graphs $H_k$ with diameter $k$ which is $G_{k-2}$-free, where $H_k$ is defined by $V(H_k) = \{v_i, x, y : 0 \leq i \leq k\}$ and $E(H_k) = \{v_i v_{i+1}, xv_{k-3}, xv_{k-2}, xy, yv_k\}$.

In [91], Kaemawichanurat and Jiarasuksakun focused on providing an upper bound on the sum of the independence number and the clique number $\omega$ of $k-\gamma_c$-critical graphs. In particular, they proved the following.

**Theorem 50 ([91])** *For $1 \leq k \leq 3$, let $G$ be a $k-\gamma_c$-critical graph of order $n$. Then $\alpha + \omega \leq n - \left\lfloor \frac{k}{2} \right\rfloor + 1$.*

The authors [91] characterized all $3-\gamma_c$-critical graphs achieving the upper bound of Theorem 50, and showed for $k \geq 4$ that there are infinitely many $k-\gamma_c$-critical graphs such that $\alpha + \omega = n - \left\lfloor \frac{k}{2} \right\rfloor + 1$. Hence the following question was posed: Does every $k-\gamma_c$-critical graph with $k \geq 4$ satisfy $\alpha + \omega \leq n - \left\lfloor \frac{k}{2} \right\rfloor + 1$. We now turn our attention to matching and hamiltonicity properties of $k-\gamma_c$-critical graphs.

**Matching Properties:** Matching properties of $k-\gamma_c$-vertex-critical graphs have been mainly investigated in [5] for $k = 3$. We recall that a matching in a graph $G$ is *perfect* if it covers all of the vertices of $G$ and it is *near-perfect* if it covers all but one of the vertices of $G$. A graph $G$ is *$t-$factor-critical* if for every set $S \subseteq V(G)$ with $|S| = t$, the graph induced by $V(G) - S$ contains a perfect matching.

We start by reminding that Theorem 44-(ii) states that a $3-\gamma_c$-critical graphs of even order contain a perfect matching. Ananchuen et al. [5] focused on $3-\gamma_c$-critical graphs. They showed that those with odd order contain a near-perfect matching. Furthermore, they established interesting properties summarized in the following results.

**Theorem 51 ([5])** *Suppose $n \geq 4$ and $G$ is a 3-connected $3-\gamma_c$-critical graph of order $2n$. Then*

**(i)** *If $\delta(G) \geq n-1$, then $G$ is $2-$factor-critical.*
**(ii)** *If $G$ is claw-free, then $G$ is $2-$factor-critical.*

It is worth noting that the authors provided an infinite family of graphs satisfying the hypotheses of Theorem 51 by constructing the following graph $H_{k,s,r,t}$ for integers $k, r, t \geq 1$ and $s \geq 2$. Let $V(H_{k,s,r,t})$ formed by four sets $X, Y, Z, W$ such that $X = \{x_1, x_2, \ldots, x_k\}$, $Y = \{y_1, y_2, \ldots, y_s\}$. Form complete graphs on $X, Y, Z$, and $W$. Also join each vertex of $Z$ to each vertex of $X \cup \{y_1\}$ and join each vertex of $W$ to every each vertex of $X \cup (Y - \{y_1\})$. Then graphs $H_{n-2,n-1,1,2}$ and $H_{2n-6,2,2,2}$ satisfy, respectively, the hypotheses of Theorem 51 and both are $2-$factor-critical.

On the other hand, the authors [5] constructed a 2-connected $3-\gamma_c$-critical claw-free graph of order $2n$ (that could satisfy in addition $\delta(G) \geq n-1$) that is not $2-$factor-critical which shows that the bound on connectivity in Theorem 51 is best possible.

**Theorem 52 ([5])** *Suppose G is a $3-\gamma_c$-critical graph of odd order n.*

**(i)** *If $n \geq 5$ and $\delta(G) \geq 2$, then G is $1-$factor-critical.*
**(ii)** *If G is 4-connected and $K_{1,4}$-free, then G is $3-$factor-critical.*

The family $\{H_{1,2n-2,1,1} : n \geq 2\}$ is an infinite family of graphs satisfying the hypotheses of Theorem 52-(i). Also, the graph $H_{n-2,n-1,1,3}$ satisfies the assumptions of Theorem 52-(ii) and hence is $3-$factor-critical.

We close this part by mentioning that Kaemawichanurat and Ananchuen [85] showed the existence of a perfect matching for $4-\gamma_c$-critical graphs of even order having connectivity one.

**Hamiltonian Properties:** Hamiltonian properties of $k-\gamma_c$-critical graphs were mainly invested by Kaemawichanurat and Caccetta in [87, 88]. The study seems to be motivated by the fact that this property has already been considered for $\gamma$-critical graphs and $\gamma_t$-critical graphs, where interesting results have been obtained but also where conjectures remained open. For $k = 3$, the following result has been shown in [87].

**Theorem 53 ([87])** *Every 2-connected $3-\gamma_c$-critical graph is Hamiltonian.*

Since connected $3-\gamma_c$-critical graphs and $3-\gamma_t$-critical graphs are the same, Theorem 53 implies that 2-connected $3-\gamma_t$-critical graphs are Hamiltonian, which improves previous results obtained on the hamiltonicity of $3-\gamma_t$-critical graphs in particular cases according to the diameter of the graph (see [136]). Furthermore, for $k \geq 4$ and $l \geq 2$, Kaemawichanurat and Caccetta constructed a class of $l$-connected $k-\gamma_c$-critical non-Hamiltonian graphs of order $n \geq (k-1)l+3$. The authors mentioned that such a class of graphs contains a claw as an induced subgraph and wondered if every connected $k-\gamma_c$-critical claw-free graph for $k \geq 4$ contains a Hamiltonian cycle. This issue was addressed in [88], where the authors showed the following.

**Theorem 54 ([88])** *Every 2-connected $4-\gamma_c$-critical claw-free graph is Hamiltonian.*

For $k \geq 5$, the authors [88] constructed a class of $k-\gamma_c$-critical claw-free non-Hamiltonian graphs of connectivity two. Moreover, for 3-connected graphs, Kaemawichanurat and Caccetta obtained the following result.

**Theorem 55 ([88])** *Let G be a 3-connected $k-\gamma_c$-critical claw-free graph. If $k \in \{3,4,5,6\}$, then G is Hamiltonian.*

Very recently, Henning, Ananchuen, and Kaemawichanurat [81] determined a connection between the traceability of a $k-\gamma_c$-critical graph and the number of cutvertices in the graph by proving the following result.

**Theorem 56 ([81])** *For $k \geq 4$ and $0 \leq \zeta \leq k-2$, every $k - \gamma_c$-critical graph with $\zeta$ cutvertices is traceable if and only if $k-3 \leq \zeta \leq k-2$.*

### 3.1.2 $(\gamma_c, r)$-critical Graphs

Ananchuen [4] extended the concept of $k-\gamma_c$-criticality to $k-(\gamma_c, r)$-criticality. For positive integers $k$ and $r \geq 2$, a graph $G$ is said to be $k - (\gamma_c, r)$-*critical* if $\gamma_c(G) = k$, but $\gamma_c(G + xy) < k$ for each pair of non-adjacent vertices $x$ and $y$ that are at distance at most $r$ apart. Clearly, every $k-\gamma_c$-critical graph is $k-(\gamma_c, r)$-critical but the converse is not true. The path $P_4$ is the simplest example of $2 - (\gamma_c, 2)$-critical graphs which is not $2 - \gamma_c$-critical. Moreover, it is also clear that $k - (\gamma_c, r)$-critical graphs with diameter at most $r$ are $k-\gamma_c$-critical. The diameter of $k-(\gamma_c, r)$-critical graphs is given by the following result shown in [4].

**Theorem 57 ([4])** *For integers $k$ and $r \geq 2$, if $G$ is a $k - (\gamma_c, r)$-critical graph, then*

$$\operatorname{diam}(G) \leq \begin{cases} k+1, & \text{for } 2 \leq r \leq k \\ k, & \text{for } r \geq k+1. \end{cases}$$

Moreover, a characterization of $k - (\gamma_c, r)$-critical graphs having diameter $k+1$ for $2 \leq r \leq k$ was given as follows. Recall that the *join* $G \vee H$ of two graphs $G$ and $H$ is the graph obtained from $G \cup H$ by joining each vertex of $G$ to each vertex of $H$.

**Theorem 58 ([4])** *For any integers $k, r$ with $2 \leq r \leq k$, $G$ is a $k - (\gamma_c, r)$-critical graph with diameter $k+1$ if and only if $G \cong K_1 \vee K_{n_1} \vee K_{n_2} \vee \ldots \vee K_{n_k} \vee K_1$ for some positive integers $n_i$, $1 \leq i \leq k$.*

For the case $k = 2$, Ananchuen [4] showed that for $r \geq 3$, $G$ is a $2 - (\gamma_c, r)$-critical graph if and only if $G$ is $2 - \gamma_c$-critical, while for $r = 2$, $G$ is a $2 - (\gamma_c, 2)$-critical graph if and only if $G$ is $2 - \gamma_c$-critical or $\overline{G}$ is a double star. For the case $k = 3$, the following characterization of $3 - (\gamma_c, r)$-critical graphs for $r \geq 3$ has also been established in [4].

**Theorem 59 ([4])** *Let $r$ be a positive integer. Then $G$ is $3-(\gamma_c, r)$-critical for $r \geq 4$ or $G$ is $3 - (\gamma_c, 3)$-critical of diameter at most 3 if and only if $G$ is $3 - \gamma_c$-critical.*

The only class of graphs not considered by Theorems 58 and 59 is the class of $3 - (\gamma_c, 2)$-critical graphs of diameter 3. For this purpose, Ananchuen, Ananchuen, and Caccetta [9] provided a characterization of $3 - (\gamma_c, 2)$-critical claw-free graphs which are not $3 - \gamma_c$-critical, and concluded that $3 - (\gamma_c, 2)$-critical graphs of diameter 3 which are not $3 - \gamma_c$-critical and contain claws do exist.

### 3.1.3 $\gamma_c$-stable Graphs

Obviously, every graph $G$ with $\gamma_c(G) = 1$ is $\gamma_c$-stable. Desormeaux et al. [55] mentioned that there exists no forbidden subgraph characterization of $\gamma_c$-stable graphs by showing that for every integer $k \geq 2$ and any graph $H$, there exists a $\gamma_c$-stable $G$ such that $\gamma_c(G) = k$ and $H$ is an induced subgraph of $G$. However, the authors gave a descriptive characterization of $\gamma_c$-stable graphs as follows. For a set of vertices $S \subseteq V$ and a vertex $v \in S$, let $\mathrm{epn}(v, S) = \{u \in V \backslash S \mid N(u) \cap S = \{v\}\}$.

**Theorem 60 ([55])** *A connected graph $G$ is $\gamma_c$-stable with $\gamma_c(G) \geq 2$ if and only if for any $\gamma_c(G)$-set $S$ and vertex $v \in S$, the following three conditions hold:*

**(i)** *If $v$ is not a cutvertex of $G[S]$, then $|\mathrm{epn}(v, S)| \geq 2$.*
**(ii)** *If $G[S \backslash \{v\}]$ has exactly two components, then $|\mathrm{epn}(v, S)| \geq 1$.*
**(iii)** *$G$ has no dominating set $D$ of cardinality at most $\gamma_c(G) - 1$, where $G[D]$ has exactly two components and the distance in $G$ between any two vertices in different components of $G[D]$ is at least three.*

Note that if $G$ is a graph of diameter two, then Condition (iii) in Theorem 60 is trivially satisfied. Also, applying Theorem 60, the authors obtained a simple characterization of $\gamma_c$-stable trees. In particular, a tree of order at least 4 does not have a degree two vertex, otherwise adding an edge between these two vertices decreases the connected domination number of the tree.

**Theorem 61 ([55])** *A tree $T$ is $\gamma_c$-stable if and only if $\gamma_c(T) = 1$ or $T$ has no degree two vertices.*

The next result gives an upper bound on the connected domination number of $\gamma_c$-stable graphs in terms of their order.

**Theorem 62 ([55])** *If $G$ is a $\gamma_c$-stable graph of order $n$, then $\gamma_c(G) \leq (n-2)/2$.*

We note that the bound of Theorem 62 is sharp as shown in [55].

## 3.2 *Edge Removal*

The first thing to note is that the study of the behavior of the connected domination parameters with respect to the deletion of edges remains little explored. Our literature search resulted in only two papers, one is due to Zelinka [153] and the other to Lemańska [103].

In [153], Zelinka was interested in graphs whose connected domatic number decreases by deleting an arbitrary edge, after showing the following result. Recall that the *connected domatic number* $d_c(G)$ of a connected graph $G$ is the maximum number of pairwise disjoint, connected dominating sets in $V(G)$. It is easy to see that the connected domatic number of a connected graph cannot increase by the deletion of any edge which is not a bridge. But it can be decreased by at most two as shown by Zelinka.

**Theorem 63 ([153])** *Let $G$ be a connected graph with at least three vertices, and let $e$ be an edge of $G$ which is not a bridge. If $e$ joins two vertices of maximum degree, then $d_c(G - e) \geq d_c(G) - 2$, otherwise $d_c(G - e) \geq d_c(G) - 1$.*

Note that for the complete graph of order at least three, $d_c(G - e) = d_c(G) - 2$. Zelinka defined a graph $G$ to be *connectively domatically critical* if for any edge $e \in E(G)$ different from a bridge, $d_c(G-e) < d_c(G)$. Moreover, the following two results have been established.

**Theorem 64 ([153])** *Let $G$ be a connectively domatically critical graph and let $d_c(G) = d$. Then $V(G)$ is the union of pairwise disjoint sets $D_1, D_2, \ldots, D_d$ such that:*

**(i)** *the subgraph $G_i$ of $G$ induced by $D_i$ is a tree for each $i \in \{1, \ldots, d\}$.*
**(ii)** *the subgraph $G_{ij}$ of $G$ with the vertex set $D_i \cup D_j$ and with the edge set consisting of all edges joining a vertex of $D_i$ with a vertex of $D_j$ is a forest, each of whose connected components is a star or a complete graph with two vertices for any $i, j$ from the set $\{1, \ldots, d\}, i \neq j$.*

Any Cartesian product of a complete graph with a tree is an example of a graph fulfilling the assumptions of Theorem 64.

**Theorem 65 ([153])** *Let $G$ be a connectively domatically critical graph with $d_c(G) = d$. If $G$ is regular of degree $d - 1$, then $G \cong K_d$. If $G$ is regular of degree $d$, then $G_i = K_2$ for each $i \in \{1, \ldots, d\}$ and $G_{ij}$ consists of two connected components isomorphic to $K_2$ for any $i, j$ from the set $\{1, \ldots, d\}, i \neq j$.*

In [103], Lemańska studied the effect on the connected domination number of a connected graph of the deletion of a set of edges. Her main result is the following.

**Theorem 66 ([103])** *Let $H_p$ be a connected subgraph of order $p$ in $G$, let $E_p$ be the edge set of $H_p$ and let $G - E_p$ be the graph obtained from $G$ by deleting edges of $E_p$. If $G$ and $G - E_p$ are connected, then $\gamma_c(G) \leq \gamma_c(G - E_p) \leq \gamma_c(G) + 2p - 2$.*

As consequence, the next result follows immediately.

**Corollary 67** *If $e$ is an edge of $G$ and if $G$ and $G - e$ are connected, then $\gamma_c(G) \leq \gamma_c(G - e) \leq \gamma_c(G) + 2$.*

## 3.3 Vertex Removal

The study of the effect of the deletion of any vertex on the connected domination number was initiated by Ananchuen, Ananchuen, and Plummer in [6]. A graph $G$ is said to be $k$ *connected domination vertex critical*, or $k - \gamma_c$*-vertex critical*, if $\gamma_c(G) = k$ and $\gamma_c(G-v) < \gamma_c(G)$ for any vertex $v$ of $G$. Clearly, since disconnected graphs do not have a connected dominating set, only the 2-connected graphs are concerned with the study of $k-\gamma_c$-vertex critical graphs. Moreover, we note that the only $k-\gamma_c$-vertex critical graphs for $k \in \{1,2\}$ are, respectively, the trivial graph $K_1$ and the complete graph $K_{2p}$, for $p \geq 2$, minus a perfect matching. We will also refer to $k-\gamma$-vertex-critical and $k-\gamma_t$-vertex-critical graphs that are defined similarly to $k-\gamma_c$-vertex-critical graphs.

It was shown in [7] that a 2-connected graph $G$ is $3-\gamma_c$-vertex critical if and only if it is $3-\gamma_t$-vertex critical. This remains true when $\gamma_c(G)=4$ as proved by Kaemawichanurat, Caccetta, and Ananchuen [89] who have shown in addition that for $k \geq 5$, the classes of 2-connected $k-\gamma_c$-vertex critical graphs and 2-connected $k-\gamma_t$-vertex critical graphs do not need to be the same.

It should be noted that for $k \geq 3$, there is no complete characterization of $k-\gamma_c$-vertex critical graphs. For the case $k=3$, it has been shown in [6] that if $G$ is a $3-\gamma_c$-vertex critical graph, then either $G$ is a cycle $C_5$ or $G$ is 3-connected. In addition, if $S$ is a vertex cutset of size 3 in a $3-\gamma_c$-vertex critical graph $G$, then the subgraph induced by $S$ contains at most one edge and $G-S$ consists of precisely two components.

Kaemawichanurat, Caccetta, and Ananchuen [90] established upper and lower bounds of the order of $k-\gamma_c$-vertex critical graphs in terms of $k$ and the maximum degree given by the following result.

**Theorem 68 ([90])** *Let $G$ be a $k-\gamma_c$-vertex critical graph of order $n$ with $k \geq 2$. Then*

$$\Delta + k \leq n \leq (\Delta - 1)(k-1) + 3,$$

*and the upper bound is sharp for all $k \geq 2$ when $\Delta$ is even.*

Similar upper bounds on the order of $k-\gamma$-vertex-critical and $k-\gamma_t$-vertex-critical graphs have been obtained and in both cases, these upper bounds are attained only if the graph is $\Delta$-regular (see Fulman et al. [66], Wang et al. [146], and Mojdeh et al. [120]). This is not the case for the upper bound of Theorem 68. It is shown in [90] that if $k \in \{2,3,4\}$, then the only graphs achieving the upper bound are $\Delta$-regular but they do not need to be $\Delta$-regular when $k \in \{5,6\}$ and $\Delta = 3$. Concerning the lower bound of Theorem 68, Kaemawichanurat et al. [90] proved that for $k \geq 5$, a graph $G$ is $k-\gamma_c$-vertex critical of order $\Delta + k$ if and only if $G = C_{k+2}$.

In [7], Ananchuen, Ananchuen, and Plummer specifically studied the case of $3-\gamma_c$-vertex-critical graphs. It has been shown that with the exception of the cycle $C_5$ these graphs must be 3-connected. Moreover, matching properties of $3-\gamma_c$-vertex-critical graphs were also explored. We begin by the following properties of $3-\gamma_c$-vertex-critical graphs with connectivity three.

**Theorem 69 ([7])** *Let $G$ be a $3-\gamma_c$-vertex-critical graph and $S$ be a vertex cutset $S$ with $|S|=3$. Then*

**(i)** *the subgraph induced by $S$ contains at most one edge.*
**(ii)** *$G-S$ consists in two components.*

**(iii)** *G is* $1-$*factor-critical.*
**(iv)** *G contains a vertex of degree* 3.

**Theorem 70 ([7])** *Let G be a* $K_{1,7}$*-free* $3 - \gamma_c$*-vertex-critical graph of order* $n$*. Then*

**(i)** *If* $n$ *is even, then G contains a perfect matching.*
**(ii)** *If* $n$ *is odd, then G contains a near-perfect matching.*

**Theorem 71 ([7])** *Let G be a* $3 - \gamma_c$*-vertex-critical graph of even order. If G is either* $K_{1,4}$*-free or* $K_{1,5}$*-free and* 5*-connected, then G is* $2-$*factor-critical.*

**Theorem 72 ([7])** *Let G be a* $K_{1,6}$*-free* $3 - \gamma_c$*-vertex-critical graph of odd order. Then G is* $1-$*factor-critical.*

**Theorem 73 ([7])** *Let G be a* $K_{1,3}$*-free* $3 - \gamma_c$*-vertex-critical graph of odd order and minimum degree at least four. Then G is* 3*-factor-critical.*

It should be noted that the authors [7] constructed several examples of graphs showing that the hypotheses of the previous theorems are the best possible.

In a very recent work, Kaemawichanurat [84] studied the 2-connected graphs that are both $k - \gamma_c$-edge critical and $k - \gamma_c$-vertex critical, and called them *maximal* $k - \gamma_c$*-vertex critical.* He proved that every maximal $3 - \gamma_c$-vertex critical graph $G$ satisfies $\alpha(G) \leq \kappa(G)$. Moreover if $\alpha(G) = \kappa(G)$, then $\kappa(G) = \delta(G)$ and if $\kappa(G) < \delta(G)$, then every two vertices of $G$ are joined by a Hamiltonian path. Finally, $\alpha(G) + \omega(G) \leq n - 1$, and the graphs achieving this upper bound are characterized.

## 3.4 Identifying Vertices

The study of the effect that identifying two vertices has on the connected domination number was investigated by Chellali, Maffray, and Tablennehas [37]. The *identification* of two vertices $u, v$ (not necessarily adjacent) in a graph $G$ is the graph $G_{uv}$ obtained from $G$ by deleting $u$ and $v$ and adding a new vertex adjacent to every vertex of $G \setminus \{u, v\}$ that is adjacent to $u$ or $v$. It was mentioned in [37] that identifying two vertices $u$ and $v$ in a connected graph $G$ cannot increase its connected domination number. But it can decrease the connected domination number by at most one if $uv \in E(G)$ or by at most three if $uv \notin E(G)$. The authors [37] define a graph $G$ to be *connected domination dot-critical*, abbreviated *cdd-critical*, if identifying any two adjacent vertices decreases $\gamma_c(G)$; and $G$ is *totally connected domination dot-critical*, abbreviated *tcdd-critical*, if identifying any two vertices decreases $\gamma_c(G)$. We present characterizations of tcdd-critical graphs obtained in [37] for the classes of block graphs, split graphs, and unicyclic graphs as well as a characterization of cdd-critical cacti.

**Theorem 74 ([37])** *Let $G$ be a connected block graph. The following statements are equivalent:*

*(a) $G$ is a tcdd-critical graph.*
*(b) $G$ is a cdd-critical graph.*
*(c) Every block $H$ of $G$ satisfies:*

- *If $H$ is an end block, then $H = P_2$.*
- *If $H$ is not an end block, then every vertex of $H$ is a cutvertex.*
- *Every support vertex belongs to exactly two blocks.*

For the class of trees, Theorem 74 leads to the following corollary.

**Corollary 75** *Let $T$ be a tree of order $n \geq 4$. The following statements are equivalent:*

*(a) $T$ is tcdd-critical.*
*(b) $T$ is cdd-critical.*
*(c) Every support vertex of $T$ has degree two.*

**Theorem 76 ([37])** *Let $G$ be a connected split graph. Then the following statements are equivalent:*

**(a)** *$G$ is tcdd-critical.*
**(b)** *$G$ is cdd-critical.*
**(c)** *$G$ is a corona $K_k \circ K_1$ with $k \geq 2$.*

**Theorem 77 ([37])** *Let $G$ be a connected unicyclic graph with cycle $C$, and let $A_C = \{x \in C : d_G(x) = 2\}$. Then $G$ is a tcdd-critical graph if and only if the following holds:*

**(a)** *Every support vertex in $C$ has degree three and every support vertex not in $C$ has degree two.*
**(b)** *If $C$ contains a support vertex, then:*

- *$A_C$ is an independent set of size different from 1 and 2.*
- *If $A_C \neq \emptyset$, then every support vertex of $C$ is adjacent to a vertex of $A_C$.*

**(c)** *If $C$ does not contain any support vertex, then the subgraph induced by $A_C$ has either zero or at least two edges. Moreover if $A_C$ is independent, then $|A_C| \geq 3$.*

**Theorem 78 ([37])** *Let $G$ be a connected cactus. Then $G$ is a cdd-critical graph if and only if the following holds:*

*(i) Every support vertex that belongs to a cycle of $G$ has degree three, and every support vertex that does not belong to any cycle has degree two.*
*(ii) For every cycle $C$ of $G$, the set $A_C = \{x \in C : d_G(x) = 2\}$ has size different from 1 and:*

*(ii.a) If $C$ contains a support vertex, then $A_C$ is an independent set and if moreover $|A_C| \geq 2$, then every support vertex of $C$ is adjacent to a vertex of $A_C$.*

*(ii.b) If $C$ does not contain any support vertex, then the subgraph induced by $A_C$ has either zero or at least two edges.*

## 3.5 Subdividing Edges

The study of the connected domination subdivision number was initiated in 2008 by Favaron, Karami, and Sheikholeslami [63]. The *subdivision* of an edge $e = uv$ is the replacement of $e$ with a new vertex $w$ and two new edges $uw$ and $wv$. The *connected domination subdivision number* $sd_{\gamma_c}(G)$ of a graph $G$ is the minimum number of edges that must be subdivided in order to increase the connected domination number. The motivation for introducing this kind of concept comes from the domination subdivision number first introduced in Velammal's PhD thesis [143]. Clearly, since the connected domination number of the graph $K_2$ does not change when its only edge is subdivided, the graph is supposed to be of order at least 3.

The connected domination subdivision number of a graph can be arbitrarily large as shown in [63] by giving for every integer $k \geq 2$ a connected graph $G_k$ for which $sd_{\gamma_c}(G_k) = k$. The graph $G_k$ is obtained from a complete graph $K_{3(k-1)}$, where for every $k$ elements subset $S$ of the vertices of $V(K_{3(k-1)})$ a new vertex $x_S$ is added as well as all the edges $x_S u$ for all $u \in S$. For example, the graph $G_2$ is illustrated in Figure 5.

Moreover, Favaron et al. [63] have established some upper bounds relating the connected domination subdivision number to the edge connectivity number of a graph, the order and the connected domination number. Recall that the *edge connectivity number* $\kappa'(G)$ of $G$ is the minimum number of edges whose removal results in a disconnected graph. It is well known that for every graph, $\kappa'(G) \leq \delta$.

**Theorem 79 ([63])** *Let $G$ be a connected simple graph $G$ of order $n \geq 3$. Then*

**(i)** $sd_{\gamma_c}(G) \leq \kappa'(G) \leq \delta$.
**(ii)** $sd_{\gamma_c}(G) \leq \lfloor n/2 \rfloor$.
**(iii)** $sd_{\gamma_c}(G) \leq \gamma_c(G) + 1$.

Under some conditions on $\gamma_c(G)$, the bound in Theorem 79-(iii) was slightly improved as follows.

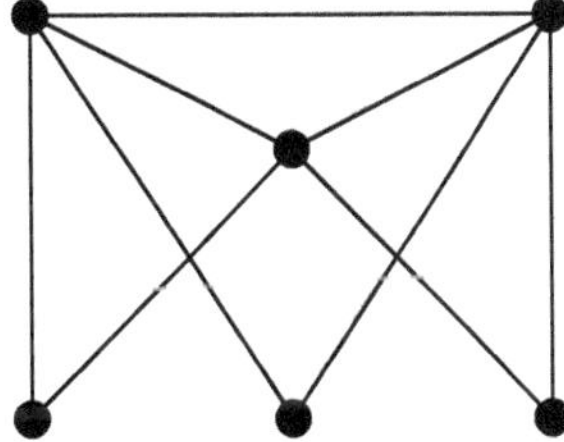

**Fig. 5** Graph $G_2$

**Theorem 80 ([63])** *Let $G$ be a graph of order $n \geq 3$. If $\gamma_c(G) \geq \delta$ or $\gamma_c(G) \geq \lfloor \frac{n}{2} \rfloor$ or if there exists a $\gamma_c$-set $S$ of $G$ such that each vertex of $S$ has an $S$-private neighbor, then $sd_{\gamma_c}(G) \leq \gamma_c(G)$.*

Moreover, the authors [63] established an upper bound on the sum $sd_{\gamma_c}(G) + \gamma_c(G)$ in terms of the order of $G$ and characterized the extremal graphs attaining this bound.

**Theorem 81 ([63])** *Every connected graph $G$ of order $n \geq 3$ satisfies $\gamma_c(G) + sd_{\gamma_c}(G) \leq n - 1$, with equality if and only if $G$ is a path or a cycle.*

We finish by mentioning that the exact value of the connected domination subdivision number was determined for grids by showing that $sd_{\gamma_c}(P_n \Box P_m) = 1$ for each two positive integers $m, n$ with $m + n \geq 4$.

# 4 Variants of Connected Domination

The purpose of this section is to present some variants of connected domination. Although quite a few variants have been considered in the literature, we only discuss seven which we feel are the most significant.

## 4.1 Connected k-domination

In 1985, Fink and Jacobson [65] gave a generalization of dominating sets in graphs as follows. For a positive integer $k$, a subset $S$ of vertices in a graph $G = (V, E)$ is *$k$-dominating* if every vertex of $V - S$ is adjacent to at least $k$ vertices in $S$. In 2009, Volkmann [144] initiated the study of connected $k$-dominating sets, that are $k$-dominating sets whose induced subgraphs are connected. The *connected $k$-domination number* $\gamma_k^c(G)$ is the minimum cardinality among the connected $k$-dominating sets of $G$, and the *$k$-domination number* $\gamma_k(G)$ is the minimum cardinality of a $k$-dominating set of $G$. Clearly, it is interesting to study $\gamma_k^c(G)$ only for graphs $G$ with $k \leq \Delta$, otherwise $\gamma_k^c(G)$ trivially equals the order of $G$. Connected $k$-domination has been studied in [15, 72, 144] and elsewhere.

Obviously, for every connected graph $G$ of order $n$, $\gamma_k^c(G) \leq n$. Volkmann [144] showed that the equality is attained if and only if all vertices of $G$ are either cutvertices or vertices of degree less than $k$. Restricted to graphs with minimum degree at least $k$, a characterization of all connected graphs $G$ of order $n$ with $\gamma_k^c(G) = n - 1$ was given in [144].

**Theorem 82 ([144])** *Let $k \geq 2$ be an integer, and let $G$ be a connected graph of order $n$ and minimum degree $\delta$.*

1. *If $\delta \geq 2$, then $\gamma_2^c(G) = n - 1$ if and only if $G$ is a cycle.*
2. *If $\delta \geq k \geq 3$, then $\gamma_k^c(G) = n - 1$ if and only if $G$ is isomorphic to the complete graph $K_{k+1}$.*

Attalah and Chellali [15] were interested in graphs $G$ of order $n$ with $\gamma_k^c(G) = n - 2$, and proved the following results.

**Proposition 83 ([15])** *Let $k \geq 2$ be an integer and $G$ a connected graph such that $\gamma_k^c(G) = n - 2$. Then*

1. *$\delta \leq k + 1$.*
2. *If $\delta \geq k$, then for every independent set $S$ of cardinality at least three, $G - S$ is a disconnected graph.*
3. *If $\delta = k + 1$, then $G$ contains no bridge.*
4. *If $\delta = k + 1$, then for every pair of adjacent vertices $x, y$, $V(G) - \{x, y\}$ is a minimum connected $k$-dominating set of $G$.*

For $k \in \{2, 3\}$, the authors [15] provided a complete characterization of connected cubic graphs $G$ with $\gamma_2^c(G) = n - 2$ and connected 4-regular claw-free graphs with $\gamma_3^c(G) = n - 2$.

**Theorem 84 ([15])** *Let $G$ be a connected cubic graph of order $n$. Then $\gamma_2^c(G) = n - 2$ if and only if $G = K_4$, $K_{3,3}$ or $G$ is the complement graph of $C_6$.*

**Theorem 85 ([15])** *Let $G$ be a connected 4-regular claw-free graph of order $n$. Then $\gamma_3^c(G) = n - 2$ if and only if $G$ is isomorphic to $K_5$ or $K_{2,2,2}$.*

Hansberg [72] established a relation between the connected $k$-domination and the connected domination numbers.

**Theorem 86 ([72])** *Let $G$ be a connected graph and $k$ an integer with $1 \leq k \leq \delta(G)$. Then $\gamma_k^c(G) \geq \gamma_c(G) + k - 2$.*

In [144], families of graphs are given showing that Theorem 86 is sharp for $k \geq 4$. Moreover, for $k \in \{2, 3\}$ an improvement of Theorem 86 was given by Volkmann [144] as follows.

**Theorem 87 ([144])** *If $G$ is a connected graph of order $n \geq 3$, then $\gamma_2^c(G) \geq \gamma_c(G) + 1$ and $\gamma_3^c(G) \geq \gamma_c(G) + 2$.*

Additional lower bounds relating the connected $k$-domination and connected domination numbers have been obtained in [72]. For any graph $G$, let $\kappa_{\max}(G)$ denote the maximum number of components of $G - x$ among all vertices $x \in V(G)$.

**Theorem 88 ([72])** *Let $G$ be a connected graph and $k$ an integer with $2 \leq k \leq \delta(G)$. Then $\gamma_k^c(G) \geq \gamma_c(G) + (k - 2)\kappa_{\max}(G)$.*

For $k = 2$ the following slightly better bound is given.

**Theorem 89** **([72])** *Let G be a connected graph on $n \geq 2$ vertices. Then*

$$\gamma_2^c(G) \geq \gamma_c(G) + \kappa_{\max}(G).$$

Volkmann [144] established two lower bounds on the connected $k$-domination number of a graph $G$ in terms of the order, size, and maximum degree of $G$.

**Theorem 90** **([144])** *Let $k \geq 2$ be an integer, and let $G$ be a connected graph of order $n$, size $m$, and maximum degree $\Delta$. Then*

1. $\gamma_k^c(G) \geq n + \frac{n-m-1}{k-1}$.
2. *If* $\gamma_k^c(G) \leq n-1$, *then* $\gamma_k^c(G) \geq \left\lceil \frac{kn-2}{\Delta+k-2} \right\rceil$.

Upper bounds for the connected $k$-domination number were also obtained, among which we present one due to Hansberg relating the connected $k$-domination number to the $k$-domination number. Additional upper bounds on the connected $k$-domination number in terms of some other domination parameters can be found in [74] and [72].

**Theorem 91** **([72])** *Let $G$ be a connected graph and $k$ an integer with $2 \leq k \leq \Delta(G)$. Then*

$$\gamma_k^c(G) \leq 2\gamma_k(G) - k + 1.$$

## 4.2 Connected Distance k-domination

In this subsection, we survey some results on a generalization of connected dominating sets using the concept of distances between vertices. Let us first give some useful definitions. For any integer $k \geq 1$, the *closed k-neighborhood* of a vertex $v \in V$ is the set $N_k[v] = \{w \in V : d(v,w) \leq k\}$, and the *open k-neighborhood* of $v$ is the set $N_k(v) = N_k[v] - \{v\}$. If $S$ is a subset of $V$, then *open (closed) k-neighborhood* of $S$, denoted by $N_k(S)$ ($N_k[S]$) is the union of *open (closed) k-neighborhoods* of vertices of $S$. A vertex $x \in X \subseteq V$ is said to be *distance k-redundant* in $X$ if $N_k[x] - N_k[X - \{x\}] = \emptyset$.

In 1975, Meir and Moon [119] introduced the concept of distance $k$-domination, while the concept of distance $k$-irredundance was introduced by Hattingh and Henning [77] in 1995. For a positive integer $k$, a set $D \subseteq V$ is a *distance k-dominating* set in $G$ if every vertex of $V(G) - D$ is within distance $k$ from some vertex of $D$. The *distance k-domination number* $\gamma_{\leq k}(G)$ is the minimum cardinality of a distance $k$-dominating set in $G$. A set $X \subseteq V$ is called *distance k-irredundant* if it has no distance $k$-redundant vertex. The *distance k-irredundance number* $ir_{\leq k}(G)$ is the minimum cardinality taken over all maximal distance $k$-irredundant sets of $G$. A *connected distance k-dominating set* (resp. *total distance k-dominating set*) is a distance $k$-dominating set of $G$ whose induced subgraph is connected (resp. has no isolated vertices). The *connected distance k-domination number* of $G$, denoted by $\gamma^c_{\leq k}(G)$, is the minimum cardinality of a connected distance $k$-dominating set and the *total distance k-domination number* $\gamma^t_{\leq k}(G)$ is the minimum cardinality of a total distance $k$-dominating set of $G$. Note that the distance 1-irredundance number $ir_{\leq 1}(G)$ and the total distance 1-domination number $\gamma^t_{\leq 1}(G)$ are the classical irredundance and total domination numbers $ir(G)$ and $\gamma_t(G)$, respectively.

We begin by the following results of Hansberg et al. [73] bounding $\gamma^c_{\leq k}(G)$ in terms of $\gamma_{\leq k}(G)$ and $\gamma^t_{\leq k}(G)$ for any connected graph $G$, which generalize for $k > 1$ the results of Theorems 25 and 27-2.

**Theorem 92 ([73])** *Let $G$ be a connected graph. Then*

*(i)* $\gamma^c_{\leq k}(G) \leq (2k+1)\gamma_{\leq k}(G) - 2k$.
*(ii)* $\gamma^c_{\leq k}(G) \leq \frac{(3k+1)}{2}\gamma^t_{\leq k}(G) - 2k$.

The next result due to Xu et al. [151] establishes a relationship between $\gamma^c_{\leq k}(G)$ and $ir_{\leq k}(G)$ for any graph $G$ and integer $k \geq 2$. Note that if $ir_{\leq k}(G) = 1$, then obviously $\gamma^c_{\leq k}(G) = 1$.

**Theorem 93 ([151])** *If a graph $G$ is connected, $k \geq 2$ and $ir_{\leq k}(G) \geq 2$, then*

*(i)* $\gamma^c_{\leq k}(G) \leq \frac{5}{2}ir_{\leq k}(G)k - 3k + 2$ *if $ir_{\leq k}(G)$ is even.*
*(ii)* $\gamma^c_{\leq k}(G) \leq \max\{(2k+1)ir_{\leq k}(G) - 2k, \frac{5}{2}ir_{\leq k}(G)k - \frac{7}{2}k + 2\}$ *if $ir_{\leq k}(G)$ is odd.*

For the class of trees, a better bound was obtained by Hansberg, Meierling, and Volkmann [73], which exactly generalizes the bound of Theorem 27-1.

**Theorem 94 ([73])** *If $T$ is a tree, then $\gamma^c_{\leq k}(T) \leq (2k+1)ir_{\leq k}(T) - 2k$.*

Let $d^c_{\leq k}(G)$ denotes the connected distance $k$-domatic number of a graph $G$ defined as the maximum number of classes in a partition of $V$ into connected distance $k$-dominating sets. Note that for $k = 1$, $d^c_{\leq 1}(G) = d_c(G)$. We recall that it was shown in [152] that $\gamma_c(G) \leq 3d_c(\overline{G})$ (see Theorem 39). In [108], Li gives an extension of this result for all $k$.

**Theorem 95 ([108])** *Let $k \geq 2$ and both $G$ and $\overline{G}$ be connected. Then $\gamma^c_{\leq k}(G) \leq (2k+1)d^c_{\leq k}(\overline{G})$.*

In [141], Tian and Xu first observed that for every nontrivial connected graph $G$ and positive integer $k$, $\gamma^c_{\leq k}(G) = \min \gamma^c_{\leq k}(T)$, where the minimum is taken over all spanning trees $T$ of $G$. It was also observed that for every connected graph $G$ of order $n$, $\gamma^c_{\leq k}(G) \leq \max\{1, 2n - 2k - \Delta + 2\}$. Moreover, Tian and Xu provided two upper bounds on $\gamma^c_{\leq k}(G)$ that generalize those given by Caro et al. [33] for $k = 1$ (see Theorem 5). The first upper bound we present is obtained by using probabilistic methods.

**Theorem 96 ([141])** *Let $G$ be a nontrivial connected graph of order $n$ with minimum degree $\delta$. Then*

$$\gamma^c_{\leq k}(G) \leq n\frac{72k + 20km^2 + 17 + 0.5\sqrt{\ln q} + \ln q}{q},$$

*where* $q = m(\delta + 1) + 2 - t$, $m = \lceil \frac{k}{3} \rceil$ *and* $t = 3\lceil \frac{k}{3} \rceil - k$.

Note that for $k = 1$, Theorem 96 yields $\gamma_c(G) \leq n\frac{109+0.5\sqrt{\ln(\delta+1)}+\ln(\delta+1)}{\delta+1}$ which slightly improves the bound $\gamma_c(G) \leq n\frac{145+0.5\sqrt{\ln(\delta+1)}+\ln(\delta+1)}{\delta+1}$ given in [33].

**Theorem 97 ([141])** *Let G be a nontrivial connected graph of order n with minimum degree δ. Then*

$$\gamma^c_{\leq k}(G) \leq (1 + o_\delta(1))n\frac{\ln q}{q},$$

*where* $q = m(\delta + 1) + 2 - t$, $m = \lceil \frac{k}{3} \rceil$ *and* $t = 3\lceil \frac{k}{3} \rceil - k$.

Note that for $k = 1$, Theorem 97 yields $\gamma_c(G) \leq (1 + o_\delta(1))n\frac{\ln(\delta+1)}{\delta+1}$, already obtained in [33].

The following upper bound on the eccentricity of $G$, due to Dankelmann and Osaye [48], is a generalization of Theorem 34.

**Theorem 98 ([48])** *Let G be a connected graph of order n and eccentricity* $ecc(G)$. *Then* $\mathrm{ecc}(G) \leq n(\gamma^c_{\leq k}(G) + 2k - 1) - k(\gamma^c_{\leq k}(G) + k - 1) + \frac{\gamma^c_{\leq k}(G)}{2} - \frac{(\gamma^c_{\leq k}(G))^2}{4}$.

## 4.3 k-connected Domination

To avoid the breakdown of a computer network due to the failure of a single node or edge, it may be interesting to increase the connectivity of its virtual backbone. For a $k$-connected (resp. $k$-edge-connected) graph $G$, let $\gamma_{kc}(G)$ (resp. $\gamma_{kec}(G)$) be the minimum cardinality of a minimum $k$-connected (resp. $k$-edge-connected) dominating set of $G$. Clearly, $\gamma_{kec}(G) \leq \gamma_{kc}(G)$, $\gamma_{(k-1)c}(G) \leq \gamma_{kc}(G)$, and $\gamma_{1c}(G) = \gamma_{1ec}(G) = \gamma_c(G)$.

For big values of $\delta$, Caro and Yuster established an asymptotical upper bound on $\gamma_{2c}$ similar as the bound on $\gamma_c$ given in Theorem 5.

**Theorem 99 ([34])** *If G is a 2-connected graph with n vertices and minimum degree δ, then* $\gamma_{2c}(G) \leq n(\ln \delta/\delta)(1 + o_\delta(1))$.

The authors conjecture that this result can be generalized and that if $G$ is $r$-connected, then $\gamma_{rc}(G) \leq n(\ln \delta/\delta)(1 + o_\delta(1))$. They also observed that, contrary to the ratio $\gamma_c(G)/\gamma(G)$ which is bounded above by 3 in any connected graph by Theorem 25, $\gamma_{2c}(G)/\gamma(G)$ may be arbitrarily large in 2-connected graphs. This can be seen, for instance, by a cycle $C_n = x_1x_2 \cdots x_n$ plus the edges $x_1x_i$ for $i \neq$

$3, n-1$. To get an upper bound on this ratio, Li et al. added the hypothesis that $G$ is triangle-free. They also established an upper bound on the ratio $\gamma_{2ec}(G)/\gamma(G)$.

**Theorem 100 ([110])** *Let $G$ be a graph such that $\gamma(G) \geq 2$ and let $S$ be a dominating set of $G$.*

1. *If $G$ is 2-connected and triangle-free, there exists a set $X \in V(G)$ such that $|X| \leq 10|S| - 13$ and $G[S \cup X]$ is 2-connected. Hence $\gamma_{2c}(G) \leq 11\gamma(G) - 13$.*
2. *If $G$ is 2-edge-connected, there exists a set $X \in V(G)$ such that $|X| \leq 4|S| - 4$ and $G[S \cup X]$ is 2-edge-connected. Hence $\gamma_{2ec}(G) \leq 5\gamma(G) - 4$.*

The chain obtained from $k$ cycles $C_6 = x_1^i x_2^i \cdots x_6^i$ by identifying $x_4^i$ and $x_1^{i+1}$ for $1 \leq i \leq k-1$ satisfies $\gamma_{2ec}(G) = 5k+1 = 5\gamma - 4$ which shows the sharpness of Item 2. However, the authors think that the upper bound for $\gamma_{2c}(G)$ in Theorem 100-1 can be improved and conjecture that under the same conditions, $\gamma_{2c}(G) \leq 6\gamma(G)$.

Considering the independence number, Li, Yang, and Wu also extended the second bound $\gamma_c(G) \leq 2\alpha(G) - 1$ of Theorem 25.

**Theorem 101 ([111])** *Let $G$ be a graph with independence number $\alpha$.*

1. *If $G$ is 2-connected, then $\gamma_{2c}(G) \leq 6\alpha - 3$.*
   *More precisely, if $G$ contains an induced cycle $C_k$, $\gamma_{2c}(G) \leq 6\alpha - 2k$ if $k$ is even, $\gamma_{2c}(G) \leq 6\alpha - 2k + 3$ if $k$ is odd.*
2. *If $G$ is 2-connected and triangle-free, then $\gamma_{2c}(G) \leq 5\alpha - 5$.*
   *More precisely, if $G$ contains an induced cycle $C_k$, $\gamma_{2c}(G) \leq 5\alpha - 3k/2$ if $c$ is even, $\gamma_{2c}(G) \leq 5\alpha - (3k-1)/2 + 2$ if $c$ is odd.*
3. *If $G$ is 2-edge-connected, then $\gamma_{2ec} \leq 4\alpha - 1$.*
   *More precisely, if $G$ contains an induced cycle $C_k$, $\gamma_{2ec}(G) \leq 4\alpha - k$ if $k$ is even, $\gamma_{2ec}(G) \leq 4\alpha - k + 2$ if $k$ is odd.*

The previous bounds are attained by cycles $C_n$, and by cliques $K_n$ for items 1 and 3. The authors of [111] also proposed the following problems: Let $S$ be a maximal independent set of a graph $G$. Is it true that if $G$ is 2-connected, there exists a set $S'$ in $V - S$ such that $|S'| \leq 6|S|$ and $G[S \cup S']$ is 2-connected, and if $G$ is 2-edge-connected, there exists a set $S'$ in $V - S$ such that $|S'| \leq 4|S|$ and $G[S \cup S']$ is 2-edge-connected?

In not necessarily 2-connected graphs, Paulraj Joseph and Arumugam [125] had previously given some properties of two variants of the 2-connected dominating sets, namely the dominating sets without cutvertex and the block dominating sets.

## 4.4 Doubly Connected Domination

Cyman, Lemańska, and Raczek [46] defined a doubly connected dominating set of a connected graph $G$ of order $n \geq 2$ as a connected dominating set $D$ such that the induced subgraph $G[V \setminus D]$ is connected. The doubly connected domination number $\gamma_{cc}(G)$ is the minimum cardinality of a doubly connected dominating set of

$G$. Clearly, $\gamma_c(G) \leq \gamma_{cc}(G)$. The differences $\gamma_{cc}(G) - \gamma_c(G)$ and $\gamma_{cc}(G) - \alpha(G)$ where $\alpha$ is the independence number can be arbitrarily large as shown, for instance, by the corona $K_{n/2} \circ K_1$. For any triple $a, b, c$ of positive integers such that $4 \leq a \leq b \leq c$ and $b \leq 2a - 3$ if $a$ is odd, $b \leq 2a - 2$ if $a$ is even, Arriola and Canoy [11] constructed a graph $G$ such that $\gamma_t(G) = a$, $\gamma_c(G) = b$, $\gamma_{cc}(G) = c$. They also determined the doubly connected dominating number of the corona and of the lexicographic product of two graphs.

In [46] it is proved that, given a connected graph $G$ and a positive integer $k$, the problem of knowing whether $\gamma_{cc}(G) \leq k$ is NP-complete even in the class of bipartite graphs.

We gather below the principal bounds on $\gamma_{cc}(G)$ in terms of the order, size, maximum degree, and connectivity of $G$. A graph $G$ belongs to Family $\mathcal{U}$ if it is obtained from two disjoint trees $T_1$ and $T_2$ by adding $n(T_2)$ edges, one edge joining each vertex of $T_2$ to one arbitrarily chosen vertex of $T_1$.

**Theorem 102 ([46])** *Let $G$ be a connected graph with $n \geq 2$.*

1. $\gamma_{cc}(G) \leq \min\{n - 1, 2m - n + 1\}$.
   $\gamma_{cc}(G) = 2m - n + 1$ *if and only if $G$ is a tree.*
   $\gamma_{cc}(G) = n - 1$ *if and only if $n = 2$ or $G - \{u, v\}$ is disconnected for each pair $u, v$ of adjacent non-cutvertices.*
2. $\gamma_{cc}(G) \geq 2n - m - 2$ *and if* $\gamma_{cc}(G) > 1$*, then* $\gamma_{cc}(G) \geq n/\Delta$*. Moreover* $\gamma_{cc}(G) = 2n - m - 2$ *if and only if $G$ belongs to Family $\mathcal{U}$.*
3. $\gamma_{cc}(G) \leq n - \kappa + 1$.

Cyman et al. also studied the effect of an edge subdivision and of a vertex deletion on $\gamma_{cc}(G)$. We denote by $G \circledast w_{uv}$ the graph obtained from $G$ by subdividing the edge $uv$ by a new vertex $w$.

**Theorem 103 ([46])** *Let $G$ be a connected graph of order $n \geq 2$ and $x$ a non-cutvertex of $G$.*

1. $\gamma_{cc}(G) \leq \gamma_{cc}(G \circledast w_{uv})$ *for any edge subdivision of $G$ and* $\gamma_{cc}(G \circledast w_{uv}) - \gamma_{cc}(G)$ *can be arbitrarily large.*
2. $\gamma_{cc}(G) - \gamma_{cc}(G - x)$ *and* $\gamma_{cc}(G - x) - \gamma_{cc}(G)$ *can be arbitrarily large.*

In view of Theorem 103-1, Karami et al. [92] studied the doubly connected domination subdivision number $\mathrm{sd}_{\gamma_{cc}}(G)$ which is the minimum number of edges that must be subdivided in order to increase the doubly connected domination number. They gave a new upper bound on $\gamma_{cc}(G)$ and bounds on $\mathrm{sd}_{\gamma_{cc}}(G)$ in terms of the order, the minimum degree, the edge connectivity $\kappa'$, the maximum matching number $\alpha'$ or the doubly connected dominating number.

**Theorem 104 ([92])** *Let $G$ be a connected graph of order $n \geq 3$.*

1. $\gamma_{cc}(G) \leq n - \delta$.
2. $sd_{\gamma_{cc}}(G) \leq \max\{1, \delta - 1\}$.
3. $sd_{\gamma_{cc}}(G) \leq \min\{\lceil \frac{n-1}{2} \rceil, n - \delta, \kappa', \alpha', \gamma_{cc} + 1\}$.

4. $sd_{\gamma_{cc}}(G) \leq \min\{\ell \mid G contains\ an\ odd\ cycle\ of\ length\ \ell\}$.
5. $sd_{\gamma_{cc}}(G) \leq n - \gamma_{cc}(G)$ *with equality if and only if* $G - \{u, v\}$ *is disconnected for all pairs* $u, v$ *of adjacent non-cutvertices of* $G$.

They also proved that for a special graph $H_k$ with order $\binom{3(k-1)}{k} + 3(k-1)$ and $\mathrm{sd}\gamma_c(H_k) = k$ [63], $\mathrm{sd}_{\gamma_{cc}}(H_k) = 3$. This shows that $\mathrm{sd}\gamma_c(G) - \mathrm{sd}\gamma_{cc}(G)$ can be arbitrarily large. From Items 2 and 4 of Theorem 104, $\mathrm{sd}_{\gamma_{cc}}(G) \leq 3$ if $G$ is planar. The authors of [92] conjecture that the good bound is 2.

Akhbari et al. established a Nordhaus–Gaddum type inequality for the doubly connected dominating number. For $n \geq 4$, let $L_n$ be the graph of order $n$ consisting of a clique $K_{n-2}$ plus two pendant edges attached at two different vertices of the clique. Note that the complement $\overline{L_n}$ is connected.

**Theorem 105 ([1])** *Let* $G$ *be a connected graph of order* $n \geq 4$ *whose complement* $\overline{G}$ *is connected. Then* $\gamma_{cc}(G) + \gamma_{cc}(\overline{G}) \leq n + 3$ *with equality if and only if* $G \in \{L_n, \overline{L_n}\}$ *for* $n \geq 5$.

## 4.5 Weakly Connected Domination

The concept of weakly connected domination was introduced by Dunbar et al. in [60]. A dominating set $S$ of vertices of $G = (V, E)$ is *weakly connected* if the subgraph generated by the edges of $G$ with at least one endvertex in $S$ is connected. The *weakly connected domination number* $\gamma_w(G)$ of $G$ is the minimum cardinality of a weakly connected dominating set of $G$. Clearly, every connected dominating set is weakly connected, thus implying $\gamma(G) \leq \gamma_w(G) \leq \gamma_c(G)$. In [57], Domke et al. defined a graph $G$ to be $(\gamma, \gamma_w)$-perfect (resp. $(\gamma_w, \gamma_c)$-perfect) if $\gamma(H) = \gamma_w(H)$ (resp. $\gamma_w(H) = \gamma_c(H)$) for every induced subgraph $H$ of $G$. Domke et al. characterized the graphs which are $(\gamma, \gamma_w)$-perfect or $(\gamma_w, \gamma_c)$-perfect. A *kite* is obtained from a cycle $C_4$ and a path $P_2$ by joining one vertex of the path and one vertex of the cycle with an edge.

**Theorem 106 ([57])** *Let* $G$ *be a connected graph. Then*

1. $G$ *is* $(\gamma, \gamma_w)$*-perfect if and only if* $G$ *is* $(P_6, C_6, kite)$*-free.*
2. $G$ *is* $(\gamma_w, \gamma_c)$*-perfect if and only if* $G$ *is* $(P_5, C_5)$*-free.*

Since $\gamma_w(P_5) = \gamma_w(C_5) = 2$ and $\gamma_c(P_5) = \gamma_c(C_5) = 3$, Item 2 of Theorem 106 is a consequence of Theorem 22. The following theorems give inequalities between $\gamma_w$ and $\gamma_c$, $\gamma$ or other graph parameters.

**Theorem 107 ([60])** *Let* $G$ *be a connected graph of order* $n$.

1. $\gamma_c(G) \leq 2\gamma_w(G) - 1$. *Hence* $\gamma_w(G) \geq \lceil \frac{diam}{2} \rceil$.
2. $\gamma_w(G) \leq 2\gamma(G) - 1$.

Moreover, given two integers $b, r$ with $0 \leq r \leq b-1$, there exists a tree $T_1$ with $\gamma_w(T_1) = b$ and $\gamma_c(T_1) = b + r$ and a tree $T_2$ with $\gamma(T_2) = b$ and $\gamma_w(T_2) = b + r$. This proves that the two bounds are attained.

Hattingh and Henning [78] related $\gamma_w$ to $\gamma_t$ and to the matching number $\alpha'$, Chen and Shiu [39] to the irredundance number $ir(G)$, and Sanchis [132] to the size $m$.

**Theorem 108 ([78])** *Let G be a connected graph of order $n \geq 2$.*

1. *If $\Delta < n-1$, $\gamma_w(G) \geq \frac{\gamma_t(G)+1}{2}$.*
2. *$\gamma_w(G) \leq \frac{3\gamma_t(G)-2}{2}$.*

The description of two families of extremal graphs for the first bound and a constructive characterization of the extremal trees for the second bound are given in [78].

**Theorem 109 ([39])** *Let G be a connected graph with irredundance number $ir(G)$. Then $\gamma_w(G) \leq \frac{5}{2}ir(G) - 2$ and if $\gamma_w(G) = \frac{5}{2}ir(G) - 2$, then $ir(G) = 2$.*

**Theorem 110 ([132])** *Let G be a connected graph of order n, size m, and weakly connected domination number $\gamma_w \geq 3$. Then $m \leq \binom{n-\gamma_w+1}{2}$. The extremal graphs are described.*

**Theorem 111 ([78])** *Let G be a connected graph of order $n \geq 2$ and matching number $\alpha'(G)$. Then $\gamma_w(G) \leq \alpha'(G)$.*

Clearly, every vertex cover of a connected graph $G$ is a weakly connected dominating set and thus $\gamma_w(G) \leq \beta(G) = n - \alpha(G)$, where $\beta$ is the vertex covering number of $G$. This bound on $\gamma_w(G)$ is improved by Theorem 111 since $\alpha'(G) \leq \beta(G)$ for every graph.

When the graph is a tree $T$, every weakly connected dominating set is a vertex cover and thus $\gamma_w(T) = \beta(T) = n - \alpha(T)$. This remark leads to the equality $\gamma_w(G) = n - \max\{\alpha(T) | T \text{ is a spanning tree of } G\}$ and to its corollary $\gamma_w(G) \leq n/2$ for every connected graph $G$ of order $n$ [60]. Some results on $\gamma_w$ are specific for trees. Recall that $\gamma_w(G) \leq \gamma_c(G) \leq n - \Delta$ for any connected graph. Dunbar et al. [60] showed that a tree $T$ satisfies $\gamma_w(T) = n - \Delta$ if and only if $T$ is obtained from a star $K_{1,k}$ by subdividing at most $k-1$ of its edges. Lemańska [104] proved that $\gamma_w(T) \geq (n - n_1 + 1)/2$ in every tree of order $n$ with $n_1$ leaves and described the family of trees attaining the bound. Domke et al. [57] studied the $\gamma_w$-excellent trees (every vertex of $T$ belongs to a $\gamma_w(T)$-set) and the trees such that $\gamma_w(T)$ is equal to $\gamma_c(T)$ or to $\gamma(T)$.

**Theorem 112 ([57])**

1. *A tree is $T$ is $\gamma_w$-excellent if and only if $\alpha(T) = n/2$.*
2. *For a tree $T$, both conditions $\gamma_w(T) = \gamma_c(T)$ and $\gamma(T) = \gamma_c(T)$ are equivalent to the property that every vertex of $T$ is a leaf or a support.*
3. *There exists a constructive characterization of the trees satisfying $\gamma_w(T) = \gamma(T)$.*

Clearly, the deletion of a non cut-edge $e$ of a connected graph $G$ cannot decrease $\gamma_w(G)$. Lemańska showed that $\gamma_w(G)$ does not increase by more than 1. More generally

**Theorem 113 ([103])** *Let $G$ be a connected graph and $E_p$ the edge set of a connected subgraph of order $p$ of $G$ such that $G - E_p$ is connected. Then $\gamma_w(G) \leq \gamma_w(G - E_p) \leq \gamma_w(G) + p - 1$.*

In the other direction, an edge addition cannot increase $\gamma_w(G)$ and by Theorem 113, $\gamma_w(G + e) \geq \gamma_w(G) - 1$ for all $e \in E(\overline{G})$. A connected graph $G$ is weakly connected domination critical, or $\gamma_w$-critical, if $\gamma_w(G + e) < \gamma_w(G)$ for all $e \in E(\overline{G})$. The graph $G$ is $k$-$\gamma_w$-critical if it is $\gamma_w$-critical and $\gamma_w(G) = k$. Lemańska and Patyk proved that a cycle $C_n$ is $k$-$\gamma_w$-critical if and only if $n$ is even, and that no tree is $k$-$\gamma_w$-critical. They observed that $G$ is 2-$\gamma_w$-critical if and only if it is 2-$\gamma_c$-critical or, equivalently, 2-$\gamma$-critical (cf Section 3.1.1) and studied the $k$-$\gamma_w$-criticality for $k = 3$.

**Theorem 114 ([105])** *There exist 3-$\gamma_w$-critical graphs of any order $n \geq 6$, and every 3-$\gamma_w$-critical graph has diameter at most 4.*

A connected graph $G$ is $\gamma_w$-stable if $\gamma_w(G + e) = \gamma_w(G)$ for all $e \in E(\overline{G})$. Lemańska and Raczek [106] showed that a tree is $\gamma_w$-stable if and only if it admits a unique minimum weakly connected dominating set, and described the family of stable trees.

Dunbar et al. [60] proved that the problem of determining whether the weakly connected dominating number of a connected graph $G$ is at most a given integer $k$ is NP-complete even if $G$ is chordal or bipartite.

## 4.6 Connected Domination Game

Very recently, Borowiecki, Fiedorowicz and Sidorowicz [27] introduced the connected domination game defined on a nontrivial connected graph $G$ as follows. The game is played by two players, $\mathcal{D}$ (called Dominator) and $\mathcal{S}$ (called Staller) who alternate taking turns choosing a vertex of $G$. A move of a player by choosing a vertex $v$ is legal, if (1) the vertex $v$ dominates at least one additional vertex that was not dominated by the set of previously chosen vertices, and (2) the set of all chosen vertices induces a connected subgraph of $G$. Player $\mathcal{D}$ starts the game and the game ends when none of the players has a legal move, that is all vertices are dominated. The aim of $\mathcal{D}$ is to finish as soon as possible, while the Staller $\mathcal{S}$ tries to delay the end of the game as much as possible. If $X$ is the set of played vertices obtained at the end of the connected domination game, then *connected game domination number* $\gamma_{cg}(G)$ is the minimum cardinality of $X$, when both players played optimally on $G$.

It was first noticed that for every connected graph $G$, $\gamma_c(G) \leq \gamma_{cg}(G) \leq 2\gamma_c(G) - 1$. Moreover, for all nontrivial trees $T$, $\gamma_{cg}(T) = \gamma_c(T)$. The authors extended their study to the class of 2-trees, where the following upper bound has been shown. Recall that a graph $G$ is a $k$-tree if either $G$ is the complete graph on $k$ vertices, or $G$ has a vertex $u$ whose neighborhood is a clique of order $k$ and the graph obtained by deleting $u$ from $G$ is also a $k$-tree.

**Theorem 115 ([27])** *If $G$ is a 2-tree of order $n \geq 4$, then*

$$\gamma_{cg}(G) \leq \left\lceil \frac{2(n-4)}{3} \right\rceil + 1.$$

Exact values of the connected domination game have been established for the special 2-trees, called 2-paths, defined inductively as follows [27]: the unique 2-path of order 4 is $K_4 - e$. If $G$ is a 2-path and $x$ and $y$ are adjacent vertices of $G$ such that $d_G(x) = 2$ and $d_G(y) = 3$, then add a new vertex by joining it to $x$ and $y$.

**Proposition 116 ([27])** *If $G$ is a 2-path of order $n \geq 5$, then $\gamma_{cg}(G) = \left\lceil \frac{2(n-4)}{3} \right\rceil + 1$ if $n \equiv 1(\frac{mod}{3})$ and $\gamma_{cg}(G) = \left\lceil \frac{2(n-4)}{3} \right\rceil$ otherwise.*

The connected domination game has been particularly studied on the Cartesian product, first in [27] and then in [30]. General upper and lower bounds have been established in [30]. In particular, the upper bound generalizes the one obtained in [27] when one of the factors is a complete graph.

**Theorem 117 ([30])** *Let $G$ and $H$ be connected graphs. Then*

1. $\gamma_{cg}(G \Box H) \leq \min\{2\gamma_c(G)n(H), 2\gamma_c(H)n(G)\} - 1$.
2. *If* $\min\{n(G), n(H)\} \geq 2$, *then* $\gamma_{cg}(G \Box H) \geq \begin{cases} 2\gamma_c(G) & \text{if } n(H) = 2 \\ 2\gamma_c(G) + 1 & \text{if } n(H) \geq 3. \end{cases}$

Moreover, Bujtás et al. showed that for the product $K_k \Box G$ with sufficiently large $k$, the upper bound of Theorem 117 can be attained for a large class of graphs.

**Theorem 118 ([30])** *If $G$ is a connected graph and $k \geq \min\{4\Delta(G) + \alpha(G), 2n(G) - 1\}$, then*

$$\gamma_{cg}(K_k \Box G) = 2n(G) - 1.$$

Exact values of the connected game domination number are also obtained in special graph products.

**Theorem 119 ([27])** *Let $G = K_n \Box K_{1,m-1}$ for $n \geq 2$ and $m \geq 4$. Then*

1. *If* $m \neq n$, *then* $\gamma_{cg}(G) = \min\{2n - 1, 2m - 1\} = 2\gamma_c(G) - 1$.
2. *If* $m = n$, *then* $\gamma_{cg}(G) = 2n - 3$.

**Theorem 120 ([30])** *Let $G = K_{1,n}$ with $n \geq 3$, and $H \in \{P_m, C_m\}$. Then*

$$\gamma_{cg}(K_{1,n} \Box H) = 2m - 1.$$

**Theorem 121 ([27])** *For* $m \geq 4$, $\gamma_{cg}(K_2 \square P_m) = 2m - 4$; $\gamma_{cg}(K_3 \square P_m) = 2m - 3$; $\gamma_{cg}(K_n \square P_m) = 2m - 1$ *for* $n \geq 4$. *Moreover, for short paths, we have* $\gamma_{cg}(K_2 \square P_2) = 2$; $\gamma_{cg}(K_n \square P_2) = 3$ *for* $n \geq 3$; $\gamma_{cg}(K_2 \square P_3) = \gamma_{cg}(K_3 \square P_3) = 3$ *and* $\gamma_{cg}(K_n \square P_3) = 5$ *for* $n \geq 4$.

**Theorem 122 ([30])** *If* $m \geq 4$, *then* $\gamma_{cg}(P_3 \square P_m) = \gamma_{cg}(K_3 \square C_m) = 2m - 2$.

The authors of [30] gave a list of open problems, in particular the determination of the exact value of the connected game domination number for $P_n \square P_m$, $P_n \square C_m$, and $C_n \square C_m$ for every $m$ and $n$.

## 4.7 Dominating Sets with at Most k Components

The connected dominating sets with at most $k$ components were studied by Hartnell and Vestergaard [76] and by Kouider and Vestergaard [99]. For a graph $G$ with at most $k$ components and a positive integer $k$, $\gamma_c^k(G)$ is the minimum cardinality of a dominating set of $G$ with at most $k$ components. Clearly $\gamma(G) \leq \gamma_c^k(G)$ and if $G$ is connected, $\gamma_c^k(G) \leq \gamma_c^{k-1}(G) \leq \gamma_c(G)$. The main results concern general graphs and trees.

**Theorem 123** *Let G be a connected graph of order n and k a positive integer.*

1. *Hartnell and Vestergaard [76]* $\gamma_c^k(G) > \gamma_c(G) - 2(k-1)$.
2. *Kouider and Vestergaard [99]* $\gamma_c^k(G) \leq n - (\mathrm{diam}(G) - 1)(\delta - 2)/3 - 2k$ *if* $\delta \geq 3$ *and* $\mathrm{diam}(G) \geq 3k - 1$.
3. *Kouider and Vestergaard [99]* $\gamma_c^k(G) \leq n - \max_{x \in V}\{d(x) + 2\min\{k - 1, (ecc(x) - 2)/3\}\}$.
4. *Kouider and Vestergaard [99]* $\gamma_c^k(G) \leq n - 2(\Delta - 1) - 2\min\{k-1, (d_\Delta - 2)/3\}$, *where* $d_\Delta$ *is the maximum distance between two vertices of degree* $\Delta$ *if such a pair exists.*

**Theorem 124** *Let T be a tree of order n and k a positive integer.*

1. *Hartnell and Vestergaard [76] If* $k \geq 2$ *and if* $d(x) \geq d \geq 3$ *for every vertex x different from a leaf or a support, then* $\gamma_c^k(T) \geq \gamma_c(T) - \frac{k-2}{d-2}$.
2. *Hartnell and Vestergaard [76] If* $k \leq (n-1)/2$, *then* $\gamma_c^k(T) \leq n - k - 1$. *For* $k = 1$, *the bound is attained by paths.*
3. *Kouider and Vestergaard [99] If* $2 \leq k \leq (n-1)/2$, $\gamma_c^k(T) = n - k - 1$ *if and only if* $(n-3)/2 \leq k \leq (n-1)/2$ *and* $\gamma(T) = \lfloor \frac{n}{2} \rfloor$.

The trees satisfying $\gamma(T) = \lfloor \frac{n}{2} \rfloor$ belong to two families described in [128] and in [20].

## 5 Complexity and Algorithmic Aspects

Given a connected graph $G = (V, E)$ and a positive integer $k \leq |V|$, the connected domination problem, to which we shall refer as C-DOM problem, asks whether there exists a connected dominating set of $G$ of cardinality at most $k$. C-DOM problem has been shown to be NP-complete for several classes of graphs. As mentioned in Section 1, Hedetniemi and Laskar [79] have shown that finding a minimum connected dominating set in a graph $G$ is equivalent to the problem of finding a spanning tree in $G$ that maximizes the number of leaves over all spanning trees, to which we shall refer as MLST problem. Based on this Gallai-type result, C-DOM problem is NP-hard for any class of graphs for which the MLST problem is NP-hard. The first result on the NP-completeness of C-DOM problem was obtained in 1983 and is due to Pfaff, Laskar, and Hedetniemi [127] who showed it for bipartite graphs. In a technical report [107], Lemke showed in 1988 the NP-hardness of the MLST problem for cubic graphs. Polynomial-time algorithms for computing the connected domination number have been designed for specific families of graphs. The first one was due to Corneil and Perl [45] in 1984 for cographs. We summarize in Table 1 some complexity results concerning some classes of graphs, where the containment relations between some of them are given below. Also, we will write NP-c (resp. P) instead of NP-complete (resp. polynomial-time solvable).

Colbourn and Stewart [44] showed in 1990 that for permutation graphs, C-DOM can be solved in polynomial time. We also note that Chang [35] was the first to propose in 1998 an $O(n + m)$ time algorithm for computing the connected domination number for circular-arc graphs but this was improved to an $O(n)$ algorithm in 2004 by Hung and Chang [82].

Since the C-DOM problem is NP-hard, several researchers explored approximation algorithms for it. It should be noted that even if C-DOM and MLST problems are equivalent from the point of view of the optimization, this is not the case in terms of approximation algorithms. For instance, Lu and Ravi [113] gave a polynomial-time 3-approximation algorithm for the MLST problem while Guha and Khuller [71] showed that a polynomial-time constant-approximation algorithm cannot exist for the C-DOM problem. More precisely, they showed that the C-DOM problem has no polynomial-time $\varepsilon \ln |V|$-approximation for $0 < \varepsilon < 1$ unless $NP \subseteq$DTIME($|V|^{O(\log\log |V|)}$). However, they proposed an algorithm with a $ln\Delta + 3$ approximation factor, which has been improved to $ln\Delta + 2$ by Ruan et al. [130]. Restricted to special classes of graphs, this approximation has been much improved. For example, for cubic graphs a $4/3$-approximation algorithm is given in [25]. But Bonsma [24] proved that in this class, there exists $\epsilon > 0$ such that no polynomial $(1 + \epsilon)$-approximation algorithm is possible unless P=NP. This proves that the C-DOM problem is APX-hard in the class of cubic graphs.

In [135], Schaudt and Schrader studied the algorithmic complexity of the problem to decide whether a given graph has a connected dominating set whose induced subgraph belongs to a given class of graphs. By considering a wide variety of graph classes (including, for example: perfect graphs, planar bipartite graphs,

**Table 1** Complexity results for the connected domination number.

| Graph class | Complexity | Citation, year |
|---|---|---|
| Bipartite graphs | NP-c | [127], 1983 |
| Split graphs | NP-c | [102], 1983 |
| Planar bipartite graphs | NP-c | [149], 1985 |
| Chordal bipartite graphs | NP-c | [123], 1987 |
| Cubic graphs | NP-c | [107], 1988 |
| Circle graphs | NP-c | [96], 1993 |
| Undirected path graphs | NP-c | [97], 1994 |
| Planar cubic graphs | NP-c | [129], 2016 |
| Grid graphs | NP-c | [43], 1990 |
| Cographs | P | [45], 1984 |
| Series-parallel graphs | P | [149], 1985 |
| Strongly chordal graphs | P | [149], 1985 |
| 2-trees | P | [149], 1985 |
| Distance-hereditary graphs | P | [49], 1988 |
| k-polygon graphs | P | [62], 1990 |
| Doubly chordal | P | [121], 1993 |
| Cocomparability graphs | P | [29, 100], 1993 |
| Trapezoid graphs | P | [112], 1995 |
| Circular-arc graphs | P | [82], 2004 |

chordal bipartite $\subset$ bipartite $\subset$ comparability,
split $\subset$ chordal,
cograph $\subset$ distance-hereditary graphs,
interval $\subset$ strongly chordal $\subset$ chordal,
permutation $\subset$ $k$-polygon $\subset$ circle,
permutation $\subset$ cocomparability,
interval $\subset$ trapezoid,
grid graphs (induced subgraphs of grids) $\subset$ Unit disk graphs.

distance-hereditary graphs, cacti, ...), they showed that this problem is NP-hard even when the instances are restricted to be $K_{1,5}$-free graphs.

We close this section by mentioning that the decision problem corresponding to the problem of computing the connected domatic number $d_c(G)$ is NP-hard. A complete proof of this result can be found in the book of Du and Wan [58].

## 6 Conjectures and Open Problems

We first list some conjectures about connected domination collected from published papers or produced by computer programs.

**1.** [Griggs, Kleitman, Shastri] [69] If $G$ is a connected cubic $\{C_3, C_4\}$-free graph, then $\gamma_c(G) \leq 3n/5 - c$ for some constant $c$.

**2.** [Comby, Schaudt] [32] Let $H$ be obtained from a triangle $abc$ by adding a pendant edge at $a$ and a pendant path of length 3 at each of $b, c$. If $G$ is connected $\{P_9, C_9, H\}$-free, then $\gamma_c(G) \leq 2\gamma(G)$.

**3.** [Conjecture Graffiti 174] [52] (see Theorem 35) In every connected graph, $\gamma_c(G) \leq b - \max_v \alpha(N(v)) + 1$, where $b$, the bipartite number equals the maximum order of an induced bipartite subgraph of $G$.

**4.** [Graffiti 177] [52] (see Theorem 35) In every connected graph of order $n$, $\gamma_c(G) \leq n - 2\alpha + b - \delta'$, where $\delta'$ is the second smallest minimum degree of $G$.

**5.** [Graffiti 179] [52] In every connected graph of order $n$, $\gamma_c(G) \leq n - \Delta + b - \max_v \alpha(N(v)) - \gamma$.

**6.** [Graffiti 2] [52] (see Theorem 3 for triangle-free graphs). In every connected graph of order $n$, $\gamma_c(G) \leq n - 2\frac{\Sigma_v \alpha(N(v))}{n} + 2$.

**7.** [Graffiti 190] [52] (see Theorem 36) Let $G$ be a connected graph. If $\delta' \geq (n - \gamma_c + 1)/2$, then $G$ is traceable.

A weaker conjecture 190a was also proposed saying: if $\delta' \geq n - \gamma_c - 1$, then $G$ is traceable [148].

**8.** [Kaemawichanurat] [84] (see Section 3.3). Let $G$ be a maximal $3 - \gamma_c$-vertex critical graph with $\alpha = \kappa = \delta$. Then $G$ is either $C_5$ or any two vertices of $G$ are joined by a Hamiltonian path.

**9.** [Li, Wu, Yang] [110] (see Theorem 100). Let $G$ be a 2-connected triangle-free graph and let $S$ be a dominating set of $G$ with $|S| \geq 2$. There exists a subset $T$ of vertices such that $|T| \leq 5|S|$ and $G[S \cup T]$ is 2-connected.

**10.** [Caro, Yuster] [34] (see Theorem 99). Let $k$ be a fixed integer and let $G$ be a $k$-connected graph of minimum degree $\delta$. Then $\gamma_{kc}(G) \leq n\frac{\ln\delta}{\delta}(1 + o_\delta)$.

**11.** [Karami, Khoeilar, Sheikholeslami] [92] (see Section 4.4). The doubly connected domination subdivision number of a connected planar graph is at most 2.

We now give some open problems unaddressed by the various studies. Problems 14 and 15 have been noticed by S.T. Hedetniemi in a private communication.

**12.** How many edges should be added to $G$ without changing $\gamma_c(G)$?

**13.** How many edges should be added to $G$ to decrease $\gamma_c(G)$?

**14.** Determine $\gamma_c(P_n \Box P_m)$ for every $m$ and $n$. Note that in [109], the exact values of $\varepsilon_T(P_n \Box P_m)$ are given when $m = 2, 3, 4, 6$.

**15.** Determine the 2-connected domination number of $P_n \Box P_m$ for every $m$ and $n$.

**16.** [Bujtás, Dokyeesun, Iršič, Klavžar] [30] (see Section 4.6). Determine $\gamma_{cg}(P_n \Box P_m)$, $\gamma_{cg}(P_n \Box C_m)$, and $\gamma_{cg}(C_n \Box C_m)$ for every $m$ and $n$.

**Acknowledgments** The authors are grateful to Dr Nacéra Meddah for her help in drawing the figures.

## References

1. M.H. Akhbari, R. Hasni, O. Favaron, H. Karami and S.M. Sheikholeslami, Inequalities of Nordhaus-Gaddum type for doubly connected domination number. *Discrete Applied. Math.* 158 (2010) 1465–1470.
2. J.D. Alvarado, S. Dantas and D. Rautenbach, Complexity of comparing the domination number to the independent domination, connected domination, and paired domination numbers. *Matemática Contemporânea* 44 (2016) 1–8.
3. N. Ananchuen, On domination critical graphs with cut vertices having connected domination number 3. *Int. Math. Forum* 2 (2007) 3041–3052.
4. N. Ananchuen, On local edge connected domination critical graphs. *Utilitas Math.* 82 (2010) 11–23.
5. N. Ananchuen, W. Ananchuen and M.D. Plummer, Matching properties in connected domination critical graphs. *Discrete Mathematics* 308 (2008) 1260–1267.
6. W. Ananchuen, N. Ananchuen and M.D. Plummer, Vertex criticality for connected domination. *Utilitas Math.* 86 (2011) 45–64.
7. W. Ananchuen, N. Ananchuen and M.D. Plummer, Connected domination: vertex criticality and matchings. *Utilitas Math.* 89 (2012) 141–159.
8. N. Ananchuen and M.D. Plummer, Some results related to the toughness of 3-domination-critical graphs. *Discrete Math.*, 272 (2003) 5–15.
9. W. Ananchuen, N. Ananchuen and L. Caccetta, A characterization of $3-(\gamma_c, 2)$-critical claw-free graphs which are not $3-\gamma_c$-critical. *Graphs and Combinatorics* 26 (2010) 315–328.
10. M. Aouchiche and P. Hansen, A survey of Nordhaus-Gaddum type relations. *Discrete Appl. Math.* 161(4–5) (2013) 466–546.
11. B.H. Arriola and S.R. Canoy Jr., Doubly connected domination in the corona and lexicographic product of graphs. *Applied Mathematical Sciences* 8 (31) (2014) 1521–1533.
12. L. Arseneau, A. Finbow, B. Hartnell, A. Hynick, D. MacLean and L. O'Sullivan, On minimal connected dominating sets. *J. Combin. Math. Combin. Comput.* 24 (1997) 185–191.
13. S. Arumugam and J. Paulraj Joseph, On graphs with equal domination and connected domination numbers. *Discrete Math.* 206 (1999) 45–49.
14. S. Arumugam and S. Velammal, Maximum size of a connected graph with given domination parameters. *Ars Combin.* 52 (1999) 221–227.
15. K. Attalah and M. Chellali, On connected $k$-domination in graphs. *Australasian Jour. of Combin.* 55 (2013) 289–298.
16. G. Bacsó, Complete description of forbidden subgraphs in the structural domination problem. *Discrete Math.* 309 (2009) 2466–2472.
17. G. Bacsó and Z. Tuza, Graphs without induced $P_5$ and $C_5$. *Discuss. Math. Graph Theory* 24 (2004) 503–507.
18. A.V. Bankevich, Bounds on the number of leaves of spanning trees in graphs without triangles. *J. Math. Sci.* 184 (5) (2012) 557–563.
19. A.V. Bankevich and D.V. Karpov, Bounds of the number of leaves of spanning trees. *J. Math. Sci.* 184(5) (2012) 564–572.
20. X. Baogen, E.J. Cockayne, T.W. Haynes, S.T. Hedetniemi and Z. Shangchao, Extremal graphs for inequalities involving domination parameters. *Discrete Math.* 216 (2000) 1–10.
21. D. Bauer, F. Harary, J. Nieminen and C. Suffel, Domination alteration sets in graphs. *Discrete Math.* 47 (1983) 153–161.
22. C. Bo and B. Liu, Some inequalities about connected domination number. *Discrete Math.* 159 (1996) 241–245.

23. P.S. Bonsma, Spanning trees with many leaves in graphs with minimum degree three. *SIAM J. Discrete Math.* 22(3) (2008) 920–937.
24. P.S. Bonsma, Max-leaves spanning trees is APX-hard for cubic graphs. *J. Discrete Algorithms* 12 (2012) 14–23.
25. P. Bonsma and F. Zickfeld, A 3/2-approximation algorithm for finding spanning trees with many leaves in cubic graphs. In H. Broersma (Ed.), WG, in: Lecture Notes in Computer Science, vol.5344, Springer, 2008, pp.66–77.
26. P. Bonsma and F. Zickfeld, Improved bounds for spanning trees with many leaves. *Discrete Math.* 312 (6) (2012) 1178–1194.
27. M. Borowiecki, A. Fiedorowicz and E. Sidorowicz, Connected domination game. *Appl. Anal. Discrete Math.* 13 (2019) 261–289.
28. R.C. Brigham, P.Z. Chinn and R.D. Dutton, Vertex domination-critical graphs. *Networks* 18 (1988) 173–179.
29. H. Breu and D.G. Kirkpatrick, Algorithms for dominating and Steiner set problems in cocomparability graphs. Manuscript, 1993.
30. C. Bujtás, P. Dokyeesun, V. Iršič and S. Klavžar, Connected domination game played on Cartesian products. *Open Math.* 17 (2019) 1269–1280.
31. T. Burton and D.P. Sumner, Domination dot-critical graphs. *Discrete Mathematics* 306 (2006) 11–18.
32. E. Camby and O. Schaudt, The price of connectivity for dominating set: Upper bounds and complexity. *Discrete Appl. Math.* 177 (2014) 53–59.
33. Y. Caro, D. West and R. Yuster, Connected domination and spanning trees with many leaves. *SIAM J. Discrete Math.* 13 (2) (2000) 202–211.
34. Y. Caro and R. Yuster, 2-connected graphs with small 2-connected dominating sets. *Discrete Math.* 269 (2003) 265–271.
35. M.S. Chang, Efficient algorithms for the domination problems on interval and circular-arc graphs. *SIAM J. Comput.* 27 (1998) 1671–1694.
36. M. Chellali, O. Favaron, A. Hansberg and L. Volkmann, On the $p$-domination, the total domination and the connected domination numbers of graphs. *J. Combin. Math. Combin. Comput.* 73 (2010) 65–75.
37. M. Chellali, F. Maffray and K. Tablennehas, Connected domination dot-critical graphs. *Contributions to Discrete Mathematics* 5 (2) (2010) 11–25.
38. X. Chen, On graphs with equal total domination and connected domination numbers. *Applied Math. Letters* 19 (2006) 472–477.
39. X. Chen and W.C. Shiu, A note on weakly connected domination number in graphs. *Ars Combin.* 97 (2010) 193–201.
40. X.G. Chen, L. Sun and D.X. Ma, Connected domination critical graphs. *Appl. Math. Lett.* 17 (5) (2004) 503–507.
41. X. Chen, L. Sun and H. Xing, Characterization of graphs with equal domination and connected domination numbers. *Discrete Math.* 289 (2004) 129–135.
42. Z. Chengye and C. Feilong, The diameter of connected domination critical graphs. *Ars Combinatoria* 107 (2012) 537–541.
43. B.N. Clark, C.J. Colbourn and D.S. Johnson, Unit Disk Graphs. *Discrete Math.* 86 (1990) 165–177.
44. C.J. Colbourn and L.K. Stewart, Permutation graphs: connected domination and Steiner trees. *Discrete Math.* 86 (1990) 179–189.
45. D.G. Corneil and Y. Perl, Clustering and domination in perfect graphs. *Discrete Applied Math.* 9 (1984) 27–39.
46. J. Cyman, M. Lemańska and J. Raczek, On the doubly connected domination number of a graph. *Central European Science Journals* 4(1) (2006) 34–45.
47. P. Dankelmann and S. Mukwembi, Upper bounds on the average eccentricity. *Discrete Appl. Math.* 167 (2014) 72–79.
48. P. Dankelmann and F.J. Osaye, Average eccentricity, $k$-packing and $k$-domination in graphs. *Discrete Math.* 342 (5) (2019) 1261–1274.

49. A. D'atri and M. Moscarini, Distance-hereditary graphs, Steiner trees and connected domination. *SIAM J. Comput.*, 17 (1988) 521–538.
50. E. DeLaViña and E. Waller, Spanning trees with many leaves and average distance. *Electron. J. Combin.* 15 (2008) #R33.
51. E. DeLaViña, S. Fajtlowicz and E. Waller, On some conjectures of Griggs and Graffiti. DIMACS volume "Graphs and Discovery: Proceedings of the 2001 Working Group on computer-generated conjectures from graph theoretic and chemical databases, 69 (2005) 119–125.
52. E. DeLaViña http://cms.uhd.edu/faculty/delavinae/research/wowII/index.htm
53. W.J. Desormeaux, T.W. Haynes and M.A. Henning, Bounds on the connected domination number of a graph. *Discrete Appl. Math.* 161 (2013) 2925–2931.
54. W.J. Desormeaux, T.W. Haynes and M.A. Henning, Domination parameters of a graph and its complement. *Discuss. Math. Graph Theory* 38 (2018) 203–215.
55. W.J. Desormeaux, T.W. Haynes and L. van der Merwe, Connected domination stable graphs upon edge addition. *Quaestiones Mathematicae* 38 (2015) 841–848.
56. G. Ding, T. Johnson and P. Seymour, Spanning trees with many leaves. *J. Graph Theory* 37(4) (2001) 189–197.
57. G.S. Domke, J.H. Hattingh and L.R. Markus, On weakly connected domination in graphs II. *Discrete Math.* 305 (2005) 112–122.
58. D.-Z. Du and P.-J. Wan, Connected dominating set: theory and applications. Springer, New York, 2013.
59. P. Duchet and H. Meyniel, On Hadwiger's number and the stability number. Graph theory (Cambridge, 1981), North-Holland Math. Stud., vol. **62**, North- Holland, Amsterdam, 71–73 (1982).
60. J.E. Dunbar, J.W. Grossman, J.H. Hattingh, S.T. Hedetniemi and A.A. McRae, On weakly connected domination in graphs. *Discrete Math.* 167/168 (1997) 261–269.
61. R.D. Dutton and R.C. Brigham, An extremal problem for edge domination in insensitive graphs. *Discrete Appl. Math.* 20 (1988) 113–125.
62. E.S. Elmallah and L.K. Stewart, Domination in polygon graphs. *Congr. Numer.* 77 (1990) 63–76.
63. O. Favaron, H. Karami and S.M. Sheikholeslami, Connected domination subdivision numbers of graphs. *Utilitas Math.* 77 (2008) 101–111.
64. O. Favaron, D. Kratsch, Ratios of domination parameters. Advances in Graph Theory, ed. V. R. Kulli, Vishwa Intern. Publications, 173–182 (1991).
65. J.F. Fink and M.S. Jacobson, $n$-domination in graphs. Graph Theory with Applications to Algorithms and Computer Science. John Wiley and Sons. New York 283–300 (1985)
66. J. Fulman, D. Hanson and G. MacGillivray, Vertex domination-critical graphs. *Networks* 25 (1995) 41–43.
67. J. Ghoshal, R. Laskar and D. Pillone, Connected domination and c-irredundance. *Congr. Numer.* 107 (1995) 161–171.
68. W. Goddard and M.A. Henning, Clique/connected/total domination perfect graphs. *Bulletin of the ICA* 41 (2004) 20–21.
69. J.R. Griggs, D.J. Kleitman and A. Shastri, Spanning trees with many leaves in cubic graphs. *J. Graph Theory* 13(6) (1989) 669–695.
70. J.R. Griggs and M. Wu, Spanning trees in graphs of minimum degree 4 or 5. *Discrete Math.*104(2) (1992) 167–183.
71. S. Guha and S. Khuller, Approximation algorithms for connected dominating sets. *Algorithmica* 20 (4) (1998) 374–387.
72. A. Hansberg, Bounds on the connected $k$-domination number. *Discrete Appl. Math.* 158 no. 14 (2010) 1506–1510.
73. A. Hansberg, D. Meierling and L. Volkmann, Distance domination and distance irredundance in graphs. *Electronic Journal of Combin.* 14 (2007) #R35.
74. A. Hansberg, D. Meierling and L. Volkmann, A general method in the theory of domination in graphs. *Int. J. Comput. Math.* 87 (2010) 2915–2924.

75. B.L. Hartnell and D.F. Rall, Connected domatic number in planar graphs. *Czechoslovak Math. J.* 51(1) (2001) 173–179.
76. B.L. Hartnell and P.D. Vestergaard, Dominating sets with at most $k$ components. *Ars Combinatoria* 74 (2005) 223–229.
77. J.H. Hattingh and M.A. Henning, Distance irredundance in graphs. *Graph Theory, Combinatorics, and Applications, John Wiley & Sons, Inc.* 1 (1995) 529–542.
78. J.H. Hattingh and M. A. Henning, Bounds relating the weakly connected domination number to the total domination number and the matching number. *Discrete Appl. Math.* 157 (2009) 3086–3093.
79. S.T. Hedetniemi and R. Laskar, Connected domination in graphs. *Graph Theory and Combinatorics,* Eds. B. Bollobàs, Academic Press (London, 1984) 209–218.
80. M.A. Henning, A survey of selected recent results on total domination in graphs. *Discrete Math.* 309 (2009) 32–63.
81. M.A. Henning, N. Ananchuen and P. Kaemawichanurat, Traceability of connected domination critical graphs. Appl. Math. Comput. 386 (2020) 125455.
82. R.-W. Hung and M.-S. Chang, A simple linear algorithm for the connected domination problem in circular-arc graphs. *Discuss. Math. Graph Theory* 24 (2004 ) 137–145.
83. J.P. Joseph and S. Arumugam, Domination and connectivity in graphs. *Intern. J. Management Systems* 8(3) (1992) 233–236.
84. P. Kaemawichanurat, Inequalities of independence number, clique number and connectivity of maximal connected domination critical graphs. *arXiv:1906.07619v1 [math.CO].*
85. P. Kaemawichanurat and N. Ananchuen, On $4-\gamma_c$-critical graphs with cut vertices. *Utilitas Math.* 82 (2010) 253–268.
86. P. Kaemawichanurat and N. Ananchuen, Connected domination critical graphs with cut vertices. *Discuss. Math. Graph Theory,* 40 (2020) 1035–1055.
87. P. Kaemawichanurat and L. Caccetta, Hamiltonicity of connected domination critical graphs. *Ars Combinatoria* 136 (2018) 127–151.
88. P. Kaemawichanurat and L. Caccetta, Hamiltonicity of domination critical claw-free graphs. *J. Combin. Math. Combin. Comput.* 103 (2017) 39–62.
89. P. Kaemawichanurat, L. Caccetta and N. Ananchuen, Critical graphs with respect to total domination and connected domination. *Australas. J. Combin.* 65 (2016) 1–13.
90. P. Kaemawichanurat, L. Caccetta and N. Ananchuen, Bounds on the order of connected domination vertex critical graphs. *J. Combin. Math. Combin. Comput.* 107 (2018) 73–96.
91. P. Kaemawichanurat and T. Jiarasuksakun, Some results on the independence number of connected domination critical graphs. *AKCE International Journal of Graphs and Combinatorics* 15 (2018) 190–196.
92. H. Karami, R. Khoeilar and S. M. Sheikholeslami, Doubly connected domination subdivision numbers of graphs. *Matematicki Vesnik* 3 (2012) 232–239.
93. H. Karami, S.M. Sheikholeslami, A. Khodkar and D. West, Connected domination number of a graph and its complement. *Graphs Combin.* 28(1) (2012) 123–131.
94. D.V. Karpov, Spanning trees with many leaves: new lower bounds in terms of the number of vertices of degree 3 and at least 4. *J. Math. Sci.* 196(6) (2014) 747–767.
95. D.V. Karpov, Spanning trees with many leaves: lower bounds in terms of the number of vertices of degree 1, 3 and at least 4. *J. Math. Sci.* 196(6) (2014) 768–783.
96. J.M. Keil, The complexity of domination problems in circle graphs. *Discrete Applied Math.,* 42 (1993) 51–63.
97. J.M. Keil, R. Laskar and P.D. Manuel, The vertex clique cover problem and some related problems in chordal graphs. *SIAM conference on Discrete Algorithms*, Albuquerque, New Mexico, June 1994.
98. D.J. Kleitman and D.B. West, Spanning trees with many leaves. *SIAM J. Discrete Math.* 4(1) (1991) 99–106.
99. M. Kouider and P.D. Vestergaard, Generalized connected domination in graphs. *Discrete Math. Theoretical Comput. Sc.* 8 (2006) 57–64.

100. D. Kratsch and L. Stewart, Domination on cocomparability graphs. *SIAM J. Discrete Math.*, 6(3) (1993) 400–417.
101. R. Laskar and K. Peters, Vertex and edge domination parameters in graphs. *Congr. Numer.* 48 (1985) 291–305.
102. R. Laskar and J. Pfaff, Domination and irredundance in split graphs. Technical report 430, Dept. Mathematical Sciences, Clemson Univ., 1983.
103. M. Lemańska, Domination numbers in graphs with removed edge or set of edges. *Discuss. Math. Graph Theory* 25 (2005 ) 51–56.
104. M. Lemańska, Lower bound on the weakly connected domination number of a tree. *Australas. J. Combin.* 37 (2007) 67–71.
105. M. Lemańska and A. Patyk, Weakly connected domination critical graphs. *Opuscula Mathematica* 28(3) (2008) 325–330.
106. M. Lemańska and J. Raczek, Weakly connected domination stable trees. *Czechoslovak Math. J.* 59(134) (2009) 95–100.
107. P. Lemke, The maximum leaf spanning tree problem for cubic graphs is NP-complete. *IMA Preprint Series* #428, Minneapolis, 1988.
108. S. Li, On connected $k$-domination numbers of graphs. *Discrete Mathematics* 274 (2004) 303–310.
109. P.C. Li and M. Toulouse, Maximum leaf spanning tree problem for grid graphs. *J. Combin. Math. Combin. Comput.* 73 (2010) 181–193.
110. H. Li, Y. Yang and B. Wu, Making a dominating set of a graph connected. *Discuss. Math. Graph Theory* 38 (2018) 947–962.
111. H. Li, Y. Yang and B. Wu, 2-edge-connected dominating sets and 2-connected dominating sets of a graph. *J. Combinat. Optimization* 31(2) (2016) 713–724.
112. Y.D. Liang, Steiner set and connected domination in trapezoid graphs. *Inform. Process. Lett.* 56 (1995) 101–108.
113. H. Lu and R. Ravi, Approximating maximum leaf spanning trees in almost linear time. *J. Algorithms* 29 (1) (1998) 132–141.
114. P. Mafuta, Leaf number and hamiltonian $C_4$-free graphs. *Afrika Matematika* 28(7–8) (2017) 1067–1074.
115. P. Mafuta, S. Mukwembi and S. Munyira, Spanning paths in graphs. *Discrete Appl. Math.* 255 (2019) 278–282.
116. P. Mafuta, J.P. Mazorodze, J. Mushanyu and G. Nhawu, Graphs with forbidden subgraphs and leaf number. *Afrika Matematika* 29 (2018) 1073–1080.
117. P. Mafuta, S. Mukwembi, S. Munyira and B.G. Rodrigues, Lower bounds on the leaf number in graphs with forbidden subgraphs. *Quaestiones Math.* 40(1) (2017) 139–149.
118. P. Mafuta, S. Mukwembi, S. Munyira and T. Vetrik, Hamiltonicity, minimum degree and leaf number. *Acta Math. Hungar.* 152(1) (2017) 217–226.
119. A. Meir and J.W. Moon, Relations between packing and covering numbers of trees. *Pacific J. Math.* 61 (1) (1975) 225–233.
120. D.A. Mojdeh and N.J. Rad, On an open problem concerning total domination critical graphs. *Exp. Math.* 25 (2007) 175–179.
121. M. Moscarini, Doubly chordal graphs, Steiner trees and connected domination. *Networks* 23 (1993) 59–69.
122. S. Mukwembi, Size, order, and connected domination. *Canad. Math. Bull.* 57(1) (2014) 141–144.
123. H. Müller and A. Brandstädt, The NP-completeness of Steiner tree and dominating set for chordal bipartite graphs. *Theoretical Computer Science* 53 (1987) 257–265.
124. J. Paulraj Joseph and S. Arumugam, On the connected domatic number of a graph. *J. Ramanujan Math. Soc.* 9 (1994) 69–77.
125. J. Paulraj Joseph and S. Arumugam, On connected cutfree domination in graphs. *Indian J. pure appl. Math.* 23(9) (1992) 643–647.
126. C. Payan, M. Tchuente and N.H. Xuong, Arbres avec un nombre maximum de sommets pendants. *Discrete Math.* 49 (1984) 267–273.

127. J. Pfaff, R. Laskar and S.T. Hedetniemi, NP-completeness of total and connected domination, and irredundance for bipartite graphs. Technical Report 428, Dept. Mathematical Sciences, Clemson Univ., 1983.
128. B. Randerath and L. Volkmann, Characterization of graphs with equal domination and covering number. *Discrete Math.* 191 (1998) 159–169.
129. A. Reich, Complexity of the maximum leaf spanning tree problem on planar and regular graphs. *Theoretical Computer Science* 626 (2016) 134–143.
130. L. Ruan, H. Du, X. Jia, W. Wu, Y. Li, and K. Ko, A greedy approximation for minimum connected dominating sets. *Theoret. Comput. Sci.*, 329 (2004) 325–330.
131. L. Sanchis, On the number of edges of a graph with a given connected domination number. *Discrete Math.* 214 (1–3) (2000) 193–210.
132. L. Sanchis, On the number of edges in graphs with a given weakly connected domination number. *Discrete Math.* 257(1) (2002) 111–124.
133. E. Sampathkumar and H.B. Walikar, The connected domination number of a graph. *J. Math. Phys. Sci.* 13(6) (1979) 607–613.
134. O. Schaudt, On graphs for which the connected domination number is at most the total domination number. *Discrete Appl. Math.* 160 (2012) 1281–1284.
135. O. Schaudt and R. Schrader, The complexity of connected dominating sets and total dominating sets with specified induced subgraphs. *Information Processing Letters* 112 (2012) 953–957.
136. J. Simmons, Closure operations and hamiltonian properties of independent and total domination critical graphs. Ph.D. Thesis, University of Victoria (2005).
137. J.A. Storer, Constructing full spanning trees for cubic graphs. *Inform. Process. Lett.* 13(1) (1981) 8–11.
138. L. Sun, Some results on connected domination in graphs. *Mathematica Applicata* (1992)(1) 29–34 (in Chinese).
139. J.D. Taylor and L.C. van der Merwe, A note on connected domination critical graphs. *J. Combin. Math. Combin. Comput.* 100 (2017) 3–8.
140. S. Thomassé and A. Yeo, Total domination of graphs and small transversals of hypergraphs. *Combinatorica* 27(4) (2007) 473–487.
141. F. Tian and J.-. Xu, On distance connected domination numbers of graphs. *Ars Combinatoria* 84 (2007) 357–367.
142. Z. Tuza, Hereditary domination in graphs: characterization with forbidden induced subgraphs. *SIAM J. Discrete Math.* 22(3) (2008) 849–853.
143. S. Velammal, Studies in Graph Theory: Covering, Independence, Domination and Related Topics. Ph.D. Thesis, Manonmaniam Sundaranar University, Tirunelveli, (1997).
144. L. Volkmann, Connected $p$-domination in graphs. *Util. Math.* 79 (2009) 81–90.
145. H.B. Walikar and B.D. Acharya, Domination critical graphs. *Nat. Acad. Sci. Lett.* 2 (1979) 70–72.
146. C. Wang, Z. Hu and X. Li, A constructive characterization of total domination vertex critical graphs. *Discrete Math.* 309 (2009) 991–996.
147. J. Wang, An inequality between the connected domination number and the irredundance number for a graph. *Mathematica Applicata* (1995)–04 (in Chinese).
148. D. West https://faculty.math.illinois.edu/~west/regs/graffiti.html
149. K. White, M. Farber and W. Pulleyblank, Steiner trees, connected domination and strongly chordal graphs. *Networks* 15 (1985) 109–124.
150. W. Wu, H. Du, X. Jia, Y. Li and S. C.-H. Huang, Minimum connected dominating sets and maximal independent sets in unit disk graphs. *Theoret. Comput. Sci* 352 (2006) 1–7.
151. J.-M. Xu, F. Tian and J. Huang, Distance irredundance and connected domination numbers of a graph. *Discrete Mathematics* 306 (2006) 2943–2953.
152. H. Yu and T. Wang, An inequality on connected domination parameters. *Ars Combin.* 50 (1998) 309–315.
153. B. Zelinka, Connected domatic number of a graph. *Mathematica Slovaca* 36 (4) (1986) 387–392.

154. W. Zhuang, Connected domination in maximal outerplanar graphs. *Discrete Appl. Math.,* 283 (2020) 533–541.
155. I.E. Zverovich, Perfect connected-dominant graphs. *Discuss. Math. Graph Theory* 23 (2003) 159–162.

# Restrained and Total Restrained Domination in Graphs

**Johannes H. Hattingh and Ernst J. Joubert**

## 1 Introduction and Terminology

For a simple undirected graph $G = (V, E)$, we denote by $n(G) = |V|$ its order, by $m(G) = |E|$ its size, by $\delta(G)$ ($\Delta(G)$, respectively) its minimum degree (maximum degree, respectively), and by $N(v)$ the neighborhood of the vertex $v$. We refer the reader to the glossary for any definitions not given here. For a set $S \subseteq V$, the *subgraph induced by* $S$ in $G$ is denoted by $G[S]$. If $H$ is a subgraph of $G$, then $G - H$ will denote the induced graph $G[V(G) - V(H)]$. A one regular spanning subgraph of a graph $G$ is called a *one factor* of $G$. A *leaf* in a graph is a vertex of degree one, while a *support vertex* is a vertex adjacent to a leaf. A *strong support vertex* is a support vertex which is adjacent to two or more leaves. A graph $G$ is said to be *claw-free* if for any vertex $u$ of degree at least three we have that if $v, w, x \in N(u)$, then $G[\{v, w, x, u\}]$ is not isomorphic to $K_{1,3}$.

A set $S \subseteq V$ is a *dominating set* of $G$, denoted by **DS**, if every vertex not in $S$ is adjacent to a vertex in $S$. The *domination number* of $G$, denoted by $\gamma(G)$, is the minimum cardinality of a **DS**. A *total dominating set*, denoted by **TDS**, of a graph $G$ without isolated vertices is a set $S$ of vertices of $G$ such that every vertex in $G$ is adjacent to a vertex in $S$ (other than itself). The *total domination number* of $G$, denoted by $\gamma_t(G)$, is the minimum cardinality of a **TDS** of $G$. The concept of domination in graphs, with its many variations, is now well studied in graph theory.

---

J. H. Hattingh (✉)
Department of Mathematics and Statistics, University of North Carolina at Wilmington, Wilmington, NC, USA
e-mail: hattinghj@uncw.edu

E. J. Joubert
Department of Mathematics, University of Johannesburg, Auckland Park 2006, South Africa



T. W. Haynes et al. (eds.), *Topics in Domination in Graphs*, Developments in Mathematics 64, https://doi.org/10.1007/978-3-030-51117-3_5

A **DS** $S \subseteq V$ is a *restrained dominating set*, denoted by **RDS**, if every vertex in $V - S$ is adjacent to a vertex in $V - S$. Every graph has a restrained dominating set, since $S = V$ is such a set. The *restrained domination number* of $G$, denoted by $\gamma_r(G)$, is the minimum cardinality of an **RDS** of $G$. Restrained domination was introduced by Telle and Proskurowski in 1997 [44], albeit indirectly, as a vertex partitioning problem (see also [45], [46], [47], [48], and [49]). However, the parameter was formally defined by Domke, Hattingh, Hedetniemi, Laskar, and Markus in their 1999 paper [11] on restrained domination in graphs, and also studied by Henning in his 1999 paper [26]. Subsequently over the past twenty or so years, the restrained domination number and its variations have been extensively studied in the literature.

As explained in the introductory 1999 paper [11], the definition of restrained domination is application driven. One application given is that of prisoners and guards. In this situation, each vertex in a graph $G = (V, E)$ represents either a guard (a vertex in an **RDS** $S$) or a prisoner (a vertex in $V - S$); every prisoner must be observed by at least one guard (be adjacent to at least one vertex in $S$) and every prisoner must be able to see at least one other prisoner (be adjacent to at least one vertex in $V - S$). The associated optimal placement of guards corresponds to an **RDS** of minimum cardinality. Since every **RDS** of a graph $G$ is a **DS** of $G$, it follows that $\gamma(G) \leq \gamma_r(G)$.

A *total restrained dominating set*, denoted by **TRDS**, of a graph $G$ without isolated vertices is a **TDS** $S$ such that every vertex of $V - S$ is adjacent to a vertex in $V - S$. The *total restrained domination* number of $G$, denoted by $\gamma_{tr}(G)$, is the smallest cardinality of a **TRDS** of $G$. In our application of prisoners and guards, each guard must now be observed by another guard. The concept of total restrained domination was introduced by Chen, Ma, and Sun in 2005 [2]. Since every **TRDS** of a graph $G$ is an **RDS** (**TDS**, respectively) of $G$, it follows that $\gamma_r(G) \leq \gamma_{tr}(G)$ ($\gamma_t(G) \leq \gamma_{tr}(G)$, respectively).

Let $G = (V, E)$ be a graph and let $\{S, V - S\}$ be a partition of $V$ into two non-empty sets. Consider vertices $u$ in $S$ and $v$ in $V - S$. One can specify a variety of conditions on the number of neighbors that $u$ must have in $S$, denoted by $\deg_S(u)$, and in $V - S$, denoted by $\deg_{V-S}(u)$, and the number of neighbors that a vertex $v$ in $V - S$ must have in $S$, denoted by $\deg_S(v)$, and in $V - S$, denoted by $\deg_{V-S}(v)$. Many dominating concepts, including restrained and total restrained domination, can be defined in terms of various combinations of these four values. In what follows, X will denote that the particular value does not matter.

| $\forall u \in S$ | | $\forall v \in V - S$ | | |
|---|---|---|---|---|
| $\deg_S(u)$ | $\deg_{V-S}(u)$ | $\deg_S(v)$ | $\deg_{V-S}(v)$ | Type of domination |
| $= 0$ | X | X | X | $S$ is an independent set |
| X | X | $\geq 1$ | X | $S$ is a dominating set |
| X | X | $\geq 1$ | $\geq 1$ | $S$ is a restrained dominating set |
| $\geq 1$ | X | $\geq 1$ | X | $S$ is a total dominating set |
| $\geq 1$ | X | $\geq 1$ | $\geq 1$ | $S$ is a total restrained dominating set |
| $= 0$ | X | $\geq 1$ | X | $S$ is an independent dominating set |
| $= 0$ | X | $\geq 1$ | $\geq 1$ | $S$ is an independent restrained dominating set |
| $= 0$ | X | $= 1$ | X | $S$ is an efficient dominating set |
| X | X | $= 1$ | X | $S$ is a perfect dominating set |
| X | $\geq 1$ | $\geq 1$ | X | two disjoint dominating sets |
| $\geq 1$ | $\geq 1$ | $\geq 1$ | $\geq 1$ | two disjoint restrained dominating sets or two disjoint total dominating sets |

In this chapter, we will survey results on restrained and total restrained domination in graphs. We begin by surveying results in the literature on restrained domination in graphs.

# 2 Restrained Domination

## 2.1 Exact Results

Let $K_n$, $C_n$, and $P_n$ denote, respectively, the complete graph, the cycle, and the path of order $n \geq 1$. Also, let $K_{n_1,\ldots,n_t}$ denote the complete multipartite graph where $n_i \geq 1$ for $1 \leq i \leq t$. We call $K_{1,n-1}$ a *star*. Domke, Hattingh, Hedetniemi, Laskar, and Markus determined the exact value for the restrained domination number for several simple graph families.

**Proposition 1** ([11]) *If $n \neq 2$ is a positive integer, then $\gamma_r(K_n) = 1$.*

**Proposition 2** ([11]) *If $n_1$ and $n_2$ are integers such that $\min\{n_1, n_2\} \geq 2$, then $\gamma_r(K_{n_1,n_2}) = 2$.*

**Proposition 3** ([11]) *If $t \geq 3$ is an integer, then*

$$\gamma_r(K_{n_1,\ldots,n_t}) = \begin{cases} 1 & \text{if } \min\{n_1, \ldots, n_t\} = 1 \\ 2 & \text{otherwise} \end{cases}.$$

**Proposition 4** ([11]) *If $n \geq 1$ is an integer, then $\gamma_r(P_n) = n - 2\lfloor \frac{n-1}{3} \rfloor$.*

**Proposition 5** ([11]) *If $n \geq 3$, then $\gamma_r(C_n) = n - 2\lfloor \frac{n}{3} \rfloor$.*

## 2.2 Lower Bounds

The following result is due to Hattingh, Joubert, Loizeaux, Plummer, and Van der Merwe.

**Theorem 6** ([24]) *If $G$ is a connected graph of order $n$ and size $m$, then $\gamma_r(G) \geq n - \frac{2m}{3}$.*

As an immediate consequence we obtain the following result due to Domke, Hattingh, Henning, and Markus.

**Corollary 7** ([12]) *If $T$ be a tree of order $n$, then $\gamma_r(T) \geq \lceil \frac{n+2}{3} \rceil$.*

Moreover, these authors also characterized the extremal trees $T$ of order $n$ achieving this lower bound. A simpler constructive characterization of the extremal trees $T$ of order $n$ achieving this lower bound was found by Hattingh and Plummer in [25].

We denote the set of leaves of a tree $T$ by $L(T)$. For $v \in V(T)$ and $\ell \in L(T)$, the path $v, x_1, \ldots, x_k, \ell$ is called a $v - L$ *endpath* if $\deg x_i = 2$ for each $i$.

In order to state the characterization, we define three simple operations on a tree $T$.

- **O1.** Join a leaf or a remote vertex, or a vertex $v$ or $x$ of $T$ on an endpath $v, x, y, z$ to a vertex of $K_1$, where $n(T) \equiv 1 \bmod 3$.
- **O2.** Join a remote vertex, or a vertex $v$ of $T$ which lies on an endpath $v, x, z$ to a leaf of $P_2$, where $n(T) \equiv 0 \bmod 3$ or $n(T) \equiv 1 \bmod 3$.
- **O3.** Join a leaf of $T$ to $\ell$ disjoint copies of $P_3$ for some $\ell \geq 1$.

Let $\mathcal{T}$ be the family of all trees obtained from $P_2$ or $P_4$ by a finite sequence of Operations **O1–O3**.

**Theorem 8** ([25]) *A tree $T$ of order $n$ has $\gamma_r(T) = \lceil (n+2)/3 \rceil$ if and only if $T \in \mathcal{T}$.*

Let $C_1, \ldots, C_\ell$ be pairwise disjoint cycles such that $\sum_{i=1}^{\ell} n(C_i) \equiv n_C \equiv 0 \bmod 3$. Let $G$ be the cubic graph obtained from the disjoint union of $\frac{n_C}{3}$ isolated vertices and the cycles $C_1, \ldots, C_\ell$ by joining each isolated vertex to exactly three vertices of the disjoint union of $C_1, \ldots, C_\ell$. As $G$ is cubic, the added edges partition the vertices of the cycles. Note that $n(G) = 4n_C/3$. All graphs of order $n$ constructed in this way will be denoted by $\mathcal{G}_n$.

The following result is due to Hattingh and Joubert.

**Theorem 9** ([20]) *If $G$ is a cubic graph of order $n$, then $\gamma_r(G) \geq n/4$. Moreover, $\gamma_r(G) = n/4$ if and only if $G \in \mathcal{G}_n$.*

## 2.3 Upper Bounds

Domke, Hattingh, Hedetniemi, Laskar, and Markus established the following upper bounds on the restrained domination of a connected graph.

**Proposition 10** ([11]) *Let $G$ be a connected graph of order $n$. Then $\gamma_r(G) = n$ if and only if $G$ is a star.*

**Proposition 11** ([11]) *If $G$ is a connected graph of order $n$ and $G$ is not a star, then $\gamma_r(G) \leq n - 2$.*

**Theorem 12** ([11]) *If $T$ is a tree of order $n \geq 3$, then $\gamma_r(T) = n - 2$ if and only if $T$ is obtained from $P_4$, $P_5$ or $P_6$ by adding zero or more leaves to the support vertices of the path.*

**Theorem 13** ([11]) *Let $G$ be a connected graph of order $n$ containing a cycle. Then $\gamma_r(G) = n-2$ if and only if $G$ is $C_4$, $C_5$ or $G$ can be obtained from $C_3$ by attaching $n - 3$ leaves to at most two of the vertices of the cycle.*

The following two results are due to Ulatowski.

**Theorem 14** ([50]) *If $T$ is a tree of order $n \geq 4$, then $\gamma_r(T) = n-3$ if and only if $T$ is obtained from $P_5$ ($P_6$, respectively) by adding zero or more leaves to the support vertices of $P_5$ ($P_6$, respectively) and adding either at least one leaf or exactly one support vertex to exactly one vertex of the center of $P_5$ ($P_6$, respectively).*

Let $\mathcal{G}$ be the family of graphs as depicted in Figure 1 of [50].

**Theorem 15** ([50]) *Let $G$ be a connected graph of order $n \geq 4$ containing a cycle. Then $\gamma_r(G) = n - 3$ if and only if $G \in \mathcal{G}$.*

Let $\mathcal{B}$ be the family of graphs depicted in Figure 1. The following result is due to Domke, Hattingh, Henning, and Markus.

**Theorem 16** ([12]) *Let $G$ be a connected graph of order $n$ with $\delta \geq 2$. If $G \notin \mathcal{B} = \{B_1, \ldots, B_8\}$, then $\gamma_r(G) \leq \frac{n-1}{2}$.*

A graph $G$ of order $n$ is an $(n-1)/2$-*minimal graph* if $G$ is edge-minimal with respect to satisfying the following three conditions:

1. $\delta(G) \geq 2$
2. $G$ is connected, and
3. $\gamma_r(G) \geq (n-1)/2$.

Let $\mathcal{B}^* = \{B_1, \ldots, B_5\}$, let $\mathcal{F} = \{F_1, \ldots, F_{22}\}$ be the collection of graphs shown in Figure 2. Construct a collection $\mathcal{H}$ of graphs as follows: Let $H_{1,m}$ be constructed from $m$ disjoint 5-cycles by identifying a set of $m$ vertices, one from each cycle, into one vertex. Let $\mathcal{H}_1 = \{H_{1,m} \mid m \geq 2\}$. For $i = 2, 3, \ldots, 7$, let $\mathcal{H}_i = \{H_{i,m} \mid m \geq 1\}$ where $H_{i,m}$ is the graph in Figure 3. For $i = 8, 9, 10$, let $\mathcal{H}_i = \{H_{i,m,\ell} \mid m \geq \ell \geq 1\}$ where $H_{i,m,\ell}$ is the graph shown in Figure 3. Let $\mathcal{H} = \cup_{i=1}^{10} \mathcal{H}_i$.

Henning characterized the $(n-1)/2$-minimal graphs as follows:

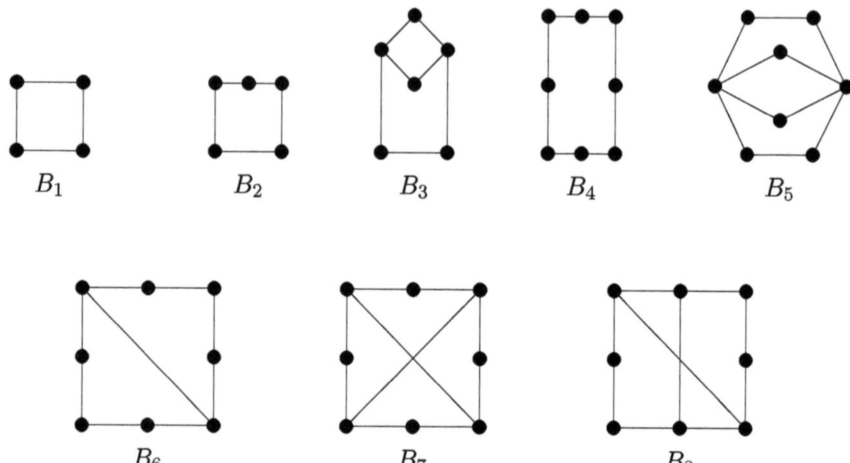


**Fig. 1** The collection $\mathcal{B}$ of graphs.

**Theorem 17** ([26]) *A graph $G$ of order $n$ is $(n-1)/2$-minimal if and only if $G \in \mathcal{B}^* \cup \mathcal{F} \cup \mathcal{H}$.*

Let $\mathcal{K}$ be the set of all even order complete graphs of order at least six with a one factor removed.

Hattingh and Joubert generalized Theorem 16 as follows:

**Theorem 18** ([18]) *$G$ be a connected graph of order $n$ with $\delta \geq 2$. If $G \notin \{B_1, \ldots, B_8\} \cup \mathcal{K}$, then $\gamma_r(G) \leq \frac{n-\delta+1}{2}$.*

Hattingh and Joubert established the following upper bound on the restrained domination number of a connected claw-free graph:

**Theorem 19** ([19]) *Let $G$ be a connected claw-free graph of order $n$ with $\delta \geq 2$. If $G \notin \{C_4, C_7\} \cup \{C_5, C_8, \ldots, C_{17}\}$, then $\gamma_r(G) \leq \frac{2n}{5}$. Moreover, this bound is best possible.*

To see that this bound is best possible, construct the graph $G$ by joining vertices $u$ and $v$ to each vertex of the disjoint union of $K_2$ and an isolated vertex. Then $G$ is a claw-free connected graph with minimum degree two and $\gamma_r(G) = 2 = \frac{2\times 5}{5} = \frac{2n}{5}$. Other examples include $C_{10}$ and $C_{20}$.

Using probabilistic methods, Cockayne established the following upper bound on the restrained domination number of a graph.

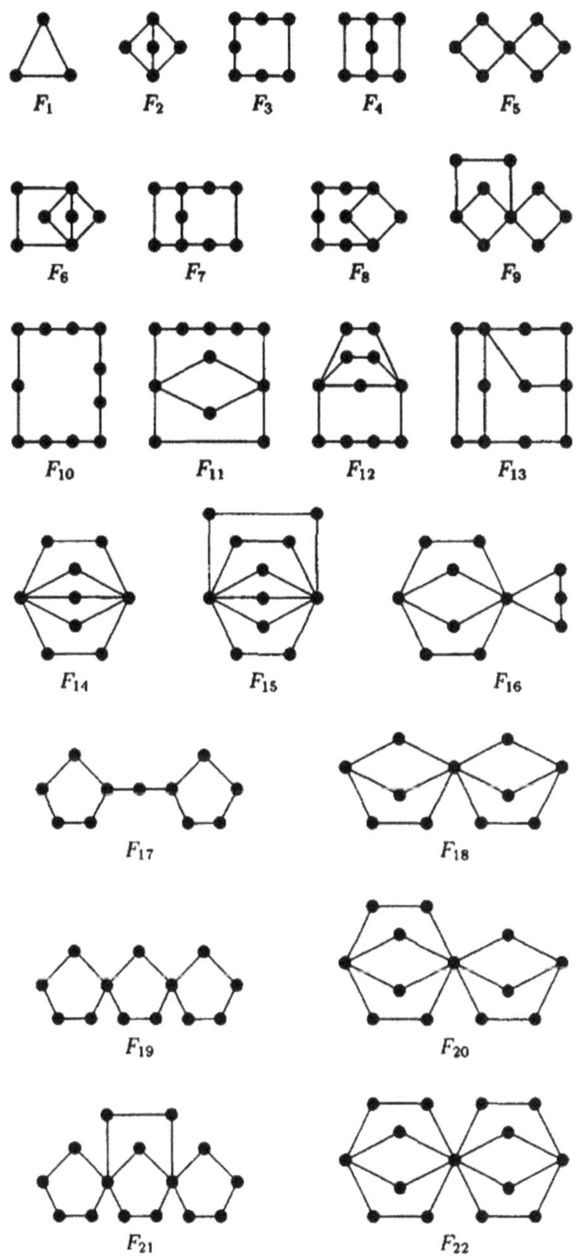


**Fig. 2** The collection $\mathcal{F}$ of graphs.

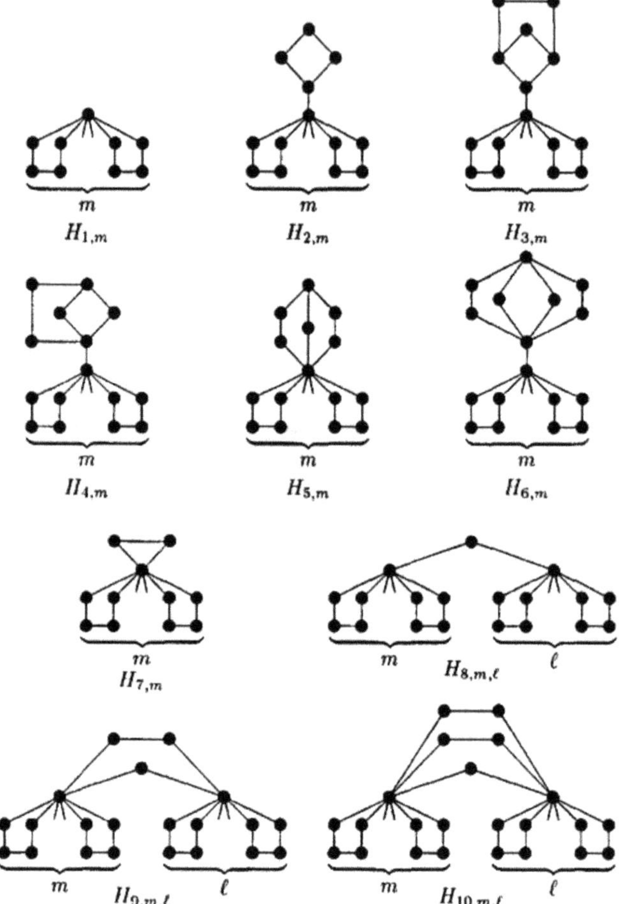


**Fig. 3** The collection $\mathcal{H}$ of graphs.

**Theorem 20** ([5]) *If G has order n and minimum degree δ, then*

$$\gamma_r(G) \leq \begin{cases} n\left(1 - \frac{2\delta}{(2\delta+2)^{1+\frac{1}{\delta}}}\right) & \text{if } \delta \geq 4 \\ .895n & \text{if } \delta = 2 \\ .668n & \text{if } \delta = 3 \end{cases}.$$

Dankelmann, Day, Hattingh, Henning, Markus, and Swart established an upper bound for the restrained domination in terms of order and maximum degree.

**Theorem 21** ([8]) *If G is a graph of order n with maximum degree* $\Delta$ *and minimum degree* $\delta \geq 2$*, then* $\gamma_r(G) \leq n - \Delta$.

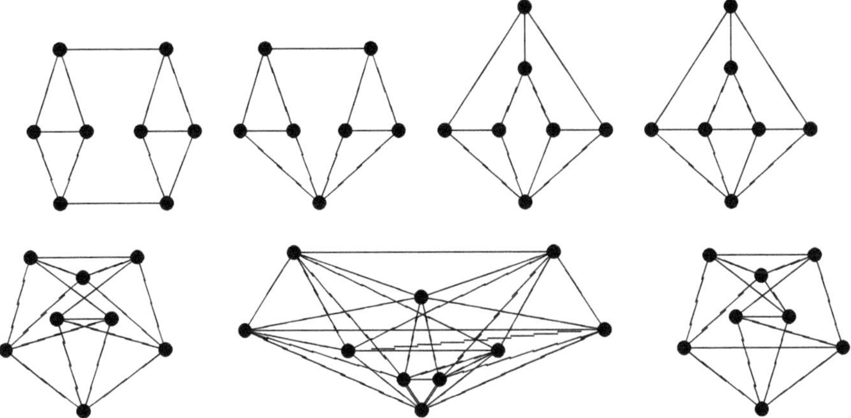

**Fig. 4** From top left to bottom right: The graphs $H_2$ to $H_8$.

Moreover, in the same paper,

- the connected triangle-free graphs of order $n$ with minimum degree $\delta \geq 2$ for which $\gamma_r(G) = n - \Delta$ are characterized; and
- for connected graphs $G$ that are both triangle-free and $C_5$-free, a particularly simple characterization is obtained.

Let $\mathcal{K} = \{K_n \mid n \geq 4\}$. Let $K_{r,s}$ be the complete bipartite graph where $r \geq 3$ and $2 \leq s \leq 3$, and suppose $\{V_1, V_2\}$ is the bipartition of $K_{r,s}$. Let $B_{r,s}$ be the graph obtained from $K_{r,s}$ by joining vertices in $V_1$ so that $\delta(B_{r,s}) \geq r$. Let $\mathcal{B} = \{B_{r,s} \mid r \geq 3 \text{ and } 2 \leq s \leq 3\}$. Let $K_{1,r}$ be the star where $r \geq 3$. Let $D_r$ be the graph obtained from $K_{1,r} \cup K_2$ by joining vertices so that $\delta(D_r) \geq r$. Let $\mathcal{D} = \{D_r \mid r \geq 3\}$. Let $H_1 \cong C_4 \times K_2$ and for $2 \leq i \leq 8$, let $H_i$ be the graphs depicted in Figure 4. Let $\mathcal{H} = \cup_{i=1}^{8}\{H_i\} \cup \mathcal{K} \cup \mathcal{B} \cup \mathcal{D}$.

The following result is due to Hattingh, Jonck, and Joubert.

**Theorem 22** ([14]) *Let $G$ be a connected graph of order $n$ with $\delta \geq 3$. If $G \notin \mathcal{H}$, then $\gamma_{tr}(G) \leq n - \delta - 2$.*

As $\gamma_r(G) \leq \gamma_{tr}(G)$ for a graph $G$, we have the following result.

**Corollary 23** *Let $G$ be a connected graph of order $n$ with $\delta \geq 3$. If $G \notin \mathcal{H}$, then $\gamma_r(G) \leq n - \delta - 2$.*

The restrained domination number of a cubic graph in terms of its order was established by Hattingh and Joubert.

**Theorem 24** ([20]) *If $G$ is a cubic graph of order $n$, then $\gamma_r(G) \leq \frac{5n}{11}$.*

An upper bound on the size of a graph when the restrained domination number is fixed was bounded by Joubert [33] as follows:

**Theorem 25** ([33] ) *If $G$ is a graph order $n \geq 5$ with $\gamma_r(G) = k \in \{3, \ldots, n-2\}$, then*

$$m(G) \leq \begin{cases} \frac{n(n-2)-1}{2} & \text{if } 2 = k \leq n \text{ and } n \geq 5 \text{ is odd,} \\ \binom{n-k}{2} + n + \frac{n-k}{2} - 3 & \text{if } 3 \leq k \leq n-3 \text{ and } n-k \text{ is even,} \\ \binom{n-k}{2} + n + \frac{n-k+1}{2} - 3 & \text{if } 3 \leq k \leq n-3 \text{ and } n-k \text{ is odd,} \\ n & \text{if } 3 \leq k = n-2. \end{cases}$$

In [22], Hattingh and Joubert extended Theorem 25 to include the cases $k = 1, 2, n$, and characterized graphs $G$ of order $n \geq 2$ and restrained domination number $k \in \{1, \ldots, n-2, n\}$ for which $m(G)$ achieves the upper bound of Theorem 25.

Consider a bipartite graph $G$ of order $n \geq 4$ and let $k \in \{2, \ldots, n-2\}$. In [34], Joubert showed that if $\gamma_r(G) = k$, then $m(G) \leq ((n-k)(n-k+6) + 4k - 8)/4$. Moreover, this bound was shown to be best possible.

## 2.4 $(\gamma, \gamma_r)$-graphs

A graph $G$ is called *$\gamma$-excellent* if every vertex of $G$ belongs to some dominating set of cardinality $\gamma(G)$. A subset $S \subseteq V$ is a *packing* in $G$ if the vertices of $S$ are pairwise at distance at least three apart in $G$. The *packing number* $\rho(G)$ is the maximum cardinality of a packing in $G$. It is immediately obvious that $\gamma(G) \leq \gamma_r(G)$ for a graph $G$. A graph $G$ such that $\gamma(G) = \gamma_r(G)$ is called a *$(\gamma, \gamma_r)$-graph.* Dankelmann, Hattingh, Henning, and Swart characterized the trees with equal domination and restrained domination numbers in [9]. In order to state the characterization we introduce two graph operations and a family of trees generated by these operations.

A *labeling* of a tree $T$ is a function $S\colon V(T) \to \{A, B\}$. The label of a vertex $v$ is also called its *status*, denoted sta$(v)$. By a *labeled* $K_1$ we shall mean a $K_1$ whose vertex is labeled with status $B$. Next we define two operations:

- **Operation $\mathcal{O}_1$.** Attach to a vertex $v$ of status $A$ a path $v, x, y$ where sta$(x) = A$ and sta$(y) = B$.

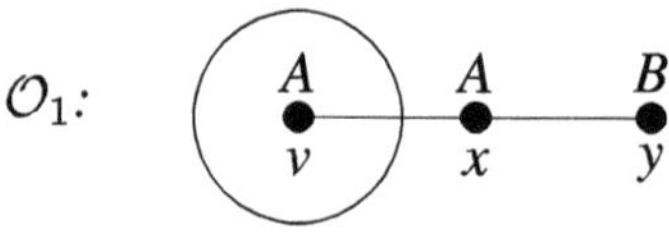

- **Operation $\mathcal{O}_2$.** Attach to a vertex $v$ of status $B$ a path $v, x, y, z$ where sta$(x)$ = sta$(y) = A$ and sta$(z) = B$.

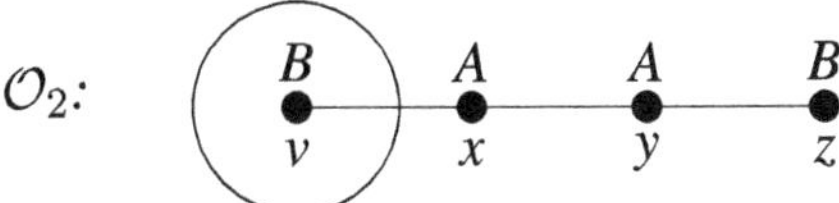


Let $\mathcal{T}$ be the family of trees that can be labeled so that the resulting family of labeled trees contains a labeled $K_1$ and is closed under the two operations $\mathcal{O}_1$ and $\mathcal{O}_2$ listed above, which extend the tree $T$ by attaching a tree to the vertex $v \in V(T)$.

**Theorem 26** ([9]) *$T$ be a tree. Then the following statements are equivalent:*

1. *$T \in \mathcal{T}$;*
2. *$T$ has a unique $\rho(T)$-set and this set is a dominating set of $T$;*
3. *$T$ is a $(\gamma, \gamma_r)$-tree;*
4. *$T$ is $\gamma$-excellent and $T \neq K_2$.*

Let $\mathcal{F}$ be the family of self-complementary graphs $G$ such that $G$ is the cycle $C_5$ or $G$ can be constructed from the disjoint union of a path $P_4$ and a self-complementary graph $H$, by adding all possible edges between the two support vertices of the path $P_4$ and the vertices of $H$.

Desormeaux, Haynes, and Henning characterized the self-complementary $(\gamma, \gamma_r)$-graphs as follows.

**Theorem 27** ([10]) *$G$ be a self-complementary graph. Then $G$ is a $(\gamma, \gamma_r)$-graph if and only if $G \notin \mathcal{F}$.*

## 2.5 Nordhaus-Gaddum Results

Nordhaus and Gaddum [41] presented best possible bounds on the sum and product of the chromatic number of a graph and its complement. The corresponding results for the domination number were presented by Jaeger and Payan in [30].

Domke, Hattingh, Hedetniemi, Laskar, and Markus obtained bounds on the sum of the restrained domination numbers of a graph and its complement.

**Theorem 28** ([11]) *Let $G$ be a graph of order $n \geq 2$ such that $G \notin \{P_3, \overline{P}_3\}$. Then $4 \leq \gamma_r(G) + \gamma_r(\overline{G}) \leq n + 2$.*

The extremal graphs achieving the lower and upper bounds in Theorem 28 were characterized by Hattingh, Jonck, Joubert, and Plummer.

Let $\mathcal{H}$ be the family of graphs $G$ of order $n$ where $G$ or $\overline{G}$ is one of the following four types:

**Type 1.** $V(G) = \{x, y, z\} \cup X$. Moreover:

- $x$ is adjacent to each vertex of $\{y, z\} \cup X$,
- each vertex of $\{y, z\} \cup X$ is adjacent to some vertex of $\{y, z\} \cup X$, and
- each vertex of $X$ is non-adjacent to some vertex of $\{y, z\}$ and non-adjacent to some vertex in $X$.

**Type 2.** $V(G) = \{x, y\} \cup X$. Moreover:

- each vertex of $X$ is adjacent to exactly one vertex of $\{x, y\}$ and also non-adjacent to exactly one vertex of $\{x, y\}$,
- each vertex of $X$ is non-adjacent to some vertex of $X$, and
- each vertex of $X$ is adjacent to some vertex of $X$.

**Type 3.** $V(G) = \{u, v, y\} \cup X$. Moreover:

- each vertex of $X \cup \{y\}$ is adjacent to some vertex of $\{u, v\}$,
- each vertex of $X \cup \{u\}$ is non-adjacent to some vertex of $\{v, y\}$,
- each vertex of $X \cup \{y\}$ is adjacent to some vertex of $X \cup \{y\}$, and
- each vertex of $X \cup \{u\}$ is non-adjacent to some vertex of $X \cup \{u\}$.

**Type 4.** $V(G) = \{x, y, u, v\} \cup X$. Moreover:

- each vertex in $\{x, y\} \cup X$ is adjacent to some vertex of $\{u, v\}$,
- each vertex in $\{u, v\} \cup X$ is non-adjacent to some vertex of $\{x, y\}$,
- each vertex in $\{x, y\} \cup X$ is adjacent to some vertex of $\{x, y\} \cup X$, and
- each vertex in $\{u, v\} \cup X$ is non-adjacent to some vertex of $\{u, v\} \cup X$.

**Theorem 29** ([17]) *If $G$ be a graph of order $n \geq 2$, then $\gamma_r(G) + \gamma_r(\overline{G}) = 4$ if and only if $G$ or $\overline{G} \in \mathcal{H}$.*

Let $\mathcal{G} = \{G \mid G$ or $\overline{G}$ is a disjoint union of non-trivial stars with $|V(G)| \neq 3\}$, $\mathcal{S} = \{G \mid G$ or $\overline{G} \cong K_1 \cup S$ where $S$ is a star of order at least three$\}$. Lastly, let $\mathcal{E} = \mathcal{G} \cup \mathcal{S}$.

**Theorem 30** ([17]) *Let $G$ be a graph of order $n = 2$ or $n \geq 4$. Then $\gamma_r(G) + \gamma_r(\overline{G}) = n + 2$ if and only if $G \in \mathcal{E}$.*

In [21], Hattingh and Joubert obtained bounds on the product of the restrained domination numbers of a graph and its complement. Let $\mathcal{E}' = \mathcal{E} - \{K_2\}$.

**Theorem 31** ([21]) *Let $G$ be a graph such that $n \geq 4$. Then $\gamma_r(G)\gamma_r(\overline{G}) \leq 2n$ with equality holding if and only if $G \in \mathcal{E}'$.*

**Sketch of Proof:** Using Theorems 16, 18, 21, and an idea which appeared in a paper by Karami, Khodkar, Sheikholeslami, and West [37] one can show:

**Lemma 32** ([21]) *If $n \geq 4$ and* $\text{diam}(G) = \text{diam}(\overline{G}) = 2$, *then $\gamma_r(G)\gamma_r(\overline{G}) < 2n$.*

**Lemma 33** ([21]) *Let $G$ be a graph of order $n \geq 4$. If $G$ or $\overline{G}$ is disconnected, then $\gamma_r(G)\gamma_r(\overline{G}) \leq 2n$ with equality if and only if $G \in \mathcal{E}'$.*

**Lemma 34** ([21]) *Let $G$ be a graph of order $n \geq 4$ such that $G$ and $\overline{G}$ are both connected. If $\delta(G) = 1$ or $\delta(\overline{G}) = 1$, then $\gamma_r(G)\gamma_r(\overline{G}) < 2n$.*

**Lemma 35** ([21]) *Let $G$ be a graph of order $n \geq 4$ with $\min\{\delta(G), \delta(\overline{G})\} \geq 2$. Moreover, suppose $G$ and $\overline{G}$ are connected. Then $\gamma_r(G)\gamma_r(\overline{G}) \leq \frac{3}{2}n$.*

The last three Lemmas establish Theorem 31.

# 3 Total Restrained Domination

Next we survey results in the literature on total restrained domination in graphs.

## 3.1 Exact Results

Chen, Ma, and Sun [2] determined the exact value for the total restrained domination number for several simple graph families.

**Proposition 36** ([2]) *If $n \geq 2$ is an integer, then*

$$\gamma_{tr}(K_n) = \begin{cases} n & \text{if } n = 2, 3, \\ 2 & \text{if } n \geq 4. \end{cases}$$

**Proposition 37** ([2]) *If $n \geq 2$ is an integer, then $\gamma_{tr}(K_{1,n-1}) = n$.*

**Proposition 38** ([2]) *If $n_1$ and $n_2$ are integers such that $\min\{n_1, n_2\} \geq 2$, then $\gamma_{tr}(K_{n_1,n_2}) = 2$.*

**Proposition 39** ([2]) *If $t \geq 3$ is an integer, then*

$$\gamma_{tr}(K_{n_1,\ldots,n_t}) = \begin{cases} 3 & \text{for } t = 3 \text{ and } n_1 = n_2 = n_3 = 1, \\ 2 & \text{otherwise.} \end{cases} .$$

**Proposition 40** ([2]) *If $n \geq 3$ is an integer, then $\gamma_{tr}(C_n) = n - 2\lfloor \frac{n}{4} \rfloor$.*

**Proposition 41** ([2]) *If $n \geq 2$ is an integer, then $\gamma_{tr}(P_n) = n - 2\lfloor \frac{n-2}{4} \rfloor$.*

## 3.2 Lower Bounds

We now turn our attention to lower bounds on the total restrained domination number of a graph.

The following result is due to Cyman and Raczek.

**Theorem 42** ([6]) *Let $G$ be a connected graph of order $n$ and size $m$. Then* $\gamma_{tr}(G) \geq \frac{3n}{2} - m$.

As an immediate consequence we obtain the following result due to Hattingh, Jonck, and Joubert.

**Theorem 43** ([16]) *If $T$ is a tree of order $n \geq 2$, then $\gamma_{tr}(T) \geq \lceil (n+2)/2 \rceil$.*

A constructive characterization of trees attaining the bound of Theorem 43 was also obtained in [16]. Let $\mathcal{T}$ be the family of all trees $T$ of order $n$ such that $\gamma_{tr}(T) = \lceil \frac{n+2}{2} \rceil$. In order to state the characterization, define four simple operations on a tree $T$.

**O1.** Join a leaf or a remote vertex of $T$ to a vertex of $K_1$, where $n(T)$ is even.
**O2.** Join a vertex $v$ of $T$ which lies on an endpath $v, x, z$ to a leaf of $P_3$, where $n(T)$ is even.
**O3.** Join a vertex $v$ which lies on an endpath $v, x_1, x_2, z$ to a leaf of $P_3$, where $n(T)$ is even.
**O4.** Join a remote vertex or a leaf of $T$ to $\ell$ disjoint copies of $P_4$ for some $\ell \geq 1$.

Let $\mathcal{C}$ be the family of all trees obtained from $P_2$ by a finite sequence of Operations **O1–O4**.

**Theorem 44** ([16]) *$T \in \mathcal{C}$ if and only if $T \in \mathcal{T}$.*

As observed in [16], the bound of Theorem 43 can be improved if the order of a tree is a multiple of 4.

**Theorem 45** ([17]) *Let $T$ be a tree of order $n$. If $n \equiv 0 \bmod 4$, then $\gamma_{tr}(T) \geq \lceil \frac{n+2}{2} \rceil + 1$.*

A characterization of trees attaining this improved bound was also given in [16].

Let $\mathcal{T}^* = \{T \mid T$ is a tree of order $n \equiv 0 \bmod 4$ such that $\gamma_{tr}(T) = \lceil \frac{n+2}{2} \rceil + 1\}$. In order to state a constructive characterization of trees in $\mathcal{T}^*$, define the following operations on a tree $T$:

**O5.** Join a leaf or a remote vertex $v$ to a vertex of $K_1$, where $n(T) \equiv 3 \bmod 4$.
**O6.** Join a vertex $v$ which lies on an endpath $v, x, z$ to a vertex of $K_2$, where $n(T) \equiv 2 \bmod 4$.
**O7.** Join a vertex $v$ which lies on an endpath $v, x_1, x_2, z$ to a vertex of $K_2$, where $n(T) \equiv 2 \bmod 4$.
**O8.** Join a vertex $v$ which lies on an endpath $v, x, z$ to a leaf of $P_3$, where $n(T) \equiv 1 \bmod 4$.
**O9.** Join a vertex $v$ which lies on an endpath $v, x_1, x_2, z$ to a leaf of $P_3$, where $n(T) \equiv 1 \bmod 4$.

Let $\mathcal{I} = \{T \mid T$ is a tree obtained by applying one of the Operations **O5–O9** to a tree $T' \in \mathcal{C}$ exactly once$\}$. Let $\mathcal{C}^* = \{T \mid T$ is a tree obtained from a tree $T' \in \mathcal{I}$ by applying the Operation **O4** to $T'$ zero or more times$\}$.

**Theorem 46** ([16]) $\mathcal{T}^* = \mathcal{C}^*$.

The following result is due to Cyman and Raczek. Let $\ell(T)$ denote the number of leaves of $T$.

**Theorem 47** ([7]) *If $T$ is a tree of order $n \geq 3$, then $3\gamma_{tr}(T) \geq n + 2 + 2\ell(T)$.*

Cyman and Raczek also provided a constructive characterization of the trees $T$ of order $n$ for which $3\gamma_{tr}(T) = n + 2 + 2\ell(T)$. To state the characterization, we introduce some additional notation. If $T_1$ and $T_2$ are vertex disjoint trees and $u$ and $v$ are strong support vertices in $T_1$ and $T_2$, respectively, then $T_1 \oplus_{uv} T_2$ will denote a tree obtained from $T_1$ and $T_2$ by adding an edge incident with a leaf adjacent to $u$ and incident with a leaf adjacent to $v$. Let $\mathcal{R}$ denote the family of trees such that:

1. Every star $K_{1,n}$ belongs to $\mathcal{R}$ where $n \geq 2$;
2. $T_1 \oplus_{uv} T_2$ belongs to $\mathcal{R}$ if only $T_1$ and $T_2$ belong to $\mathcal{R}$, where $u$ and $v$ are strong support vertices in $T_1$ and $T_2$, respectively.

**Theorem 48** ([7]) *If $T$ is a tree of order $n \geq 3$, then $3\gamma_{tr}(T) = n + 2 + 2\ell(T)$ if and only if $T$ belongs to the family $\mathcal{R}$.*

## 3.3 Upper Bounds

Let $S$ be the set of leaves and support vertices of $G$. The following result of Chen, Liu, and Meng characterized the graphs $G$ of order $n$ for which $\gamma_{tr}(G) = n$.

**Theorem 49** ([4]) *Let $G$ be a connected graph of order $n \geq 4$. Then $\gamma_{tr}(G) = n$ if and only if $G - S$ is an empty graph.*

The following result is due to Chen, Liu, and Meng. Let $B$ constructed from two copies of $K_2$ and joining a new vertex to each of the vertices of the two copies of $K_2$.

**Theorem 50** ([4]) *Let $G$ be a connected graph of order $n \geq 4$. Then $\gamma_{tr}(G) = n-2$ if and only if one of the following holds:*

1. *$G \cong K_4$ or $G \cong K_4 - e$ or $G \cong \overline{B}$ or $G \cong C_i$ where $i \in \{4, 5, 6, 7\}$.*
2. *$G - S$ is the disjoint union of $C_4$ and isolated vertices. Furthermore, either one vertex or two adjacent vertices of $C_4$ have neighbors in $S$.*
3. *$G - S$ is the disjoint union of $K_4 - e$ and isolated vertices. Moreover, only one vertex of $K_4 - e$ whose degree is two has neighbors in $S$.*
4. *$G - S$ is the disjoint union of $C_3$ and isolated vertices. Furthermore, at most two vertices of $C_3$ have neighbors in $S$.*
5. *$G - S$ is the disjoint union of a $C_3$ with an edge attached to one vertex of $C_3$ and isolated vertices. Furthermore, either both ends of the edge or one end of the edge which does not coincide with the vertex of $C_3$ has neighbors in $S$.*
6. *$G - S$ is the disjoint union of a $C_3$ with a path $P_3$ attached to one vertex of $C_3$ and isolated vertices. Furthermore, only the end vertex of $P_3$ which does not coincide with the vertex of $C_3$ has neighbors in $S$.*

7. *$G - S$ is the disjoint union of $\overline{B}$ and isolated vertices. Furthermore, only one vertex whose degree is two in $\overline{B}$ has neighbors in $S$.*
8. *$G - S$ is the disjoint union of a path of order at most 5 and isolated vertices. Furthermore, only the ends of path are adjacent to vertices of $S$.*

In [15], Hattingh, Jonck, and Joubert obtained an upper bound on the total restrained domination number of a tree, and constructively characterized the trees attaining this bound. For a tree $T$, let $s(T)$ and $\ell(T)$ denote the number of support vertices and leaves, respectively.

**Theorem 51** ([15]) *If $T$ is a tree of order $n \geq 3$, then $\gamma_{tr}(T) \leq \left\lfloor \frac{n+2s(T)+\ell(T)-1}{2} \right\rfloor$.*

Let $\mathcal{T}$ be the family of all trees $T$ of order $n \geq 3$ such that $\gamma_{tr}(T) = \left\lfloor \frac{n+2s(T)+\ell(T)-1}{2} \right\rfloor$. For a tree $T$, let $k(T) = n(T) + 2s(T) + \ell(T) - 1$. In order to state the characterization, define five simple operations on a tree $T$.

**O1.** If $k(T)$ is even, and $T$ has a remote vertex adjacent to at least two leaves, join one of these leaves to a vertex of $K_1$.
**O2.** If $T$ has a remote vertex adjacent to at least two leaves, join one of these leaves to a leaf of a $K_2$.
**O3.** Join a remote vertex of $T$ to a vertex of $K_1$.
**O4.** Suppose $k(T)$ is even. Join a leaf of $K_2$ to $w$, where $w$ is a vertex such that $\deg(w) \geq 2$ and adjacent to a vertex $u$ which is either a leaf or a remote vertex whose other neighbors only consist of leaves.
**O5.** Join the vertex $v$ of $T$ which lies on an endpath $u, w, v$, where $\deg(u) \geq 2$, to a leaf of $P_4$.

Let $\mathcal{C}$ be the family of all trees obtained from $P_3$ by a finite sequence of Operations **O1–O5**.

**Theorem 52** ([15]) *$T \in \mathcal{T}$ if and only if $T \in \mathcal{C}$.*

Theorem 22, above, of Hattingh, Jonck, and Joubert provides an upper bound on the total restrained domination number of a graph in terms of its order and minimum degree.

Hattingh, Jonck, and Joubert showed that:

**Theorem 53** ([14]) *Let $G$ be an $r$-regular graph of order $n$, where $4 \leq r \leq n - 3$. Then $\gamma_{tr}(G) \leq n - \operatorname{diam}(G) - r + 1$, and this bound is best possible.*

The following two results are due to Henning and Maritz.

**Theorem 54** ([28]) *If $G$ is a connected graph of order $n \geq 4$ with $\delta \geq 2$ and $\Delta \leq n - 2$, then $\gamma_{tr}(G) \leq n - \Delta/2 - 1$. Moreover, this bound is best possible.*

**Theorem 55** ([28]) *If $G$ is a connected bipartite graph of order $n \geq 5$, maximum degree $\Delta$ where $3 \leq \Delta \leq n-2$ and $\delta \geq 2$, then $\gamma_{tr}(G) \leq n - \frac{2}{3}\Delta - \frac{2}{9}\sqrt{3\Delta - 8} - \frac{7}{9}$. Moreover, this bound is best possible.*

There are several results involving upper bounds on the total restrained domination number in terms of the order of a graph. Let $B$ be a graph obtained by attaching the two vertices of a $K_2$ to exactly one vertex of a $C_3$. Let $\mathcal{K} = \{B, C_3, C_5, C_6, C_7, C_{10}, C_{11}, C_{15}, C_{19}\}$. Joubert showed that:

**Theorem 56** ([32]) *If $G$ is a claw-free graph with $\delta \geq 2$ and $G \notin \mathcal{K}$, then $\gamma_{tr}(G) \leq 4n/7$.*

Koh, Maleki, and Omoomi used probabilistic methods to establish the following bound on the total restrained domination number.

**Theorem 57** ([39]) *Let $G$ be a graph such that $n \geq 4$ and $\delta \geq 2$. Then $\gamma_{tr}(G) \leq n - \sqrt[3]{\frac{n}{4}}$.*

Koh, Maleki, and Omoomi also conjectured the following.

**Conjecture 58** ([39]) *Let $G$ be a graph such that $n \geq 4$ and $\delta \geq 2$. Then $\gamma_{tr}(G) \leq n - \theta(\sqrt{n})$.*

This conjecture was settled by Joubert.

**Theorem 59** ([35]) *Let $G$ be a graph without any $C_3$ components. Then $\gamma_{tr}(G) \leq n - \sqrt{\frac{n}{2}}$.*

The total restrained domination number of a cubic graph in terms of its order was established by Jiang, Kang, and Shan.

**Theorem 60** ([31]) *If $G$ is a cubic graph of order $n$, then $\gamma_{tr}(G) \leq \frac{13n}{19}$.*

In the same paper, Jiang, Kang, and Shan also showed that if adding the restriction that $G$ is claw-free, then $\gamma_{tr}(G) = \gamma_t(G)$, and thus some results on total domination in claw-free cubic graphs are valid for total restrained domination.

Henning and Southey [29] established the following improved upper bound on the total restrained domination number of a cubic graph.

**Theorem 61** ([29]) *If $G$ is a cubic graph of order $n$, then $\gamma_{tr}(G) \leq (n + 4)/2$.*

They also provided two infinite families of connected cubic graphs $G$ with $\gamma_{tr}(G) = n/2$ showing that the upper bound of Theorem 61 is essentially best possible.

## 3.4 $(\gamma_t, \gamma_{tr})$-graphs and $(\gamma_r, \gamma_{tr})$-graphs

A graph $G$ such that $\gamma_t(G) = \gamma_{tr}(G)$ is called a $(\gamma_t, \gamma_{tr})$-*graph*. Cyman and Raczek characterized the trees with equal total domination and total restrained domination numbers in [7].

Let $\mathcal{T}$ be the family of trees $T$ that can be obtained from the sequence $T_1, \ldots, T_j (j \geq 1)$ of trees such that $T_1$ is the path $P_2$ and $T = T_j$, and, if $j > 1$, then $T_{i+1}$ can be obtained recursively from $T_i$ by one of the two operations $\mathcal{O}_1$ and $\mathcal{O}_2$ below.

- **Operation $\mathcal{O}_1$.** The tree $T_{i+1}$ is obtained from $T_i$ by adding a path $y, x_1, x_2, x_3, x_4$ where $y \in V(T_i)$ belongs to some $\gamma_{tr}(T_i)$-set.
- **Operation $\mathcal{O}_2$.** The tree $T_{i+1}$ is obtained from $T_i$ by adding a path $y, x_1, x_2, x_3$ where $y \in V(T_i)$ belongs to none of the $\gamma_{tr}(T_i)$-sets.

**Theorem 62** ([7]) *A tree $T$ is $(\gamma_t, \gamma_{tr})$-graph if and only if $T$ belongs to the family $\mathcal{T}$.*

Let $K_{1,r,4}$ be the graph obtained by subdividing each edge of the star $K_{1,r}$ twice. The vertex of degree $r$ is called the *central vertex* of $K_{1,r,4}$. Let $\mathcal{K} = \{K_{1,r,4} \mid r$ is a positive integer$\}$.

Consider the following two operations on a tree $T$:

- **Operation $\mathcal{O}_1$.** Let $x$ be a vertex of $T$ which is either a leaf or support vertex of $T$. Join $x$ to the central vertex of one or more of the trees in $\mathcal{K}$.
- **Operation $\mathcal{O}_2$.** Let $x$ be a non-leaf of $T$ which is adjacent to a support vertex of $T$. Join $x$ to one leaf of one or more copies of $P_3$.

Let $\mathcal{T}' = \{T \mid T$ is obtained from $P_6$ by a finite sequence of operations $\mathcal{O}_1$ or $\mathcal{O}_2\} \cup \{P_2, P_6\}$.

We are now ready to state an alternative constructive characterization of trees with equal total domination and total restrained domination numbers, which was obtained by Chen, Shui, and Chen.

**Theorem 63** ([3]) *A tree $T$ is $(\gamma_t, \gamma_{tr})$-graph if and only if $T$ belongs to the family $\mathcal{T}'$.*

A graph $G$ such that $\gamma_r(G) = \gamma_{tr}(G)$ is called a *$(\gamma_r, \gamma_{tr})$-graph*. Raczek characterized the trees with equal restrained domination and total restrained domination numbers in [43].

Consider the following two operations on a tree $T$:

- **Operation $\mathcal{O}_1$.** Let $x$ be a support vertex of $T$. Join $x$ to a new vertex $y$.
- **Operation $\mathcal{O}_2$.** Let $x$ be a support vertex of $T$. Add a path $y_1, y_2, y_3, y_4$ and join $x$ to $y_1$.

Let $\mathcal{T}'' = \{T \mid T$ is obtained from $P_3$ by a finite sequence of operations $\mathcal{O}_1$ or $\mathcal{O}_2\} \cup \{P_2, P_6\}$.

**Theorem 64** ([43]) *A tree $T$ is $(\gamma_r, \gamma_{tr})$-graph if and only if $T$ belongs to the family $\mathcal{T}''$.*

## 3.5 Nordhaus-Gaddum Results

In [17], Hattingh, Jonck, and Joubert obtained bounds on the sum of the total restrained domination of a graph and its complement, and characterized the graphs attaining these bounds.

Construct the graph $K$ by matching the vertices of $\overline{K}_2$ to distinct vertices of $K_3$.

**Theorem 65** ([17]) *If $G$ is a graph of order $n \geq 2$ such that neither $G$ nor $\overline{G}$ contains isolated vertices or is isomorphic to $K$, then $4 \leq \gamma_{tr}(G)+\gamma_{tr}(\overline{G}) \leq n+4$.*

Let $n \geq 5$ be an integer and suppose $\{x, y, u, v\}$ and $X$ are disjoint sets of vertices such that $|X| = n - 4$. Let $\mathcal{L}$ be the family of graphs $G$ of order $n$ where $V(G) = \{x, y, u, v\} \cup X$ and with the following properties:

**P1:** $x$ and $y$ are non-adjacent, while $u$ and $v$ are adjacent,
**P2:** each vertex in $\{x, y\} \cup X$ is adjacent to some vertex of $\{u, v\}$,
**P3:** each vertex in $\{u, v\} \cup X$ is non-adjacent to some vertex of $\{x, y\}$,
**P4:** each vertex in $\{x, y\} \cup X$ is adjacent to some vertex of $\{x, y\} \cup X$,
**P5:** each vertex in $\{u, v\} \cup X$ is non-adjacent to some vertex of $\{u, v\} \cup X$.

**Theorem 66** ([17]) *If $G$ be a graph of order $n \geq 2$ such that neither $G$ nor $\overline{G}$ contains isolated vertices, then $\gamma_{tr}(G) + \gamma_{tr}(\overline{G}) = 4$ if and only if $G \in \mathcal{L}$.*

Let $\mathcal{U} = \{G \mid G$ is a graph of order $n$ which can be obtained from a $P_4$ with consecutive vertices labeled $u, v_1, v_2, v$ by joining vertices $v_1$ and $v_2$ to each vertex of $K_{n-4}$ where $n \geq 6\}$.

**Theorem 67** ([17]) *If $G$ is a graph of order $n \geq 2$ such that neither $G$ nor $\overline{G}$ contains isolated vertices or is isomorphic to $K$, then $\gamma_{tr}(G) + \gamma_{tr}(\overline{G}) = n + 4$ if and only if $G \in \mathcal{U}$ or $\overline{G} \in \mathcal{U}$ or $G \cong P_4$.*

In [23], Hattingh and Joubert obtained bounds on the the product of the total restrained domination of a graph and its complement, and characterized the graphs attaining these bounds.

Let $\mathcal{L}$ be the family of all graphs constructed in the following way: Let $u$ and $v$ be two distinct isolates and consider the complete graph $K_n$, where $n = 2$ or $n \geq 4$. Let $u'$ and $v'$ be two distinct vertices of $K_n$. Join $u$ to $u'$, and join $v$ to $v'$. Let $K$ be the graph as defined earlier.

**Theorem 68** ([23]) *Let $G$ be a graph of order $n \geq 4$, and suppose neither $G$ nor $\overline{G}$ contains isolated vertices or is isomorphic to $K$. Then $\gamma_{tr}(G)\gamma_{tr}(\overline{G}) \leq 4n$ with equality holding if and only if either $G \in \mathcal{L}$ or $\overline{G} \in \mathcal{L}$.*

## 3.6 Partitions

A classical result in domination theory is that if $S$ is a minimal dominating set of a graph $G = (V, E)$ without isolates, then $V - S$ is also a dominating set of $G$. Thus, the vertex set of every graph without any isolates can be partitioned into two dominating sets. However, it is not the case that the vertex set of every graph can be partitioned into two total restrained dominating sets. For example, the vertex set of $C_5$ cannot be partitioned into two total restrained dominating sets.

A partition of the vertex set can also be thought of as a coloring. In particular, a partition into two total restrained dominating sets is a 2-coloring of the graph such that no vertex has a monochromatic neighborhood. As an example of such a 2-coloring in $K_n$ with $n \geq 4$, take any 2-coloring with at least two vertices of each color, while in $K_{m,n}$ with $m, n \geq 2$ take any 2-coloring where neither partite set is monochromatic.

We now consider the question of how many edges must be added to $G$ to ensure that there is a partition of $V$ into two total restrained dominating sets. We denote this minimum number by $rd(G)$.

The following result is due to Broere, Dorfling, Goddard, Hattingh, Henning, and Ungerer.

**Theorem 69** ([1]). *If $T$ is a tree with $\ell$ leaves, then $\ell/2 \leq rd(T) \leq \ell/2 + 1$.*

The following result is due to Goddard, Hattingh, and Henning.

**Theorem 70** ([13]) *If $G$ is a graph of order $n \geq 4$ and minimum degree at least 2, then $rd(G) \leq (n - 2\sqrt{n})/4 + O(\log n)$, and this bound is best possible.*

## 4 Open Problems

1. Is it true that if $G$ is a cubic graph of order $n$, then $\gamma_{tr}(G) \leq n/2$?
2. Characterize the trees $T$ with $\ell$ leaves for which $rd(T) = \ell/2$.
3. Is it true that if $G$ is a graph of order $n \geq 2$ and $\delta \geq 2$, then $rd(G) \leq \lceil (n - \gamma_t(G))/2 \rceil$?

## 5 Concluding Remarks

Many variations of restrained domination have been introduced in the past 20 years. We discussed one such variation in this survey paper, namely total restrained domination in graphs. Other variations of restrained domination in graphs include: secure restrained domination [42], $k$-tuple restrained domination [27], $k$-tuple total restrained domination [38], restrained double domination [36], and inverse restrained domination [40]. Restrained domination and its variations continue to be a fruitful area to explore in graph theory, and a quick survey of the literature reveals at least a hundred papers on restrained domination and its variations.

## References

1. I. Broere, M. Dorfling, W. Goddard, J.H. Hattingh, M.A. Henning and E. Ungerer, Augmenting trees to have two disjoint total dominating sets. *Bull. Inst. Combin. Appl.* **42** (2004) 12–18.
2. X.G. Chen, D.X. Ma and L. Sun, On total restrained domination in graphs. *Czechoslovak Math. J.* **55(130)** (2005) 165–173.
3. X. G. Chen, W. C. Shiu and H. Y. Chen, Trees with equal total domination and total restrained domination numbers. *Discuss. Math. Graph Theory* **28** (2008) 59–66.
4. X. Chen, J. Liu and J. Meng, Total restrained domination in graphs. *Comput. Math. App.* **62** (2011) 2892–2898
5. E. J. Cockayne, An upper bound for the restrained domination number of a graph. Unpublished Manuscript circa 1999.
6. J. Cyman and J. Raczek, On the total restrained domination number of a graph. *Australas. J. Combin.* **36** (2006), 91–100.
7. J. Cyman and J. Raczek, Total restrained domination numbers of trees. *Discrete Math.* **308** (2008) 44–50.
8. P. Dankelmann, D. Day, J.H. Hattingh, M. A. Henning, L.R. Markus and H.C. Swart, On equality in an upper bound for the restrained and total domination numbers of a graph. *Discrete Math.* **307** (2007) 2845–2852.
9. P. Dankelmann, J.H. Hattingh, M.A Henning, and H.C. Swart, Trees with equal domination and restrained domination numbers. *J. Glob. Optim.* **34** (2006) 597–607.
10. W. Desormeaux, T. Haynes and M.A. Henning, A note on restrained domination in self-complementary graphs. *Discuss. Math. Graph Theory* (2019) 10.7151/dmgt.2222.
11. G.S. Domke, J.H. Hattingh, S.T. Hedetniemi, R.C. Laskar and L.R. Markus, Restrained domination in graphs. *Discrete Math.* **203** (1999) 61–69.
12. G.S. Domke, J.H. Hattingh, M.A. Henning, and L.R. Markus, Restrained domination in graphs with minimum degree two. *J. Combin. Math. Combin. Comput.* **35** (2000) 239–254.
13. W. Goddard, J.H. Hattingh and M.A. Henning, Augmenting a graph of minimum degree 2 to have two disjoint total dominating sets. *Discrete Math.* **300** (2005) 82–90.
14. J.H. Hattingh, E. Jonck, and E.J. Joubert, Bounds on the total restrained domination number of a graph. *Graphs Combin.* **26** (2010) 77–93.
15. J.H. Hattingh, E. Jonck, and E.J. Joubert, An upper bound on the total restrained domination number of a tree. *J. Comb. Optim.* **20** (2010) 205–223.
16. J.H. Hattingh, E. Jonck, E.J. Joubert and A.R. Plummer, Total restrained domination in trees. *Discrete Math.* **307** (2007) 1643–1650
17. J.H. Hattingh, E. Jonck, E.J. Joubert and A.R. Plummer, Nordhaus-Gaddum results for restrained domination and total restrained domination in graphs. *Discrete Math.* **308** (2008) 1080–1087.
18. J.H. Hattingh and E.J. Joubert, An upper bound for the restrained domination number of a graph with minimum degree at least two in terms of order and minimum degree. *Discrete Appl. Math.* **157** (2009) 2846–2858.
19. J.H. Hattingh and E.J. Joubert, Restrained domination in claw-free graphs with minimum degree at least two. *Graphs Combin.* **25** (2009) 693–706.
20. J.H. Hattingh and E.J. Joubert, Restrained domination in cubic graphs. *J. Combin. Optim.* **22** (2011) 166–179.
21. J.H. Hattingh and E.J. Joubert, The product of the restrained domination numbers of a graph and its complement. *Acta Math. Sin.* **30** (2014) 445–452.
22. J.H. Hattingh and E.J. Joubert, Equality in a bound that relates the size and the restrained domination number of a graph. *J. Comb. Optim.* **31** (2016) 1586–1608.
23. J.H. Hattingh and E.J. Joubert, The product of the total restrained domination numbers of a graph and its complement. *Australasian J. Comb.* **70** (2018) 297–308.
24. J.H. Hattingh, E.J. Joubert, M. Loizeaux, A.R. Plummer and L. van der Merwe, Restrained domination in unicyclic graphs. *Discuss. Math. Graph Theory* **29** (2009) 71–86.

25. J.H.Hattingh and A.R. Plummer, A note on restrained domination in trees. *Ars Combin.* **94** (2010) 477–483.
26. M.A. Henning, Graphs with large restrained domination number. *Discrete Math.* **197/198** (1999) 415–429.
27. M.A. Henning and A.P. Kazemi, $k$-tuple restrained domination in graphs. Manuscript circa 2016.
28. M.A. Henning and J.E. Maritz, Total restrained domination in graphs with minimum degree two. *Discrete Math.* **308** (2008) 1909–1920.
29. M.A. Henning and J. Southey, An improved upper bound on the total restrained domination number in cubic graphs. *Graphs Combin.* **28** (2012) 547–554.
30. F. Jaeger and C. Payan, Relations du type Nordhaus-Gaddum pour le nombre d' absorption d' un graphe simple. *C. R. Acad. Sci Ser A* **274** (1972) 728–730.
31. H. Jiang, L. Kang and E. Shan, Total restrained domination in cubic graphs. *Graphs Combin.* **25** (2009) 341–350.
32. E.J. Joubert, Total restrained domination in claw-free graphs with minimum degree at least two. *Discrete Applied Math.* **159** (2011) 2078–2097.
33. E.J. Joubert, Maximum sizes of graphs with given restrained domination numbers. *Discrete Applied Math.* **161** (2013) 829–837.
34. E.J. Joubert, An inequality that relates the size of a bipartite graph with its order and restrained domination number. *J. Comb. Optim.* **31** (2016) 44–51.
35. E.J. Joubert, On a conjecture involving a bound for the total restrained domination number of a graph. *Discrete Applied Math.* **258** (2019) 177–187.
36. R. Kala and T.R. Nirmala Vasantha, Restrained double domination number of a graph. *AKCE Int. J. Graphs Comb.* **5** (2008) 73–82.
37. H. Karami, A. Khodkar, S.M. Sheikholeslami and D.B. West, Inequalities of Nordhaus-Gaddum type for connected domination. *Graphs Combin.* **28** (2012) 123–131.
38. A.P. Kazemi, $k$-tuple total restrained domination in graphs. *Bull. Iranian Math. Soc.* **40** (2014) 751–763.
39. K.M. Koh, Z. Maleki and B. Omoomi, An upper bound for the total restrained domination number of graphs. *Graphs and Combinatorics* **29** (2013) 1443–1452.
40. V.R. Kulli, Inverse and Disjoint Restrained Domination in Graphs. *J. Fuzzy Mathematical Archive* **11** (2016) 9–5.
41. E.A. Nordhaus and J.W. Gaddum, On complementary graphs. *Amer. Math. Monthly* **63** (1956) 175–177.
42. P.R.L. Pushpam and C. Suseendran, Secure Restrained Domination in Graphs. *Math. Comput. Sci.* **9** (2015) 239–247.
43. J.Raczek, Trees with equal restrained domination and total restrained domination numbers. *Discuss. Math. Graph Theory* **27** (2007) 83–91.
44. J.A. Telle, Characterization of domination-type parameters in graphs, in Proceedings of the 24th South Eastern Conference on Combinatorics, Graph Theory and Computing, *Congr. Numer.* **94** (1993) 9–16.
45. J.A. Telle, Vertex Partitioning Problems: Characterization, Complexity and Algorithms on Partial k-Trees, PhD thesis, University of Oregon CIS Department Technical Report TR-94-18 (1994).
46. J.A. Telle and A. Proskurowski, Practical algorithms on partial k-trees with an application to domination-type problems, in Proceedings WADS 93, LNCS **709** (1993) 610–621.
47. J.A. Telle and A. Proskurowski, Efficient sets in partial $k$-trees. *Discrete Appl. Math.* **44** (1993) 109–117.
48. J.A. Telle, Complexity of domination-type problems in graphs. *Nordic J. Comput.* **1** (1994), 157–171.
49. J. A. Telle and A. Proskurowski, Algorithms for vertex partitioning problems on partial $k$-trees. *SIAM J. Discrete Math.* **10** (1997) 529–550.
50. W. Ulatowski, All graphs with restrained domination number three less than their order.Australas. J. Combin. **48** (2010) 73–86.

# Multiple Domination

**Adriana Hansberg and Lutz Volkmann**

## 1 Introduction

### *1.1 Main Concepts*

Throughout this chapter, we will assume $k$ to be a positive integer. In 1985, Fink and Jacobson [63, 64] generalized the concept of dominating sets. We say that a subset $D$ of $V(G)$ is *$k$-dominating* if every vertex of $V(G) \setminus D$ has at least $k$ neighbors in $D$. A $k$-dominating set $D$ is *minimal* if, for every vertex $v \in D$, the set $D \setminus \{v\}$ is not $k$-dominating in $G$. The *$k$-domination number* $\gamma_k(G)$ and the *upper $k$-domination number* $\Gamma_k(G)$ are, respectively, the minimum cardinality of a $k$-dominating set and the maximum cardinality of a minimal $k$-dominating set of $G$. When $k = 1$, we set $\gamma(G)$ and $\Gamma(G)$ instead of $\gamma_1(G)$ and $\Gamma_1(G)$.

Similarly, Harary and Haynes [83, 84] introduced, in two papers published in 1996 and 2000, the concept of double domination and, more generally, of $k$-tuple domination. A subset $D \subseteq V(G)$ is said to be *$k$-tuple dominating* if the closed neighborhood of every vertex $v \in V(G)$ intersects with $D$ in at least $k$ elements. Of course, this definition requires that the graph in question has minimum degree at least $k-1$. A $k$-tuple dominating set $D$ is minimal if, for every $v \in D$, the set $D \setminus \{v\}$ is not $k$-tuple dominating. The minimum and, respectively, maximum cardinality of a minimal $k$-tuple dominating set of $G$ is called the *$k$-tuple domination number* $\gamma_{\times k}(G)$ and the *upper $k$-tuple domination number* $\Gamma_{\times k}(G)$.

A. Hansberg (✉)
Instituto de Matemáticas, UNAM Juriquilla, Blvd. Juriquilla 3001, 76230 Querétaro, Mexico
e-mail: ahansberg@im.unam.mx

L. Volkmann
Lehrstuhl II für Mathematik, RWTH Aachen University, Templergraben 55, D-52056 Aachen, Germany
e-mail: volkm@math2.rwth-aachen.de



T. W. Haynes et al. (eds.), *Topics in Domination in Graphs*, Developments in Mathematics 64, https://doi.org/10.1007/978-3-030-51117-3_6

Clearly, the 1-dominating sets and the 1-tuple dominating sets are the same as the classical dominating sets. Hence, $\gamma_1(G) = \gamma_{\times 1}(G) = \gamma(G)$ and $\Gamma_1(G) = \Gamma_{\times 1}(G) = \Gamma(G)$.

We have the following general properties that follow directly from the definitions. Observe that, when we talk about a $k$-tuple dominating set of a graph $G$, we are assuming implicitly that $\delta(G) \geq k - 1$.

- Every $k$-dominating set contains all vertices of degree less than $k$, and every $k$-tuple dominating set contains all vertices of degree equal to $k - 1$.
- Every $k$-tuple dominating set is a $k$-dominating set and thus $\gamma_k(G) \leq \gamma_{\times k}(G)$.
- Every maximum minimal $k$-tuple dominating set is also a maximum minimal $k$-dominating set, so $\Gamma_k(G) \leq \Gamma_{\times k}(G)$.
- Every $k$-dominating set and every $k$-tuple dominating set of a graph $G$ with $n(G) \geq k$ contain at least $k$ vertices and so $\gamma_{\times k}(G) \geq \gamma_k(G) \geq k$.
- Clearly, $\gamma_k(G) \leq \Gamma_k(G)$ and $\gamma_{\times k}(G) \leq \Gamma_{\times k}(G)$.
- Every $(k + 1)$-dominating set is also a $k$-dominating set and every $(k + 1)$-tuple dominating set is also a $k$-tuple dominating set. Hence, $\gamma_k(G) \leq \gamma_{k+1}(G)$ and $\gamma_{\times k}(G) \leq \gamma_{\times (k+1)}(G)$.
- The vertex set $V$ of a graph $G$ is the only $(\Delta + 1)$-dominating set and the only $(\Delta + 1)$-tuple dominating set, but it is not a minimal $\Delta$-dominating set nor a minimal $\Delta$-tuple dominating set. Thus,

$$\Gamma_\Delta(G) < n, \text{ and } \Gamma_{\times \Delta}(G) < n.$$

- The vertex set $V$ of a graph $G$ is the only $(\Delta + 1)$-dominating set and the only $(\Delta + 1)$-tuple dominating set. Hence,

$$\gamma(G) = \gamma_1(G) \leq \gamma_2(G) \leq \ldots . \leq \gamma_\Delta(G) < \gamma_{\Delta+1}(G) = n, \text{ and}$$

$$\gamma(G) = \gamma_{\times 1}(G) \leq \gamma_{\times 2}(G) \leq \ldots . \leq \gamma_{\times \Delta}(G) < \gamma_{\times (\Delta+1)}(G) = n.$$

- The properties for a subset of $V$ to be $k$-dominating or $k$-tuple dominating are both superhereditary.

Necessary and sufficient conditions for a set to be a minimal $k$-dominating or a minimal $k$-tuple dominating are given below and are straightforward from the definitions.

**Proposition 1.1 ([64])** *Let $G$ be a graph and $D \subseteq V(G)$ a $k$-dominating set of $G$. Then $D$ is minimal if and only if, for every vertex $v \in D$,*

*(i) $|N(v) \cap D| \leq k - 1$, or*
*(ii) there exists a vertex $u \in N(v) \setminus D$ such that $|N(u) \cap D| = k$.*

**Proposition 1.2** *Let $G$ be a graph and $D \subseteq V(G)$ a $k$-tuple dominating set of $G$. Then $D$ is minimal if and only if, for every vertex $v \in D$,*

*(i)* $|N(v) \cap D| = k - 1$, *or*
*(ii)* *there exists a vertex* $u \in N(v) \setminus D$ *such that* $|N(u) \cap D| = k$, *or*
*(iii)* *there exists a vertex* $u \in N(v) \cap D$ *such that* $|N(u) \cap D| = k - 1$.

A very similar concept to $k$-tuple domination is the so-called total $k$-domination. In this case, the vertices inside the set are required to have at least $k$ neighbors inside the set, not only $k - 1$ as with $k$-tuple domination. To be precise, a subset $D \subseteq V(G)$ is *total k-dominating* if $|N(v) \cap D| \geq k$ for all $v \in V(G)$ and we denote with $\gamma_k^t(G)$ the cardinality of a minimum total $k$-dominating set of $G$. Clearly, a total 1-dominating set is the same as a total dominating set and $\gamma_t^1(G) = \gamma_t(G)$. Total $k$-dominating sets were introduced by Caro [29] in 1990 and by Kulli [111] in 1991 under the name of *k-total dominating sets*. They were reintroduced later, in 2006, under the term of *total k-tuple dominating sets* by Dorbec, Gravier, Klavžar, and Spacapan [51] who denoted $\gamma_k^t(G)$ by $\gamma_t^{(\times k)}(G)$. Clearly, every total $k$-dominating set is a $k$-tuple dominating set, and so

$$\gamma_{\times k}(G) \leq \gamma_k^t(G).$$

Since $k$-tuple domination and total $k$-domination differ only in one unit with respect of the number of neighbors of vertices inside the corresponding dominating set, the results concerning these parameters are very similar and, because of that, we will talk about the two parameters in the same section, giving the results that we think are the most representative of the state of the art in the topic. Of course, there is a characterization of a minimal total $k$-dominating set analogous to Proposition 1.2 (see [105] for an analogous formulation).

**Proposition 1.3** *Let* $G$ *be a graph and* $D \subseteq V(G)$ *a total* $k$*-dominating set of* $G$. *Then* $D$ *is minimal if and only if, for every vertex* $v \in D$, *there exists a vertex* $u \in N(v)$ *such that* $|N(u) \cap D| = k$.

Observe that, in this case, there is no condition on the neighbors of $v$ inside $D$ as it is necessary in Proposition 1.2 because, in this case, every vertex inside $D$ has at least $k$ neighbors there.

We note at this point that there are many other different multiple domination parameters. However, most research has been done precisely on $k$-domination, $k$-tuple domination, and total $k$-domination. For this reason, and because we also think that these three parameters are the best representatives in the area, we will not include results concerning other types of multiple domination. For the interested reader, we refer to [3, 4, 18, 38, 40, 44, 55, 67, 129, 130, 135, 151, 153].

## 1.2 First Observations and Global Results

Recall that, if $n(G) \geq k$, then $k \leq \gamma_k(G) \leq n(G)$. Also we know that $\gamma_{\times k}(G)$ exists if and only if $\delta \geq k - 1$ and, similarly, $\gamma_k^t(G)$ exists if and only if $\delta \geq k$. Moreover, in each such case, $k \leq \gamma_{\times k}(G) \leq n(G)$ and $k + 1 \leq \gamma_k^t(G) \leq n(G)$.

The following propositions give a characterization of the extremal graphs in these inequalities. For this purpose, we define, for a non-negative integer $q$, a *$q$-join* of a graph $G$ to a graph $H$ of order at least $q$ to be the graph obtained from the disjoint union of $G$ and $H$ by joining, with edges, each vertex of $G$ to at least $q$ vertices of $H$. We denote a $q$-join of $G$ to $H$ by $G \circ_q H$.

**Proposition 1.4** *Let $G$ be a graph of order $n \geq k$.*

*(i) $G$ satisfies $\gamma_k(G) = k$ if and only if $n = k$ or $G = F \circ_k H$, for any graphs $F, H$ with $n(H) = k$ and $n(F) = n - k$.*
*(ii) Let $\delta(G) \geq k - 1$. Then $\gamma_{\times k}(G) = k$ if and only if $G = K_k$ or $G = F \circ_k K_k$ for some graph $F$.* (**[84]** for $k = 2$, **[35]** in general.)
*(iii) Let $\delta(G) \geq k$. Then $\gamma_k^t(G) = k+1$ if and only if $G = K_{k+1}$ or $G = F \circ_k K_{k+1}$ for some graph $F$.* (**[89]**)

**Proof.**

(i) If $G$ is such that $n(G) = k$ or $G = F \circ_k H$ for graphs $F, H$ with $n(H) = k$ and $n(F) = n - k > 0$, it clearly holds that $\gamma_k(G) = k$. Conversely, suppose that $G$ satisfies $\gamma_k(G) = k$ and let $D$ be a minimum $k$-dominating set of $G$. Then $|D| = k$. If $V(G) = D$, we are done. Otherwise, every vertex $v \in V(G) \setminus D$ is adjacent to every vertex in $D$, implying that $G = F \circ_k H$ with $H = G[D]$.
(ii) That $\gamma_{\times k}(K_k) = k$, $\gamma_{\times k}(F \circ_k K_k) = k$, $\gamma_t^k(K_{k+1}) = k+1$ and $\gamma_t^k(F \circ_k K_{k+1}) = k+1$, for some arbitrary graph $F$, is easy to check. Now let $G$ be a graph such that $\gamma_{\times k}(G) = k$ and let $S$ be a minimum $k$-tuple dominating set. Then $|S| = k$ and every vertex $v \in S$ has at least $k - 1$ neighbors in $S$. This implies that $G[S] = K_k$. If $V(G) \setminus S = \emptyset$, then $G = K_k$ and we are done. Otherwise let $F = G[V(G) \setminus S]$. Since $S$ is $k$-tuple dominating, every vertex $u \in V(F)$ has at least $k$ neighbors in $S$. Hence, we conclude that $G = F \circ_k K_k$.
(iii) To show that $\gamma_k^t(G) = k + 1$ holds only if $G = K_{k+1}$ or $G = F \circ_k K_{k+1}$ for some graph $F$ is completely analogous to the previous case. ■

On the other hand, we will characterize the graphs $G$ attaining the upper bound $n(G)$ for each of the three parameters.

**Proposition 1.5** *Let $G$ be a graph of order $n$ and minimum degree $\delta$.*

*(i) $G$ satisfies $\gamma_k(G) = n$ if and only if $\delta \leq k - 1$.*
*(ii) Let $\delta \geq k - 1$. Then $\gamma_{\times k}(G) = n$ if and only if every vertex of $G$ has either degree $k - 1$ or has a neighbor of degree $k - 1$.*
*(iii) Let $\delta \geq k$. Then $\gamma_k^t(G) = n$ if and only if every vertex of $G$ has a neighbor of degree $k$.*

**Proof.**

(i) If $G$ has $\delta \leq k - 1$, then clearly $\gamma_k(G) = n$. Conversely, if $\gamma_k(G) = n$, then $V(G)$ is the only minimum $k$-dominating set, and, by Proposition 1.1, it follows that $d(v) \leq k - 1$ for all $v \in V(G)$ and we are done.

(ii) Suppose that $G$ is such that $\gamma_{\times k}(G) = n$. Then $V(G)$ is the only minimum $k$-tuple dominating set, and Proposition 1.2 yields that every vertex of $G$ has either degree $k-1$ or has a neighbor of degree $k-1$. On the other hand, the converse can be easily seen.

(iii) This is straightforward from Proposition 1.3. ■

The most known result about the domination number is due to Ore [121].

**Theorem 1.6 ([121])** *If $G$ is a graph of order $n$ without isolates, then $\gamma(G) \leq \frac{n}{2}$.*

An also very well-known result on the domination number, which was proved independently by Arnautov [5] in 1974 and, in 1975, by Lovász [117] and by Payan [122], is the following.

**Theorem 1.7 ([5, 117, 122])** *Let $G$ be a graph on $n$ vertices and minimum degree $\delta \geq 1$. Then*

$$\gamma(G) \leq \frac{\ln(\delta+1)+1}{\delta+1}n.$$

In 1990, Alon [2] showed that this result is asymptotically optimal for $G$ ranging among all graphs on $n$ vertices and minimum degree $\delta$, where $\delta$ tends to infinity. In [29], Caro derived a bound with the same characteristics for the $k$-domination number (see Theorem 2.17), and showed that the growing rate for $\gamma_k$ with $\delta$ tending to infinity is $(1+o_\delta(1))\frac{\ln\delta}{\delta}n$, solving an open problem stated in [139]. Using the concept of *(F,k)-cores*, Caro and Yuster [31] generalized all these results. If $F = \{G_1, G_2, \ldots, G_t\}$ is a family of graphs on the same vertex set $V$, a subset $D \subset V$ is called an *(F, k)-core* if $D$ is a *total k-dominating set* of each graph in $F$, i.e., if $|N_{G_i}(x) \cap D| \geq k$ for every vertex in $V$, $1 \leq i \leq t$. We denote with $c(k, F)$ the minimum cardinality of an $(F, k)$-core. Evidently, if $F = \{G\}$, an $(F, k)$-core is precisely a total $k$-dominating set in $G$ and vice versa.

**Theorem 1.8 ([31])** *Let $k$, $t$, and $\delta$ be positive integers satisfying $k < \sqrt{\ln\delta}$ and $t < \ln\ln\delta$. Let $F$ be a family of graphs on the same $n$-vertex set. Assume that every graph in $F$ has minimum degree at least $\delta$. Then:*

$$c(k, F) \leq \frac{\ln\delta}{\delta}n(1+o_\delta(1)).$$

Note that, from the definition of an $(F, k)$-core, Theorem 1.8 yields the following bound for the $k$-domination, the $k$-tuple, and the total $k$-domination numbers for graphs with large minimum degree $\delta$.

$$\gamma_k(G) \leq \gamma_{\times k}(G) \leq \gamma_k^t(G) \leq \frac{\ln\delta}{\delta}n(1+o_\delta(1)).$$

Caro and Yuster's result [31] is stronger since they prove the bound actually for connected $(F, k)$-cores, where, given $F = \{G_1, G_2, \ldots, G_t\}$, the underlying induced graphs $G_i[D]$ are connected for $1 \leq i \leq t$.

## *1.3 Complexity*

In 1989, Jacobson and Peters [97] showed that the decision problem to determine if a graph has a $k$-dominating set of size at most $K$ for some fixed integer $K$ is NP-complete in general. However, for trees and series-parallel graphs, they were able to provide linear time algorithms to compute the $k$-domination number [97]. In 1994, Bean, Henning, and Swart [8] proved that the problem remains NP-complete in bipartite or chordal graphs. Lan and Chang [112] showed the NP-completeness in split graphs, which is a subfamily of chordal graphs. In 2004, Klasing and Laforest [108] showed that the $k$-tuple domination problem is even hard to approximate. Later, in 2013, Cicalese, Milanič, and Vaccaro [44] proved the same result for the $k$-domination number. More precisely, they showed that, unless $\text{NP} \subseteq \text{DTIME}(n^{\mathcal{O}(\log logn)})$, for any $\epsilon > 0$, there is neither a polynomial time algorithm approximating the $k$-domination problem [44] nor the $k$-tuple domination problem [108] within a factor of $(1 - \epsilon)\ln n$, where $n$ is the order of the input graph. On the other hand, they also demonstrated that the $k$-domination and the $k$-tuple domination numbers can be approximated in polynomial time by a factor of $\ln(2\Delta) + 1$ [44] and $\ln(\Delta + 1) + 1$ [108], respectively, where $\Delta$ is the maximum degree of the input graph. Analogous results for the total $k$-domination number are obtained in [44], together with the proof of its NP-completeness. In fact, the results provided in [44] deal actually with more complex domination variants, namely *vector domination* (also called *f-domination* in [151] or *threshold ordinary domination* in [67]) and *α-domination*, among others. In [115] Liao and Chang proved that the $k$-tuple domination problem is NP-Complete even for bipartite and split graphs. They also gave a linear time algorithm to compute $\gamma_{\times k}$, and with some adaptation $\gamma_k^t$, in strongly chordal graphs. In [108], Klasing and Laforest gave an $(\ln n + 1)$-approximation algorithm for computing $\gamma_{\times k}(G)$ in general graphs. For more information on domination complexity, we refer the reader to Part 3.3 of [85] and to the references [3, 4, 7, 43, 50, 113, 114].

## 2 $k$-Domination

### 2.1 Relations Between k-Domination Numbers for Different k's

We saw in the introduction that the sequence $(\gamma_k)$ is nondecreasing. In [63, 64], Fink and Jacobson raised the question of the rate at which the $k$-domination number increases with $k$. They proved that $\gamma_3(G) > \gamma(G)$ for graphs with $\Delta \geq 3$ (by Theorem 2.4 below) and their first conjecture in [63] was $\gamma_{2k+1}(G) > \gamma_k(G)$ if $\delta \geq k$. This strict inequality was proved for the case $k = 2$ by Chen and Jacobson:

**Theorem 2.1 ([41])** *For every graph $G$ with minimum degree $\delta \geq 2$, $\gamma_2(G) < \gamma_5(G)$.*

Theorem 2.1 is best possible in the sense that there exist infinitely many graphs $G$ with minimum degree at least 2 having $\gamma_2(G) = \gamma_4(G)$ [41]. However, Schelp (unpublished) disproved Fink and Jacobson's conjecture exhibiting the graph $G = \overline{K}_{k+1} + (k+1)K_k$, which has minimum degree $2k$, and $\gamma_{2k}(G) = \gamma_{k(k+1)}(G) = k(k+1)$. The problem of the rate of growth of the sequence $(\gamma_k)$ remains open under the following form.

**Problem 2.2 ([64])** *Find a function $f$ such that*

$$\gamma_k(G) < \gamma_{f(k)}(G)$$

*for every graph $G$ with $\delta(G) \geq k$.*

By Schelp's counterexample, if $f$ exists, then $f(k) > k^2/4$. We give below some particular classes of graphs for which a function $f$ has been determined.

**Theorem 2.3** *Let $k$ be a positive integer and let $G$ be a graph with maximum degree $\Delta \geq k$.*

*(i) If $k \geq 2$ and $G$ is claw-free, then $\gamma_k(G) < \gamma_{2k}(G)$* **[64]**.
*(ii) If $G$ is $\{K_{1,3}, K_{1,3}+e\}$-free, then $\gamma_k(G) < \gamma_{k+2}(G)$* **[58]**.
*(iii) If $k \geq 2$ and $G$ is $\{K_{1,3}, K_{1,3}+e, C_4+\overline{K_2}\}$-free, then $\gamma_k(G) < \gamma_{k+1}(G)$* **[58]**.

Let $G$ be a graph such that $2 \leq k \leq \Delta(G)$ and let $D$ be a minimum $k$-dominating set of $G$. Then $V(G) \setminus D$ is not empty and we can take a vertex $x \in V(G) \setminus D$. Consider a set $X \subseteq N_G(x) \cap D$ such that $|X| = k-1$. We can easily see that $(D \setminus X) \cup \{x\}$ is a dominating set of $G$. This implies the following theorem of Fink and Jacobson, which establishes a relation between the usual domination number $\gamma(G)$ and the $k$-domination number $\gamma_k(G)$.

**Theorem 2.4 ([63])** *If $G$ is a graph with $\Delta(G) \geq k \geq 2$, then $\gamma_k(G) \geq \gamma(G) + k - 2$.*

Several authors have studied when this inequality is sharp and for which graph classes it can be improved. For the general case, we can cite the following result given by Hansberg.

**Theorem 2.5 ([69])** *Let $G$ be a connected graph and $k$ an integer with $\Delta(G) \geq k \geq 2$. If $\gamma_k(G) = \gamma(G) + k - 2$, then the following statements hold.*

*(i) Every vertex of $G$ lies on an induced cycle of length 4.*
*(ii) $G$ contains at least $(\gamma(G) - 1)(k - 1)$ induced cycles of length 4.*

Let $r$ and $k$ be two positive integers, where $k \geq 2$. Let $G$ be a graph consisting of a complete graph $H$ on $k - 1$ vertices and of vertices $u_i$, $v_i$, $w_i$, for $1 \leq i \leq r$, such that every $u_i$ and $w_i$ is adjacent to every vertex of $H$ and to $v_i$, see Figure 1. Then it is easy to see that $\gamma_k(G) = k-1+r$, $\gamma(G) = r+1$ and thus $\gamma_k(G) = \gamma(G)+k-2$. Since $G$ contains exactly $r(k - 1) = (\gamma(G) - 1)(k - 1)$ induced cycles of length 4, it follows that the bound on the number of induced $C_4$'s given in Theorem 2.5 (ii) can be attained.

Theorem 2.5 improves or extends previous results by Chellali, Favaron, Hansberg, and Volkmann [34]. In particular, it implies that $\gamma_k(T) \geq \gamma(T) + k - 1$ for any tree $T$. In the same paper [34], the authors characterize, for $k \geq 3$, the trees attaining equality in this bound, extending a previous result of Volkmann for the case $k = 2$ [143]. We state these results together in the following theorem, where, by a *subdivided star $SS_t$*, we mean a graph arising from the star $K_{1,t}$ by subdividing each of its edges, while by a *subdivided double star $SS_{s,t}$* we mean a graph formed by means of two stars $K_{1,s-1}$ and $K_{1,t-1}$, where their centers are joined by an edge and then every edge is subdivided.

**Theorem 2.6** *Let $T$ be a tree such that $\Delta(T) \geq k \geq 2$ for an integer $k$.*

*(i) $\gamma_2(T) = \gamma(T)+1$ if and only if $T$ is a subdivided star $SS_t$ or a subdivided star $SS_t$ minus a leaf or a subdivided double star $SS_{s,t}$* **[143]**.
*(ii) Let $k \geq 3$. Then $\gamma_k(T) = \gamma(T) + k - 1$ if and only if $T$ is isomorphic to a subdivided star $SS_k$ minus $p$ leaves for an integer $1 \leq p \leq k$* **[34]**.

Hansberg and Volkmann considered the case $k = 2$ [74–76]. They proved that for nontrivial connected graphs, $\gamma_2(G) \geq \gamma(G) + 1$ if $G$ is a block graph or a unicyclic graph different from $C_4$ and characterized the graphs of these two families satisfying

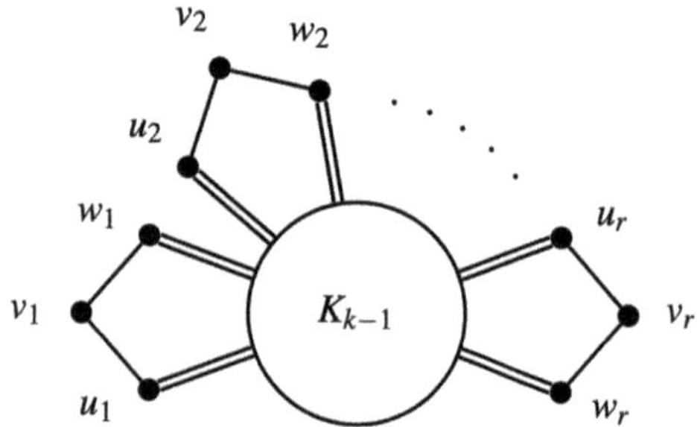


**Fig. 1** Example of a graph $G$ with $\gamma_k(G) = \gamma(G) + k - 2$ and exactly $(\gamma(G) - 1)(k - 1)$ induced cycles of length 4. A double line connecting a vertex $u_i$ or $w_i$ to the complete graph $K_{k-1}$ in the middle means that it is adjacent to all vertices of $K_{k-1}$.

$\gamma_2(G) = \gamma(G) + 1$ (comprising thus Theorem 2.6). In [76] they proved that any graph $G$ with $\gamma_2(G) = \gamma(G)$ cannot have leaves and characterized the block-cactus graphs fulfilling this equality. In [73], Hansberg, Randerath, and Volkmann considered claw-free graphs and were able to characterize all claw-free graphs and all line-graphs (which are also claw-free) with equal domination and 2-domination numbers.

In a recent paper [19], Gülnaz Borunzanı, and Bujtás proved that, in general, it is NP-hard to decide whether a graph has equal domination and 2-domination numbers and that there is no forbidden subgraph characterization for such graphs. They consider also a large class of graphs in which those with equal domination and 2-domination numbers can be characterized via a family of forbidden subgraphs. Although the family of such forbidden subgraphs is infinite, they prove that the recognition problem is solvable in polynomial time. Moreover, they study the so-called *$(\gamma, \gamma_2)$-perfect graphs*, which are those for which all induced subgraphs with minimum degree at least 2 have equal domination and 2-domination numbers. Observe that the condition on the minimum degree is natural because it has been noted that graphs $G$ with $\gamma(G) = \gamma_2(G)$ cannot have vertices of degree 1 [76]. The following characterization of $(\gamma, \gamma_2)$-perfect graphs was obtained in [19].

**Theorem 2.7** *Let $G$ be a connected graph with minimum degree at least 2. Then $G$ is a $(\gamma, \gamma_2)$-perfect graph if and only if $G$ contains no subgraph isomorphic to any $K_2 \circ 2K_1$, $P_8$ or $C_n$ for any $n \neq 4$.*

In [34], the authors showed that the inequality in Theorem 2.6(ii) can be improved to $\gamma_k(G) \geq \gamma_t(G) + k - 2$ for block graphs. Moreover, for $k \geq 3$, they showed that $\gamma_k(T) \geq \gamma_t(T) + k - 1$ and that $\gamma_k(T) \geq \gamma_c(T) + k - 1$ for any tree $T$ with $\Delta(T) \geq k$ and characterized both equalities.

That $\gamma_{k+1}(G)$ cannot be too large with respect to $\gamma_k(G)$ is nevertheless shown by the following two theorems. The first one was proved by Favaron [59] and Volkmann [142] independently, the second by Favaron, Hansberg, and Volkmann [60].

**Theorem 2.8 ([59], [142] p. 195)** *Let $k$ be a positive integer. If $G$ is a graph of order $n$ and $\delta(G) \geq k + 1$, then $\gamma_{k+1}(G) \leq \frac{n+\gamma_k(G)}{2}$ and the bound is sharp.*

**Proof.** Let $S$ be a minimum $k$-dominating set of $G$. Consider the set $A \subseteq V(G) \setminus S$ of vertices $v$ with $|N(v) \cap S| = k$. If $A = \emptyset$, then $\gamma_k(G) = \gamma_{k+1}(G)$ and the inequality follows easily. If $A \neq \emptyset$, then $|N(x) \cap S| \geq k + 1$ for all vertices $x$ in $V(G) \setminus (A \cup S)$. Consider now the graph $H = G[N[A] \cap (V(G) \setminus S)]$ and let $D$ be a minimum dominating set of $H$. Clearly, $\delta(H) \geq 1$ and Ore's Theorem (Theorem 1.6) implies that

$$|D| \leq \frac{1}{2}|N[A] \cap (V(G) \setminus S)| \leq \frac{1}{2}|V(G) \setminus S| = \frac{1}{2}(n - \gamma_k(G)).$$

It is straightforward to check that $D \cup S$ is a $(k+1)$-dominating set of $G$. Hence,

$$\gamma_{k+1}(G) \le |D \cup S| \le \frac{1}{2}(n - \gamma_k(G)) + \gamma_k(G) = \frac{n + \gamma_k(G)}{2}.$$

That the bound is sharp may be seen by the graph $G$ obtained from a clique $B$ with $|B| = k$ and $q \ge k$ cliques $A_i$ each of order two by adding all edges between $B$ and all $A_i$'s. Then $n = k + 2q, \delta = k + 1, \gamma_k(G) = k$ and $\gamma_{k+1}(G) = k + q = (n + \gamma_k(G))/2$. ■

The special case $k = 2$ of Theorem 2.8 can be found in [15]. In [146], Volkmann characterized the connected $P_4$-free graphs $G$ with the property that $\overline{G}$ is $(K_4 - e)$-free, which attain equality in Theorem 2.8.

**Theorem 2.9 ([60])** *Every graph of order $n$ and minimum degree $\delta$ satisfies*

$$\gamma_k(G) + \frac{(k' - k + 1)}{2k' - k}\gamma_{k'}(G) \le n$$

*for all integers $k$ and $k'$ with $1 \le k \le k' \le \delta$.*

## 2.2 Bounds in Terms of Order and Degrees

We begin by the lower bounds on the $k$-domination number given by Fink and Jacobson in their paper introducing $k$-domination [63]. Note that a bipartite graph is called *$k$-semiregular* if every vertex in one of the two partite sets has degree $k$.

**Theorem 2.10 ([63])** *Let $G$ be a graph $G$ of order $n$ with maximum degree $\Delta$ and $m$ edges and let $k \le \Delta$ be a positive integer.*

*(i)* $\gamma_k(G) \ge \dfrac{kn}{k + \Delta}$.
*(ii)* $\gamma_k(G) \ge n - \frac{m}{k}$. *Furthermore, if $m \ne 0$, then $\gamma_k(G) = n - \frac{m}{k}$ if and only if $G$ is a bipartite $k$-semiregular graph.*

**Proof.** Let $D$ be a minimum $k$-dominating set of $G$. Since $|N(u) \cap D| \ge k$ for all $u \in V \setminus D$ and $|N(v) \cap (V \setminus D)| \le \Delta$ for all $v \in D$, we derive

$$k(n - \gamma_k(G)) \le m(D, V \setminus D) \le \Delta\gamma_k(G).$$

Hence, rearranging the terms on both sides of the inequality chain, (i) follows easily. For the bound in (ii), the left inequality combined with $m(D, V \setminus D) \le m$ yields the result. To have equality in (ii), $D$ and $V \setminus D$ have to be independent sets and each vertex in $V \setminus D$ has to have exactly $k$ neighbors in $D$, hence $G$ is a bipartite $k$-semiregular graph. On the other side, let $G$ be a $k$-semiregular bipartite graph with bipartition $(S, V \setminus S)$ such that every vertex in $S$ has degree $k$. Let $m$ be the edge number of $G$ and $n$ its order. Clearly, $V \setminus S$ is a $k$-dominating set and $m = |S|k$. Hence, $\gamma_k(G) \leq |V \setminus S| = n - \frac{m(G)}{k}$ and, with the inequality in (ii), we obtain $\gamma_k(G) = n - \frac{m}{k}$. ■

In [77], Hansberg and Volkmann extended the result in Theorem 2.10(ii) introducing a parameter $\mu_0(G)$ that denotes the minimum number of edges that can be removed from a graph $G$ such that the remaining graph is bipartite.

**Theorem 2.11 ([77])** *If $G$ is a graph of order $n$ and edge number $m$ and $\mu_0 = \mu_0(G)$, then*

$$\gamma_k(G) \geq n - \frac{m - \mu_0}{k}.$$

*Additionally, if $m \neq 0$, then $\gamma_k(G) = \lceil n - \frac{m-\mu_0}{k} \rceil$ if and only if $G$ contains a bipartite $k$-semiregular factor $H$ with $m(H) = m - \mu_0 - r$, where $r$ is an integer such that $0 \leq r \leq k - 1$ and $m - \mu_0 - r \equiv 0 \pmod{k}$.*

Since a tree $T$ of order $n$ has $n - 1$ edges, it follows from Theorem 2.10.2 that $\gamma_k(T) \geq \dfrac{(k-1)n + 1}{k}$. In [143], Volkmann provided a characterization of the trees achieving equality in this bound. Related results involving the number of leaves and support vertices were given by Lu, Hou, and Xu in [118] for trees and by Chellali [33] for graphs with at most one cycle.

We consider now upper bounds on $\gamma_k(G)$. The first upper bound on the $k$-domination number generalizing Ore's upper bound (Theorem 1.6) for the domination number was given by Cockayne, Gamble, and Shepherd in 1985 [46].

**Theorem 2.12 ([46])** *If $G$ is a graph of order $n$ and minimum degree $\delta \geq k$, then $\gamma_k(G) \leq \frac{k}{k+1} n$.*

Observe that, with $k = k'$ in Theorem 2.9, we immediately obtain Theorem 2.12. Using Theorem 2.9, Favaron, Hansberg, and Volkmann [60] characterized the extremal graphs attaining equality in Theorem 2.12. This characterization generalizes the one of graphs $G$ without isolated vertices realizing $\gamma(G) = n/2$ [62, 123]. To state the result, we need to give the following definition. The *corona* of two graphs $G_1$ and $G_2$ is the graph $G_1 \circ G_2$ formed from one copy of $G_1$ and $|V(G_1)|$ copies of $G_2$ where the $i$th vertex of $G_1$ is adjacent to every vertex in the $i$th copy of $G_2$.

**Theorem 2.13 ([60])** *Let $G$ be a connected graph of order $n$ and minimum degree $\delta$. Then $G$ satisfies $\gamma_k(G) = \frac{k}{k+1} n$ for some integer $k$ with $1 \leq k \leq \delta$ if and only if $G$ is the corona $J \circ K_k$, when $k \geq 2$, and $J \circ K_1$ or $G \cong C_4$, when $k = 1$, where $J$ is any connected graph.*

The upper bound in Theorem 2.12 has been extended later by different authors, among them (given in chronological order) Caro and Roditty [28], Stracke and Volkmann [135], Chen and Zhou [40] and Favaron, Hansberg, and Volkmann [60].

Since the bounds given in [40, 135] are comprised by the other bounds, we only state those in [28, 60].

**Theorem 2.14 ([28])** *Let $r, k$ be positive integers and $G$ a graph of order $n$ and minimum degree $\delta \geq \frac{r+1}{r}k - 1$. Then $\gamma_k(G) \leq \frac{r}{r+1}n$.*

**Proof.** Let $V_1, V_2, \ldots, V_r$ be a partition of $V(G)$ such that the number of edges in $E' = \cup_{i=1}^{r} E_i$ is minimum, where $E_i = E(G[V_i])$. By a classical Erdős argument [53], the graph $H = G - E'$ has minimum degree at least $\left\lceil \frac{r\delta}{r+1} \right\rceil$. Hence,

$$\delta(H) \geq \left\lceil \frac{r\delta}{r+1} \right\rceil \geq \left\lceil \frac{r}{r+1}\left(\frac{r+1}{r}k - 1\right) \right\rceil = k.$$

Assume, without loss of generality, that $|V_1| \geq |V_2| \geq \ldots \geq |V_r|$. Then the set $D = \cup_{i=2}^{r} V_i$ is a $k$-dominating set since every vertex in $V_1$ has at least $k$ neighbors in $D$. It follows that

$$\gamma_k(G) \leq |D| = n - |V_1| \leq n - \frac{n}{r} = \frac{r+1}{r}n.$$

■

Observe that the condition $\delta \geq \frac{r+1}{r}k - 1$ is equivalent to $r \geq \frac{k}{\delta+1-k}$. Hence, the smallest $r$ we can take in Theorem 2.14 is $\lceil k/(\delta + 1 - k) \rceil$. Hence, the bound of Theorem 2.14 can be stated in the following equivalent but more explicit form.

**Theorem 2.15 ([60])** *If $G$ is a graph of order $n$ and minimum degree $\delta$, then, for every positive integer $k \leq \delta$,*

$$\gamma_k(G) \leq \frac{\lceil k/(\delta + 1 - k) \rceil}{\lceil k/(\delta + 1 - k) \rceil + 1} n .$$

Using the fact that $\lceil \frac{k}{\delta+1-k} \rceil \leq \frac{\delta}{\delta+1-k}$ and that the function $\frac{x}{x+1}$ is monotonically increasing for positive $x$, the following simple bound follows from Theorem 2.15.

**Corollary 2.16 ([60])** *If $G$ is a graph of order $n$ and minimum degree $\delta$ and $k \leq \delta$ is a positive integer, then*

$$\gamma_k(G) \leq \frac{\delta}{2\delta + 1 - k} n.$$

As mentioned in the introduction in the general setting of $(F, k)$-cores, when the minimum degree is large, one can expect better bounds on the corresponding domination parameter. Ten years before Caro and Yuster [31] provided the general bound for $(F, k)$-cores, Caro [29] proved, in 1990, the corresponding result for $k$-dominating sets, hence generalizing the bound on the usual domination number by Arnautov [5], Lovázs [117], and Payan [122] that we stated in Theorem 1.7. We will state Caro's theorem together with its proof, since it contains the essence of all these results that use probabilistic arguments.

**Theorem 2.17 ([29])** *Let $k$ and $\delta$ be positive integers satisfying $k < \sqrt{\ln \delta}$ and let $G$ be a graph on $n$ vertices with minimum degree at least $\delta$. Then*

$$\gamma_k(G) \leq n\frac{\ln \delta}{\delta}(1 + o_\delta(1)).$$

**Proof.** Fix $0 < \epsilon < \frac{1}{2}$ and $p = \left(1 + \frac{\epsilon}{2}\right)\frac{\ln \delta}{\delta}$. Let $X \subseteq V(G)$ be a random subset of vertices, where each vertex is selected independently with probability $p$. Let $Y \subseteq V(G)$ be the set of vertices $v$ with $|N(v) \cap X| \leq k - 1$. Clearly, $X \cup Y$ is a $k$-dominating set and $E[|X|] = pn$. We will calculate now $E[|Y|]$. Observe that, since $k < \sqrt{\ln \delta}$, for each $0 \leq i \leq k - 1$ and $t \geq \delta$, we have

$$\frac{i}{t} \leq \frac{k-1}{\delta} \leq \left(1 + \frac{\epsilon}{2}\right)\frac{\ln \delta}{\delta} = p,$$

and thus

$$\binom{t}{i}(1-p)^{t-i} \leq \binom{t-1}{i}(1-p)^{t-1-i}. \tag{1}$$

Hence,

$$P(v \in Y) = \sum_{i=1}^{k-1}\binom{d(v)}{i}p^i(1-p)^{d(v)-i} \leq \sum_{i=1}^{k-1}\binom{\delta}{i}p^i(1-p)^{\delta-i}. \tag{2}$$

Considering the terms inside the sum, we see that, for $0 \leq i \leq k - 1$,

$$\begin{aligned}
\binom{\delta}{i}p^i(1-p)^{\delta-i} &\leq (\delta p)^i e^{-p(\delta-i)} \\
&= \left(1 + \frac{\epsilon}{2}\right)^i (\ln \delta)^i \delta^{-(1+\frac{\epsilon}{2})(1-\frac{i}{\delta})} \\
&< \left(1 + \frac{\epsilon}{2}\right)^{k-1} (\ln \delta)^{k-1} \delta^{-(1+\frac{\epsilon}{2})(1-\frac{k}{\delta})} \\
&= \mathcal{O}(2^k(\ln \delta)^k \delta^{-(1+\frac{\epsilon}{2})}).
\end{aligned}$$

Hence, $P(v \in Y) = \mathcal{O}\left(k2^k(\ln \delta)^k\delta^{-(1+\frac{\epsilon}{2})}\right)$, which is at most $\mathcal{O}\left(\delta^{-(1+\frac{\epsilon}{4})}\right)$ for $k < \sqrt{\ln \delta}$, and thus

$$E[|Y|] = \mathcal{O}\left(n\delta^{-(1+\frac{\epsilon}{4})}\right) = o\left(\frac{n}{\delta}\right).$$

By the linearity of expectation, we have $E[|X \cup Y|] \le E[|X| + |Y|] = E[|X|] + E[|Y|]$ and we conclude that

$$\begin{aligned} E[|X \cup Y|] &\le np + o\left(\frac{n}{\delta}\right) \\ &= n\left(1 + \frac{\epsilon}{2}\right)\frac{\ln \delta}{\delta} + o\left(\frac{n}{\delta}\right) \\ &= n\frac{\ln \delta}{\delta}\left(1 + \frac{\epsilon}{2} + o(1)\right), \end{aligned}$$

implying that there is a $k$-dominating set of size at most $n\frac{\ln \delta}{\delta}\left(1 + \frac{\epsilon}{2} + o(1)\right)$. Therefore, for sufficiently large $\delta$, $\gamma_k(G) \le n\frac{\ln \delta}{\delta}(1 + \epsilon)$ and the result follows. ■

Observe that previous theorem is useful when we are dealing with large graphs and large minimum degree. For example, if $k = 2$, the assumption $2 < \sqrt{\ln \delta}$ forces us to take graphs with minimum degree at least 55. Theorem 2.17 tells us that a graph on $n$ vertices and minimum degree at least 55 will have a 2-dominating set of cardinality at most $\frac{\ln 55}{55}n$. Hence, if $n = 100$, the graph will have a 2-dominating set of at most 7 vertices. If we increase the minimum degree, say to 80, and keep $n = 100$, then we could find a 2-dominating set on at most 5 vertices.

Caro [29] showed that the bound in Theorem 2.17 is asymptotically sharp by means of the construction of a graph derived from a certain $d$-uniform hypergraph $H$ with transversal number at least $(n + m)(1 + o(1))\frac{\ln d}{d}$, where $n$ is the order of $H$ and $m$ its edge number, which was given by Alon in [2].

Weakening considerably the condition on the minimum degree, Rautenbach and Volkmann [128] gave, using similar probabilistic arguments, another upper bound on the $k$-domination number $\gamma_k$. Hansberg and Volkmann [78] improved later their technique to obtain the following bound. Due to the weaker conditions, these bounds are, as expected, not as strong as Caro's bound.

**Theorem 2.18 ([78])** *Let $G$ be a graph on $n$ vertices with minimum degree $\delta \ge 1$ and let $k \in \mathbb{N}$. If $\frac{\delta+1}{\ln(\delta+1)} \ge 2k$, then*

$$\gamma_k(G) \le \frac{n}{\delta + 1}\left(k \ln(\delta + 1) + \sum_{i=0}^{k-1} \frac{\delta^i}{i!\,(\delta + 1)^{k-1}}\right).$$

The proof of this theorem uses similar probabilistic arguments as Caro's proof of Theorem 2.17. Since $\sum_{i=0}^{k-1} \delta^i \le (\delta + 1)^{k-1}$, it is easy to see that Theorem 2.18 implies that, for $\frac{\delta+1}{\ln(\delta+1)} \ge 2k$, $\gamma_k(G) \le \frac{n}{\delta+1}(k \ln(\delta + 1) + 1)$. Observe that this statement generalizes the bound for the domination number given in Theorem 1.7. With a less strong assumption on the minimum degree (by a factor of 2), Przybyło obtained the same bound for the $k$-domination number.

**Theorem 2.19 ([126])** *Let $G$ be a graph on $n$ vertices with minimum degree $\delta \geq 1$. If $\frac{\delta+1}{\ln(\delta+1)} \geq k$, then*

$$\gamma_k(G) \leq \frac{n}{\delta+1}(k\ln(\delta+1)+1).$$

For the case $k = 1$, these theorems (and also Rautenbach and Volkmann's bound in [128]) yield directly the well-known bound for the usual domination number $\gamma$, see Theorem 1.7. Weakening a bit more the assumption on the minimum degree, other similar results were obtained in [78].

We would like to finish this subsection mentioning some results concerning upper bounds on the 2-domination number in terms of the order of the graph and where the minimum degree is small. First note that the bounds in Theorems 2.14 and 2.15 give that $\gamma_2(G) \leq \frac{2}{3}n$ for $\delta \geq 2$ and $\gamma_2(G) \leq \frac{1}{2}n$ for $\delta \geq 3$ and this is the best one can get, even if the minimum degree is increased. On the other hand, the bounds in Theorems 2.18 and 2.19 yield a bound of less than $\frac{1}{2}n$ only if $\delta \geq 11$. Hence, better bounds for small $\delta$ are desirable to find. This was achieved by Bujtás and Jaskó [23] for $6 \leq \delta \leq 21$. Their result relies on an algorithmic method to construct a 2-dominating set, where weights are assigned to the vertices, which in turn change according to some rules during the greedy 2-domination procedure. In particular, the following bounds are obtained.

**Theorem 2.20** *Let $G$ be a graph of order $n$ and minimum degree $\delta$.*

*(i) If $\delta = 6$, then $\gamma_2(G) < 0.498n$.*
*(ii) If $\delta = 7$, then $\gamma_2(G) < 0.467n$.*
*(iii) If $\delta = 8$, then $\gamma_2(G) < 0.441n$.*
*(iv) If $\delta \geq 9$, then $\gamma_2(G) < 0.418n$.*

## 2.3 Relationship to Other Graph Parameters

One of the graph parameters most studied in connection with the multiple domination number is the $k$-independence number. We say that a subset $S$ of $V$ is *$k$-independent* if the maximum degree of the subgraph induced by the vertices of $S$ is less than $k$. The cardinality of a $k$-independent set of maximum cardinality in a graph $G$ is denoted with $\alpha_k(G)$. Clearly, a 1-independent set is an independent set and $\alpha_1(G) = \alpha(G)$. Note that a $k$-independent set is sometimes called $(k-1)$-independent [30, 109], $(k-1)$-dependent [63, 64], $k$-dependent [98], or $(k-1)$-small [96]. The property for a subset of $V$ to be $k$-independent is hereditary. A $k$-independent set $S$ of $G$ is *maximal* if for every vertex $v \in V - S$, $S \cup \{v\}$ is not $k$-independent. We denote with $i(G)$ the minimum cardinality of an *independent dominating set*, i.e., a set that is both independent and dominating.

It is well-known that an independent set is maximal if and only if it is also dominating. So we can say that domination, which is defined even for non-independent sets, is the property which makes an independent set maximal. Moreover, every set which is both independent and dominating is a minimal dominating set of $G$. This observation leads to the well-known inequality chain:

$$\gamma(G) \leq i(G) \leq \alpha(G) \leq \Gamma(G) \quad \text{for all graphs } G. \tag{3}$$

In [63], Fink and Jacobson proved that $\gamma_2(G) \leq \alpha_2(G)$ and conjectured that for every graph $G$ and any positive integer $k$, $\gamma_k(G) \leq \alpha_k(G)$. This inequality is not obvious because for $k \geq 2$, a maximal $k$-independent set is not necessarily $k$-dominating as it is for $k = 1$. The conjecture proved to be true by means of the following result due to Favaron [57].

**Theorem 2.21 ([57])** *For any graph $G$, every $k$-independent set $D$ such that $k|D| - m(G[D])$ is maximum is a $k$-dominating set of $G$.*

**Proof.** Let $D$ be a $k$-independent set such that $k|D| - m(G[D])$ is maximum. If $D = V(G)$, then $\Delta(G) \leq k - 1$ and $D = V(G)$ is a $k$-dominating set and we are done. Hence, we may assume that $V(G) \setminus D \neq \emptyset$. Suppose now that $D$ is not $k$-dominating. Then there is a vertex $v \in V(G) \setminus D$ with $|N(v) \cap D| \leq k - 1$. Let $B = N(v) \cap D$, $A = \{a \in B : |N(a) \cap D| = k - 1\}$ and let $S$ be a maximal independent set of $G[A]$. Then $S \subseteq A \subseteq B \subseteq D$ and $|B| < k$ (see Figure 2). Observe also that $A$, and hence $S$, may be empty, but, if $A$ is not empty, then neither $S$. Also note that, because of the maximality of $S$, the vertices in $A$, if any, have one or more neighbors in $S$. We will show that $C = (D \setminus S) \cup \{v\}$ is again a $k$-independent set. This is indeed true, as

$$\begin{aligned}
&|N(v) \cap C| \leq |B| < k, \\
&|N(x) \cap C| \leq |N(x) \cap D| < k, \ \text{for all } x \in D \setminus B \\
&|N(b) \cap C| \leq |N(b) \cap D| + 1 < k, \ \text{for all } b \in B \setminus A \\
&|N(a) \cap C| \leq |N(a) \cap D| < k, \ \text{for all } a \in A \setminus S.
\end{aligned}$$

However,

$$\begin{aligned}
k|C| - m(G[C]) &= k|C| - (m(G[D]) - |S|(k-1) + |B| - |S|)) \\
&= k(|C| + |S|) - m(G[D]) - |B| \\
&= k(|D| + 1) - m(G[D]) - |B| \\
&= k|D| - m(G[D]) + k - |B| \\
&> k|D| - m(G[D]),
\end{aligned}$$

which is a contradiction to the choice of $D$. Hence, $D$ is a $k$-dominating set. ■

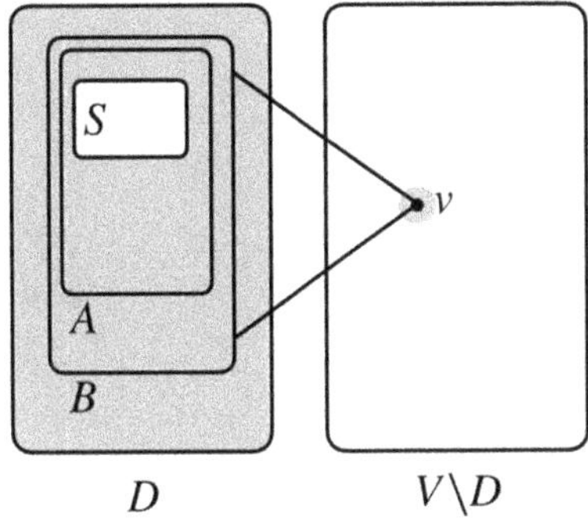


**Fig. 2** Sketch of the proof of Theorem 2.21: The set $C = (D \setminus S) \cup \{v\}$ is marked with grey color.

The proof of Theorem 2.21 allows one to construct a $k$-independent $k$-dominating set from a $k$-independent one, and thus also from any independent set (which is trivially $k$-independent). However, the algorithmic aspect is not developed in [57].

By Theorem 2.21, any graph $G$ admits a set $S$ that is both $k$-independent and $k$-dominating. Since such a set is a minimal $k$-dominating set and a maximal $k$-independent set, $\gamma_k(G) \leq |S| \leq \Gamma_k(G)$ and $i_k(G) \leq |S| \leq \alpha_k(G)$. Therefore we have the following corollary.

**Corollary 2.22 ([57])** *For any graph G and positive integer k,*

$$\gamma_k(G) \leq \alpha_k(G) \quad \text{and} \quad i_k(G) \leq \Gamma_k(G).$$

Chellali and Meddah [39] provided a constructive characterization of trees with equal 2-domination and 2-independence numbers. Their characterization is in terms of global properties of a tree, and involves properties of minimum 2-dominating and maximum 2-independent sets in the tree at each stage of the construction. Brause, Henning, and Krzywkowski [20] provided later a constructive characterization that relies only on local properties of the tree at each stage of the construction.

The inequality chain (3) is now partially generalized. We can wonder whether a complete generalization of (3) is possible for every positive integer $k$. The answer is negative as noticed in [58]. The following examples show that each of the four inequalities $\gamma_k(G) > i_k(G)$, $i_k(G) > \gamma_k(G)$, $\Gamma_k(G) > \alpha_k(G)$, and $\alpha_k(G) > \Gamma_k(G)$ are possible.

For a double star $S_{k-1,k-1}$ $(k \geq 2)$ we have $\gamma_k(G) = n-1 > i_k(G) = n-2$ while for the graph $G$ constructed from three subdivided stars $SS_p$, $p \geq 2$, with centers $x$, $y$, and $z$ by adding edges $xy$ and $xz$, $i_2(G) = 4p+2$ and $\gamma_2(G) = 3p+3$.

Now let us consider the graph $H_k$ obtained from $k \geq 2$ disjoint stars $F_i \simeq K_{1,k}$ with centers $c_i$ and leaves $u_{i,1}, u_{i,2}, \cdots, u_{i,k}$ by adding a new vertex $x$ and the $k$ edges $xc_i$, $1 \leq i \leq k$ (see Figure 3). Then $n = k^2 + k + 1$ and, since $k < \Delta$, $\Gamma_k(H_k) < n$ follows. Moreover, $\bigcup_{i=1}^{k} V(F_i)$ is a minimal $k$-dominating set of $H_k$. Therefore $\Gamma_k(H_k) = k^2 + k$. On the other hand, let $S$ be a $k$-independent set of $H_k$. Clearly, every maximal $k$-independent set has at most $k$ vertices in each star $F_i$ and $\alpha_k(H_k) \leq k^2 + 1$. Since $V(H_k) - \{c_1, c_2, \cdots, c_k\}$ is a maximal $k$-independent set,

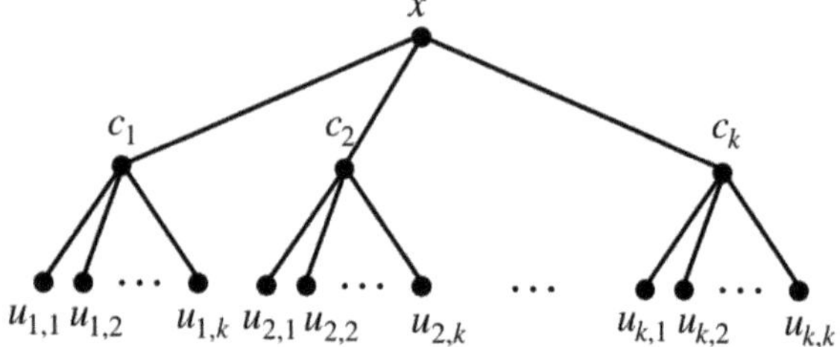


**Fig. 3** The graph $H_k$.

$\alpha_k(H_k) = k^2+1$ follows. Thus $\Gamma_k(H_k) > \alpha_k(G)$. The same graph $H_k$ also provides an example for the opposite inequality since $\Gamma_{k+1}(H_k) = k^2 + 1 < k^2 + k = \alpha_{k+1}(H_k)$.

However, some inequalities can be proved in particular situations as shown by Theorems 2.23 and 2.24.

**Theorem 2.23 ([98])** *If $\gamma_{k+1}(G) = \gamma_k(G)$ for some positive integer $k$, then $\gamma_k(G) \ge i_k(G)$.*

**Theorem 2.24 ([35])** *Every graph $G$ satisfies $\alpha_\Delta(G) \ge \Gamma_\Delta(G)$.*

**Proof.** Let $S$ be a minimal $\Delta$-dominating set of $G$ with maximum order. Then the maximum degree in $G[S]$ is at most $\Delta - 1$ for otherwise a vertex $v \in S$ of degree $\Delta$ has all its neighbors in $S$ and then $S - \{v\}$ is a $\Delta$-dominating set, contradicting the minimality of $S$. Thus $S$ is a $\Delta$-independent set of $G$ and $\alpha_\Delta(G) \ge |S| = \Gamma_\Delta(G)$. ■

In [12], Blidia, Chellali, and Favaron show that, if a graph $G$ is claw-free, then $\gamma_2(G) \ge \alpha(G)$. This result can be extended to $K_{1,k+1}$-free graphs.

**Theorem 2.25** *Let $G$ be a $K_{1,k+1}$-free graph. Then $\gamma_k(G) \ge \alpha(G)$.*

**Proof.** Let $D$ be a minimum $k$-dominating set and $S$ a maximum independent set of $G$. If $S \subseteq D$, we are done. So we may assume that $S \setminus D \neq \emptyset$. Now every vertex in $S \setminus D$ has at least $k$ neighbors in $D \setminus S$. Moreover, since $G$ is $K_{1,k+1}$-free, every vertex in $D \setminus S$ has at most $k$ neighbors in $S \setminus D$. We obtain

$$k(|S| - |S \cap D|) = k|S \setminus D| \le m(S \setminus D, D \setminus S) \le k|D \setminus S| = k(|D| - |S \cap D|),$$

which yields $\gamma_k(G) \ge \alpha(G)$. ■

**Theorem 2.26 ([79])** *If $G$ is a connected nontrivial block-cactus graph, then $\gamma_2(G) \ge i(G)$. Moreover $\gamma_2(G) = i(G)$ if and only if $G$ is a $C_4$-cactus.*

Since trees are contained in the class of block-cactus graphs, clearly $\gamma_2(T) \ge i(T) + 1$ for any nontrivial tree $T$. Using the characterization of trees $T$ such that $\gamma_2(T) = \gamma(T) + 1$ given in Theorem 2.6 (i), Hansberg and Volkmann [79] characterized all trees $T$ with $\gamma_2(T) = i(T) + 1$.

**Theorem 2.27 ([79])** *Let $T$ be a nontrivial tree. Then $\gamma_2(T) = i(T)+1$ if and only if $T$ is a subdivided star $SS_t$ or a subdivided star $SS_t$ minus a leaf or a subdivided double star $SS_{s,t}$ or $T$ is isomorphic to the tree of order 6 with two adjacent vertices of order 3 and 4 leaves.*

Jacobson, Peters, and Rall [98] observed that if $S$ is a $k$-independent set of a graph $G$ with minimum degree $\delta \geq k$, then $V - S$ is a $(\delta - k + 1)$-dominating set. Similarly, Favaron [59] showed that if $S$ is a $k$-dominating set of a graph $G$ of maximum degree $\Delta \geq k$, then $V - S$ is a $(\Delta - k + 1)$-independent set. Therefore we have the following.

**Theorem 2.28** *Let $G$ be a graph with minimum degree $\delta$ and maximum degree $\Delta$.*

*(i) If $\delta \geq k$, then $\alpha_k(G) + \gamma_{\delta-k+1}(G) \leq n$.* **[98]**
*(ii) If $\Delta \geq k$, then $\alpha_k(G) + \gamma_{\Delta-k+1}(G) \geq n$.* **[59]**

These theorems were extended by Pepper [124] and Hansberg and Pepper [72] to the following shape.

**Theorem 2.29** *Let $G$ be a graph of order $n$, let $k, j, m$ be positive integers such that $m = k + j - 1$, and let $G_{\geq m}$ and $G_{\leq m}$ denote, respectively, the subgraphs induced by the vertices of degree at least $m$ and by the vertices of degree at most $m$. Then*

*(i) $\gamma_k(G) + \alpha_j(G_{\geq m}) \leq n$,* **[124]**
*(ii) $\gamma_k(G_{\leq m}) + \alpha_j(G) \geq n(G_{\leq m})$.* **[72]**

**Proof.** Let $I$ be a maximum $j$-independent set of $G_{\geq m}$ and $D$ a minimum $k$-dominating set of $G_{\leq m}$. We will show that $V(G) - I$ is a $k$-dominating set of $G$ and that $V(G_{\leq m}) - D$ is a $j$-independent set of $G$.

Let $u \in I$. Then $d(u) \geq m$ and $|N(u) \cap I| \leq j - 1$. This implies

$$|N(u) \cap (V(G) - I)| = |N(u)| - |N(u) \cap I| \geq m - (j - 1) = k$$

and hence $V(G) - I$ is a $k$-dominating set of $G$. If $V(G_{\leq m}) - D$ is empty, then $n(G_{\leq m}) - |D| = n(G_{\leq m}) - \gamma_k(G_{\leq m}) = 0$ and the statement is trivial. Thus suppose there is a vertex $v \in V(G_{\leq m}) - D$. Then $deg(v) \leq m$ and $|N(v) \cap D| \geq k$. This implies

$$|N(v) \cap (V(G_{\leq m}) - D)| = |N(v) \cap V(G_{\leq m})| - |N(v) \cap D| \leq m - k = j - 1,$$

and hence $V(G_{\leq m}) - D$ is a $j$-independent set of $G$.

Altogether we obtain

$$\gamma_k(G) \leq |V(G) - I| = n - \alpha_j(G_{\geq m})$$

and

$$\alpha_j(G) \geq |V(G_{\leq m}) - D| = n(G_{\leq m}) - \gamma_k(G_{\leq m}),$$

which completes the proof. ■

Favaron[59] noted that regular graphs attain the bound in Theorem 2.28.1. Similarly, the inequalities in Theorem 2.29 take the following form for regular graphs.

**Corollary 2.30 ([72])** *Let $G$ be an $r$-regular graph and $k$ and $j$ two positive integers. Then*

$$\gamma_k(G) + \alpha_j(G) \geq n, \text{ if } r \leq k + j - 1 \text{ and}$$

$$\gamma_k(G) + \alpha_j(G) \leq n, \text{ if } r \geq k + j - 1.$$

*In particular, $\gamma_k(G) + \alpha_j(G) = n$ when $r = k + j - 1$.*

**Proof.** It is evident that $H_m = G$ for $m \leq r$ and $G_m = G$ for $m \geq r$. Thus with Theorem 2.29, we obtain $\gamma_k(G)+\alpha_j(G) \leq n$ for $r \geq k+j-1$ and $\gamma_k(G)+\alpha_j(G) \geq n$ for $r \leq k + j - 1$, which implies the desired result. ■

It was also noted in [72] that if $\gamma_k(G)+\alpha_j(G) = n$ for every pair of integers $k$, $j$ such that $k + j - 1 = \delta$ and $\gamma_{k'}(G) + \alpha_{j'}(G) = n$ for every pair of integers $k'$, $j'$ such that $k' + j' - 1 = \Delta$, then, in fact, $G$ is regular.

Clearly, Theorem 2.29 is useful when details about the graph are known like, for instance, the degree sequence or facts about the number of vertices with large degree ($\geq m$) or small degree ($\leq m$). In this line, combining upper bounds on the independence number given in [22, 56, 87, 134] and the inequality $\gamma_\delta(G) \leq n - \alpha(G)$, the authors in [72] concluded the following.

**Corollary 2.31 ([72])** *Let $G$ be a graph on $n$ vertices with maximum degree $\Delta$ and minimum degree $\delta$. Then the following hold:*

*(i) If $G$ is $r$-regular and different from the complete graph or a cycle of odd length, then $\gamma_r(G) \leq \frac{r-1}{r}n$.*
*(ii) If $G$ is triangle-free and planar with $\Delta = 3$, then $\gamma_\delta(G) \leq \frac{5}{8}n$.*
*(iii) If $G$ is $K_q$-free, then $\gamma_\delta(G) \leq \frac{\Delta+q-2}{\Delta+q}\, n$.*
*(iv) If $G$ is cubic and triangle-free, then $\gamma_3(G) \leq \frac{9}{14}n$.*

*Moreover, all bounds are sharp.*

Considering the independence number $\alpha(G)$ and the chromatic number $\chi(G)$, we can cite the following results of Hansberg, Meierling, and Volkmann [71] and Fujisawa, Hansberg, Kubo, Saito, Sugita, and Volkmann [65].

**Theorem 2.32 ([71])** *Let $G$ be a graph of order $n$. Then*

$$\gamma_k(G) \leq \frac{(\chi(G) - 1)n + n(G_{\leq k-1})}{\chi(G)}.$$

**Proof.** Let $S = V(G_{\leq k-1})$ and $r = \chi(G)$. Clearly, $S$ is contained in every $\gamma_k(G)$-set. In the case that $|S| = |V(G)| = n$, we are done. In the remaining case that $|S| < |V(G)|$, let $V_1, V_2, \ldots, V_r$ be a partition of $V(G) \setminus S$ derived from a partition of $V(G)$ into $\chi(G)$ chromatic classes and such that $|V_1| \geq |V_2| \geq \ldots \geq |V_r|$. Observe that $V_i = \emptyset$ is possible for any $i \geq 2$. Then every vertex of $V_1$ has degree at least $k$ and all its neighbors are in $V(G) \setminus V_1$. Thus $V(G) \setminus V_1$ is a $k$-dominating set of $G$ such that

$$|V_1| \geq \frac{|V_1| + |V_2| + \ldots + |V_r|}{r} = \frac{n - |S|}{r}$$

and thus

$$\gamma_k(G) \leq n - |V_1| \leq n - \frac{n - |S|}{r} = \frac{(r-1)n + n(G_{\leq k-1})}{r}.$$

■

Clearly, if we assume that $G$ is a graph with $\delta(G) \geq k$, then Theorem 2.32 gives $\gamma_k(G) \leq \frac{\chi(G)-1}{\chi(G)}n$. This is interesting when considering graphs with chromatic number at most $k$ and minimum degree at least $k$, since then this gives a better bound than the inequality $\gamma_k(G) \leq \frac{k}{k+1}n$ of Theorem 2.12. In particular, we mention the following corollary for bipartite graphs.

**Corollary 2.33** *If $G$ is a bipartite graph of order $n$ and $\delta(G) \geq k$, then $\gamma_k(G) \leq \frac{1}{2}n$.*

On the other hand, Brooks' theorem [22] states that any connected graph $G$ different from the complete graph and from a cycle of odd length has $\chi(G) \leq \Delta(G)$. Moreover, if we also assume that $\delta(G) \geq k$, then Theorem 2.32 together with Brooks' Theorem and the fact that $\alpha(G) \geq \frac{n(G)}{\Delta(G)}$ yield

$$\gamma_k(G) \leq \frac{(\chi(G)-1)n + n(G_{\leq k-1})}{\chi(G)} \leq \frac{(\chi(G)-1)n}{\chi(G)} \leq \frac{(\Delta(G)-1)n}{\Delta(G)} \leq (\Delta(G)-1)\alpha(G).$$

This is the statement of the following theorem, where the authors also characterize the non-regular graphs attaining equality.

**Theorem 2.34 ([71])** *Let $G$ be a connected nontrivial graph with maximum degree $\Delta$ and minimum degree $\delta(G) \geq k$. If $G$ is neither isomorphic to a cycle of odd length when $k = 2$ nor to the complete graph $K_{k+1}$, then*

$$\gamma_k(G) \leq (\Delta - 1)\alpha(G).$$

*Moreover, if $G$ is non-regular, then $\gamma_k(G) = (\Delta - 1)\alpha(G)$ if and only if $G \cong K_2 \circ K_k$.*

The next theorem, stated also in [71], establishes a relation between the $k$-domination number $\gamma_k(G)$, the independence number $\alpha(G)$, and the chromatic number $\chi(G)$ of a graph $G$.

**Theorem 2.35 ([71])** *Let G graph be a graph with $\Delta(G) \geq k$. Then*

$$\gamma_k(G) \leq \alpha(G)\left(\chi(G) - 1 + \frac{k-1}{\chi(G)}\right).$$

**Proof.** For $k = 1$, use the fact that the complement of a maximum independent set is dominating, which yields $\gamma(G) \leq n - \alpha(G)$, together with the inequality $\alpha(G) \geq \frac{n}{\chi(G)}$. Now suppose that $k \geq 2$. Let $S = V(G_{\leq k-1})$. Since $G$ is connected and $V(G) - S$ is not empty, every component $Q$ of $G[S]$ fulfills $\delta(Q) \leq k-2$ and $\Delta(Q) \leq k-1$. From Brook's Theorem, it follows that $\alpha(Q) \geq \frac{n(Q)}{\chi(Q)} \geq \frac{n(Q)}{k-1}$. Thus, if $Q_1, Q_2, \ldots, Q_s$ are the components of $G[S]$, then we obtain

$$\alpha(G) \geq \alpha(G[S]) \geq \sum_{i=1}^{s} \alpha(Q_i) \geq \frac{n(G[S])}{k-1} = \frac{|S|}{k-1}.$$

Hence, together with $n(G) \leq \chi(G)\alpha(G) = r\alpha(G)$, we have, with Theorem 2.32,

$$\begin{aligned}\gamma_k(G) &\leq \frac{(r-1)n(G) + |S|}{r} \\ &\leq \frac{(r-1)r\alpha(G) + (k-1)\alpha(G)}{r} \\ &= \alpha(G)\left(\chi(G) - 1 + \frac{k-1}{\chi(G)}\right).\end{aligned}$$

■

If $G$ is a graph of order $n$ and $\Delta(G) \geq k$ with $\chi(G) \leq k-1$, then the right-hand side of the inequality of previous theorem would give, by means of $\alpha(G)\chi(G) \geq n$,

$$\alpha(G)\left(\chi(G) - 1 + \frac{k-1}{\chi(G)}\right) = \alpha(G)\chi(G) + \alpha(G)\left(\frac{k-1}{\chi(G)} - 1\right) \geq n.$$

Hence, the equality is only meaningful for $k \leq \chi(G)$. In particular, this implies that, for bipartite graphs, the only case worth of studying is $k = 2$. This is precisely what was done previously by Fujisawa, Hansberg, Kubo, Saito, Sugita, and Volkmann [65], who showed that $\gamma_2(G) \leq \frac{3}{2}\alpha(G)$ for bipartite graphs $G$ with maximum degree at least 2 and who characterized the equality. This inequality had been previously shown for trees by Blidia, Chellali, and Favaron [11]. We recall here that, in [62, 123], the graphs $G$ satisfying equality in Ore's bound (Theorem 1.6), i.e., such that $\gamma(G) = \frac{1}{2}n(G)$, were shown to be those isomorphic to a corona graph $H \circ K_1$ or to the cycle of length 4.

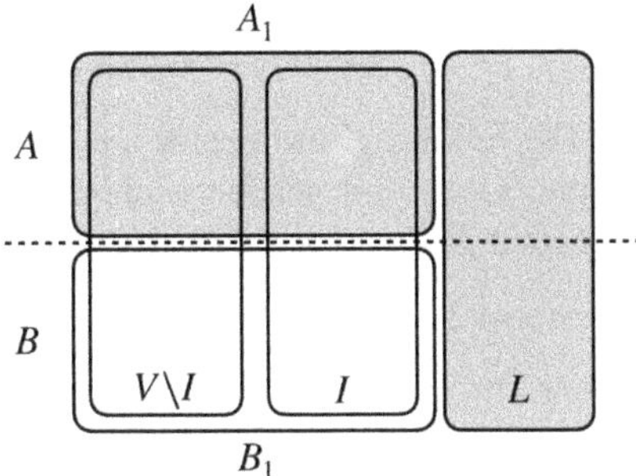


**Fig. 4** Sketch of the proof of Theorem 2.36: The set $A_1 \cup L$ is a 2-dominating set.

**Theorem 2.36 ([65])** *If $G$ is a connected bipartite graph of order at least 3, then $\gamma_2(G) \le \frac{3}{2}\alpha(G)$ and equality holds if and only if $G \cong (J \circ K_1) \circ K_1$ for some connected bipartite graph $J$ or $G \cong C_4 \circ K_1$.*

**Proof.** Let $G$ be a connected bipartite graph with $n(G) \ge 3$. Let $L$ be the set of leaves in $G$, and let $I$ be a maximum independent set of $G$. Since $n(G) \ge 3$, we can assume, without loss of generality, that $L \subseteq I$ and thus $|L| \le \alpha(G)$. Since $G$ is bipartite, we have $2\alpha(G) \ge n(G)$.

Let $A$ and $B$ be the partition sets of $G$. Define $A_1 := A \setminus L$ and $B_1 := B \setminus L$ and assume, without loss of generality, that $|A_1| \le |B_1|$ (see Figure 4). Then $|A_1| \le \frac{1}{2}(n(G) - |L|)$. Since every vertex in $B_1$ has at least two neighbors in $A_1 \cup L$,

$$\gamma_2(G) \le |A_1 \cup L| \le \frac{1}{2}(n(G) - |L|) + |L| = \frac{1}{2}(n(G) + |L|).$$

Combining this inequality with $|L| \le \alpha(G)$ and $n(G) \le 2\alpha(G)$, we obtain the desired bound

$$\gamma_2(G) \le \frac{1}{2}(n(G) + |L|) \le \frac{2\alpha(G) + \alpha(G)}{2} = \frac{3}{2}\alpha(G).$$

Moreover, if $\gamma_2(G) = \frac{3}{2}\alpha(G)$ holds, we necessarily have $n(G) = 2\alpha(G)$, $|L| = \alpha(G)$, and $\gamma_2(G) = \frac{1}{2}(n(G) + |L|)$. Hence, $G$ is such that $|L| = \alpha(G)$ and $n(G) = 2\alpha(G) = 2|L|$, implying that $G \cong H \circ K_1$. Furthermore, the identities $\gamma_2(G) = \frac{1}{2}(n(G) + |L|)$ and $n(G) = 2|L|$ yield $\gamma_2(G) = \frac{3}{4}n(G)$. Consider now a minimum dominating set $D$ of $H = G - L$. Since $G \cong H \circ K_1$, $D \cup L$ is a minimum 2-dominating set of $G$ and we obtain, with Ore's inequality (Theorem 1.6),

$$\gamma_2(G) = \gamma(H) + |L| \le \frac{1}{2}n(G - L) + |L| = \frac{3}{4}n(G).$$

In view of the characterization of the graphs attaining Ore's bound mentioned before this theorem, equality holds if and only if $H \cong J \circ K_1$ for some connected bipartite graph $J$ or if $H \cong C_4$.

Conversely, if $G \cong (J \circ K_1) \circ K_1$ for some connected bipartite graph $J$ or $G \cong C_4 \circ K_1$, then it is straightforward to check that $\gamma_2(G) = \frac{3}{2}\alpha(G)$. ■

The following results give a lower bound on the $k$-domination number in terms of the $(k-1)$-independent number for special classes of graphs. The case $k = 2$ of Theorem 2.37 was given in [11] previously.

**Theorem 2.37 ([17])** *If $T$ is a tree, then $\gamma_k(T) \geq \alpha_{k-1}(T)$ for every integer $k \geq 2$.*

**Proof.** If $\Delta(T) \leq k - 1$, then $\gamma_k(T) = n = \alpha_{k-1}(T)$. So suppose that $\Delta(T) \geq k$. Let $B = V(T_{\geq k})$. Observe that the condition on the maximum degree lets us assume that $B \neq \emptyset$. We will proceed by induction on $|B|$. If $|B| = 1$, then clearly $\gamma_k(T) = n - 1 = \alpha_{k-1}(T)$. Let now $|B| \geq 2$ and assume that $\gamma_k(T') \geq \alpha_{k-1}(T')$ for every tree $T'$ with $|V(T'_{\geq k})| < |B|$. Root $T$ at a vertex $r$ of maximum eccentricity and let $w$ be a vertex of degree at least $k$ of maximum distance from $r$. Since $r$ is a leaf, $r \neq w$. Let $T_w$ be the subtree of $T$ consisting of $w$ and all its descendants, and let $T' = T - T_w$. Let $v$ be the unique neighbor of $w$ in $V(T')$. If $S$ is a maximum $(k-1)$-independent set of $T'$, then $S \cup (V(T_w) \setminus \{w\})$ is a $(k-1)$-independent set of $T'$. It follows that

$$\alpha_{k-1}(T) \geq |S \cup (V(T_w) \setminus \{w\})| = \alpha_{k-1}(T') + n(T_w) - 1.$$

On the other hand, if $D$ is a minimum $k$-dominating set of $T$, then $(D \setminus V(T_w)) \cup \{v\}$ is a $k$-dominating set in $T'$. Hence, we have

$$\gamma_k(G) \leq |(D \setminus V(T_w)) \cup \{v\}| \leq \gamma_k(T) - n(T_w) + 1.$$

Since by the induction hypothesis we have $\gamma_k(T') \geq \alpha_{k-1}(T')$, we obtain, together with both inequalities above, that $\gamma_k(T) \geq \alpha_{k-1}(T)$. ■

Equivalent conditions for trees to satisfy $\gamma_k(T) = \alpha_{k-1}(T)$ are also given in [17], among them an explicit construction of the family via two different operations. For $k = 2$, the authors of Theorem 2.37 extended their result to block graphs showing that $\gamma_2(G) \geq \alpha(G)$ for any block graph $G$ and established the bound $\gamma_2(G) \geq \alpha(G) - c(G) + 1$ for connected cactus graphs with $c(G) \geq 1$ cycles [16]. In [35] this last inequality was refined to $\gamma_2(G) \geq \Gamma(G) - c(G) + 1$.

## *2.4 Nordhaus–Gaddum Bounds*

Using Ore's upper bound (Theorem 1.6), Theorems 2.12, 2.14, and two further results in the paper by Stracke and Volkmann [135], Volkmann [144] derived the following Nordhaus–Gaddum bound for the 2-domination number.

**Theorem 2.38 ([144])** *If G is a graph of order n, then*

$$\gamma_2(G) + \gamma_2(\overline{G}) \le n + 2.$$

In 2008, Prince [125] extended the above theorem in the case where $k \ge 2$. However, the bound in Theorem 2.38 is one unit below the bound in Theorem 2.39 for $k = 2$. In [125], a lower bound for the sum as well as lower and upper bounds for the bounds are given, too.

**Theorem 2.39 ([125])** *If G is a graph of order $n \ge n_0(k)$. Then*

*(i)* $\frac{8}{3}k \le \gamma_k(G) + \gamma_k(\overline{G}) \le n + 2k - 1$, *and*
*(ii)* $\frac{16}{9}k^2 \le \gamma_k(G)\gamma_k(\overline{G}) \le (2k-1)(n-k+2)$.

Note that, in order to prove the above bounds, Prince [125] proved first a series of lemmas, some of which are themselves inequalities of Nordhaus–Gaddum-type under certain conditions. The function $n_0(k)$ appearing in Theorem 2.39 is small for three of the inequalities. Indeed, the lower bounds hold already for $n \ge \frac{25}{9}k$, and the upper bound for $\gamma_k(G) + \gamma_k(\overline{G})$ is valid for all $n$. However, the argument used to prove the upper bound for $\gamma_k(G)\gamma_k(\overline{G})$ requires $n$ to be at least $\exp(\Omega(k \log k))$. This requirement comes from a probabilistic upper bound on the $k$-domination number in which the number of vertices with low degree plays a particular role (also given in [125]). We also point out that many of the results in [125] are proven via the interpretation of $k$-dominating sets of a graph in the neighborhood hypergraph. In this context, the so-called $\Delta$-systems (also called sunflowers) play an important role (see also [54, 138]). As to the sharpness, all inequalities in Theorem 2.39 are proved to be sharp. However, the sharpness of the inequality $\gamma_k(G) + \gamma_k(\overline{G}) \le n + 2k - 1$ is given only for small $n$'s.

In [120], Mojdeh, Samadi, and Volkmann gave the following Nordhaus–Gaddum-type result involving the connectivity $\kappa(G)$ of the graph, i.e., the minimum size of a vertex cut.

**Theorem 2.40 ([120])** *If G is a graph with $\gamma(G), \gamma(\overline{G}) \ge k + 2$, then*

$$\gamma_k(G) + \gamma_k(\overline{G}) \le \kappa(G) + \kappa(\overline{G}) - (\gamma(G) + \gamma(\overline{G})) + 2k + 4.$$

Observe that, since the connectivity of a graph is upper-bounded by the minimum degree of the graph and since we are assuming $\gamma(G), \gamma(\overline{G}) \ge k + 2$, the bound of Theorem 2.40 implies that

$$\gamma_k(G) + \gamma_k(\overline{G}) \le \delta(G) + \delta(\overline{G}) - (\gamma(G) + \gamma(\overline{G})) + 2k + 4 \le \delta(G) + \delta(\overline{G}).$$

# 3 $k$-Tuple Domination and Total $k$-Domination

## 3.1 Bounds in Terms of Order and Degrees

### 3.1.1 The General Case

In their paper introducing the $k$-tuple domination and analogous to the bounds of Fink and Jacobson for the $k$-domination number (Theorem 2.10), Harary and Haynes [84] gave two lower bounds on $\gamma_{\times k}(G)$ in terms of the order, the edge number, and the maximum degree of the graph $G$.

**Theorem 3.1 ([84])** *Let $G$ be a graph of order $n$, edge number $m$, maximum degree $\Delta$, and minimum degree at least $k-1$. Then*

$$\gamma_{\times k}(G) \geq \frac{kn}{\Delta + 1} \quad \text{and} \quad \gamma_{\times k}(G) \geq \frac{2kn - 2m}{k+1}$$

*and the two bounds are sharp.*

The proofs of these bounds are analogous to those of Theorem 2.10. Similarly, one can see that

$$\gamma_k^t(G) \geq \frac{kn}{\Delta} \quad \text{and} \quad \gamma_k^t(G) \geq \frac{2kn - 2m}{k}.$$

Some variations of these bounds for the total $k$-domination number were given in [10]. Also in [10], Bermudo, Hernández, and Sigarreta gave the following upper bound on the total $k$-domination number.

**Theorem 3.2 ([10])** *Let $G$ be a graph of order $n$, with maximum degree $\Delta$ and minimum degree $\delta \geq k+1$. Then*

$$\gamma_k^t(G) \leq \frac{\Delta(k+1)}{\Delta(k+1)+1} n.$$

**Proof.** Let $D$ be a minimum total $k$-dominating set. Let $X = \{x \in D \mid N(x) \subseteq D\}$ be the set of vertices in $D$ having all neighbors inside $D$. Then $V(G) \setminus D$ is a dominating set of $G - X$. By Theorem 2.10 with $k = 1$, we deduce

$$n - |D| = |V(G) \setminus D| \geq \frac{n - |D|}{\Delta(G - X) + 1} \geq \frac{n - |D|}{\Delta + 1},$$

which yields $|X| \geq |D|(\Delta + 1) - n\Delta$.

Let $S = \{v \in D \mid |N(v) \cap D| = k\}$. Since $\delta \geq k+1$, $S \cap X = \emptyset$. Moreover, by the minimality of $D$, we know by Proposition 1.3 that, for all $x \in X$, $S \cap N(x) \neq \emptyset$. Hence,

$$|X| \le m(X, S) \le k|S| \le k|D \setminus X|,$$

which gives $|X| \le \frac{k}{k+1}|D|$.

Combining both obtained inequalities, it follows that

$$|D|(\Delta + 1) - n\Delta \le |X| \le \frac{k}{k+1}|D|,$$

which gives $\gamma_k^t(G) = |D| \le \frac{\Delta(k+1)}{\Delta(k+1)+1}n$. ∎

Recall Caro and Yuster's result (Theorem 1.8) stating that, for large minimum degree $\delta$, more precisely for $k < \sqrt{\ln \delta}$,

$$\gamma_{\times k}(G) \le \gamma_k^t(G) \le \frac{\ln \delta}{\delta} n(1 + o_\delta(1)).$$

As for the $k$-domination number, there has been made much effort in upper bounding the $k$-tuple domination and the total $k$-domination numbers in terms of the order and the degrees of the graph by means of less strong conditions on the minimum degree as those assumed in Caro and Yuster's bound. Similar bounds for the $k$-tuple number and for the total $k$-domination number are given in [47] and, respectively, [89]. Building up on results by Harant and Henning [80] (case $k = 2$), Rautenbach and Volkmann [128] (case $k = 3$) and Gagarin and Zverovich [66], Chang [32], Xu, Kang, Shan, and Yan [149], and Zverovich [152] proved, independently, the following upper bound of the $k$-tuple domination number in terms of the order and the degree sequence of the graph. This bound was stated in [128] as a conjecture. Observe that the condition on the minimum degree is the weakest reasonable assumption.

**Theorem 3.3 ([32, 149, 152])** *For a graph G with minimum degree $\delta \ge k - 1$,*

$$\gamma_{\times k}(G) \le \frac{n}{\delta - k + 2}\left(\ln(\delta - k + 2) + \ln\left(\frac{1}{n}\sum_{v \in V}\binom{d(v)+1}{k-1}\right) + 1\right).$$

As it can be expected, if we set a stronger assumption on the minimum degree, one can obtain better bounds on the $k$-tuple domination and the total $k$-domination numbers. Along this line, we can cite the following results.

**Theorem 3.4 ([128])** *Let G be an n-order graph of minimum degree $\delta$. If $\frac{\delta+1}{2\ln(\delta+1)} \ge k$, then*

$$\gamma_{\times k}(G) \le \frac{n}{\delta + 1}\left(k\ln(\delta + 1) + \sum_{i=0}^{k-1}\frac{k-i}{i!(\delta + 1)^{k-1-i}}\right).$$

The following upper bound on $\gamma_{\times k}$ was obtained by Przybyło [126] using a probabilistic approach, in which the construction of the $k$-tuple dominating set is made by selecting its elements randomly in $k$ rounds, in contrast to the one round that is usually performed.

**Theorem 3.5 ([126])** *For a graph $G$ with minimum degree $\delta \geq k-1$ such that $k \leq \frac{\delta+2-k}{\ln(\delta+2-k)+1}$,*

$$\gamma_{\times k}(G) \leq \left(\sum_{i-1}^{k} \frac{\ln(\delta+2-i)+1}{(\delta+2-i)}\right) n.$$

Considering the function $g(x) = \frac{\ln(\delta+2-x)+1}{(\delta+2-x)}$, which is convex and increasing for $x \leq \delta+1$, it is not difficult to prove the following facts:

$$\frac{g(1+j)+g(k-j)}{2} \leq \frac{g(1)+g(k)}{2}, \text{ for } 0 \leq j \leq \left\lfloor \frac{k}{2} \right\rfloor, \text{ and}$$

$$\sum_{i=1}^{k} g(i) \leq \int_{1}^{k+1} g(x)dx = \ln\left(\frac{\delta+1}{\delta+1-k}\right)(1+\ln(\sqrt{(\delta+1)(\delta+1-k)})).$$

Having this, the following corollary is immediate.

**Corollary 3.6 ([126])** *Let $G$ be a graph with minimum degree $\delta \geq k-1$ such that $k \leq \frac{\delta+2-k}{\ln(\delta+2-k)+1}$. Then we have the following bounds.*

*(i)* $\gamma_{\times k}(G) \leq \frac{k}{2}\left(\frac{\ln(\delta+1)+1}{(\delta+1)} + \frac{\ln(\delta+2-k)+1}{(\delta+2-k)}\right) n.$

*(ii)* $\gamma_{\times k}(G) \leq \ln\left(\frac{\delta+1}{\delta+1-k}\right)\left(1+\ln\left(\sqrt{(\delta+1)(\delta+1-k)}\right)\right) n.$

Analogous results for the total $k$-domination number were also obtained in [126]. Other probabilistic bounds for both the $k$-tuple domination and the total $k$-domination numbers can be found in [99].

### 3.1.2 The Case $k = 2$

Much research on $k$-tuple and total $k$-domination deals with the case $k = 2$. In this case, the 2-tuple domination is called *double domination* and instead of $\gamma_{\times 2}(G)$, many authors use the parameter $\mathrm{dd}(G)$, which was as Harary and Haynes denoted it originally in their seminal paper [84].

Following ideas from [82], Harant and Henning [80] show that, for a graph $G$ on the vertex set $\{1, 2, \ldots, n\}$,

$$\gamma_{\times 2}(G) = \min f(\mathbf{p}),$$

where the minimum is taken within all points $\mathbf{p}$ of the $n$-dimensional cube

$$C^n = \{\mathbf{p} = (p_1, p_2, \dots, p_n) \mid p_i \in [0,1], 1 \le i \le n\}$$

and the function $f$ is the following:

$$f(\mathbf{p}) = \sum_{i=1}^{n} p_i + \left(\sum_{i=1}^{n} p_i \prod_{j\in N(i)} (1-p_j)\right) + \left(2\sum_{i=1}^{n}(1-p_i) \prod_{j\in N(i)} (1-p_j)\right)$$
$$+ \sum_{i=1}^{n}(1-p_i)\left(\sum_{j\in N(i)} p_j \prod_{\ell\in N(i)\setminus\{j\}} (1-p_\ell)\right).$$

With this approach, they also obtain the bound $\gamma_{\times 2}(G) \le \left(\frac{\ln(1+\mathrm{d})+\ln\delta+1}{\delta}\right) n$ for a graph $G$ of order $n$, minimum degree $\delta \ge 1$, and average degree d, result that was later generalized for general $k$ (Theorem 3.3). A few years later, Harant and Henning [81] presented an efficient algorithm that yields, for a given $\mathbf{p} \in C^n$, a double dominating set of cardinality at most $f(\mathbf{p})$. We observe in passing that analogous results were obtained for $k = 3$ in [82], too.

There are some interesting results on the double domination and the total 2-domination numbers in terms of the order of the graph. Henning gave in [88] a $\frac{3}{4}n$-bound for the double domination number of connected $n$-vertex graphs with minimum degree $\delta \ge 2$, with $C_5$ being the only exception. A characterization of the extremal family was also achieved. Let $F_8$ be the graph consisting of a $C_8$ together with an extra edge joining two of its vertices at distance 4. Let $\mathcal{H}$ be the family of graphs of order $n \ge 4$ divisible by 4 and consisting of an arbitrary connected graph $H$ with $V(H) = \{v_1, v_2, \dots, v_{n/4}\}$ and $\frac{n}{4}$ disjoint $C_4$'s, say $C^1, C^2, \dots, C^{n/4}$, such that each vertex $v_i$ is identified with one vertex of $C^i$, $1 \le i \le \frac{n}{4}$. See Figure 5 for a drawing of the family $\mathcal{H}$ and the graph $F_8$.

**Theorem 3.7 ([88])** *Let $G$ be a connected graph of order $n$ and minimum degree $\delta \ge 2$. If $G \ne C_5$, then $\gamma_{\times 2}(G) \le \frac{3}{4}n$ with equality if and only if $G \in \{C_8, F_8\} \cup \mathcal{H}$.*

The proof of this theorem is interesting but a bit long to be displayed here. It makes a careful analysis of the structure of the minimal connected graphs (in terms of edges) of minimum degree $\delta \ge 2$ and such that its double domination number

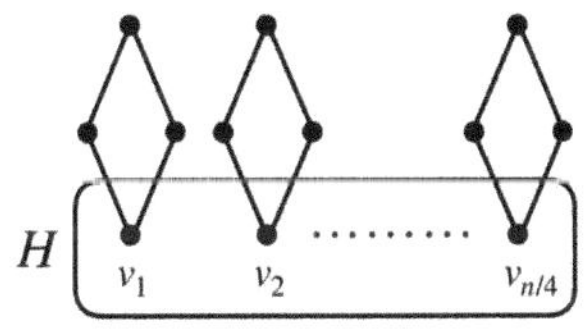


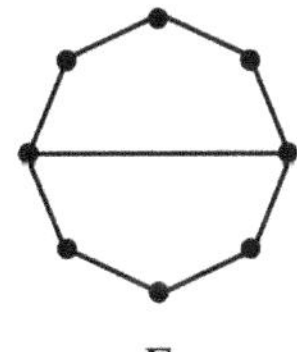


**Fig. 5** Family $\mathcal{H}$ and the graph $F_8$.

is at least $\frac{3}{4}$ its order, which yields finally the bound and the characterization of the desired extremal family of Theorem 3.7.

Concerning the total 2-domination number and building up on ideas that involve the total domination number (see [92, 93, 137]), an interesting relation to transversals in hypergraphs has been established in [94]. This relation can be extended to the total $k$-domination number. Given a hypergraph $H$ with all edges of size at least $k+1$, we say that a subset $S$ of the vertex set is a *$k$-transversal* if, for every hyperedge $e \in E(H)$, $|e \cap S| \geq k$. The minimum cardinality of a $k$-transversal in $H$ is called the *$k$-transversal number* of $H$ and denoted by $\tau_k(H)$. Consider now the neighborhood hypergraph $H_G$ of a graph $G$ on minimum degree $\delta \geq k$, which has vertex set $V(H_G) = V(G)$ and edge set consisting of all open neighborhoods $N(x)$ of $G$, $x \in V(G)$. It is now easy to see that $S$ is a $k$-transversal in $H_G$ if and only if $S$ is a total $k$-dominating set of $G$, and so we have

$$\gamma_k^t(G) = \tau_k(H_G).$$

Henning and Yeo showed in [94] that, if $H$ is a connected hypergraph on $n$ vertices with only 2-edges and 3-edges, and not belonging to a certain family of "bad" hypergraphs, then

$$12\tau_2(H) \leq 6n + 4e_3(H) + 2e_2(H), \tag{4}$$

where $e_j(H)$ is the number of edges of size $j$ in $H$. For the neighborhood hypergraph of a given connected graph $G$, the following useful facts have been established.

**Theorem 3.8 ([92])** *Let $G$ be a connected graph.*

*(i) If $G$ is non-bipartite, then $H_G$ is connected.*
*(ii) If $G$ is bipartite, then $H_G$ consists of exactly two connected components which are induced by the partite sets of $G$.*

Observe that the neighborhood hypergraph $H_G$ of a graph $G$ on $n$ vertices has at most $n$ hyperedges, since sometimes it happens that different vertices have the same neighborhood in $G$. Consider, for instance, the graph $K_{3,3}$ whose neighborhood hypergraph $H_{K_{3,3}}$ has only 2 disjoint hyperedges of size 3. By this reason we also have $d_{H_G}(x) \leq d_G(x)$ and $\Delta(H_G) \leq \Delta(G)$. (We remark at this point that in [94] it was erroneously assumed that $\Delta(H_G) = \Delta(G)$ and so the proof of Theorem 3.10 in that paper contains a flaw.) However, if, for certain graph $G$, a component $C$ of $H_G$ (out of at most 2) is isomorphic to the hypergraph $F_7$ depicted in Figure 6, then $G$ cannot have vertices with equal neighborhoods that represent a hyperedge in $C$. This is due to the fact that any two hyperedges in $F_7$ have exactly one vertex in common. Indeed, if there were vertices $x, y \in V(G)$ such that $N_G(x) = N_G(y) = e \in E(C)$, then $\{x, y\} \in N(u) \cap N(v)$ for every pair $u, v \in e$. This is not possible, since we know that $|N(u) \cap N(v)| = 1$. In fact, if $F_7$ is a component of $H_G$, then $G$ is necessarily the Heawood graph, which we call here $G_{14}$, see Figure 6. This is because of the following reasoning. First, we can assume that $H_G$ consists of two

**Fig. 6** Hypergraphs $H_3$ and $F_7$ (the Fano-plane) and the Heawood graph $G_{14}$.

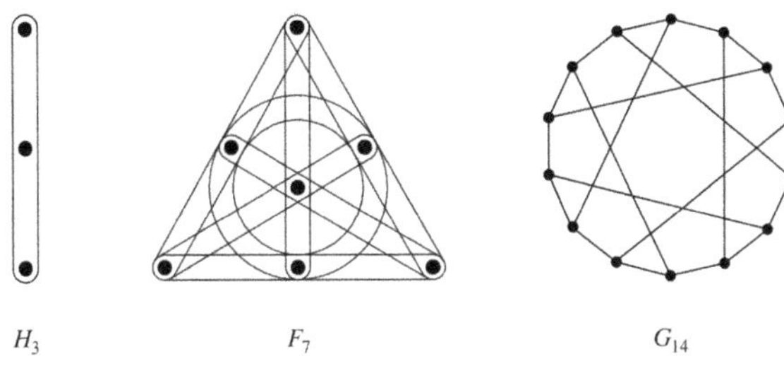


components. If not, then the unique component would be isomorphic to $F_7$ and $G$ would be a cubic graph on 7 vertices, which is impossible. Now, each of the 7 hyperedges of the $F_7$-component represents the neighborhood of a vertex from the second component (and vice versa). Then, $G$ is precisely the incidence graph of the Fano-plane, which it is well-known to be the Heawood graph $G_{14}$.

When dealing with cubic graphs, it is straightforward to see that the only possible "bad" components (as defined in [94]) of the neighborhood hypergraph are the hypergraphs $H_3$ and $F_7$ depicted in Figure 6, where $F_7$ is the Fano-plane.

Having this, the inequality (4) can be stated for the special case that the hypergraph in question is a component of the neighborhood graph $H_G$ of a connected cubic graph $G$.

**Theorem 3.9 (see [94] for the general case)** *Let $G$ be a connected cubic graph. If $C$ is a component of $H_G$ such that $C \neq H_3, F_7$, then*

$$\tau_2(C) \leq \frac{1}{2}n(C) + \frac{1}{3}e_3(C).$$

Now we have enough elements to prove the following theorem.

**Theorem 3.10 ([94])** *Let $G$ be a connected cubic graph of order $n$ such that $G$ is not the Heawood graph. Then*

$$\gamma_2^t(G) \leq \frac{5}{6}n.$$

**Proof.** Let $H_G$ be the neighborhood hypergraph of $G$. If $G$ is non-bipartite, then $H_G$ has only one component by Theorem 3.8. Clearly, $H_3$ cannot be a neighborhood graph and so $H_G \neq H_3$. From what we discussed previous to this theorem, we know that $H_G$ cannot be isomorphic to $F_7$. Now Theorem 3.9 yields

$$\gamma_2^t(G) = \tau_2(H_G) \leq \frac{1}{2}n(H_G) + \frac{1}{3}e_3(C) \leq \frac{1}{2}n + \frac{1}{3}n = \frac{5}{6}n.$$

Hence, we can assume that $G$ is bipartite. By Theorem 3.8, $H_G$ has two components, say $C_1$ and $C_2$, induced by the partite sets of $G$. Since $G$ is cubic, we have $n(C_1) = n(C_2) = \frac{n}{2}$. If $C_1, C_2 \neq H_3, F_7$, then we have with Theorem 3.9

$$\begin{aligned}\gamma_2^t(G) = \tau_2(G) = \tau_2(C_1) + \tau_2(C_2) \\ \leq \frac{1}{2}(n(C_1) + n(C_2)) + \frac{1}{3}(e_3(C_1) + e_3(C_2)) \\ \leq \frac{1}{2}n + \frac{1}{3}n = \frac{5}{6}n,\end{aligned}$$

and we are done. Hence we may assume that $C_1 = H_3$ or $C_1 = F_7$. If $C_1 = H_3$, then $C_2 = H_3$ and we have $G = K_{3,3}$, which has $\gamma_2^t(G) = 4 < 5 = \frac{5}{6}n$. If $C_1 = F_7$, then $C_2 = F_7$ and $G = G_{14}$ since the latter is the incidence graph of $F_7$. This is a contradiction since $G$ was assumed not to be the Heawood graph $G_{14}$. ■

Henning and Yeo also considered more generally graphs of minimum degree $\delta \geq 3$ and, with a much more involved proof, they were able to give the following theorem.

**Theorem 3.11 ([94])** *Let G be a connected graph of order n and minimum degree $\delta \geq 3$ such that G is not the Heawood graph. Then*

$$\gamma_2^t(G) \leq \frac{11}{13}n.$$

In [89], some results on the $k$-total domination number for certain bipartite graphs are presented. Among them one can find the following one, which bounds the total 2-domination number of cubic bipartite graphs. We mention in passing that the sharpness is not discussed in [89].

**Theorem 3.12 ([89])** *If G is a cubic bipartite graph of order n, then $\gamma_2^t(G) \leq \frac{8}{9}n$.*

We finish this section mentioning that results about special graph classes for which either the precise multiple domination number or bounds in terms of its order are given, may be found in [9, 10, 25, 89, 90, 101–103, 105, 119].

## 3.2 Relationship to Other Graph Parameters

Klasing and Laforest [108] gave a lower bound on $\gamma_{\times k}$ in terms of the independence number in graphs which are $K_{1,r}$-free.

**Theorem 3.13 ([108])** *If G is a $K_{1,r}$-free graph with minimum degree $\delta \geq k - 1$, then*

$$\gamma_{\times k}(G) \geq \frac{k}{r-1}\alpha(G).$$

Other lower bounds on the $k$-tuple domination number in terms of the independence number and maximum degree were found by Harant and Henning [81].

For the case $k = 2$, Blidia, Chellali, and Favaron proved the following relation between the double domination number, the independent domination number $i(G)$, and the independence number $\alpha(G)$.

**Theorem 3.14 ([12])** *Let G be a graph without isolates. Then*

$$\gamma_{\times 2}(G) \leq i(G) + \alpha(G).$$

**Proof.** Let $S$ be a minimum independent set of $G$ and $S'$ a maximum independent set of $G[V(G) \setminus S]$. Let $B$ be the set of isolated vertices of $G[S \cup S']$. Since $S$ is dominating, clearly $B \subseteq S$ and $S' \cup B$ is an independent set, so that we have $|S' \cup B| \leq \alpha(G)$. Since $G$ has no isolates, each vertex $x \in B$ has at least one neighbor in $V(G) \setminus (S \cup S')$. Let $A \subseteq V(G) \setminus (S \cup S')$ be a selection of one neighbor for each $x \in B$. Then $|A| \leq |B|$. Observe that the set $D = S \cup S' \cup A$ is a double dominating set. Indeed, $G[D]$ has no isolates and every vertex in $V(G) \setminus D$ has at least a neighbor in $S$ and a neighbor in $S'$. Now we have

$$\gamma_{\times 2}(G) \leq |S \cup S' \cup A| = |S| + |S'| + |A| \leq |S| + |S'| + |B| = i(G) + \alpha(G).$$

■

Observe that, if $G$ is claw-free, Theorem 2.25 yields $\gamma_{\times 2}(G) \geq \gamma_2(G) \geq \alpha(G)$, as was also noted in [12].

In [10], Bermudo, Hernández, and Sigarreta considered relations between the total $k$-domination number and the chromatic number, the diameter and the girth of the graph.

**Theorem 3.15 ([10])** *Let G be a graph of minimum degree $\delta \geq k$. Then*

$$\gamma_k^t(G) \geq \frac{k\chi(G)}{\chi(G) - 1}.$$

**Proof.** Let $\chi = \chi(G)$ and let $V_1, V_2, \ldots, V_\chi$ be a partition of $V(G)$ into $\chi$ chromatic classes. Let $D$ be a minimum total $k$-dominating set of $G$ and define $D_i = D \cap V_i$. Then, for each $i = 1, 2, \ldots, \chi$ and each vertex $x \in D_i$, $|D \setminus D_i| \geq |N(x)| \geq k$. Hence,

$$k\chi \leq \sum_{i=1}^{\chi} |D \setminus D_i| = |D|(\chi - 1) = \gamma_k^t(G)(\chi - 1),$$

and the bound follows. ■

The sharpness of previous theorem can be seen with a complete $\chi$-partite graph $H = K_{n_1, n_2, \ldots, n_\chi}$ such that $n_i \geq \frac{k}{\chi - 1}$, provided that $k$ is divisible by $\chi - 1$. In such a graph, a total $k$-dominating set can be formed by taking $\frac{k}{\chi - 1}$ vertices from each partition set, and so $\gamma_k^t(H) \le \frac{k\chi(G)}{\chi(G)-1}$, which by previous theorem has to be an equality.

**Theorem 3.16 ([10])** *Let $G$ be a graph of order $n$ and minimum degree $\delta \ge k+1$. Then we have the following bounds.*

*(i)* $\gamma_k^t(G) \le n - \left\lceil \frac{\text{diam}(G)}{3} \right\rceil$;
*(ii) If $G$ has finite girth $g$ and $k \ge 2$, then $\gamma_k^t(G) \le (g-2)(k-1)+2$;*

When $k = 2$, the following results have been established.

**Theorem 3.17 ([107])** *Let $G$ be a connected graph with minimum degree $\delta$ and maximum degree $\Delta$.*

*(i) If* $\text{diam}(G) = 2$, *then* $\gamma_{\times 2}(G) \le \delta + \Delta$.
*(ii)* $\gamma_{\times 2}(G) \ge 2\lceil \frac{2\text{diam}(G)+1}{3} \rceil$.

The bounds (i) and (ii) in Theorem 3.17 are proved to be sharp [107]. The same authors also considered graphs with finite girth and obtained the following results.[1]

**Theorem 3.18 ([107])** *Let $G$ be a connected graph with finite girth $g$, minimum degree $\delta$ and maximum degree $\Delta$.*

*(i) If $g = 4$, then $\gamma_{\times 2}(G) \ge 3$ with equality if and only if $G \cong K_{2,n-2}$.*
*(ii) If $g \in \{5, 6\}$, then $\gamma_{\times 2}(G) \ge 2\delta$; if $g \ge 7$, then $\gamma_{\times 2}(G) \ge 2\delta + 1$.*
*(iii) If $g \ge 5$, then $\gamma_{\times 2}(G) \ge \Delta + \lceil \frac{2g-7}{3} \rceil$.*
*(iv) If $g \ge 7$ and $\delta \ge 2$, then $\gamma_{\times 2}(G) \ge 2\Delta + 1$.*

## 3.3 Nordhaus–Gaddum Bounds

Clearly, $\gamma_{\times k}(G) \ge k$, and it is easy to verify that $\gamma_{\times k}(G) = k$ if and only if $G$ contains a clique $K_k$, each vertex of which has degree $n-1$. To study Nordhaus–Gaddum-type bounds on $\gamma_{\times k}$, we assume $\delta(G) \ge k-1$ and $\delta(\overline{G}) \ge k-1$. Therefore, $\Delta(G), \Delta(\overline{G}) \le n-k \le n-2$ when $k \ge 2$ and by the observation above $\gamma_{\times k}(G), \gamma_{\times k}(\overline{G}) \ge k+1$. Hence

$$2k+2 \le \gamma_{\times k}(G) + \gamma_{\times k}(\overline{G}) \le 2n \quad \text{and} \quad 4(k+1)^2 \le \gamma_{\times k}(G)\gamma_{\times k}(\overline{G}) \le n^2$$

for $k \ge 2$. The previous bounds were given in [83] for $k = 2$, with the determination of $P_4$ as the unique extremal graph for the upper bounds.

Other Nordhaus–Gaddum bounds can of course be obtained by using known bounds on $\gamma_{\times k}(G)$. For instance, the inequality $\gamma_{\times k}(G) \le n - \delta(G) + k - 1 = \Delta(\overline{G}) + k$ of Corollary 4.4 gives

[1]The results here are altered from their original shape to display them cleaner.

$$\gamma_{\times k}(G) + \gamma_{\times k}(\overline{G}) \leq \Delta(G) + \Delta(\overline{G}) + 2k$$

(observed in [37] for $k = 2$). When $k$ is odd, let $G$ be a $(k - 1)$-regular graph of order $2k - 1$. When $k$ is even, let $G$ be obtained from a $(k - 1)$-regular graph of order $2k$ by adding a matching of $n/2$ edges. These graphs are extremal for the two upper bounds $2n$ and $\Delta(G) + \Delta(\overline{G}) + 2k$ on $\gamma_{\times k}(G) + \gamma_{\times k}(\overline{G})$. The following theorem gives a clear improvement on this bound (note that, in [132], the $k$-tuple domination number $\gamma_{\times k}$ is denoted $\gamma_k$). The same statement was conjectured in [83] for $k = 2$ after proving a slightly weaker result, namely that $\gamma_{\times 2}(G)+\gamma_{\times 2}(\overline{G}) \leq \delta(G)+\delta(\overline{G})$ for any graph $G$ with $\gamma(G), \gamma(\overline{G}) \geq 5$.

**Theorem 3.19 ([132])** *For any integer $k \geq 1$, if a graph $G$ has $\gamma(G), \gamma(\overline{G}) \geq k + 2$, then*

$$\gamma_{\times k}(G) + \gamma_{\times k}(\overline{G}) \leq \delta(G) + \delta(\overline{G}).$$

The next result by Mojdeh, Samadi, and Volkmann [120] improves the upper bound in Theorem 3.19.

**Theorem 3.20 ([120])** *For any integer $k \geq 1$, if a graph $G$ has $\gamma(G), \gamma(\overline{G}) \geq k + 2$, then*

$$\gamma_{\times k}(G) + \gamma_{\times k}(\overline{G}) \leq \delta(G) + \delta(\overline{G}) - (\gamma(G) + \gamma(\overline{G})) + 2k + 4.$$

For the particular case $k = 2$, Chen and Sun proved the following Nordhaus–Gaddum bound involving conditions on the diameter.

**Theorem 3.21 ([42])** *If $G$ is a graph with $\delta(G) \geq 1$, $\delta(\bar{G}) \geq 1$ and $diam(G) > 2$ or $diam(\overline{G}) > 2$, then $\gamma_{\times 2}(G) + \gamma_{\times 2}(\overline{G}) \leq n + 4$.*

It is also known that, if $\mathrm{diam}(G) \geq 4$, then $\gamma_{\times 2}(\overline{G}) \leq 4$, and if moreover $G$ is triangle-free or $\mathrm{diam}(G) \geq 6$, then $\gamma_{\times 2}(\overline{G}) = 3$ [107].

## 3.4 Cartesian Products of Graphs

The *cartesian product* of two graphs $G$ and $H$ is the graph $G \Box H$ with vertex set $V(G) \times V(H)$ and vertices $(u_1, u_2)$ and $(v_1, v_2)$ are adjacent if and only if either $u_1 = v_1$ and $u_2v_2 \in E(H)$ or $u_2 = v_2$ and $u_1v_1 \in E(G)$. Arguably the most important conjecture in domination theory is Vizing's Conjecture [140] that claims that, for any two graphs $G$ and $H$

$$\gamma(G \Box H) \leq \gamma(G)\gamma(H).$$

This conjecture has been subject of study for over 50 years, and has been confirmed for several graph classes, and many partial results have been obtained, see [21] for a comprehensive survey on the subject. A significant breakthrough was obtained in [45], where it was shown that $\gamma(G\Box H) \leq 2\gamma(G)\gamma(H)$, which led to the discovery of analogous inequalities for several other domination parameters, including $\gamma_t(G\Box H) \leq 2\gamma_t(G)\gamma_t(H)$ [91, 95]. There are some results on $k$-tuple and total $k$-domination concerning cartesian products of graphs, including some Vizing-type inequalities. The following inequalities, in which the packing number is involved, were both proven independently in [9] and [106]. Recall that a *packing* of a graph $G$ is a set of vertices in $G$ that are pairwise at distance more than two, while the *packing number* $\rho(G)$ of $G$ is the size of a largest packing in $G$.

**Theorem 3.22 ([9, 106])** *Let $G$ and $H$ be two graphs such that $k \leq \delta(G)$. Then we have*

$$\gamma_k^t(G)\rho(H) \leq \gamma_k^t(G\Box H) \leq \gamma_k^t(G)n(H).$$

It was shown in [10] that, if $H$ is a graph with minimum degree $\delta(H) \geq k$, then $k\rho(H) \leq \gamma_k^t(H)$. However, it is not clear how good this inequality is, since only one particular example for $k = 3$ is given in [10]. Let $f(k) \geq k$ be a function such that $\gamma_k^t(H) \leq f(k)\rho(H)$. Then

$$\gamma_k^t(G)\gamma_k^t(H) \leq f(k)\gamma_k^t(G\Box H).$$

If $f(k) = 2k$, this would imply that $\gamma_k^t(G)\gamma_k^t(H) \leq 2k\gamma_k^t(G\Box H)$ [106], while if $f(k) = k$, then $\gamma_k^t(G)\gamma_k^t(H) \leq k\gamma_k^t(G\Box H)$ would follow [9]. Neither of these two conditions is shown to be realizable for a $k$ different from 1 in [9, 106].

We remark also that in [106] the total 2-domination of $K_n\Box K_m$ for any $n, m \geq 1$ is determined, while in [24] the $k$-tuple domination of $K_n\Box K_n$, i.e., the rook's graph, is obtained. Other results related to $k$-tuple or total $k$-domination in cartesian products of graphs (including grid graphs) can be found in [9, 10, 48, 106, 131]. Finally, results on cross products or tensor products of graphs are given in [48, 90].

# 4 Relations Involving Multiple Domination and Other Domination Parameters

## 4.1 The General Case

Favaron, Henning, Puech, and Rautenbach defined in 2001 the following domination parameter which generalizes both $k$-domination, $k$-tuple domination, and total $k$-domination.

**Definition 4.1 ([61])** *For two integers $l \geq 0$ and $k > 0$, the subset $S$ of vertices of $G$ is $l$-total $k$-dominating if every vertex $x$ has at least $l$ neighbors in $S$ if $x \in S$ and $k$ neighbors in $S$ if $x \in V - S$. The $l$-total $k$-dominating number $\gamma_{l,k}(G)$ is the minimum cardinality of an $l$-total $k$-dominating set of $G$.*

Clearly, $\gamma_{0,k}(G) = \gamma_k(G)$, $\gamma_{k-1,k}(G) = \gamma_{\times k}(G)$ and $\gamma_{k,k}(G) = \gamma_k^t(G)$. The first question raised by this definition is that of the existence of $l$-total $k$-dominating sets. If such a set $S$ exists, every vertex of degree $\delta$ belongs to $S$ or to $V - S$, thus implying $\delta \geq \min\{l, k\}$. Conversely, if $\delta \geq l$, then $V$ itself is a $l$-total $k$-dominating set. Therefore if $l \leq k$, then $l$-total $k$-dominating sets exist if and only if $\delta \geq l$. In particular, $\gamma_{k-1,k}(G)$ $(= \gamma_{\times k}(G))$ is defined if and only if $k \leq \delta + 1$ and $\gamma_{k,k}(G)$ $(= \gamma_k^t(G))$ is defined if and only if $k \leq \delta$. But if $k \leq \delta < l$, $l$-total $k$-dominating sets may or not exist. Another generalization of these domination parameters, which also contemplates restrained domination, is given in [130].

In the following theorem, the lower bound on $\gamma_{l,k}(G)$ is an immediate consequence of the definition.

**Theorem 4.2 ([61])** *If $\delta \geq \max\{l, k-1\}$, then*

$$\max\{l+1, k\} \leq \gamma_{l,k}(G) \leq \max\{n - \delta + l, n - \delta + k - 1\}.$$

Let $S$ be a minimum $l$-total $k$-dominating set and let $0 \leq l' \leq l$ and $1 \leq k' \leq k$. If $Y \subseteq S$ with $|Y| = \min\{l - l', k - k'\}$, then $S - Y$ is an $l'$-total $k'$-dominating set. Therefore

**Theorem 4.3 ([61])** *If $\gamma_{l,k}(G)$ exists, then $\gamma_{l',k'}(G)$ exists for every $l', k'$ with $0 \leq l' \leq l$, $1 \leq k' \leq k$ and*

$$\gamma_{l',k'}(G) \leq \gamma_{l,k}(G) - \min\{l - l', k - k'\}.$$

The following bounds on the $k$-tuple domination number are immediate from Theorems 4.2 and 4.3.

**Corollary 4.4** *Let $1 \leq k' \leq k$. If $G$ is a graph of order $n$ and with minimum degree $\delta \geq k$, then*

*(i)* $\gamma_{k'}(G) \leq \gamma_{\times k}(G) - k + k'$,
*(ii)* $\gamma_{\times k'}(G) \leq \gamma_{\times k}(G) - k + k'$,
*(iii)* $\gamma_{k'}^t(G) \leq \gamma_{\times k}(G) - k + k' + 1$, *if* $k' \neq k$,
*(iv)* $\gamma_{k'}^t(G) \leq \gamma_k^t(G) - k + k'$,
*(v)* $k \leq \gamma_k(G) \leq \gamma_{\times k}(G) \leq n - \delta + k - 1 \leq n$,
*(vi)* $k + 1 \leq \gamma_k^t(G) \leq n - \delta + 1 \leq n - k + 2$.

The following theorem, whose proof is very similar to the one of Theorem 2.8, was given by Kazemi in [104] and later again in [10]. The case $k = 2$ was given previously in [37].

**Theorem 4.5 ([10, 104])** *Let $G$ be a graph of order $n$ and minimum degree $\delta \geq k+1$. Then*

$$\gamma_{\times(k+1)}(G) \leq \frac{1}{2}(\gamma_k^t(G) + n).$$

A characterization of the graphs achieving equality was also given in [104]. The latter depends, however, on having determined, in certain cases, a minimum total $k$-dominating set with some particular structure.

We have also the following relations between the three different multiple domination parameters, all considered with the same multiplicity $k$.

**Theorem 4.6 ([10])** *Let $G$ be a graph with minimum degree $\delta \geq k+1$. Then we have the following upper bounds for the total $k$-domination number.*

*(i)* $\gamma_k^t(G) \leq (k+1)\gamma_k(G) - k(k-1)$;
*(ii)* $\gamma_k^t(G) \leq 2\gamma_{\times k}(G) - k + 1$.

The inequalities given in previous theorem were also proved in [18] for the case $k = 2$.

The following lemma is helpful in proving inequalities between the domination number of a graph and other domination parameters in claw-free graphs.

**Lemma 4.7 ([61])** *Let $X$ be a $k$-dominating set of a claw-free graph $G$ and let $D'$ be a dominating set of $G[X]$. Then there is a dominating set $D$ of $G$ with*

$$D' \subseteq D \quad \text{and} \quad |D| \leq \frac{|X| + (k-1)|D'|}{k}.$$

**Proof.** Let $A$ be a maximum independent set in $G[V(G) \setminus (X \cup N(D'))]$. Observe that, since $G$ is claw-free, the vertices in $X \setminus D'$ have at most one neighbor in $A$: indeed, if $x \in X \setminus D'$, $x$ has at least one neighbor in $D'$, which by construction is not adjacent to any vertex in $A$, and so it is easily seen that $x$ cannot have more than one neighbor in $A$. Together with the fact that every vertex in $A$ has at least $k$ neighbors in $X \setminus D'$, it follows that

$$k|A| \leq m(A, X \setminus D') \leq |X \setminus D'|.$$

Hence, $|A| \leq \frac{1}{k}|X \setminus D'|$. Now we see that $D = A \cup D'$ is a dominating set of $G$ with

$$|D| = |A \cup D'| \leq \frac{1}{k}|X \setminus D'| + |D'| = \frac{|X| + (k-1)|D'|}{k}.$$

■

**Corollary 4.8** *Suppose that, for all graphs $H$ of minimum degree $\delta(H) \geq l$, there is a bound $\gamma(H) \leq c \cdot n$ for some constant $c = c(l)$ depending on $l$. Then, given a claw-free graph $G$ of minimum degree $\delta \geq l$, we have*

$$\gamma(G) \leq \frac{1 + c \cdot (k-1)}{k} \gamma_{l,k}(G).$$

**Proof.** Take a minimum $l$-total $k$-dominating set $X$ of $G$, a minimum dominating set $D'$ of $H = G[X]$ and apply Lemma 4.7 together with the bound $|D'| = \gamma(H) \leq c \cdot |X| = c \cdot \gamma_{l,k}(G)$. ■

## *4.2 The Case k = 2*

As usual, there are many more results for $k$-tuple and total $k$-domination concerning the special case $k = 2$.

In [18], Bonomo, Brešar, Grippo, Milanič, and Safe compared many different kinds of double domination numbers that include the 2-domination, the double domination, and the total 2-domination. Among many other results, they gave the following relation between the double domination and the 2-domination numbers. The corresponding inequalities between total 2-domination and 2-domination and between total 2-domination and double domination were also given in [18] and are special cases of Theorem 4.6.

**Theorem 4.9 ([18])** *If G is a graph with minimum degree $\delta \geq 1$, then $\gamma_{\times 2}(G) \leq 2\gamma_2(G) - 1$.*

Observe that the inequality of the above theorem was proved previously for trees in [36]. Harary and Haynes [84] gave bounds on the double domination related to the domination number. They showed that if a graph has no isolates, then $\gamma_{\times 2}(G) \geq \gamma(G) + 1$. We will refine this result by considering the set of vertices of degree one. To see this, consider a connected graph $G$ of order $n \geq 3$ and with $l$ vertices of degree 1. Let $L$ be the set of vertices of degree 1 and let $D$ be a minimum double dominating set of $G$. Then, clearly, $L \cup N(L) \subseteq D$. Since $G$ is connected and $n \geq 3$, we have that $N(L) \cap L = \emptyset$, and we deduce that $D \setminus L$ is a dominating set. Hence, $\gamma(G) \leq \gamma_{\times 2}(G) - l$. To see that this bound is sharp, take any corona graph $J \circ K_1$. Observe that we are forced to assume that the graph is connected, because if there are $K_2$-components, we cannot distinguish between vertices of degree 1 and their neighbors as two disjoint sets. We state this fact, together with an upper and a lower bound for the double domination number of graphs of minimum degree at least 2 in terms of their domination number, which was also found in [84].

**Theorem 4.10** *Let G be a graph on $n \geq 3$ vertices and minimum degree $\delta \geq 1$.*

*(i) If G is connected and has l vertices of degree 1, then $\gamma_{\times 2}(G) \geq \gamma(G) + l$.*

*(ii) If $\delta \geq 2$, then* $\gamma(G) + 1 \leq \gamma_{\times 2}(G) \leq \begin{cases} \lfloor \frac{n}{2} \rfloor + \gamma(G), & \text{if } n = 3, 5 \\ \lfloor \frac{n}{2} \rfloor + \gamma(G) - 1, & \text{otherwise.} \end{cases}$ **([84])**

*Furthermore, all the bounds are sharp.*

Observe that the lower bound in Theorem 4.10 (ii) can be also obtained from Corollary 4.4 (i).

Harant and Henning [81] gave also probabilistic upper bounds for the double domination number that involve the domination and the total domination numbers.

Recall that the *paired domination number* $\gamma_{pr}(G)$ is the minimum cardinality of a dominating set whose induced subgraph admits a perfect matching. Clearly, a paired dominating set is also a total dominating set and thus the inequality $\gamma_t(G) \leq \gamma_{pr}(G)$ follows trivially. Similarly, we have the inequality $\gamma_t(G) \leq \gamma_{\times 2}(G)$. In [37], the double domination number is compared to the total domination number as well as to the paired domination number.

**Theorem 4.11 ([37])** *Let G be a graph of order n and minimum degree δ. Then we have the following bounds.*

*(i) If* $\gamma_{\times 2}(G) \neq 2$*, then* $\gamma_{\times 2}(G) \geq \frac{1}{2}(\gamma_{pr}(G) + 4) \geq \frac{1}{2}(\gamma_t(G) + 4)$.
*(ii) If* $\delta \geq 2$*, then* $\gamma_{\times 2}(G) \leq \frac{1}{2}(n + \gamma_t(G)) \leq \frac{1}{2}(n + \gamma_{pr}(G))$.

*Moreover, all inequalities are sharp.*

Clearly, the second inequality from item (i) and the second inequality from item (ii) follow from the trivial inequality $\gamma_t(G) \leq \gamma_{pr}(G)$. Observe also that the first inequality in item (ii) is the special case of Theorem 4.5 where $k = 2$.

The class of claw-free graphs has particularly interesting properties in domination theory. For these graphs it is known that

$$\gamma(G) = i(G)[1] \text{ and } \alpha(G) \leq 2i(G).[136] \tag{5}$$

This, combined with Theorem 3.14, gives $\gamma_{\times 2}(G) \leq 3\gamma(G)$ for claw-free graphs without isolates.

In the following theorem, we will state this together with other known relations between the double domination number and the domination the total domination and the paired domination numbers in claw-free graphs.

**Theorem 4.12** *Let G be a claw-free graph without isolates. Then we have the following inequalities.*

*(i)* $\frac{1}{3}\gamma_{\times 2}(G) \leq \gamma(G) \leq \frac{3}{4}\gamma_{\times 2}(G)$ (second inequality in [61]).
*(ii)* $\frac{1}{2}\gamma_{\times 2}(G) \leq \gamma_t(G) \leq \gamma_{pr}(G) \leq \gamma_{\times 2}(G)$ (first inequality in [12], third inequality in [37]).

**Proof.** (i) The first inequality was mentioned before stating this theorem. The second inequality follows from Corollary 4.8 using $l = 1$, $k = 2$, and Ore's inequality $\gamma(H) \leq \frac{1}{2}n(H)$ for graphs $H$ without isolates (Theorem 1.6).

(ii) Let $S$ be a minimum total dominating set of $G$. Let $X$ be the set of vertices in $V(G) \setminus S$ with exactly one neighbor in $S$. Let $S \cap N(X) = \{s_1, s_1, \ldots, s_r\}$ and let $X_i = N(s_i) \cap X$. Clearly, $X_1 \cup X_2 \ldots \cup X_r$ is a partition of $X$. Since $G$ is claw-free and every vertex $s_i$ has at least a neighbor in $S$, which by the definition of $X$ cannot have neighbors in $S$, it follows that $X_i$ is a clique, for

$1 \leq i \leq r$. Select a vertex $x_i \in X_i$ for each $i$. Then $S \cup \{x_1, x_2, \ldots, x_r\}$ is a double dominating set of $G$ and we have

$$\gamma_{\times 2}(G) \leq |S \cup \{x_1, x_2, \ldots, x_r\}| \leq 2|S| = 2\gamma_t(G),$$

so the bound follows.

To show that $\gamma_{pr}(G) \leq \gamma_{\times 2}(G)$, we consider a minimum double dominating set $D$ of $G$. Let $M$ be a maximum matching in $G[D]$ and let $S = D \setminus V(M)$. Clearly, $S$ is an independent set. Let $Q_1$ be the set of vertices in $x \in V(G) \setminus D$ with at least one neighbor in $V(M)$ and let $Q_2 = V \setminus (D \cup Q_1)$. Now consider a maximal independent set $I$ in $G[Q_2]$. Since $D$ is a double dominating set, every vertex in $S$ has a neighbor in $V(M)$ and every vertex in $I$ has two neighbors in $S$. Suppose there is a vertex $s \in S$ having two neighbors $x_1, x_2$ in $I$. Let $y \in V(M) \cap N(s)$. But we know that, since $x_1, x_2 \in I \subseteq Q_2$, neither $x_1$ nor $x_2$ can be adjacent to $y$. It follows that the set $\{s, x_1, x_2, y\}$ induces a claw, a contradiction. Hence, every vertex in $S$ has at most one neighbor in $I$. For each $x \in I$, select one neighbor $s_x \in S$. Then we have

$$|I| = |\{s_x \mid x \in I\}| \leq \frac{1}{2}|S|. \tag{6}$$

We define the set $D' = V(M) \cup \{s_x \mid x \in I\} \cup I$. We will show that $D'$ is a paired dominating set of $G$. By construction, $G[D']$ has a perfect matching. Let $v \in V(G) \setminus D'$. If $v \in Q_1 \cup (S \setminus D')$, then $v$ has a neighbor in $V(M) \subseteq D'$. If $v \in Q_2 \setminus I$, then $v$ has a neighbor in $I$ because $I$ is maximal independent in $G[Q_2]$. Hence, $D'$ is a paired dominating set and, using inequality (6), we obtain

$$\gamma_{pr}(G) \leq |D'| = |V(M)|+|\{s_x \mid x \in I\}|+|I| \leq |D \setminus S|+\frac{1}{2}|S|+\frac{1}{2}|S| = |D| = \gamma_{\times 2}(G).$$

■

Concerning the inequality $\gamma(G) \leq \frac{3}{4}\gamma_{\times 2}(G)$, the authors of [61] prove it actually for a wider family of graphs, namely the family of graphs whose blocks are all claw-free. For the inequality $\gamma_{pr}(G) \leq \gamma_{\times 2}(G)$ in claw-free graphs there is also the following extension to $K_{1,r}$-free graphs.

**Theorem 4.13 ([52])** *Let $r \geq 2$ be an integer. If $G$ is a connected $K_{1,r}$-free graph of order at least 2, then*

$$\gamma_{pr}(G) \leq \frac{2r^2 - 6r + 6}{r(r-1)}\gamma_{\times 2}(G).$$

There are also many special results concerning trees. It has been shown that, like for claw-free graphs, the double domination number is at least as large as the paired domination number in trees [13]. In the following theorem, we gather such comparisons between different domination parameters in trees.

**Theorem 4.14** *Let $T$ be a tree on $n \geq 2$ vertices, within $\ell$ leaves and $s$ support vertices. The following inequalities hold.*

*(i)* $\gamma(T) + \ell \leq \gamma_{\times 2}(T) \leq 2\gamma(T) + \ell - 1$. **[14]**
*(ii)* $\gamma_t(T) \leq \gamma_2(T)$. **[86]**
*(iii)* $\gamma_{pr}(T) \leq \gamma_{\times 2}(T)$. **[13]**
*(iv)* $2i(T) \leq \gamma_{\times 2}(T)$. **[14]**
*(v)* *If $n \geq 4$, then* $\gamma_{\times 2}(T) \leq 2\gamma_2(T) - 2$. **[36]**
*(vi)* *If $n \geq 3$, then* $\gamma_t(T) \leq \gamma_{\times 2}(T) - 1$. **[110]**

From the obvious $\gamma(T) \leq i(T)$, (iii) implies $\gamma_{\times 2}(T) \leq 2\gamma(T)$. It was shown in [14] that $\gamma_{\times 2}(T) = 2i(T)$ if and only if $T$ has two disjoint minimum independent dominating sets and that $\gamma_{\times 2}(T) = 2\gamma(T)$ if and only if $T$ has two disjoint minimum dominating sets. In [110], the trees $T$ with $\gamma_{\times 2}(T) = \gamma_t(T) + 1$ are characterized.

# 5 Open Problems and Final Remarks

In this final section, we discuss the state of the art of several problems concerning $k$-domination, $k$-tuple domination, and total $k$-domination, and gather some open problems as well as ideas for future research.

## *5.1 Bounds on Order and Degree*

One of the most interesting questions concerning domination parameters is to find bounds in terms of the order of the graph, say $n$. With the weakest reasonable condition on the minimum degree, namely that $\delta \geq k$, the bound $\gamma_k(G) \leq \frac{k}{k+1}n$ was given by Cockayne, Gamble, and Shepherd [46] (Theorem 2.12) and an infinite family attains the equality (Theorem 2.13). It would be interesting to find similar bounds on the $k$-tuple domination and the total $k$-domination numbers.

**Problem 5.1** *Find functions $f(k)$, $g(k)$ such that $\gamma_{\times k}(G) \leq f(k)n$ and $\gamma_k^t(G) \leq g(k)n$ for any graph $G$ of order $n$ and minimum degree $\delta \geq k$.*

As we have mentioned, letting grow the minimum degree has the effect of obtaining better upper bounds on these domination parameters. A testimony of this fact is all the probabilistic bounds that we have cited in this work, from which Theorem 1.8 is the one representative for all three parameters $\gamma_k$, $\gamma_{\times k}$, and $\gamma_k^t$. Although these probabilistic bounds give a good shade of the behavior of these domination parameters for graphs of large order and large minimum degree, they do not give light on the proportion of vertices that is needed for a $k$-dominating set, a $k$-tuple dominating set or a total $k$-dominating set when dealing with smaller but yet growing minimum degrees or when general structural constraints of the graph may be taken into account. Results in this direction are those on $k$-domination given in Corollary 4.4 (that are consequences of known lower bounds for the independence number) and Theorem 2.20, the $\frac{3}{4}n$-bound on the double domination number given for graphs $G \neq C_5$ with minimum degree at least 2 (Theorem 3.7) and the bounds on the total 2-domination number giving a $\frac{5}{6}n$-upper bound for cubic graphs $G$ different from the Heawood graph (Theorem 3.10), and a $\frac{11}{13}n$-upper bound for connected graphs $G$ with minimum degree at least 3 and which are different from the Heawood graph (3.11). However, there are still interesting research paths to pursue. Most intriguing could be to find sharp upper bounds for the 2-domination number in terms of the order of the graph when the minimum degree is quite small, which for strange reasons has not been studied yet.

**Problem 5.2** *Find constants $c_1, c_2$ such that*

*(i) $\gamma_2(G) \leq c_1 n$ for any graph $G$ on $n$ vertices and minimum degree $\delta \geq 3$, and*
*(ii) $\gamma_2(G) \leq c_2 n$ for any cubic graph $G$ on $n$ vertices.*

Of course, the problem given above can be extended to graphs of minimum degree at least 4 or 5 (for $6 \leq \delta \leq 21$, we have already the result in Theorem 2.20, although this may be improved, too), regular graphs and other graph families like claw-free graphs, triangle-free graphs and others. The family of bipartite graphs has also shown to have nice properties. In this context, Corollary 2.33, stating that $\gamma_k(G) \leq \frac{1}{2}n$ for any bipartite graph $G$ of order $n$ and $\delta(G) \geq k$, seems to be a good candidate for the study of the extremal family.

**Problem 5.3** *Characterize all bipartite graphs $G$ of order $n$ and $\delta(G) \geq k$ with $\gamma_k(G) = \frac{1}{2}n$.*

Also very interesting are the following two conjectures stated in [94].

**Conjecture 5.4 ([94])** *Let $G$ be a connected graph of order $n$ and minimum degree $\delta \geq 3$ such that $G$ is not the Heawood graph. Then $\gamma_2^t(G) \leq \frac{5}{6}n$.*

**Conjecture 5.5 ([94])** *Let $G$ be a connected graph of order $n$ and minimum degree $\delta \geq 3$. If $n$ is sufficiently large, then $\gamma_2^t(G) \leq \frac{3}{4}n$.*

## 5.2 Multiple Domination Numbers for Different k's

We started Section 2 on $k$-domination talking about the problem of finding a function $f$ such that $\gamma_k(G) < \gamma_{f(k)}(G)$ for every graph $G$ with $\delta(G) \geq k$ (see Problem 2.2), and some partial results have been found (see Theorems 2.1 and 2.3). The same question can be stated for $\gamma_k^t(G)$ and $\gamma_{\times k}(G)$.

**Problem 5.6** *Find functions $f, g, h$ such that*

(i) $\gamma_k(G) < \gamma_{f(k)}(G)$ *for every graph $G$ with $\delta(G) \geq k$;*
(ii) $\gamma_k^t(G) < \gamma_{g(k)}(G)$ *for every graph $G$ with $\delta(G) \geq k - 1$;*
(iii) $\gamma_{\times k}(G) < \gamma_{\times h(k)}(G)$ *for every graph $G$ with $\delta(G) \geq k$;*

In Section 2.1, we proved the inequality $\gamma_k(G) \geq \gamma(G) + k - 2$ (Theorem 2.4) and gave a series of different results discussing when equality is attained. From Corollary 4.4, we can deduce the corresponding inequalities for $k$-tuple and total $k$-domination by setting $k' = 1$, which are $\gamma_{\times k}(G) \geq \gamma(G) + k - 1$ and $\gamma_k^t(G) \geq \gamma_t(G)+k-1$, from which nothing is known concerning conditions for their equality. The special case $k = 2$ for double domination was already discussed in [84], where it is shown that the inequality is attained for the complete bipartite graph $K_{2,t}$. However, there is no characterization provided, which could be an interesting problem to settle. We gather these questions below.

**Problem 5.7**

(i) *Find further families of graphs $G$ with $\gamma_2(G) = \gamma(G)$ (for instance, outerplanar graphs, diamond-free graphs, etc.).* **[73]**
(ii) *Characterize the $\{K_{1,4}, K_{1,3}+e\}$-free graphs $G$ with $\gamma_3(G) = \gamma(G) + 1$.* **[73]**
(iii) *Characterize the families of graphs $G$ such that $\gamma_{\times 2}(G) = \gamma(G) + 1$ or such that $\gamma_2^t(G) = \gamma_t(G) + 1$.*
(iv) *Find necessary and/or sufficient conditions for a graph $G$ having $\gamma_k(G) = \gamma(G) + k - 2$* **[73]**, *$\gamma_{\times k}(G) = \gamma(G) + k - 1$, or $\gamma_k^t(G) = \gamma_t(G) + k - 1$.*

## 5.3 *Relations Between Parameters of Different Kind*

There are some results that have been given for the case that $k = 2$ that may be generalized to arbitrary $k$. An intriguing relation between double domination and independence was the one given in Theorem 3.14, stating that $\gamma_{\times 2}(G) \leq i(G) + \alpha(G)$ for any graph $G$ without isolates. Also, the Nordhaus–Gaddum bound $\gamma_{\times 2}(G) + \gamma_{\times 2}(\overline{G}) \leq n + 4$ given in Theorem 3.21, for graphs $G$ with $\delta(G) \geq 1$, $\delta(\overline{G}) \geq 1$ and $\mathrm{diam}(G) > 2$ or $\mathrm{diam}(\overline{G}) > 2$, could have an extension for any $k$, or could have its analogon for $\gamma_k$. Finally, a generalization to arbitrary $k$ of Theorem 4.9 that states that $\gamma_{\times 2}(G) \leq 2\gamma_2(G) - 1$ for graphs $G$ with no isolates could be interesting, too.

**Problem 5.8** *Give an upper bound for $\gamma_{\times k}(G)$ in terms of $\gamma_k(G)$ for any graph $G$ of minimum degree $\delta \geq k - 1$.*

Recall Theorem 2.25 stating that $\gamma_k(G) \geq \alpha(G)$ for a $K_{1,k+1}$-free graph $G$, and, on the other hand, look at Theorem 3.13 that gives the bound $\gamma_{\times k}(G) \geq \frac{k}{r-1}\alpha(G)$ for a $K_{1,r}$-free graph $G$ with $\delta(G) \geq k - 1$. So one can naturally ask if the first inequality can be extended for $K_{1,r}$-free graphs, too.

**Problem 5.9** *Find a function $f(k, r)$ such that $\gamma_k(G) \geq f(k, r)\alpha(G)$ for any $K_{1,r}$-free graph $G$.*

Another inequality that involves $k$-domination and independence is $\gamma_k(G) \leq (\Delta - 1)\alpha(G)$ for a connected nontrivial graph $G$ with maximum degree $\Delta$ and $\delta(G) \geq k$, where $G$ is neither isomorphic to a cycle of odd length when $k = 2$ nor to the complete graph $K_{k+1}$, that was given in Theorem 2.34. It is somehow peculiar that the proof of this result can be only used to show that the non-regular graphs attaining equality are isomorphic to $K_2 \circ K_k$, while for the regular graphs nothing is concluded. However, it is easy to see that even cycles attain equality for $k = 2$ as well as the complete graph $K_{k+2}$ for every $k \geq 2$. It would be interesting to know if there are other $r$-regular graphs $G$ with $\gamma_k(G) = (r-1)\alpha(G)$. Moreover, the bound does not seem to be very good since it appears to be attained only by a finite family of graphs for each $k$. Hence, an improvement of this bound would be desirable.

Finally, we would like to say a word about vertex covers. A *vertex cover $S$* in a graph $G$ is a set of vertices such that every edge of $G$ is incident to at least one vertex of $S$. The minimum cardinality of a vertex cover in $G$ is denoted $\beta(G)$ and is called the *covering number* of $G$. If $G$ has no isolates, then it is straightforward to see that $\gamma(G) \leq \beta(G)$, since the complement of a vertex cover is an independent set. The graphs $G$ with $\gamma(G) \leq \beta(G)$ have been characterized in [141] and studied again in [116]. If, moreover, $\delta \geq k$, then any minimum vertex cover is a $k$-tuple dominating set, and we have $\gamma_{\times k}(G) \leq \beta(G)$. On the other hand, if $\delta \geq k+1$, then any minimum vertex cover is a total $k$-dominating set, implying that $\gamma_k^t(G) \leq \beta(G)$.The following problems arise naturally.

**Problem 5.10**

*(i) Characterize the family of graphs $G$ with minimum degree $\delta \geq k$ and $\gamma_{\times k}(G) = \beta(G)$.*
*(ii) Characterize the family of graphs $G$ with minimum degree $\delta \geq k + 1$ and $\gamma_k^t(G) = \beta(G)$.*

## 5.4 Cartesian Products of Graphs

We discussed in Section 3.4 results on $k$-tuple and total $k$-domination of cartesian products of graphs. The upper bounds in Theorem 3.22 are interesting but should be subject to further study. In particular, it would be interesting to know if there is a function $f(k) \geq k$ and an infinite family of graphs $H$ of minimum degree $\delta \geq k$ with $\gamma_k^t(H) \leq f(k)\rho(H)$, since, as mentioned in the discussion after Theorem 3.22, this would give an upper bound $\gamma_k^t(G)\gamma_k^t(H) \leq f(k)\gamma_k^t(G\Box H)$.

As to $k$-domination on cartesian products, not many results can be cited up to now. In [127], the 2-domination number of grid graphs is determined, while in [26] the $k$-domination number of the rook's graph $K_n \Box K_n$ is determined. Further development on the theory of $k$-domination of cartesian products of graphs would be desirable to pursue. In particular, 2-domination would be of considerable interest in all manner of chessboard problems, where the basic objective is to not only dominate squares of the board, but dominate them more than once. Thus, for example, queens 2-domination would be a very interesting problem to consider.

## 5.5 *Miscellaneous*

There are many other variants of multiple domination that we do not mention in this chapter like connected $k$-(tuple/total) domination [68, 70, 145], dynamical $k$-domination [55], (total) $\{k\}$-domination [3, 4, 18], $f$-domination [135, 151], $(k, l)$-domination and $[r, s]$-domination [153]. As it is extensively studied for standard domination, one could consider multiple independent domination, as well as multiple distance domination parameters, too.

A *greedy coloring* is defined by repeatedly removing from a graph maximal independent set. This produces a proper coloring into, say, $c$ classes, $V_1, V_2, \ldots, V_c$. The *Grundy number* $\Gamma(G)$ of a graph $G$ is the largest $c$ such that $G$ has a greedy $c$-coloring (see, for instance, [6, 27]). Now, observe that $V_1 \cup V_2 \cup \ldots \cup V_k$ is a $k$-dominating set. Hence, there surely are nice relations between the Grundy number and the $k$-domination number of a graph $G$.

A *domatic partition* of a graph $G$ is a partition of its vertices into dominating sets. The maximum number of sets in a domatic partition of $G$ is its *domatic number*, denoted by $d(G)$ (see, for instance, [49, 150]). Domatic partitions could also have interesting relations with $k$-domination, since any domatic partition of a graph $G$ provides $\binom{d(G)}{k}$ $k$-dominating sets for any $k < d(G)$. Also note that $k$-domatic partitions, i.e., such that every set on the partition is $k$-dominating, together with other variants, have been studied in [100, 133, 147], to cite some.

Finally, we would like to mention that there is still little research done on multiple domination and random graphs. In [148], Wang and Xiang were able to prove that the 2-tuple domination number of the random graph $G(n, p)$ with fixed $p \in (0, 1)$ has a.a.s. a two-point concentration. It would be interesting to delve deeper into the matter with respect to $k$-domination, $k$-tuple, or total $k$-domination.

**Acknowledgments** We would like to thank the anonymous referee, whose comments and suggestions helped improve considerably the contents and style of this chapter. The first author was partially supported by PAPIIT IN111819 and CONACYT project 282280.

## References

1. Robert B. Allan and Renu Laskar, *On domination and independent domination numbers of a graph*, Discrete Math. **23** (1978), no. 2, 73–76. MR 523402
2. Noga Alon, *Transversal numbers of uniform hypergraphs*, Graphs Combin. **6** (1990), no. 1, 1–4. MR 1058542

3. G. Argiroffo, V. Leoni, and P. Torres, *Complexity of $k$-tuple total and total $\{k\}$-dominations for some subclasses of bipartite graphs*, Inform. Process. Lett. **138** (2018), 75–80. MR 3826535
4. Gabriela Argiroffo, Valeria Leoni, and Pablo Torres, *On the complexity of $\{k\}$-domination and $k$-tuple domination in graphs*, Inform. Process. Lett. **115** (2015), no. 6–8, 556–561. MR 3327090
5. V. I. Arnautov, *Estimation of the exterior stability number of a graph by means of the minimal degree of the vertices*, Prikl. Mat. i Programmirovanie (1974), no. 11, 3–8, 126. MR 0340103
6. Marie Asté, Frédéric Havet, and Claudia Linhares-Sales, *Grundy number and products of graphs*, Discrete Math. **310** (2010), no. 9, 1482–1490. MR 2601252
7. Sambhu Charan Barman, Sukumar Mondal, and Madhumangal Pal, *Minimum 2-tuple dominating set of permutation graphs*, J. Appl. Math. Comput. **43** (2013), no. 1–2, 133–150. MR 3096395
8. Timothy J. Bean, Michael A. Henning, and Henda C. Swart, *On the integrity of distance domination in graphs*, Australas. J. Combin. **10** (1994), 29–43. MR 1296938
9. S. Bermudo, J. L. Sanchéz, and J. M. Sigarreta, *Total $k$-domination in Cartesian product graphs*, Period. Math. Hungar. **75** (2017), no. 2, 255–267. MR 3718519
10. Sergio Bermudo, Juan C. Hernández-Gómez, and José M. Sigarreta, *On the total $k$-domination in graphs*, Discuss. Math. Graph Theory **38** (2018), no. 1, 301–317. MR 3743968
11. Mostafa Blidia, Mustapha Chellali, and Odile Favaron, *Independence and 2-domination in trees*, Australas. J. Combin. **33** (2005), 317–327. MR 2170368
12. ———, *Ratios of some domination parameters in graphs and claw-free graphs*, Graph theory in Paris, Trends Math., Birkhäuser, Basel, 2007, pp. 61–72. MR 2279167
13. Mostafa Blidia, Mustapha Chellali, and Teresa W. Haynes, *Characterizations of trees with equal paired and double domination numbers*, Discrete Math. **306** (2006), no. 16, 1840–1845. MR 2251565
14. Mostafa Blidia, Mustapha Chellali, Teresa W. Haynes, and Michael A. Henning, *Independent and double domination in trees*, Util. Math. **70** (2006), 159–173. MR 2238438
15. Mostafa Blidia, Mustapha Chellali, and Lutz Volkmann, *Bounds of the 2-domination number of graphs*, Util. Math. **71** (2006), 209 216. MR 2278833
16. ———, *Bounds of the 2-domination number of graphs*, Util. Math. **71** (2006), 209–216. MR 2278833
17. ———, *Some bounds on the $p$-domination number in trees*, Discrete Math. **306** (2006), no. 17, 2031–2037. MR 2251821
18. Flavia Bonomo, Boštjan Brešar, Luciano N. Grippo, Martin Milanič, and Martín D. Safe, *Domination parameters with number 2: interrelations and algorithmic consequences*, Discrete Appl. Math. **235** (2018), 23–50. MR 3732593
19. Gülnaz Boruzanlı Ekinci and Csilla Bujtás, *On the equality of domination number and 2-domination number*, https://arxiv.org/abs/1907.07866.
20. Christoph Brause, Michael A. Henning, and Marcin Krzywkowski, *A characterization of trees with equal 2-domination and 2-independence numbers*, Discrete Math. Theor. Comput. Sci. **19** (2017), no. 1, Paper No. 1, 14. MR 3626982
21. Boštjan Brešar, Paul Dorbec, Wayne Goddard, Bert L. Hartnell, Michael A. Henning, Sandi Klavžar, and Douglas F. Rall, *Vizing's conjecture: a survey and recent results*, J. Graph Theory **69** (2012), no. 1, 46–76. MR 2864622
22. R. L. Brooks, *On colouring the nodes of a network*, Proc. Cambridge Philos. Soc. **37** (1941), 194–197. MR 12236
23. Csilla Bujtás and Szilárd Jaskó, *Bounds on the 2-domination number*, Discrete Applied Mathematics **242** (2018), 4–15.
24. Paul A. Burchett, *On the border queens problem and $k$ tuple domination on the rook's graph*, Proceedings of the Forty-Second Southeastern International Conference on Combinatorics, Graph Theory and Computing, vol. 209, 2011, pp. 179–187. MR 2856346
25. ———, *$k$-tuple domination on the bishop's graph*, Util. Math. **101** (2016), 351–358. MR 3585669

26. Paul A. Burchett, David Lane, and Jason A. Lachniet, *k-tuple and k-domination on the rook's graph and other results*, Proceedings of the Fortieth Southeastern International Conference on Combinatorics, Graph Theory and Computing, vol. 199, 2009, pp. 187–204. MR 2584370
27. Victor Campos, András Gyárfás, Frédéric Havet, Claudia Linhares Sales, and Frédéric Maffray, *New bounds on the Grundy number of products of graphs*, J. Graph Theory **71** (2012), no. 1, 78–88. MR 2959289
28. Y. Caro and Y. Roditty, *A note on the k-domination number of a graph*, Internat. J. Math. Math. Sci. **13** (1990), no. 1, 205–206. MR 1038667
29. Yair Caro, *On the k-domination and k-transversal numbers of graphs and hypergraphs*, Ars Combin. **29** (1990), no. C, 49–55, Twelfth British Combinatorial Conference (Norwich, 1989). MR 1412836
30. Yair Caro and Adriana Hansberg, *New approach to the k-independence number of a graph*, Electron. J. Combin. **20** (2013), no. 1, Paper 33, 17. MR 3035043
31. Yair Caro and Raphael Yuster, *Dominating a family of graphs with small connected subgraphs*, Combin. Probab. Comput. **9** (2000), no. 4, 309–313. MR 1786921
32. Gerard J. Chang, *The upper bound on k-tuple domination numbers of graphs*, European J. Combin. **29** (2008), no. 5, 1333–1336. MR 2419234
33. Mustapha Chellali, *Bounds on the 2-domination number in cactus graphs*, Opuscula Math. **26** (2006), no. 1, 5–12. MR 2272278
34. Mustapha Chellali, Odile Favaron, Adriana Hansberg, and Lutz Volkmann, *On the p-domination, the total domination and the connected domination numbers of graphs*, J. Combin. Math. Combin. Comput. **73** (2010), 65–75. MR 2657315
35. ———, *k-domination and k-independence in graphs: a survey*, Graphs Combin. **28** (2012), no. 1, 1–55. MR 2863534
36. Mustapha Chellali, Odile Favaron, Teresa W. Haynes, and Dalila Raber, *Ratios of some domination parameters in trees*, Discrete Math. **308** (2008), no. 17, 3879–3887. MR 2418091
37. Mustapha Chellali and Teresa W. Haynes, *On paired and double domination in graphs*, Util. Math. **67** (2005), 161–171. MR 2137931
38. Mustapha Chellali, Abdelkader Khelladi, and Frédéric Maffray, *Exact double domination in graphs*, Discuss. Math. Graph Theory **25** (2005), no. 3, 291–302. MR 2232995
39. Mustapha Chellali and Nacéra Meddah, *Trees with equal 2-domination and 2-independence numbers*, Disc. Math. Graph Theory **32** (2012), no. 2, 263.
40. Beifang Chen and Sanming Zhou, *Upper bounds for f-domination number of graphs*, Discrete Math. **185** (1998), no. 1–3, 239–243. MR 1614254
41. Guantao Chen and Michael S. Jacobson, *On a relationship between 2-dominating and 5-dominating sets in graphs*, J. Combin. Math. Combin. Comput. **39** (2001), 139–145. MR 1865852
42. Xue-gang Chen and Liang Sun, *Some new results on double domination in graphs*, J. Math. Res. Exposition **25** (2005), no. 3, 451–456. MR 2163723
43. Nina Chiarelli, Tatiana Romina Hartinger, Valeria Alejandra Leoni, Maria Inés Lopez Pujato, and Martin Milanič, *Improved algorithms for k-domination and total k-domination in proper interval graphs*, Combinatorial optimization, Lecture Notes in Comput. Sci., vol. 10856, Springer, Cham, 2018, pp. 290–302. MR 3835950
44. Ferdinando Cicalese, Martin Milanič, and Ugo Vaccaro, *On the approximability and exact algorithms for vector domination and related problems in graphs*, Discrete Appl. Math. **161** (2013), no. 6, 750–767. MR 3027964
45. W. Edwin Clark and Stephen Suen, *An inequality related to Vizing's conjecture*, Electron. J. Combin. **7** (2000), Note 4, 3. MR 1763970
46. E. J. Cockayne, B. Gamble, and B. Shepherd, *An upper bound for the k-domination number of a graph*, J. Graph Theory **9** (1985), no. 4, 533–534. MR 890244
47. E. J. Cockayne and A. G. Thomason, *An upper bound for the k-tuple domination number*, J. Combin. Math. Combin. Comput. **64** (2008), 251–254. MR 2389082
48. Arnel Marino Cuivillas and Sergio R. Canoy, Jr., *Double domination in the Cartesian and tensor products of graphs*, Kyungpook Math. J. **55** (2015), no. 2, 279–287. MR 3367944

49. Peter Dankelmann and Neil Calkin, *The domatic number of regular graphs*, Ars Combin. **73** (2004), 247–255. MR 2098768
50. M. P. Dobson, V. Leoni, and G. Nasini, *The multiple domination and limited packing problems in graphs*, Inform. Process. Lett. **111** (2011), no. 23–24, 1108–1113. MR 2893943
51. Paul Dorbec, Sylvain Gravier, Sandi Klavžar, and Simon Špacapan, *Some results on total domination in direct products of graphs*, Discuss. Math. Graph Theory **26** (2006), no. 1, 103–112. MR 2230822
52. Paul Dorbec, Bert Hartnell, and Michael A. Henning, *Paired versus double domination in $K_{1,r}$-free graphs*, J. Comb. Optim. **27** (2014), no. 4, 688–694. MR 3181748
53. P. Erdős, *On some extremal problems in graph theory*, Israel J. Math. **3** (1965), 113–116. MR 0190027
54. P. Erdős and R. Rado, *Intersection theorems for systems of sets*, J. London Math. Soc. **35** (1960), 85–90. MR 111692
55. Ch. Eslahchi, H. R. Maimani, R. Torabi, and R. Tusserkani, *Dynamical 2-domination in graphs*, Ars Combin. **134** (2017), 339–350. MR 3677172
56. S. Fajtlowicz, *The independence ratio for cubic graphs*, (1977), 273–277. Congressus Numerantium, No. XIX. MR 0491297
57. Odile Favaron, *On a conjecture of Fink and Jacobson concerning k-domination and k-dependence*, J. Combin. Theory Ser. B **39** (1985), no. 1, 101–102. MR 805459
58. ———, *k-domination and k-independence in graphs*, Ars Combin. **25** (1988), no. C, 159–167, Eleventh British Combinatorial Conference (London, 1987). MR 943386
59. ———, *Graduate course in the University of Blida*, unpublished (2005).
60. Odile Favaron, Adriana Hansberg, and Lutz Volkmann, *On k-domination and minimum degree in graphs*, J. Graph Theory **57** (2008), no. 1, 33–40. MR 2370182
61. Odile Favaron, Michael A. Henning, Joël Puech, and Dieter Rautenbach, *On domination and annihilation in graphs with claw-free blocks*, Discrete Math. **231** (2001), no. 1–3, 143–151, 17th British Combinatorial Conference (Canterbury, 1999). MR 1821955
62. J. F. Fink, M. S. Jacobson, L. F. Kinch, and J. Roberts, *On graphs having domination number half their order*, Period. Math. Hungar. **16** (1985), no. 4, 287–293. MR 833264
63. John Frederick Fink and Michael S. Jacobson, *n-domination in graphs*, Graph theory with applications to algorithms and computer science (Kalamazoo, Mich., 1984), Wiley-Intersci. Publ., Wiley, New York, 1985, pp. 283–300. MR 812671
64. ———, *On n-domination, n-dependence and forbidden subgraphs*, Graph theory with applications to algorithms and computer science (Kalamazoo, Mich., 1984), Wiley-Intersci. Publ., Wiley, New York, 1985, pp. 301–311. MR 812672
65. Jun Fujisawa, Adriana Hansberg, Takahiro Kubo, Akira Saito, Masahide Sugita, and Lutz Volkmann, *Independence and 2-domination in bipartite graphs*, Australas. J. Combin. **40** (2008), 265–268. MR 2381434
66. Andrei Gagarin and Vadim E. Zverovich, *A generalised upper bound for the k-tuple domination number*, Discrete Math. **308** (2008), no. 5–6, 880–885. MR 2378922
67. Wayne Goddard and Michael A. Henning, *Restricted domination parameters in graphs*, J. Comb. Optim. **13** (2007), no. 4, 353–363. MR 2308225
68. Adriana Hansberg, *Bounds on the connected k-domination number in graphs*, Discrete Appl. Math. **158** (2010), no. 14, 1506–1510. MR 2659165
69. ———, *On the k-domination number, the domination number and the cycle of length four*, Util. Math. **98** (2015), 65–76. MR 3410883
70. Adriana Hansberg, Dirk Meierling, and Lutz Volkmann, *A general method in the theory of domination in graphs*, Int. J. Comput. Math. **87** (2010), no. 13, 2915–2924. MR 2754237
71. ———, *Independence and k-domination in graphs*, Int. J. Comput. Math. **88** (2011), no. 5, 905–915. MR 2775086
72. Adriana Hansberg and Ryan Pepper, *On k-domination and j-independence in graphs*, Discrete Appl. Math. **161** (2013), no. 10–11, 1472–1480. MR 3043099
73. Adriana Hansberg, Bert Randerath, and Lutz Volkmann, *Claw-free graphs with equal 2-domination and domination numbers*, Filomat **30** (2016), no. 10, 2795–2801. MR 3583404

74. Adriana Hansberg and Lutz Volkmann, *Characterization of block graphs with equal 2-domination number and domination number plus one*, Discuss. Math. Graph Theory **27** (2007), no. 1, 93–103. MR 2321425
75. ———, *Characterization of unicyclic graphs with equal 2-domination number and domination number plus one*, Util. Math. **77** (2008), 265–276. MR 2462645
76. ———, *On graphs with equal domination and 2-domination numbers*, Discrete Math. **308** (2008), no. 11, 2277–2281. MR 2404554
77. ———, *Lower bounds on the p-domination number in terms of cycles and matching number*, J. Combin. Math. Combin. Comput. **68** (2009), 245–255. MR 2494123
78. ———, *Upper bounds on the k-domination number and the k-Roman domination number*, Discrete Appl. Math. **157** (2009), no. 7, 1634–1639. MR 2510244
79. ———, *On 2-domination and independence domination numbers of graphs*, Ars Combin. **101** (2011), 405–415. MR 2838924
80. Jochen Harant and Michael A. Henning, *On double domination in graphs*, Discuss. Math. Graph Theory **25** (2005), no. 1–2, 29–34. MR 2152046
81. ———, *A realization algorithm for double domination in graphs*, Util. Math. **76** (2008), 11–24. MR 2428932
82. Jochen Harant, Anja Pruchnewski, and Margit Voigt, *On dominating sets and independent sets of graphs*, Combin. Probab. Comput. **8** (1999), no. 6, 547–553. MR 1741406
83. Frank Harary and Teresa W. Haynes, *Nordhaus-Gaddum inequalities for domination in graphs*, vol. 155, 1996, Combinatorics (Acireale, 1992), pp. 99–105. MR 1401362
84. ———, *Double domination in graphs*, Ars Combin. **55** (2000), 201–213. MR 1755232
85. Teresa W. Haynes, Stephen T. Hedetniemi, and Michael A. Henning, *Domination in graphs: Major topics*, Springer, 2020.
86. Teresa W. Haynes, Stephen T. Hedetniemi, Michael A. Henning, and Peter J. Slater, *H-forming sets in graphs*, Discrete Math. **262** (2003), no. 1–3, 159–169. MR 1951385
87. Christopher Carl Heckman and Robin Thomas, *Independent sets in triangle-free cubic planar graphs*, J. Combin. Theory Ser. B **96** (2006), no. 2, 253–275. MR 2208354
88. Michael A. Henning, *Graphs with large double domination numbers*, Discuss. Math. Graph Theory **25** (2005), no. 1–2, 13–28. MR 2152045
89. Michael A. Henning and Adel P. Kazemi, *k-tuple total domination in graphs*, Discrete Appl. Math. **158** (2010), no. 9, 1006–1011. MR 2607047
90. ———, *k-tuple total domination in cross products of graphs*, J. Comb. Optim. **24** (2012), no. 3, 339–346. MR 2970502
91. Michael A. Henning and Douglas F. Rall, *On the total domination number of Cartesian products of graphs*, Graphs Combin. **21** (2005), no. 1, 63–69. MR 2136709
92. Michael A. Henning and Anders Yeo, *Hypergraphs with large transversal number and with edge sizes at least 3*, J. Graph Theory **59** (2008), no. 4, 326–348. MR 2463184
93. ———, *Total domination in graphs with given girth*, Graphs Combin. **24** (2008), no. 4, 333–348. MR 2438865
94. ———, *Strong transversals in hypergraphs and double total domination in graphs*, SIAM J. Discrete Math. **24** (2010), no. 4, 1336–1355. MR 2735927
95. Pak Tung Ho, *A note on the total domination number*, Util. Math. **77** (2008), 97–100. MR 2462630
96. Glenn Hopkins and William Staton, *Vertex partitions and k-small subsets of graphs*, Ars Combin. **22** (1986), 19–24. MR 867729
97. Michael S. Jacobson and Ken Peters, *Complexity questions for n-domination and related parameters*, vol. 68, 1989, Eighteenth Manitoba Conference on Numerical Mathematics and Computing (Winnipeg, MB, 1988), pp. 7–22. MR 995851
98. Michael S. Jacobson, Kenneth Peters, and Douglas F. Rall, *On n-irredundance and n-domination*, Ars Combin. **29** (1990), no. B, 151–160, Twelfth British Combinatorial Conference (Norwich, 1989). MR 1412871
99. Nader Jafari Rad, *Upper bounds on the k-tuple domination number and k-tuple total domination number of a graph*, Australas. J. Combin. **73** (2019), 280–290. MR 3900544

100. Karsten Kämmerling and Lutz Volkmann, *The k-domatic number of a graph*, Czechoslovak Math. J. **59(134)** (2009), no. 2, 539–550. MR 2532389
101. A. P. Kazemi, *k-tuple total domination and Mycieleskian graphs*, Trans. Comb. **1** (2012), no. 1, 7–13. MR 3145625
102. Adel P. Kazemi, *On the total k-domination number of graphs*, Discuss. Math. Graph Theory **32** (2012), no. 3, 419–426. MR 2974027
103. ———, *k-tuple total domination in inflated graphs*, Filomat **27** (2013), no. 2, 341–351. MR 3287383
104. ———, *A note on the k-tuple total domination number of a graph*, Tbilisi Math. J. **8** (2015), no. 2, 281–286. MR 3441142
105. ———, *Upper k-tuple total domination in graphs*, Pure Appl. Math. Q. **13** (2017), no. 4, 563–579. MR 3903060
106. Adel P. Kazemi, Behnaz Pahlavsay, and Rebecca J. Stones, *Cartesian product graphs and k-tuple total domination*, Filomat **32** (2018), no. 19, 6713–6731. MR 3899305
107. Abdollah Khodkar, S. M. Sheikholeslami, and H. Hasanzadeh, *Bounds on double domination numbers of graphs*, vol. 177, 2005, 36th Southeastern International Conference on Combinatorics, Graph Theory, and Computing, pp. 77–87. MR 2198652
108. Ralf Klasing and Christian Laforest, *Hardness results and approximation algorithms of k-tuple domination in graphs*, Inform. Process. Lett. **89** (2004), no. 2, 75–83. MR 2024938
109. Shimon Kogan, *New results on k-independence of graphs*, Electron. J. Combin. **24** (2017), no. 2, Paper 2.15, 19. MR 3650264
110. Marcin Krzywkowski, *On trees with double domination number equal to total domination number plus one*, Ars Combin. **102** (2011), 3–10. MR 2847954
111. V. R. Kulli, *On n-total domination number in graphs*, Graph theory, combinatorics, algorithms, and applications (San Francisco, CA, 1989), SIAM, Philadelphia, PA, 1991, pp. 319–324. MR 1132914
112. James K. Lan and Gerard Jennhwa Chang, *Algorithmic aspects of the k-domination problem in graphs*, Discrete Appl. Math. **161** (2013), no. 10–11, 1513–1520. MR 3043104
113. V. Leoni and G. Nasini, *Limited packing and multiple domination problems: polynomial time reductions*, Discrete Appl. Math. **164** (2014), no. part 2, 547–553. MR 3159142
114. Chung-Shou Liao and Gerard J. Chang, *Algorithmic aspect of k-tuple domination in graphs*, Taiwanese J. Math. **6** (2002), no. 3, 415–420. MR 1921604
115. ———, *k-tuple domination in graphs*, Inform. Process. Lett. **87** (2003), no. 1, 45–50. MR 1979859
116. Andrzej Lingas, Mateusz Miotk, Jerzy Topp, and PawełŻyliński, *Graphs with equal domination and covering numbers*, J. Comb. Optim. **39** (2020), no. 1, 55–71. MR 4047096
117. L. Lovász, *On the ratio of optimal integral and fractional covers*, Discrete Math. **13** (1975), no. 4, 383–390. MR 0384578
118. You Lu, Xinmin Hou, Jun-Ming Xu, and Ning Li, *Trees with unique minimum p-dominating sets*, Util. Math. **86** (2011), 193–205. MR 2884786
119. Javad Mehri, MirKamal Mirnia, and S. M. Sheikholeslami, *3-tuple domination number in complete grid graphs*, Int. Math. Forum **1** (2006), no. 21–24, 1099–1112. MR 2256668
120. Doost Ali Mojdeh, Babak Samadi, and Lutz Volkmann, *Nordhaus-Gaddum type inequalities for multiple domination and packing parameters in graphs*, Contrib. Discrete Math. **15** (2020), no. 1, 154–162
121. Oystein Ore, *Theory of graphs*, Third printing, with corrections. American Mathematical Society Colloquium Publications, Vol. XXXVIII, American Mathematical Society, Providence, R.I., 1967. MR 0244094
122. C. Payan, *Sur le nombre d'absorption d'un graphe simple*, Cahiers Centre Études Recherche Opér. **17** (1975), no. 2–4, 307–317, Colloque sur la Théorie des Graphes (Paris, 1974). MR 0401551
123. Charles Payan and Nguyen Huy Xuong, *Domination-balanced graphs*, J. Graph Theory **6** (1982), no. 1, 23–32. MR 644738

124. Ryan Pepper, *Implications of some observations about the k-domination number*, Proceedings of the Forty-First Southeastern International Conference on Combinatorics, Graph Theory and Computing, vol. 206, 2010, pp. 65–71. MR 2762480
125. Noah B. Prince, *Delta-system methods in contemporary graph theory*, Ph.D. thesis, Graduate College of the University of Illinois at Urbana-Champaign, 2008.
126. Jakub Przybyło, *On upper bounds for multiple domination numbers of graphs*, Discrete Appl. Math. **161** (2013), no. 16–17, 2758–2763. MR 3101756
127. Michaël Rao and Alexandre Talon, *The 2-domination and Roman domination numbers of grid graphs*, Discrete Math. Theor. Comput. Sci. **21** (2019), no. 1, Paper No. 9, 14. MR 3962627
128. Dieter Rautenbach and Lutz Volkmann, *New bounds on the k-domination number and the k-tuple domination number*, Appl. Math. Lett. **20** (2007), no. 1, 98–102. MR 2273616
129. Babak Samadi, Hamid R. Golmohammadi, and Abdollah Khodkar, *Bounds on several versions of restrained domination numbers*, Contrib. Discrete Math. **12** (2017), no. 1, 14–19. MR 3710021
130. Babak Samadi, Abdollah Khodkar, and Hamid R. Golmohammadi, *$(k, k', k'')$-domination in graphs*, J. Combin. Math. Combin. Comput. **98** (2016), 343–349. MR 3560493
131. Ramy S. Shaheen, *Bounds for the 2-domination number of toroidal grid graphs*, Int. J. Comput. Math. **86** (2009), no. 4, 584–588. MR 2514154
132. Erfang Shan, Chuangyin Dang, and Liying Kang, *A note on Nordhaus-Gaddum inequalities for domination*, vol. 136, 2004, Discrete mathematics and theoretical computer science (DMTCS), pp. 83–85. MR 2043928
133. S. M. Sheikholeslami and L. Volkmann, *k-tuple total domatic number of a graph*, Util. Math. **95** (2014), 189–197. MR 3243930
134. William Staton, *Some Ramsey-type numbers and the independence ratio*, Trans. Amer. Math. Soc. **256** (1979), 353–370. MR 546922
135. Christoph Stracke and Lutz Volkmann, *A new domination conception*, J. Graph Theory **17** (1993), no. 3, 315–323. MR 1220992
136. David P. Sumner, *Critical concepts in domination*, Discrete Math. **86** (1990), no. 1–3, 33–46. MR 1088558
137. Stéphan Thomassé and Anders Yeo, *Total domination of graphs and small transversals of hypergraphs*, Combinatorica **27** (2007), no. 4, 473–487. MR 2359829
138. Ioan Tomescu, *Sunflower hypergraphs are chromatically unique*, Discrete Math. **285** (2004), no. 1–3, 355–357. MR 2062863
139. Zsolt Tuza, Ph.D. thesis, Eötvös Univ. Hungary, 1983.
140. V. G. Vizing, *Some unsolved problems in graph theory*, Uspehi Mat. Nauk **23** (1968), no. 6 (144), 117–134. MR 0240000
141. Lutz Volkmann, *On graphs with equal domination and covering numbers*, vol. 51, 1994, 2nd Twente Workshop on Graphs and Combinatorial Optimization (Enschede, 1991), pp. 211–217. MR 1279636
142. ———, *Graphen an allen Ecken und Kanten*, Rheinisch-Westfälische Technische Hochschule. Lehrstuhl II für Mathematik, Aachen, 2006. MR 2229730
143. ———, *Some remarks on lower bounds on the p-domination number in trees*, J. Combin. Math. Combin. Comput. **61** (2007), 159–167. MR 2322211
144. ———, *A Nordhaus-Gaddum-type result for the 2-domination number*, J. Combin. Math. Combin. Comput. **64** (2008), 227–235. MR 2389080
145. ———, *Connected p-domination in graphs*, Util. Math. **79** (2009), 81–90. MR 2527288
146. ———, *A bound on the k-domination number of a graph*, Czechoslovak Math. J. **60(135)** (2010), no. 1, 77–83. MR 2595071
147. ———, *Bounds on the l-total k-domatic number of a graph*, Util. Math. **104** (2017), 103–113. MR 3701848
148. Bin Wang and Kai-Nan Xiang, *On k-tuple domination of random graphs*, Appl. Math. Lett. **22** (2009), no. 10, 1513–1517. MR 2561727
149. Guangjun Xu, Liying Kang, Erfang Shan, and Hong Yan, *Proof of a conjecture on k-tuple domination in graphs*, Appl. Math. Lett. **21** (2008), no. 3, 287–290. MR 2433743

150. Bohdan Zelinka, *Domatic numbers of graphs and their variants: a survey*, Domination in graphs, Monogr. Textbooks Pure Appl. Math., vol. 209, Dekker, New York, 1998, pp. 351–377. MR 1605698
151. Sanming Zhou, *Invariants concerning f-domination in graphs*, Bull. Malays. Math. Sci. Soc. (2) **37** (2014), no. 4, 1047–1055.
152. Vadim Zverovich, *The k-tuple domination number revisited*, Appl. Math. Lett. **21** (2008), no. 10, 1005–1011. MR 2450630
153. ———, *On general frameworks and threshold functions for multiple domination*, Discrete Math. **338** (2015), no. 11, 2095–2104. MR 3357796

# Distance Domination in Graphs

**Michael A. Henning**

**AMS Subject Classification** 05C65, 05C69

## 1 Introduction

The notion of distance domination in graphs has been studied a great deal; a rough estimate says that it occurs in more than 100 papers to date. In this chapter, we survey selected results on distance domination in graphs. First we recall the fundamental concepts of a dominating set in a graph and of distance in a graph. Thereafter, we combine the concepts of both distance and domination in graphs to define distance domination in graphs.

A *dominating set* of a graph $G$ is a set $S$ of vertices of $G$ such that every vertex not in $S$ has a neighbor in $S$. The *domination number* of $G$, denoted $\gamma(G)$, is the minimum cardinality of a dominating set.

The *distance* between two vertices $u$ and $v$ in a connected graph $G$, denoted $d_G(u, v)$ or simply $d(u, v)$ if the graph $G$ is clear from context, is the minimum length of a $(u, v)$-path in $G$. The *eccentricity* $\text{ecc}_G(v)$ of a vertex $v$ in $G$ is the distance between $v$ and a vertex farthest from $v$ in $G$. The minimum eccentricity among all vertices of $G$ is the *radius* of $G$, denoted by $\text{rad}(G)$, while the maximum eccentricity among all vertices of $G$ is the *diameter* of $G$, denoted by $\text{diam}(G)$. Thus, the diameter of $G$ is the maximum distance among all pairs of vertices of $G$. The *distance* from a vertex $v$ to the set $S$ in $G$, denoted by $d_G(v, S)$, is the minimum distance from $v$ to a vertex of $S$; that is, $d_G(v, S) = \min\{d(u, v) \mid u \in S\}$. In particular, if $v \in S$, then $d(v, S) = 0$. The *eccentricity* of the set $S$ in $G$ is the maximum distance of a vertex from $S$; that is, $\text{ecc}_G(S) = \max\{d(v, S) \mid v \in V(G)\}$. If $S = \{v\}$, we simply write $\text{ecc}_G(v)$ rather than $\text{ecc}_G(\{v\})$.

M. A. Henning (✉)
Department of Mathematics and Applied Mathematics, University of Johannesburg, Auckland Park, 2006, South Africa
e-mail: mahenning@uj.ac.za



T. W. Haynes et al. (eds.), *Topics in Domination in Graphs*, Developments in Mathematics 64, https://doi.org/10.1007/978-3-030-51117-3_7

A *neighbor* of a vertex $v$ in $G$ is a vertex adjacent to $v$. The *open neighborhood* of a vertex $v$ in $G$, denoted $N_G(v)$, is the set of all neighbors of $v$ in $G$, while the *closed neighborhood* of $v$ is the set $N_G[v] = N_G(v) \cup \{v\}$. If the graph $G$ is clear from the context, we write $N(v)$ and $N[v]$ rather than $N_G(v)$ and $N_G[v]$, respectively. We denote the *degree* of a vertex $v$ in $G$ by $d_G(v) = |N_G(v)|$. The minimum and maximum degrees among all vertices of $G$ are denoted by $\delta(G)$ and $\Delta(G)$, respectively.

For $k \geq 1$ an integer, the *closed k-neighborhood* of $v$ in $G$, denoted $N_k[v; G]$, is the set of all vertices within distance $k$ from $v$; that is, $N_k[v; G] = \{u \mid d_G(u, v) \leq k\}$. The *open k-neighborhood* of $v$, denoted $N_k(v; G)$, is the set of all vertices different from $v$ and at distance at most $k$ from $v$ in $G$; that is, $N_k(v; G) = N_k[v; G] \setminus \{v\}$. The *k-degree* of a vertex $v$ in $G$, denoted $d_k(v; G)$, is the number of vertices different from $v$ within distance $k$ from $v$ in $G$. Thus, $d_k(v; G) = |N_k(v; G)|$. The minimum and maximum $k$-degrees among all vertices of $G$ are denoted by $\delta_k(G)$ and $\Delta_k(G)$, respectively.

If the graph $G$ is clear from context, we omit the subscript $G$ from the above notational definitions. For example, we simply write $N(v)$, $N[v]$, $N_k(v)$, and $N_k[v]$ rather than $N_G(v)$, $N_G[v]$, $N_k(v; G)$, and $N_k[v; G]$, respectively. When $k = 1$, the set $N_k[v] = N[v]$ and the set $N_k(v) = N(v)$. In what follows, for $k \geq 1$ an integer, we use the standard notation $[k] = \{1, \dots, k\}$ and $[k]_0 = [k] \cup \{0\} = \{0, 1, \dots, k\}$.

Let $k \geq 1$ be an integer and $G$ be a graph. In 1975 Meir and Moon [62] combined the concepts of distance and domination in graphs, and introduced the concept of distance domination (also called a "$k$-covering" in [62]) in a graph. A set $S$ is a *distance k-dominating set* of $G$ if every vertex is within distance $k$ from some vertex of $S$; that is, for every vertex $v$ of $G$, we have $d(v, S) \leq k$. The *distance k-domination number* of $G$, denoted $\gamma_k(G)$ (and also $\gamma_{\leq k}(G)$ in the literature), is the minimum cardinality of a distance $k$-dominating set of $G$. The *upper k-domination number* $\Gamma_k(G)$ of $G$ is the maximum cardinality taken over all minimal distance $k$-dominating sets of $G$. A distance $k$-dominating set of $G$ of cardinality $\gamma_k(G)$ (respectively, $\Gamma_k(G)$) is called a *$\gamma_k$-set of $G$* (respectively, *$\Gamma_k$-set of $G$*). When $k = 1$, we note that a dominating set is a distance $k$-dominating set and $\gamma(G) = \gamma_k(G)$. If $X$ and $Y$ are subsets of vertices of $G$, then the set $X$ *distance k-dominates* the set $Y$ if every vertex of $Y$ is within distance $k$ from some vertex of $X$. In particular, if $X$ distance $k$-dominates the set $V(G)$, then $X$ is a distance $k$-dominating set of $G$.

For example, if $G$ is the graph shown in Figure 1, then $\gamma(G) = \gamma_1(G) = 4$ and the four darkened vertices of $G$ illustrated in Figure 1(a) form a $\gamma$-set of $G$. Moreover, $\gamma_2(G) = \gamma_3(G) = 2$ and the two darkened vertices of $G$ illustrated in Figure 1(b) form a $\gamma_2$-set and a $\gamma_3$-set of $G$. Furthermore, $\gamma_4(G) = 1$ and the darkened vertex of $G$ illustrated in Figure 1(c) forms a $\gamma_4$-set of $G$.

Throughout the remainder of this survey chapter, for notational simplicity we write "$k$-dominating set" and "$k$-domination number" rather than the more accurate terminology "distance $k$-dominating set" and "distance $k$-domination number," respectively. We emphasize that in this chapter a $k$-dominating set is therefore different from multiple domination introduced in 1984 by Fink and Jacobson [31], where they define a $k$-dominating set of a graph $G$ as a set $S$ of vertices of $G$

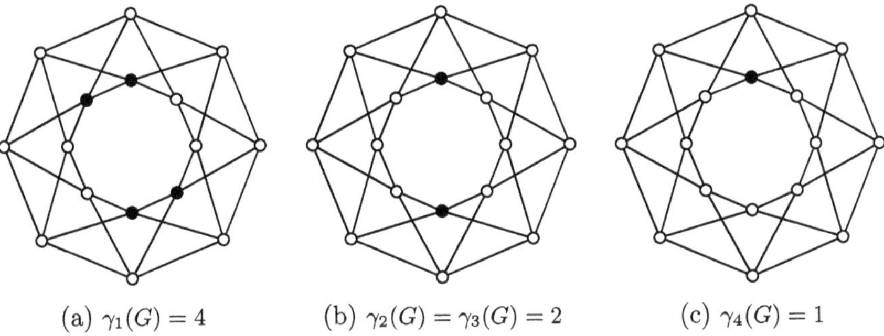

(a) $\gamma_1(G) = 4$ (b) $\gamma_2(G) = \gamma_3(G) = 2$ (c) $\gamma_4(G) = 1$

**Fig. 1** A graph $G$

such that every vertex in $V(G) \setminus S$ is dominated by at least $k$ vertices of $S$; that is, $|N_G(v) \cap S| \geq k$ for every vertex $v \in V(G) \setminus S$. By a $k$-dominating set in this chapter we will always mean a distance $k$-dominating set, and by the $k$-domination number of a graph we will always mean the distance $k$-domination number of the graph.

## 2 Applications of Distance Domination

This concept of distance domination in graphs finds applications in many situations and structures which give rise to graphs, including communication networks, geometric problems, and facility location problems in operations research.

In 1976 Slater [68] continued the study of distance domination in graphs. He referred to a $k$-dominating set of minimum cardinality in a graph as a "$k$-basis" of the graph. Slater's study was application driven and related to communication networks. More precisely, he considered a graph associated with a collection of cities where the vertices correspond to the cities and where two vertices are joined by an edge if there is a communication link between the corresponding cities. Slater considered the problem of selecting a minimum number of cities as sites for transmitting stations so that every city either contains a transmitter or can receive messages from at least one of the transmitting stations through the links. If communication over paths of $k$ links (but not of $k+1$ links) is adequate in quality and speed, then as remarked by Slater [68] the problem becomes that of determining a minimum $k$-dominating set in the associated graph. Slater [68] introduced a more general problem in which each vertex $v_i$ has an associated value $a_i$, where he required that there be a transmitting station within distance $a_i$ of $v_i$.

In 1998, the author in [45] gave a discussion on the concept of distance domination in graphs. Consider, for instance, the following illustration given in [45]. Let $G$ be the graph associated with the road grid of a city where the vertices of $G$ correspond to the street intersections and where two vertices are adjacent if and only if the corresponding street intersections are a block apart. A minimum $k$-dominating set in $G$ may be used to locate a minimum number of facilities (such as utilities, police stations, waste disposal dumps, hospitals, blood banks, transmission towers) such that every intersection is within $k$ city blocks of a facility.

In 1977 Lichtenstein [60] considered the following geometric problem: For a given (finite) subset $P$ of points in $\mathbb{Z} \times \mathbb{Z}$ and a positive integer $k$, what is the smallest cardinality of a subset $P'$ of $P$ such that every point of $P \setminus P'$ is within Euclidean distance $k$ of some point in $P'$ given that the graph with vertex set $P'$ in which two points of $P'$ are adjacent if and only if they are within Euclidean distance $k$ of each other is connected? As noted in [60], this problem is defined only if the graph with vertex set $P$ and such that two points of $P$ are joined by an edge if and only if they are within distance $k$ of each other is itself connected. It was shown [60] that this optimization problem appears to be computationally difficult by showing that a corresponding decision problem is $NP$-complete. This geometric problem suggests several related graph problems, including the concept of distance domination, which have since then been introduced and studied.

## 3 Properties of Minimal $k$-Dominating Sets

In this section, we present various properties of $k$-dominating sets in graphs. We begin with properties of minimal $k$-dominating sets. The following proposition from [47] generalizes a classical result of Ore [65] concerning dominating sets.

**Proposition 1 ([47])** *For $k \ge 1$, let $D$ be a $k$-dominating set of a graph $G$. Then $D$ is a minimal $k$-dominating set of $G$ if and only if every vertex $v \in D$ has at least one of the following two properties.*

$P_1$: *There exists a vertex $u \in V(G) \setminus D$ such that $N_k(u) \cap D = \{v\}$;*
$P_2$: *The vertex $v$ is at distance at least $k+1$ from every other vertex of $D$ in $G$.*

The $k$-th power of the graph $G$, denoted $G^k$, is the graph with the same vertex set as $G$ and where there is an edge between two vertices in $G^k$ if and only if the distance between them is at most $k$ in $G$; that is, $V(G^k) = V(G)$ and $E(G^k) = \{uv \mid u, v \in V(G) \text{ and } d_G(u,v) \le k\}$. We state next a useful observation, where $G^k$ denotes the $k$-th power of the graph $G$.

**Observation 2** *If $G$ is a connected graph, then $\gamma_k(G) = \gamma(G^k)$.*

Bollobás and Cockayne [13] established the following result. By a nontrivial graph we mean a graph on at least two vertices.

**Theorem 3 ([13])** *If $G$ is a connected nontrivial graph, then there exists a $\gamma$-set $D$ of $G$ such that for every vertex $v \in D$, there exists a vertex $w \in V(G) \setminus D$ with $N(v) \cap D = \{v\}$.*

An immediate consequence of Observation 2 and Theorem 3, if $G$ is a connected nontrivial graph, then there exists a $\gamma_k$-set $D$ of $G$ such that for every vertex $v \in D$, there exists a vertex $w \in V(G) \setminus D$ with $N_k(v) \cap D = \{v\}$. The following stronger result is proved in [49].

**Theorem 4 ([49])** *For $k \geq 1$, if $G$ is a connected graph of order at least $k+1$ with* $\text{diam}(G) \geq k$*, then there exists a $\gamma_k$-set $D$ of $G$ such that for every vertex $v \in D$, there is a vertex $w \in V(G) \setminus D$ at distance exactly $k$ from $v$ in $G$ with $N_k(v) \cap D = \{v\}$.*

We next present a proof of Theorem 4 given in [49]. For this purpose, we introduce some additional notation. Let $S$ be a set of vertices of a connected graph $G$. We will call a nondecreasing sequence $\ell_1, \ell_2, \ldots, \ell_{|S|}$ of integers the *distance sequence* of $S$ in $G$ if the vertices of $S$ can be labeled $v_1, v_2, \ldots, v_{|S|}$ so that $\ell_i = d_G(v_i, S \setminus \{v_i\})$ for all $i \in [|S|]$. For example, if $k \geq 1$ is an arbitrary integer and $G$ is the graph obtained from a connected graph $H$ by attaching a path of length $k$ to each vertex of $H$ so that the resulting paths are vertex disjoint, then the set $S = V(H)$ is a $\gamma_k$-set of $G$ and has distance sequence $1, 1, \ldots, 1$ of length $|S|$ in $G$.

Suppose $s_1 \colon a_1, a_2, \ldots, a_p$ and $s_2 \colon b_1, b_2, \ldots, b_q$ are two nondecreasing sequences of positive integers. We say that the sequence $s_1$ *precedes* the sequence $s_2$ in *dictionary order* if either $p \leq q$ and $a_i = b_i$ for $i \in [p]$ or if there exists an integer $i$ where $i \in [\min\{p, q\}]$ such that $a_i < b_i$ and $a_j = b_j$ for $j < i$. For example, for the tree $T$ given in Figure 2, the set $\{v_2, v_5, v_8\}$ has distance sequence $s_1 \colon 3, 3, 3$ in $T$, while the distance sequence of the set $\{u_1, v_5, u_4\}$ in $G$ is $s_2 \colon 5, 5, 5$. Although both sets $\{v_2, v_5, v_8\}$ and $\{u_1, v_5, u_4\}$ are 2-dominating sets of $T$, the distance sequence $s_1$ of the set $\{v_2, v_5, v_8\}$ precedes the distance sequence $s_2$ of the set $\{u_1, v_5, u_4\}$ in dictionary order.

We are now in a position to present a proof of Theorem 4.

**Proof of Theorem 4** Among all $\gamma_k$-sets of $G$, let $D$ be one that has the smallest distance sequence in dictionary order. Let $s = \gamma_k(G)$ and let the distance sequence of $D$ be given by $\ell_1, \ell_2, \ldots, \ell_s$, where $D = \{v_1, v_2, \ldots, v_s\}$ and $\ell_i = d_G(v_i, D \setminus \{v_i\})$ for $i \in [s]$.

We show first that each vertex of $D$ has property $P_1$. If this is not the case, then let $i$ be the smallest integer such that the vertex $v_i$ does not have property $P_1$. By Proposition 1, the vertex $v_i$ has property $P_2$, and so $\ell_i \geq k+1$. Now let $v_i' \in N_k(v_i)$ and consider the set $D' = (D \setminus \{v_i\}) \cup \{v_i'\}$. Necessarily, the set $D'$ is a $\gamma_k$-set of

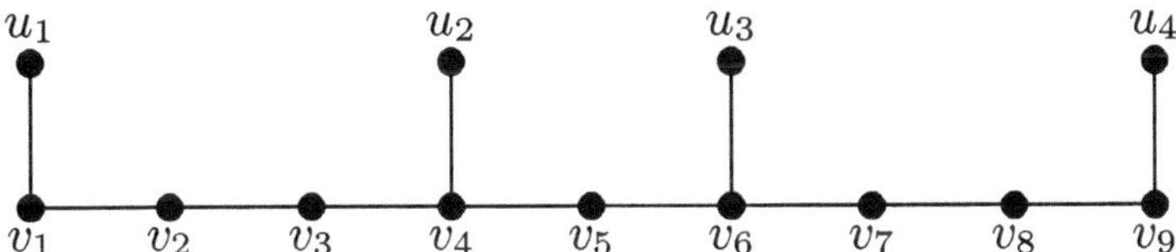


**Fig. 2** A tree $T$

$G$. Furthermore, the vertex $v_i'$ is within distance $k$ from some vertex of $D \setminus \{v_i\}$. Consequently, $\ell_i' = d_G(v_i', D' \setminus \{v_i'\}) < \ell_i$. Now let $j$ be the largest integer for which $\ell_j < \ell_i$, and consider the value $\ell_t' = d_G(v_t, D' \setminus \{v_t\})$ for each $t$ with $t \in [j]$. Since $\ell_t < \ell_i$, a shortest path from the vertex $v_t$ to a vertex of $D \setminus \{v_i\}$ does not contain $v_i$. It follows, therefore, that $\ell_t' \le \ell_t$ for all $t \in [j]$. This, together with the observation that $\ell_i' < \ell_r$ for all $r > j$, implies that the distance sequence of $D'$ precedes that of $D$ in dictionary order. This produces a contradiction. Hence every vertex of $D$ has property $P_1$.

For each vertex $v_i$ of $D$, let $w_i$ be a vertex of $V(G) \setminus D$ at maximum distance from $v_i$ in $G$ satisfying $N_k(w_i) \cap D = \{v_i\}$ for $i \in [s]$. We show that $d(v_i, w_i) = k$ for all $i \in [s]$. If this is not the case, then let $i$ be the smallest integer for which $d(v_i, w_i) < k$. Observe that every vertex of $V(G) \setminus D$ at distance greater than $k - 1$ from $v_i$ is within distance $k$ from some vertex of $D \setminus \{v_i\}$. Consider a shortest path from the vertex $v_i$ to a vertex of $D \setminus \{v_i\}$ in $G$. Let $v_i^*$ denote the vertex adjacent to $v_i$ on such a path. Further, let $D^* = (D \setminus \{v_i\}) \cup \{v_i^*\}$. Necessarily, the set $D^*$ is a $\gamma_k$-set of $G$. Now let $j$ be the largest integer for which $\ell_j < \ell_i$, and consider the value $\ell_t^* = d_G(v_t, D^* \setminus \{v_t\})$ for each $t \in [j]$. Necessarily, $\ell_t^* \le \ell_t$ for all $t \in [j]$. Furthermore, $d_G(v_i^*, D^* \setminus \{v_i^*\}) = \ell_i - 1 < \ell_r$ for all $r > j$. It follows, therefore, that the distance sequence of $D^*$ precedes that of $D$ in dictionary order, a contradiction. Hence, $d(v_i, w_i) = k$ for all $i \in [s]$. This completes the proof of Theorem 4. □

## 4 Algorithmic and Complexity Results

From a computational point of view the problem of finding $\gamma_k(G)$ for a general graph $G$ appears to be difficult. To date there is no known efficient algorithm for solving this problem and the corresponding decision problem is NP-complete (see [16]). Even if we restrict the graph $G$ to belong to certain special classes of graphs, including bipartite or chordal graphs of diameter $2k + 1$, the problem remains $NP$-hard (see [16]). Algorithmic and complexity results to compute $\gamma_k(G)$ are given, for example, in [14, 16, 27, 68].

In 2012 Datta, Larmore, Devismes, Heurtefeux, and Rivierre [23] present a distributed self-stabilizing algorithm to compute a $k$-dominating set. They show that for unit disk graphs the size of the resulting $k$-dominating set is at most $7.2552k + O(1)$ times the minimum possible size. Turau and Köhler [74] present a distributed algorithm to compute a minimum $k$-dominating set of a tree $T$. Their algorithm terminates in $O(height(T))$ rounds and requires $O(\log k)$ storage in their distributed model.

In 2018 Jaffke, Kwon, Strømme, and Telle [52] studied the complexity of generalized distance domination problems on graphs of bounded mim-width. We do not define the *maximum induced matching width*, or *mim-width* for short, in this survey chapter suffice it to say that it is a structural graph parameter described over *decomposition trees* (also called *branch decompositions* in the literature), similar to graph parameters such as *rank-width* and *module-width*. The concept of mim-width was introduced in the 2012 Ph.D. thesis of Vatshelle [76], and used implicitly by Belmonte and Vatshelle [9]. Subsequently, it has been used of late in several important algorithmic and complexity papers (see, for example, [12, 32, 38, 53–55, 79]).

Jaffke et al. [52] raise the problem to a higher level and study distance domination in a more general setting of distance-$k$ $(\sigma, \rho)$-domination problems, which they define as follows. Let $\sigma$ and $\rho$ be finite or co-finite subsets of the natural numbers $\mathbb{N}$, and so $\sigma, \rho \subseteq \mathbb{N}$ where $\mathbb{N} = \{0, 1, \ldots\}$. For $k \geq 1$ an integer, a set $S$ is a *distance-$k$ $(\sigma, \rho)$-dominating set* in a graph $G$ if the following holds.

(a) $|N_k(v) \cap S| \in \sigma$ for each vertex $v \in S$.
(b) $|N_k(v) \cap S| \in \rho$ for each vertex $v \in V(G) \setminus S$.

For instance, when $\sigma = \mathbb{N}$ and $\rho = \mathbb{N}^+ = \{1, 2 \ldots\}$, then a distance-$k$ $(\sigma, \rho)$-dominating set is a $k$-dominating set since in this case each vertex in $V(G) \setminus S$ has at least one vertex in its open $k$-neighborhood that belongs to the set $S$; that is, each vertex in $V(G) \setminus S$ is within distance $k$ from at least one vertex in $S$.

The main result of Jaffke et al. [52] is that for any positive integer $k$ the mim-width of the $k$-th power $G^k$ of a graph $G$ is at most twice the mim-width of $G$. This key structural result implies that we can reduce the distance-$k$ $(\sigma, \rho)$-domination problem to the standard version by taking the graph power $G^k$, while preserving small mim-width. In particular, this implies that the $k$-domination problem is XP parameterized by min-width if a decomposition tree is given. As explained in [52], the $k$-domination problem "is therefore solvable in polynomial time for many interesting graph classes where mim-width is bounded and quickly computable, such as $k$-trapezoid graphs, Dilworth $k$-graphs, (circular) permutation graphs, interval graphs and their complements, convex graphs and their complements, $k$-polygon graphs, circular arc graphs, complements of $d$-degenerate graphs, and $H$-graphs if given an $H$-representation." Thus for the above classes of graphs, Jaffke et al. [52] obtain the first polynomial-time algorithms to compute the $k$-domination number. This is a significant breakthrough on the algorithmic and complexity results for $k$-domination in graphs.

Having said that, we remark that the meta-algorithms due to Jaffke et al. [52] have runtimes which are most likely not optimal on a particular graph class. For instance, as pointed out by Jaffke et al. [52], applying their results to solve the $k$-domination problem on permutation graphs of order $n$ yields an algorithm that runs in time $O(n^8)$. However, there is an algorithm presented in a 2016 paper due to Rana, Pal, and Pal [66] that solves the $k$-domination problem in time $O(n^2)$; a much faster runtime. This example illustrates that the meta-algorithms due to Jaffke et al. [52] will have runtimes which can likely be improved significantly for particular graph class.

Jaffke et al. [52] comment that the downside to showing results using the parameter mim-width, is that it is not yet known if there is an XP approximation algorithm for computing mim-width. As remarked in [52], "computing a decomposition tree with optimal mim-width is NP-complete in general and W[1]-hard parameterized by itself. Determining the optimal mim-width is not in APX unless NP = ZPP, making it unlikely to have a polynomial-time constant-factor approximation algorithm [67], but saying nothing about an XP algorithm."

## 5 A Duality Lemma

In this section, we present a relationship between the concept of distance domination in graphs and the $p$-center problem studied in operations research. For an integer $p \geq 1$, the *p-radius* of a connected graph $G$ is minimum eccentricity among all $p$-element subsets of $G$; that is,

$$\text{rad}_p(G) = \min\{\text{ecc}(S) \colon S \subseteq V(G) \text{ and } |S| = p\}.$$

We note that if $p = 1$, then $\text{rad}(G) = \text{rad}_1(G)$. If $S$ is a subset of $V(G)$ such that $|S| = p$ and $\text{ecc}_G(S) = \text{rad}_p(G)$, then $S$ is the *p-center* of $G$. An excellent survey paper on network location problems with an emphasis on the $p$-center, $p$-median, and $p$-radius of a graph is given in the 1983 paper by Tansel, Francis, and Lowe [70]. Motivated by applications to location problems, $p$-centers in graphs have been extensively studied in the literature. Recent studies on algorithms to find a $p$-center of a graph can be found, for example, in [3, 17, 26, 78].

To illustrate the concept of the $p$-radius of a graph, consider again the graph $G$ shown in Figure 1. We note that $\text{rad}_4(G) = 1$ and the set $S$ of four darkened vertices of $G$ illustrated in Figure 1(a) satisfies $\text{rad}_4(G) = \text{ecc}(S) = 1$. Moreover, $\text{rad}_3(G) = 2$ and the set $S$ of three darkened vertices of $G$ illustrated in Figure 3 satisfies $\text{rad}_3(G) = \text{ecc}(S) = 2$. We also note that $\text{rad}_2(G) = 2$ and the set $S$ of two darkened vertices of $G$ illustrated in Figure 1(b) satisfies $\text{rad}_2(G) = \text{ecc}(S) = 2$. Furthermore, $\text{rad}_1(G) = 4$ and the set $S$ consisting of an arbitrary vertex of $G$ satisfies $\text{rad}_1(G) = 4$. For example, the set $S$ consisting of the singleton darkened vertex illustrated in Figure 1(c) satisfies $\text{rad}_1(G) = \text{ecc}(S) = 4$.

The problem of determining the $p$-radius of a graph is in a sense a dual problem to that of determining the $k$-domination number of a graph as observed by Tansel, Francis, and Lowe [70] and others.

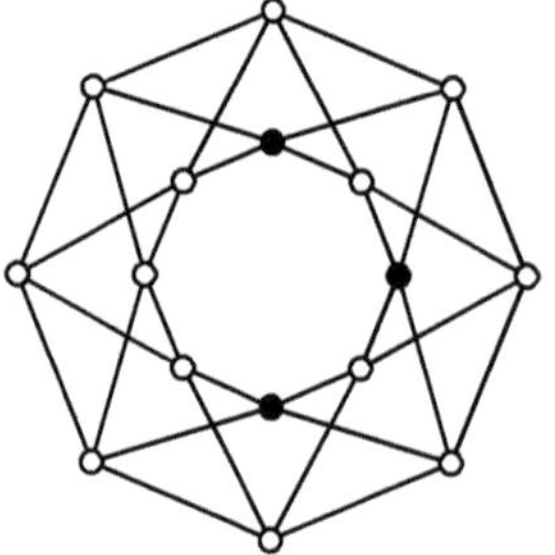

**Fig. 3** A graph $G$ with $\text{rad}_3(G) = 2$

**Lemma 5 (Duality Lemma)** *For a connected graph $G$ and positive integers $k$ and $p$, we have*

$$\mathrm{rad}_p(G) \le k \quad \text{if and only if} \quad \gamma_k(G) \le p.$$

**Proof.** Suppose that $\mathrm{rad}_p(G) \le k$. Let $S$ be a $p$-center of $G$, and so $S$ is a subset of $V(G)$ satisfying $|S| = p$ and $\mathrm{ecc}_G(S) = \mathrm{rad}_p(G) \le k$. The set $S$ is a $k$-dominating set of $G$, and so $\gamma_k(G) \le p$. Conversely, suppose that $\gamma_k(G) \le p$. Let $S$ be $\gamma_k$-set of $G$, and so $S$ is a $k$-dominating set of $G$ and $|S| = \gamma_k(G) \le p$. Adding vertices to the set $S$ if necessary, there therefore exists a subset $S'$ of $V(G)$ such that $S \subseteq S'$, $|S'| = p$, and $\mathrm{rad}_p(G) \le \mathrm{ecc}_G(S') \le k$. □

## 6 Lower Bounds on the Distance Domination Number

In this section, we discuss some lower bounds on the $k$-domination number of a graph for $k \ge 1$ an integer. For this purpose, we first present a property of $k$-dominating sets in spanning subgraphs of a graph. Since every $k$-dominating set of a spanning subgraph of a graph $G$ is a $k$-dominating set of $G$, we have the following observation due to Zelinka [80].

**Observation 6 ([80])** *For $k \ge 1$, if $H$ is a spanning subgraph of a graph $G$, then $\gamma_k(G) \le \gamma_k(H)$.*

Tian and Xu [72] showed next that every connected graph has a spanning tree with equal $k$-domination number. The proof we present is from [24].

**Theorem 7 ([24])** *For $k \ge 1$, every connected graph $G$ has a spanning tree $T$ such that $\gamma_k(T) = \gamma_k(G)$.*

**Proof.** Let $S$ be a $\gamma_k$-set of $G$. Thus, $|S| = \gamma_k(G)$. For $i \in [k]_0$, let $D_i(S) = \{v \in V(G) \setminus S \mid d_G(v, S) = i\}$. Since $S$ is a $k$-dominating set of $G$, every vertex $v$ in $V(G) \setminus S$ is within distance $k$ from some vertex of $S$ for some $i \in [k]_0$, and therefore the sets $D_0(S), D_1(S), \ldots, D_k(S)$ form a weak partition of $V(G) \setminus S$ (where in a weak partition some of the sets may be empty). Furthermore for $i \in [k]$, every vertex $v \in D_i(S)$ is adjacent to at least one vertex of $D_{i-1}(S)$, and possibly to vertices in $D_i(S)$ and $D_{i+1}(S)$. For all $i \in [k]$ and for each vertex $v \in D_i(S)$, we delete all but one edge that joins $v$ to a vertex of $D_{i-1}(S)$. Further, for all $i \in [k]_0$ we delete all edges, if any, that join $v$ to vertices in $D_i(S)$. Let $F$ denote the resulting spanning subgraph of the graph $G$.

We claim that $F$ is a forest. Suppose, to the contrary, that $F$ contains a cycle $C$. Let $v$ be a vertex in such a cycle $C$ at maximum distance from a vertex of $S$ in $G$, and let $v_1$ and $v_2$ be the two neighbors of $v$ on $C$. Suppose that $v \in D_p(S)$ for some $p \in [k]$. Thus, $d_G(v, S) = p$ and $d_G(w, S) \le p$ for every vertex $w$ of $C$ different from $v$. If $v_1$ or $v_2$ belongs to $D_p(S)$, then this contradicts the way in which $F$ was constructed, noting that no edge in $F$ joins two vertices in the same set $D_i(S)$. Thus, both $v_1$ and $v_2$ belong to $D_{p-1}(S)$. Once again, this contradicts the way in which $F$ was constructed, noting that exactly one edge in $F$ joins a vertex in $D_i(S)$ to a vertex in $D_{i-1}(S)$. Therefore, $F$ is a forest.

If $F$ is a tree, then we let $T = F$; otherwise, if the forest $F$ has $\ell \geq 2$ components, then we let $T$ be obtained from $F$ by adding to it $\ell - 1$ edges in such a way that the resulting subgraph is connected. We note that $T$ is a tree. By construction, if $v \in D_i(S)$ for some $i \in [k]$, then there is a path from $v$ to $S$ of length $i$ in $T$, and so $d_T(v, S) \leq d_G(v, S)$. Since $T$ is a spanning tree of $G$, $d_G(v, S) \leq d_T(v, S)$ for every vertex $v \in V(G)$. Consequently, the spanning tree $T$ of $G$ is distance-preserving from the set $S$ in the sense that $d_G(v, S) = d_T(v, S)$ for every vertex $v \in V(G)$. Since $S$ is a $k$-dominating set of $G$, the set $S$ is therefore a $k$-dominating set of $T$, and so $\gamma_k(T) \leq |S| = \gamma_k(G)$. However, by Observation 6, $\gamma_k(G) \leq \gamma_k(T)$. Consequently, $\gamma_k(T) = \gamma_k(G)$. □

Davila, Fast, Henning, and Kenter [24] established the following lower bound on the $k$-domination number of a graph in terms of its diameter.

**Theorem 8 ([24])** *For $k \geq 1$, if $G$ is a connected graph with diameter $d$, then*

$$\gamma_k(G) \geq \frac{d+1}{2k+1}.$$

That the lower bound of Theorem 8 is tight may be seen by taking $G$ to be a path, $v_1 v_2 \dots v_n$, of order $n = \ell(2k+1)$ for some $\ell \geq 1$. Let $d = \text{diam}(G)$, and so $d = n - 1 = \ell(2k+1) - 1$. By Theorem 8, $\gamma_k(G) \geq (d+1)/(2k+1) = \ell$. The set

$$S = \bigcup_{i=0}^{\ell-1} \{v_{k+1+i(2k+1)}\}$$

is a $k$-dominating set of $G$, and so $\gamma_k(G) \leq |S| = \ell$. Consequently, $\gamma_k(G) = \ell = (d+1)/(2k+1)$. We state this formally as follows.

**Observation 9 ([24])** *If $G = P_n$ where $n \equiv 0 \bmod (2k+1)$, then*

$$\gamma_k(G) = \frac{\text{diam}(G) + 1}{2k+1}.$$

More generally, by applying Theorem 8, the $k$-domination number of a path $P_n$ on $n \geq 3$ vertices is easy to compute.

**Observation 10 ([24])** *For $k \geq 1$ and $n \geq 3$, $\gamma_k(P_n) = \lceil \frac{n}{2k+1} \rceil$.*

For $k \geq 1$ and $n \geq 3$, every vertex of a cycle $C_n$ $k$-dominates exactly $2k + 1$ vertices. Thus, if $S$ is a minimum $k$-dominating set of $G$, then the set $S$ $k$-dominates at most $|S|(2k+1)$ vertices of $P$, implying that $|S|(2k+1) \geq n$, or, equivalently, $\gamma_k(C_n) = |S| \geq n/(2k+1)$. Conversely, by Observation 6 and Proposition 10, $\gamma_k(C_n) \leq \gamma_k(P_n) = \lceil \frac{n}{2k+1} \rceil$. Consequently, the $k$-domination number of a cycle is determined by the following result.

**Observation 11 ([24])** *For $k \geq 1$ and $n \geq 3$, $\gamma_k(C_n) = \lceil \frac{n}{2k+1} \rceil$.*

For $k \geq 1$ and $n \geq 3$, where $n \equiv 0 \bmod (2k+1)$, consider a path $P\colon v_1 v_2 \dots v_n$. By replacing each vertex $v_i$, for $2 \leq i \leq n-1$, on the path $P$ with a clique $V_i$ of size at least $\delta \geq 1$, and adding all edges between $v_1$ and vertices in $V_2$, adding all edges between $v_n$ and vertices in $V_{n-1}$, and adding all edges between vertices in $V_i$ and $V_{i+1}$ for $2 \leq i \leq n-2$, we obtain a graph with minimum degree at least $\delta$ achieving the lower bound of Theorem 8.

From Theorem 8, we have the following lower bound on the $k$-domination number of a graph in terms of its radius.

**Corollary 12 ([24])** *For $k \geq 1$, if $G$ is a connected graph with radius $r$, then*

$$\gamma_k(G) \geq \frac{2r}{2k+1}.$$

**Proof.** By Theorem 7, the graph $G$ has a spanning tree $T$ such that $\gamma_k(T) = \gamma_k(G)$. Since adding edges to a graph cannot increase its radius, $\text{rad}(G) \leq \text{rad}(T)$. Since $T$ is a tree, we note that $\text{diam}(T) \geq 2\text{rad}(T) - 1$. Applying Theorem 8 to the tree $T$, we have that

$$\gamma_k(G) = \gamma_k(T) \geq \frac{\text{diam}(T)+1}{2k+1} \geq \frac{2\text{rad}(T)}{2k+1} \geq \frac{2\text{rad}(G)}{2k+1}.$$

□

That the lower bound of Corollary 12 is tight, may be seen by taking $G$ to be a path, $P_n$, of order $n = 2\ell(2k+1)$ for some integer $\ell \geq 1$. Let $d = \text{diam}(G)$ and let $r = \text{rad}(G)$, and so $d = 2\ell(2k+1) - 1$ and $r = \ell(2k+1)$. In particular, we note that $d = 2r - 1$. By Observation 10, $\gamma_k(G) = \frac{d+1}{2k+1} = \frac{2r}{2k+1}$. As before, by replacing each internal vertex on the path with a clique of size at least $\delta \geq 1$, we can obtain a graph with minimum degree at least $\delta$ achieving the lower bound of Corollary 12.

In 2005 Meierling and Volkmann [61] and in 2006 Cyman, Lemańska, and Raczek [21] studied lower bounds for the distance $k$-domination number of a tree. The lower bound presented in [21] is in terms of the order and number of leaves of the tree.

**Theorem 13** ([21]) *For $k \geq 1$, if $T$ is a tree of order $n$ with $\ell$ leaves, then*

$$\gamma_k(G) \geq \frac{n + 2k - k\ell}{2k+1}.$$

The following lower bound on the $k$-domination number of a graph in terms of its girth is established in [24], where the girth of a graph is the length of a shortest cycle in the graph.

**Theorem 14 ([24])** *For $k > 1$, if $G$ is a connected graph with girth $g < \infty$, then*

$$\gamma_k(G) \geq \frac{g}{2k+1}.$$

The lower bound of Theorem 14 is tight, as may be seen by taking $G$ to be a cycle $C_n$, where $n \equiv 0 \bmod (2k+1)$. We note that $G$ has girth $g = n$ and, by Observation 11, $\gamma_k(G) = \frac{n}{2k+1} = \frac{g}{2k+1}$.

## 7 Upper Bounds on the Distance Domination Number

In this section, we discuss several upper bounds on the $k$-domination number of a graph for $k \geq 1$ an integer. In 1975 Meir and Moon [62] established the following upper bound for the $k$-domination number of a graph in terms of its order. The proof we provide follows readily from Theorem 4.

**Theorem 15 ([62])** *For $k \geq 1$, if $G$ is a connected graph of order $n \geq k+1$, then*

$$\gamma_k(G) \leq \frac{n}{k+1}.$$

**Proof.** Let $D = \{v_1, \ldots, v_r\}$ be a $\gamma_k$-set of $G$ satisfying the statement of Theorem 4. For each $i \in [r]$, let $w_i$ be a vertex in $V(G) \setminus D$ at distance exactly $k$ from $v_i$ in $G$ and at distance greater than $k$ from every vertex of $D \setminus \{v_i\}$; that is, $d_G(w_i, v_i) = k$ and $d_G(w_i, D \setminus \{v_i\}) > k$. Further, let $Q_i$ be a shortest $(v_i, w_i)$-path in $G$ from $v_i$ to $w_i$ for $i \in [r]$, and so $Q_i \cong P_{k+1}$. We note that the paths $Q_1, \ldots, Q_r$ are vertex disjoint, implying that

$$\gamma_k(G) \cdot (k+1) = r \cdot (k+1) = \sum_{i=1}^{r} |V(Q_i)| = |\bigcup_{i=1}^{r} V(Q_i)| \leq |V(G)| = n,$$

or, equivalently, $\gamma_k(G) \leq n/(k+1)$. □

We remark that in the original proof of Theorem 15, Meir and Moon [62] first proved that for $k \geq 1$, if $T$ is a tree of order $n \geq k+1$, then $\gamma_k(T) \leq n/(k+1)$. From this result, they immediately deduced the result of Theorem 15 noting that if $G$ is a connected graph of order $n \geq k+1$ and $T$ is an arbitrary spanning tree of $G$, then since adding edges to a graph cannot increase its $k$-domination number, we have $\gamma_k(G) \leq \gamma_k(T) \leq n/(k+1)$.

The proof of Theorem 15 given in [47] suggests an algorithm that finds, for a connected graph $G$ of order $n$, a $k$-dominating set of cardinality at most $n/(k+1)$. If $G$ is a connected graph of order $n \leq k+1$, then any vertex of $G$ forms a $k$-dominating set of $G$, and in this case $\gamma_k(G) = 1 \leq n/(k+1)$. For $n \geq k+2$, we have the following algorithm from [47].

**Algorithm 16 ([47])** *For $k \geq 1$ an integer and $G$ a connected graph of order $n \geq k+1$, perform the following steps.*

(a) *Find a spanning tree $T$ of $G$. Set $D_k \leftarrow \emptyset$.*

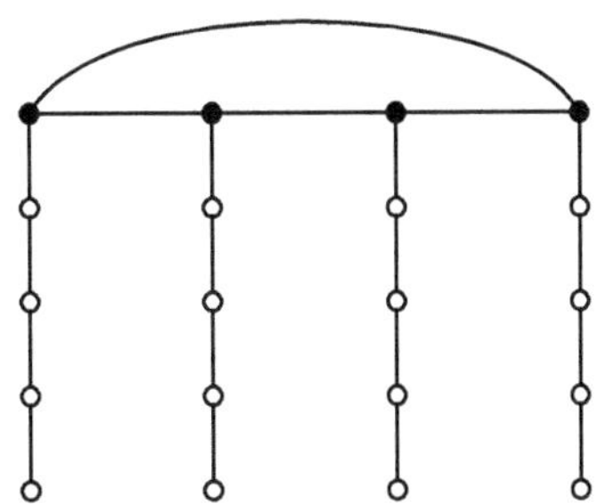

**Fig. 4** The 4-corona $C_4 \circ P_5$ of a 4-cycle

(b) *If* rad$(T) \leq k$, *then let $v$ be a central vertex of $T$, output $D_k \cup \{v\}$, and stop. Otherwise continue.*

(c) *Let $d =$* diam$(T)$ *and find a path $u_0, u_1, \ldots, u_d$ of length $d$ in $T$. Let $T_k$ be the component of $T - u_k u_{k+1}$ that contains the vertex $u_{k+1}$. Set $T \leftarrow T_k$ and $D_k \leftarrow D_k \cup \{u_k\}$, and return to Step 2.*

For $k \geq 1$ an integer, the *$k$-corona* $H \circ P_k$ of a graph $H$ is the graph of order $(k+1)|V(H)|$ obtained from $H$ by attaching a path of length $k$ to each vertex of $H$ so that the resulting paths are vertex disjoint. For example, if $H \cong C_4$ and $k = 4$, then the $k$-corona $H \circ P_k$ of $H$ is illustrated in Figure 4. We note in this example, the four darkened vertices (that induce the cycle $C_4 \cong H$) form a $\gamma_5$-set of $C_4 \circ P_4$.

If $G = H \circ P_k$ and $H$ is a connected graph of order at least 2, then the set $V(H)$ is a $k$-dominating set of $G$, and so $\gamma_k(G) \leq |V(H)|$. However in order to distance $k$-dominate the set of $|V(H)|$ leaves of $G$, every $k$-dominating set of $G$ contains at least one vertex from each of the added $|V(H)|$ paths, and so $\gamma_k(G) \geq |V(H)|$. Consequently, $\gamma_k(G) = |V(H)| = n/(k+1)$, showing that the upper bound of Theorem 15 is tight. In 1991, Topp and Volkmann [73] gave a complete characterization of the graphs achieving equality in the upper bound of Theorem 15.

**Theorem 17 ([73])** *For $k \geq 1$, if $G$ is a connected graph of order $n \geq k+1$ satisfying $\gamma_k(G) \leq \frac{n}{k+1}$, then one of the following holds.*

(a) $n = k + 1$.
(b) $G = C_{2k+2}$.
(c) $G = H \circ P_k$ *for some connected graph $H$ of order at least* 2.

For positive integers $k$ and $\ell$, if $G$ is the graph of order $n = \ell(k+1)$ obtained from an arbitrary connected graph $H$ of order $\ell$ by attaching a path of length $k$ (equivalently, of order $k+1$) to each vertex of $H$ so that the resulting paths are vertex disjoint, then the set $V(H)$ is a $k$-dominating set of $G$, and so $\gamma_k(G) \leq |V(H)| = \ell$. However in order to distance $k$-dominate the set of $\ell$ leaves of $G$, every $k$-dominating set of $G$ contains at least one vertex from each of the added $\ell$ paths, and so $\gamma_k(G) \geq \ell$. Consequently, $\gamma_k(G) = \ell = n/(k+1)$, showing that the upper bound of Theorem 15 is tight.

In 2009 Tian and Xu [71] established the following upper bound for the $k$-domination number of a graph in terms of its order and maximum degree.

**Theorem 18 ([71])** *For $k \geq 1$, if $G$ is a connected graph of order $n \geq k+1$ with maximum degree $\Delta$, then*

$$\gamma_k(G) \leq \frac{n - \Delta + k - 1}{k}.$$

In 2009 Tian and Xu [71] also established the following upper bound on the $k$-domination number of a graph in terms of its minimum degree.

**Theorem 19 ([71])** *For $k \geq 1$, if $G$ is a connected graph of order $n$ with minimum degree $\delta$, then*

$$\gamma_k(G) \leq \left( \frac{1 + \ln(m(\delta+1) + 2 - t)}{m(\delta+1) + 2 - t} \right) n,$$

*where $m = \left\lceil \frac{k}{3} \right\rceil$ and $t = 3 \left\lceil \frac{k}{3} \right\rceil - k$.*

In 2017 Henning and Lichiardopol [46] proved the following stronger results.

**Theorem 20 ([46])** *For $k \geq 2$, if $G$ is a connected graph with minimum degree $\delta \geq 2$ and maximum degree $\Delta$ and of order $n \geq \Delta + k - 1$, then*

$$\gamma_k(G) \leq \frac{n + \delta - \Delta}{\delta + k - 1}.$$

We remark that for $k \geq 2$, if $G$ is a connected graph with minimum degree $\delta \geq 2$, maximum degree $\Delta$ and of order $n < \Delta + k - 1$, then $(n + \delta - \Delta)/(\delta + k - 1) \leq (\delta + k - 2)/(\delta + k - 1) < 1$. However, in this case a vertex of maximum degree, $\Delta$, in $G$ is a $k$-dominating set, which implies that $\gamma_k(G) = 1$. Hence, the requirement that $n \geq \Delta + k - 1$ in the statement of Theorem 20 is essential.

As an immediate consequence of Theorem 20, we have the following general upper bound on the distance domination number of a graph in terms of its order and minimum degree.

**Theorem 21 ([46])** *For $k \geq 2$, if $G$ is a connected graph with minimum degree $\delta \geq 2$ and of order $n \geq \delta + k - 1$, then*

$$\gamma_k(G) \leq \frac{n}{\delta + k - 1}.$$

If $\delta = 2$ and $\Delta > 2$, then Theorem 20 implies strict inequality in the bound of Theorem 21. If $\delta \geq 3$, then Theorem 20 is a significant improvement on the bound of Theorem 15. As shown in [46], the result in Theorem 20 is also a stronger result than that of Theorem 18. If $\delta \geq \frac{1}{m}\left(e^{m-1} - m + i - 2\right)$, then as shown in [46], the inequality

$$\frac{1}{\delta + k - 1} < \frac{1 + \ln(m(\delta+1) + 2 - t)}{m(\delta+1) + 2 - t} \tag{1}$$

holds, implying that in this case the upper bound in Theorem 20 is an improvement on the upper bound of Theorem 19. In particular, if $k \leq 9$, then Inequality (1) holds (recalling that $\delta \geq 2$), and therefore the upper bound in Theorem 20 is an improvement on the upper bound of Theorem 19 for any value of $\delta \geq 2$. Further, as shown in [46] for any given $\delta \geq 2$ and for sufficiently large $k$, the upper bound in Theorem 20 is an improvement on the upper bound of Theorem 19.

However, none of the bounds in Theorem 18 to Theorem 21 is sharp or close to being sharp. In 2019 Dankelmann and Erwin [22] presented the following upper bound on the $p$-radius of a graph.

**Theorem 22 ([22])** *If $G$ is a connected graph of order $n$ with minimum degree $\delta$ and maximum degree $\Delta$, and $p$ is an integer satisfying $1 \leq p \leq n-1$, then*

$$\mathrm{rad}_p(G) \leq \frac{3(n+\delta-\Delta-1)}{(p+1)(\delta+1)} + 2.$$

As an immediate consequence of Theorem 22, we have the following upper bound on the $p$-radius of a graph in terms of its order and minimum degree.

**Theorem 23 ([22])** *If $G$ is a connected graph of order $n$ with minimum degree $\delta$ and $p$ is an integer satisfying $1 \leq p \leq n-1$, then*

$$\mathrm{rad}_p(G) \leq \frac{3(n-1)}{(p+1)(\delta+1)} + 2.$$

As a consequence of the Duality Lemma 5 and Theorem 23, Dankelmann and Erwin [22] obtained the following asymptotically sharp upper bound on the $k$-domination number of a graph in terms of its order and minimum degree, which significantly improves the earlier bounds.

**Theorem 24 ([22])** *For $3 \leq k \leq n-1$, if $G$ is a connected graph of order $n$, minimum degree $\delta$, and maximum degree $\Delta$, then*

$$\gamma_k(G) \leq \left\lceil \frac{3(n+\delta-\Delta-1)}{(k-2)(\delta+1)} \right\rceil - 1.$$

**Proof.** Let

$$p = \left\lceil \frac{3(n+\delta-\Delta-1}{(k-2)(\delta+1)} \right\rceil - 1.$$

By Theorem 22, we have

$$\mathrm{rad}_p(G) \leq \frac{3(n+\delta-\Delta-1)}{(p+1)(\delta+1)} + 2 \leq k.$$

The desired result now follows immediately from the Duality Lemma 5. □

As an immediate consequence of Theorem 24, we have the following upper bound on the distance domination number of a graph in terms of its order and minimum degree.

**Theorem 25 ([22])** *For $k \geq 3$ an integer, if $G$ is a connected graph of order $n$ and minimum degree $\delta$ where $k \leq n-1$, then*

$$\gamma_k(G) \leq \left\lceil \frac{3(n-1)}{(k-2)(\delta+1)} \right\rceil - 1.$$

To show that the upper bound of Theorem 25 is asymptotically sharp, Dankelmann and Erwin [22] construct the following family of graphs. Let $k$, $p$, and $\delta$ be fixed positive integers where $k \equiv 2 \pmod 3$ and $k \geq 5$, and where $\delta \geq 2$. Let $G_{k,p,\delta}$ be the graph constructed as follows. Let

$$V(G_{k,p,\delta}) = \bigcup_{i=1}^{p} \bigcup_{j=0}^{k} V_{ij},$$

where

$$|V_{ij}| = \begin{cases} \delta & \text{if } j = k-1 \\ \delta - 1 & \text{if } j \equiv 1 \pmod 3 \text{ and } j \neq k-1 \\ 1 & \text{otherwise.} \end{cases}$$

If $|V_{ij}| = 1$, then we let $V_{ij} = \{v_{ij}\}$ for $i \in [p]$ and $j \in [k]_0$. The edge set of $G_{k,p,\delta}$ is defined as follows. The subgraph induced by $\{v_{1,0}, \ldots, v_{p,0}\}$ is the path $P\colon v_{1,0}v_{2,0}\ldots v_{p,0}$ on $p$ vertices. For $i \in [p]$ and $j \equiv 1 \pmod 3$, the set $V_{ij}$ is an independent set and every vertex in $V_{ij}$ is adjacent to both vertices $v_{i,j-1}$ and $v_{i,j+1}$. To complete the construction of $G_{k,p,\delta}$ we add the edge $v_{ij}v_{i,j+1}$ for all $i \in [p]$ and $j \equiv 2 \pmod 3$. When $k = 8$ and $p = 4$, the graph $G_{k,p,\delta}$ is illustrated in Figure 5.

The graph $G_{k,p,\delta}$ satisfies

$$\gamma_k(G_{k,p,\delta}) = p = \frac{3n}{3 + (k+1)(\delta+1)} \geq \frac{3(n-1)}{(k+2)(\delta+1)}$$

since $\delta \geq 2$. Hence, the upper bound of Theorem 25 is asymptotically sharp. For fixed positive integers $k$, $p$, $\delta$, and $\Delta$ where $k \equiv 2 \pmod 3$ and $k \geq 5$, and where $2 \leq \delta < \Delta$, let $H_{k,p,\delta,\Delta}$ be the graph constructed in the same way as the graph $G_{k,p,\delta}$ except that $|V_{1,k-1}| = \Delta$. The resulting graph $H_{k,p,\delta,\Delta}$ has minimum degree $\delta$, maximum degree $\Delta$, and satisfies

$$\gamma_k(H_{k,p,\delta,\Delta}) = p = \frac{3(n+\delta-\Delta)}{3 + (k+1)(\delta+1)} \geq \frac{3(n+\delta-\Delta-1)}{(k+2)(\delta+1)},$$

implying that the bound given in Theorem 24 is asymptotically sharp. We close this section on upper bounds on the $k$-domination number of a graph with small girth.

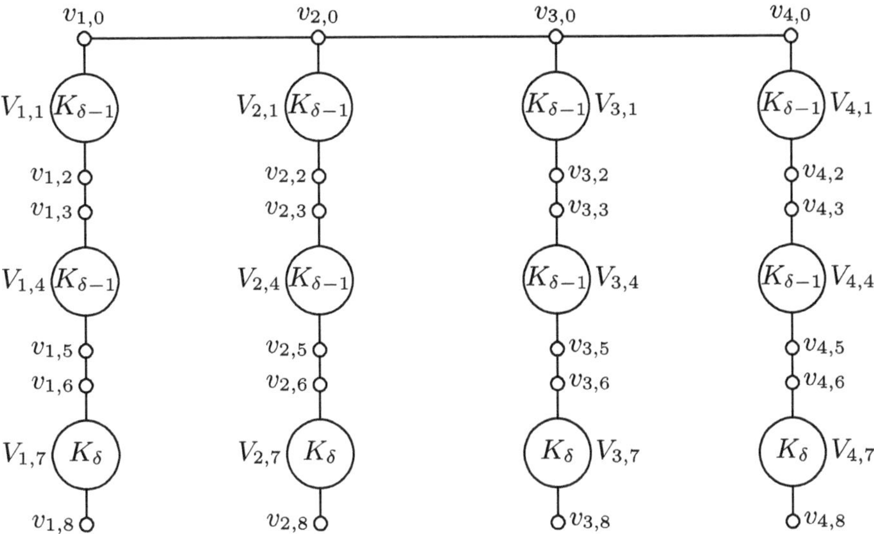


**Fig. 5** The graph $G_{8,4,\delta}$

**Theorem 26 ([22])** *If $G$ is a triangle-free connected graph of order $n$ and minimum degree $\delta$ and $k$ is an integer satisfying $5 \le k \le n-1$, then*

$$\gamma_k(G) \le \left\lceil \frac{2(n-1)}{\delta(k-4)} \right\rceil - 1.$$

**Theorem 27 ([22])** *If $G$ is a $C_4$-free connected graph of order $n$ and minimum degree $\delta$ and $k$ is an integer satisfying $5 \le k \le n-1$, then*

$$\gamma_k(G) \le \left\lceil \frac{5(n-1)}{(k-4)(1+\delta^2-2\lfloor \delta/2 \rfloor} \right\rceil - 1.$$

Both bounds in Theorems 26 and 27 are shown to be asymptotically sharp in [22].

**Theorem 28 ([46])** *For $k \ge 2$, if $G$ is a graph of girth at least 5, minimum degree $\delta$, maximum degree $\Delta$, and order $n$ where $n \ge \Delta(G^2) + k - 1$, then*

$$\gamma_{2k}(G) \le \frac{n+\delta^2-\delta\Delta}{\delta^2+k-1}.$$

## 8 Distance Domination in Graph Products

The *direct product graph*, $G \times H$, of graphs $G$ and $H$ is the graph with vertex set $V(G) \times V(H)$ and with edges $(g_1, h_1)(g_2, h_2)$, where $g_1g_2 \in E(G)$ and $h_1h_2 \in E(H)$. Davila et al. [24] proved the following result on the $k$-domination number of the direct product graph of two graphs.

**Theorem 29 ([24])** *If G and H are connected graphs, then*

$$\gamma_k(G \times H) \geq \gamma_k(G) + \gamma_k(H) - 1.$$

The *Cartesian product* $G \square H$ of graphs $G$ and $H$ is the graph whose vertex set is $V(G) \times V(H)$. Two vertices $(g_1, h_1)$ and $(g_2, h_2)$ are adjacent in $G \square H$ if either $g_1 = g_2$ and $h_1 h_2$ is an edge in $H$, or $h_1 = h_2$ and $g_1 g_2$ is an edge in $G$.

For $k \geq 2$, the $k$-domination number of the Cartesian product of graphs does not appear to have been studied much in the literature in comparison with the domination number of the Cartesian product of graphs. We present here some results on distance domination in $m \times n$ grids, or equivalently in the Cartesian product $P_m \square P_n$ of $P_m$ and $P_n$ (also denoted by $G_{m,n}$ in the literature). By Observation 10, we have the following result.

**Observation 30** *For $k \geq 1$ and $n \geq 2$,*

$$\gamma_k(P_1 \square P_n) = \gamma_k(P_n) = \left\lceil \frac{n}{2k+1} \right\rceil.$$

In 2005 Klobučar [59] determined the $k$-domination of a $2 \times m$ grid and $3 \times m$ grid.

**Theorem 31 ([59])** *For $k \geq 1$ and $n \geq 2$,*

$$\gamma_k(P_2 \square P_n) = \begin{cases} \frac{n}{2k} + 1 & \text{if } n \equiv 0 \,(mod\ 2k) \\ \lceil \frac{n}{2k} \rceil & \text{otherwise.} \end{cases}$$

To illustrate Theorem 31, Figure 6 shows an optimal solution for a 2-by-$n$ grid in the case when $n = 17$ and $k = 3$. In this case, $\gamma_3(P_2 \square P_{17}) = \lceil \frac{17}{6} \rceil = 3$.

**Theorem 32 ([59])** *For $k \geq 2$ and $n \geq 2$,*

$$\gamma_k(P_3 \square P_n) = \left\lceil \frac{n}{2k-1} \right\rceil.$$

To illustrate Theorem 57, Figure 7 shows an optimal solution for a 3-by-$n$ grid in the case when $n = 15$ and $k = 3$. In this case, $\gamma_3(P_3 \square P_{15}) = \left\lceil \frac{15}{5} \right\rceil = 3$.

It remains an open problem to determine the exact value of the $k$-domination of the $m \times n$ grid for all $k \geq 2$ and for $n \geq m \geq 3$. Klobučar [59] determined

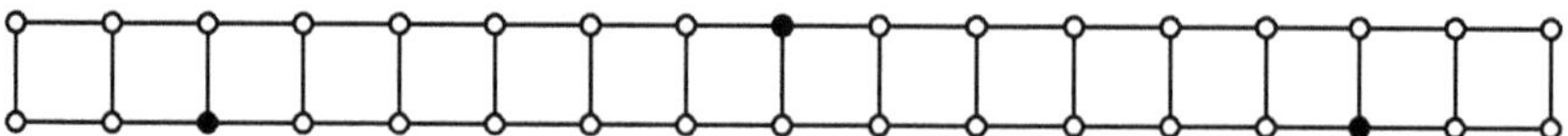

**Fig. 6** $\gamma_3(P_2 \square P_{17}) = 3$

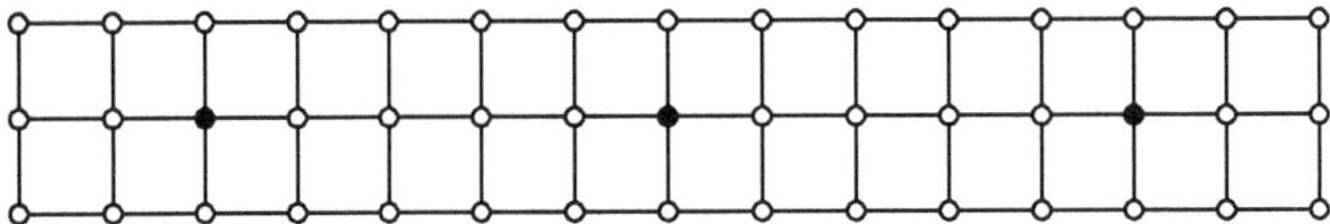

**Fig. 7** $\gamma_3(P_3 \square P_{15}) = 3$

the exact value of $\gamma_2(P_m \square P_n)$ when $m \in \{4, 5, 6, 7\}$. An upper bound on the 2-domination number $\gamma_2(P_n \square P_n)$ of a (balanced) $n \times n$ grid is given by Hemalatha and Jeyanthi [44].

In 2013, Fata, Smith, and Sundaram [29] gave a polynomial-time construction of a $k$-distance dominating set (using what they called $k$-diagonalization and $k$-projection arguments) in an $m \times n$ grid. As a consequence of their distributed algorithm, they obtained the following upper bound on the $k$-distance domination number of an $m \times n$ grid.

**Theorem 33 ([29], Theorem V.10)** *For $k \geq 1$ and $n \geq m \geq 2$,*

$$\gamma_k(P_m \square P_n) \leq \left\lceil \frac{(m+2k)(n+2k)}{2k^2+2k+1} + \frac{2k^2+2k+1}{4} \right\rceil.$$

When $k = 2$, the bound for the 2-domination number given in Theorem 33 is

$$\gamma_2(P_m \sqcap P_n) \leq \left\lceil \frac{(m+4)(n+4)}{13} + \frac{13}{4} \right\rceil.$$

In 2014, Blessing, Insko, Johnson, and Mauretour [10] improved this upper bound slightly for sufficiently large $m$ and $n$.

**Theorem 34 ([10], Theorem 3.8)** *For sufficiently large $m$ and $n$.*

$$\gamma_2(P_m \square P_n) \leq \left\lceil \frac{(m+4)(n+4)}{13} \right\rceil - 4.$$

Using algebraic and geometric arguments, in 2016 Farina and Grez [28] improved the upper bounds established in Theorem 33 by proving the following result.

**Theorem 35 ([28], Theorem 1.1)** *For $k \geq 1$ and $n \geq m \geq 2(2k^2 + 2k + 1)$,*

$$\gamma_k(P_m \square P_n) \leq \left\lceil \frac{(m+2k)(n+2k)}{2k^2+2k+1} \right\rceil - 4.$$

Klobučar [59] established the following result for $m \times n$ grids as $m$ and $n$ approach infinity. We note that one vertex can $k$-dominate at most $2k^2 + 2k + 1$ vertices. Therefore, every $\gamma$-set of $P_m \square P_n$ must contain at least $\frac{mn}{2k^2+2k+1}$ vertices.

Klobučar [59] constructed a $k$-dominating set that contains at most $\lceil \frac{mn}{2k^2+2k+1} \rceil + 2m + 2n$ vertices. From these observations, we have the following result.

**Theorem 36 ([59])** *For integers m and n,*

$$\lim_{m,n\to\infty} \frac{\gamma_k(P_m \,\square\, P_n)}{mn} = \frac{1}{2k^2+2k+1}.$$

Vizing's conjecture [77] from 1968 asserts that the domination number of the Cartesian product of two graphs is at least as large as the product of their domination numbers. We state next a distance version of Vizing's conjecture.

**Conjecture 1** *For $k \geq 1$ an integer and for every pair of finite graphs G and H,*

$$\gamma_k(G \,\square\, H) \geq \gamma_k(G)\,\gamma_k(H).$$

When $k = 1$, Conjecture 1 is Vizing's Conjecture on the domination number of Cartesian product of graphs stating that for every pair of finite graphs $G$ and $H$, $\gamma(G \,\square\, H) \geq \gamma(G)\,\gamma(H)$. This 50+ year old conjecture has yet to be settled and remains one of the outstanding unsolved problems in graph theory. If $k \geq \max\{\mathrm{rad}(G), \mathrm{rad}(H)\}$, then by Observation 2 we have $\gamma_k(G) = \gamma(G^k) = 1$ and $\gamma_k(H) = \gamma(H^k) = 1$, and so in this case $\gamma_k(G \,\square\, H) \geq 1 = \gamma_k(G)\,\gamma_k(H)$. Also if $k \geq \mathrm{rad}(G \,\square\, H)$, then by Observation 2 we note that

$$\gamma_k(G \,\square\, H) = \gamma((G \,\square\, H)^k) = 1 = \gamma_k(G)\,\gamma_k(H).$$

The distance version of Vizing's conjecture, namely Conjecture 1, therefore trivially holds for sufficiently large $k$. However it would be interesting to find the smallest value of $k$ for which Conjecture 1 holds. Vizing's Conjecture is that $k = 1$ suffices.

## 9 Nordhaus–Gaddum Type Results

In 1956 Nordhaus and Gaddum [63] established sharp bounds on the sum and product of the chromatic numbers of a graph $G$ and its complement $\overline{G}$. Since then such results have been given for many parameters, as discussed in the 81-page survey by Aouchiche and Hansen [2] on Nordhaus–Gaddum type relations. They include a 1972 result by Jaeger and Payan [64] on the domination number that if $G$ is a graph of order $n \geq 2$, then $2 \leq \gamma(G) + \gamma(\overline{G}) \leq n + 1$ and $1 \leq \gamma(G)\gamma(\overline{G}) \leq n$, and these bounds are sharp. This result was generalized to the distance domination number in 1991 by Henning, Oellermann, and Swart [47].

**Theorem 37 ([47])** *For $k \geq 2$, if G is a graph of order $n \geq k + 1$, then*

$$2 \le \gamma_k(G) + \gamma_k(\overline{G}) \le n + 1 \quad \textit{and} \quad 1 \le \gamma_k(G)\gamma_k(\overline{G}) \le n,$$

*and these bounds are sharp.*

The lower bounds of Theorem 37 are reached for any graph $G$ such that $\text{diam}(G) \le 2$ and $\text{diam}(\overline{G}) \le 2$. For example, one could take $G$ to be a self-complementary graph of diameter 2 and order $n$ (mod 4) $\in \{0, 1\}$. For such graphs $G$, we have $\gamma_k(G) = \gamma_k(\overline{G}) = 1$. The upper bounds of Theorem 37 are achieved if $G$ or $\overline{G}$ is the complete graph $K_n$. We remark that the $k$-domination number of a disconnected graph is by linearity the sum of the $k$-domination numbers of its components. If we restrict both $G$ and $\overline{G}$ to be connected, then the upper bounds of Theorem 37 can be improved, as shown in [47].

**Theorem 38 ([47])** *For $k \ge 2$, if both $G$ and $\overline{G}$ are connected graphs of order $n \ge k + 1$, then*

$$2 \le \gamma_k(G) + \gamma_k(\overline{G}) \le \left\lfloor \frac{n}{k+1} \right\rfloor + 1 \quad \textit{and} \quad 1 \le \gamma_k(G)\gamma_k(\overline{G}) \le \left\lfloor \frac{n}{k+1} \right\rfloor,$$

*and these bounds are sharp.*

That the upper bounds in Theorem 38 are sharp may be seen by taking the $k$-corona $G = H \circ P_k$ of a graph connected $H$ for $k \ge 2$ as defined in Section 7. In this case, recall that $G$ has order $n = (k+1)|V(H)|$ and satisfies $\gamma_k(G) = n/(k+1)$ and $\gamma_k(\overline{G}) = 1$. Thus, the graph $G$ achieves the upper bounds in Theorem 38.

## 10 Distance Domination Versus Iterated Domination

In 2012 Bacsó and Tuza [7] studied the structure of $k$-dominating subgraphs, and looked for conditions under which a graph surely admits a $k$-dominating set that induces a subgraph belonging to a prescribed graph class $\mathcal{D}$. Among the requirements, they show that connectivity plays a central role both for dominating subgraphs and for the graphs to be dominated. In this section, we present selected results from their paper [7].

Given a class $\mathcal{D}$ of finite simple graphs closed under connected induced subgraphs, they completely characterize those graphs $G$ in which every connected induced subgraph has a connected $k$-dominating subgraph isomorphic to some $D \in \mathcal{D}$. They apply this result to prove that the class of graphs hereditarily $\mathcal{D}$-dominated within distance $k$ is the same as the one obtained by iteratively taking the class of graphs hereditarily dominated by the previous class in the iteration chain.

In order to precisely state their results, we shall need the following notation. The class $\mathcal{D}$ is *compact* if it is closed under taking connected induced subgraphs, and the class $\mathcal{D}$ is *concise* if it is compact and contains connected graphs only. A graph $G$ is defined as *minimal not-in-$\mathcal{D}$* if it is connected, $G \notin \mathcal{D}$, and all proper connected induced subgraphs of $G$ are in the class $\mathcal{D}$. Moreover, a graph $G$ is *$\mathcal{D}$-dominated* if there exists a dominating connected induced subgraph $D \in \mathcal{D}$ in G. A graph $G$ is *hereditarily dominated* by $\mathcal{D}$ if each of its connected induced subgraphs is $\mathcal{D}$-dominated. Further, a connected graph $G$ is defined as *minimal non-$\mathcal{D}$*-dominated if it is not $\mathcal{D}$-dominated but all of its proper connected induced subgraphs are.

The class $Dom_k \mathcal{D}$ is defined to consists of the graphs $G$ for which every connected induced subgraph $H$ of $G$ is $k$-dominated by some connected graph $D \in \mathcal{D}$. We remark that $Dom_k$ is an operator acting on graph classes. For simplicity, they omit the subscript "1" and write $Dom_1\mathcal{D} = Dom\,\mathcal{D}$.

For any operator $\phi$, operating on a set $X$ and having its values in $X$, for an arbitrary element $x \in X$ and integer $k \geq 1$, the notation $\phi^k(x)$ is defined to mean the element obtained from $x$ by applying the operation $\phi$ $k$ times. They also write $\phi^0(x) = x$, and denote by $\mathcal{M}_k\mathcal{D}$ the set of minimal connected forbidden induced subgraphs for the class of graphs $Dom^k(\mathcal{D})$. We remark that $\mathcal{M}_k\mathcal{D}$ is well-defined because membership in $Dom^k(\mathcal{D})$ is an additive induced hereditary property for all $\mathcal{D}$ and all $k \geq 1$.

The *leaf-graph* of a connected graph $H$, denoted $F(H)$, is defined as the graph obtained from $H$ by attaching a leaf (or, equivalently, a pendant edge) to each non-cut vertex of $H$. Let

$$\mathcal{F}_k := \{F^k(M) : M \text{ minimal not-in-}\mathcal{D}\}.$$

For a class $\mathcal{C}$ of graphs, they denote by $\Theta(\mathcal{C})$ the minimum element of the set $\{j : P_j \notin \mathcal{C}\}$, where $P_j$ denotes an induced path on $j$ vertices, also referred to as a chordless path on $j$ vertices. Further, the classes are grouped into two types of classes, namely,

*Type 1.* All chordless paths are elements of $\mathcal{D}$.
*Type 2.* Some paths are not in $\mathcal{D}$.

We are now in a position to state the main results of Bacsó and Tuza [7].

**Theorem 39 ([7])** *If $\mathcal{D}$ is a compact class of graphs, then for all integers $k \geq 1$ we have*

$$Dom_k\mathcal{D} = Dom^k(\mathcal{D}).$$

We remark that the equation in the statement of Theorem 39 is not true in general, and they show that for any $k \geq 2$, there exists some class $\mathcal{D}$ of graphs for which the equation $Dom_k\mathcal{D} = Dom^k(\mathcal{D})$ is not valid.

In order to prove Theorem 39, they first prove the following lemma, which they call the *Legged Cycle Lemma*, which characterizes the class $\mathcal{M}_k\mathcal{D}$ of minimal connected forbidden induced subgraphs of $Dom^k(\mathcal{D})$. Recall that $C_n$ denotes a (chordless) cycle on $n$ vertices.

**Lemma 40 ([7])** *If $\mathcal{D}$ is a compact class of graphs, then the following holds.*

(a) *If $\mathcal{D}$ is of Type 1, then $\mathcal{M}_k\mathcal{D} = \mathcal{F}_k$.*
(b) *If $\mathcal{D}$ is of Type 2 and $\theta := \Theta(\mathcal{G})$ where $\mathcal{G}$ is the class of connected graphs in $Dom^{k-1}(\mathcal{D})$, then*

$$\mathcal{M}_k\mathcal{D} = \mathcal{F}_k \cup \{F^i(C_{\theta+2-2i}) : i \in [k-1]_0\}.$$

## 11 Distance Domination and Average Distance

In 2009, Tian and Xu [71] studied tight upper bounds on the average distance of a graph in terms of its order and distance domination number. The *average* (or *mean*) *distance* $\mu(G)$ of a graph $G$ is the average value of distance over all pairs of vertices of $G$; that is, if $G$ has order $n$, then

$$\mu(G) = \frac{1}{n(n-1)} \sum_{u,v \in V(G)} d_G(u, v).$$

For $k \geq 1$, Tian and Xu [71] provide sharp upper bounds on the average distance of a graph $G$ with given order $n$ and $k$-domination number $\gamma_k(G)$. Recall that by Theorem 15, for $k \geq 1$ if $G$ is a connected graph of order $n \geq k+1$, then $\gamma_k(G) \leq n/(k+1)$. The result of Tian and Xu [71] on the average distance differs depending on whether $\gamma_k(G) \leq \lceil n/(2k+1) \rceil$ or $\gamma_k(G) > \lceil n/(2k+1) \rceil$.

We first consider the case when $\gamma_k(G) \leq \lceil n/(2k+1) \rceil$. Recall that a *double star* is a tree with exactly two (adjacent) non-leaf vertices. Further if one of these vertices is adjacent to $\ell_1$ leaves and the other to $\ell_2$ leaves, then we denote the double star by $S(\ell_1, \ell_2)$. For example, the double star $S(1, 1)$ is the path $P_4$. For positive integers $n$ and $\gamma_k$, let $G_{n,\gamma_k,k}$ be the graph obtained as follows. If $\gamma_k = \lceil n/(2k+1) \rceil$, let $G_{n,\gamma_k,k}$ consist of a path $P_n$. If $\gamma_k \leq n/(2k+1)$, let $G_{n,\gamma_k,k}$ be the graph obtained from a double star

$$S\left(\left\lfloor \frac{1}{2}(n-(2k+1)\gamma_k+2) \right\rfloor, \left\lceil \frac{1}{2}(n-(2k+1)\gamma_k+2) \right\rceil\right)$$

by subdividing the edge joining the two central vertices of the star exactly $(2k+1)\gamma_k - 4$ times. For example, when $n = 20$, $\gamma_2 = 3$ and $k = 2$, the graph $G_{20,3,2}$ is shown in Figure 8. We note if $G = G_{20,3,2}$, then $\gamma_2(G) = 3$, and the darkened vertices in Figure 8 form a $\gamma_2$-set of $G$.

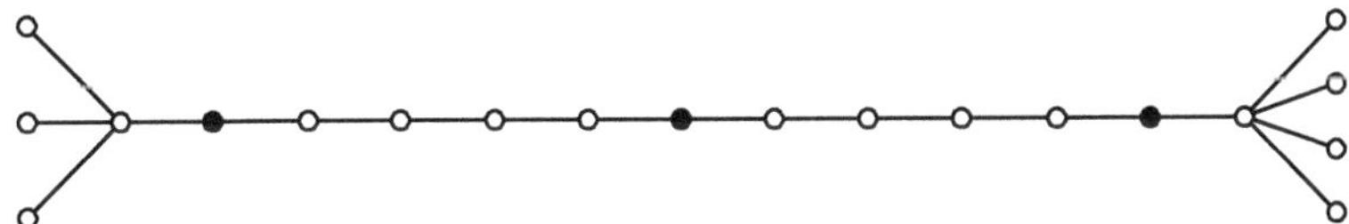

**Fig. 8** The graph $G_{20,3,2}$

For positive integers $n$ and $\gamma_k$ where $\gamma_k \le \lceil \frac{n}{2k+1} \rceil$, let

$$\Phi(n,k) = (n-(2k+1)\gamma_k)(n-(2k+1)\gamma_k+2)(2n+(2k+1)\gamma_k-7).$$

We are now in a position to state the result due to Tian and Xu [71] in the case when $\gamma_k(G) \le \lceil \frac{n}{2k+1} \rceil$.

**Theorem 41 ([71])** *If $G$ is a connected graph of order $n$ with $k$-domination number $\gamma_k \le \lceil \frac{n}{2k+1} \rceil$, then*

$$\mu(G) \le \begin{cases} \dfrac{n+1}{3} - \dfrac{\Phi(n,k)}{6n(n-1)} & \text{if } \gamma_k \le \dfrac{n}{2k+1} \text{ and } n-\gamma_k \text{ is even} \\ \dfrac{n+1}{3} - \dfrac{\Phi(n,k)-3((2k+1)\gamma_k-3)}{6n(n-1)} & \text{if } \gamma_k \le \dfrac{n}{2k+1} \text{ and } n-\gamma_k \text{ is odd} \\ \dfrac{n+1}{3} & \text{if } \gamma_k = \left\lceil \dfrac{n}{2k+1} \right\rceil. \end{cases}$$

*Further, equality holds if and only if $G = G_{n,\gamma_k,k}$.*

We consider next the second case when $\lceil \frac{n}{2k+1} \rceil < \gamma_k(G) \le \frac{n}{k+1}$.

Let $s$ and $t$ be the quotient and the reminder of the division of $(2k+1)\gamma_k - n$ by $k$; that is, $(2k+1)\gamma_k - n = sk + t$ where $s \ge 0$ and $t \in [k-1]_0$. Further, let

$$\begin{aligned} \Psi_1(n,k) &= n - s(k+1) - t \\ \Psi_2(n,k) &= s(k+1) \\ \Psi_3(n,k) &= 3n - s(k+1) \\ \Psi_4(n,k) &= 2n - s(k+1) - 2t. \end{aligned}$$

Let $H_{n,\gamma_k,k}$ be the graph obtained from a path $P_{2n-(2k+1)\gamma_k}$ given by $v_1 v_2 \dots v_{2n-(2k+1)\gamma_k}$ by applying the following:

- Attaching a path of length $k$ to each vertex $v_i$ for $1 \le i \le \lceil s/2 \rceil$.
- Attaching a path of length $k$ to each vertex $v_{2n-(2k+1)\gamma_k+1-j}$ for $1 \le j \le \lceil s/2 \rceil$.
- Attaching a path of length $t$ to the vertex $v_{\lceil s/2 \rceil + \Psi_1(n,k)+t-k}$.

For example, when $n = 25$ and $k = 3$ the graph $H_{25,3}$ is shown in Figure 9 (here $s = 5$, $t = 2$, and $\Psi_1(n,k) = 3$). We note if $G = H_{25,6,3}$, then $\gamma_3(G) = 6$.

For positive integers $n$ and $\gamma_k$ where $\gamma_k > \lceil \frac{n}{2k+1} \rceil$, let

$$\begin{aligned} \Theta_1(n,k) &= ((2k+1)\gamma_k - n - t - 2k)(\Psi_3(n,k) - 2(k+1)) + 3t(\Psi_4(n,k) - 2) \\ \Theta_2(n,k) &= \tfrac{1}{n(n-1)}(\Psi_1(n,k) + t - k - 1) \\ \Theta_3(n,k) &= ((2k+1)\gamma_k - n - t - 3k)(\Psi_3(n,k) - (k+1)) + 3t(\Psi_4(n,k) + 2k) \\ &\quad +3(k\Psi_4(n,k) + (k-1)t - k(k+1)) \\ \Theta_4(n,k) &= \tfrac{1}{n(n-1)}(\Psi_1(n,k) + t). \end{aligned}$$

**Fig. 9** The graph $H_{25,6,3}$

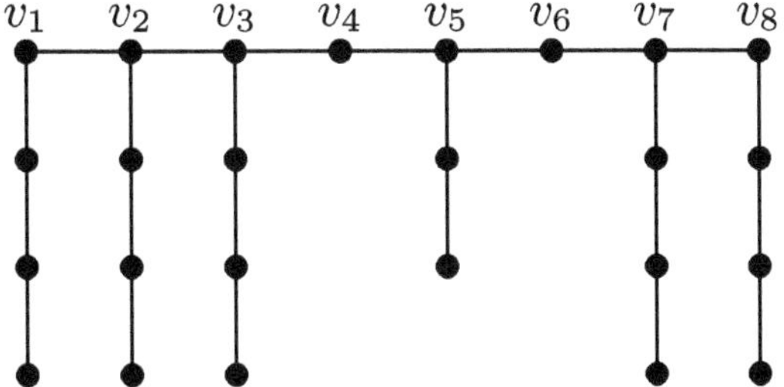


We are now in a position to state the result due to Tian and Xu [71] in the case when $\gamma_k(G) > \lceil \frac{n}{2k+1} \rceil$.

**Theorem 42 ([71])** *If $G$ is a connected graph of order $n$ with $k$-domination number $\gamma_k > \lceil \frac{n}{2k+1} \rceil$, then*

$$\mu(G) \leq \begin{cases} \dfrac{n+1}{3} - \dfrac{\Psi_2(n,k)\Theta_1(n,k)}{6n(n-1)} - 2t(k-t)\Theta_2(n,k) & \text{if } \dfrac{\gamma_k - n - t}{k} \text{ is even} \\ \dfrac{n+1}{3} - \dfrac{(\Psi_2(n,k) - k - 1)\Theta_3(n,k)}{6n(n-1)} - 2t(k-t)\Theta_4(n,k) & \text{if } \dfrac{\gamma_k - n - t}{k} \text{ is odd.} \end{cases}$$

*Further, equality holds if and only if $G = H_{n,\gamma_k,k}$.*

## 12 Well-$k$-Dominated Graphs

Finbow, Hartnell, and Nowakowski [30] introduced the concept of a well-dominated graph. In [30], a graph is defined to be *well-dominated* if every minimal dominating set has the same cardinality. We extend here the definition of well-dominated graphs to distance well-dominated. Let $k \geq 1$ be an integer, and let $G$ be a graph. Recall that the *upper k-domination number* $\Gamma_k(G)$ of a graph $G$ is the maximum cardinality taken over all minimal $k$-dominating sets of $G$. In 1993 Hattingh and Henning [41] defined a graph to be *well-k-dominated* if every minimal $k$-dominating set of the graph has the same cardinality. Hence, $G$ is well-$k$-dominated if and only if $\gamma_k(G) = \Gamma_k(G)$.

A parameter of interest here is the $k$-packing number defined by Meir and Moon [62]. A set $S$ of vertices of a graph $G$ is a *k-packing* of $G$ if $d_G(x, y) > k$ for all pairs of distinct vertices $x$ and $y$ in $S$. The *k-packing number* $\rho_k(G)$ of $G$ is the maximum cardinality of a $k$-packing set in $G$. In 1988 Domke, Hedetniemi, and Laskar [25] established the following important relationship between the $2k$-packing number and the $k$-domination number of a connected block graph, where we recall that a *block graph* is a graph in which each block is complete. We note that a tree is a block graph where each block is $K_2$.

**Proposition 43 ([25])** *For $k \geq 1$ an integer, if $G$ is a connected block graph, then $\rho_{2k}(G) = \gamma_k(G)$.*

In 1993 Hattingh and Henning [41] characterized block graphs that are well-$k$-dominated. Since a graph is well-$k$-dominated if and only if each of its components is well-$k$-dominated, we restrict ourselves to connected graphs. The following result of [41] extends a 1991 result due to Topp and Volkmann [73] from trees to connected block graphs.

**Theorem 44 ([41])** *If $G$ is a connected block graph, then the following statements are equivalent.*

(1) $\gamma_k(G) = \rho_{2k}(G) = r$.
(2) *One of the following statements hold.*
  (a) $\mathrm{diam}(G) \leq k$ *and* $r = 1$.
  (b) *There exists a decomposition of $G$ into $r$ subgraphs $G_1, \ldots, G_r$ in such a way that the following hold.*
    (i) *The graph $G_i$ is a connected block graph and* $\mathrm{diam}(G_i) = k$ *for* $i \in [r]$.
    (ii) *For each $i \in [r]$, there exists a vertex $u_i \in V(G_i) \setminus V(G_0)$ such that $d_G(u_i, V(G_0)) = k$, where $G_0$ is the subgraph of $G$ induced by those edges that do not belong to any of the subgraphs $G_1, \ldots, G_r$.*
    (iii) *There is at most one edge with one end in $V(G_i)$ and the other end in $V(G_j)$ for $1 \leq i < j \leq r$.*
(3) *The graph $G$ is well-$k$-dominated.*

If $G$ is any connected graph, then the conditions given in Theorem 44(b) are easily seen to be sufficient for $G$ to be well-$k$-dominated. That the conditions are not necessary for any connected graph $G$, may be seen by considering the graph $H_k$ constructed as follows. Let $T$ be a complete binary tree of height $k$ in which every leaf is at level $k$ (and so $T$ has order $2^{k+1} - 1$). Let $T_1$ and $T_2$ be two (disjoint) copies of $T$. Finally, let $H_k$ be obtained from $T_1$ and $T_2$ by inserting a 1-factor between the corresponding leaves of $T_1$ and $T_2$. The graphs $H_1$ and $H_2$, for example, are shown in Figure 10(a) and 10(b), respectively. The resulting graphs $H_k$ is well-$2k$-dominated with $\gamma_{2k}(H_k) = 2$, but $H_k$ does not satisfy the conditions given in Theorem 44(b).

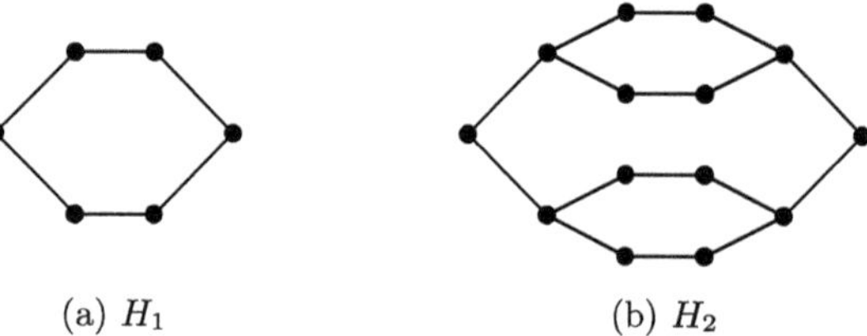


**Fig. 10** The graphs $H_1$ and $H_2$

**Corollary 45 ([41])** *If $G$ is a connected block graph, then $\gamma_{2k}(G) = \rho_k(G)$ if and only if well-2k-dominated.*

## 13 Distance Domination Critical Graphs

As remarked by Sumner [69], "graphs which are minimal or critical with respect to a given property frequently play an important role in the investigation of that property. Not only are such graphs of considerable interest in their own right, but also a knowledge of their structure often aids in the development of the general theory." In this subsection we consider graphs which are critical with respect to their $k$-domination number. We examine the effects on $\gamma_k(G)$ when $G$ is modified by deleting a vertex. Unless otherwise stated, the results of this subsection are from [51].

Brigham, Chinn, and Dutton [15] define a vertex $v$ of a graph $G$ to be *critical* if $\gamma(G - v) < \gamma(G)$. The graph $G$ is *vertex domination-critical* (or *$\gamma$-critical*) if each vertex is critical. For $k \geq 1$ an integer, a vertex $v$ of a graph $G$ is defined in [51] to be *k-critical* if $\gamma_k(G - v) < \gamma_k(G)$. The graph $G$ is *vertex k-domination-critical* (or *$\gamma_k$-critical*) if each vertex of $G$ is $k$-critical. If $G$ is $\gamma_k$-critical and $\gamma_k(G) = \ell$, we say $G$ is *$(\gamma_k, \ell)$-critical.* For example, the graphs $G_2$ and $G_3$ of Figure 11 are $(\gamma_2, 2)$-critical and $(\gamma_3, 2)$-critical, respectively. Further, for integers $k \geq 1$ and $\ell \geq 2$, the cycle $C_{(\ell-2)(2k+1)+1}$ is $(\gamma_k, \ell)$-critical. Note that $\gamma_1$-critical graphs are vertex domination-critical graphs.

As pointed out in [15], vertex domination-critical graphs can be used to model multiprocessor networks. Similarly, $\gamma_k$-critical graphs can serve as models for multiprocessor networks in which a subset of processors (represented by an $k$-dominating set) can transmit messages to all remaining processors in at most $k$ time units (where a time unit is the time it takes for a message to be sent between adjacent processors). These $\gamma_k$-critical networks have the desirable characteristics that any processor can be in a minimum set of "$k$-dominating" processors and the failure of any processor leaves a network which requires one fewer "dominating" processors.

**Fig. 11** The graphs $G_2$ and $G_3$

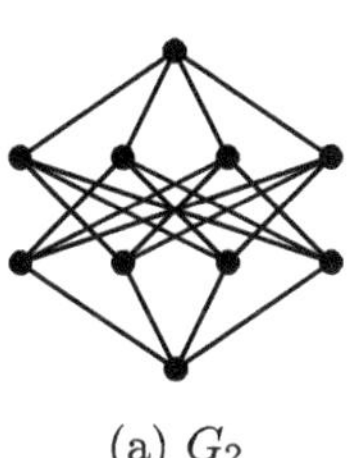

(a) $G_2$

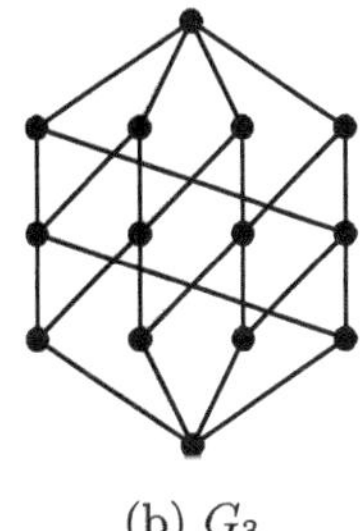

(b) $G_3$

We present next some basic properties of $\gamma_k$-critical graphs.

**Lemma 46 ([51])** *If $G$ is a connected $\gamma_k$-critical graph of order $n \geq 2$, then the following hold.*

(a) $N_k[u] \not\subseteq N_k[v]$ *for every pair of distinct vertices $u$ and $v$ of $G$.*
(b) $\delta_k(G) \geq 2k$.
(c) $n \leq (\Delta_k(G)+1)(\gamma_k(G)-1)+1$.

We remark that the upper bound on the order of a graph $G$ given in Lemma 46(c) in fact holds if $G$ contains at least one $k$-critical vertex. That this upper bound on the order of a connected $\gamma_k$-critical graph is best possible may be seen by considering the infinite class of $\gamma_k$-critical graphs $G_{s,r}$, where $r \geq 2$, $s = 2\ell$ for some positive integer $\ell$ and $G_{s,r} \cong C^{k\ell}_{(r-1)(sk+1)+1}$ (that is, the $k\ell$th power of the cycle on $(r-1)(sk+1)+1$ vertices). Then $\gamma_k(G_{s,r}) = r$ while $\gamma_k(G_{s,r} - v) = r - 1$ for every vertex $v$ of $G_{s,r}$. This example serves to illustrate the existence of $(\gamma_k, r)$-critical graphs of connectivity $s$ for every integer $s \geq 2$.

**Lemma 47 ([51])** *For $k \geq 2$, if $G$ is a $(\gamma_k, 2)$-critical graph of order $n$, then the following hold.*

(a) $\delta_k(G) = \Delta_k(G)$.
(b) $n = \delta_k(G) + 2 \geq 2k + 2$.

The following question is posed in [51]: "For integers $k \geq 2$ and $\ell \geq 2$, is it true that if $G$ is a $(\gamma_k, \ell)$-critical graph of order $n$, then $n \geq (\delta_k(G)+1)(\gamma_k(G)-1)+1$?" If this question can be answered in the affirmative, then this result and Lemma 46(b), imply that $n \geq (2k+1)(\ell-1)+1$ for every connected $(\gamma_k, \ell)$-critical graph $G$ of order $n \geq 2$. Moreover, the cycle $C_{(\ell-1)(2k+1)+1}$ shows that this bound is attainable. Lemma 47(b) solves this problem in the special case of $(\gamma_k, 2)$-critical graphs.

**Lemma 48 ([51])** *For integers $k \geq 2$ and $\ell \geq 2$, a graph $G$ is $(\gamma_k, \ell)$-critical graph if and only if $G^k$ is a $(\gamma, \ell)$-critical graph.*

We next describe a construction technique which can be employed to produce large classes of $\gamma_k$-critical graphs. We note that a graph is $\gamma_k$-critical if and only if each of its components is $\gamma_k$-critical. A similar statement holds for the blocks of $G$. Suppose $H$ and $G$ are nonempty graphs. Let $u$ and $w$ be two non-isolated vertices of $H$ and $G$, respectively. By $(H \cdot G)(u, w : v)$ we mean the graph obtained from $H$ and $G$ by identifying the vertices $u$ and $w$ in a vertex labeled $v$.

**Theorem 49 ([51])** *If $u$ and $w$ are two non-isolated vertices of two nonempty graphs $H$ and $G$, respectively, then*

$$\gamma_k(H) + \gamma_k(G) - 1 \leq \gamma_k((H \cdot G)(u, w : v)) \leq \gamma_k(H) + \gamma_k(G).$$

*Furthermore, the following hold.*

(a) *If $v$ is a critical vertex of $(H \cdot G)(u, w : v)$, then $\gamma_k((H \cdot G)(u, w : v)) = \gamma_k(H) + \gamma_k(G) - 1$ and $u$ and $w$ are critical vertices of $H$ and $G$, respectively.*

(b) *If $u$ and $w$ are critical vertices of $H$ and $G$, respectively, then $\gamma_k((H \cdot G)(u, w : v)) = \gamma_k(H) + \gamma_k(G) - 1$ and $v$ is a critical vertex of $(H \cdot G)(u, w : v)$.*

**Theorem 50 ([51])** *If $u$ and $w$ are two non-isolated vertices of two nonempty graphs $H$ and $G$, respectively, then $(H \cdot G)(u, w : v)$ is $\gamma_k$-critical if and only if $H$ and $G$ are both $\gamma_k$-critical.*

The following result establishes a relationship between the $k$-domination number of a graph and the $k$-domination number of its blocks.

**Theorem 51 ([51])** *A graph $G$ is $\gamma_k$-critical if and only if each block of $G$ is $\gamma_k$-critical. Further, if $G$ is $\gamma_k$-critical with blocks $G_1, \ldots, G_b$, then*

$$\gamma_k(G) = \sum_{i=1}^{b} \gamma_k(G_i) \ - (b-1).$$

As an illustration of Theorem 51, let $B_1, B_2, \ldots, B_{2k+3}$ be $(2k+2)$-cycles. For each $i \in [2k+2]$, let $u_i$ be a vertex of $B_i$ and let $w_1, w_2, \ldots, w_{2k+2}$ be the vertices of $B_{2k+3}$. Let $G$ be obtained by identifying $u_i$ and $w_i$ for all $i \in [2k+2]$. Then, $B_1, B_2, \ldots, B_{2k+3}$ are the blocks of $G$. Since each $B_i$ is $(\gamma_k, 2)$-critical, the graph $G$ is $\gamma_k$-critical by Theorem 51. Furthermore,

$$\gamma_k(G) = \sum_{i=1}^{2k+3} \gamma_k(B_i) \ - (2k+2) = 2k+4.$$

This example serves to illustrate the existence of $\gamma_k$-critical graphs that contain cut-vertices. We remark that attempts to date to characterize $\gamma_k$-critical graphs have been unsuccessful. The following result shows that it is not possible to do so in terms of forbidden subgraphs.

**Proposition 52 ([51])** *For $k \geq 1$ and for any graph $G$ there is a $(\gamma_k, \ell)$-critical graph $H$ containing $G$ as an induced subgraph.*

## 14 The Distance Domatic Number of a Graph

For $k \geq 1$, the *distance $k$-domatic number* of a graph $G$, denoted $\text{dom}_k(G)$, is the maximum order of a partition on the vertex set $V(G)$ into classes each of which is a distance $k$-dominating set of $G$. The concept of distance domatic number of a graph was introduced and first studied in 1983 by Zelinka [80], and further studied, for example, by Kämmerling and Volkmann [56]. In this introductory paper, he made the following observations.

**Observation 53 ([80])** *For positive integers $k$ and $\ell$, if $G$ is a graph of order $n$, then the following hold.*

(a) *If* $\mathrm{diam}(G) \le k$, *then* $\mathrm{dom}_k(G) = n$.
(b) *If* $k \ge \ell \ge 1$, *then* $\mathrm{dom}_\ell(G) \le \mathrm{dom}_k(G)$.
(c) $\mathrm{dom}_k(G) \le \frac{n}{\gamma_k(G)}$.

We present here some selected results on the distance 2-domatic number. The *total domatic number* of a graph $G$, denoted by $\mathrm{tdom}(G)$ and first defined by Cockayne, Dawes, and Hedetniemi [19], is the maximum number of total dominating sets into which the vertex set of $G$ can be partitioned. The parameter $\mathrm{tdom}(G)$ is equivalent to the maximum number of colors in a (not necessarily proper) coloring of the vertices of a graph where every color appears in every open neighborhood. Chen, Kim, Tait, and Verstraete [18] called this the *coupon coloring problem*. Kiser [57] shows that the distance 2-domatic number of a graph is at least twice its total domatic number.

**Theorem 54 ([57])** *If $G$ is a graph with no isolated vertex, then* $\mathrm{dom}_2(G) \ge 2\mathrm{tdom}(G)$.

As a consequence of Theorem 54, we have the following result on the distance 2-domatic number of a regular graph.

**Theorem 55 ([18])** *If $G$ is a $d$-regular graph, then for $d$ sufficiently large we have*

$$\mathrm{dom}_2(G) \ge \frac{2(1 - o(1))d}{\log d}.$$

**Theorem 56 ([40])** *There exist planar graphs $G$ with* $\mathrm{dom}_2(G) \ge 8$, *and there exist toroidal graphs $G$ with* $\mathrm{dom}_2(G) \ge 10$.

Kiser and Haynes [58] studied distance-2 domatic numbers of grid graphs.

**Theorem 57 ([58])** *For $n \ge m \ge 2$, the following holds.*

(a) $\mathrm{dom}_2(P_2 \square P_n) = 5$ *for* $n \ge 5$.
(b) *For* $m \in \{3, 4, 5\}$ *and* $n \ge 6$, *we have* $\mathrm{dom}_2(P_m \square P_n) = 6$.
(c) *For the infinite grid when* $m, n \to \infty$, *we have* $\mathrm{dom}_2(P_m \square P_n) = 13$.

The distance $k$-domatic number problem was studied from a slightly different angle by Alon et al. [1] who consider factor distance $k$-domatic coloring in graph. For $r \ge 2$, an $r$-factorization of a graph $G = (V, E)$ is a collection $S_1, S_2, \ldots, S_r$ of connected (not necessarily edge-disjoint) spanning subgraphs (the factors) of $G$, whose union is $G$. A *$k$-domatic coloring of $G$* is a coloring of $V$ such that each color class is a distance $k$-dominating set of $G$. The coloring is an *all-factor $k$-domatic coloring* of $G$ with respect to $S_1, S_2, \ldots, S_r$ if each color class is a $k$-dominating set of every factor. Given integers $t$ and $r$, $\alpha(t, r)$ is the minimum $k$ such that every $r$-factorization of every graph on at least $t$ vertices has an all-factor $k$-domatic coloring with $t$ colors. Alon et al. [1] determine upper and lower bounds on $\alpha(t, r)$. Surprisingly, they show that the upper bound is finite and does not depend on the order of the graph.

**Theorem 58 ([1])** *The following holds.*

(a) *For $r \geq 2$ and $t \geq r$, we have $\alpha(t, r) \leq \lceil \frac{3}{2}(rt - 1) \rceil$.*
(b) *For $t \geq 2$ and $r \geq 4$, we have $\alpha(t, r) \geq \Omega(t \log r)$.*

**Theorem 59 ([1])** *For every $r \geq 1$ and $t \geq 1$, $\alpha(t, r) \leq O(t \log(rt))$.*

For a survey on distance $k$-domatic number we refer the reader to Chapter 13.3.3 of Zelinka [81] on the domatic number of graphs and their variants.

## 15 Fractional Distance Domination

In 1993 Hattingh, Henning, and Walters [42] introduced and first studied the concept of fractional distance domination. For a graph $G = (V, E)$ with vertex set $V$ and edge set $E$ and for a real-valued function $f : V \to [0, 1]$, the *weight* of $f$ is $\mathrm{w}(f) = \sum_{v \in V} f(v)$. Further, for $S \subseteq V$ we define $f(S) = \sum_{v \in S} f(v)$. In particular, we note that $\mathrm{w}(f) = f(V)$. A real-valued function $f : V \to [0, 1]$ is a *fractional dominating function* of $G$ if $f(N[v]) \geq 1$ for each $v \in V$. The minimum weight of a fractional dominating function of graph $G$ is the *fractional domination number* $\gamma_f(G)$ of $G$. Thus,

$$\gamma_f(G) = \min \{\mathrm{w}(f) \mid f \text{ is a fractional dominating function for } G\}.$$

Fractional domination in graphs was formally defined in 1987 by Stephen Hedetniemi reporting on results in [43] at the Eighteenth Southeastern Conference, and in 1988 by Domke, Hedetniemi, and Laskar [25] and in 1990 by Grinstead and Slater [37]. For $k \geq 1$ an integer, a real-valued function $f : V \to [0, 1]$ is a *fractional $k$-dominating function* of $G$ if $f(N_k[v]) \geq 1$ for each $v \in V$. The minimum weight of a fractional $k$-dominating function of graph $G$ is the *fractional $k$-domination number* $\gamma_g(G)$ of $G$. Thus,

$$\gamma_{kf}(G) = \min \{\mathrm{w}(f) \mid f \text{ is a fractional } k\text{-dominating function for } G\}.$$

A fractional $k$-dominating function $f$ of a graph $G$ is *minimal* if there does not exist a fractional $k$-dominating function $g$ of $G$, $f \neq g$, for which $g(v) \leq f(v)$ for every $v \in V$. The following property of a minimal fractional $k$-dominating function is established in [42].

**Lemma 60 ([42])** *Let $f$ be a fractional $k$-dominating function of a graph $G$. Then, $f$ is a minimal fractional $k$-dominating function of $G$ if and only if whenever $f(v) > 0$ for some vertex $v \in V$, there exists some vertex $u \in N_k[v]$ such that $f(N_k[u]) = 1$.*

The fractional $k$-domination number is readily viewed as a linear program. Thus we can talk of minimum, rather than infimum. The dual of the fractional $k$-domination number is the *fractional k-packing number*. This is the linear programming relaxation of the packing number.

A real-valued function $f: V \to [0, 1]$ is called a *fractional k-packing function* of $G$ if $f(N_k[v]) \le 1$ for every $v \in V$. A fractional $k$-packing function $f$ is maximal if there does not exist a fractional $k$-packing function $g: V \to [0, 1]$, $f \neq g$, for which $g(v) \ge f(v)$ for every $v \in V$. This is equivalent to saying that a fractional $k$-packing function $f$ is maximal if for every vertex $v$ with $f(v) < 1$, there exists a vertex $u \in N_k[v]$ such that $f(N_k[u]) = 1$. The maximum weight of a fractional $k$-packing function of graph $G$ is the *fractional k-packing number* $\rho_{kf}(G)$ of $G$. We have:

**Fractional $k$-domination $\gamma_{kf}$**

minimize $\quad \mathrm{w}(f) = \sum_{v \in V} f(v)$

$$\text{subject to:} \quad \begin{cases} \sum_{u \in N_k[v]} f(u) \ge 1 \\ 0 \le f(v) \le 1 \text{ for all } v \in V \end{cases}$$

**Fractional $k$-packing $\rho_{kf}$**

maximize $\quad \mathrm{w}(f) = \sum_{v \in V} f(v)$

$$\text{subject to:} \quad \begin{cases} \sum_{u \in N_k[v]} f(u) \le 1 \\ 0 \le f(v) \le 1 \text{ for all } v \in V \end{cases}$$

Thus, by the fundamental theorem of linear programming, it follows that:

**Theorem 61** *For $k \ge 1$ an integer and any graph $G$, it holds that*

$$\gamma_{kf}(G) = \rho_{kf}(G).$$

Hence by Theorem 61 if there exists a minimal fractional $k$-dominating function $f$ of a graph $G$ and a maximal fractional $k$-packing function $g$ of $G$ with $\mathrm{w}(f) = \mathrm{w}(g)$, then $\gamma_{kf}(G) = \mathrm{w}(f) = \mathrm{w}(g) = \rho_{kf}(G)$.

We note that if $k \ge \mathrm{rad}(G)$, then $\gamma_{kf}(G) = \gamma_k(G) = 1$. Hence, it is only of interest to consider graphs $G$ for which $k < \mathrm{rad}(G)$. Further we note that $\gamma_{kf}(G) \ge 1$ for every graph $G$. If $f$ is the characteristic function of a $\gamma_k$-set $S$ of $G$, and so $f(v) = 1$ if $v \in S$ and $f(v) = 0$ otherwise, then $f$ is a fractional $k$-dominating function of $G$, and so $\gamma_{kf}(G) \le \mathrm{w}(f) = \gamma_k(G)$. Thus, as an immediate consequence of Theorem 15, we have the following upper bound on the fractional $k$-domination number.

**Theorem 62 ([62])** *For $k \ge 1$, if $G$ is a connected graph of order $n \ge k + 1$, then*

$$\gamma_{fk}(G) \le \frac{n}{k+1}.$$

Recall that $G^k$ denotes the $k$-th power of a graph $G$.

**Observation 63** *For $k \geq 1$, if $G$ is a connected graph, then $\gamma_{kf}(G) = \gamma_f(G^k)$.*

Recall that the minimum and maximum $k$-degrees among all vertices of $G$ are denoted by $\delta_k(G)$ and $\Delta_k(G)$, respectively. Assigning the value of $1/(\Delta_k(G)+1)$ on every vertex of $G$ is a fractional $k$-packing function of $G$ and assigning the value of $1/(\delta_k(G)+1)$ on every vertex is a fractional $k$-dominating function of $G$. Thus for all graphs, we have the following bounds involving the minimum and maximum $k$-degrees.

**Observation 64** *For $k \geq 1$, if $G$ is a graph of order $n$ with minimum $k$-degree $\delta_k(G) = \delta_k$ and maximum $k$-degree $\Delta_k(G) = \Delta_k$, then*

$$\frac{n}{\Delta_k + 1} \leq \rho_{kf}(G) = \gamma_{kf}(G) \leq \frac{n}{\delta_k + 1}.$$

As a consequence of Observation 64, we immediately determine the fractional $k$-domination number of a graph in which every vertex has the same $k$-degree.

**Theorem 65** *For $k \geq 1$, if $G$ is a graph of order $n$ in which every vertex has $k$-degree equal to $r$, then $\gamma_{kf}(G) = \frac{n}{r+1}$.*

As an application of Theorem 65, we have the following result first observed in 2012 Arumugam, Mathew, and Karuppasamy [6].

**Proposition 66 ([6])** *For $k \geq 1$ and for a hypercube $Q_n$, we have*

$$\gamma_{kf}(G) = \frac{2^n}{\binom{n}{0} + \binom{n}{0} + \cdots + \binom{n}{k}}.$$

**Proof.** For any two vertices $x = (x_1, x_2, \ldots, x_n)$ and $y = (y_1, y_2, \ldots, y_n)$ in $Q_n$, $d(x, y) \leq k$ if and only if $x$ and $y$ differ in at most $k$ coordinates. This implies that every vertex of $Q_n$ has the same $k$-degree, namely $r = \binom{n}{0} + \binom{n}{0} + \cdots + \binom{n}{k}$. The desired result now follows from Theorem 65 noting that $|V(Q_n)| = 2^n$. □

In 2012 Arumugam, Mathew, and Karuppasamy [6] determined the fractional $k$-domination number of a $2 \times n$ grid and the fractional 2-domination number of a $3 \times n$ grid.

**Theorem 67 ([6])** *For $k \geq 1$ and $n \geq 2$,*

$$\gamma_{kf}(P_2 \square P_n) = \begin{cases} \frac{n(n+2k)}{2k(n+k)} & \text{if } n \equiv 0 \ (mod\ 2k) \\ \lceil \frac{n}{2k} \rceil & \text{otherwise.} \end{cases}$$

**Theorem 68 ([6])** *For $n \geq 2$, $\gamma_{2f}(P_3 \square P_n) = \gamma_2(P_3 \square P_n) = \lceil \frac{n}{3} \rceil$.*

We say a function $f : V \to [0, 1]$ is an *efficient fractional $k$-dominating function* if for every vertex $v$ it holds that $f(N_k[v]) = 1$. Such a function $f$ is also a fractional

$k$-packing function of $G$, implying by Theorem 61 that $\gamma_{kf}(G) = \rho_{kf}(G)$. We state this formally as follows.

**Theorem 69** *For $k \geq 1$, if a graph $G$ has an efficient fractional $k$-dominating function $f$, then $\gamma_{kf}(G) = \rho_{kf}(G) = \mathrm{w}(f)$.*

To illustrate Theorem 69, if $G$ is a cycle $C_n$ where $n \geq 2k+1$ and $k \geq 1$, then the function $f$ that assigns a weight of $\frac{1}{2k+1}$ to every vertex of $G$ is an efficient fractional $k$-dominating function of $G$, implying by Theorem 69 that $\gamma_{kf}(G) = \rho_{kf}(G) = \mathrm{w}(f) = \frac{n}{2k+1}$.

For $k \geq 1$, an *efficient $k$-dominating set* of a graph $G$ is a $k$-dominating set $S$ of $G$ such that $|N_k[v] \cap S| = 1$ for every vertex $v \in V(G)$. If a graph $G$ has an efficient $k$-dominating set $S$, then the characteristic function of $S$ is an efficient fractional $k$-dominating function of $G$. However, the converse is not true: having an efficient fractional $k$-dominating function does not imply an efficient $k$-dominating set. Consider, for example, a cycle $C_n$ where $n > 2k+1$ and $n \not\equiv 0 \pmod{2k+1}$. As observed earlier, the function $f$ that assigns a weight of $\frac{1}{2k+1}$ to every vertex of $G$ is an efficient fractional $k$-dominating function of $G$. However, $G$ does not have an efficient $k$-dominating set $S$.

More generally, if $G$ is a graph of order $n$ in which every vertex has the same $k$-degree, say $r$, and $n > r+1$ and $n \not\equiv 0 \pmod{r+1}$, then the function $f$ that assigns a weight of $\frac{1}{r+1}$ to every vertex of $G$ is an efficient fractional $k$-dominating function of $G$. However, $G$ does not have an efficient $k$-dominating set $S$. As an example, the 5-prism, $C_5 \square K_2$, which is the Cartesian product of a 5-cycle with a copy of $K_2$, is a graph of order $n = 10$ in which every vertex has the same 2-degree, namely 7. The function $f$ that assigns a weight of $\frac{1}{8}$ to every vertex of $G$ is an efficient fractional 2-dominating function of $G$, as illustrated in Figure 12. However, $G$ does not have an efficient 2-dominating set.

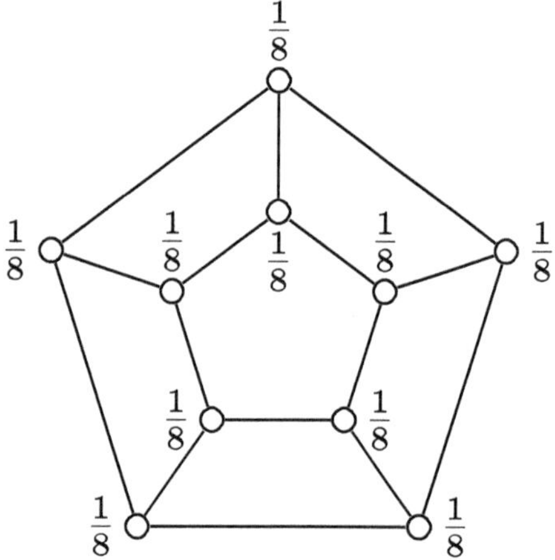


**Fig. 12** An efficient fractional 2-dominating function of $C_5 \square K_2$.

# 16 Distance Independent Domination in Graphs

Recall that an *independent dominating set* of a graph $G$ is a set $S$ of vertices of $G$ that is both independent and dominating. Equivalently, $S$ is a dominating set and no two vertices of $S$ are adjacent. The *independent domination number* $i(G)$ of $G$ is the minimum cardinality of an independent dominating set of $G$. In this section, we consider two extensions of the definition of independent domination in graphs to distance independent domination.

## *16.1 Independent k-Domination in Graphs*

The first extension we present is that due to Beineke and Henning [8] in 1994. For $k \geq 1$ an integer, an *independent k-dominating set* of a graph $G$ is defined in [8] as a $k$-dominating set of $G$ that has the additional property of being independent. The *independent k-domination number* $id(k, G)$ of $G$ is the minimum cardinality among all independent $k$-dominating sets of $G$. In particular, we note that an independent 1-dominating set of $G$ is precisely an independent dominating set of $G$, and $id(1, G) = i(G)$.

In [8], it is shown that the decision problem corresponding to the problem of computing $id(k, G)$ is $NP$-complete, even when restricted to bipartite graphs, by demonstrating a polynomial-time reduction from the decision problem **Independent Dominating Set** (**IDOM**). The decision problem **IDOM** for the independent domination number of a graph is known to be NP-complete (see Garey and Johnson [36]), and remains NP-complete for the class of bipartite graphs, as shown by Corneil and Perl [20]. Since the problem of computing $id(k, G)$ appears to be a difficult one, it is desirable to find good upper bounds on this parameter. For $k \geq 2$, Beineke and Henning [8] established the following upper bound on $id(k, G)$ for a connected graph $G$.

**Theorem 70 ([8])** *For $k \geq 2$, if $G$ is a connected graph of order $n \geq k + 1$, then $id(k, G) \leq \frac{n}{k}$, and this bound is asymptotically best possible.*

That the bound given in Theorem 70 is in a sense best possible, may be seen by considering the connected graph $G$ constructed as follows: For positive integers $\ell$ and $b$, let $G$ be obtained from a complete graph on $b$ vertices by attaching to each of its vertices $\ell$ disjoint paths of length $k$. The graph $G$ is illustrated in Figure 13. Then, $id(k, G) = (b - 1)\ell + 1$ and $n = |V(G)| = b(k\ell + 1)$, and so

$$\frac{id(k, G)}{n} = \frac{b\ell - \ell + 1}{bk\ell + b} = \frac{1 - \frac{1}{b} + \frac{1}{b\ell}}{k + \frac{1}{\ell}} \overset{b,\ell\to\infty}{\longrightarrow} \frac{1}{k}.$$

If we restrict our attention to trees, then the upper bound on the independent $k$-domination number given in Theorem 70 can be improved. In this case, the upper

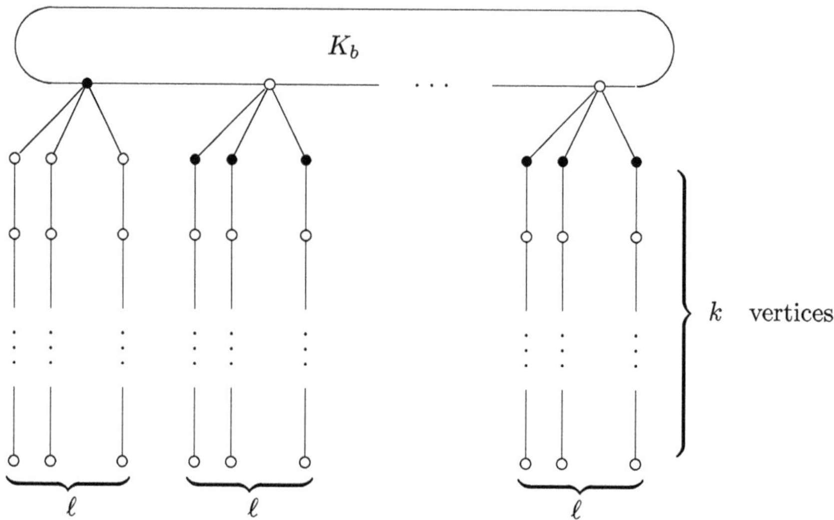


**Fig. 13** A graph $G$.

bound on the $k$-domination number established by Meir and Moon [62] also holds for the independent $k$-domination number. We state this formally as follows.

**Theorem 71 ([62])** *For $k \geq 1$, if $G$ is a tree of order $n \geq k+1$, then $id(k, G) \leq \frac{n}{k+1}$.*

The upper bound in Theorem 71 is tight, may be seen by taking $T$ to be the tree obtained from a path $P_b$ on $b \geq 1$ vertices by attaching a path of length $k$ to each vertex of the path. The resulting tree $T$ satisfies $id(k, T_k) = b = \frac{n}{k+1}$.

In 1996 Gimbel and Henning [39] improved the upper bound in Theorem 71. Their proof is constructive in that it presents an algorithm that constructs an independent $k$-dominating set of sufficiently small cardinality.

**Theorem 72 ([39])** *For $k \geq 1$, if $G = (V, E)$ is a connected graph of order $n \geq k+1$, then*

$$id(k, G) \leq \frac{n + k + 1 - 2\sqrt{n}}{k}$$

*and this bound is sharp.*

## 16.2 k-Independent k-Domination in Graphs

The second extension we present is that due to Henning, Oellermann, and Swart [47] in 1991. For $k \geq 1$ an integer, a set $S$ of vertices of a graph $G$ is defined in [47] to be *k-independent* in $G$ if every vertex of $S$ is at distance at least $k + 1$ from every other vertex of $S$ in $G$. Further, a *k-independent k-dominating set* of $G$ is defined in [47] as a set that is both $k$-independent and $k$-dominating set. The *k-independent k-domination number* $i_k(G)$ of $G$ is the minimum cardinality among all $k$-independent $k$-dominating sets of $G$. In particular, we note that a 1-independent 1-dominating set of $G$ is precisely an independent dominating set of $G$, and $i_1(G) = i(G)$. For the tree $T$ shown in Figure 2, the set $S = \{v_3, v_7\}$ is a 3-independent 3-dominating set of $T$ of minimum cardinality 2, and so $i_3(T) = 4$.

This concept of independent distance domination in graphs finds applications in many situations and structures which give rise to graphs. Consider, for example, the facility location problem discussed in Section 2. To avoid interference and contamination, it may be required that no two facilities be within $k$ blocks of each other, and facilities should then be sited at points corresponding to vertices in a minimum $k$-independent $k$-dominating set.

With this extension of the independent domination number, we have the following observation, where as before $G^k$ denotes the $k$-th power of the graph $G$.

**Observation 73** *If $G$ is a connected graph, then $i_k(G) = i(G^k)$.*

We present next a series of results from Fricke, Hedetniemi, and Henning [33] in 1995.

**Lemma 74 ([33])** *For each positive integer $k$, there exists a connected graph $G$ so that every spanning tree $T$ of $G$ satisfies $i_k(T) < i_k(G)$.*

**Lemma 75 ([33])** *The tree $T'$ obtained from a tree $T$ by joining a new vertex to some vertex of $T$ satisfies $i_k(T') \geq i_k(T)$ for $k \in [2]$.*

It is somewhat surprising that Lemma 75 is not true for $k \geq 3$ as shown in [33]. In [33] it is shown that the decision problem corresponding to the problem of computing $i_k(G)$ is NP-complete, even when restricted to bipartite graphs, by describing a polynomial transformation from the known NP-complete decision problem *One-In-Three 3SAT* (see [36]). Since the problem of computing $i_k(G)$ appears to be a difficult one, it is desirable to find good bounds on this parameter. The following lower and upper bounds on $i_k(G)$ in terms of the maximum $k$-degree $\Delta_k(G)$ are presented in [33].

**Theorem 76 ([33])** *For $k \geq 1$, if $G$ is a graph of order $n$ and maximum $k$-degree $\Delta_k \geq 2k$, then*

$$i_k(G) \geq \frac{n}{(\frac{k+1}{k})\Delta_k - 1}.$$

*Furthermore, we have equality if and only if all components of $G$ are either paths or cycles on $\ell \equiv 0 \pmod{2k+1}$ vertices, or have order exactly $2k + 1$.*

**Theorem 77 ([33])** *For $k \geq 1$, if $G$ is a graph of order $n$ and maximum $k$-degree $\Delta_k \geq 2k$, then $i_k(G) \leq n - \Delta_k(G)$, and this bound is sharp.*

To see that the bound in Theorem 77 is sharp, consider the graph $G$ obtained from a star $K_{1,\ell}$, $\ell \geq 2$, by subdividing $\ell - 1$ of the edges $k$ times and one edge $k - 1$ times. Then $i_k(G) = \ell$, $n = |V(G)| = \ell(k + 1)$ and $\Delta_k(G) = k\ell$; consequently, $i_k(G) = n - \Delta_k(G)$. The following upper bound on $i_k(G)$ in terms of the order of the graph is given in [35].

**Theorem 78 ([35])** *For $k \geq 1$, if $G$ is a connected graph of order $n > \lceil \frac{1}{2}(k + 1) \rceil$, then*

$$i_k(G) < \frac{n}{\lceil \frac{1}{2}(k + 1) \rceil}.$$

The bound given in Theorem 78 is shown in [35] to be asymptotically best possible. One such construction is the following. For $k \geq 2$ an even integer and for $\ell \geq 1$ an integer, let $T'_{k,\ell}$ be a complete $\ell$-ary tree of height $\lceil (k+1)/2 \rceil$ in which every leaf is at level $\lceil (k + 2)/2 \rceil$. Further, let $T_{k,\ell}$ be the tree obtained from $T'_{k,\ell}$ by attaching a path of length $\lfloor (k+1)/2 \rfloor$ to every leaf of $T'_{k,\ell}$. For $k \geq 3$ an odd integer and for $\ell \geq 1$ an integer, let $T'_{k,\ell}$ be a complete $\ell$-ary tree of height $(k + 1)/2$ in which every leaf is at level $(k+1)/2$. Further, let $T_{k,\ell}$ be the tree of order $n$ obtained from $T'_{k,\ell}$ by attaching a path of length $(k + 1)/2$ to every leaf of $T'_{k,\ell}$. In both cases, the resulting tree $T_{k,\ell}$ of order $n$ is such that

$$i_k(T_{k,\ell}) \to \frac{n}{\lceil \frac{1}{2}(k + 1) \rceil} \text{ as } \ell \to \infty.$$

**Theorem 79 ([35])** *For $k \geq 1$, if $i_k(n) = \max\{ i_k(G) \mid G$ is a connected graph of order $n\}$, then*

$$\frac{i_k(n)}{n} \to \frac{1}{\lceil \frac{1}{2}(k + 1) \rceil} \textit{ as } n \to \infty.$$

The result of Theorem 78 is extended to $\ell$-connected graphs in [35].

**Theorem 80 ([35])** *For $k \geq 1$, if $G$ is an $\ell$-connected graph of order $n > \lfloor \frac{1}{2}k \rfloor \ell + 1$, then*

$$i_k(G) < \frac{n}{\lfloor \frac{k}{2} \rfloor \ell + 1},$$

*and this bound is asymptotically best possible.*

We next present results on bounds relating $i_k$ and $\gamma_k$. Since every $k$-independent $k$-dominating set of a graph $G$ is a $k$-dominating set of $G$, we have the following relation.

**Observation 81** *For $k \geq 1$ an integer and for every graph $G$, $\gamma_k(G) \leq i_k(G)$.*

We note that strict inequality may occur in Observation 81. For example, for integers $\ell_1, \ell_2 \geq 1$, let $G$ be the graph obtained from a double star $S(\ell_1, \ell_2)$ by subdividing each edge $k - 1$ times. The resulting graph $G$ satisfies $\gamma_k(G) = 2$ and $i_k(G) = 3$.

In order to present the next two results, we first define a generalization of $K_{1,r}$ for $r \geq 3$. Let $G$ be a graph that contains a $k$-independent set $I_r$ of $r$ vertices and a vertex $v$ of $G$ that is within distance $k$ from every vertex of $I_r$. We refer to a connected subgraph of $G$ of minimum size that contains all the vertices in $I_r \cup \{v\}$ as a $k$-generalized $K_{1,r}$ in $G$. We note that if $k = 1$ and $r = 3$, then a $k$-generalized $K_{1,r}$ is an induced copy of $K_{1,3}$, also called a claw in the literature. The following result established a sufficient condition for the $k$-independent $k$-domination number of a graph to equal its $k$-domination number. We remark that this result generalizes that of Allan and Laskar [4].

**Theorem 82 ([49])** *For $k \geq 1$, if a graph $G$ contains no $k$-generalized $K_{1,3}$, then $\gamma_k(G) = i_k(G)$.*

The following result generalizes a result due to Bollobás and Cockayne [11].

**Theorem 83 ([49])** *For integers $k \geq 1$ and $r \geq 2$, if a graph $G$ contains no $k$-generalized $K_{1,r+1}$, then $i_k(G) \leq (r - 1)\gamma_k(G) - (r - 2)$.*

The following relation between $i_k$ and $\gamma_k$ is established in [49].

**Theorem 84 ([49])** *For $k \geq 1$, if $G$ is a connected graph of order $n \geq k + 1$, then $i_k(G) + k\gamma_k(G) \leq n$.*

That the bound given in Theorem 88 is best possible, may be seen by considering a graph $G$ of order $n$ obtained from a connected graph $H$ by attaching a path of length $k$ to each vertex of $H$. The resulting graph $G$ satisfies $i_k(G) = \gamma_k(G) = |V(H)|$ and $i_k(G) + k\gamma_k(G) = (k + 1)|V(H)| = n$.

## 17 Distance Total Domination in Graphs

Let $k \geq 1$ be an integer and let $G = (V, E)$ be a graph. Henning, Oellermann, and Swart [47] defined a set $S$ of vertices of $G$ to be a *distance total $k$-dominating set* of $G$ if every vertex is within distance $k$ from some vertex of $S$ other than itself; that is, for every vertex $v \in V$, we have $d_G(v, S \setminus \{v\}) \leq k$. The *distance total $k$-domination number* of $G$, denoted, $\gamma_k^t(G)$, is the minimum cardinality of a distance total $k$-dominating set of $G$. Throughout this section, for notational simplicity we write "total $k$-dominating set" and "total $k$-domination number" rather than the more accurate terminology "distance total $k$-dominating set" and "distance total $k$-domination number," respectively.

A total $k$-dominating set of $G$ of cardinality $\gamma_k^t(G)$ is called a *$\gamma_k^t$-set of $G$*. We note that the parameter $\gamma_k^t(G)$ is defined only for graphs with no isolated vertex. When $k = 1$, we note that a total dominating set is a distance total 1-dominating set and $\gamma_t(G) = \gamma_k^t(G)$, where recall that $\gamma_t(G)$ denotes the total domination number of $G$. For the tree $T$ shown in Figure 2, the set $S = \{v_2, v_3, v_7, v_8\}$, for example, is a distance total 2-dominating set of $T$ of minimum cardinality 4, and so $\gamma_2^t(T) = 4$.

As remarked in [45, 47], the concept of distance total domination in graphs finds applications in many situations and structures which give rise to graphs. Consider, for example, the facility location problem discussed in Section 2. For practical reasons it may be desirable that each facility be sited within $k$ blocks of some other facility (for instance, to cope with emergencies and breakdowns), in which case the use of a total $k$-dominating set of minimum cardinality is indicated. Corresponding applications to the design of computer networks and defense systems exist. For example, the problem of finding total $k$-dominating sets has potential applications to storage location problems in a computer network. Suppose $G$ is a graph that models a multiprocessor computer network where the vertices of $G$ represent processors and an edge of $G$ indicates that the processors corresponding to its end vertices can communicate directly. The same data are to be stored at each member of a subset $S$ of these processors so that any processor in the rest of the network can be sent this information in at most $k$ time units (where a time unit is the time it takes for the data to be sent between adjacent processors). Furthermore, we wish to select $S$ in such a way so that if a processor should lose its data due to failure, then it can obtain the data from another element of $S$ in at most $k$ time units. (We assume here that at most one of the elements of $S$ will fail at any one time.) The problem of finding such a set $S$ corresponds to the problem of finding a total $k$-dominating set of vertices of $G$, and an optimal solution to the problem has cardinality $\gamma_k^t(G)$.

As is the case for the $k$-domination number of a graph, it appears to be a computationally difficult problem to determine the total $k$-domination number of a graph. There is no known efficient algorithm for this purpose. The following result in [47] provides a tight upper bound on the total $k$-domination number of a connected graph.

**Theorem 85 ([47])** *For an integer $k \geq 1$, if $G$ is a connected graph of order $n \geq 2$, then $\gamma_k^t(G) = 2$ if $n \leq 2k$ and*

$$\gamma_k^t(G) \leq \frac{2n}{2k+1}$$

*if $n \geq 2k + 1$.*

If $2 \leq n \leq 2k$, then any central vertex of $G$ together with any other vertex of $G$ forms a total $k$-dominating set of $G$. For $n \geq 2k + 1$, an algorithm for finding a total $k$-dominating set $D_k$ of vertices of a connected graph $G$ of order $n$ such that $|D_k| \leq \frac{2n}{2k+1}$ is given in [50]. This algorithm is based on the proof of Theorem 85 given in [47]. That the bound of Theorem 85 is tight may be seen as follows. For integers $k \geq 1$ and $\ell \geq 1$, let $G$ be obtained from an arbitrary connected graph $H$ of order $\ell$

by attaching a path of length $2k$ to each vertex of $H$ so that the added paths of vertex disjoint. The resulting graph $G$ has order $n = (2k+1)\ell$ and $\gamma_k^t(G) = 2\ell = \frac{2n}{2k+1}$.

We close this section with the following Nordhaus-Gaddum type results with respect to the total $k$-domination number obtained in [47].

**Theorem 86 ([47])** *For integers $n \geq k+1 \geq 2$, if $G$ and $\overline{G}$ are connected graphs of order $n$, then the following holds.*

(a) $\gamma_k^t(G) + \gamma_k^t(\overline{G}) = 4$ *and* $\gamma_k^t(G) \cdot \gamma_k^t(\overline{G}) = 4$ *if* $n \leq 2k+1$.

(b) $4 \leq \gamma_k^t(G) + \gamma_k^t(\overline{G}) \leq \frac{2n}{2k-1} + 2$ *and* $4 \leq \gamma_k^t(G) \cdot \gamma_k^t(\overline{G}) \leq \frac{4n}{2k+1}$ *if* $n \geq 2k+2$.

**Theorem 87 ([47])** *For integers $n \geq k+1 \geq 2$, if $G$ and $\overline{G}$ are graphs of order $n$ with no isolated vertices, then*

$$\gamma_k^t(G) + \gamma_k^t(\overline{G}) \leq n+2 \quad \textit{and} \quad \gamma_k^t(G)\gamma_k^t(\overline{G}) \leq 2n.$$

The bounds given in Theorem 86 are best possible in the following sense. For integers $n$, $k$, and $\ell$ where $2 \leq \ell \leq \frac{2n}{2k+1}$, let $G = J(n,k,\ell)$ be the graph obtained from a complete graph $K_{n-2k(\ell-1)}$ by selecting one vertex, $v$ say, from the complete graph and attaching to it $\ell - 1$ paths of length $2k-1$ so that the resulting paths have only the vertex $v$ in common. We note that the resulting $G$ has order $n$ and has a unique $\gamma_k^t$-set that consists of the vertex $v$ and the $\ell - 1$ vertices at distance $k$ from $v$ that belong to the $\ell - 1$ paths attached to $v$. In particular, $\gamma_k^t(G) = \ell$. Further, we note that $\gamma_k^t(\overline{G}) = 2$. Thus, $\gamma_k^t(G) + \gamma_k^t(\overline{G}) = \ell + 2$ and $\gamma_k^t(G) \cdot \gamma_k^t(\overline{G}) = 2\ell$. We note that if $\ell = \frac{2n}{2k+1}$, then $\gamma_k^t(G) \cdot \gamma_k^t(\overline{G}) = \frac{4n}{2k+1}$.

We next discuss bounds relating $\gamma_k^t$ and $i_k$. In 1996 Fricke, Henning, Oellermann, and Swart [34] established the following relationship between the total $k$-domination number and $k$-independent $k$-domination number of a graph. We remark that in the special case when $k = 1$, this result was first proven by Allan, Laskar, and Hedetniemi [5] and simplifies to the following statement: If $G$ is a connected graph of order $n \geq 3$, then $i(G) + \gamma_t(G) \leq n$.

**Theorem 88 ([34])** *For $k \geq 1$ if $G$ is a connected graph of order $n \geq 2k+1$, then*

$$i_k(G) + k\,\gamma_k^t(G) \leq n.$$

We remark that the proof of Theorem 88 presented in [34] is algorithmic in nature. That the bound in Theorem 88 is best possible may be seen by letting $G$ be the graph obtained from a star $K_{1,r}$ where $r \geq 1$, by subdividing each edge $2k$ times. The resulting graph $G$ has order $n = (2k+1)r + 1$ and satisfies $\gamma_k^t(G) = 2r$ and $i_k(G) = r+1$, whence $i_k(G) + k\,\gamma_k^t(G) = n$.

We discuss next bounds relating $\gamma_k$ and $\gamma_k^t$. Since $\gamma_k(G) \leq i_k(G)$ for all graphs $G$, as an immediate consequence of Theorem 88 we have the following result.

**Corollary 89 ([34])** *For $k \geq 1$ if $G$ is a connected graph of order $n \geq 2k+1$, then*

$$\gamma_k(G) + k\,\gamma_k^t(G) \leq n.$$

We remark that Tuza [75] provided an elegant proof of Corollary 89. That the bound given in Corollary 89 is best possible, may be seen by considering the graph $G$ obtained from a connected graph $H$ by attaching a path of length $k$ to each vertex of $H$. The resulting graph $G$ satisfies $\gamma_k(G) = \gamma_k^t(G) = |V(H)|$ and $\gamma_k(G) + k\gamma_k^t(G) = (k+1)|V(H)| = |V(G)|$. However, as pointed out by Tuza [75], the bound is best possible in a much stronger sense as well; namely, its left-hand side cannot be replaced by $(1-\epsilon)\gamma_k(G) + (k+\epsilon)\gamma_k^t(G)$, for any $\epsilon > 0$. To see this, take $r-1$ (where $r \geq 1$) vertex-disjoint paths $T_1, \ldots, T_{r-1}$ of length $k$ and one path of length $2k-1$. Joining a new vertex $v$ with one end-vertex of each path $T_i$, we obtain a tree $T$ of order $n = (r+1)(k+1) - 1$, with $\gamma_k(G) = r$ and $\gamma_k^t(G) = r+1$, hence $\gamma_k(T) + k\,\gamma_k^t(T) = n$ while $(1-\epsilon)\gamma_k(T) + (k+\epsilon)\gamma_k^t(T) = n + \epsilon$.

As a consequence of Corollary 89, we have the following result which was first established in [48].

**Theorem 90 ([48])** *For $k \geq 1$ if $G$ is a connected graph of order $n \geq 2k+1$, then*

$$\gamma_k(G) + \gamma_k^t(G) \leq \frac{2n}{k+1},$$

*and this bound is best possible.*

## 18 Concluding Remarks

In this chapter, we combine the concepts of both distance and domination in graphs to define distance domination in graphs. Since the inception of this concept of distance domination in graphs in 1975 by Meir and Moon [62], it has been extensively studied in the literature. We describe a relationship between the concept of distance domination in graphs and the $p$-center problem studied in operations research. Among other results we provide best known lower and upper bounds on the distance domination number, present algorithmic and complexity results, discuss distance domination in graph products, study the structure of distance dominating subgraphs, examine well-$k$-dominated graphs, explore the concept of distance domination critical graphs, present results on fractional distance domination, and study distance independent domination and distance total domination in graphs. The selected results we present in this chapter are by no means exhaustive and serve to provide numerous interesting theoretical and computational questions.

## References


1. N. Alon, G. Fertin, A. Liestman, T. C. Shermer, and L. Stacho, Factor $d$-domatic colorings of graphs. *Discrete Math.* **262** (2003), no. 1–3, 17–25.
2. M. Aouchiche and P. Hansen, A survey of Nordhaus-Gaddum type relations. *Discrete Appl. Math.* **161** (2013), No. 4–5, 466–546.
3. H. Ahn, H. Kim, S. Kim, and W. Son, Computing $k$-centers over streaming data for small $k$. *Internat. J. Comput. Geom. Appl.* **24** (2014), No. 2, 107–123.
4. R. B. Allan and R. Laskar, On domination and independent domination numbers of a graph. *Discrete Math.* **23** (1978), 73–76.
5. R. B. Allan, R. Laskar and S. T. Hedetniemi, A note on total domination. *Discrete Math.* **49** (1984), 7–13.
6. S. Arumugam, V. Mathew, and K. Karuppasamy, Fractional distance domination in graphs. *Discuss. Math. Graph Theory* **32** (2012), No. 3, 449–459.
7. G. Bacsó and Zs. Tuza, Distance domination versus iterated domination. *Discrete Math.* **312** (2012), no. 17, 2672–2675.
8. L. Beineke and M. A. Henning, Some extremal results on independent distance domination in graphs. *Ars Combin.* **37** (1994), 223–233.
9. R. Belmonte and M. Vatshelle. Graph classes with structured neighborhoods and algorithmic applications. *Theor. Comput. Sci.* **511** (2013), 54–65.
10. D. Blessing, K. Johnson, C. Mauretour, and E. Insko, On $(t, r)$ broadcast domination numbers of grids. *Discrete Appl. Math.* **187** (2015), 19–40.
11. B. Bollobás and E.J. Cockayne, Graph-theoretic parameters concerning domination, independence, and irredundance. *J. Graph Theory* **3** (1979), 241–249.
12. B. Bui-Xuan, J. A. Telle, and M. Vatshelle, Fast dynamic programming for locally checkable vertex subset and vertex partitioning problems. *Theor. Comput. Sci.* **511** (2013), 66–76.
13. B. Bollobás and E. J. Cockayne, Graph-theoretic parameters concerning domination, independence, and irredundance. *J. Graph Theory* **3** (1979), 241–249.
14. K. S. Booth and J. H. Johnson, Dominating sets in chordal graphs. *SIAM J. Comput.* **11** (1982), 191–199.
15. R. C. Brigham, P. Z. Chinn, and R. D. Dutton, Vertex domination-critical graphs. *Networks* **18** (1988), 173–179.
16. G. J. Chang and G. L. Nemhauser, The $k$-domination and $k$-stability problem on graphs. *Tech. Report* 540, School of Operations Res. and Industrial Eng., Cornell Univ., 1982.
17. D. Z. Chen, J. Li, and H. Wang, Efficient algorithms for the one-Řdimensional $k$-center problem. *Theor. Comput. Sci.* **592** (2015), 135–142.
18. B. Chen, J. H. Kim, M. Tait, and J. Verstraete, On coupon colorings of graphs. *Discrete Appl. Math.* **193** (2015), 94–101.
19. E. J. Cockayne, R. M. Dawes, and S. T. Hedetniemi, Total domination in graphs. *Networks* **10** (1980), 211–219.
20. D. G. Corneil and Y. Perl, Clustering and domination in perfect graphs. *Discrete Appl. Math.* **9** (1984), 27–40.
21. J. Cyman, M. Lemańska and J. Raczek, Lower bound on the distance $k$-domination number of a tree. *Math. Slovaca* **56**(2) (2006), 235—243.
22. P. Dankelmann and D. Erwin, Distance domination and generalized eccentricity in graphs with given minimum degree. *J. Graph Theory* **94**(1) (2020), 5–19. https://doi.org/10.1002/jgt.22503
23. A. Datta, L. Larmore, S. Devismes, K. Heurtefeux, and Y. Rivierre. Competitive self-stabilizing $k$-clustering. *IEEE 32nd International Conference on Distributed Computing Systems ICDCS* (2012), 476–485.
24. R. Davila, C. Fast, M. A. Henning, and F. Kenter, Lower bounds on the distance domination number of a graph. *Contrib. Discrete Math.* **12**(2017), No.2, 11–21.
25. G. S. Domke, S. T. Hedetniemi, and R. Laskar, Generalized packings and coverings of graphs. *Congr. Numer.* **62** (1988), 159–270.

26. H. Du, Y. Xu, and B. Zhu, An incremental version of the $k$-center problem on boundary of a convex polygon. *J. Comb. Optim.* **30** (2015), No. 4, 1219–1227.
27. M. Farber, Applications of linear programming duality to problems involving independence and domination. *Tech. Report* 81–13, Department of Computer Science, Simon Fraser University, Canada (1981).
28. M. Farina and A. Grez, New upper bounds on the distance domination numbers of grids. *Rose-Hulman Undergrad. Math. J.* **17** (2016), no. 2, Art. 7, 133–145.
29. E. Fata, S. L. Smith, and S. Sundaram, Distributed dominating sets on grids. *2013 American Control Conference*, Washington, DC, 2013, 211–216.
30. A. Finbow, B. Hartnell, and R. Nowakowski, Well-dominated graphs: a collection of well-covered ones. *Ars Combin.* **25** (1988), 5–10.
31. J. F. Fink and M. Jacobson, $n$-domination in graphs. *Graph theory with applications to algorithms and computer science (Kalamazoo, Mich., 1984)*, 283–300, Wiley-Intersci. Publ., Wiley, New York, 1985.
32. F. V. Fomin, P. A. Golovach, and J. F. Raymond, On the tractability of optimization problems on $H$-graphs. *26th European Symposium on Algorithms*, Art. No. 30, 14 pp., *LIPIcs. Leibniz Int. Proc. Inform.*, **112**, Schloss Dagstuhl. Leibniz-Zent. Inform., Wadern, 2018.
33. G. Fricke, S. T. Hedetniemi, and M. A. Henning, Distance independent domination in graphs. *Ars Combin.* **40** (1995), 1–12.
34. G. Fricke, M. A. Henning, O. R. Oellermann, and H. C. Swart, An efficient algorithm to compute the sum of two distance domination parameters. *Discrete Appl. Math.* **68** (1996), 85–91.
35. G. Fricke, S. T. Hedetniemi, and M. A. Henning, Asymptotic results on distance independent domination in graphs. *J. Combin. Math. Combin. Comput.* **17** (1995), 160–174.
36. M. R. Garey and D. S. Johnson, *Computers and Intractability: A Guide to the Theory of NP-Completeness*, W. H. Freeman and Company, New York (1979).
37. D. L. Grinstead and P. J. Slater, Fractional domination and fractional packing in graphs. *Congr. Numer.* **71** (1990), 153–172.
38. E. Galby, A. Munaro, and B. Ries, Semitotal domination: New hardness results and a polynomial-time algorithm for graphs of bounded mim-width. *Theor. Comput. Sci.* **814** (2020), 28–48.
39. J. Gimbel and M. A. Henning, Bounds on an independent distance domination parameter. *J. Combin. Math. Combin. Comput.* **20** (1996), 193–205.
40. W. Goddard and M. A. Henning, Thoroughly dispersed colorings. *J. Graph Theory* **88** (2018), 174–191.
41. J. H. Hattingh and M. A. Henning, A characterization of block graphs that are well-$k$-dominated. *J. Combin. Math. Combin. Comput.* **13** (1993), 33–38.
42. J. H. Hattingh, M. A. Henning, and J. L. Walters, On the computational complexity of upper distance fractional domination. *Australas. J. Combin.* **7** (1993), 133–144.
43. S. M. Hedetniemi, S. T. Hedetniemi, T. V. Wimer, Linear time resource allocation for trees. *Technical Report URI-014*, Dept. Mathematical Sciences, Clemson Univ. (1987).
44. G. Hemalatha and P. Jeyanthi, Results on distance-2 domination subdivision number of Cartesian product graph. *Int. J. Appl. Adv. Sci. Res.* Special Issue, February, 87–93, 2017.
45. M. A. Henning, Distance domination in graphs. *Domination in graphs: Advanced Topics*, 321–349, *Monogr. Textbooks Pure Appl. Math.*, **209**, Marcel Dekker, Inc. New York, 1998.
46. M. A. Henning and N. Lichiardopol, Distance domination in graphs with given minimum and maximum degree. *J. Combin. Optim.* **34** (2017), 545–553.
47. M. A. Henning, O. R. Oellermann, and H. C. Swart, Bounds on distance domination parameters. *J. Combin. Comput. Inf. Sys. Sciences* **16** (1991), 11–18.
48. M. A. Henning, O. R. Oellermann, and H. C. Swart, Relationships between distance domination parameters. *Mathematica Pannonica* **5**(1) (1994), 69–79.
49. M. A. Henning, O. R. Oellermann, and H. C. Swart, Relating pairs of distance domination parameters. *J. Combin. Math. Combin. Comput.* **18** (1995), 233–244.

50. M. A. Henning, O. R. Oellermann, and H. C. Swart, The diversity of domination. *Discrete Math.* **161** (1996), 161–173.
51. M. A. Henning, O. R. Oellermann, and H.C. Swart, Distance domination critical graphs. *J. Combin. Math. Combin. Comput.* **44** (2003), 33–45.
52. J. Jaffke, O. Kwon, T. J. F. Strømme, and J. A. Telle, Generalized distance domination problems and their complexity on graphs of bounded mim-width. *13th International Symposium on Parameterized and Exact Computation*, Art. No. 6, 14 pp., *LIPIcs. Leibniz Int. Proc. Inform.*, **115**, Schloss Dagstuhl. Leibniz-Zent. Inform., Wadern, 2019.
53. J. Jaffke, O. Kwon, and J. A. Telle, Mim-width I. Induced path problems. *Discrete Appl. Math.* **278** (2020), 153–168
54. J. Jaffke, O. Kwon, and J. A. Telle, Mim-width II. The feedback vertex set problem. *Algorithmica* **82**(1) (2020), 118–145.
55. J. Jaffke, O. Kwon, T. J. F. Strømme, and J. A. Telle, Mim-width III. Graph powers and generalized distance domination problems. *Theor. Comput. Sci.* **796** (2019), 216–236.
56. K. Kämmerling and L. Volkmann, The $k$-domatic number of a graph. *Czech. Math. J.* **59** (2009), No. 2, 539–550.
57. D. Kiser, Distance-2 domatic numbers of graphs. Masters Thesis, East Tennessee State University, May 2015.
58. D. Kiser and T. W. Haynes, Distance-2 domatic numbers of grid graphs. *Congr. Numer.* **225** (2015), 55–62.
59. A. Klobučar, On the $k$-dominating number of Cartesian products of two paths. *Math. Slovaca* **55** (2005), No. 2, 141–154.
60. D. Lichtenstein, Planar satisfiability and its uses. *SIAM J. Comput.* **11** (1982), 329–343.
61. D. Meierling and L. Volkmann, A lower bound for the distance $k$-domination number of trees. *Result. Math.* **47** (2005), 335–339.
62. A. Meir and J. W. Moon, Relations between packing and covering number of a tree. *Pacific J. Math.* **61** (1975), 225–233.
63. E. A. Nordhaus and J. W. Gaddum, On complementary graphs. *Amer. Math. Monthly* **63** (1956), 175–177.
64. F. Jaeger and C. Payan, Relations du type Nordhaus Gaddum pour le nombre d'absorption d'un graphe simple. *C.R. Acad. Sci. Ser. A* **274** (1972), 728–730.
65. O. Ore, *Theory of Graphs. Amer. Math. Soc. Colloq. Publ.* Vol. XXXVIII, American Mathematical Society, Providence, R.I. (1962), 270 pp.
66. A. Rana, A. Pal, and M. Pal, An efficient algorithm to solve the distance $k$-domination problem on permutation graphs. *J. Discrete Math. Sci. Cryptography* **19**(2) (2016), 241–255.
67. S. H. Sæther and M. Vatshelle, Hardness of computing width parameters based on branch decompositions over the vertex set. *Theor. Comput. Sci.* **615** (2016), 120–125.
68. P. J. Slater, R-domination in graphs. *J. Assoc. Comp. Mach.* **23**(3) (1976), 446–450.
69. D. P. Sumner, Critical concepts in domination. *Discrete Math.* **86** (1990), 33–46.
70. B. C. Tansel, R. L. Francis, and T. J. Lowe, Location on networks: a survey. I. The $p$-center and $p$-median problems. *Management Sci.* **29** (1983), 282–297.
71. F. Tian and J. M. Xu, A note on distance domination numbers of graphs. *Austral. J. Combin.* **43** (2009), 181–190.
72. F. Tian and J. M. Xu, Average distances and distance domination numbers. *Discrete Appl. Math.* **157** (2009), 1113–1127.
73. J. Topp and L. Volkmann, On packing and covering numbers of graphs. *Discrete Math.* **96** (1991), 229–238.
74. V. Turau and S. Köhler, A distributed algorithm for minimum distance-$k$ domination in trees. *J. Graph Algorithms Appl.* **19** (2015), no. 1, 223–242.
75. Zs. Tuza, Small $n$-dominating sets. *Mathematica Pannonica* **5** (1994), No. 2, 271–273.
76. M. Vatshelle. New width parameters of graphs. PhD thesis, University of Bergen, 2012. https://www.ii.uib.no/~martinv/Papers/MartinThesis.pdf
77. V. G. Vizing. Some unsolved problems in graph theory. *Uspehi Mat. Nauk* **23** no. 6(144) (1968), 117–134.

78. H. Wang and J. Zhang, An $O(n \log n)$-time algorithm for the $k$-center problem in trees. *34th International Symposium on Computational Geometry, LIPIcs*. Leibniz International Proceedings in Informatics, vol. 99, Schloss Dagstuhl. Leibniz-ŘZent. Inform., Wadern, 2018, pp. Art. No. 72, 15.
79. K. Yamazaki, Inapproximability of rank, clique, Boolean, and maximum induced matching-widths under small set expansion hypothesis. *Algorithms (Basel)* **11** (2018), no. 11, paper No. 173, 10 pp.
80. B. Zelinka, On the $k$-domatic numbers of graphs. *Czech. Math. J.* **33**(2) (1983), 309–313.
81. B. Zelinka, Domatic numbers of graphs and their variants: a survey. *Domination in graphs*, 351–377. *Monogr. Textbooks Pure Appl. Math.* **209**, Dekker, New York, 1998.

# Locating-Domination and Identification

**Antoine Lobstein, Olivier Hudry, and Irène Charon**

## 1 Introduction

Locating-dominating codes were introduced by Slater in 1983 [168], but for more easily accessible sources, see Rall & Slater [155], or Colbourn, Slater, & Stewart [60]. The term "identifying code" is used in the 1998 paper [134] by Karpovsky, Chakrabarty, & Levitin, which certainly marks the starting point for the blossoming of works on this topic, but the concept is already contained in [158] (Rao, 1993). For both locating-dominating and identifying codes, see the ongoing bibliography at [142].

The graphs $G = (V, E)$ that we shall consider will usually be finite, undirected, and connected. Before we proceed, and since we consider domination at distance $r$, we extend the notion of neighborhood: for any integer $r \geqslant 1$, the *open r-neighborhood* of a vertex $v \in V$ is the set $N_r(v) = \{u : 0 < d(u, v) \leqslant r\}$. The set $N_r[v] = N_r(v) \cup \{v\}$ is called the *closed r-neighborhood* of $v$. A *code* is simply a set of vertices, whose elements are called *codewords*.

Formally, an *r-locating-dominating code* $C \subseteq V$, abbreviated $r$-LD code, is a distance-$r$ dominating code such that

$$\forall v_1 \in V \setminus C, \forall v_2 \in V \setminus C, v_1 \neq v_2 : N_r(v_1) \cap C \neq N_r(v_2) \cap C.$$

A. Lobstein (✉)
Centre National de la Recherche Scientifique, Laboratoire de Recherche en Informatique, Orsay Cedex, France
e-mail: lobstein@lri.fr

O. Hudry
Télécom Paris, Institut polytechnique de Paris, Palaiseau, France
e-mail: hudry@enst.fr

I. Charon
Télécom Paris, Institut polytechnique de Paris, (idem), France



T. W. Haynes et al. (eds.), *Topics in Domination in Graphs*, Developments in Mathematics 64, https://doi.org/10.1007/978-3-030-51117-3_8

An $r$-locating-dominating code always exists (e.g., $C = V$, or $|C| = |V| - 1 > 0$ whenever $|E| > 0$).

An *r-identifying code* $C \subseteq V$, abbreviated $r$-Id code, is a distance-$r$ dominating code such that

$$\forall v_1 \in V, \forall v_2 \in V, v_1 \neq v_2 : N_r[v_1] \cap C \neq N_r[v_2] \cap C.$$

Identifying codes are sometimes called *differentiating*, mostly when $r = 1$. We can see that it is possible to retrieve a vertex $v$ simply by knowing which codewords $r$-dominate $v$. Thus, if each codeword sends an alarm to a central controller whenever there is a malfunctioning vertex in its closed neighborhood, then, only knowing which codewords gave the alarm, the controller can unambiguously retrieve that vertex (if there is at most one). The same is true for a non-codeword in the case of $r$-LD codes. See also [149] for an illustration with smoke detectors.

Two distinct vertices $v_1 \in V$, $v_2 \in V$, are said to be *r-twins* if $N_r[v_1] = N_r[v_2]$. It is easy to see that a graph $G$ admits an $r$-identifying code if and only if it has no $r$-twins, i.e.,

$$\forall v_1 \in V, \forall v_2 \in V, v_1 \neq v_2 : N_r[v_1] \neq N_r[v_2]. \tag{1}$$

A graph satisfying (1) is usually called *r-identifiable*, *r-twin-free* or *r-distinguishable*, or also *point-distinguishing* ([175], for $r = 1$). If three vertices $x, y, z$ are such that $z \in N_r[x]$ and $z \notin N_r[y]$, we say that *$z$ r-separates $x$ and $y$* in $G$ (note that $z = x$ is possible). A set of vertices $r$-separates $x$ and $y$ if at least one of its elements does. So we can rephrase the definitions: an $r$-LD code (respectively, an $r$-Id code) $r$-dominates every vertex and $r$-separates every pair of non-codewords (respectively, of vertices).

Usually, one is interested in finding the smallest possible size of a code, LD or Id, in a given graph. Such codes are called *optimal*. Several notations exist; in this chapter, we denote by $LD_r(G)$ (respectively, $Id_r(G)$) the smallest possible cardinality of an $r$-locating-dominating code (respectively, an $r$-identifying code when $G$ is $r$-twin-free), and we call these numbers the *r-locating-domination number* (respectively, the *r-identification number*) of $G$. They are abbreviated as $r$-LD and $r$-Id numbers, respectively. We may drop $r$ when $r = 1$ or when it is irrelevant.

**Example** See Figure 1.

$r = 1$: (a) $\{v_1, v_4\}$ is locating-dominating, not identifying. (b) $\{v_1, v_3\}$ is neither locating-dominating nor identifying. (c) $\{v_1, v_2, v_4\}$ is both locating-dominating and identifying. We have: $LD(G_1) = 2$, $Id(G_1) = 3$. (d) $v_1$ and $v_3$ are twins; $\{v_1, v_4\}$ is locating-dominating, and $LD(G_2) = 2$.

$r = 4$: (e) $v_4$ and $v_5$ are 4-twins; $\{v_1, v_3, v_5, v_7\}$ is 4-locating-dominating, and $LD_4(G_3) = 4$.

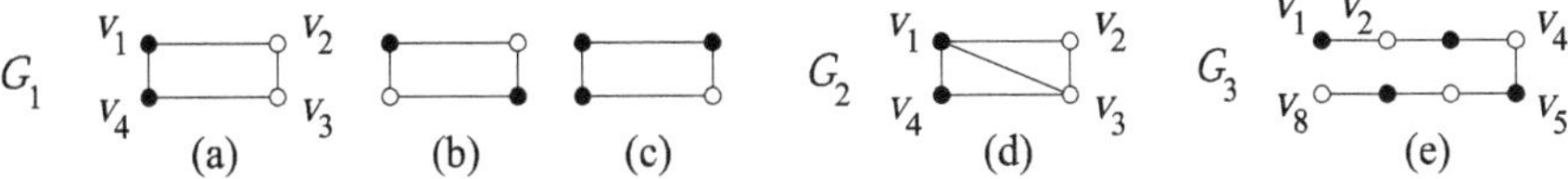


**Fig. 1** Different graphs and codes. Black vertices represent codewords. (**a**)–(**d**) $r = 1$; (**e**) $r = 4$.

The decision problems classically associated with the search of optimal $r$-LD and $r$-Id codes are NP-complete for all $r \geqslant 1$, see Section 8 on complexity. The following inequalities hold for any $r \geqslant 1$:

$$\gamma_r(G) \leqslant LD_r(G) \leqslant Id_r(G). \tag{2}$$

Almost all of this chapter is devoted to undirected graphs, but the above definitions can be extended to *digraphs*, by considering, e.g., *insets* instead of neighborhoods; see [39, 88], or [59] for an illustration.

The following three theorems are probably as old as the definitions of $r$-dominating, $r$-locating-dominating or $r$-identifying codes.

**Theorem 1** *Let $r \geqslant 2$ be any integer and $G$ be any graph. A code is 1-dominating (respectively, 1-locating-dominating, 1-identifying) in $G^r$, the $r$-th power of $G$, if and only if it is $r$-dominating (respectively, $r$-locating-dominating, $r$-identifying) in $G$.*

**Theorem 2** *Let $r \geqslant 1$ be any integer. Any connected $r$-twin-free graph has order $n = 1$ or $n \geqslant 2r + 1$. The only connected $r$-twin-free graph with order $n = 2r + 1$ is the path $P_{2r+1}$.*

*Any cycle with order $n \geqslant 2r + 2$ is $r$-twin-free.*

**Theorem 3** *Let $r \geqslant 1$ be any integer. Every $r$-dominating code is $(r+1)$-dominating. If $C$ is $r$-dominating (respectively, $r$-locating-dominating, $r$-identifying), then so is any superset $C^* \supseteq C$.*

## 2 Possible Values for LD and Id Numbers

Almost everyone's first results on LD and Id numbers are the following two theorems, which give three easy bounds, whereas Theorem 6 is more difficult.

**Theorem 4**

*(a) For any integer $r \geqslant 1$ and any graph $G = (V, E)$ of order $n$, we have*

$$LD_r(G) \geqslant \lceil \log_2(n - LD_r(G) + 1) \rceil. \tag{3}$$

(b) *For any integer* $r \geqslant 1$ *and any* $r$*-twin-free graph* $G = (V, E)$ *of order* $n$*, we have*

$$Id_r(G) \geqslant \lceil \log_2(n+1) \rceil.$$

**Proof.** (a) Let $C$ be any $r$-LD code in $G$. All the $n - |C|$ non-codewords $v \in V \setminus C$ must be given nonempty and distinct sets $N_r(v) \cap C$, constructed with the $|C|$ codewords, so $2^{|C|} - 1 \geqslant n - |C|$, from which (3) follows when $C$ is optimal; (b) the argument is the same, but we have to consider all the $n$ vertices $v \in V$, so $2^{|C|} - 1 \geqslant n$. □

**Theorem 5** *For any* $r \geqslant 1$ *and connected graph* $G$ *of order* $n \geqslant 2$*, we have* $LD_r(G) \leqslant n - 1$.

**Theorem 6** ([15]) *For any* $r \geqslant 1$ *and connected* $r$*-twin-free graph* $G$ *of order* $n \neq 1$*, we have* $Id_r(G) \leqslant n - 1$.

**Proof.** We give the short elegant proof by Gravier & Moncel [98] in 2007, six years after the first proof in [15]. Using Theorem 1, we can see that it is sufficient to prove the case $r = 1$. We assume that $G = (V, E)$ is a connected 1-twin-free graph, of order $n \geqslant 3$. Let $C_1 = V \setminus \{a\}$ for $a \in V$ with maximum degree in $G$. If $C_1$ is identifying, we are done, so we assume that it is not. Because $G$ is connected, $C_1$ is a dominating code. If $u$ and $v$ belong to $N[a]$ and because $N[u] \neq N[v]$, we have $N[u] \setminus \{a\} \neq N[v] \setminus \{a\}$, so $C_1$ separates $u$ and $v$, and the same is true for $u \notin N[a]$ and $v \notin N[a]$. Therefore, there must be $u \in N[a]$ and $v \notin N[a]$ which are not separated by $C_1$, i.e., $N[u] \cap C_1 = N[v] \cap C_1$, or $N[u] \setminus \{a\} = N[v]$.

We claim that $C_2 = V \setminus \{v\}$ is an identifying code. By the previous discussion with $a$, which did not use the maximum degree assumption yet, it is sufficient to check that every vertex in $N[v]$ is separated from every vertex not in $N[v]$. And indeed: (i) since $N[u] = N[v] \cup \{a\}$, each vertex in $V \setminus (N[v] \cup \{a\})$ is separated from each vertex in $N[v]$ by $u$; (ii) the vertex $a$ is separated from $v$ by itself; (iii) the vertex $a$ is separated from each vertex $a' \in N[v] \setminus \{v\}$, for otherwise, $N[a'] = N[a] \cup \{v\}$, i.e., $a'$ has degree greater than $a$, a contradiction. □

The previous result actually holds for all graphs, finite or infinite, with bounded degree. There exist (infinite) graphs (with unbounded degree) such that the only 1-identifying code is the whole vertex set [15]: take two copies, $G_1$ and $G_2$, of the infinite complete graph with vertex set $\mathbb{Z}$, and link $i \in V(G_1)$ to $j \in V(G_2)$ if and only if $i \geqslant j$.

## 2.1 Reaching the Bounds

Graphs exist that meet the lower and upper bounds given previously. Theorem 9(a) even characterizes the graphs $G$ of order $n$ such that $LD(G) = n - 1$, as will Theorem 12 for 1-identifying codes.

**Theorem 7** ([43]) *Let $r \geqslant 1$ and $n$ be integers such that $n \geqslant 2^{2r+1} + 2r + 1$. There exists a connected graph of order $n$ admitting an $r$-locating-dominating code achieving the lower bound (3).*

**Theorem 8** ([134, for $r$=1]), ([64]), ([43]) *Let $r \geqslant 1$ and $n$ be integers such that $n \geqslant 2^{2r}$. There exists a connected graph of order $n$ admitting an $r$-identifying code of size $\lceil \log_2(n+1) \rceil$.*

For $r = 1$, Theorem 8 holds for all $n \geqslant 3$; for $r = 2$, it holds if and only if $n \geqslant 6$.

**Theorem 9**

*(a)* ([169]), ([170]) *A connected graph $G$ of order $n \geqslant 2$ is such that $LD(G) = n - 1$ if and only if $G$ is the star $K_{1,n-1}$ or the complete graph $K_n$.*
*(b) For all $r \geqslant 1$ and $n \geqslant 2$, we have $LD_r(K_n) = n - 1$.*

**Theorem 10** *For all $n \geqslant 3$, there exists a connected graph $G$ of order $n$ such that $Id(G)$ achieves the upper bound $n - 1$.*

An easy example of such a graph is the star, see Figure 2(a). We give two more examples, which will be used for the construction of graphs $G$ such that $Id_r(G) = n - 1$ for all $r$.

In the following, a vertex is *$r$-universal* if it $r$-dominates all the vertices in the graph. Let $p \geqslant 2$.

(a) Take $G^*_{2p} = (V^*_{2p}, E^*_{2p})$, with $V^*_{2p} = \{v_0, v_1, \dots, v_{2p-1}\}$, $E^*_{2p} = \{v_i v_j : v_i \in V^*_{2p}, v_j \in V^*_{2p}, i \neq j, i \neq j + p \bmod 2p\}$: the graph $G^*_{2p}$ has even order and is the complete graph $K_{2p}$ minus the edges of a perfect matching.
(b) Take $G^*_{2p+1}$, obtained from $G^*_{2p}$ simply by adding one 1-universal vertex; its order is odd.

In view of Section 3, the following theorem gives the domination, locating-domination, and identification numbers of these two classes of graphs, as well as for the star.

**Theorem 11** *Let $p \geqslant 2$ and $n \geqslant 3$. Then*

*(a)* $\gamma(G^*_{2p}) = 2$, $LD(G^*_{2p}) = p$, *and* $Id(G^*_{2p}) = 2p - 1$;
*(b)* $\gamma(G^*_{2p+1}) = 1$, $LD(G^*_{2p+1}) = p$, *and* $Id(G^*_{2p+1}) = 2p$;
*(c)* $\gamma(K_{1,n-1}) = 1$, $LD(K_{1,n-1}) = n - 1$, *and* $Id(K_{1,n-1}) = n - 1$.

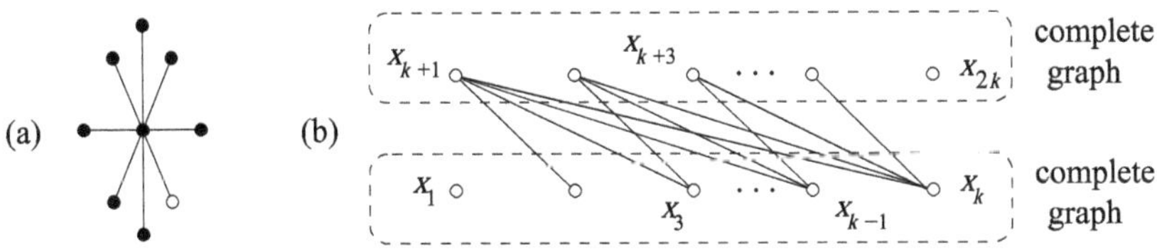


**Fig. 2** (a) The star $K_{1,8}$, with an optimal identifying code (of size 8), which is also an optimal locating-dominating code. (b) The graph $A_k$.

It is an interesting fact, published in 2010 [82], that it is possible to characterize the graphs $G$ with $Id(G) = n - 1$. For two graphs $G = (V, E)$ and $G' = (V', E')$, let $G \bowtie G' = (V \cup V', E \cup E' \cup \{vv' : v \in V, v' \in V'\})$ be their join graph. Let $A_k = (V_k, E_k)$, with $V_k = \{x_1, \ldots, x_{2k}\}$ and $E_k = \{x_i x_j : |i-j| \leqslant k-1\}$, see Figure 2(b). Note that $A_1$ consists of two isolated vertices, and for $k \geqslant 2$, $A_k$ is the $(k-1)$-th power of the path $P_{2k}$. Finally, let $\mathcal{A}$ be the closure of all the graphs $A_i$ for $\bowtie$.

**Theorem 12** (Foucaud *et al.* [82]), (Foucaud *et al.* [83]) *A connected graph $G$ of order $n \geqslant 3$ is such that $Id(G) = n - 1$ if and only if $G$ is the star or belongs to the set of graphs $\mathcal{A} \cup \{A \bowtie K_1 : A \in \mathcal{A}\}$.*

Note that $G_4^* \simeq A_1 \bowtie A_1$, $G_6^* \simeq G_4^* \bowtie A_1$, ..., $G_{2p}^* \simeq G_{2p-2}^* \bowtie A_1$, and that $G_{2p+1}^* \simeq G_{2p}^* \bowtie K_1$. We now turn to the case $r \geqslant 2$: are there graphs $G$ such that $Id_r(G) = n - 1$? (as we have already seen, complete graphs give a trivial positive answer for $r$-LD codes). A crucial fact is that the two graphs $G_{2p}^*$ and $G_{2p+1}^*$ admit $r$-th roots for any $r$, if $p$ is sufficiently large. More precisely:

**Theorem 13** ([43]) *Let $r \geqslant 2$ and $p \geqslant 2$ be integers.*

*(a) If $2p \geqslant 3r^2$, then there exists a graph $G_{2p}$ of order $2p$ such that $(G_{2p})^r = G_{2p}^*$.*
*(b) If $2p \geqslant 3r^2$, then there exists a graph $G_{2p+1}$ of order $2p+1$ such that $(G_{2p+1})^r = G_{2p+1}^*$.*

**Corollary 1**

*(a) For $n \geqslant 3r^2$, there exists a graph $G_n$ of even order $n$ such that $\gamma_r(G_n) = 2$, $LD_r(G_n) = \frac{n}{2}$ and $Id_r(G_n) = n - 1$.*
*(b) For $n \geqslant 3r^2 + 1$, there exists a graph $G_n$ of odd order $n$ such that $\gamma_r(G_n) = 1$, $LD_r(G_n) = \frac{n-1}{2}$ and $Id_r(G_n) = n - 1$.*

**Proof.** Use Theorems 11(a)–(b), 13 and 1. □

## 2.2 *Reaching All Intermediate Values*

As for the lower and upper bounds, constructions show how to achieve all intermediate values.

**Theorem 14** ([42]) *Let $r \geqslant 1$ and $c \geqslant 5r^2 + 5r + 1$. For $n$ between $c + 1$ and $2^c + c - 1$, there exists a connected graph $G$ of order $n$ such that $LD_r(G) = c$.*

For Id codes, the case $r = 1$ is easy and one can even address the special case of bipartite graphs.

**Theorem 15** ([41]) *For $n \geqslant 3$ and for any integer $c$ between $\lceil \log_2(n+1) \rceil$ and $n - 1$, there exists a connected bipartite graph $G$ with $n$ vertices such that $Id(G) = c$.*

**Theorem 16** ([41]) *Let $r \geqslant 1$ and $c \geqslant 5r^2 + 5r + 1$. For $n$ between $c + 1$ and $2^c - 1$, there exists a connected graph $G$ of order $n$ such that $Id_r(G) = c$.*

## 3 The Cost of Locating-Domination and Identification

The inequalities $\gamma_r(G) \leqslant LD_r(G) \leqslant Id_r(G)$ in (2) express that locating-domination is more "expensive" than domination, and identification is more expensive than locating-domination. In this section, we compare the respective "costs" for these three notions. More precisely, denoting, for graphs of order $n$,

$$\mathcal{G}_{r,n} = \{G : r\text{-twin-free, connected, } n \geqslant 2r + 1\} \text{ and}$$

$$\mathcal{G}_{r,n}^{tw} = \{G : \text{ with } r\text{-twins, connected, } n \geqslant 2\},$$

we study the following maximum and minimum differences:

- $F_{Id,LD}(r, n) = \max_{G\in\mathcal{G}_{r,n}}\{Id_r(G) - LD_r(G)\}$,
  $f_{Id,LD}(r, n) = \min_{G\in\mathcal{G}_{r,n}}\{Id_r(G) - LD_r(G)\}$,
- $F_{Id,\gamma}(r, n) = \max_{G\in\mathcal{G}_{r,n}}\{Id_r(G) - \gamma_r(G)\}$,
  $f_{Id,\gamma}(r, n) = \min_{G\in\mathcal{G}_{r,n}}\{Id_r(G) - \gamma_r(G)\}$.

In order to see the influence of the twin-free property on locating-domination and domination, we distinguish, for dominating and locating-dominating codes, between two cases, and study the following functions:

- $F_{LD,\gamma}(r, n) = \max_{G\in\mathcal{G}_{r,n}}\{LD_r(G) - \gamma_r(G)\}$,
  $f_{LD,\gamma}(r, n) = \min_{G\in\mathcal{G}_{r,n}}\{LD_r(G) - \gamma_r(G)\}$;

these two functions are considered on the same set of graphs (the twin-free graphs) as the four functions involving identification, unlike the two functions below:

- $F_{LD,\gamma}^{tw}(r, n) = \max_{G\in\mathcal{G}_{r,n}^{tw}}\{LD_r(G) - \gamma_r(G)\}$,
  $f_{LD,\gamma}^{tw}(r, n) = \min_{G\in\mathcal{G}_{r,n}^{tw}}\{LD_r(G) - \gamma_r(G)\}$.

Finally, if we want to consider all the connected graphs of order $n$, twin-free or not, the result is obviously obtained by taking $\max\{F_{LD,\gamma}(r, n), F_{LD,\gamma}^{tw}(r, n)\}$ and $\min\{f_{LD,\gamma}(r, n), f_{LD,\gamma}^{tw}(r, n)\}$.

Note that [20] characterizes the *trees* $T$ such that $LD_1(T) = \gamma_1(T)$ or $Id_1(T) = \gamma_1(T)$. Most results in this section on cost (namely, Theorems 19–26) are due to Hudry & Lobstein [124] in 2020.

### 3.1 Preliminary Results

**Theorem 17** ([147]), ([102, p. 41]) *If $G$ has order $n$ and no isolated vertices, then $\gamma(G) \leqslant \frac{n}{2}$.*

**Theorem 18** ([96]) *If $G$ is 1-twin-free, then $Id(G) \leqslant 2LD(G)$.*

Thanks to Theorem 1 on the powers of graphs, the previous two results are true also for $\gamma_r(G)$, $Id_r(G)$ (for $r$-twin-free graphs) and $LD_r(G)$, for any $r \geqslant 2$.

## 3.2 Identification vs Domination

First, we construct an infinite family of graphs $G_n^*$ of order $n$ satisfying $Id_r(G_n^*) = \gamma_r(G_n^*)$. These graphs have order $n = k(r+1)$, $k \geqslant 2r+2$, and are informally given by Figure 3(a).

**Theorem 19**

*(a) For all $r \geqslant 1$, $k \geqslant 2r+2$ and $n = k(r+1)$, we have $\gamma_r(G_n^*) = Id_r(G_n^*) = k$.*
*(b) For all $r \geqslant 1$ and $n \geqslant (2r+2)(r+1)$, we have $f_{Id,\gamma}(r,n) = 0$, and obviously $f_{Id,LD}(r,n) = f_{LD,\gamma}(r,n) = 0$.*

**Proof.**

(a) The $k$ vertices $v_{i,r}$ must be $r$-dominated by at least one codeword, and no vertex can $r$-dominate two such vertices, so $\gamma_r(G_n^*) \geqslant k$. On the other hand, the code $C = V_0$ represented by the black vertices on Figure 3(a) has cardinality $k$, and it is straightforward to check that it is $r$-identifying. Note in particular that vertices in $\{v_{1,0}, v_{2,0}, \ldots, v_{k,0}\}$ are $r$-dominated by exactly $2r+1$ codewords (this is where the assumption $k \geqslant 2r+2$ is crucial), and more generally, vertices $v_{i,j} \in \{v_{i,1}, v_{i,2}, \ldots, v_{i,r}\}$ are $r$-dominated by an *odd* number, namely $2r-2j+1$, of codewords, for all $i \in \{1,2,\ldots,k\}$ and $j \in \{1,2,\ldots,r\}$. So $k \leqslant \gamma_r(G_n^*) \leqslant Id_r(G_n^*) \leqslant k$.

(b) Then it is easy to see that we can reach all intermediate values between $k(r+1)$ and $(k+1)(r+1)-1$. Figure 3(b) illustrates the case $k(r+1)+1$, with one additional vertex, $w_1$, which is $r$-dominated by an *even* number of codewords in $G_{n+1}$. □

For $r = 1$, Theorem 19 starts at $n = 8$, for $k = 4$, but we can fill in the holes: $f_{Id,\gamma}(1,n) = 1$ for $n \in \{3,4,5\}$, and for $n \in \{6,7\}$, $f_{Id,\gamma}(1,n) = 0 = f_{Id,LD}(1,n) = f_{LD,\gamma}(1,n)$.

Now how large can the difference $Id_r(G) - \gamma_r(G)$ be? We know that it is at most $n-2$, obtained by graphs $G$ with $Id_r(G) = n-1$ and $\gamma_r(G) = 1$.

**Theorem 20**

*(a) For all $n \geqslant 3$, we have $F_{Id,\gamma}(1,n) = n-2$.*
*(b) For all $r \geqslant 2$ and even $n \geqslant 3r^2$, we have $F_{Id,\gamma}(r,n) = n-3$.*
*(c) For all $r \geqslant 2$ and odd $n \geqslant 3r^2+1$, we have $F_{Id,\gamma}(r,n) = n-2$.*

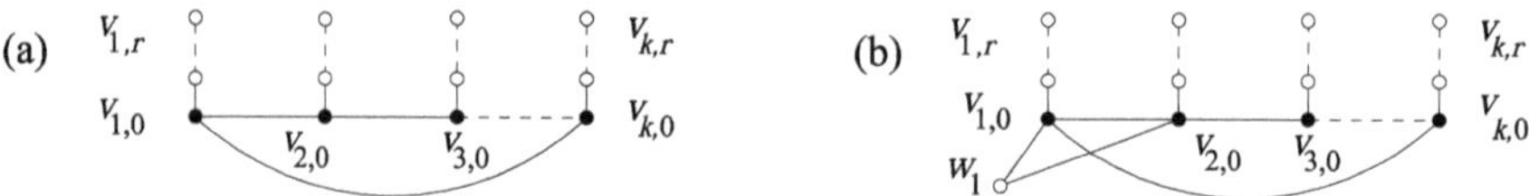


**Fig. 3** The $k$ black vertices represent codewords. (a) The graph $G_n^*$. (b) The graph $G_{n+1}$.

**Proof.** Use the star for (a), Corollary 1(b) for (c). For (b), observe that, among all the graphs $G$ of even order $n$ such that $Id(G) = n - 1$ (see Theorem 12), none of them, except the star, contains a 1-universal vertex, i.e., is such that $\gamma(G) = 1$. But the star cannot be the power of any graph. Therefore, for $r \geqslant 2$, there can exist no graph $G$ with even order $n$ such that $Id_r(G) = n - 1$ and $\gamma_r(G) = 1$, since the $r$-th power of this graph would contradict our previous observation; consequently the difference $Id_r(G) - \gamma_r(G)$ is at most $n - 3$. On the other hand, Corollary 1(a) gives an example achieving $n - 3$. □

### 3.3 Identification vs Locating-Domination

**Theorem 21**

*(a) (= Theorem 19(b)) For all $r \geqslant 1$ and $n \geqslant (2r + 2)(r + 1)$, we have $f_{Id,LD}(r, n) = 0$.*
*(b) For all $n \geqslant 3$, we have $f_{Id,LD}(1, n) = 0$.*

**Theorem 22**

*(a) For all $r \geqslant 1$ and $n \geqslant 3r^2 + 1$, we have $F_{Id,LD}(r, n) = \lceil \frac{n}{2} \rceil - 1$.*
*(b) We have $F_{Id,LD}(1, 3) = 0$.*

**Proof.** We know (cf. Theorem 18) that any (connected) $r$-twin-free graph $G$ is such that $Id_r(G) \leqslant 2LD_r(G)$. Therefore, $Id_r(G) - LD_r(G) \leqslant Id_r(G) - \frac{Id_r(G)}{2} \leqslant \frac{n-1}{2}$, leading to $Id_r(G) - LD_r(G) \leqslant \lceil \frac{n}{2} \rceil - 1$. On the other hand, Corollary 1(a)–(b) provides graphs reaching $\lceil \frac{n}{2} \rceil - 1$. □

### 3.4 Domination vs Locating-Domination

The two cases, without and with twins, show quite a difference between $F_{LD,\gamma}$ and $F^{tw}_{LD,\gamma}$.

#### 3.4.1 Domination vs Locating-Domination in Twin-Free Graphs

Using Theorem 19(b), the sentence following its proof, and the paths $P_4$ and $P_5$, we obtain the following.

**Theorem 23**

*(a) For all $n \geqslant 4$, we have $f_{LD,\gamma}(1, n) = 0$; $f_{LD,\gamma}(1, 3) = 1$.*
*(b) For all $r \geqslant 2$ and $n \geqslant (2r + 2)(r + 1)$, we have $f_{LD,\gamma}(r, n) = 0$.*

**Theorem 24**

*(a)* *For all* $n \geqslant 3$, *we have* $F_{LD,\gamma}(1,n) = n-2$.
*(b)* *For all* $r \geqslant 2$ *and* $n \geqslant 3r^2+1$, *we have* $F_{LD,\gamma}(r,n) \geqslant \frac{n}{2}-2$.
*(c)* *For all* $r \geqslant 2$ *and* $n \geqslant 2r+1$, *we have* $F_{LD,\gamma}(r,n) \leqslant n-3$.

**Proof.** (a) The star shows that $F_{LD,\gamma}(1,n) = n-2$. (b) For any $r$, Corollary 1(a)–(b) immediately gives examples proving that $F_{LD,\gamma}(r,n) \geqslant \frac{n}{2}-2$. (c) The characterization of the graphs $G$ of order $n$ such that $Id(G) = n-1$ (Theorem 12) gives graphs which, apart from the star which is not the power of any graph, are such that $LD(G) \leqslant n-2$. This allows to conclude that $F_{LD,\gamma}(r,n) \leqslant n-3$, using the powers of graphs as in the proof of Theorem 20. □

Improvements are possible on Theorem 24(b) when $r=2$.

**Theorem 25**

*(a)* *Let* $n = 8t \geqslant 24$. *Then* $F_{LD,\gamma}(2,n) \geqslant 5t-3$.
*(b)* *Let* $n = 8t+i \geqslant 24$, *with* $1 \leqslant i \leqslant 7$. *Then* $F_{LD,\gamma}(2,n) \geqslant 5t+i-6$.

### 3.4.2 Domination vs Locating-Domination in Graphs with Twins

**Theorem 26**

*(a)* *For* $r \geqslant 1$ *and* $n \geqslant 2$, *we have* $F^{tw}_{LD,\gamma}(r,n) = n-2$.
*(b)* *For* $n \in \{2,5\}$ *or* $n \geqslant 7$, *we have* $f^{tw}_{LD,\gamma}(1,n) = 0$; *for* $n \in \{3,4,6\}$, *we have* $f^{tw}_{LD,\gamma}(1,n) = 1$.
*(c)* *For all* $r \geqslant 2$ *and* $n \geqslant (2r+2)(r+1)+1$, *we have* $f^{tw}_{LD,\gamma}(r,n) = 0$.

To conclude this section on compared costs, we can see that for $r=1$, we have exact values for all $n$ and functions $f$ and $F$. For $r \geqslant 2$, most results are exact but valid for $n$ large. One open problem is to establish results for all $n$, another to reduce the gap between lower and upper bounds for $F_{LD,\gamma}(r,n)$, cf. Theorems 24(b)–(c) and 25. Note that all open problems and conjectures mentioned throughout this chapter are gathered at its end.

# 4 Specific Families of Graphs

We survey some well-known and some not so well-known families of graphs.

## *4.1 Infinite Grids and Strips*

The four infinite 2-dimensional grids $G_S$ (square), $G_T$ (triangular), $G_K$ (king), and $G_H$ (hexagonal), partially represented in Figure 4, have been much studied in the

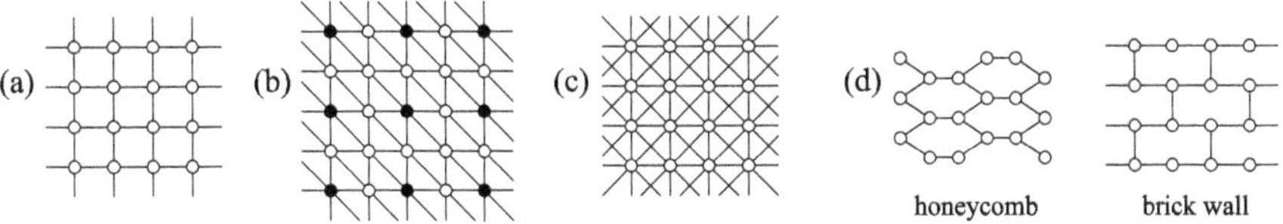


**Fig. 4** Partial representations of the four grids: (a) the square grid; (b) the triangular grid: black vertices are codewords (cf. Theorem 32); (c) the king grid; (d) the hexagonal grid (with two possible representations).

**Fig. 5** A periodic 5-identifying code in the square grid $G_S$, of density 2/25. Codewords are in black.

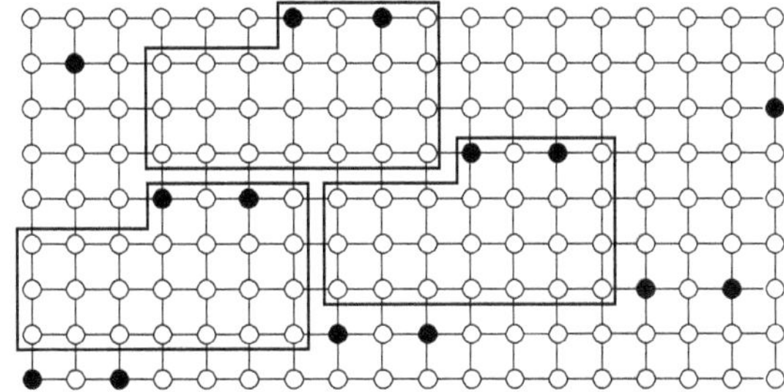

literature, especially with respect to identifying codes, and the densities $\partial_r^{LD}(G_\pi)$ and $\partial_r^{Id}(G_\pi)$ of optimal $r$-LD and $r$-Id codes investigated for $\pi \in \{S, T, K, H\}$. One can also consider the infinite strips $G_\pi^{[k]}$ of height $k \geqslant 1$ (the case $k = 1$ gives the infinite path, see also below, Section 4.2). We remind that the *density* of a code $C$ in $G_\pi^{[k]}$ can be defined for every $k \geqslant 1$ by $\partial(C, G_\pi^{[k]}) = \limsup_{n\to\infty} \frac{|C \cap N_n[v]|}{|N_n[v]|}$, where $v$ is an arbitrary vertex.

Some constructions of codes are obtained by heuristics searching for small subcodes inside tiles that will be repeated periodically [38], see Figure 5 for a first example.

We give, as far as we know, the best lower and upper bounds, for grids and strips. Note how few results there are for $r$-LD codes when $r \neq 1$.

### 4.1.1 The Square Grid

The square grid, $G_S$, has vertex set $V_S = \mathbb{Z} \times \mathbb{Z}$ and edge set $E_S = \{uv : u - v \in \{(1, 0), (0, 1)\}\}$. Figure 6 will give constructions proving the upper bounds for Theorems 27(a) and 28, Figure 7 for Theorems 27(c) and 30(b).

- **Locating-dominating codes**

**Theorem 27** ([171]), ([172])

*(a) We have* $\partial_1^{LD}(G_S) = \frac{3}{10}$.
*(b) We have* $\partial_1^{LD}(G_S^{[1]}) = \frac{2}{5}$.
*(c) We have* $\partial_1^{LD}(G_S^{[2]}) = \frac{3}{8} = 0.375$, $\partial_1^{LD}(G_S^{[3]}) = \frac{1}{3}$.

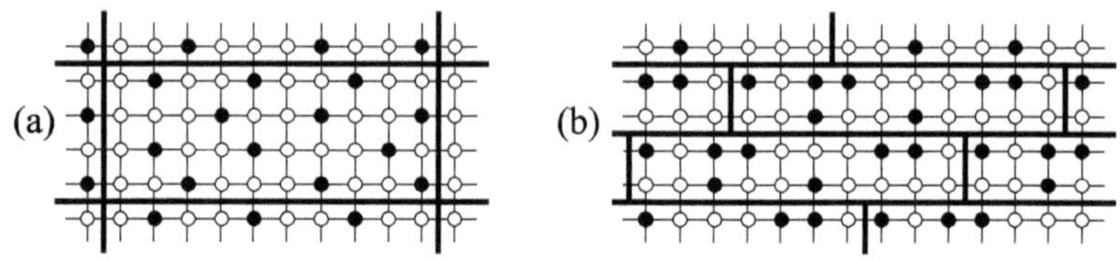


**Fig. 6** Black vertices are the elements of a periodic code in $G_S$, which is (a) 1-LD with density $\frac{3}{10}$, (b) 1-Id with density $\frac{7}{20}$.

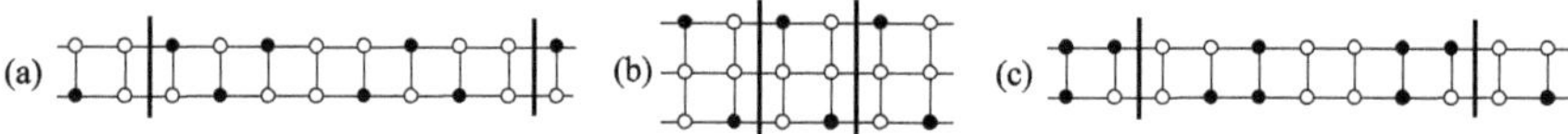


**Fig. 7** Black vertices are the elements of a periodic code which is (a) 1-LD with density $\frac{3}{8}$ in $G_S^{[2]}$, (b) 1-LD with density $\frac{1}{3}$ in $G_S^{[3]}$, (c) 1-Id with density $\frac{3}{7}$ in $G_S^{[2]}$.

- **Identifying codes**

The following upper bound $\partial_1^{Id}(G_S) \leqslant \frac{7}{20}$ is from [54] in 1999, the lower bound $\partial_1^{Id}(G_S) \geqslant \frac{7}{20}$ by Ben-Haim & Litsyn [13] in 2005.

**Theorem 28** *We have* $\partial_1^{Id}(G_S) = \frac{7}{20}$.

The following general lower bounds come from [33], the lower bound in (b) from [128], and all the upper bounds from [117].

**Theorem 29**

*(a) For every* $r \geqslant 1$, *we have* $\frac{3}{8r+4} \leqslant \partial_r^{Id}(G_S) \leqslant \begin{cases} \frac{2}{5r} & : r \text{ even} \\ \frac{2r}{5r^2-2r+1} & : r \text{ odd} \end{cases}$. *When* $r$ *increases, these bounds are close to* $\frac{3}{8r} = \frac{0.375}{r}$ *and* $\frac{2}{5r}$.

*(b) We have* $\frac{6}{35} \approx 0.17143 \leqslant \partial_2^{Id}(G_S) \leqslant \frac{5}{29} \approx 0.17241$.

The previous upper bounds have been improved in [38], using heuristics, for $r \in \{3, 4, 5, 6\}$. The strips of all heights have also been studied, for $r = 1$; in (d) below, the upper bound is from [22], the lower bound from [126].

**Theorem 30**

*(a)* ([94]) *We have* $\partial_1^{Id}(G_S^{[1]}) = \frac{1}{2}$.
*(b)* ([64]), ([65]) *We have* $\partial_1^{Id}(G_S^{[2]}) = \frac{3}{7} \approx 0.42857$.
*(c)* ([22]) *We have* $\partial_1^{Id}(G_S^{[3]}) = \frac{7}{18} \approx 0.38889$.
*(d) We have* $\partial_1^{Id}(G_S^{[4]}) = \frac{11}{28} \approx 0.39286$.
*(e)* ([126]) *We have* $\partial_1^{Id}(G_S^{[5]}) = \frac{19}{50} = 0.38$.
*(f)* ([23]) *For* $k \geqslant 6$, *we have* $\frac{7}{20} + \frac{1}{20k} \leqslant \partial_1^{Id}(G_S^{[k]}) \leqslant \frac{7}{20} + \frac{3}{10k}$.

### 4.1.2 The Triangular Grid

The triangular grid, or square grid with one diagonal, $G_T$, has vertex set $V_T = \mathbb{Z}\times\mathbb{Z}$ and edge set $E_T = \{uv : u - v \in \{(1,0), (0,1), (1,-1)\}\}$. Figure 8 will give constructions proving the upper bounds in $G_T^{[3]}$ for Theorems 31(b) and 34(a).

- **Locating-dominating codes**

**Theorem 31**

*(a)* ([107]) *We have* $\partial_1^{LD}(G_T) = \frac{13}{57} \approx 0.22807$.
*(b)* ([22]) *We have* $\partial_1^{LD}(G_T^{[2]}) = \frac{1}{3}$, $\partial_1^{LD}(G_T^{[3]}) = \frac{3}{10}$.

- **Identifying codes**

**Theorem 32** (Karpovsky, Chakrabarty & Levitin [134]) *We have* $\partial_1^{Id}(G_T) = \frac{1}{4}$.

**Proof.** The upper bound is proved by the construction given in Figure 4(b), which has the property (P) that every codeword is dominated by exactly one codeword (itself) and every non-codeword is dominated by exactly two codewords (either horizontally or vertically or diagonally). This proves that the code is indeed 1-identifying, and moreover that it is best possible. □

Property (P) is at the root of Theorem 70 when $r = 1$. Another example of a graph and a code having this property is the cycle of length $2p \geqslant 6$, with $p$ pairwise nonadjacent vertices forming the code, see Theorem 44. Property (P) cannot be true for $r > 1$ [57].

**Theorem 33**

*(a)* ([33]) *For every* $r \geqslant 1$, *we have* $\frac{2}{6r+3} \leqslant \partial_r^{Id}(G_T) \leqslant \begin{cases} \frac{1}{2r+2} & : r \in \{1,2,3\} \bmod 4 \\ \frac{1}{2r+4} & : r = 0 \bmod 4 \end{cases}$.

*When r increases, these bounds are close to* $\frac{1}{3r}$ *and* $\frac{1}{2r}$.

*(b)* ([38]) *We have* $\partial_3^{Id}(G_T) \leqslant \frac{2}{17} \approx 0.11765$ *and* $\partial_5^{Id}(G_T) \leqslant \frac{1}{13} \approx 0.07692$.

Results on the different triangular strips are available when $r = 1$.

(b)

**Fig. 8** Black vertices are the elements of a periodic code in $G_T^{[3]}$ which is (a) 1-LD with density $\frac{3}{10}$, (b) 1-Id with density $\frac{1}{3}$.

**Theorem 34** ([66])

*(a) We have* $\partial_1^{Id}(G_T^{[2]}) = \frac{1}{2}$, $\partial_1^{Id}(G_T^{[3]}) = \partial_1^{Id}(G_T^{[4]}) = \frac{1}{3}$, $\partial_1^{Id}(G_T^{[5]}) = \frac{3}{10}$, *and* $\partial_1^{Id}(G_T^{[6]}) = \frac{1}{3}$.

*(b) For odd* $k \geqslant 7$, *we have* $\partial_1^{Id}(G_T^{[k]}) = \frac{1}{4} + \frac{1}{4k}$.

*(c) For even* $k \geqslant 8$, *we have* $\frac{1}{4} + \frac{1}{4k} \leqslant \partial_1^{Id}(G_T^{[k]}) \leqslant \frac{1}{4} + \frac{1}{2k}$.

### 4.1.3 The King Grid

The king grid, or square grid with two diagonals, $G_K$, has vertex set $V_K = \mathbb{Z} \times \mathbb{Z}$ and edge set $E_K = \{uv : u - v \in \{(1,0), (0,1), (1,-1), (1,1)\}\}$. Its name comes from the fact that, on an infinite chessboard, the $r$-neighborhood of a vertex $v$ is the set of vertices that a king, starting from $v$, can reach in at most $r$ moves. Figure 9 will give constructions proving the upper bounds for Theorems 35, 37 and 38(b).

- **Locating-dominating codes**

**Theorem 35** ([113]) *We have* $\partial_1^{LD}(G_K) = \frac{1}{5}$.

For $r > 1$, there are good bounds on $\partial_r^{LD}(G_K)$, and the first two strips are solved for $r = 1$.

**Theorem 36** ([152]) *For every* $r \geqslant 1$, *we have* $\frac{1}{4r+2} \leqslant \partial_r^{LD}(G_K) \leqslant \begin{cases} \frac{1}{4r} & : r \text{ even} \\ \frac{r+1}{4r(r+1)+2} & : r \text{ odd} \end{cases}$. *When r increases, these bounds are all equivalent to* $\frac{1}{4r}$.

**Theorem 37** ([22]) *We have* $\partial_1^{LD}(G_K^{[2]}) = \frac{1}{2}$, $\partial_1^{LD}(G_K^{[3]}) = \frac{4}{15} \approx 0.26667$.

- **Identifying codes**

It is remarkable that the best density is known for all $r \geqslant 1$ for identification [34] (2004). In the following theorem, the lower bound on $\partial_1^{Id}(G_K)$ comes from [58], the upper bound from [38].

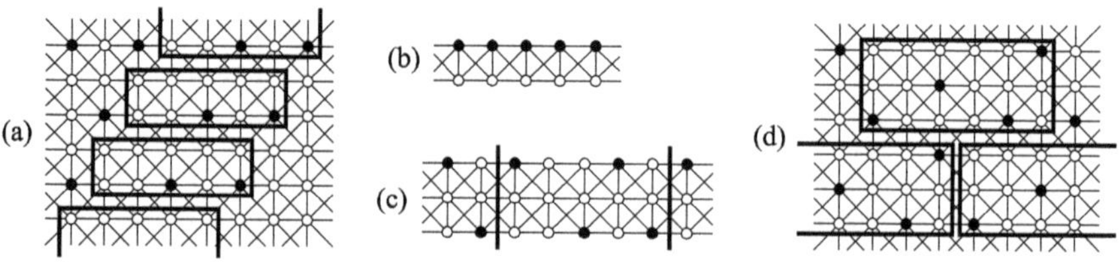


**Fig. 9** Black vertices are the elements of a periodic code which is (a) 1-LD with density $\frac{1}{5}$ in $G_K$, (b) 1-LD with density $\frac{1}{2}$ in $G_K^{[2]}$, (c) 1-LD with density $\frac{4}{15}$ in $G_K^{[3]}$, (d) 1-Id with density $\frac{2}{9}$ in $G_K$.

**Theorem 38**

*(a)* (Charon *et al.* [34]) *For every* $r > 1$, *we have* $\partial_r^{Id}(G_K) = \frac{1}{4r}$.
*(b) We have* $\partial_1^{Id}(G_K) = \frac{2}{9} \approx 0.22222$.

For $r = 1$, the strips have been investigated. Note that $G_K^{[2]}$ is not 1-twin-free.

**Theorem 39** ([67])

*(a) We have* $\partial_1^{Id}(G_K^{[3]}) = \frac{1}{3}$; $\partial_1^{Id}(G_K^{[4]}) = \frac{5}{16} = 0.3125$; $\partial_1^{Id}(G_K^{[5]}) = \frac{4}{15} \approx 0.26667$; $\partial_1^{Id}(G_K^{[6]}) = \frac{5}{18} \approx 0.27778$.

*(b) For* $k \geqslant 7$, *we have* $\frac{2}{9} + \frac{8}{81k} \leqslant \partial_1^{Id}(G_K^{[k]}) \leqslant \begin{cases} \frac{2}{9} + \frac{6}{18k} : k = 0 \bmod 3 \\ \frac{2}{9} + \frac{8}{18k} : k = 1 \bmod 3 \\ \frac{2}{9} + \frac{7}{18k} : k = 2 \bmod 3 \end{cases}$.

### 4.1.4 The Hexagonal Grid

The hexagonal grid, $G_H$, has vertex set $V_H = \mathbb{Z} \times \mathbb{Z}$ and edge set $E_H = \{uv : u = (i, j) \text{ and } u - v \in \{(0, (-1)^{i+j+1}), (1, 0)\}\}$. It is the grid for which one has the sparsest and weakest results. Figure 10 will give constructions proving the upper bounds for Theorems 40 and 41.

- **Locating-dominating codes**

**Theorem 40** ([113]) *We have* $\partial_1^{LD}(G_H) = \frac{1}{3}$.

- **Identifying codes**

The following upper bound is from [56], the lower bound from [61].

**Theorem 41** *We have* $\frac{5}{12} \approx 0.41667 \leqslant \partial_1^{Id}(G_H) \leqslant \frac{3}{7} \approx 0.42857$.

In the following theorem, the lower bounds in (a) and (b) come from [131] and [144], respectively, and both upper bounds from [38]; the general lower bounds in (c) come from [33], and the upper bounds from [174].

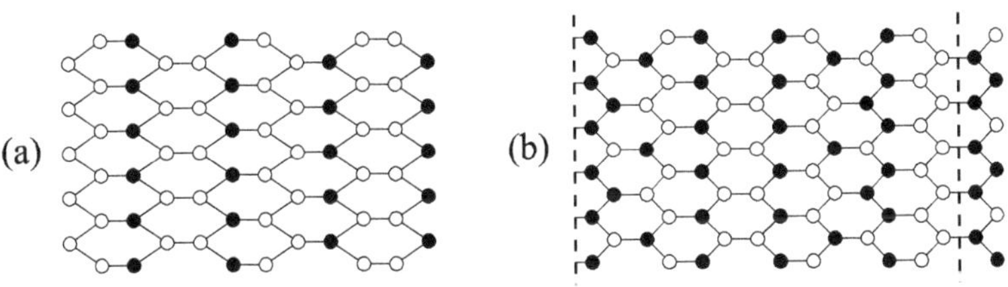


**Fig. 10** Black vertices are the elements of a periodic code in $G_H$, which is (a) 1-LD with density $\frac{1}{3}$, (b) 1-Id with density $\frac{3}{7}$.

**Theorem 42**

*(a) We have* $\partial_2^{Id}(G_H) = \frac{4}{19} \approx 0.21053$.
*(b) We have* $\frac{3}{25} = 0.12 \leqslant \partial_3^{Id}(G_H) \leqslant \frac{1}{6} \approx 0.16667$.
*(c) For* $r \geqslant 8$, *we have* $\left.\begin{array}{ll}\frac{2}{5r+3} & : r \text{ even} \\ \frac{2}{5r+2} & : r \text{ odd}\end{array}\right\} \leqslant \partial_r^{Id}(G_H) \leqslant \left\{\begin{array}{ll}\frac{5r+3}{6r(r+1)} & : r \text{ even} \\ \frac{5r^2+10r-3}{(6r-2)(r+1)^2} & : r \text{ odd}\end{array}\right.$.
*When* $r$ *increases, these bounds are close to* $\frac{2}{5r}$ *and* $\frac{5}{6r} \approx \frac{0.83333}{r}$.

There are many better upper bounds, obtained by the use of heuristics, in [38] for $r \leqslant 30$.

## 4.2 Paths and Cycles

For $r = 1$, the values of $LD(P_n)$, $LD(C_n)$, $Id(P_n)$, and $Id(C_n)$ have been completely determined.

**Theorem 43** ([169]), ([170]) *For every path or cycle* $G_n$ *of order* $n \geqslant 1$, *we have* $LD(G_n) = \left\lceil \frac{2n}{5} \right\rceil$.

**Proof.** For the lower bound, we give an alternative by Bertrand *et al.* [16] in 2004 to the 1988 proof by Slater. This proof uses the following counting argument, which can be applied to graphs other than paths or cycles, and can be adapted to Id codes, cf. proof of Theorem 32. Let $C$ be a 1-LD code in $G_n$. Now $n - |C|$ non-codewords must be dominated by $C$, at most $|C|$ of these are dominated by one codeword, and the remaining are dominated by at least two codewords. Therefore,

$$1 \times |C| + 2 \times (n - 2|C|) \leqslant \sum_{c \in C} |N(c)|.$$

For paths and cycles, $|N(c)| \leqslant 2$, which leads to $|C| \geqslant \lceil 2n/5 \rceil$. This lower bound is met with equality, see Figure 11(a) for paths; for cycles, simply link the leftmost and rightmost vertices in the same figure. □

**Theorem 44** ([94]) *For every path* $P_n$ *of order* $n \geqslant 3$, *we have* $Id(P_n) = \lceil \frac{n+1}{2} \rceil$. *For every cycle* $C_n$ *of length* $n \geqslant 6$, *we have* $Id(C_n) = 3\lceil \frac{n}{2} \rceil - n$, *and* $Id(C_4) = Id(C_5) = 3$.

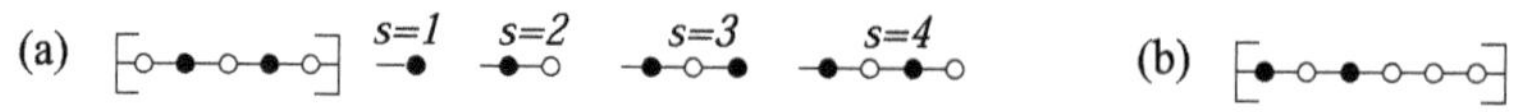

**Fig. 11** Black vertices in the path with $n$ vertices are codewords. (a) $n = 5k + s$, $0 \leqslant s < 5$: repeat the pattern between brackets $k - 1$ times to the left and paste the appropriate tail to the right, to obtain a 1-LD code. (b) $n = 6k + s$, $0 < s \leqslant 6$: repeat the pattern $k - 1$ times to the left, then paste appropriate tails, not given here, to the left and to the right, to obtain a 2-LD code.

It is then immediate that the best density in an infinite path is 2/5 for 1-LD codes and 1/2 for 1-Id codes (cf. Theorems 27(b) and 30(a), respectively). For $r > 1$, it is not difficult to establish that these densities do not depend on $r$ and are 1/3 and 1/2, respectively [16]. But the general case $r > 1$ for finite paths and cycles is surprisingly difficult and the problem is not settled yet for LD codes.

#### 4.2.1 Paths with $r > 1$

- **Locating-dominating codes**

**Theorem 45** ([16]) *Let $r \geqslant 2$. For all $n \geqslant 1$, we have $\frac{n+1}{3} \leqslant LD_r(P_n)$; for all $n \geqslant 2r+1$, we have $LD_r(P_n) \leqslant \lceil \frac{n+7r+6}{3} \rceil$.*

For infinitely many values of $n$, the above upper bound can be improved to $\frac{n+r+2}{3}$, see [16] and [108]; the latter also completely solves the case $r = 2$. Figure 11(b) gives the pattern for the upper bound in the following result.

**Theorem 46** ([108]) *For all $n \geqslant 1$, we have $LD_2(P_n) = \lceil \frac{n+1}{3} \rceil$.*

In [71], the exact values of $LD_3(P_n)$ and $LD_4(P_n)$ are given for $n \geqslant 1$, the exact value of $LD_r(P_n)$ is given for $r \geqslant 5$, $1 \leqslant n \leqslant 7r+3$, and the following is proved.

**Theorem 47** ([71]) *For $r \geqslant 5$ and $n \geqslant 3r+2+6r((r-3)(2r+1)+r)$, we have $LD_r(P_n) = \lceil \frac{n+1}{3} \rceil$.*

- **Identifying codes**

**Theorem 48** ([16]) *Let $r \geqslant 2$. For all $n \geqslant 2r+1$, we have $\frac{n+1}{2} \leqslant Id_r(P_n)$, and this bound can be achieved for infinitely many values of $n$.*

The case $r = 2$ is completely settled in [160]. And in 2011–12, the complete results were given independently by Chen, Lu, & Miao [52] and by Junnila & Laihonen [130].

**Theorem 49** ([52]), ([130]) *Let $r \geqslant 2$ and $n = (2r+1)p+q$, $p \geqslant 1$, $0 \leqslant q < 2r+1$.*

*(a) If $q = 0$, then $Id_r(P_n) = \frac{(2r+1)p}{2} + 1$ if $p$ is even; $Id_r(P_n) = \frac{(2r+1)(p-1)}{2} + 2r$ if $p$ is odd.*
*(b) If $1 \leqslant q \leqslant r+1$, then $Id_r(P_n) = \frac{(2r+1)p}{2} + q$ if $p$ is even; $Id_r(P_n) = \frac{(2r+1)(p-1)}{2} + 2r + 1$ if $p$ is odd.*
*(c) If $r+2 \leqslant q \leqslant 2r$, then $Id_r(P_n) = \frac{(2r+1)p}{2} + q - 1$ if $p$ is even; $Id_r(P_n) = \frac{(2r+1)(p-1)}{2} + 2r + 1$ if $p$ is odd.*

This means that $Id_r(P_n)$ lies between $\frac{n+1}{2}$ and $\frac{n}{2} + r$, according to the values of $p$ and $q$.

### 4.2.2 Cycles with $r > 1$

- **Locating-dominating codes**

**Theorem 50** ([16]) *Let $r \geqslant 2$. For all $n \geqslant 1$, we have $\frac{n}{3} \leqslant LD_r(C_n)$, and this bound can be achieved for infinitely many values of $n$.*

In [52], the exact value of $LD_2(C_n)$ is given for $n \geqslant 1$, and in [72], the same is done for $LD_3(C_n)$ and $LD_4(C_n)$; furthermore, the exact values of $LD_r(C_n)$ are given for $r \geqslant 5$, $n \in [3, 2r+3] \cup \{3r, 3r+3\}$, and the following is proved.

**Theorem 51** ([72]) *Let $r \geqslant 5$ and $n \geqslant 12r+5+6r((r-3)(2r+1)+r-1)$. Then we have $LD_r(C_n) = \lceil \frac{n}{3} \rceil$ if $n \neq 3 \bmod 6$, and $\frac{n}{3} \leqslant LD_r(C_n) \leqslant \frac{n}{3}+1$ otherwise.*

The previous result is improved in [70]: the bound on $n$ becomes a polynomial in $r^2$ instead of $r^3$.

- **Identifying codes**

The crucial following lower bound is from [64] and solves the case $n$ even.

**Theorem 52** ([64]), ([65]) *For all $r \geqslant 2$ and $n \geqslant 2r+2$, we have $Id_r(C_n) \geqslant \gcd(2r+1, n) \times \lceil \frac{n}{2\gcd(2r+1,n)} \rceil$. If $n \geqslant 2r+4$ is even, then $Id_r(C_n) = \frac{n}{2}$, and $Id_r(C_{2r+2}) = 2r+1$.*

Then [99] provides cases when the exact value of $Id_r(C_n)$ is known. The case $r = 2$ is completely settled in [160]. More cases giving the exact value for $Id_r(C_n)$ are given in [178], further results of this type are given in [52], and finally in 2012, Junnila & Laihonen [130] closed the case: for all $r \geqslant 2$ and $n \geqslant 2r+2$, $Id_r(C_n)$ is known; but the many cases do not allow to give the results in a compact way. Let us simply mention here that $Id_r(C_n)$ lies between $\frac{n+1}{2}$ and $\frac{n}{2}+r$.

## 4.3 Trees

The following upper bound comes from [20], the lower bound from [166].

**Theorem 53** *Let $T$ be a tree of order $n$ with $\ell$ leaves and $s$ support vertices.*

*(a) We have $\frac{n+2(\ell-s)+1}{3} \leqslant LD_1(T) \leqslant \frac{n+\ell-s}{2}$.*
*(b) ([153]) We have $LD_1(T) \leqslant \frac{2n+3\ell-2}{5}$.*

Moreover, the trees achieving the above bounds can be characterized. When $\ell = s$, the lower bound reads $LD_1(T) > n/3$. The star $K_{1,n-1}$ has $s = 1$, $\ell = n-1$ and $LD_1(K_{1,n-1}) = n-1$. All the intermediate values between $\lceil \frac{n+1}{3} \rceil$ and $n-1$ can be reached [166].

**Theorem 54** ([26]) *For any tree $T$ of order $n$, we have $LD_2(T) > \frac{n}{4}$, and $LD_3(T) > \frac{n}{5}$.*

From this it is conjectured that for any tree $T$ of order $n$, $LD_r(T) > \frac{n}{r+2}$; if true, the conjecture is sharp [26]. The following upper bound comes from [94], the lower bound from [20].

**Theorem 55** *For any tree $T$ of order $n$ with $\ell$ leaves and $s$ support vertices, we have*

*(a)* $\frac{n+2\ell-2}{2} \geqslant Id_1(T) \geqslant \frac{3(n+\ell-s+1)}{7}$. *Both bounds are sharp.*
*(b)* ([154]) $\frac{3n+2\ell-1}{5} \geqslant Id_1(T) \geqslant \frac{2n-s+3}{4}$.

The trees achieving the above bounds in (b) can be characterized. All the intermediate values between $\lceil\frac{3(n+1)}{7}\rceil$ and $n-1$ can be reached [41].

The following result *corrects* and completes the study of complete (or balanced) $q$-ary trees for 1-Id codes in [134], and also settles the case of 1-LD codes. It is illustrated by Figure 12.

**Theorem 56** ([17]) *Let $h \geqslant 1$, $q \geqslant 2$, $CT_h^q$ be the complete $q$-ary tree of depth $h$, and $\xi_h^q = \frac{q^h-1}{q-1}$ be its order.*

*(a) We have* $LD(CT_h^q) = \left\lceil \frac{q^2\xi_h^q}{q^2+q+1} \right\rceil$.

*(b) We have* $Id(CT_h^2) = \left\lceil \frac{20\xi_h^q}{31} \right\rceil$; *if $q > 2$, then* $Id(CT_h^q) = \left\lceil \frac{q^2\xi_h^q}{q^2+q+1} \right\rceil$.

Comparing (a) and (b) above, one can see that the complete nonbinary trees give yet another example of graphs in which 1-identification costs the same as 1-locating-domination—see Theorem 21(b).
Although the computing problems of finding optimal $r$-LD and $r$-Id codes are generally NP-hard, even for some restricted classes of graphs (see Section 8 below), trees (and forests) form a class for which polynomial and even linear algorithms exist that output an optimal 1-LD or 1-Id code: see [169] for LD codes, [4] for Id codes, and [32] for Id codes in oriented trees.

**Fig. 12** Codewords are in black. (a) The complete binary tree of depth 4, with an optimal 1-LD code (of size 9) which is not 1-Id. (b) The complete ternary tree of depth 3, with an optimal 1-Id code (of size 9). This code is also an optimal 1-LD code.

### 4.4 The q-ary n-Cube

Here, the standard notation $n$ is not for the order of the cube, but for its dimension: the *$q$-ary $n$-cube* is denoted $\mathbb{F}_q^n$, where $\mathbb{F}_q = \{0, 1, \ldots, q-1\}$; it has $q^n$ $q$-ary vectors $v_1v_2\ldots v_n$, and the distance is usually the *Hamming distance* $d_H$ defined by $d_H(u, v) = |\{i : u_i \neq v_i, 1 \leqslant i \leqslant n\}|$. In this setting, $r$-dominating codes are rather called *$r$-covering codes*, and constitute an important topic inside coding theory. Their strong algebraic structure makes them very particular: for instance, a code may be a vector subspace of dimension $k$ in $\mathbb{F}_q^n$, in which case it has size $q^k$ and is said to be *linear*. Since a generator matrix with $k$ rows suffices to describe the code, this affects the size of an instance when considering complexity issues, see, e.g., [91, 118, 119, 143]. See also [55] for an overview of covering codes, with tables giving bounds on the sizes of optimal codes, linear or not, for the first values of $r$ and $n$. We shall restrict ourselves to the case $q = 2$, although there are some works on the nonbinary cube, and denote $\mathbb{F}_2^n$ simply $\mathbb{F}^n$. Almost all the results given below concern identification. There are strong links between Id codes and $\mu$-fold coverings (where each vertex is dominated by at least $\mu$ codewords), but we do not have enough space to discuss them here; see, e.g., [76]. The following theorem connects the 1-LD number in $\mathbb{F}^n$ and the 2-domination number in $\mathbb{F}^{n-1}$.

**Theorem 57** ([116]) *For all $n \geqslant 5$, we have* $\frac{n^2 2^{n+1}}{n^3+2n^2+3n-2} \leqslant LD_1(\mathbb{F}^n) \leqslant (2n-3)\gamma_2(\mathbb{F}^{n-1})$.

The exact values of $LD_1(\mathbb{F}^n)$ are known for $1 \leqslant n \leqslant 6$; for instance, $\{u \in \mathbb{F}^5 : d_H(u, 00000) = 1 \text{ or } 4\}$ is an optimal 1-LD code (of size 10) [116]. See also [73, 133]. The following result gives the complete answer for linear codes.

**Theorem 58** ([116]) *Let $n = 3 \times 2^k - 5 + s$, for $k \geqslant 1$ and $0 \leqslant s < 3 \times 2^k$. Then the size of an optimal 1-LD linear code in $\mathbb{F}^n$ is $2^{n-k}$.*

We now turn to Id codes. The following theorem links the $r$-Id and $(2r)$-domination numbers.

**Theorem 59** ([134]) *For $n \geqslant 3$, we have* $\frac{n\,2^{n+1}}{n(n+1)+2} \leqslant Id(\mathbb{F}^n)$ *and* $2^{n+1}/(1 + \sum_{j=0}^{r}\binom{n}{j}) \leqslant Id_r(\mathbb{F}^n)$. *If $r < n/2$, then* $Id_r(\mathbb{F}^n) \leqslant \binom{n}{r}\gamma_{2r}(\mathbb{F}^n)$.

The above lower bound on $Id_r(\mathbb{F}^n)$ is improved in [77] (but both bounds coincide for $r$ fixed and $n$ large enough), then in [75]. The next result uses the direct sum construction (DSC) $C \oplus \mathbb{F}^s = \{c|u : c \in C \subseteq \mathbb{F}^n, u \in \mathbb{F}^s\}$, where | stands for concatenation. This is a classical tool, often used in coding theory, allowing to go from $\mathbb{F}^n$ to $\mathbb{F}^{n+s}$, $s \geqslant 1$.

**Theorem 60** ([18]) *For $n \geqslant 2$, we have $Id(\mathbb{F}^{n+2}) \leqslant 4Id(\mathbb{F}^n)$. If $C$ is an optimal 1-Id code in $\mathbb{F}^n$ such that $d_H(c, C \setminus \{c\}) = 1$ for every $c \in C$, then $Id(\mathbb{F}^{n+1}) \leqslant 2Id(\mathbb{F}^n)$.*

Since $\mathbb{F}^s$ is a 0-Id code of size $2^s$, a generalization of the previous result would be the following conjecture: $Id_{r_1+r_2}(\mathbb{F}^{n_1+n_2}) \leqslant Id_{r_1}(\mathbb{F}^{n_1}) \times Id_{r_2}(\mathbb{F}^{n_2})$. The case $r_1 = r_2 = 1$ is proved in [76], which by iteration leads to

$$Id_r(\mathbb{F}^{\sum_{i=1}^r n_i}) \leqslant \prod_{i=1}^r Id_1(\mathbb{F}^{n_i}).$$

Also, by refining the condition on $C$ in Theorem 60, one can widen the possibilities of the DSC, see, e.g., [31, 74, 76]. But $Id(\mathbb{F}^{n+1}) \leqslant 2Id(\mathbb{F}^n)$ remains a conjecture; to our knowledge, the closest result obtained so far is the following.

**Theorem 61** ([74]) *For* $n \geqslant 2$, *we have* $Id(\mathbb{F}^{n+1}) \leqslant \left(2 + \frac{1}{n+1}\right) Id(\mathbb{F}^n)$.

There are also asymptotic results (with $n$ going to infinity) in [74, 118, 134], or [125].

In 2010, Charon *et al.* [31] gave tables with bounds on $Id_r(\mathbb{F}^n)$, $1 \leqslant r \leqslant 5$, $n \leqslant 21$; the same was done for $r \in \{2, 3\}$ and $n \leqslant 30$ by Exoo *et al.* [75] in the same year; all sources are given. The lower bounds stem from the bounds discussed above, and from more ad hoc methods using the topology of the cube and studying local situations, possibly with the help of a computer. The upper bounds use widely the DSC, often enhanced by computer, or heuristics such as simulated annealing or the noising method.

In the linear case, the complete answer is known for $r = 1$ (see [157] for $r > 1$), and is very similar to Theorem 58 for 1-LD codes. In particular, it uses the second part of Theorem 60, which always works with linear codes and allows to go from $\mathbb{F}^n$ to $\mathbb{F}^{n+1}$ by doubling the size of the code.

**Theorem 62** ([156]) *Let* $n = 3 \times 2^k - 3 + s$, *for* $k \geqslant 1$ *and* $0 \leqslant s < 3 \times 2^k$. *Then the size of an optimal 1-Id linear code in* $\mathbb{F}^n$ *is* $2^{n-k}$.

In conclusion, we offer an open problem about the monotonicity of $Id_r(\mathbb{F}^n)$. It is true that for all $n \geqslant 2$, we have $Id_1(\mathbb{F}^n) \leqslant Id_1(\mathbb{F}^{n+1})$ (Moncel [145] in 2006), but this is not the case for all $r$. For instance, $Id_5(\mathbb{F}^6) = 63$, $Id_5(\mathbb{F}^7) \in \{31, 32\}$, $Id_5(\mathbb{F}^8) \in [19, 21]$, $Id_5(\mathbb{F}^9) \in [12, 17]$. We conjecture however that for a fixed $r > 1$, there exists $n(r)$ such that for all $n \geqslant n(r)$, we have $Id_r(\mathbb{F}^n) \leqslant Id_r(\mathbb{F}^{n+1})$.

## 4.5 *Bipartite Graphs and Discriminating Codes*

There are scattered results regarding bipartite graphs (see Theorems 15 and 100(c), Section 6.2.3, and the second sentence after Theorem 107), but these graphs are also of interest to us because they lend themselves to the natural definition of *discriminating codes*: consider a bipartite graph $G = (I \cup A, E)$ where $I$ represents individuals and $A$ their attributes (hair color, age, glasses, ...). A code $C \subseteq A$ is discriminating if every individual has at least one attribute in $C$, and no two individuals have the same set of attributes in $C$ [30].

**Theorem 63** ([28]) *If $C$ is an optimal discriminating code in $G$, then $\lceil \log_2(|I| + 1) \rceil \leqslant |C| \leqslant |I|$. Both bounds are sharp.*

Discriminating codes may be seen as a generalization of 1-Id codes, since a 1-Id code in a graph $G = (V, E)$ is clearly a discriminating code in the bipartite graph consisting of $V$ on the one hand, and vertices representing the closed neighborhoods in $G$ on the other hand, with edges linking each closed neighborhood to its members.

The definition can be extended to any odd $r \geqslant 1$, by asking that for $i \in I$, $i_1 \in I$, $i_2 \in I$, $i_1 \neq i_2$, we have $N_r(i) \cap C \neq \emptyset$ and $N_r(i_1) \cap C \neq N_r(i_2) \cap C$. The associated decision problem is NP-complete for any fixed $r \geqslant 1$ [28], but for $r = 1$ there is a linear algorithm in the case of trees [30].

Special cases of bipartite graphs have been investigated: the infinite square and hexagonal grids [28], bipartite planar graphs [30], and the binary $n$-cube, for which it can be proved that for any odd $r$, there is a bijection between the set of $r$-Id codes in $\mathbb{F}^n$ and the set of $r$-discriminating codes in $\mathbb{F}^{n+1}$ [29].

## 4.6 Line Graphs and Edge Identification

We can *identify an edge with edges* in a graph $G = (V, E)$: for $e \in E$, we denote $N(e)$ the set of edges which are adjacent to $e$, and $N[e] = N(e) \cup \{e\}$. A code $C_E \subseteq E$ is edge-identifying if for every $e \in E$, we have $N[e] \cap C_E \neq \emptyset$, and for every $e_1 \in E$, $e_2 \in E$, $e_1 \neq e_2$, we have $N[e_1] \cap C_E \neq N[e_2] \cap C_E$. Clearly, $C_E$ is edge-identifying in $G$ if and only if, in the line graph $L(G)$, the vertices corresponding to the edges in $C_E$ form a 1-Id code. This motivates the study of Id codes in the class of line graphs. For instance, one has the following results.

**Theorem 64** ([81])

(a) *If $G$ is a 1-twin-free line graph of order $n$, then $Id(G) > \frac{3\sqrt{2}}{4}\sqrt{n} \approx 1.06066\sqrt{n}$.*
(b) *For $n \geqslant 4$, the minimum size of an edge-identifying code in the binary cube $\mathbb{F}^n$ is $2^{n-1}$.*

The decision problem associated with edge identification is NP-complete [148].

### 4.7 Miscellaneous

- **Planar and outerplanar graphs**

**Theorem 65** ([155]) *For any planar graph G of order $n \geqslant 18$, we have $LD(G) \geqslant \frac{n+10}{7}$; for any outerplanar graph G of order n, we have $LD(G) \geqslant \frac{2n+3}{7}$. Both bounds are sharp.*

- **Split graphs**

Split graphs have been studied in several works; in particular, complete suns have been intensively investigated, both for 1-LD and 1-Id codes: see, e.g., [2, 122].

- In the context of LD and Id codes, attention has been given in the existing literature to the following families: graph products (mostly Cartesian, but also corona, lexicographic, Kronecker, or direct), cubic graphs, rotagraphs and fasciagraphs, finite rectangles extracted from the infinite grids ($k_1$ rows, $k_2$ columns), $n$-dimensional infinite grids, Kneser graphs, circulant networks, complementary prisms, the $n$-cube endowed with the Lee metric, triangle-free graphs, cographs, Sierpiński graphs, interval graphs, permutation graphs, triangular graphs, vertex-transitive graphs, fractal graphs, block graphs, chordal graphs, series parallel graphs, and others. Random graphs have also been studied.

See also Tables 1.3–1.5 in [79] for a survey of the different lower and upper bounds, and their tightness, for different classes of graphs.

## 5 Relationships with Other Parameters

We study the different relationships linking the LD and Id numbers of a graph to different parameters such as maximum and minimum degree, girth, domination number, independent domination numbers, or diameter, mostly for $r = 1$.

### 5.1 Locating-Dominating Codes

**Theorem 66** ([170]) *For any graph G of order $n \geqslant 2$ and maximum degree $\Delta$, we have $n \leqslant LD_1(G) + \sum_{i=1}^{\Delta} \binom{LD_1(G)}{i}$.*

For $\Delta = 2$ (paths and cycles), this can be improved to $n \leqslant \frac{5}{2} LD_1(G)$, see Theorem 43.

Theorem 67 below gives upper bounds on $LD_1(G)$, with conditions on minimum degree and girth, then the best known constructions, for $n$ small or arbitrarily large.

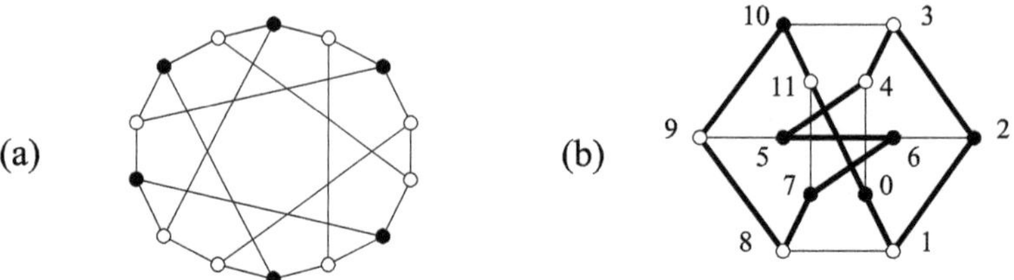


**Fig. 13** Black vertices form optimal codes. (a) The Heawood graph $H_{14}$; (b) The graph $G_{12}$.

**Theorem 67** *Let G be a graph of order n, minimum degree $\delta$, and girth $g \geqslant 5$.*

*($a_1$)* ([50]) *If $\delta \geqslant 2$, then $LD_1(G) \leqslant \frac{n}{2}$.*
*($a_2$)* ([10]) *If $\delta = 3$, then $LD_1(G) \leqslant \frac{22n}{45} \approx 0.4889\,n$.*
*(b)* ([10]) *The cycle $C_6$ has $n = 6$, $\delta = 2$, $g = 6$, and $LD_1(C_6) = 3 = \frac{n}{2}$; there are infinitely many connected graphs G of order n with $\delta = 2$, $g = 5$, and $LD_1(G) = \frac{n-1}{2}$.*

*The Heawood graph $H_{14}$ (see Figure 13(a)) has $n = 14$, $\delta = 3$, $g = 6$, and $LD_1(H_{14}) = 6 = \frac{3n}{7} \approx 0.4286\,n$; there are infinitely many connected graphs G of order n with $\delta = 3$, $g = 5$, and $LD_1(G) = \frac{4(n-1)}{11} \approx 0.3636\,n$.*

We recall that $i(G)$ (respectively, $\beta(G)$) denotes the minimum (respectively, maximum) cardinality of an independent dominating set in a graph $G$.

**Theorem 68** ([50])

*(a) If T is a nontrivial tree, then $i(T) \leqslant LD_1(T) \leqslant \beta(T)$.*
*(b) For every nontrivial tree T, we have $LD_1(T) + \gamma_1(T) \leqslant n$, with equality if and only if T is a tree consisting only of leaves and support vertices.*

**Theorem 69** ([25]) *Let G be a graph of order n and diameter $D \geqslant 3$. Then $LD_1(G) \leqslant n - \lceil \frac{3D-1}{5} \rceil$, and the bound is tight. Consequently, if $LD_1(G) \in \{n-2, n-1\}$, then G has diameter at most 3.*

## 5.2 *Identifying Codes*

We first give a lower bound involving the maximum degree, then discuss different upper bounds, first in terms of maximum degree, then of minimum degree and girth.

**Theorem 70** ([134]) *Let $r \geqslant 1$. If G is an r-twin-free graph of order n and maximum degree $\Delta$, then $Id_r(G) \geqslant \frac{2n}{\Delta+2}$.*

For $r = 1$, [79] gives a full characterization of the graphs reaching the previous lower bound.

**Theorem 71** ([83]) *Let $r \geqslant 1$ and $G$ be a connected $r$-twin-free graph of order $n$ and maximum degree $\Delta$. Then $Id_r(G) \leqslant n - \frac{n(\Delta-2)}{\Delta(\Delta-1)^{5r}-2} = n - \frac{n}{\Theta(\Delta^{5r})}$.*

It is conjectured in [84] that there exists a constant $c$ such that, for every nontrivial connected 1-twin-free graph $G$, we have $Id_1(G) \leqslant n - \frac{n}{\Delta} + c$. Graphs such that $Id_1(G) = n - \frac{n}{\Delta}$ exist (e.g., $K_{\Delta,\Delta}$ with $\Delta \geqslant 3$). The complete $q$-ary trees $T$ have $\Delta = q + 1$ and, by Theorem 56(b), we have $Id_1(T) = \lceil n - \frac{n}{\Delta-1+1/\Delta} \rceil$. The conjecture holds for graphs with very high $\Delta$, and for $\Delta = 2$, and the following results are intended to further substantiate it.

**Theorem 72** ([84]) *Let $G$ be a connected, 1-twin-free, triangle-free graph with order $n$ and maximum degree $\Delta \geqslant 3$.*

*(a) We have $Id_1(G) \leqslant n - \frac{n}{\Delta + \frac{3\Delta}{\ln \Delta - 1}} = n - \frac{n}{\Delta + o(\Delta)}$.*
*(b) If moreover $G$ is a nontrivial planar or bipartite graph, then $Id_1(G) \leqslant n - \frac{n}{\Delta+9}$.*

Other graphs can be exhibited, with $Id_1(G) \leqslant n - \frac{n}{o(\Delta)}$. In the following theorem, a vertex $z$ is said to be *forced* if there are two vertices $u$ and $v$ with symmetric difference $N[u]\Delta N[v]$ equal to $\{z\}$, which implies that $z$ belongs to any 1-Id code in $G$.

**Theorem 73** ([89])

*(a) Let $G$ be a connected, 1-twin-free graph, with order $n$ and maximum degree $\Delta \geqslant 3$. Then $Id_1(G) \leqslant n - \frac{nf^2}{103\Delta}$, where $f$ is the ratio over $n$ of the number of non-forced vertices. In particular, if $G$ is $\Delta$-regular, then $Id_1(G) \leqslant n - \frac{n}{103\Delta} = n - \frac{n}{\Theta(\Delta)}$.*
*(b) There exists an integer $\Delta_0$ such that for every connected, 1-twin-free graph with order $n$ and maximum degree $\Delta \geqslant \Delta_0$, we have $Id_1(G) \leqslant n - \frac{n}{103\Delta(\Delta+1)^2} = n - \frac{n}{\Theta(\Delta^3)}$.*

The following result can be compared to Theorem 67.

**Theorem 74** ([10]) *Let $G$ be a twin-free graph of order $n$, minimum degree $\delta$, and girth $g \geqslant 5$.*

*($a_1$) If $\delta \geqslant 2$, then $Id_1(G) \leqslant \frac{5n}{7} \approx 0.7143\, n$.*
*($a_2$) If $\delta = 3$, then $Id_1(G) \leqslant \frac{31n}{45} \approx 0.6889\, n$.*
*(b) The cycle $C_7$ has $n = 7$, $\delta = 2$, $g = 7$, and $Id_1(C_7) = 5 = \frac{5n}{7}$; there are infinitely many connected graphs $G$ of order $n$ with $\delta = 2$, $g = 5$, and $Id_1(G) = \frac{3(n-1)}{5}$.*

*The graph $G_{12}$ with vertex set $\{x_i : 0 \leqslant i \leqslant 11\}$ and the edges of the Hamiltonian cycle $x_0x_1 \dots x_{11}x_0$ plus the six edges $x_0x_4$, $x_1x_8$, $x_2x_6$, $x_3x_{10}$, $x_5x_9$, and $x_7x_{11}$ (see Figure 13(b)) has $n = 12$, $\delta = 3$, $g = 5$, and $Id_1(G_{12}) = 6 = \frac{n}{2}$; there are infinitely many connected graphs $G$ of order $n$ with $\delta = 3$, $g = 5$, and $Id_1(G) = \frac{5(n-1)}{11} \approx 0.4545\, n$.*

## 5.3 Twin-Free Graphs and Parameters

From now on and until the end of this section, we investigate the extremal values that the following parameters: size, order, maximum and minimum degree, size of a maximum clique, radius, diameter, size of a maximum independent set, and identification number can achieve in a connected $r$-twin-free graph. If $\pi$ stands for such a parameter, we fix $r$ and search for the smallest value $f_r(\pi)$ that $\pi$ can reach, or we fix $r$ and $n$ and search for the smallest and largest values $f_{r,n}(\pi)$ and $F_{r,n}(\pi)$. Therefore, $f_r(\pi) = \min\{\pi(G) : G$ connected, $r$-twin-free with at least $2r+1$ vertices$\}$, $f_{r,n}(\pi) = \min\{\pi(G) : G$ connected, $r$-twin-free with $n \geqslant 2r+1\}$, and $F_{r,n}(\pi)$ is defined similarly. The first case, when $\pi = \varepsilon$ is the size of the graph, will help understand easily these notions.

- **Number of edges** $\varepsilon$

For any connected graph of order $n$, the size is between $n-1$ (trees) and $\frac{n(n-1)}{2}$ (the complete graph). The size of a connected $r$-twin-free graph of order at least $2r+1$ is at least $2r$, and if the order is $n \geqslant 2r+1$, then this number is at least $n-1$; moreover, the paths $P_{2r+1}$ and $P_n$ meet these bounds. So $f_r(\varepsilon) = 2r$ and $f_{r,n}(\varepsilon) = n-1$. The maximum number of edges possible for $r$-twin-free graphs of order $n$ is known for $r=1$, and can be achieved only by the complete graphs minus a maximum matching [64]. For $r>1$, we are close to the exact value.

**Theorem 75** ([64]) *For all $n \geqslant 3$, we have $F_{1,n}(\varepsilon) = \frac{n(n-1)}{2} - \lfloor \frac{n}{2} \rfloor$.*

**Theorem 76** ([5])

*(a) For $n$ large enough, we have $\frac{n^2}{2} - 2n\log_2 n \lesssim F_{2,n}(\varepsilon) \lesssim \frac{n^2}{2} - \frac{1}{2}n\log_2 n$.*
*(b) For $r>2$ and $n$ large enough with respect to $r$, we have $\frac{n^2}{2} - rn\log_2 n \lesssim F_{r,n}(\varepsilon) \lesssim \frac{n^2}{2} - 0.63(r-0.915)n\log_2 n$.*

For $r>2$, the gap between the above lower and upper bounds is about $(0.37r + 0.58)n\log_2 n$. For $r=10$ and $n$ around 60000, it represents 0.2%.

- **Number of vertices** $n$

Obviously, $f_{r,n}(n) = F_{r,n}(n) = n$ for all $r \geqslant 1$, $n \geqslant 2r+1$, and the following comes from Theorem 2.

**Theorem 77** *For all $r \geqslant 1$, we have $f_r(n) = 2r+1$.*

- **Maximum degree** $\Delta$

In any connected graph of order $n$, the maximum degree is between 1 (if $n=2$) or 2 (the path, the cycle, $n \geqslant 3$), and $n-1$ (the complete graph, the star, every graph $G$ with $\gamma_1(G) = 1$).

**Theorem 78** ([45])

*(a)* *For all* $r \geqslant 1$, *we have* $f_r(\Delta) = 2$; *for all* $r \geqslant 1$ *and* $n \geqslant 2r + 1$, *we have* $f_{r,n}(\Delta) = 2$.
*(b)* *For all* $n \geqslant 3$, *we have* $F_{1,n}(\Delta) = n - 1$.
*(c)* *For all* $p \geqslant 2$ *and* $n \geqslant 5$, *if* $2^{p-1} + p - 1 < n \leqslant 2^p + p$, *then* $F_{2,n}(\Delta) = n - p - 1$.
*(d)* *For all* $r \geqslant 3$ *and* $n \geqslant 2r + 1$, *we have* $F_{r,n}(\Delta) \leqslant k$, *where* $k$ *is the largest integer such that* $k + (r - 2)\lceil \log_3(k + 1)\rceil + \lceil \log_2(k + 1)\rceil \leqslant n - 1$.

When $r > 2$, there are too many cases for the lower bounds on $F_{r,n}(\Delta)$ to be given here. Let us only mention that: (i) we have the exact value of $F_{r,n}(\Delta)$ for infinitely many values of $n$ and $r$—for instance, if $r = 100$ and $60\,050 \leqslant n \leqslant 60\,143$, then $F_{100,n}(\Delta) = 59\,048$; (ii) if $n$ is large with respect to $r$, the lower bound on $F_{r,n}(\Delta)$ approximately behaves like $n - r \log_3 n$, whereas the upper bound in Case (d) above behaves like $n - (r - 2 + \log_2 3) \log_3 n$, which gives a gap between the bounds roughly equal to $0.415 \times \log_3 n$ (independent of $r$).

- **Minimum degree** $\delta$

In any connected graph of order $n$, the minimum degree is between 1 and $n - 1$ (the complete graph). The first part of Case (b) below relies on the fact that $K_n$ minus a maximum matching is 1-twin-free, cf. Theorem 75.

**Theorem 79** ([5])

*(a)* *For all* $r \geqslant 1$, *we have* $f_r(\delta) = 1$; *for all* $r \geqslant 1$ *and* $n \geqslant 2r + 1$, *we have* $f_{r,n}(\delta) = 1$.
*(b)* *For all* $n \geqslant 3$, *we have* $F_{1,n}(\delta) = n-2$; *for all* $n \geqslant 5$, *we have* $F_{2,n}(\delta) = \lfloor \frac{n-2}{2} \rfloor$.
*(c)* *For all* $r \geqslant 3$, *we have* $F_{r,2r+1}(\delta) = 1$.
*(d)* *Let* $r \geqslant 3$, $n \geqslant 2r + 2$ *and* $k = \lfloor \frac{n-2}{r} \rfloor$.

*(d₁)* *We have* $k - 1 \leqslant F_{r,n}(\delta)$ *if* $k$ *is odd, and* $k \leqslant F_{r,n}(\delta)$ *if* $k$ *is even.*
*(d₂)* *For* $r \in \{3, 4, 5, \ldots, 24\}$, *we have* $F_{r,n}(\delta) \leqslant \frac{n}{\lfloor \frac{r}{2} \rfloor + 1} - 1$.
*(d₃)* *For* $r \geqslant 25$, *we have* $F_{r,n}(\delta) \leqslant \min\left\{\frac{n}{\lfloor \frac{r}{2} \rfloor + 1} - 1, \frac{3n-r+2}{2(r-5)}\right\}$.

We conjecture that $F_{r,n}(\delta)$ is close, possibly equal, to $\lfloor \frac{n-2}{r} \rfloor$.

- **Size** $\omega$ **of a maximum clique**

In any connected graph of order $n$, a maximum clique has size between 2 and $n$. The clique $K_n$ is far from being $r$-twin-free; actually any graph whose $r$-th power is the clique has $r$-twins. However, a twin-free graph may contain quite a large clique, and it is remarkable that we can exactly determine its size. Case (b) below is connected to Theorem 95. Case (c) depends on some conditions on $k$ and $n$ that we do not give here.

**Theorem 80** (Charon, Hudry & Lobstein 2011 [44])

*(a) For all $r \geqslant 1$, we have $f_r(\omega) = 2$; for all $r \geqslant 1$ and $n \geqslant 2r + 1$, we have $f_{r,n}(\omega) = 2$.*
*(b) For all $n \geqslant 3$, we have $F_{1,n}(\omega) = k$, where $k$ is the largest integer such that $k + \lceil \log_2 k \rceil \leqslant n$.*
*(c) For all $r \geqslant 2$ and $n \geqslant 2r + 1$, let $k$ be the largest integer such that $k + r\lceil \log_2 k \rceil \leqslant n-1$. Then, according to conditions on $n$ and $k$, we have $F_{r,n}(\omega) = k$ or $k + 1$.*

If $n$ is large with respect to $r$, then $F_{r,n}(\omega)$ behaves like $n - r \log_2 n$. If $r$ is a fraction of $n$, then $F_{r,n}(\omega)$ is bounded by above by a constant; the extremal case is when $r = \frac{n-1}{2}$ and $F_{r,n}(\omega) = 2$ (path $P_{2r+1}$).

- **Radius $\rho$**

In any connected graph of order $n$, the radius is between 1 (the complete graph, the star) and $\lfloor \frac{n}{2} \rfloor$ (the path, the cycle). The study of the radius in twin-free graphs is easy, the results are complete.

**Theorem 81** ([5])

*(a) For all $r \geqslant 1$, we have $f_r(\rho) = r$; for all $r \geqslant 1$ and $n \geqslant 2r + 1$, we have $f_{r,n}(\rho) = r$.*
*(b) For all $r \geqslant 1$ and $n \geqslant 2r + 1$, we have $F_{r,n}(\rho) = \lfloor \frac{n}{2} \rfloor$.*

- **Diameter $D$**

In any connected graph of order $n$, the diameter is between 1 (the complete graph) and $n - 1$ (the path). The results on the diameter in twin-free graphs are complete. Figure 14 illustrates Case (c).

**Theorem 82** ([5])

*(a) For all $r \geqslant 1$, we have $f_r(D) = r + 1$.*
*(b) For all $r \geqslant 1$, we have $f_{r,2r+1}(D) = 2r$.*
*(c) For all $r \geqslant 1$ and $n \geqslant 2r + 2$, we have $f_{r,n}(D) = r + 1$.*
*(d) For all $r \geqslant 1$ and $n \geqslant 2r + 1$, we have $F_{r,n}(D) = n - 1$.*

- **Size $\alpha$ of a maximum independent set**

In any connected graph of order $n$, $\alpha$ lies between 1 (the complete graph) and $n - 1$ (the star). Theorem 89 contributes to Case (a) in the following theorem.

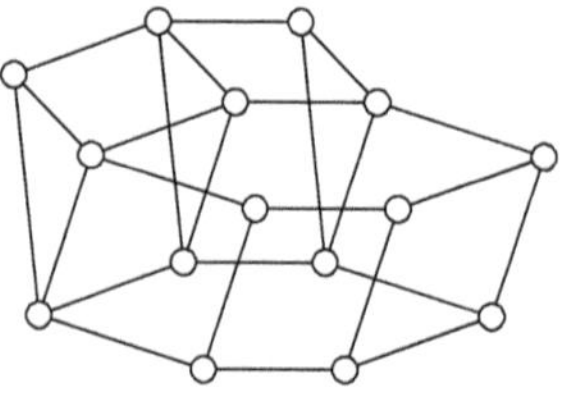

**Fig. 14** The case $n = 15$, $r = 3$: the graph is 3-twin-free and has diameter $r + 1 = 4$.

**Theorem 83** ([5])

*(a) For all $r \geqslant 1$, we have $f_r(\alpha) = r + 1$.*
*(b) For all $n \geqslant 3$, we have $f_{1,n}(\alpha) = 2$.*
*(c) For all $r \geqslant 2$, we have $f_{r,2r+1}(\alpha) = r + 1$.*
*(d) Let $r \geqslant 2$, $n \geqslant 2r + 2$ and $k = \lfloor \frac{n-2}{r} \rfloor$. We have* $r + 1 \leqslant f_{r,n}(\alpha) \leqslant \begin{cases} \frac{2n}{k+2} : k \text{ even} \\ \frac{2n}{k+1} : k \text{ odd} \end{cases}$.

The star gives Case (a) in the following theorem.

**Theorem 84** ([5])

*(a) For all $n \geqslant 3$, we have $F_{1,n}(\alpha) = n - 1$.*
*(b) For all $r \geqslant 2$ and $n \geqslant 2r + 1$, we have* $\max\left\{\lceil \frac{n}{2} \rceil, k + \lceil \log_2 k \rceil \lfloor \frac{r}{2} \rfloor\right\} \leqslant F_{r,n}(\alpha) \leqslant n - r$, *where $k$ is the largest integer such that $k + r\lceil \log_2 k \rceil \leqslant n - 1$.*

If $n$ is large with respect to $r$, then $k + \lceil \log_2 k \rceil \lfloor \frac{r}{2} \rfloor$ behaves approximately like $n - \frac{r}{2} \log_2 n$.

- **Identification number $Id_r$**

We start with $F_{r,n}(Id_r)$ and $f_{r,n}(Id_r)$, and conclude with $f_r(Id_r)$, the most interesting of the three functions. From Section 2.1, in particular Theorems 8 and 13, we derive the following results.

**Theorem 85**

*(a) For all $r \geqslant 1$, if $n = 2r + 1$, or if $n \geqslant 3r^2$ is even, or if $n \geqslant 3r^2 + 1$ is odd, we have $F_{r,n}(Id_r) = n - 1$. In particular, for all $n \geqslant 3$, we have $F_{1,n}(Id_1) = n - 1$.*
*(b) For all $r \geqslant 1$ and $n \geqslant 2^{2r}$, we have $f_{r,n}(Id_r) = \lceil \log_2(n + 1) \rceil$.*

**Theorem 86** ([44])

*(a) We have $f_1(Id_1) \geqslant 2$ and for all $r \geqslant 2$, $f_r(Id_r) \geqslant \lceil \log_2(2r + 4) \rceil$.*
*(b) For all $r \geqslant 1$, we have $f_r(Id_r) \leqslant r + 1$.*

The constructive upper bound is illustrated by Figure 15.

The lower and upper bounds in Theorem 86 coincide for $r \in \{1, 2, 3\}$. For $r = 4$, ad hoc arguments show that no graph admits a 4-Id code of size 4. Therefore,

**Theorem 87** ([44]) *For $r \in \{1, 2, 3, 4\}$, we have $f_r(Id_r) = r + 1$.*

Open problem: for $r \geqslant 5$, find the exact value of $f_r(Id_r)$, knowing that it lies between $\lceil \log_2(2r + 4) \rceil$ and $r + 1$.

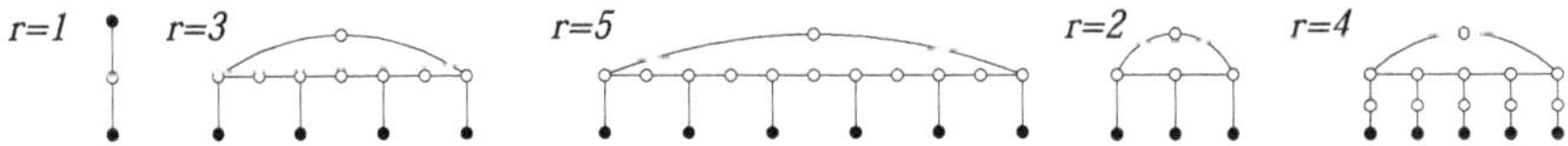


**Fig. 15** Different graphs with optimal $r$-Id codes of size $r + 1$. Codewords are in black. For $r \geqslant 3$ odd, there is a cycle $C_{2r+2}$ and $n = (2r + 2) + r + 1$. For $r$ even, there is a cycle $C_{r+2}$ and $n = (r + 2) + \frac{r}{2}(r + 1)$.

# 6 Structural Issues

This section is mostly devoted to identifying codes, either because identifying codes lead to the problem of their very existence, i.e., to the important issue of when a graph is twin-free, or simply because in the literature, more attention has been devoted to these topics for identifying codes, compared to locating-dominating codes.

We shall study problems such as the existence of a chordless path in twin-free graphs, whether a twin-free graph deprived of a vertex remains twin-free, the consequence of adding or deleting an edge or a vertex on the identification number, or graphs that are critical, in some sense, with respect to locating-domination or identification.

Locating-dominating and identifying codes do not lend themselves well to Nordhaus–Gaddum-type bounds, and there are very few results. We give two examples.

**Theorem 88** ([106]) *For any graph $G$ of order $n \geqslant 1$, we have $|LD_1(G) - LD_1(\overline{G})| \leqslant 1$.*

Also, because $P_4 = \overline{P}_4$ and $Id_1(P_4) = 3$, the trivial bounds $2\lceil \log_2(n+1) \rceil \leqslant Id_1(G) + Id_1(\overline{G}) \leqslant 2(n-1)$ and $\lceil \log_2(n+1) \rceil^2 \leqslant Id_1(G) \times Id_1(\overline{G}) \leqslant (n-1)^2$ are tight.

## 6.1 *Structural Properties of Twin-Free Graphs*

The first part of Theorem 2 is an obvious consequence of the following result, due to Auger in 2008 [3], which is not obvious.

**Theorem 89** *Let $r \geqslant 1$ and $G$ be a connected $r$-twin-free graph. Then $G$ admits $P_{2r+1}$, the path on $2r+1$ vertices, as an induced subgraph, i.e., there is a chordless path of length $2r$ in $G$.*

### 6.1.1 Deletion of a Vertex in a Twin-Free Graph

Theorems 90–94, due to Charon *et al.* [35] in 2007, give results which vary widely with $r$. If $G = (V, E)$ is $r$-twin-free, we say that $G$ is *$r$-terminal* if for all $v \in V$, $G - v$ is not $r$-twin-free (so $G$ is not $r$-terminal if there exists $v \in V$ such that $G - v$ is also $r$-twin-free). An alternative would be *$r$-twin-free critical*, see also the discussion before Theorem 96.

If $n = 2r + 1$, the only connected $r$-twin-free graph is the path $P_{2r+1}$ for $r \geqslant 1$, and the only $r$-terminal graph is $P_{2r+1}$ for $r > 1$ (the case of $P_3$ is particular, because removing the middle vertex yields two isolated vertices, constituting a 1-twin-free graph). We address the following questions: (a) are there 1-terminal graphs? (b) For $r > 1$, are there $r$-terminal graphs other than $P_{2r+1}$?

The answer to (a) is negative, and the answer to (b) is multifold: it is negative if we restrict ourselves to trees; it is positive if $r \geqslant 3$. The case $r = 2$ remains open.

- **The case** $r = 1$

**Theorem 90** ([35]) *Let $n \geqslant 3$ and $G = (V, E)$ be any connected 1-twin-free graph of order $n$. Then $G$ is not 1-terminal.*

**Proof.** If $n = 3$, then $G = P_3$ is not 1-terminal, as already seen; so we can assume that $n \geqslant 4$. By Theorem 6, there is a 1-Id code $C$ of size $n-1$ in $G$. Consider $G - v$, where $\{v\} = V \setminus C$ ($G - v$ may be connected or not); then $C$ is still 1-identifying in $G - v$, because removing $v$ did not cut connexions of length $r (= 1)$ between pairs of vertices not containing $v$ itself—this explains why the cases $r = 1$ and $r > 1$ are different. Therefore, $G - v$ is 1-twin-free. □

The following theorem sharpens the previous one with respect to connectivity.

**Theorem 91** ([35]) *Let $n \geqslant 4$ and $G = (V, E)$ be any connected 1-twin-free graph of order $n$. Then there exists $v \in V$ such that $G - v$ is 1-twin-free and connected.*

- **The case of trees**

**Theorem 92** ([35]) *Let $r \geqslant 1$, $n \geqslant 2r + 2$ and $T = (V, E)$ be any (connected) $r$-twin-free tree of order $n$. Then there exists a leaf $v \in V$ such that $T - v$ is $r$-twin-free (and connected). Consequently, the only $r$-terminal trees are the paths $P_{2r+1}$, for $r > 1$.*

- **The case** $r \geqslant 3$

We now consider general graphs, for $r \geqslant 3$.

**Theorem 93** ([35]) *For each integer $r \geqslant 3$, there is a graph $G$, $G \neq P_{2r+1}$, which is $r$-terminal.*

**Sketch of Proof.** We search for a connected $r$-twin-free graph $G = (V, E)$, with $|V| \geqslant 2r + 2$, $r \geqslant 3$, such that for all $v \in V$, $G - v$ is not $r$-twin-free. One possible construction is the following: take a cycle of length $2r$ with vertices $c_i$ ($i \in \{0, 1, \ldots, 2r - 1\}$), and add one vertex $s_i$ together with the edge $c_i s_i$ for every value of $i$ but one. □

- **Other values of** $r$ **and open problems**

The construction of Theorem 93 does not work for $r = 2$. Other constructions have been tried and failed, and the problem remains open: Apart from $P_5$, do 2-terminal graphs exist?

**Theorem 94** ([35]) *For each integer $r \geqslant 6$, there are infinitely many $r$-terminal graphs.*

Now, another open problem is the situation for $r \in \{3, 4, 5\}$: there exist $r$-terminal graphs, but are they in finite or infinite number? Conversely, there is a bound on the number of vertices that need to be added in order to obtain a 1-twin-free graph.

**Theorem 95** ([94]) *If $G$ has order $n$, then $G$ embeds as an induced subgraph in a 1-twin-free graph with order at most $n + \lceil \log_2 n \rceil$. The bound is tight.*

### 6.1.2 Deletion of an Edge in a Twin-Free Graph

The graphs that we have called terminal in the previous section might as well have been called *vertex-terminal*, and the $r$-twin-free graphs $G = (V, E)$ such that for all $e \in E$, $G - e$ is not $r$-twin-free could then have been called *r-edge-terminal*. Back to 1974 however, the expression "line-critical point distinguishing graph" was coined in [68] (for $r = 1$ only) and the following result was proved.

**Theorem 96** ([68]) *A nontrivial connected graph $G$ is line-critical point distinguishing [or 1-edge-terminal] if and only if $G = P_3$.*

*A nonempty graph is line-critical point distinguishing if and only if it is the disjoint union of paths of length two and isolated vertices.*

To our knowledge, everything remains to be done for $r > 1$.

## 6.2 *Adding or Deleting Edges or Vertices Can Save Codewords*

The question is to determine how the identification or locating-domination numbers change when we add or delete edges or vertices. We first deal with identifying codes.

### 6.2.1 Identifying Codes: Adding or Deleting an Edge

The problem is the following: given an $r$-twin-free graph $G = (V, E)$ and an edge $e \in E$, and assuming that the graph $G - e$ is $r$-twin-free, what can be said about the relationship between $Id_r(G)$ and $Id_r(G - e)$? The answer widely depends on $r$, as show the following two theorems by Charon *et al.* [37] in 2014.

- **The case $r = 1$**

**Theorem 97** *If $G$ and $G - e$ are 1-twin-free, then $Id(G) - Id(G - e) \in \{-2, -1, 0, 1, 2\}$. Moreover, pairs of connected graphs $G$ and $G - e$ such that $Id(G) - Id(G - e) = 0$, $Id(G) - Id(G - e) = \pm 1$ or $Id(G) - Id(G - e) = \pm 2$ exist.*

- **The case $r \geqslant 2$**

Now the differences $Id_r(G) - Id_r(G - e)$ and $Id_r(G - e) - Id_r(G)$ can be large, and we obtain results which slightly vary with $r$.

**Theorem 98** *Let $k \geqslant 2$ be an arbitrary integer.*

*(a1) Let $r \geqslant 2$. There exist two $r$-twin-free graphs, $G$ and $G-e$, with $(r+1)k + r\lceil \log_2(k+2)\rceil + 2r$ vertices, such that $Id_r(G) \geqslant k$ and $Id_r(G-e) \leqslant r\lceil \log_2(k+2)\rceil + r + 3$.*
*(a2) Let $r \geqslant 5$. There exist two $r$-twin-free graphs, $G$ and $G-e$, with $(2r-2)k + r\lceil \log_2(k+2)\rceil + r + 3$ vertices, such that $Id_r(G) \geqslant k$ and $Id_r(G-e) \leqslant r\lceil \log_2(k+2)\rceil + r + 1$.*
*(b1) There exist two 2-twin-free graphs, $G$ and $G-e$, with $3k+2\lceil \log_2(k+2)\rceil+5$ vertices, such that $Id_2(G-e) \geqslant k$ and $Id_2(G) \leqslant 2\lceil \log_2(k+2)\rceil + 5$.*
*(b2) Let $r \geqslant 3$. There exist two $r$-twin-free graphs, $G$ and $G-e$, with $(2r-2)k + r\lceil \log_2(k+2)\rceil + r + 2$ vertices, such that $Id_r(G-e) \geqslant k$ and $Id_r(G) \leqslant r\lceil \log_2(k+2)\rceil + r + 1$.*

Very close results can be obtained with pairs of connected graphs. Whether these inequalities can be substantially improved is an open problem.

Note that $k$ can be taken arbitrarily and is linked to the order $n$ of $G$ and $G-e$ by the relation $n = (c_1r+c_2)k+r\lceil \log_2(k+2)\rceil+(c_3r+c_4)$ where the integer quadruple $(c_1, c_2, c_3, c_4)$ takes different values in (a1), (a2), (b1), and (b2) above. This means, roughly speaking, that $k$ is a fraction, depending on $r$, of $n$; therefore, Theorem 98 implies that, given $r \geqslant 2$, there is an infinite sequence of graphs $G$ of order $n$ and two positive constants $\alpha$ and $\beta$ such that $Id_r(G) \geqslant \alpha n$ and, after deletion of a suitable edge $e$, $Id_r(G-e) \leqslant \beta \log_2 n$ (or the other way round: $Id_r(G) \leqslant \beta \log_2 n$ and $Id_r(G-e) \geqslant \alpha n$). We can see that adding or deleting one edge can lead to quite a drastic difference for the identification numbers. On this topic, see also [90] for a probabilistic approach.

### 6.2.2 Identifying Codes: Adding or Deleting One Vertex or More

- **The case $r = 1$**

If $S \subset V$, we denote by $G - S$ the graph obtained from $G$ by deleting the vertices of $S$.

**Theorem 99** ([83]) *If $G$ and $G-S$ are 1-twin-free, then $Id(G)-Id(G-S) \leqslant |S|$. In particular, if $v \in V$, we have $Id(G) - Id(G-v) \leqslant 1$; moreover, the two inequalities are tight.*

**Theorem 100** ([151]) *Let $n$ be the order of $G$, $S \subset V$ and $v \in V$.*

*(a) If $n \geqslant 2^{|S|-1}$, then $Id(G-S) - Id(G) \leqslant n - 2|S| - \lfloor \frac{n-|S|}{2^{|S|}} \rfloor$; the inequality is tight for $n$ sufficiently large with respect to $|S|$.*
*(b) We have $Id(G-v) - Id(G) \leqslant \frac{n}{2} - \varepsilon$, with $\varepsilon = 2$ if $n \in \{2, 4, 5, 6, 8\}$, $\varepsilon = 1$ otherwise; the inequality is tight.*
*(c) If $G$ is bipartite, then $Id(G-v) - Id(G) \leqslant \frac{n-\log_2(n-\log_2 n)}{2} - 1$; the inequality is tight.*

So, if one deletes a vertex, the identification number cannot drop by more than one, and can increase by a quantity close to $n/2$.

- **The case** $r \geqslant 2$

Now both $Id_r(G-v)-Id_r(G)$ and $Id_r(G)-Id_r(G-v)$ can be large: the difference $Id_r(G-v)-Id_r(G)$ can increase to approximately $\frac{n}{4}$ (for even $r$) and $\frac{n(3r-1)}{12r}$ (for odd $r$), or $\frac{n(2r-2)}{2r+1}$, according to whether we want the graphs to be connected or not, and $Id_r(G)-Id_r(G-v)$ to approximately $\frac{n(r-1)}{r}$.

**Theorem 101** ([36]) *Let $n$ be the order of $G$ and $v \in V$.*

*(a) Let $r \geqslant 2$ be even. There exist two connected graphs, $G$ and $G-v$, such that $Id_r(G-v)-Id_r(G) \geqslant \frac{n}{4}-(r+1)$.*
*(b) Let $r \geqslant 3$ be odd. There exist two connected graphs, $G$ and $G-v$, such that $Id_r(G-v)-Id_r(G) \geqslant \frac{n(3r-1)}{12r}-r$.*
*(c) Let $r \geqslant 2$. There exist two graphs, $G$ and $G-v$, such that $Id_r(G-v)-Id_r(G) \geqslant \frac{(n-1)(2r-2)}{2r+1}-2r$.*

**Theorem 102** ([36]) *Let $n$ be the order of $G$ and $v \in V$. Then there exist two connected graphs, $G$ and $G-v$, such that $Id_r(G)-Id_r(G-v) \geqslant \frac{(n-3r-1)(r-1)+1}{r}$.*

Whether all these inequalities can be substantially improved is an open problem.
Figure 16 illustrates Theorem 102 for $n=17, r=3$.

### 6.2.3 Criticality Concepts

In this section, $r=1$. Let $\pi$ stand for $LD=LD_1$ or $Id=Id_1$. A graph $G=(V,E)$ is said to be $\pi^+$*-edge removal critical*, or $\pi^+$-ER-*critical*, if $\pi(G-e)>\pi(G)$ for all $e \in E$, and $\pi^-$-ER-*critical* if $\pi(G-e)<\pi(G)$ for all $e \in E$. Similarly, one can define $\pi^+$*-vertex removal critical* graphs and $\pi^-$*-vertex removal critical* graphs.

For instance, it is possible to characterize all graphs that are LD$^+$-ER-critical. Let $H=(V_1 \cup V_2, E)$ be a connected bipartite graph such that: for every $w \in V_2$ and for every nonempty subset $V_1' \subseteq N(w)$ there exists a unique $w' \in V_2$ such that $N(w')=V_1'$. Let $\mathcal{H}$ be the set of all such graphs, see Figure 17 for an example.

**Theorem 103** ([21]) *A nontrivial connected graph $G$ is $LD^+$-ER-critical if and only if $G \in \mathcal{H}$.*

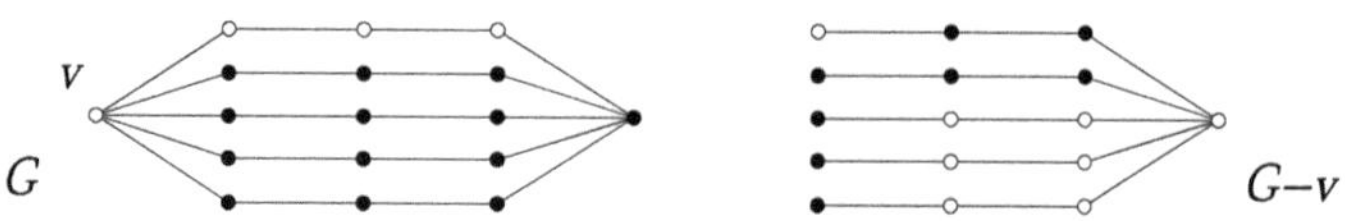


**Fig. 16** In each graph, the black vertices form an optimal 3-Id code, and $Id_3(G)-Id_3(G-v)=13-8=5$.

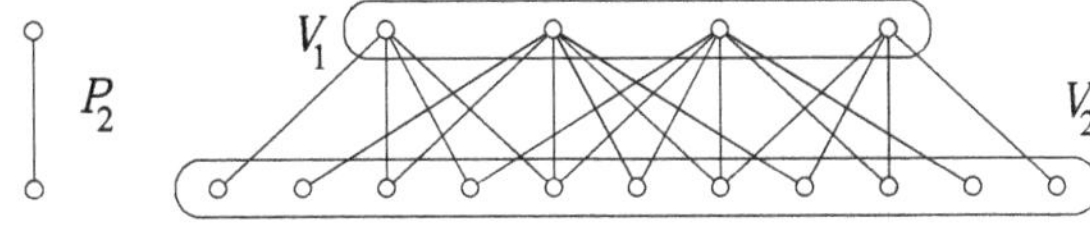


**Fig. 17** Two graphs belonging to $\mathcal{H}$ [21].

*A nonempty graph is $LD^+$-ER-critical if and only if it is the disjoint union of independent sets and of graphs in $\mathcal{H}$.*

For more results, see, e.g., [92] for graphs which are critical for identification, or [62] and [63] for locating-domination.

In Section 6.1, we mostly showed what a twin-free graph can become when we delete vertices; so far, in Section 6.2 we studied what the identification and locating-domination numbers can become when adding or deleting vertices or edges; but we can also be interested in what an *existing* (optimal) code becomes when edges or vertices are deleted or added—this leads naturally to the notion of *robustness*: an $r$-Id code $C$ is *$t$-edge-robust* [111] in $G$ if $C$ remains $r$-Id in all the graphs obtained from $G$ by adding or deleting edges, with a total amount of additions and deletions at most $t$; see also, among others, [114, 139]. Different definitions exist for $t$-vertex-robust codes, see [111, 159].

## 7 Number of Optimal Codes

In this section, we are interested in graphs with many optimal codes—or only one.

In order to obtain a large number of optimal 1-Id codes, Honkala, Hudry, & Lobstein [110] considered in 2015 the binary 3-cube and used it to build a graph with $8k + 1$ vertices, $k \geqslant 1$, admitting at least $56^k + 32k \times 56^{k-1}$ optimal 1-Id codes, of size $4k$, then they built graphs of higher orders and numbers of optimal codes, until they obtained the following result.

**Theorem 104** ([110]) *There exist infinitely many connected graphs of order $n$ admitting approximately $2^{0.770 \times n}$ different optimal 1-identifying codes.*

For $r > 1$, they used trees admitting many optimal codes, and combined them.

**Theorem 105** ([110]) *Let $r \geqslant 1$ be an integer and $\varepsilon > 0$ be a real. There exist infinitely many connected graphs of order $n$ admitting $2^{\left(\frac{1+\log_2 5}{2} - \varepsilon\right)n}$ different optimal $r$-identifying codes.*

Note that $\frac{1+\log_2 5}{2} \approx 0.664$. An obvious upper bound is $\binom{n}{\lceil \frac{n}{2} \rceil}$, which, using Stirling's formula, can be approximated above by $2^{n - \frac{1}{2}\log_2 n}$; to reduce the gap between these lower and upper bounds remains an open problem.

In [95] there is the description of graphs $G$ where every $k$-set of vertices is a 1-Id code, but the codes are not necessarily optimal, that is, $k$ may be different from $Id_1(G)$. See a small example in Figure 18(a).

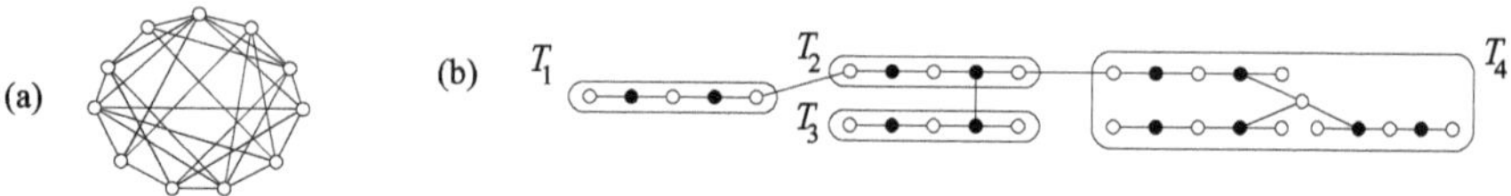


**Fig. 18** (a) A graph where all sets with 7 vertices are 1-identifying codes (not optimal). (b) Codewords are in black. Each code inside $T_1$, $T_2$, $T_3$, $T_4$, $T_1 \cup T_2$, $T_2 \cup T_3$, $T_2 \cup T_4$, $T_1 \cup T_2 \cup T_3$, $T_1 \cup T_2 \cup T_4$, $T_2 \cup T_3 \cup T_4$, $T_1 \cup T_2 \cup T_3 \cup T_4$ is a unique optimal 1-locating-dominating code.

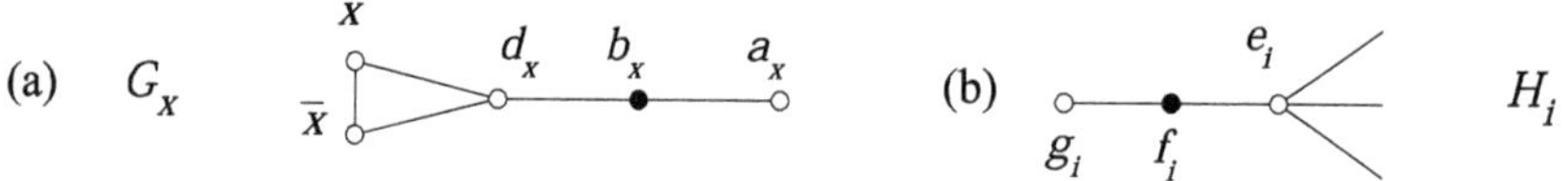


**Fig. 19** Black vertices will be part of a LD code. (a) The graph $G_x$. (b) The graph $H_i$.

On the other hand, a characterization of trees admitting a unique optimal 1-LD code is given in [19], see examples of such trees in Figure 18(b).

For complexity issues related to the uniqueness of optimal codes, see Section 8.2.

# 8 Complexity

## 8.1 How Hard Is It to Find Optimal Codes?

For any integer $r \geqslant 1$, we consider the following two decision problems.

| $r$-**LD CODE** | $r$-**Id CODE** |
|---|---|
| **Instance:** A graph $G = (V, E)$ and an integer $k \leqslant \lvert V \rvert$. | **Instance:** A graph $G = (V, E)$ and an integer $k \leqslant \lvert V \rvert$. |
| **Question:** Does $G$ admit an $r$-locating-dominating code of size at most $k$? | **Question:** Does $G$ admit an $r$-identifying code of size at most $k$? |

**Theorem 106** ([60]) *The problem 1-**LD CODE** is NP-complete.*

**Proof.** We simplify the 1987 proof by Colbourn, Slater, & Stewart in [60]. We take an arbitrary instance of 3-SAT, that is, a set $X = \{x_1, \ldots, x_n\}$ of variables and a set $\mathcal{C} = \{c_1, \ldots, c_m\}$ of clauses of size 3. For each variable $x$, we construct the 5-vertex graph $G_x = (V_x, E_x)$ given in Figure 19(a), in which additional edges may be incident only with $x$ or $\overline{x}$.

For each clause $c_i = \{\ell_{i,1}, \ell_{i,2}, \ell_{i,3}\}$, we construct the 3-vertex graph $H_i = (W_i, F_i)$ given by Figure 19(b), where the 3 edges incident with $e_i$ and not with $f_i$, are incident with $\ell_{i,1}$, $\ell_{i,2}$, and $\ell_{i,3}$. We claim that the resulting graph $G$, whose order $5n + 3m$ is polynomial with respect to the size of the instance of 3-SAT, admits a LD code of size $k = 2n + m$ if and only if the set of clauses $\mathcal{C}$ is satisfiable.

(a) If $\mathcal{C}$ is satisfiable, take as codewords the $n$ vertices $b_x$, the $m$ vertices $f_i$, $1 \leqslant i \leqslant m$, then for each variable $x \in X$, whichever of $x$ and $\overline{x}$ is true. This code, of size $k$, is locating-dominating, in particular because $e_i$ and $g_i$ are separated by at least one vertex corresponding to a true literal, i.e., are separated by a codeword.
(b) Let $C$ be a LD code in $G$, of size $k$. Obviously, $|C \cap V_x| \geqslant 2$ and $|C \cap W_i| \geqslant 1$; therefore, thanks to the choice of $k$, we have $|C \cap V_x| = 2$ and $|C \cap W_i| = 1$. Also, because $g_i$ must be dominated by $C$, we have $|C \cap \{g_i, f_i\}| = 1$ and none of the vertices $e_i$ is in $C$. This implies that $|C \cap \{x, \overline{x}\}| = 1$, because if $C \cap \{x, \overline{x}\} = \emptyset$, then the non-codewords $x$ and $\overline{x}$ are not separated by $C$ (the vertex $d_x$ cannot do it), and if $|C \cap \{x, \overline{x}\}| = 2$, then $b_x$ and $a_x$ are not dominated by $C$. So we can define a valid truth assignment of the variables by setting $x$ true if and only if $x \in C$.

(i) If $C \cap \{g_i, f_i\} = \{g_i\}$, then $e_i$ must be dominated by a codeword of type $x$ or $\overline{x}$; (ii) the same is true if $C \cap \{g_i, f_i\} = \{f_i\}$, because $g_i$ and $e_i$ must be separated by the code. This shows that every clause $c_i$ is satisfied by this assignment.

□

The same result for $r \geqslant 2$ was stated in [26] in 1995. We indicate the 2003 reference [40], which gives a proof.

**Theorem 107**

*(a)* ([26]), ([40]) *For any $r > 1$, the problem $r$-**LD CODE** is NP-complete.*
*(b)* ([57] for $r = 1$), ([40]) *For any $r \geqslant 1$, the problem $r$-**Id CODE** is NP-complete.*

This means that, given $G$ and $r$, determining $LD_r(G)$ or $Id_r(G)$, or finding an optimal $r$-LD or $r$-Id code, is NP-hard.

Note that Theorem 107 holds even if the instances are restricted to bipartite graphs; it also exists for bipartite oriented graphs [39]. For more on complexity, see [4, 6] for identification in planar graphs with arbitrarily high girth or low maximal degree, or [87] for interval and permutation graphs; Table 1 in [80] extends Table 1.8 in [79] and summarizes what was known around 2015 about the complexity of certain classes of graphs, for LD and Id codes, as well as dominating codes, for comparison.

See [118, 119] for identification in the binary $n$-cube, and [121] for problems closely related, such as the search for an $r$-Id code containing a prescribed vertex subset. For concepts close to locating-domination and identification that will be surveyed in Section 9: complexity results on $(r, \leqslant \ell)$-identifying codes are given in [6], on watching systems in [7], on open neighborhood locating-dominating codes in [164], on metric bases in [93, p. 204]; usually, the decision problems are NP-complete, even for some restricted classes of graphs, but linear for trees (when $r = 1$). The problem 1-**LD CODE** is also linear for series parallel graphs [60], and so is 1-**Id CODE** for block graphs [1].

For approximability issues, see, e.g., [80] (where Table 2 gives a state of play) or [87].

Complexity can bring what may look like surprises [79]: there are classes of graphs in which 1-**LD CODE** is polynomial and 1-**Id CODE** is NP-complete, and, if we define the decision problem 1-**DOM CODE** similarly to 1-**LD CODE** or 1-**Id CODE**, there are even classes of graphs in which 1-**LD CODE** and 1-**Id CODE** are polynomial, whereas 1- **DOM CODE** is NP-complete.

### *8.2 Uniqueness of Optimal Codes*

We consider the following decision problems, stated for fixed $r \geqslant 1$.

| **UNIQUE OPTIMAL** $r$**-LD CODE** | **UNIQUE OPTIMAL** $r$**-Id CODE** |
|---|---|
| **Instance:** A graph $G = (V, E)$. | **Instance:** A graph $G = (V, E)$. |
| **Question:** Does $G$ admit a *unique optimal* $r$-locating-dominating code? | **Question:** Does $G$ admit a *unique optimal* $r$-identifying code? |

In [123], it is proved, among other results, that for every $r \geqslant 1$, these two problems are NP-hard and belong to the class $L^{NP}$ (also denoted by $P^{NP[O(\log n)]}$ or $\Theta_2$), which contains the decision problems which can be solved by applying, with a number of calls which is logarithmic with respect to the size of the instance, a subprogram able to solve an appropriate problem in NP.

## 9 Related Concepts, Generalizations

- One of the first possible generalizations is to consider that *more than one vertex* may need to be retrieved by the code, and usually one puts a limit $\ell$ on the number of these vertices. Therefore, an *$(r, \leqslant \ell)$-locating-dominating code* $C$ in a graph $G = (V, E)$ is an $r$-dominating code such that for all $X \subseteq V \setminus C$, $Y \subseteq V \setminus C$, such that $X \neq Y$, $|X| \leqslant \ell$, $|Y| \leqslant \ell$, we have

$$\bigcup_{u \in X} N_r(u) \cap C \neq \bigcup_{v \in Y} N_r(v) \cap C.$$

  One can similarly define *$(r, \leqslant \ell)$-identifying codes*. Both generalizations, and other similar ones, have been studied mostly in the $n$-cube and in the grids; we refer to, e.g., [70, 73, 86, 97, 112, 116, 134, 137, 150].
- Next, considering that using $r$-LD or $r$-Id codes amounts to using $r$-neighborhoods centered at the codewords, one can think of *changing the pattern* surrounding a codeword.

One choice of pattern is, given $G = (V, E)$ and $r \geqslant 1$, to consider that every codeword $c \in C$ can check a connected *subset* of $N_r[c]$, instead of the whole set $N_r[c]$ for an $r$-identifying code, and that several codewords can be on the same vertex. This leads to the notion of *watching systems* [7], which are more complex but can be very efficient: consider, for instance, the star $K_{1,n-1}$, which requires as many as $n - 1$ codewords for 1-identification and even for 1-locating-domination, and as few as $\lceil \log_2(n + 1) \rceil$ if one puts suitable watchers at the center of the star. See also [8, 9]. When a watching system has at most one codeword on each vertex, we talk of *choice identification* [27].

Other patterns for different graphs can be thought of: cycles, paths (including paths of even order in the infinite path, as opposed to odd paths which are closed neighborhoods around their centers), squares of even side in $\mathbb{Z} \times \mathbb{Z}$, as opposed to odd squares, which are closed neighborhoods in the king grid, etc. See, e.g., [85, 120, 162].

- A slight modification in the definition of a 1-identifying code leads to *open neighborhood locating-dominating codes* (OLD codes): an OLD code $C$ is such that for every two vertices $u$, $v$ of $V(G)$, the sets $N(u) \cap C$ and $N(v) \cap C$ are nonempty and different. This was introduced in [115] (for $r \geqslant 1$ and for identification of sets of vertices) in the binary $n$-cube, then extended to all graphs in [164] (for $r = 1$). This describes a situation where an intruder at a vertex $v$ can prevent the detection device at $v$ from signaling the intrusion, that is, $v$ can only check $N(v)$. In most of our models, we assumed that a codeword correctly sent a 1 if it detected something in its neighborhood, a 0 otherwise. The definition of OLD codes is one of several definitions for the so-called *fault-tolerant codes*, where different scenarios are considered for the alarms given by the codewords.

If $OLD(G)$ denotes the minimum size of an OLD code in $G$, the only graphs with $OLD(G) = |V(G)|$ are the three graphs given by Figure 20 and the only graphs with $OLD(G) = 2$ are the complete graphs $K_2$ and $K_3$ [51]. Another representative result is that the minimum density of an OLD code in the triangular grid is $4/13 \approx 0.3077$ [136], to be compared with 0.25 for a 1-identifying code, see Theorem 32. For more on this topic, see also, e.g., [163, 165, 176].
One can also mention here the *liar's problem*, when the vertices of $C$ can tell precisely where the malfunctioning vertex $v$ is if it belongs to their closed neighborhoods, so a dominating code would be enough, except that any one codeword in the closed neighborhood of $v$ can lie, that is, either misidentify any vertex in its closed neighborhood as $v$, or fail to report any vertex. For instance, three codewords are necessary and sufficient in the complete graph $K_n$ with $n \geqslant 3$. See [173] or [161].

- Another related definition is the following: a code $C$ is *strongly r-identifying* [115] if for all $v_1 \in V$, $v_2 \in V$, $v_1 \neq v_2$, the sets $\{N_r[v_1] \cap C, N_r(v_1) \cap C\}$ and $\{N_r[v_2] \cap C, N_r(v_2) \cap C\}$ are disjoint; this can be extended

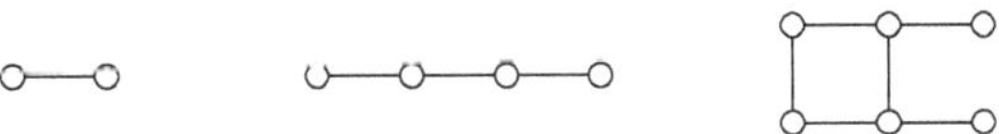

**Fig. 20** The three graphs $G = (V, E)$ with $OLD(G) = |V|$.

to codes identifying more than one vertex. See [109], where the best density for a strongly 1-Id code in the triangular grid is proved to be $6/19 \approx 0.3158$, or [138, 141].

- Using identifying codes in search of one defective vertex can be seen as asking to the codewords, all at one time: "Is there a defective vertex in your closed neighborhood?" *Adaptive identification* consists in asking the queries one by one, taking into account the previous answers. So, after the first yes answer is obtained in one neighborhood, we dichotomize this neighborhood with other neighborhoods (= new queries) so as to minimize the maximum number of queries [12, 127, 135].
- The notions of 1-LD and 1-Id codes can be modified by adding the condition that the code, instead of being dominating, must be total dominating: a total dominating code $C$ of a graph $G = (V, E)$ is a *locating-total dominating code* if for every pair of distinct vertices $u$ and $v$ in $V \setminus C$, one has $N(u) \cap C \neq N(v) \cap C$, and $C$ is a *differentiating-total dominating code*, or an *identifying-total dominating code*, if for every pair of distinct vertices $u$ and $v$ in $V$, $N[u] \cap C \neq N[v] \cap C$. See, e.g., [22, 49, 53, 103, 105, 129, 146].
- An older way of locating vertices is the following: given a graph $G = (V, E)$ and a code $C = \{c_1, c_2, \ldots, c_k\}$, the $C$-*location* of a vertex $v \in V$ is the distance vector $(d(v, c_1), d(v, c_2), \ldots, d(v, c_k))$. If no two vertices have the same $C$-location, then $C$ is said to be *resolving*, as introduced in [101], or *locating*, as introduced independently in [167], around 1975. An optimal resolving code is called a *metric basis* and its size is the *metric dimension* of the graph. For example, the only graphs with metric dimension 1 are the paths, for which each endvertex constitutes a resolving code. See also [24, 87], or [100]; see [104] for *metric-locating-domination* (where the code must also be 1-dominating), and [48] for the related concepts of *distance-location* and *external distance-location*.

Moreover, if we have a proper vertex coloring $c$ with $k$ colors, and if $\Pi = (C_1, C_2, \ldots, C_k)$ is the resulting partition of $V$, we can define, for every $v \in V$, the *color code* of $v$, that is, the distance vector $(d(v, C_1), d(v, C_2), \ldots, d(v, C_k))$. If distinct vertices have distinct color codes, then $c$ is called a *locating-coloring*. The *locating-chromatic number* of $G$ is the minimum number of colors in a locating-coloring. See, e.g., [11, 46, 47]. Other more or less distant concepts using colorings exist, such as *locally identifying colorings* [69], and many more, see, e.g., [148, 180, 181].

- For the model of *information retrieval* in associative memories, see, e.g., [132, 140, 179].
- The *d-identifying codes* [177] generalize Id codes by endowing every vertex with a positive integer cost, which will give a global cost to the code that may be different from its size, and with a positive integer weight which, together with $d$, allows and measures a degree of uncertainty in the identification.
- Another parameter, the *propagation time*, can be considered; in this model, a fault at one vertex $v$ in a weighted digraph spreads along the arcs and reaches

any out-neighbor $w$ in a time equal to the weight of the arc $vw$, then spreads from $w$, and so on. See, e.g., [14, 158].
- Other definitions: *paired*-LD or Id codes (the code must be paired-dominating), *co-isolated* LD or Id codes (the code $C$ is such that there exists at least one isolated vertex in $\langle V - C \rangle$), *independent* LD or Id codes, *connected* codes, *weak* $r$-codes and *light* $r$-codes, *self*-LD or Id codes, *solid*-LD codes.

It is also possible to generalize these concepts to *hypergraphs*, and of course, some of these definitions can be combined, e.g., connected vertex-robust identifying codes [78].

## 10 Concluding Remarks, Open Problems, Conjectures

We have tried to give an overview of the major results on locating-domination and on identification. We apologize in advance if your favorite variation or result is omitted. We simply refer people interested to know more about these concepts to [142] for more literature on the topic; they can search for keywords there, or on the web for even more references.

We conclude with two seemingly paradoxical remarks partly collected from [160]:

(a) For some graphs, *increasing* $r$, i.e., increasing the power of the codewords, can require *more* codewords; this was first noticed in [26], where a tree $T$ is given, such that $LD_2(T) = LD_3(T) = 6$ and $LD_4(T) = 5$, while $LD_7(T) > 6$. Paths provide similar cases for identification; for instance, if $k \geqslant 1$, then by Theorem 44 we have $Id_1(P_{10k+5}) = 5k+3$, whereas by Theorem 49(a) we have $Id_2(P_{10k+5}) = 5k+4$. In the case of the binary $n$-cube, we have $Id_1(\mathbb{F}^6) = 19$, $Id_2(\mathbb{F}^6) = 8$, $Id_3(\mathbb{F}^6) = 7$, $Id_4(\mathbb{F}^6) = 18$, $Id_5(\mathbb{F}^6) = 63$.

This paradox can be understood when $r$ becomes relatively large with respect to the graph: if the $r$-neighborhoods grow too much, they cannot separate anything anymore; the extremal case is when $r$ reaches the diameter of the graph, where $n - 1$ codewords are needed for an $r$-LD code, and no $r$-Id code exists. But this is not the case for the example of the path $P_{10k+5}$ above.

(b) If we take a path or a cycle and make it *longer*, it may need *fewer* codewords; for instance, for $k \geqslant 1$, one has $Id_2(C_{10k+6}) = Id_2(C_{10k+5}) - 2$.

The discussion following Theorem 62 gives another example, which however can be explained by the fact that increasing the dimension $n$ of the cube makes $r$ not too large with respect to $n$ anymore.

Finally, we recapitulate the open problems and conjectures given throughout the text.

**Open Problem 1** [2020] (see End of Section 3) For $r \geqslant 2$, most results on the costs of locating-domination vs domination are exact but valid for $n$ large.

(a) Establish results for all $n$. (b) Reduce the gap between lower and upper bounds for $F_{LD,\gamma}(r,n)$.

**Open Problem 2** [2011] (see Theorem 87) For $r \geqslant 5$, find the exact value of $f_r(Id_r)$, knowing that it lies between $\lceil \log_2(2r+4) \rceil$ and $r+1$.

**Open Problem 3** [2007] (see End of Section 6.1.1) (a) Apart from $P_5$, do 2-terminal graphs exist? (b) There exist $r$-terminal graphs for $r \in \{3,4,5\}$, but are they in finite or infinite number?

**Open Problem 4** [2013] Improve the bounds in Theorems 98, 101 and 102.

**Open Problem 5** [2015] (see Theorems 104 and 105) Reduce the gap between the lower and upper bounds on the number of different optimal identifying codes.

**Conjecture 1** [1995] (see Theorem 54) For any tree $T$ of order $n$, we have $LD_r(T) > \frac{n}{r+2}$.

**Conjecture 2** [1999] (see the discussion following Theorem 60) (a) We have $Id_{r_1+r_2}(\mathbb{F}^{n_1+n_2}) \leqslant Id_{r_1}(\mathbb{F}^{n_1}) \times Id_{r_2}(\mathbb{F}^{n_2})$. (b) We have $Id(\mathbb{F}^{n+1}) \leqslant 2Id(\mathbb{F}^n)$.

**Conjecture 3** [2010] (see the discussion after Theorem 62) For a fixed $r > 1$, there exists $n(r)$ such that for all $n \geqslant n(r)$, we have $Id_r(\mathbb{F}^n) \leqslant Id_r(\mathbb{F}^{n+1})$.

**Conjecture 4** [2012] (see the discussion following Theorem 71) There exists a constant $c$ such that, for every nontrivial connected 1-twin-free graph $G$, with maximum degree $\Delta$, we have $Id_1(G) \leqslant n - \frac{n}{\Delta} + c$.

**Conjecture 5** [2009] (see Theorem 79) The value $F_{r,n}(\delta)$ is close, possibly equal, to $\lfloor \frac{n-2}{r} \rfloor$.

## References

1. G. R. Argiroffo, S. M. Bianchi, Y. P. Lucarini and A. K. Wagler, A linear-time algorithm for the identifying code problem on block graphs. *Electron. Notes Discrete Math.* **62** (2017), 249–254.
2. G. R. Argiroffo, S. M. Bianchi, Y. P. Lucarini and A. K. Wagler, Polyhedra associated with identifying codes in graphs. *Discrete Appl. Math.* **245** (2018), 16–27.
3. D. Auger, Induced paths in twin-free graphs. *Electron. J. Combin.* **15(1)** (2008), N17.
4. D. Auger, Minimal identifying codes in trees and planar graphs with large girth. *European J. Combin.* **31** (2010), 1372–1384.
5. D. Auger, I. Charon, I. Honkala, O. Hudry and A. Lobstein, Edge number, minimum degree, maximum independent set, radius and diameter in twin-free graphs. *Adv. Math. Commun.* **3(1)** (2009), 97–114. Erratum **3(4)** (2009), 429–430.
6. D. Auger, I. Charon, O. Hudry and A. Lobstein, Complexity results for identifying codes in planar graphs. *Int. Trans. Oper. Res.* **17** (2010), 691–710.
7. D. Auger, I. Charon, O. Hudry and A. Lobstein, Watching systems in graphs: an extension of identifying codes. *Discrete Appl. Math.* **161** (2013), 1674–1685.
8. D. Auger, I. Charon, O. Hudry and A. Lobstein, Maximum size of a minimum watching system and the graphs achieving the bound. *Discrete Appl. Math.* **164** (2014), 20–33.

9. D. Auger and I. Honkala, Watching systems in the king grid. *Graphs Combin.* **29** (2013), 333–347.
10. C. Balbuena, F. Foucaud and A. Hansberg, Locating-dominating sets and identifying codes in graphs of girth at least 5. *Electron. J. Combin.* **22(2)**, (2015) P2.15.
11. A. Behtoei and M. Anbarloei, A bound for the locating chromatic number of trees. *Transactions on Combinatorics* **4(1)** (2015), 31–41.
12. Y. Ben-Haim, S. Gravier, A. Lobstein and J. Moncel, Adaptive identification in torii in the king lattice. *Electron. J. Combin.* **18(1)** (2011), P116.
13. Y. Ben-Haim and S. Litsyn, Exact minimum density of codes identifying vertices in the square grid. *SIAM J. Discrete Math.* **19** (2005), 69–82.
14. T. Y. Berger-Wolf, W. E. Hart and J. Saia, Discrete sensor placement problems in distribution networks. *Math. Comput. Modelling* **42** (2005), 1385–1396.
15. N. Bertrand, *Codes identifiants et codes localisateurs-dominateurs sur certains graphes.* Mémoire de stage de maîtrise. ENST, Paris, France, (2001), 28 pages.
16. N. Bertrand, I. Charon, O. Hudry and A. Lobstein, Identifying and locating-dominating codes on chains and cycles. *European J. Combin.* **25** (2004), 969–987.
17. N. Bertrand, I. Charon, O. Hudry and A. Lobstein, 1-identifying codes on trees. *Australas. J. Combin.* **31** (2005), 21–35.
18. U. Blass, I. Honkala and S. Litsyn, On the size of identifying codes. *Lecture Notes in Comput. Sci.* **1719** (1999), 142–147.
19. M. Blidia, M. Chellali, R. Lounes and F. Maffray, Characterizations of trees with unique minimum locating-dominating sets. *JCMCC* **76** (2011), 225–232.
20. M. Blidia, M. Chellali, F. Maffray, J. Moncel and A. Semri, Locating-domination and identifying codes in trees. *Australas. J. Combin.* **39** (2007), 219–232.
21. M. Blidia and W. Dali, A characterization of locating-domination edge critical graphs. *Australas. J. Combin.* **44** (2009), 297–300.
22. M. Bouznif, *Algorithmes génériques en temps constant pour la résolution de problèmes combinatoires dans la classe des rotagraphes et fasciagraphes. Application aux codes identifiants, dominant-localisateurs et total-dominant-localisateurs.* Thèse de Doctorat. Université de Grenoble, France, (2012), 131 pages.
23. M. Bouznif, F. Havet, M. Preissmann, Minimum-density identifying codes in square grids. *Lecture Notes in Comput. Sci.* **9778** (2016), 77–88.
24. J. Cáceres, D. Garijo, M. L. Puertas and C. Seara, On the determining number and the metric dimension of graphs. *Electron. J. Combin.* **17(1)** (2010), R63.
25. J. Cáceres, C. Hernando, M. Mora, I. M. Pelayo and M. L. Puertas, Locating-dominating codes: bounds and extremal cardinalities. *Appl. Math. Comput.* **220** (2013), 38–45.
26. D. I. Carson, On generalized location-domination. In *Graph Theory, Combinatorics, and Applications: Proceedings of the 7th Quadrennial International Conference on the Theory and Applications of Graphs*, Wiley, **1** (1995), 161–179.
27. T. P. Chang and L. D. Tong, Choice identification of a graph. *Discrete Appl. Math.* **167** (2014), 61–71.
28. E. Charbit, I. Charon, G. Cohen, O. Hudry and A. Lobstein, Discriminating codes in bipartite graphs: bounds, extremal cardinalities, complexity. *Adv. Math. Commun.* **4(2)** (2008), 403–420.
29. I. Charon, G. Cohen, O. Hudry and A. Lobstein, Links between discriminating and identifying codes in the binary Hamming space. *Lecture Notes in Comput. Sci.* **4851** (2007), 267–270.
30. I. Charon, G. Cohen, O. Hudry and A. Lobstein, Discriminating codes in (bipartite) planar graphs. *European J. Combin.* **29** (2008), 1353–1364.
31. I. Charon, G. Cohen, O. Hudry and A. Lobstein, New identifying codes in the binary Hamming space. *European J. Combin.* **31** (2010), 491–501.
See also: perso.telecom-paristech.fr/~hudry/newIdentifyingNcube.html
32. I. Charon, S. Gravier, O. Hudry, A. Lobstein, M. Mollard and J. Moncel, A linear algorithm for minimum 1-identifying codes in oriented trees. *Discrete Appl. Math.* **154** (2006), 1246–1253.

33. I. Charon, I. Honkala, O. Hudry and A. Lobstein, General bounds for identifying codes in some infinite regular graphs. *Electron. J. Combin.* **8(1)** (2001), R39.
34. I. Charon, I. Honkala, O. Hudry and A. Lobstein, The minimum density of an identifying code in the king lattice. *Discrete Math.* **276** (2004), 95–109.
35. I. Charon, I. Honkala, O. Hudry and A. Lobstein, Structural properties of twin-free graphs. *Electron. J. Combin.* **14(1)** (2007), R16.
36. I. Charon, I. Honkala, O. Hudry and A. Lobstein, Minimum sizes of identifying codes in graphs differing by one vertex. *Cryptogr. Commun.* **5** (2013), 119–136.
37. I. Charon, I. Honkala, O. Hudry and A. Lobstein, Minimum sizes of identifying codes in graphs differing by one edge. *Cryptogr. Commun.* **6** (2014), 157–170.
38. I. Charon, O. Hudry and A. Lobstein, Identifying codes with small radius in some infinite regular graphs. *Electron. J. Combin.* **9(1)** (2002), R11.
39. I. Charon, O. Hudry and A. Lobstein, Identifying and locating-dominating codes: NP-completeness results for directed graphs. *IEEE Trans. Inform. Theory* **IT-48** (2002), 2192–2200.
40. I. Charon, O. Hudry and A. Lobstein, Minimizing the size of an identifying or locating-dominating code in a graph is NP-hard. *Theoret. Comput. Sci.* **290** (2003), 2109–2120.
41. I. Charon, O. Hudry and A. Lobstein, Possible cardinalities for identifying codes in graphs. *Australas. J. Combin.* **32** (2005), 177–195.
42. I. Charon, O. Hudry and A. Lobstein, Possible cardinalities for locating-dominating codes in graphs. *Australas. J. Combin.* **34** (2006), 23–32.
43. I. Charon, O. Hudry and A. Lobstein, Extremal cardinalities for identifying and locating-dominating codes. *Discrete Math.*, **307** (2007), 356–366.
44. I. Charon, O. Hudry and A. Lobstein, Extremal values for identification, domination and maximum cliques in twin-free graphs. *Ars Combin.* **101** (2011), 161–185.
45. I. Charon, O. Hudry and A. Lobstein, Extremal values for the maximum degree in a twin-free graph. *Ars Combin.* **107** (2012), 257–274.
46. G. Chartrand, D. Erwin, M. A. Henning, P. J. Slater and P. Zhang, The locating-chromatic number of a graph. *Bull. Inst. Combin. Appl.* **36** (2002), 89–101.
47. G. Chartrand, D. Erwin, M. A. Henning, P. J. Slater and P. Zhang, Graphs of order $n$ with locating-chromatic number $n-1$. *Discrete Math.* **269** (2003), 65–79.
48. G. Chartrand, D. Erwin, P. J. Slater and P. Zhang, Distance-location numbers of graphs. *Util. Math.* **63** (2003), 65–79.
49. M. Chellali, On locating and differentiating-total domination in trees. *Discuss. Math. Graph Theory* **28** (2008), 383–392.
50. M. Chellali, M. Mimouni and P. J. Slater, On locating-domination in graphs. *Discuss. Math. Graph Theory* **30** (2010), 223–235.
51. M. Chellali, N. J. Rad, S. J. Seo and P. J. Slater, On open neighborhood locating-dominating in graphs. *Electron. J. Graph Theory and Applications* **2(2)** (2014), 87–98.
52. C. Chen, C. Lu and Z. Miao, Identifying codes and locating-dominating sets on paths and cycles. *Discrete Appl. Math.* **159** (2011), 1540–1547.
53. X. Chen and M. Y. Sohn, Bounds on the locating-total domination number of a tree. *Discrete Appl. Math.* **159** (2011), 769–773.
54. G. Cohen, S. Gravier, I. Honkala, A. Lobstein, M. Mollard, Ch. Payan and G. Zémor, Improved identifying codes for the grid. *Electron. J. Combin.* **6(1)** (1999), Comments to R19.
55. G. Cohen, I. Honkala, S. Litsyn and A. Lobstein, *Covering Codes*, Elsevier, Amsterdam, 1997.
56. G. Cohen, I. Honkala, A. Lobstein and G. Zémor, Bounds for codes identifying vertices in the hexagonal grid. *SIAM J. Discrete Math.* **13** (2000), 492–504.
57. G. Cohen, I. Honkala, A. Lobstein and G. Zémor, On identifying codes. In *Proceedings of DIMACS Workshop on Codes and Association Schemes '99, Piscataway, USA*, Barg A. and Litsyn S. (eds), American Mathematical Society, 2001, **56** 97–109.
58. G. Cohen, I. Honkala, A. Lobstein and G. Zémor, On codes identifying vertices in the two-dimensional square lattice with diagonals. *IEEE Trans. Comput.* **50** (2001), 174–176.

59. N. Cohen and F. Havet, On the minimum size of an identifying code over all orientations of a graph. *Electron. J. Combin.* **25(1)** (2018), P1.49.
60. C. J. Colbourn, P. J. Slater and L. K. Stewart, Locating dominating sets in series parallel networks. *Congr. Numer.* **56** (1987), 135–162.
61. A. Cukierman and G. Yu, New bounds on the minimum density of an identifying code for the infinite hexagonal grid. *Discrete Appl. Math.* **161** (2013), 2910–2924.
62. W. Dali and M. Blidia, Criticality indices of locating-domination of paths and cycles. *Util. Math.* **94** (2014), 199–219.
63. W. Dali and M. Blidia, On locating and locating-total domination edge addition critical graphs. *Util. Math.* **94** (2014), 303–313.
64. M. Daniel, *Codes identifiants.* Mémoire pour le DEA ROCO. Université Joseph Fourier, Grenoble, France, (2003), 46 pages.
65. M. Daniel, S. Gravier and J. Moncel, Identifying codes in some subgraphs of the square lattice. *Theoret. Comput. Sci.* **319** (2004), 411–421.
66. R. Dantas, F. Havet and R. M. Sampaio, Identifying codes for infinite triangular grids with a finite number of rows. *Discrete Math.* **340** (2017), 1584–1597.
67. R. Dantas, F. Havet and R. M. Sampaio, Minimum density of identifying codes of king grids. *Discrete Math.* **341** (2018), 2708–2719.
68. R. C. Entringer and L. D. Gassman, Line-critical point determining and point distinguishing graphs. *Discrete Math.*, **10** (1974), 43–55.
69. L. Esperet, S. Gravier, M. Montassier, P. Ochem and A. Parreau, Locally identifying coloring of graphs. *Electron. J. Combin.* **19(2)** (2012), P40.
70. G. Exoo, V. Junnila and T. Laihonen, On location-domination of set of vertices in cycles and paths. *Congr. Numer.* **202** (2010), 97–112.
71. G. Exoo, V. Junnila and T. Laihonen, Locating-dominating codes in paths. *Discrete Math.* **311** (2011), 1863–1873.
72. G. Exoo, V. Junnila and T. Laihonen, Locating-dominating codes in cycles. *Australas. J. Combin.* **49** (2011), 177–194.
73. G. Exoo, V. Junnila, T. Laihonen and S. Ranto, Locating vertices using codes. *Congr. Numer.* **191** (2008), 143–159.
74. G. Exoo, V. Junnila, T. Laihonen and S. Ranto, Upper bounds for binary identifying codes. *Adv. in Appl. Math.* **42** (2009), 277–289.
75. G. Exoo, V. Junnila, T. Laihonen and S. Ranto, Improved bounds on identifying codes in binary Hamming spaces. *European J. Combin.* **31** (2010), 813–827.
76. G. Exoo, T. Laihonen and S. Ranto, Improved upper bounds on binary identifying codes. *IEEE Trans. Inform. Theory* **IT-53** (2007), 4255–4260.
77. G. Exoo, T. Laihonen and S. Ranto, New bounds on binary identifying codes. *Discrete Appl. Math.* **156** (2008), 2250–2263.
78. N. Fazlollahi, D. Starobinski and A. Trachtenberg, Connecting identifying codes and fundamental bounds. *Proceedings of Information Theory and Applications Workshop ITA 2011, La Jolla, USA* (2011), 403–409.
79. F. Foucaud, Aspects combinatoires et algorithmiques des codes identifiants dans les graphes. Thèse de Doctorat, Université Bordeaux 1, France, (2012), 194 pages (in English).
80. F. Foucaud, Decision and approximation complexity for identifying codes and locating-dominating sets in restricted graph classes. *J. Discrete Alg.* **31** (2015), 48–68.
81. F. Foucaud, S. Gravier, R. Naserasr, A. Parreau, and P. Valicov, Identifying codes in line graphs. *J. Graph Theory* **73** (2013), pp. 425–448.
82. F. Foucaud, E. Guerrini, M. Kovše, R. Naserasr, A. Parreau and P. Valicov, Classifying graphs with minimum identifying codes of size $n - 1$, *Abstracts of the 8th French Combinatorial Conference*, Orsay, France, **151** (2010).
83. F. Foucaud, E. Guerrini, M. Kovše, R. Naserasr, A. Parreau and P. Valicov, Extremal graphs for the identifying code problem. *European J. Combin.* **32** (2011), 628–638.
84. F. Foucaud, R. Klasing, A. Kosowski and A. Raspaud, On the size of identifying codes in triangle-free graphs. *Discrete Appl. Math.* **160** (2012), 1532–1546.

85. F. Foucaud and M. Kovše, Identifying path covers in graphs, *J. Discrete Alg.* **23** (2013), 21–34.
86. F. Foucaud, T. Laihonen and A. Parreau, An improved lower bound for $(1, \leqslant 2)$-identifying codes in the king grid. *Adv. Math. Commun.* **8** (2014), 35–52.
87. F. Foucaud, G. B. Mertzios, R. Naserasr, A. Parreau and P. Valicov, Identification, location-domination and metric dimension on interval and permutation graphs: II. Algorithms and complexity, *Algorithmica* **78** (2017), 914–944.
88. F. Foucaud, R. Naserasr and A. Parreau, Characterizing extremal digraphs for identifying codes and extremal cases of Bondy's theorem on induced subsets. *Graphs Combin.* **29** (2013), 463–473.
89. F. Foucaud and G. Perarnau, Bounds for identifying codes in terms of degree parameters. *Electron. J. Combin.* **19(1)** (2012), P32.
90. F. Foucaud, G. Perarnau and O. Serra, Random subgraphs make identification affordable. *J. Comb.* **8** (2017), 57–77.
91. M. Frances and A. Litman, On covering problems of codes. *Theory Comput. Syst.* **30(2)** (1997), 113–119.
92. M. Frick, G. H. Fricke, C. M. Mynhardt and R. D. Skaggs, Critical graphs with respect to vertex identification. *Util. Math.* **76** (2008), 213–227.
93. M. R. Garey and D. S. Johnson, *Computers and Intractability, a Guide to the Theory of NP-Completeness*, Freeman, New York, 1979.
94. J. Gimbel, B. D. Van Gorden, M. Nicolescu, C. Umstead and N. Vaiana, Location with dominating sets. *Congr. Numer.* **151** (2001), 129–144.
95. S. Gravier, S. Janson, T. Laihonen and S. Ranto, Graphs where every $k$-subset of vertices is an identifying set. *Discrete Math. Theor. Comput. Sci.* **16** (2014), 73–88.
96. S. Gravier, R. Klasing and J. Moncel, Hardness results and approximation algorithms for identifying codes and locating-dominating codes in graphs. *Algorithmic Oper. Res.* **3** (2008), 43–50.
97. S. Gravier and J. Moncel, Construction of codes identifying sets of vertices. *Electron. J. Combin.* **12(1)** (2005), R13.
98. S. Gravier and J. Moncel, On graphs having a $V \setminus \{x\}$ set as an identifying code. *Discrete Math.* **307** (2007), 432–434.
99. S. Gravier, J. Moncel and A. Semri, Identifying codes of cycles. *European J. Combin.* **27** (2006), 767–776.
100. A. Hakanen and T. Laihonen, On $\{\ell\}$-metric dimensions in graphs. *Fund. Inform.* **162** (2018), 143–160.
101. F. Harary and R. A. Melter, On the metric dimension of a graph. *Ars Combin.* **2** (1976), 191–195. Addendum **4** (1977), 318.
102. T. W. Haynes, S. T. Hedetniemi and P. J. Slater, *Fundamentals of Domination in Graphs*, Marcel Dekker, New York, 1998.
103. T. W. Haynes, M. A. Henning and J. Howard, Locating and total dominating sets in trees. *Discrete Appl. Math.* **154** (2006), 1293–1300
104. M. A. Henning and O. R. Oellermann, Metric-locating-dominating sets in graphs. *Ars Combin.* **73** (2004), 129–141.
105. M. A. Henning and N. J. Rad, Locating-total domination in graphs. *Discrete Appl. Math.* **160** (2012), 1986–1993.
106. C. Hernando, M. Mora and I. M. Pelayo, Nordhaus-Gaddum bounds for locating domination. *European J. Combin.* **36** (2014), 1–6.
107. I. Honkala, An optimal locating-dominating set in the infinite triangular grid. *Discrete Math.* **306** (2006), 2670–2681.
108. I. Honkala, On $r$-locating-dominating sets in paths. *European J. Combin.* **30** (2009), 1022–1025.
109. I. Honkala, An optimal strongly identifying code in the infinite triangular grid. *Electron. J. Combin.* **17(1)** (2010), R91.

110. I. Honkala, O. Hudry and A. Lobstein, On the number of optimal identifying codes in a twin-free graph. *Discrete Appl. Math.* **180** (2015), 111–119.
111. I. Honkala, M. G. Karpovsky and L. B. Levitin, On robust and dynamic identifying codes. *IEEE Trans. Inform. Theory* **IT-52** (2006), 599–612.
112. I. Honkala and T. Laihonen, On identifying codes in the triangular and square grids. *SIAM J. Comput.* **33** (2004), 304–312.
113. I. Honkala and T. Laihonen, On locating-dominating sets in infinite grids. *European J. Combin.* **27** (2006), 218–227.
114. I. Honkala and T. Laihonen, On identifying codes that are robust against edge changes. *Inform. and Comput.* **205** (2007), 1078–1095.
115. I. Honkala, T. Laihonen and S. Ranto, On strongly identifying codes. *Discrete Math.* **254** (2002), 191–205.
116. I. Honkala, T. Laihonen and S. Ranto, On locating-dominating codes in binary Hamming spaces. *Discrete Math. Theor. Comput. Sci.* **6** (2004), 265–282.
117. I. Honkala and A. Lobstein, On the density of identifying codes in the square lattice. *J. Combin. Theory*, Ser. B **85** (2002), 297–306.
118. I. Honkala and A. Lobstein, On identifying codes in binary Hamming spaces. *J. Combin. Theory*, Ser. A **99** (2002), 232–243.
119. I. Honkala and A. Lobstein, On the complexity of the identification problem in Hamming spaces. *Acta Inform.* **38** (2002), 839–845.
120. I. Honkala and A. Lobstein, On identification in $Z^2$ using translates of given patterns. *J. UCS* **9(10)** (2003), 1204–1219.
121. O. Hudry and A. Lobstein, More results on the complexity of identifying problems in graphs. *Theoret. Comput. Sci.* **626** (2016), 1–12.
122. O. Hudry and A. Lobstein, Some results about a conjecture on identifying codes in complete suns. *Int. Trans. Oper. Res.* **26** (2019), 732–746.
123. O. Hudry and A. Lobstein, Unique (optimal) solutions: Complexity results for identifying and locating-dominating codes. *Theoret. Comput. Sci.* **767** (2019), 83–102.
124. O. Hudry and A. Lobstein, The compared costs of domination, location-domination and identification. *Discuss. Math. Graph Theory* **40(1)** (2020), 127–147.
125. S. Janson and T. Laihonen, On the size of identifying codes in binary hypercubes. *J. Combin. Theory*, Ser. A **116** (2009), 1087–1096.
126. M. Jiang, Periodicity of identifying codes in strips. *Inform. Process. Lett.* **135** (2018), 77–84.
127. V. Junnila, Adaptive identification of sets of vertices in graphs. *Discrete Math. Theor. Comput. Sci.* **14** (2012), 69–86.
128. V. Junnila, New lower bound for 2-identifying code in the square grid. *Discrete Appl. Math.* **161** (2013), 2042–2051.
129. V. Junnila, Optimal locating-total dominating sets in strips of height 3. *Discuss. Math. Graph Theory* **35** (2015), 447–462.
130. V. Junnila and T. Laihonen, Optimal identifying codes in cycles and paths. *Graphs Combin.* **28** (2012), 469–481.
131. V. Junnila and T. Laihonen, Optimal lower bound for 2-identifying codes in the hexagonal grid. *Electron. J. Combin.* **19(2)** (2012), P38.
132. V. Junnila and T. Laihonen, Codes for information retrieval with small uncertainty. *IEEE Trans. Inform. Theory* **IT-60** (2014), 976–985.
133. V. Junnila, T. Laihonen and T. Lehtilä, On regular and new types of codes for location-domination. *Discrete Appl. Math.* **247** (2018), 225–241.
134. M. G. Karpovsky, K. Chakrabarty and L. B. Levitin, On a new class of codes for identifying vertices in graphs, *IEEE Trans. Inform. Theory* **IT-44** (1998), 599–611.
135. Y. Kim, M. Kumbhat, Z. L. Nagy, B. Patkós, A. Pokrovskiy and M. Vizer, Identifying codes and searching with balls in graphs. *Discrete Appl. Math.* **193** (2015), 39–47.
136. R. Kincaid, A. Oldham and G. Yu, Optimal open-locating-dominating sets in infinite triangular grids. *Discrete Appl. Math.* **193** (2015), 139–144.

137. T. Laihonen, Sequences of optimal identifying codes. *IEEE Trans. Inform. Theory* **IT-48** (2002), 774–776.
138. T. Laihonen, Optimal codes for strong identification, *European J. Combin.* **23** (2002), 307–313.
139. T. Laihonen, Optimal $t$-edge-robust $r$-identifying codes in the king lattice. *Graphs Combin.* **22** (2006), 487–496.
140. T. Laihonen, Information retrieval and the average number of input clues. *Adv. Math. Commun.* **11** (2017), 203–223.
141. T. Laihonen and S. Ranto, Families of optimal codes for strong identification. *Discrete Appl. Math.* **121** (2002), 203–213.
142. A. Lobstein, Watching systems, identifying, locating-dominating and discriminating codes in graphs, a bibliography. https://www.lri.fr/~lobstein/debutBIBidetlocdom.pdf
143. A. McLoughlin, The complexity of computing the covering radius of a code. *IEEE Trans. Inform. Theory* **IT-30** (1984), 800–804.
144. R. Martin and B. Stanton, Lower bounds for identifying codes in some infinite grids. *Electron. J. Combin.* **17(1)** (2010), R122.
145. J. Moncel, Monotonicity of the minimum cardinality of an identifying code in the hypercube. *Discrete Appl. Math.* **154** (2006), 898–899.
146. B. N. Omamalin, S. R. Canoy, Jr., and H. M. Rara, Differentiating total domination in graphs: revisited. *Internat. J. Math. Analysis* **8** (2014), 2789–2798.
147. O. Ore, *Theory of Graphs*, American Mathematical Society, Providence, 1962.
148. A. Parreau, Problèmes d'identification dans les graphes. Thèse de Doctorat, Université de Grenoble, France, (2012), 214 pages.
149. M. Pastori, Les codes identifiants ou comment sauver le Palais des flammes ? *Découverte* **369** (2010), 56–59.
150. M. Pelto, New bounds for $(r, \leqslant 2)$-identifying codes in the infinite king grid. *Cryptogr. Commun.* **2** (2010), 41–47.
151. M. Pelto, Maximum difference about the size of optimal identifying codes in graphs differing by one vertex. *Discrete Math. Theor. Comput. Sci.* **17(1)** (2015), 339–356.
152. M. Pelto, On locating-dominating codes in the infinite king grid. *Ars Combin.* **124** (2016), 353–363.
153. N. J. Rad and H. Rahbani, Bounds on the locating-domination number and differentiating-total domination number in trees. *Discuss. Math. Graph Theory* **38** (2018), 455–462.
154. H. Rahbani, N. J. Rad and S. M. MirRezaei, Bounds on the identifying codes in trees. *Graphs Combin.* **35** (2019), 599–609.
155. D. F. Rall and P. J. Slater, On location-domination numbers for certain classes of graphs. *Congr. Numer.* **45** (1984), 97–106.
156. S. Ranto, Optimal linear identifying codes. *IEEE Trans. Inform. Theory* **IT-49** (2003), 1544–1547.
157. S. Ranto, On binary linear $r$-identifying codes. *Des. Codes Cryptogr.* **60** (2011), 81–89.
158. N. S. V. Rao, Computational complexity issues in operative diagnosis of graph-based systems. *IEEE Trans. Comput.* **42** (1993), 447–457.
159. S. Ray, R. Ungrangsi, F. De Pellegrini, A. Trachtenberg and D. Starobinski, Robust location detection in emergency sensor networks. *Proceedings of INFOCOM 2003, San Francisco, USA*, (2003), 1044–1053.
160. D. L. Roberts and F. S. Roberts, Locating sensors in paths and cycles: the case of 2-identifying codes. *European J. Combin.* **29** (2008), 72–82.
161. M. L. Roden and P. J. Slater, Liar's domination in graphs. *Discrete Math.* **309** (2009), 5884–5890.
162. P. Rosendahl, On the identification of vertices using cycles. *Electron. J. Combin.* **10(1)** (2003), R7.
163. S. J. Seo, Open-locating-dominating sets in the infinite king grid. *JCMCC* **104** (2018), 31–47.
164. S. J. Seo and P. J. Slater, Open neighborhood locating-dominating sets. *Australas. J. Combin.* **46** (2010), 109–119.

165. S. J. Seo and P. J. Slater, OLD trees with maximum degree three. *Util. Math.* **94** (2014), 361–380.
166. J. L. Sewell and P. J. Slater, A sharp lower bound for locating-dominating sets in trees. *Australas. J. Combin.* **60** (2014), 136–149.
167. P. J. Slater, Leaves of trees. *Congr. Numer.* **14** (1975), 549–559.
168. P. J. Slater, *Domination and location in graphs*. Research Report No. 93. National University of Singapore, (1983).
169. P. J. Slater, Domination and location in acyclic graphs. *Networks* **17** (1987), 55–64.
170. P. J. Slater, Dominating and reference sets in a graph. *J. Math. Phys. Sci.* **22** (1988), 445–455.
171. P. J. Slater, Locating dominating sets and locating-dominating sets. In *Graph Theory, Combinatorics, and Applications: Proceedings of the 7th Quadrennial International Conference on the Theory and Applications of Graphs*, Wiley, **2** (1995), 1073–1079.
172. P. J. Slater, Fault-tolerant locating-dominating sets. *Discrete Math.* **249** (2002), 179–189.
173. P. J. Slater, Liar's domination. *Networks* **54** (2009), 70–74.
174. B. Stanton, Improved bounds for $r$-identifying codes of the hex grid. *SIAM J. Discrete Math.* **25** (2011), 159–169.
175. D. P. Sumner, Point determination in graphs. *Discrete Math.* **5** (1973), 179–187.
176. D. B. Sweigart, J. Presnell and R. Kincaid, An integer program for open locating dominating sets and its results on the hexagon-triangle infinite grid and other graphs. *Proceedings of 2014 Systems and Information Engineering Design Symposium (SIEDS), Charlottesville, USA* (2014), 29–32.
177. Y. Xiao, C. Hadjicostis and K. Thulasiraman, The $d$-identifying codes problem for vertex identification in graphs: probabilistic analysis and an approximation algorithm. *Proceedings of COCOON 2006, 12th Annual International Computing and Combinatorics Conference, Taipei, Taiwan* (2006), 284–298.
178. M. Xu, K. Thulasiraman and X. D. Hu, Identifying codes of cycles with odd orders. *European J. Combin.* **29** (2008), 1717–1720.
179. E. Yaakobi and J. Bruck, On the uncertainty of information retrieval in associative memories. *Proceedings of the 2012 IEEE International Symposium on Information Theory* (2014), 106–110.
180. J. Yao, X. Yu, G. Wang and C. Xu, Neighbor sum distinguishing total coloring of 2-degenerate graphs. *J. Combin. Optim.* **34** (2017), 64–70.
181. Z. Zhang, X. Chen, J. Li, B. Yao, X. Lu and J. Wang, On adjacent-vertex-distinguishing total coloring of graphs. *Sci. China*, Ser. A **48** (2005), 289–299.

# Signed and Minus Dominating Functions in Graphs

**Liying Kang and Erfang Shan**

## 1 Introduction

Dominating functions in domination theory have received much attention. A purely graph-theoretic motivation is that the dominating function problem can be regarded as a proper generalization of the classical domination problem. The study of dominating functions was formally initiated in [2, 13, 16]. The literature on the topic has been surveyed and detailed in [33, 45].

For a graph $G = (V, E)$ and for any real-valued function $f : V \to \mathbf{R}$, we define $f(S) = \sum_{u \in S} f(u)$. The *weight* of $f$ is defined as $f(V)$. For a vertex $v \in V$, we denote $f(N[v])$ by $f[v]$ for notational convenience, where $N[v] = \{u : uv \in E\} \cup \{v\}$.

For an arbitrary set $Y$ of the integers, a *$Y$-dominating function* of $G$ is defined as an integer-valued function $f : V \to Y$ such that $f[v] \geq 1$ for every $v \in V$. The notion of $Y$-dominating functions can be extended in a more general situation. Let $k$ be positive integer such that $1 \leq k \leq |V|$. A *$(Y, k)$-subdominating function* of $G$ is defined as an integer-valued function $f : V \to Y$ such that $f[v] \geq 1$ for at least $k$ vertices $v \in V$. We can obtain the definitions of *total $Y$-dominating function* and *total $(Y, k)$-subdominating function* by simply changing "closed" neighborhood $N[v]$ in the definitions of $Y$-dominating function and $(Y, k)$-subdominating function to "open" neighborhood $N(v)$, respectively. Obviously, when $k = |V|$, the $(Y, k)$-subdominating function and total $(Y, k)$-subdominating function are the $Y$-dominating function and total $Y$-dominating

L. Kang
Department of Mathematics, Shanghai University, Shanghai, 200444, P.R. China
e-mail: lykang@shu.edu.cn

E. Shan (✉)
School of Management, Shanghai University, Shanghai, 200444, P.R. China
e-mail: efshan@i.shu.edu.cn



T. W. Haynes et al. (eds.), *Topics in Domination in Graphs*, Developments in Mathematics 64, https://doi.org/10.1007/978-3-030-51117-3_9

function, respectively. We say $f$ is a *minimal* $(Y, k)$-subdominating function if there does not exist a $(Y, k)$-subdominating function $g : V \to Y$, $f \neq g$, for which $g(v) \leq f(v)$ for every $v \in V$.

Variations of domination are defined by taking the different weight subsets $Y$ of integers in the definition of a $Y$-dominating function or a $(Y, k)$-subdominating function.

When $Y = \{0, 1\}$, we obtain the standard domination number. In this case, a *$Y$-dominating function* of $G$ is called a *dominating function* of a graph $G$ and the set of vertices assigned weight 1 is a *dominating set* of $G$ (see [57]). The corresponding domination parameters are, respectively, called the *domination number* $\gamma(G)$ and *upper domination number* $\Gamma(G)$ of $G$, which are now well studied in graph theory [31].

In this chapter we survey main results concerning minus and signed domination of graphs. Also, we list some conjectures and open problems which have yet to be settled or solved on this subject. For more variations on signed and minus domination, including studying these functions in digraphs, we refer the reader to the survey paper [63].

## 1.1 Terminology and Notation

In general, we consider a *simple graph* $G = (V, E)$ with *vertex set* $V$ and *edge set* $E$. Let $v$ be a vertex in $V$. The *(open) neighborhood* of $v$, $N_G(v)$, is defined as the set of vertices adjacent to $v$, i.e., $N_G(v) = \{u : uv \in E\}$. The *closed neighborhood* of $v$ is $N_G[v] = N_G(v) \cup \{v\}$. We write $d_G(v)$ for the *degree* of $v$ in $G$, and $\Delta(G)$ and $\delta(G)$ denote the *maximum degree* and the *minimum degree* of $G$, respectively. If the graph $G$ is clear from the context, we simply write $N(v)$, $N[v]$, $d(v)$, $\Delta$, and $\delta$ for $N_G(v)$, $N_G[v]$, $d_G(v)$, $\Delta(G)$, and $\delta(G)$.

As usual, $P_n$, $C_n$, and $K_n$ denote a path, a cycle, and a complete graph on $n$ vertices, respectively. A *clique* of $G$ is a complete subgraph of $G$. The *clique number* of $G$, denoted by $\omega(G)$, is the cardinality of a maximum clique of $G$. For $S \subseteq V$, denote by $G[S]$ the subgraph induced by $S$ and by $d_S(v)$ the number of vertices in $S$ adjacent to $v$. Let $F$ be a graph. we call $G$ *$F$-free* if $G$ contains no $F$ as an induced subgraph. A *triangle-free graph* is a graph containing no cycles of length three. For $A, B \subseteq V(G)$, $A \cap B = \emptyset$, let $e(A, B)$ be the number of edges between $A$ and $B$.

A *$k$-regular graph* is a graph whose each vertex has degree $k$. If $d(v) = k - 1$ or $k$ for all $v \in V$, then we call $G = (V, E)$ a *nearly $k$-regular graph*. A graph is called an *Eulerian graph* if the degree of each vertex in the graph is even.

Let $k \geq 2$ be an integer. A graph $G = (V, E)$ is called *$k$-partite* if $V$ admits a partition into $k$ classes such that every edge has its ends in different classes: vertices in the same partition class must not be adjacent. Instead of "2-partite" one usually says *bipartite*. The *complement* $\overline{G}$ of a graph $G = (V, E)$ is defined by taking $V(\overline{G}) = V(G)$ and making two vertices $u$ and $v$ adjacent in $\overline{G}$ if and only if they are nonadjacent in $G$.

Let $G_1$ and $G_2$ be two graphs with vertex sets $V(G_1)$ and $V(G_2)$ and edge sets $E(G_1)$ and $E(G_2)$, respectively. The *join* $G_1 + G_2$ of two disjoint graphs $G_1$ and $G_2$ is the graph obtained from the union of $G_1$ and $G_2$ by adding all possible edges between $V(G_1)$ and $V(G_2)$. The *Cartesian product* $G = G_1 \Box G_2$ has vertex set $V(G) = V(G_1) \times V(G_2)$, and two vertices $(u_1, u_2)$ and $(v_1, v_2)$ of $G$ are adjacent if and only if either $u_1 = v_1$ and $u_2v_2 \in E(G_2)$ or $u_2 = v_2$ and $u_1v_1 \in E(G_1)$. The Cartesian product $P_r \Box P_s$ of two paths is called a $(r \times s)$-*grid*. The Cartesian product graphs $P_r \Box C_s$ and $C_r \Box C_s$ are called a $(r \times s)$-*stacked prism* and a $(r \times s)$-*torus grid*, respectively. A *n-cube* (also called a *hypercube*) $Q_n$ is given by the Cartesian product $P_2 \Box P_2 \Box \ldots \Box P_2$ of $n$ paths $P_2$.

A graph $G = (V, E)$ is an *interval graph*, if the vertex set $V$ can be put into one-to-one correspondence with a set of intervals $I$ on the real line $R$ such that two vertices are adjacent in $G$ if and only if their corresponding intervals have a non-empty intersection. If $G$ is an interval graph, then it contains no induced cycles with more than three edges.

# 2 Signed Domination and Its Generalizations in Graphs

## 2.1 Signed Domination

The *signed dominating function* for a graph $G = (V, E)$ is defined in Dunbar et al. [16] as a two-valued function $f : V \to \{-1, 1\}$ such that $f[v] \geq 1$ for every vertex $v$ in $G$. The sum $f(V)$ is called the *weight* $w(f)$ of $f$. A *signed dominating set* of $G$ (associated with the signed dominating function $f$) is the set of vertices in $G$ that are assigned value $+1$ by the function $f$. The minimum of weights $w(f)$, taken over all signed dominating functions $f$ on $G$, is called the *signed domination number* of $G$, denoted by $\gamma_s(G)$. A signed dominating function $f : V \to \{-1, 1\}$ is *minimal* if every signed dominating function $g$ satisfies $f(v) \leq g(v)$ for every $v \in V$. The *upper signed domination number* $\Gamma_s(G)$ is the maximum weigh $w(f)$ of a minimal signed dominating function of $G$. Clearly, $\gamma_s(G) \leq \Gamma_s(G)$.

There are a variety of applications for signed dominating sets in graph theory. Assigning values $-1$ or $+1$ to the vertices of a graph can be modeled as networks of positive and negative electrical charges, networks of positive and negative spins of electrons, and networks of people or organizations in which global decisions must be made (e.g., yes–no, agree–disagree, like–dislike, etc.). In such a context, for example, the minimum number of vertices assigned value $+1$ by a signed dominating function represents the minimum number of people whose positive votes can dominate all local groups (represented by vertex neighborhoods in the graph), even though the entire network may have many negative voters. Hence this variation of domination studies situations in which, in spite of the presence of negative vertices, the neighborhoods of all vertices are required to maintain a positive sum.

The concept of signed domination in graphs was introduced by Dunbar et al. [16] and has been studied in [19, 22, 26, 32, 33, 36, 52, 58, 62, 78, 81–83]. The decision version of the problem of computing the signed domination number of a graph is NP-complete, even when the graphs are restricted to being chordal or bipartite [26], doubly chordal, chordal bipartite, or bipartite planar [47]. For a fixed $k$, the problem of determining if a graph has a signed dominating function of weight at most $k$ is also NP-complete [26]. However, they showed that there is a linear-time algorithm for finding a minimum signed dominating function on a tree. Much research on signed domination has been focused on deriving better upper and lower bounds on the signed domination numbers $\gamma_s$ of graphs.

## 2.2 Lower Bounds on the Signed Domination Number

In this subsection, we focus on lower bounds on the signed domination number $\gamma_s$ of a graph.

The following basic property of signed domination functions is frequently used when studying signed domination.

**Proposition 1.** *A signed dominating function $f$ on a graph $G$ is minimal if and only if for every vertex $v$ such that $f(v) = 1$, there exists a vertex $u \in N[v]$ for which $f[u] \in \{1, 2\}$.*

For trees, a sharp lower bound for $\gamma_s$ has been obtained as follows.

**Proposition 2 (Kang and Shan [43]).** *For any tree $T$ of order $n \geq 2$,*

$$\gamma_s(T) \geq \frac{n + 2 + n_o}{3},$$

*where $n_o$ is the number of vertices of odd-degree in $T$, and this bound is sharp.*

Another sharp lower bound for the signed domination number of a tree can be stated in terms of the independence domination number $i$.

**Proposition 3 (Dunbar et al. [16]).** *For a tree $T$ of order $n \geq 2$, $\gamma_s(T) \geq i(T)+1$.*

That the bounds given in Propositions 2 and 3 are sharp may be seen by considering the path $P_{3k+2}$, $k \geq 0$, on $3k + 2$ vertices.

A lower bound on $\gamma_s$ in terms of the degree sequence of the graph can be given.

**Proposition 4 (Dunbar et al. [16]).** *Let $G$ be a graph of order $n$ with degrees $d_1 \leq d_2 \leq \ldots \leq d_n$. If $k$ is the smallest integer for which*

$$\sum_{i=0}^{k-1} d_{n-i} \geq 2(n-k) + \sum_{i=0}^{n-k} d_i,$$

*then $\gamma_s(G) \geq 2k - n$.*

For cubic graphs, the following result is known.

**Proposition 5 (Henning and Slater [36]).** *If G is a cubic graph of order n, then* $\gamma_s(G) \geq n/2 \geq IR(G)$, *where* $IR(G)$ *is the upper irredundance number of G.*

**Proposition 6 (Henning and Slater [36]).** *If G is a cubic graph, then* $\gamma_s(G) \geq i(G)+1$ *and this bound is sharp, where* $i(G)$ *is the independent domination number of G.*

For a $k$-regular graph $G$ of order $n$, Dunbar et al. [16] showed that $\gamma_s(G) \geq n/(k+1)$ if $n$ is even and Henning and Slater [36] further showed that $\gamma_s(G) \geq 2n/(k+1)$ if $n$ is odd. For general graphs $G$, Zhang et al. [82] obtained several sharp lower bounds in terms of order $n$ and size $m$. Chen and Song [7] proved the following strengthening of the above results.

To state the result, we need more notation. For a graph $G$, let $C(G) = \{v \mid d_G(v) > 1$ and $v$ is not adjacent to 1-degree vertices$\}$. If $C(G) = \emptyset$, then clearly $\gamma_s(G) = |V(G)|$. Otherwise, we let $\delta^* = \min\{d_G(v) : v \in C(G)\}$. Then $\delta^* \geq 2$. Further, let $\Delta(G) = \Delta$ and $\delta(G) = \delta$.

**Theorem 1 (Chen and Song [7]).** *Let G be a graph of order n and size m, and* $n_o$ *is the number of odd vertices of G. Then*

$$\gamma_s(G) \geq \max\Big\{ \big((\delta^*+2-\Delta)n+2n_o\big)/(\delta^*+2+\Delta),\ (2(m+n)+n_o)/(\Delta+1)-n,$$

$$n-(2m-n_o)/(\delta^*+1),\ 2\left\lceil \left(-\delta^*+\sqrt{\delta^{*2}+8(\delta^*+2)n+8n_o}\right)/4\right\rceil - n,$$

$$2\left\lceil \left(1+\sqrt{1+8(m+n)+4n_o}\right)/4\right\rceil - 4\Big\},$$

*and these bounds are sharp.*

Zhang et al. [82] established the lower bound on $\gamma_s$ of a graph in terms of its maximum degree $\Delta$ and minimum degree $\delta$.

**Theorem 2 (Zhang et al. [82]).** *If G is a graph of order n with minimum degree* $\delta$ *and maximum degree* $\Delta$, *then*

$$\gamma_s(G) \geq \frac{\delta-\Delta+2}{\delta+\Delta+2}n.$$

Haas and Wexler [22] improved this lower bound on $\gamma_s$ given in Theorem 2, they obtained the following sharp bound on $\gamma_s$ of a graph, which also generalized the result for regular graphs.

**Theorem 3 (Haas and Wexler [22]).** *If $G = (V, E)$ is a graph of order $n$ with minimum degree $\delta \geq 2$ and maximum degree $\Delta$, then*

$$\gamma_s(G) \geq \frac{\lceil \delta/2 \rceil - \lfloor \Delta/2 \rfloor + 1}{\lceil \delta/2 \rceil + \lfloor \Delta/2 \rfloor + 1} n,$$

*and this bound is sharp.*

The following example illustrates that the lower bound given in Theorem 3 is sharp for $\delta(G) \geq 2$. Take the disjoint union of $2\lceil \Delta/\delta \rceil$ copies of the complete bipartite graph $K_{\lceil \delta/2 \rceil+1, \lfloor \Delta/2 \rfloor}$ to get a bipartite graph with partite sets $P$ and $M$ such that each vertex in $P$ is of degree $\lfloor \Delta/2 \rfloor$, and each vertex in $M$ is of degree $\lceil \delta/2 \rceil + 1$. The desired graph $G$ will be obtained from this bipartite graph by adding some edges in the subgraphs $G[P]$ and $G[M]$ induced by $P$ and $M$, respectively. Since $P$ is a set of $2\lceil \Delta/\delta \rceil (\lceil \delta/2 \rceil + 1) \geq 2\lceil \Delta/2 \rceil$ vertices, by the Erdös–Gallai graphic criterion (see, [17]), we can construct a $\lceil \Delta/2 \rceil$-regular graph on the vertices of $P$. Similarly, $M$ is a set of $2\lceil \Delta/\delta \rceil \lfloor \Delta/2 \rfloor \geq 2(\lfloor \delta/2 \rfloor - 1)$ vertices, so we construct a $(\lfloor \delta/2 \rfloor - 1)$-regular graph on the vertices of $M$. By our construction, every vertex of $P$ has exactly degree $\Delta$, every vertex of $M$ has exactly degree $\delta$, and $G$ has order $n = 2\lceil \Delta/\delta \rceil (\lceil \delta/2 \rceil + \lfloor \Delta/2 \rfloor + 1)$.

Now let $f : V \to \{-1, 1\}$ be a function by assigning $+1$ to each vertex of $P$ and $-1$ to each vertex of $M$. It is easy to check that $f$ is a signed dominating function of $G$ with weight $f(V) = 2\lceil \Delta/\delta \rceil (\lceil \delta/2 \rceil - \lfloor \Delta/2 \rfloor + 1)$. So

$$\gamma_s(G) \leq 2\left\lceil \frac{\Delta}{\delta} \right\rceil \left( \left\lceil \frac{\delta}{2} \right\rceil - \left\lfloor \frac{\Delta}{2} \right\rfloor + 1 \right) = \left( \frac{\lceil \delta/2 \rceil - \lfloor \Delta/2 \rfloor + 1}{\lceil \delta/2 \rceil + \lfloor \Delta/2 \rfloor + 1} \right) n.$$

By Theorem 3, the equality holds in the above inequality.

In 2010, Poghosyan and Zverovich [58] provided a lower bound on the signed domination number of a graph $G$ depending on its order and a parameter $\lambda$, which is determined on the basis of the degree sequence of $G$ (note that $\lambda$ may be equal to 0, in this case we put $\sum_{i=1}^{\lambda} = 0$).

**Theorem 4 (Poghosyan and Zverovich [58]).** *Let $G = (V, E)$ be a graph of order $n$ with degrees $d_1 \leq d_2 \leq \cdots \leq d_n$. Then $\gamma_s(G) \geq n - 2\lambda$, where $\lambda \geq 0$ is the largest integer such that*

$$\sum_{i=1}^{\lambda} \left( \left\lceil \frac{d_i}{2} \right\rceil + 1 \right) \leq \sum_{i=\lambda+1}^{n} \left\lfloor \frac{d_i}{2} \right\rfloor.$$

*Proof.* Let $f$ be a signed domination function of minimum weight of the graph $G$. Let us denote

$$P = \{v \in V : f(v) = 1\} \text{ and } M = \{v \in V : f(v) = -1\}.$$

Then $\gamma_s(G) = f(V) = |P| - |M| = n - 2|M|$. By definition, for any vertex $v \in V$, $f[v] = \sum_{u \in N[v]} f(u) \geq 1$. Therefore, for all $v \in V$, $|N[v] \cap P| - |N[v] \cap M| \geq 1$. This implies that

$$|N[v]| = d_G(v) + 1 = |N[v] \cap P| + |N[v] \cap M| \leq 2|N[v] \cap P| - 1.$$

Hence $|N[v] \cap P| \geq d_G(v)/2 + 1$. We deduce that $|N[v] \cap P| \geq \lceil d_G(v)/2 \rceil + 1$ and $|N[v] \cap M| = d_G(v) + 1 - |N[v] \cap P| \leq \lfloor d_G(v)/2 \rfloor$. Let us estimate the number of edges between $P$ and $M$ in two ways.

$$e(P, M) = \sum_{v \in M} |N[v] \cap P| \geq \sum_{v \in M} \left( \left\lceil \frac{d_G(v)}{2} \right\rceil + 1 \right) \geq \sum_{i=1}^{|M|} \left( \left\lceil \frac{d_i}{2} \right\rceil + 1 \right).$$

Note that if $M = \emptyset$, then we put $\sum_{i=1}^{0} g(i) = 0$. On the other hand,

$$e(P, M) = \sum_{v \in P} |N[v] \cap M| \leq \sum_{v \in P} \left\lfloor \frac{d_G(v)}{2} \right\rfloor \leq \sum_{i=n-|P|+1}^{n} \left\lfloor \frac{d_i}{2} \right\rfloor = \sum_{i=|M|+1}^{n} \left\lfloor \frac{d_i}{2} \right\rfloor.$$

Therefore, the following inequality holds:

$$\sum_{i=1}^{|M|} \left( \left\lceil \frac{d_i}{2} \right\rceil + 1 \right) \leq \sum_{i=|M|+1}^{n} \left\lfloor \frac{d_i}{2} \right\rfloor.$$

Since $\lambda \geq 0$ is the largest integer such that

$$\sum_{i=1}^{\lambda} \left( \left\lceil \frac{d_i}{2} \right\rceil + 1 \right) \leq \sum_{i=\lambda+1}^{n} \left\lfloor \frac{d_i}{2} \right\rfloor,$$

$|M| \leq \lambda$. Thus $\gamma_s(G) = n - 2|M| \geq n - 2\lambda$. This completes the proof. □

This result improves the bound given in Proposition 4 and immediately implies the result in Theorem 3. In some cases, it provides a much better lower bound. For example, let us consider a graph $G$ consisting of two vertices of degree 5 and $n-2$ vertices of degree 3. It is easy to see that $k = (5n-4)/8$ is the smallest integer for which $\sum_{i=0}^{k-1} d_{n-i} \geq 2(n-k) + \sum_{i=0}^{n-k} d_i$. Then, by Proposition 4, $\gamma_s(G) \geq 0.25n - 1$. However, since $\lambda = (n+2)/4$ is the largest integer satisfying the inequality in Theorem 4, Theorem 4 can yield $\gamma_s(G) \geq 0.5n - 1$.

**Problem 1.** Let $\lambda$ be as defined in Theorem 4. Characterize the graphs $G$ with equality $\gamma_s(G) = n - 2\lambda$.

Xu [76] gave lower bounds on $\gamma_s$ for planar graphs and triangle-free graphs. The result on triangle-free graphs can be generalized. Indeed, applying a well-known theorem of Turán, Shan, Cheng and Kang [62] established a sharp lower bound on $\gamma_s$ for a $K_{r+1}$-free graph.

If $F$ is a simple graph, we denote by $ex(n, F)$ the maximum number of edges in a graph $G$ of order $n$ which does not contain an induced copy of $F$. Such a graph $G$ is called an *extremal graph*. The unique complete $k$-partite graphs on $n \geq k$ vertices whose partition sets differ in size by at most 1 are called *Turán graphs*; we denote them by $T^k(n)$ and their number of edges by $t_k(n)$. Clearly, $T^k(n) = K_n$ for all $n \leq k$. The following theorem of Turán from extremal theory is well-known, we will make use of it in our proof.

**Theorem 5 (Turán's theorem [67]).** *For any integer $k \geq 1$, if $G = (V, E)$ is a graph on $n$ vertices and $\mathrm{ex}(n, K_{k+1})$ edges containing no $(k+1)$-cliques, then $G$ is a $T^k(n)$ and*

$$|E| = t_k(n) \leq \frac{k-1}{2k}n^2$$

*with equality if and only if $k$ divides $n$.*

By applying Turán's theorem, a sharp lower bound on $\gamma_s(G)$ for graphs $G$ containing no $(k+1)$ cliques can be obtained.

**Theorem 6 (Shan et al. [62]).** *For any integer $k \geq 2$, let $G = (V, E)$ be a graph of order $n$ with no $(k+1)$-cliques and $c = \lceil \delta^*/2 \rceil + 1$, where $\delta^* = \max\{2, \delta(G)\}$. Then*

$$\gamma_s(G) \geq \frac{k}{k-1}\left(-c + \sqrt{c^2 + 4\frac{k-1}{k}nc}\right) - n$$

*and this bound is sharp.*

*Proof.* Let $f : V \to \{+1, -1\}$ be a signed dominating function on $G$ with $f(V(G)) = \gamma_s(G)$ and let $P$ and $M$ be the sets of vertices in $V$ that are assigned the value $+1$ and $-1$, respectively, under $f$. Then $n = |P| + |M|$. For convenience, let $|P| = p$ and $|M| = m$. For each vertex $v \in M$, $v$ is adjacent to at least $c$ $(\geq 2)$ vertices in $P$ since $f[v] \geq 1$, i.e., $|N_G(v) \cap P| \geq c$. Hence,

$$e(P, M) = \sum_{v \in M} |N_G(v) \cap P| \geq c|M| = cm = c(n-p). \tag{1}$$

On the other hand, for each vertex $v \in P$, $|N_G(v) \cap M| \leq |N_G(v) \cap P|$, and so

$$e(M, P) = \sum_{v \in P} |N_G(v) \cap M| \leq \sum_{v \in P} |N_G(v) \cap P| = \sum_{v \in P} d_{G[P]}(v). \tag{2}$$

Since $G$ contains no $(k+1)$-cliques, so does $G[P]$. Applying Turán's theorem, together with inequalities (1) and (2), we have

$$c(n-p) \leq e(P, M) \leq \frac{k-1}{k}p^2, \tag{3}$$

or equivalently,

$$\frac{k-1}{k}p^2 + cp - cn \geq 0.$$

Hence,

$$p \geq \left(-c + \sqrt{c^2 + 4\frac{k-1}{k}cn}\right) / 2\left(\frac{k-1}{k}\right).$$

Therefore,

$$\gamma_s(G) = 2p - n \geq \frac{k}{k-1}\left(-c + \sqrt{c^2 + 4\frac{k-1}{k}cn}\right) - n.$$

That the bound is sharp may be seen as follows: For positive integers $k, s \geq 2$, let $F_1$ be the Turán graph $T^k(ks)$, that is, $F_1$ is a complete $k$-partite graph of order $ks$ with equal partition sets $V_1, V_2, \ldots, V_k$ and $|V_i| = s$ for $i = 1, \ldots, k$. Let $F_2$ be a $(s-2)$- or $(s-3)$ $(\geq 0)$-regular $k$-partite graph of order $k(k-1)s$ with equal partition sets $U_1, U_2, \ldots, U_k$ and $|U_i| = (k-1)s$ for $i = 1, \ldots, k$. Let $F(k, s)$ be a family of graphs obtained from the disjoint union of $F_1$ and $F_2$ by joining each vertex of $V_i$ with all the vertices of $U_i$ for each $i = 1, \ldots, k$. Let $X_i = V_i \cup U_{i+1}$, where $i + 1$ (mod $k$). Then every one of $F(k, s)$ be a $k$-partite graph of order $n = k^2 s$ with equal partition sets. Note that for all $i$, each vertex of $U_i$ in $F(k, s)$ has minimum degree $2(s-1)$ or $2(s-1) - 1$. For each graph $H \in F(k, s)$, we assign to each vertex of $F_1$ the value $+1$ and to each vertex of $F_2$ the value $-1$. It is easy to check that $f[v] = 1$ for each vertex $v \in V(F_1)$ and $f[v] = 1$ or $2$ for each vertex $v \in V(F_2)$, so we produce a signed dominating function $f$ of $H$ with weight

$$\begin{aligned} w(f) = f(V(H)) &= p - m = V(F_1) - V(F_2) \\ &= ks - k(k-1)s = ks(2-k) \\ &= \frac{k}{k-1}\left(-s + \sqrt{s^2 + 4\frac{k-1}{k}ns}\right) - n. \end{aligned}$$

Note that $s \geq 2$. If $s = 2$, then $\delta^* = 2$, and thus $c = s$; if $s \geq 3$, then $\delta^* = \delta(G) = 2(s-1)$ or $2(s-1) - 1$, and thus $c = s$. Consequently,

$$\gamma_s(G) = \frac{k}{k-1}\left(-c + \sqrt{c^2 + 4\frac{k-1}{k}nc}\right) - n.$$

This completes the proof. □

As a somewhat weak version of Theorem 6, Shan, Cheng, and Kang [62] extended a result on $\gamma_s(G)$ for bipartite graphs, due to Wang [71], to graphs containing no $(k+1)$-cliques and characterized the extremal graphs achieving this bound. For this purpose, we define a family $\mathscr{H}(k,s)$ of graphs as follows.

Let $p, m, k$, and $s$ be positive integers satisfying the following conditions:

(1) $p = ks$, $s$ is even if $k = 2$; $s \geq 1$ if $k \geq 3$.
(2) $m = (k-1)p^2/2k = \frac{1}{2}k(k-1)s^2$.

Let $F_1$ be the Turán graph $T^k(ks)$ and $F_3$ be an independent set on $m$ vertices. Let $H(k,s)$ be the family of graphs in which each graph is obtained from the disjoint union of $F_1$ and $F_3$ by adding edges as follows: if $k \geq 3$, we join each vertex of $F_1$ with exactly $(k-1)s$ vertices of $F_3$ and join each vertex of $F_3$ with exactly 2 vertices of $F_1$ (since $2m = (k-1)p^2/k = k(k-1)s^2$, such an addition of edges is possible); if $k = 2$, we can partition $V(F_3)$ into two subsets $U_1$ and $U_2$ with $|U_1| = |U_2| = s^2/2$, then join each vertex of $V_i$ with exactly $s$ vertices of $U_i$ and join each vertex of $U_i$ with exactly 2 vertices of $V_i$ for $i = 1, 2$. By our construction, every one of $H(k,s)$ contains no $(k+1)$-cliques, and each vertex of $F_1$ in $H(k,s)$ has degree $2(k-1)s$, while each vertex of $F_3$ in $H(k,s)$ has degree 2. An example of the graphs $H(3,2)$ is shown in Figure 1. Let $\mathscr{H}(k,s) = \cup H(k,s)$, where $p, m, k$, and $s$ take over all integers satisfying conditions (1) and (2).

**Theorem 7 (Shan et al. [62]).** *For any integer $k \geq 2$, if $G = (V, E)$ be a graph of order $n$ with no $(k+1)$-cliques, then*

$$\gamma_s(G) \geq \frac{2k}{k-1}\left(-1 + \sqrt{1 + \frac{2(k-1)}{k}n}\right) - n,$$

*where equality holds if and only if $G \in \mathscr{H}(k,s)$.*

*Proof.* We define

$$g(x) = \frac{k}{k-1}\left(-x + \sqrt{x^2 + 4\frac{k-1}{k}nx}\right) - n.$$

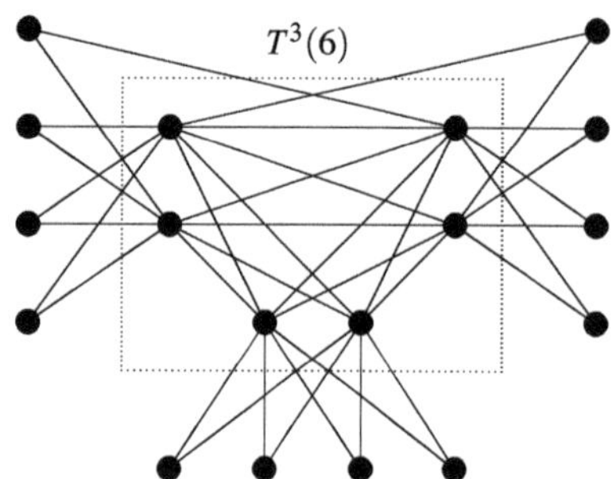


**Fig. 1** An example of the graphs $H(3,2)$.

It is easy to check that $g'(x) > 0$ when $x, n \geq 1$, so $g(x)$ is a strictly monotone increasing function when $x \geq 1$. Note that $c \geq 2$, hence

$$\gamma_s(G) \geq g(c) \geq g(2) = \frac{2k}{k-1}\left(-1+\sqrt{1+\frac{2(k-1)}{k}n}\right) - n.$$

The first part of the assertion follows.

Next we characterize the extremal graphs achieving this lower bound. First, suppose that $\gamma_s(G) = 2k/(k-1)\left(-1+\sqrt{1+2(k-1)n/k}\right) - n$ holds. Then $c = 2$ and all the equalities hold in (1), (2), and (3). Hence, we obtain

$$2|M| = e(P, M) = \sum_{v\in P} |N_G(v) \cap M| = \sum_{v\in P} |N_G(v) \cap P| = \sum_{v\in P} d_{G[P]}(v) = \frac{k-1}{k} p^2 \quad (4)$$

The equality chain implies that

$$|E(G[P])| = \frac{k-1}{2k} p^2 \text{ and } m = |M| = \frac{k-1}{2k} p^2.$$

Note the fact that $G[P]$ contains no $(k+1)$-cliques. Applying Turán's theorem, $G[P]$ is a complete $k$-partite graph with equal partition classes, and so $k$ divides $p$. Let $p = ks$. Then $G[P]$ is isomorphic to some $F_1$. The equality $\sum_{v\in P} |N_G(v) \cap M| = \sum_{v\in P} |N_G(v) \cap P|$ implies that each vertex $v$ of $P$ has exactly $(k-1)s$ neighbors in $M$. Hence $d_G(v) = 2(k-1)s \geq 2$ as $k \geq 2$. By definition, each vertex of $M$ has degree at least 2. Then $\delta(G) \geq 2$. However, since $c = \lceil \delta^*/2 \rceil + 1 = \lceil \delta(G)/2 \rceil + 1 = 2$, it follows that $\delta(G) = 2$. The equality chain (4) implies that each vertex of $M$ in $G$ is exactly adjacent to two vertices of $P$ and has minimum degree 2. Hence $M$ is an independent set of vertices in $G$, and so $M$ is isomorphic to some empty graph $F_3$ of order $m$. So $G$ is isomorphic to one of the families $H(k, s)$ of graphs. It follows that $G \in \mathscr{H}(k, s)$.

On the other hand, suppose $G \in \mathscr{H}(k, s)$. Thus, there exist integers $k$ and $s$ such that $G \in H(k, s)$. Assigning to each vertex of $F_1$ the value $+1$ and to each vertex of $F_3$ the value $-1$, we produce a signed dominating function $f$ of $G$ with weight

$$\begin{aligned} w(f) = f(V(G)) = p - m = V(F_1) - V(F_3) &= p - (k-1)p^2/2k \\ &= \frac{2k}{k-1}\left(-1+\sqrt{1+\frac{2(k-1)}{k}n}\right) - n. \end{aligned}$$

Consequently,

$$\gamma_s(G) = \frac{2k}{k-1}\left(-1+\sqrt{1+\frac{2(k-1)}{k}n}\right) - n.$$

This completes the proof. □

Let $f$ be a signed dominating function on a graph $G$ and $D$ be its signed dominating set. We define $E(G[D]) + |D|$ to be the *kernel* of $f$. In 2013, Zheng et al. [83] developed kernelization algorithms that produce small kernels for the signed dominating set problem on some graph classes, such as general graphs, planar graphs, grid graphs, $k$-partite graphs, bipartite graphs, bounded-degree graphs, and $r$-regular graphs. The kernelization results lead to tight lower bounds on the signed domination number of these graph classes, including some known results. The following result is not found in the literature.

**Theorem 8 (Zheng et al. [83]).**

(1) *For a planar graph $G$ of order $n$, $\gamma_s(G) \geq (6-n)/2$, and this bound is tight.*
(2) *For a graph $G$ with maximum degree $\Delta$,*

$$\gamma_s(G) \geq \begin{cases} \dfrac{4-\Delta}{4+\Delta}n & \text{if } \Delta \text{ is even} \\ \dfrac{5-\Delta}{3+\Delta}n & \text{if } \Delta \text{ is odd,} \end{cases}$$

*and these bounds are tight.*

Haas and Wexler [22] are the first to investigate signed domination of graph products, they provided bounds and some exact formulas for signed domination of the grids $P_r \Box P_s$, stacked prisms $P_r \Box C_s$, and torus grids $C_r \Box C_s$.

**Theorem 9 (Haas and Wexler [22]).** *For the ladder $P_2 \Box P_s$, we have*

$$\gamma_s(P_2 \Box P_s) = \begin{cases} s & s \text{ is even,} \\ s+1 & s \text{ is odd.} \end{cases}$$

*For $s \geq 3$, we have*

$$\gamma_s(P_2 \Box C_s) = \begin{cases} s & s \equiv 0 \pmod 4, \\ s+2 & s \equiv 2 \pmod 4, \\ s+1 & s \text{ is odd.} \end{cases}$$

$$\gamma_s(C_3 \Box P_s) = s+2, \quad \gamma_s(C_3 \Box C_s) = s.$$

$$(7s-8)/5 \leq \gamma_s(P_3 \Box P_s) \leq (7s+10-2c)/5, \quad \text{where } c \equiv s \pmod 5,$$

$$7s/5\gamma_s(P_3 \Box C_s \leq (7s+8)/5.$$

*In general, when $r, s \geq 4$, let $n = rs$,*

$$(n+4r+4s-24)/5 \leq \gamma_s(P_r \Box P_s) \leq (n+8r+4s)/5,$$

$$(n+4r)/5 \leq \gamma_s(C_r \Box P_s) \leq (n+4r+4s+8)/5,$$

$$n/5 \leq \gamma_s(C_r \Box C_s) \leq (n+4r+4s-4)/5.$$

However, signed domination of grids has not yet been settled, we close this section with the following open problem.

**Problem 2.** Determine the values or sharp bounds of $\gamma_s(G)$ for a grid graph $G = P_r \Box P_s$, $C_r \Box P_s$, and $C_r \Box C_s$ for all $r, s$.

## 2.3 Upper Bounds on the Signed Domination Number

This subsection is devoted to determining upper bounds on the signed domination number $\gamma_s$ and upper signed domination number $\Gamma_s$ of graphs.

Zelinka [78] showed that for every cubic graph $G$ of order $n$, $\gamma_s(G) \leq 4n/5$. Favaron [19] improved Zelinka's result. She showed that every cubic graph of order $n$ different from the Petersen graph has a signed domination number at most $3n/4$. An upper bound on the signed domination number of regular graphs was given by both Favaron [19] and Henning [32] at just about the same time and the graphs attaining the bound are described.

**Theorem 10 (Favaron [19] and Henning [32]).** *If $G$ is a $k$-regular graph of order $n$, then*

$$\Gamma_s(G) \leq \begin{cases} \dfrac{(k+1)^2}{k^2+4k-1}n & \text{if } k \text{ is odd} \\ \dfrac{k+1}{k+3}n & \text{if } k \text{ is even,} \end{cases}$$

*and these bounds are sharp.*

Wang and Mao established sharp bounds on $\Gamma_s$ for nearly regular graphs.

**Theorem 11 (Wang and Mao [70]).** *If $G$ is a nearly $(k+1)$-regular graph of order $n$, then*

$$\Gamma_s(G) \leq \begin{cases} \dfrac{k^2+3k+4}{k^2+5k+2}n & \text{if } k \text{ is odd} \\ \dfrac{(k+2)^2}{k^2+6k+4}n & \text{if } k \text{ is even,} \end{cases}$$

*and these bounds are sharp.*

However, the characterization of the extremal graphs attaining the upper bounds in Theorem 11 remains open.

In 2008, Tang and Chen [66] generalized these results in Theorems 10 and 11 to an arbitrary graph. They found a sharp upper bound on $\Gamma_s$ for a graph in terms of maximum degree $\Delta$ and minimum degree $\delta$ and constructed a class of extremal graphs achieving the upper bound.

In order to prove Theorem 12, we need the following lemma.

**Lemma 1 (Favaron [19]).** *Let $G$ be a graph and $f$ a minimal signed dominating function on $G$ such that $w(f) = \Gamma_s(G)$. Let*

$$M = \{x_1, x_2, \ldots, x_m\}$$
$$P = \{x_{m+1}, x_{m+2}, \ldots, x_n\}$$

*be the sets of vertices that are assigned the value $-1$ and $1$ under $f$, respectively. Write $\lfloor \Delta/2 \rfloor = l$. Define $A_i = \{v \in P : d_M(v) = i\}$ for $0 \le i \le l$ and set $|A_i| = a_i$. If $\delta \ge 2$, then*

*(1) $n = m + \sum_{i=0}^{l} a_i$ and*
*(2) $e(M, P) = \sum_{i=1}^{l} i a_i \le m\Delta$.*

**Theorem 12 (Tang and Chen [66]).** *If $G$ is a graph of order $n$ with minimum degree $\delta$ and maximum degree $\Delta$, then*

$$\Gamma_s(G) \le \begin{cases} \dfrac{\Delta(\delta+3)-(\delta-1)}{\Delta(\delta+3)+(\delta-1)}n & \text{for } \delta \text{ odd}, \\ \dfrac{\Delta(\delta+4)-\delta}{\Delta(\delta+4)+\delta}n & \text{for } \delta \text{ even}. \end{cases}$$

*In particular, if $G$ is an Eulerian graph, then*

$$\Gamma_s(G) \le \frac{\Delta(\delta+2)-\delta}{\Delta(\delta+2)+\delta}n.$$

*Furthermore, these bounds are sharp.*

*Proof.* Let $f$ be a minimal signed dominating function on $G$ such that $w(f) = \Gamma_s(G)$ and let $M, P, l$, and $A_i$ be as defined in Lemma 1. Write $\lfloor \delta/2 \rfloor = k$.

If $\delta = 1$, then the result is trivial. Thus we may assume $\delta \ge 2$.

If $A_0 = \emptyset$, then by Lemma 1, we have $n = m + \sum_{i=1}^{l} a_i \le m + \sum_{i=1}^{l} i a_i \le (\Delta+1)m$, which implies that $m \ge n/(\Delta+1)$, and so $\Gamma_s(G) = n - 2m \le (\Delta - 1)n/(\Delta+1)$. Noting that $(\Delta-1)n/(\Delta+1) < \min\{(\delta\Delta + 4\Delta - \delta)n/(\delta\Delta + 4\Delta + \delta), (\delta\Delta + 3\Delta - \delta + 1)n/(\delta\Delta + 3\Delta + \delta - 1)\}$, we see the conclusion holds. Thus we may assume that $A_0 \ne \emptyset$.

For any $v \in A_0$, since $f[v] = d(v) + 1 \ge \delta + 1 \ge 3$ and $f$ is minimal, by Proposition 1, there exists a vertex $u \in N(v)$ such that $u \notin A_0$ and $f[u] = 1$ or $2$. Let $Q = \{v \in N(A_0) : f[v] = 1 \text{ or } 2\}$. Noting that $f[v] \ge 3$ for any $v \in \bigcup_{i=0}^{k-1} A_i$, we see that $Q \subseteq \bigcup_{i=k}^{l} A_i$. Obviously, each $u \in Q \cap A_i$ has at most $i+1$ neighbors in $A_0$ . Thus $Q \cap A_i$ has at most $(i+1)|Q \cap A_i|$ neighbors in $A_0$. By the arguments above, we have $A_0 \subseteq \bigcup_{i=k}^{l} N(Q \cap A_i)$. This implies that

$$a_0 = |A_0| \leq \sum_{i=k}^{l} |N(Q \cap A_i)| \leq \sum_{i=k}^{l} (i+1)a_i. \tag{5}$$

By (5) and Lemma 1 (1), we have

$$n \leq m + \sum_{i=k}^{l} (i+1)a_i + \sum_{i=1}^{l} a_i. \tag{6}$$

If $k = 1$, then by (6), we have

$$n \leq m + \sum_{i=1}^{l} (i+2)a_i, \tag{7}$$

and if $k \geq 2$, then by (6), we have

$$n \leq m + \sum_{i=1}^{l-1} a_i + \sum_{i=k}^{l} (i+2)a_i. \tag{8}$$

If $\delta$ is odd, then since $(\delta+3)i/(\delta-1) \geq i+2$ for $i \geq k = (\delta-1)/2$, by (7) and (8), we have $n \leq m + [(\delta+3)/(\delta-1)] \sum_{i=1}^{l} ia_i$. By Lemma 1 (2), $n \leq m + m\Delta(\delta+3)/(\delta-1)$, which implies that $m \geq n(\delta-1)/(\delta\Delta+3\Delta+\delta-1)$, and hence

$$\Gamma_s(G) = n - 2m \leq \frac{\Delta(\delta+3)-(\delta-1)}{\Delta(\delta+3)+(\delta-1)} n.$$

If $\delta$ is even, then since $(\delta+4)i/\delta \geq i+2$ for $i \geq k = \delta/2$, by (7) and (8), we have $n \leq m + [(\delta+4)/\delta] \sum_{i=1}^{l} ia_i$. By Lemma 1 (2), $n \leq m + m\Delta(\delta+4)/\delta$, which implies that $m \geq n\delta/(\delta\Delta+4\Delta+\delta)$, and hence

$$\Gamma_s(G) = n - 2m \leq \frac{\Delta(\delta+4)-\delta}{\Delta(\delta+4)+\delta} n.$$

Finally, if $G$ is an Eulerian graph, then every vertex of $G$ has even degree. This implies that each $u \in Q \cap A_i$ has at most $i$ neighbors in $A_0$. Thus the inequality (6) can be improved as follows:

$$n \leq m + \sum_{i=k}^{l} ia_i + \sum_{i=1}^{l} a_i.$$

From the similar proof, it follows that $n \geq m + m\Delta(\delta+2)/\delta$. This implies that

$$\Gamma_s(G) \leq \frac{\Delta(\delta+2)-\delta}{\Delta(\delta+2)+\delta}n.$$

This completes the proof. □

The bounds on $\Gamma_s(G)$ given in Theorem 12 are sharp. Note that Theorem 10 is a special case of Theorem 12. Thus the sharpness of the bounds in Theorem 10 implies that the bounds in Theorem 12 are also sharp when $\Delta = \delta$. The following example illustrates that the bounds in Theorem 12 are still sharp in the case when $\Delta - \delta \geq 1$. For $r = \lfloor (s+2)/2 \rfloor$, $s \geq 2$, let $K_{2r}^*$ be a graph obtained from complete graph $K_{2r}$ by deleting a perfect matching if $s$ is odd and the edges of a Hamiltonian cycle if $s$ is even and $K_{r,r,r}^*$ a graph obtained from complete 3-partite graph $K_{r,r,r}$ by deleting the edges of a Hamiltonian cycle if $s$ is odd and the edges of a Hamiltonian cycle together with any other edge if $s$ is even. Let $t \geq s+1$. Now let $G$ be the graph, as shown in Figure 2, with vertex set $V(G) = X \cup Y \cup Z$, where $|X| = \lfloor s/2 \rfloor$, $|Y| = t$, and $|Z| = rt$, $G[X \cup Y]$ is a complete bipartite graph, $G[Z] = (t/2)K_{2r}^*$ if $t$ is even and $G[Z] = [(t-3)/2]K_{2r}^* \cup K_{r,r,r}^*$ if $t$ is odd, $d_Z(y) = r$ for any $y \in Y$, and $\bigcup_{y \in Y} N_Z(y) = Z$, and there are no edges between $X$ and $Z$. Obviously, the order of $G$ is $t + \lfloor (s+2)/2 \rfloor t + \lfloor s/2 \rfloor$, $\delta = s$ and $\Delta = t$. We define a function $f$ on $G$ by assigning $-1$ to every vertex of $X$ and $+1$ to all other vertices. It is easy to check that $f$ is a signed dominating function. Since $f[y] = 2$ for each $y \in Y$ and $Y$ is a dominating set of $G$, by Proposition 1, $f$ is minimal. Clearly, $w(f) = |V(G)| - 2|X| = t + \lfloor (s+2)/2 \rfloor t - \lfloor s/2 \rfloor$. It is easy to see that $w(f) = [(\delta+3)\Delta - (\delta-1)]n/[(\delta+3)\Delta + (\delta-1)]$ if $s$ is odd, and $w(f) = [(\delta+4)\Delta - \delta]n/[(\delta+4)\Delta + \delta)]$ if $s$ is even.

Let $k \geq 1$ be an integer, and let $G = (V, E)$ be a graph with minimum degree $\delta \geq k$. A *signed k-dominating function* of $G$ is defined by Wang [12] as a function $f : V \to \{-1, 1\}$ such that $f[v] \geq k$ for every $v \in V$. The minimum of the values

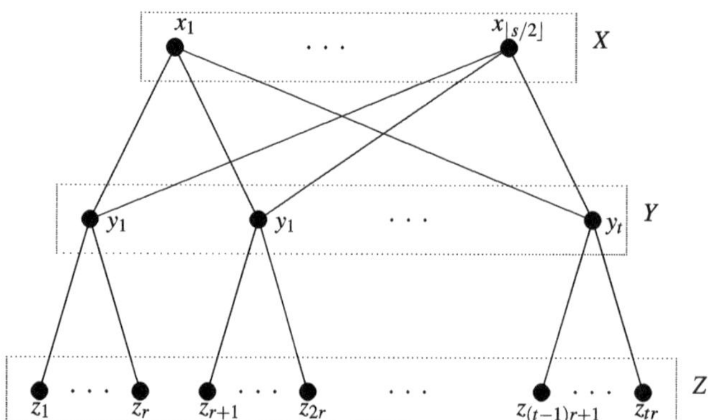


**Fig. 2** An extremal graph $G$ achieving the bounds in Theorem 12.

of $f(V)$, taken over all signed $k$-domination functions $f$, is called the *signed k-domination number* of $G$ and is denoted by $\gamma_s^k(G)$. The maximum of the values of $f(V)$, taken over all minimal signed $k$-dominating function $f$, is called the *upper signed k-domination number*, denoted by $\Gamma_s^k(G)$. In the special case $k = 1$, the signed $k$-domination is the ordinary signed domination in graphs.

In 2010, Delić and Wang generalized the two results to the signed $k$-domination number of graphs, which is based on the same ideas as described in the proof of Theorem 12.

**Theorem 13 (Delić and Wang [12]).** *If $G = (V, E)$ is a graph of order $n$ with minimum degree $\delta \geq k+1$ and maximum degree $\Delta$, then*

$$\Gamma_k^s(G) \leq \begin{cases} \dfrac{\Delta(\delta+k+2)-(\delta-k)}{\Delta(\delta+k+2)+(\delta-k)}n & \text{for } \delta-k \equiv 0 \pmod 2, \\ \dfrac{\Delta(\delta+k+3)-(\delta-(k-1))}{\Delta(\delta+k+3)+(\delta-(k-1))}n & \text{for } \delta-k \equiv 1 \pmod 2. \end{cases}$$

*If G is Eulerian, then*

$$\Gamma_k^s(G) \leq \begin{cases} \dfrac{\Delta(\delta+k+2)-(\delta-k)}{\Delta(\delta+k+2)+(\delta-k)}n & \text{for } k \equiv 0 \pmod 2, \\ \dfrac{\Delta(\delta+k+1)-(\delta-(k-1))}{\Delta(\delta+k+1)+(\delta-(k-1))}n & \text{for } k \equiv 1 \pmod 2. \end{cases}$$

*Furthermore, these bounds are sharp.*

A signed domination function of a graph $G$ can be regarded as a two-coloring of the vertices of $G$ with colors 1 and $-1$ such that the closed neighborhood of every vertex contains more 1's than $-1$'s. This concept is closely related to *combinatorial discrepancy theory* discussed by Füredi and Mubayi in their fundamental paper [21].

A function $f : V \to \{1, -1\}$ is called a *signed domination function* (SDF) of the hypergraph $H = (V, E)$ if $f(e) = \sum_{v \in e} f(v) \geq 1$ for every hyperedge $e \in E$, i.e., each hyperedge has a positive *imbalance*. The *signed discrepancy* of $H$, denoted by $SD(H)$, is defined as $SD(H) = \min_{SDF f} f(V)$ where the minimum is taken over all signed domination functions of $H$. Thus, in this version of discrepancy, the success is measured by minimizing the imbalance of the vertex set $V$ while keeping the imbalance of every hyperedge $e \in E$ positive.

Füredi and Mubayi [21] first obtained an upper bound on the signed discrepancy of a hypergraph.

**Theorem 14 (Füredi and Mubayi [21]).** *Let $H = (V, E)$ be a hypergraph of order $n$ with every hyperedge containing at least $k$ vertices, where $k \geq 100$. Then*

$$SD(H) \leq 4\sqrt{\frac{\ln k}{k}}n + \frac{1}{k}|E|.$$

This result can be easily reformulated in terms of the signed domination of graphs by considering the neighborhood hypergraph of a given graph.

**Theorem 15 (Füredi and Mubayi [21]).** *If $G$ is a graph of order $n$ with minimum degree $\delta \geq 99$, then*

$$\gamma_s(G) \leq \left(4\sqrt{\frac{\ln(\delta+1)}{\delta+1}} + \frac{1}{\delta+1}\right)n.$$

Furthermore, Füredi and Mubayi [21] also found a good upper bound on $\gamma_s$ for small minimum degree $\delta$. Using Hadamard matrices, they constructed a $\delta$-regular graph $G$ of order $4\delta$ with $\gamma_s(G) \geq 0.5\delta - O(1)$. This means that the upper bound in Theorem 15 is off from optimal by at most the factor of $\sqrt{\ln \delta}$. For this reason, they posed an interesting conjecture: there exists some constant $C$ such that $\gamma_s(G) \leq Cn/\sqrt{\delta}$. If the discrepancy conjecture is true, then, by the above construction, this is asymptotically tight. In 2000, Matoušek [52] proved the conjecture. The proof uses the partial coloring method from combinatorial discrepancy theory. However, the constant $C$ in his proof is big making the result of rather theoretical interest.

In 2010, Poghosyan and Zverovich [58] provided an upper bound on the signed domination number, which is better than the bound of Theorem 15 for "relatively small" values of $\delta$. For example, if $\delta = 99$, then, by Theorem 15, $\gamma_s(G) \leq 0.869n$, while Theorem 16 yields $\gamma_s(G) \leq 0.537n$.

**Theorem 16 (Poghosyan and Zverovich [58]).** *For any graph $G$ of order $n$ with minimum degree $\delta \geq 1$,*

$$\gamma_s(G) \leq \left(1 - \frac{2\lfloor\delta/2\rfloor}{(1+\lfloor\delta/2\rfloor)^{1+1/\lfloor\delta/2\rfloor}\widetilde{d}_{0.5}^{1/\lfloor\delta/2\rfloor}}\right)n.$$

For larger values of minimum degree $\delta$, the result can be further improved. Poghosyan and Zverovich [58] believed that Füredi–Mubayi's conjecture is true for some small constant $C$. However, as the Peterson graph shows, $C > 1$, i.e., the behavior of the conjecture is not good for relatively small values of minimum degree $\delta$. For this reason, they proposed the following refined conjecture, which consists of two functions for "small" and "large" values of $\delta$.

*Conjecture 1 (Poghosyan and Zverovich [58]).* For some $C \leq 10$ and $\alpha$, $0.18 \leq \alpha \leq 0.21$, if $G$ is a graph of order $n$ with minimum degree $\delta$, then

$$\gamma_s(G) \leq \min\left\{\frac{n}{\delta^\alpha}, \frac{Cn}{\sqrt{\delta}}\right\}.$$

Let $A(G)$ be the adjacency matrix of $G$ and $D(G) = diag(d(v_1), d(v_2), \ldots, d(v_n))$ be the diagonal matrix of vertex degrees. The Laplacian matrix of $G$ is $L(G) = D(G) - A(G)$. Clearly, $L(G)$ is a real symmetric matrix. From this fact and Geršgorin's theorem, it follows that its eigenvalues are nonnegative real numbers. The eigenvalues of an $n \times n$ matrix $M$ are denoted by $\lambda_1(M), \lambda_2(M), \ldots, \lambda_n(M)$, while for a graph $G$, we will use $\lambda_i(G) = \lambda_i$ to denote $\lambda_i(L(G))$, $i = 1, 2, \ldots, n$ and assume that $\lambda_1(G) \geq \lambda_2(G) \geq \cdots \geq \lambda_{n-1}(G) \geq \lambda_n(G)$. It is well-known that $\lambda_n(G) = 0$ and the algebraic multiplicity of zero as an eigenvalue of $L(G)$ is exactly the number of connected components of $G$ (see, [54]). In particular, the second smallest eigenvalue $\lambda_{n-1}(G) > 0$ if and only if $G$ is connected. This leads Fiedler [20] to define it as the *algebraic connectivity* of $G$. The eigenvalue $\lambda_1(G)$ is called *Laplacian spectral radius* of $G$. We have known that $\lambda_1(G) = max\lambda_1(G_i)$, $i = 1, 2, \ldots, n$ if $G_1, G_2, \ldots, G_n$ are all components of $G$ [20]. In recent years, the eigenvalues $\lambda_1(G)$ and $\lambda_{n-1}(G)$ have received a great deal of attention (see, for example, [54, 56]).

The following result is motivated by the results on graph eigenvalues involving domination of graphs [56]. Shi et al. [65] gave an upper bound on the algebraic connectivity $\lambda_{n-1}(G)$ and lower bounds on the Laplacian spectral radius $\lambda_1(G)$ for a connected graph $G$ in terms of $\gamma_s$.

**Theorem 17 (Shi et al. [65]).** *If $G$ is a connected graph of order $n \geq 2$, then*

$$\lambda_{n-1}(G) \leq \frac{n(\gamma_s(G) + n - 2)}{n - \gamma_s(G)},$$

*and this bound is sharp.*

**Theorem 18 (Shi et al. [65]).** *If $G = (V, E)$ is a connected graph of order $n \geq 2$, then*

$$\lambda_1(G) \geq \frac{4n}{\gamma_s(G) + n},$$

*with equality if and only if $G = K_3$.*

**Theorem 19 (Shi et al. [65]).** *If $G$ is a $k$-regular graph of order $n$, then*

$$\lambda_1(G) \geq \begin{cases} \dfrac{n(k+3)}{\gamma_s(G) + n} & \text{for } k \text{ odd,} \\ \dfrac{n(k+2)}{\gamma_s(G) + n} & \text{for } k \text{ even} \end{cases}$$

*with equality if and only if $G = K_n$.*

Although many bounds on the (upper) signed domination number have been found, no one has yet discovered what are bounds of the difference $\Gamma_s - \gamma_s$. We propose the following open problems.

**Problem 3.** Let $G$ be a graph of order $n$. How large can the difference $\Gamma_s(G) - \gamma_s(G)$ be? If $G$ is a cubic graph, how large can the difference $\Gamma_s(G) - \gamma_s(G)$ be?

**Problem 4.** Is it true that if $T$ is a tree of order $n$, then $\Gamma_s(T) - \gamma_s(T) \leq 4n/7$?

**Problem 5.** Characterize the cubic graphs for which $\Gamma_s(G) = \gamma_s(G)$.

## 2.4 Signed Total Domination in Graphs

A *signed total dominating function* of a graph $G$ is defined in [79] as a function $f : V \to \{+1, -1\}$ such that for every vertex $v$, $f(N(v)) \geq 1$, and the minimum cardinality of the weight $w(f) = \sum_{v \in V} f(v)$ over all such functions is called the *signed total domination number* of $G$, denoted by $\gamma_{st}(G)$. The *upper signed total domination number* of $G$, denoted by $\Gamma_{st}(G)$, is the maximum weight of a minimal signed total dominating function of $G$. A signed total dominating function of weight $\Gamma_{st}(G)$ is called a $\Gamma_{st}(G)$-function.

The motivation for considering signed total domination of graphs comes from the fact observed by Henning [34] that signed total domination differs significantly from signed domination. For example, for a path $P_n$ on $n$ vertices, $\gamma_{st}(P_n) = n$ and $\gamma_s(P_n) = n - 2\lfloor (n-2)/3 \rfloor$ (see [16]), while for a star $K_{1,n-1}$ on $n$ vertices, $\gamma_{st}(K_{1,n-1}) = 2$ if $n$ is even and 3 if $n$ is odd and $\gamma_s(K_{1,n-1}) = n$.

The decision problem for the total domination number of a graph has been known to be NP-complete, even when restricted to bipartite graphs [31]. By demonstrating a polynomial time reduction of the known NP-complete decision problem **Total Dominating Set** to the signed total dominating function problem, Henning [34] showed that the decision problem for the signed total domination number of a graph is NP-complete, even when restricted to bipartite graphs or chordal graphs.

Concerning the extremal behavior of the difference $\Gamma_{st}(G) - \gamma_{st}(G)$ for a graph $G$, many interesting questions can be raised. We first ask how large the difference $\Gamma_{st}(G) - \gamma_{st}(G)$ is, when $n$ is given or $n \to \infty$.

**Problem 6.** For $k$-regular graph $G$, what is the upper bound of the difference $\Gamma_{st}(G) - \gamma_{st}(G)$? Characterize the regular graphs for which $\Gamma_{st}(G) = \gamma_{st}(G)$.

Now we turn our attention to bounds of the signed total domination number of graphs. First, we give some sharp lower bounds on the signed total domination number of a graph. The proofs of the results rely on the following basic property of a minimal signed total dominating function.

**Proposition 7.** *A signed total dominating function $f$ on $G = (V, E)$ is minimal if and only if for $v \in V$ with $f(v) = 1$, there exists a vertex $u \in N(v)$ with $f(N(u)) \leq 2$, where $f(N(u)) = 1$ if $d_G(u)$ is odd and $f(N(u)) = 2$ otherwise.*

**Theorem 20 (Henning [34]).** *If $G$ is a graph of order $n$ and size $m$, then*

$$\gamma_{st}(G) \geq \max\{2(n-m), \sqrt{4n+1} + 1 - n\},$$

*and these bounds are sharp.*

The next result gives a lower bound on $\gamma_{st}$ in terms of the degree sequence of a graph.

**Theorem 21 (Henning [34]).** *Let $G$ be a graph of order $n$ with degrees $d_1 \leq d_2 \leq \ldots \leq d_n$. If $G$ has $n_e$ vertices of even degree, and if $k$ is the smallest integer for which*

$$d_{n-k+1} + d_{n-k+2} + \ldots + d_n - (d_1 + \ldots + d_{n-k}) \geq n + n_e,$$

*then $\gamma_{st}(G) \geq 2k - n$.*

Zelinka [79] presented a sharp lower bound on $\gamma_{st}$ for a regular graph. Henning generalized the result to a general graph in terms of its minimum degree, maximum degree and its order.

**Theorem 22 (Henning [34]).** *If $G$ is a graph of order $n$ with minimum degree $\delta \geq 2$ and maximum degree $\Delta$, then*

$$\gamma_{st}(G) \geq \left( \frac{\lceil(\delta - 1)/2\rceil - \lfloor(\Delta - 1)/2\rfloor + 1}{\lceil(\delta - 1)/2\rceil + \lfloor(\Delta - 1)/2\rfloor + 1} \right) n,$$

*and this bound is sharp.*

Chen and Song [7] gave several lower bounds on $\gamma_{st}$ for a general graph in terms of order, size, minimum degree, maximum degree, and the number of vertices of odd-degree in graphs, whose proofs are similar to those of Theorem 1.

Henning [34] showed that for a bipartite graph $G$, $\gamma_{st}(G) \geq 2\sqrt{2n} - n$. Shan and Cheng [61] extended this result to $k$-partite graphs and characterized the extremal graphs. In fact, we can continue to generalize the result to $K_{r+1}$-free graphs where $r \geq 2$ by applying Theorem 5 (Turán's theorem).

**Theorem 23 (Shan and Cheng [61]).** *If $G = (V, E)$ is a $K_{r+1}$-free graph of order $n$ with $\delta(G) \geq 1$ and $c = \lceil(\delta(G) + 1)/2\rceil$, then*

$$\gamma_{st}(G) \geq \frac{r}{r-1}\left( -(c-1) + \sqrt{(c-1)^2 + 4\frac{r-1}{r}cn} \right) - n$$

*and this bound is sharp.*

That the bound is sharp may be seen as follows: For integers $k \geq 2$, let $H_i$ be a complete bipartite graph with vertex classes $V_i$ and $U_i$, where $|V_i| = k$ and $|U_i| = k^2 - k - 1$, for $i = 1, 2, \ldots, k$. We let $H(k)$ be the graph obtained from the disjoint union of $H_1, H_2, \ldots, H_k$ by joining each vertex of $V_i$ with all the vertices of $\bigcup_{j=1, j\neq i}^{k} V_j$ and adding $(k-1)(k^2-k-1)$ edges between $U_i$ with $\bigcup_{j=1, j\neq i}^{k} U_j$ so that each vertex of $U_i$ has exactly $k-1$ neighbors in $\bigcup_{j=1, j\neq i}^{k} U_j$, while each vertex of $\bigcup_{j=1, j\neq i}^{k} U_j$ has exactly one neighbor in $U_i$ for all $i = 1, 2, \ldots, k$. Let $Y_i = V_i \cup U_{i+1}$, where $i + 1 \pmod k$. Then $H(k)$ is a $k$-partite graph of order

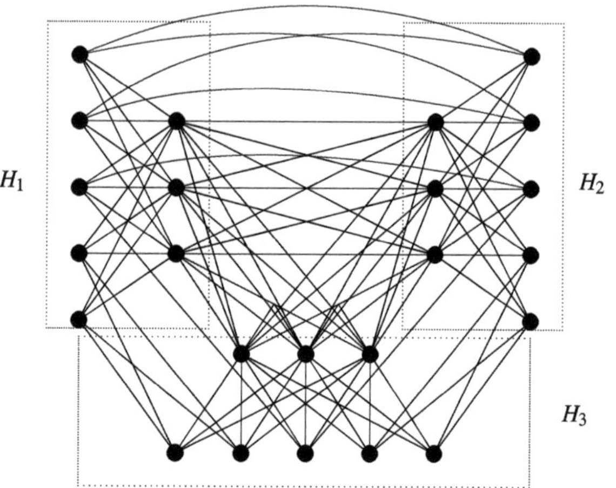


**Fig. 3** The graph $H(3)$.

$n = k(k^2 - 1)$ with vertex classes $Y_1, Y_2, \ldots, Y_k$ and $|Y_i| = k^2 - 1$. The graph $H(3)$ is shown in Figure 3. Note that each vertex of $U_i$ has minimum degree $2k - 1$. Assigning to each vertex of $\bigcup_{i=1}^{k} V_i$ the value $+1$ and to each vertex of $\bigcup_{i=1}^{k} U_i$ the value $-1$, we produce a signed total dominating function $f$ of $H$ with weight

$$\begin{aligned} f(V(H(k)) &= k^2 - k(k^2 - k - 1) \\ &= k(-k^2 + 2k + 1) \\ &= \frac{k}{k-1}\left(-(c-1) + \sqrt{(c-1)^2 + 4\frac{k-1}{k}cn}\right) - n. \end{aligned}$$

Consequently,

$$\gamma_{st}(H(k)) = \frac{k}{k-1}\left(-(c-1) + \sqrt{(c-1)^2 + 4\frac{k-1}{k}cn}\right) - n.$$

From Theorem 23, one can easily extend the result to $k$-partite graphs and characterize the extremal graphs achieving this bound. For this purpose, we recall a family $\mathscr{T}$ of graphs due to Kang et al. [46] as follows.

For integers $r \geq 1, k \geq 2$, let $H_i$ ($i = 1, 2, \ldots, k$) be the graph obtained from the disjoint union of $r$ stars $K_{1,(k-1)r-1}$ (the graph $K_{1,0}$ is regarded as $K_1$ when $r = 1$ and $k = 2$) with centers $V_i = \{x_{i,j} \mid j = 1, 2, \ldots, r\}$. Furthermore, let $U_i$ denote the set of vertices of degree 1 in $H_i$ that are not central vertices of stars and write $X_i = V_i \cup U_{i+1}$, where $i + 1 \pmod k$. We let $G_{k,r}$ be the $k$-partite graph obtained from the disjoint union of $H_1, H_2, \ldots, H_k$ by joining each center of $H_i$ ($i = 1, 2, \ldots, k$) with all the centers of $\bigcup_{j=1, j\neq i}^{k} H_j$. By construction, we know that $G_{k,r}$ is a $k$-partite graph of order $n = k(k-1)r^2$ with vertex classes

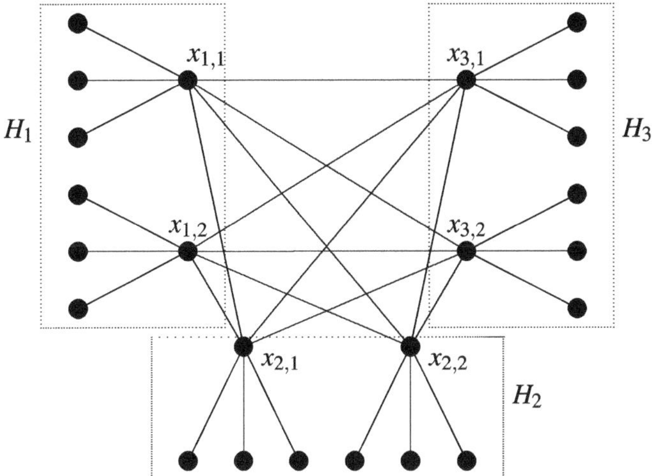


**Fig. 4** The graph $G_{3,2}$.

$X_1, X_2, \ldots, X_k$, and $|X_i| = (k-1)r^2$. Let $\mathscr{T} = \{G_{k,r} \mid r \geq 1, k \geq 2\}$. An example $G_{3,2}$ is shown in Figure 4.

**Theorem 24 (Shan and Cheng [61]).** *If $G = (V, E)$ is a $k$-partite graph of order $n$ with no isolated vertex, then*

$$\gamma_{st}(G) \geq 2\sqrt{\frac{k}{k-1}n} - n,$$

*where equality holds if and only if $G \in \mathscr{T}$.*

Let $k \geq 1$ be an integer, and let $G = (V, E)$ be a graph with minimum degree $\delta > k$. A *signed total $k$-dominating function* of $G$ is defined by Wang [69] as a function $f : V \to \{-1, 1\}$ such that $f(N(v)) \geq k$ for every $v \in V$. The minimum of the values of $f(V)$, taken over all signed total $k$-domination functions $f$, is called the *signed total $k$-domination number* of G and is denoted by $\gamma_{st}^k(G)$. By definition, the condition $\delta \geq k$ is clearly necessary. When $k = 1$, the signed total $k$-domination is the ordinary signed total domination in graphs. The upper signed total $k$-domination number $\Gamma_{st}^k(G)$ can be defined similarly.

By using an argument similar to the one used to prove in Theorem 6, Volkmann [68] (2016) generalized the result in Theorem 23 to the signed total $k$-domination number of graphs.

**Theorem 25 (Volkmann [68]).** *Let $k \geq 1$ and $r \geq 2$ be integers, and let $G$ be a $K_{r+1}$-free graph of order $n$. If $c = \lceil(\delta + k)/2\rceil$, then*

$$\gamma_{st}(G) \geq \frac{r}{r-1}\left(-(c-k)+\sqrt{(c-k)^2+4\frac{r-1}{r}cn}\right)-n,$$

*and this bound is sharp.*

We now turn to the upper bounds on the (upper) signed total domination number $\Gamma_{st}$ in graphs. For regular and nearly regular graphs, best possible upper bounds have been obtained.

**Theorem 26 (Henning [34]).** *If $G$ is a $k$-regular graph of order $n$, then*

$$\Gamma_{st}(G) \leq \begin{cases} \dfrac{k^2+1}{k^2+2k-1}n & \text{for } k \text{ odd} \\ \dfrac{k^2+k+2}{k^2+3k-2}n & \text{for } k \text{ even,} \end{cases}$$

*and these bounds are sharp.*

**Theorem 27 (Kang and Shan [44]).** *If $G$ is a nearly $(k+1)$-regular graph of order $n$, then*

$$\Gamma_{st}(G) \leq \begin{cases} \dfrac{k^2+3k+4}{k^2+5k+2}n & \text{for } k \text{ odd} \\ \dfrac{(k+1)^2+3}{k(k+4)}n & \text{for } k \text{ even,} \end{cases}$$

*and these bounds are sharp.*

Later, Shan and Cheng [60] generalized the two results above to an arbitrary graph and obtained the result similar to Theorem 12.

**Theorem 28 (Shan and Cheng [60]).** *If $G = (V, E)$ is a graph of order $n$ with minimum degree $\delta$ and maximum degree $\Delta$, then*

$$\Gamma_{st}(G) \leq \begin{cases} \dfrac{\Delta(\delta+3)-(\delta-1)}{\Delta(\delta+3)+(\delta-1)}n & \text{for } \delta \text{ odd,} \\ \dfrac{\Delta(\delta+2)-(\delta-2)}{\Delta(\delta+2)+(\delta-2)}n & \text{for } \delta \text{ even.} \end{cases}$$

*In particular, if $G$ is an odd-degree graph, then*

$$\Gamma_{st}(G) \leq \frac{\Delta(\delta+1)-(\delta-1)}{\Delta(\delta+1)+(\delta-1)}n.$$

*Furthermore, these bounds are sharp.*

## 2.5 Signed k-Subdomination and Majority Domination

Signed $k$-subdomination, a generalization of signed domination in a different direction, was initially studied in [9]. We now begin to survey the main results on this topic.

For a positive integer $k$, a signed *k-subdominating function* of $G = (V, E)$ is defined in [9] as a function $f : V \to \{-1, 1\}$ such that $f[v] \geq 1$ for at least $k$ vertices $v$ of $G$. The signed *k-subdomination number* of $G$ is

$$\gamma_{ks}(G) = \min\{f(V) : f \text{ is a signed } k\text{-subdominating function on } G\}.$$

A signed $k$-subdominating function $f$ of $G = (V, E)$ is *minimal* if there does not exist a signed $k$-subdominating $h : V \to \{-1, 1\}$ such that $h \neq f$ and $h(v) \leq f(v)$ for every $v \in V$. The *upper k-subdomination number* of $G$ is defined as

$$\Gamma_{ks}(G) = \min\{f(V) : f \text{ is a minimal signed } k\text{-subdominating function on } G\}.$$

In the special cases where $k = |V|$, $k = \lceil |V|/2 \rceil$, and $\lceil (|V| + 1)/2 \rceil$, $\gamma_{ks}(G)$ is, respectively, the *signed domination number* $\gamma_s(G)$ defined in [16], *(weak) majority domination number* $\gamma_{maj}(G)$ defined in [5], and *strict majority domination number* $\gamma_{smaj}(G)$ defined in [35]. For $\Gamma_{ks}$, we have, respectively, the upper signed domination number $\Gamma_s$, upper (weak) majority number $\Gamma_{maj}$, and upper strict majority number $\Gamma_{smaj}$.

**Proposition 8.** *For every graph $G$, $\gamma_{ks}(G) \leq \gamma_s(G)$.*

Note that $\gamma_{ks}(K_n) = \gamma_s(K_n)$ for a complete graph $K_n$ of order $n$.

The *comet* $C_{s,t}$, where $s$ and $t$ are positive integers, denotes the tree obtained by identifying the center of the star $K_{1,s}$ with an end-vertex of $P_t$, the path of order $t$. So $C_{s,1} \cong K_{1,s}$ and $C_{1,p-1} \cong P_p$. Beineke and Henning [3] computed the value of $\gamma_{ks}(C_{s,t})$ for $k = s + t$ and for $k = \lceil (s + t)/2 \rceil + 1$. In [29] the value $\gamma_{ks}(C_{s,t})$ for all possible values of $k$ where $1 \leq k \leq s + t$ was provided.

Chang et al. [6] established a lower bound on $\gamma_{ks}$ in terms of the degree sequence.

**Theorem 29 (Chang et al. [6]).** *If $G = (V, E)$ is a graph of order $n$ with degree sequence $d_1 \leq d_2 \leq \ldots \leq d_n$, then*

$$\gamma_{ks}(G) \geq -n + \frac{2}{d_n + 1} \sum_{j=1}^{k} \left\lceil \frac{d_j + 2}{2} \right\rceil.$$

*Proof.* Let $g$ be an optimal $k$-subdominating function of $G$, say, $g(N_G[v]) \geq 1$ for $k$ distinct vertices $v$ in $\{v_{j_1}, v_{j_2}, \ldots, v_{j_k}\}$. Let $f(x) = (g(x) + 1)/2$ for all vertices $x \in V$. Then $f$ is a 0-1 valued function. First, note that

$$\sum_{i=1}^{k} f(N_G[v_{j_i}]) = \sum_{i=1}^{k} \left\lceil \frac{g(N_G[v_{j_i})+d(v_{j_i})+1}{2} \right\rceil \geq \sum_{i=1}^{k} \left\lceil \frac{d(v_{j_i})+2}{2} \right\rceil \geq \sum_{i=1}^{k} \left\lceil \frac{d_i+2}{2} \right\rceil.$$

On the other hand, we have

$$\sum_{i=1}^{k} f(N_G[v_{j_i}]) \leq \sum_{i=1}^{n} f(N_G[v_i]) = \sum_{i=1}^{n} (d_i + 1) f(v_i) \leq (d_n + 1) f(V).$$

Therefore, $f(V) \geq 1/(d_n + 1) \sum_{i=1}^{k} \lceil (d_i + 2)/2 \rceil$ and so

$$\gamma_{ks}(G) = g(V) = 2f(V) - n \geq -n + \frac{2}{d_n + 1} \sum_{j=1}^{k} \left\lceil \frac{d_j + 2}{2} \right\rceil.$$

□

**Corollary 1 (Chang et al. [6]).** *If $G = (V, E)$ is a graph of order $n$ and size $m$ with maximum degree $\Delta$, then*

$$\gamma_{ks}(G) \geq k - 2n + \frac{2m + n + k}{\Delta + 1}.$$

The lower bounds on $\gamma_{ks}$ and $\gamma_{maj}$ for a general graph were, respectively, obtained in Kang et al. [42] and Kang and Shan [43] in terms of its minimum degree, maximum degree, order, and size.

**Theorem 30 (Kang et al. [42]).** *If $G = (V, E)$ is a graph of order $n$ and size $m$ with minimum degree $\delta$ and maximum degree $\Delta$, then*

$$\gamma_{ks}(G) \geq n - \frac{2m + (n - k)(\Delta + 2)}{\delta + 1}.$$

**Theorem 31 (Kang and Shan [43]).** *If $G$ is a graph of order $n$ and size $m$ with minimum degree $\delta$ and maximum degree $\Delta$, then*

$$\gamma_{maj}(G) \geq \frac{n(2\delta - \Delta) - 4m}{2(\delta + 1)}.$$

For the graphs in which each vertex is odd-vertex, the lower bound on $\gamma_{ks}$ in Theorem 30 can be slightly improved [42]. In 2008, Chen and Song [7] improved these results in Theorems 29 and 30 again.

**Theorem 32 (Chen and Song [7]).** *Let $G = (V, E)$ be a graph of order $n$ and size $m$ with degree sequence $d_1 \leq d_2 \leq \ldots \leq d_n$ and let $f_k = \sum_{j=1}^{k} \lceil (d_j + 2)/2 \rceil$, then*

$$\gamma_{ks}(G) \geq \max\left\{\frac{2f_k}{\Delta+1} - n,\ n - \frac{2(n+2m) - 2f_k}{\delta+1}\right\},$$

*and these bounds are all sharp.*

Theorem 32 generalized some previous results on $\gamma_{ks}$, $\gamma_{maj}$, and $\gamma_s$ for regular graphs. By setting $d_1 = d_2 = \cdots = d_n = r$ in Theorem 32, we immediately have

**Corollary 2 (Hattingh et al. [30]).** *For every r-regular ($r \geq 2$) graph G of order n,*

$$\gamma_{ks}(G) \geq \begin{cases} -n + \dfrac{r+3}{r+1}k & \text{if } \Delta \text{ is odd} \\ -n + \dfrac{r+2}{r+1}k & \text{if } \Delta \text{ is even.} \end{cases}$$

In the special case where $k = \lceil |V|/2 \rceil$, the following result follows directly from Theorem 32.

**Corollary 3 (Henning [32]).** *For every r-regular ($r \geq 2$) graph G of order n,*

$$\gamma_{maj}(G) \geq \begin{cases} \dfrac{(1-r)}{2(r+1)}n & \text{if } \Delta \text{ is odd} \\ \dfrac{-r}{2(r+1)}n & \text{if } \Delta \text{ is even.} \end{cases}$$

Next we consider upper bounds on $\gamma_{ks}$ of a graph. Cockayne and Mynhardt [9] proved that for any tree $T$ of order $n$, $\gamma_{ks}(T) \leq 2(k+1) - n$. This bound is clearly sharp when $k \leq n/2$ as shown by the example $K_{1,n-1}$. They then proposed a conjecture: for a tree of order $n$ with $n/2 < k \leq n$, $\gamma_{ks}(T) \leq 2k - n$. The conjecture was completely settled by the following result, which was independently obtained by Kang et al. [40] and Chang et al. [6] by different techniques.

**Theorem 33 (Kang et al. [40] and Chang et al. [6]).** *If $T$ is a tree of order $n$ and $n/2 < k \leq n$, then $\gamma_{ks}(T) \leq 2k - n$, and this bound is sharp.*

Note that $\gamma_{ks}(K_{1,n-1}) = 2k - n$ if $k > n/2$, so the bound established in Theorem 33 is sharp.

**Problem 7.** For $n/2 < k \leq n$, characterize the trees of order $n$ with equality $\gamma_{ks}(T) = 2k - n$.

Cockayne and Mynhardt [9] also posed another conjecture: for any connected graph $G$ of order $n$ and any $k$ with $n/2 < k \leq n$, $\gamma_{ks}(G) \leq 2k - n$. Zelinka [80] disproved the conjecture by a simple counterexample $Q_3$ of the three-dimensional cube and $k = 5$, for which $\gamma_{ks}(Q_3) = 4$.

However, Kang et al. [40] gave the following approximate result.

**Theorem 34 (Kang et al. [40]).** *For any connected graph $G$ of order $n$ and any $k$ with $n/2 < k \leq n$,*

$$\gamma_{ks}(G) \leq 2\left\lceil \frac{k}{n-k+1} \right\rceil (n-k+1) - n.$$

By describing a polynomial transformation from known NP-complete decision problem, i.e., dominating set problem, Broere et al. [5] showed that the decision problem corresponding to the problem of computing the majority domination number $\gamma_{maj}$ is NP-complete. They also computed the (weak) majority domination number for various classes of graphs such as $K_n$, $K_{1,n}$, $K_{m,n}$, $P_n$, $C_n$, and $K_m \cup K_m$. Beineke and Henning [3] gave bounds on $\gamma_{maj}(T)$ for any tree $T$ and determined the exact value of $\gamma_{maj}$ for the class $C_{s,t}$ of trees called comets.

In 2001, Holm [37] further showed that the decision problem corresponding to computing $\gamma_{maj}$ of an arbitrary disjoint union of complete graphs is NP-complete and determined the values of $\gamma_{maj}$ for certain families of graphs, such as $\overline{K}+G$, the complete multipartite graph $K_{n,n,\ldots,n}$, and some special union graphs.

The following result was proved by Alon (see, [5]). It is the most elegant result on the weak majority domination number.

**Theorem 35 (Alon [5]).** *The (weak) majority domination number $\gamma_{maj}$ of a connected graph is at most 2.*

*Proof.* Let $G = (V, E)$ be a connected graph of order $n$. If $n = 2k + 1$ is odd, then we partition $V$ into two sets $V_1$ and $V_2$ with $|V_1| = k$ and $|V_2| = k + 1$ such that the sum of the number of edges induced by $V_1$ and $V_2$ is as large as possible. Then each vertex $v$ in $V_2$ is adjacent to at least as many vertices in $V_2$ as in $V_1$; otherwise, we may remove $v$ from the set $V_2$ and add it to the set $V_1$ to produce a new partition of $V$ in which the sum of the number of edges induced by the resulting partite sets exceeds that of the original partition of $V$. Hence assigning to each vertex of $V_2$ the value 1 and to each vertex of $V_1$ the value $-1$, we produce a majority dominating function of weight 1. Hence if $G$ has odd order, then $\gamma_{maj}(G) \leq 1$. On the other hand, if $n = 2k$ is even, then let $v$ be any vertex in $G$ and consider the graph $G - v$ of odd order. As above, we may produce a majority dominating function of $G - v$ of weight 1. We now extend this function to a majority dominating function of $G$ of weight 2 by assigning to the vertex $v$ the value 1. Hence if $G$ has even order, then $\gamma_{maj}(G) \leq 2$. □

For the strict majority domination number, Henning and Hind [35] obtained some interesting results. They determined the values of $\gamma_{smaj}$ for certain families of graphs such as $K_n$, $K_{1,n}$, and $K_{m,n}$. In particular, $\gamma_{smaj}(K_{n,n}) = 2$ for equitable complete bipartite graphs $K_{n,n}$. However, the situation is somewhat different for complete bipartite graphs $K_{m,n}$ which are not equitable.

**Proposition 9 (Henning and Hind [35]).** *For $n > m \geq 2$ integers,*

$$\gamma_{smaj}(K_{m,n}) = \begin{cases} 2-n & \textit{for } n \textit{ is even} \\ 3-n & \textit{for } n \textit{ is odd.} \end{cases}$$

For the weak majority number, the formula of Proposition 9 also holds when $n = m \geq 2$ (see Proposition 4 in [5]). This implies the following corollary, which serves to illustrate that the strict majority number and the weak majority number of a graph may differ by an arbitrarily large amount.

**Proposition 10 (Henning and Hind [35]).** *For $n > m \geq 2$ integers,*

$$\gamma_{smaj}(K_{n,n}) - \gamma_{maj}(K_{n,n}) = \begin{cases} n & \textit{if } n \textit{ is even} \\ n-1 & \textit{if } n \textit{ is odd.} \end{cases}$$

By somewhat improving Alon's method described in Theorem 35, Henning and Hind [35] showed that every graph with odd order has the strict majority domination number $\gamma_{smaj}$ at most 1.

**Theorem 36 (Henning and Hind [35]).** *For a graph $G$ of odd order, $\gamma_{smaj}(G) \leq 1$.*

Suppose that $G = (V, E)$ has even order $2k$ and $V$ is partitioned into two sets, $W_1$ and $W_2$, each of cardinality $k$ (such a partition is an *equipartition* of $G$). Given such a partition, we define a function to measure the excess of a vertex's neighbors in its own set over the opposite set; for $v \in W_i$, define $\delta(v) = |N(v) \cap W_i| - |N(v) \cap W_{3-i}|$. An equipartition is *strong* if $\delta(v) > 0$ for all $v \in V$.

**Theorem 37 (Henning and Hind [35]).** *If $G$ is a graph, then $\gamma_{smaj}(G) \leq 4$.*

*Proof.* By Theorem 36, we may assume that $G$ has even order $2k$. If $G$ has no strong equipartition, then we claim that $\gamma_{smaj}(G) \leq 2$. Let $\{W_1, W_2\}$ be an equipartition for $G$ such that the total number of edges induced within these two sets is maximized. By this choice, $\delta(u) + \delta(v) \geq 0$ whenever $u \in W_1$ and $v \in W_2$ (otherwise we move $u$ to $W_2$ and $v$ to $W_1$ to improve the partition). Since $\{W_1, W_2\}$ is not a strong equipartition, we have $\delta(u) \leq 0$ for some vertex $u$ , say $u \in W_i$ . Thus we have $\delta(v) \geq 0$ for all $v \in W_{3-i}$. Giving a positive opinion $+1$ to $u$ and to all of $W_{3-i}$ now yields a strict majority function of weight 2. We next assume that $G$ has a strong equipartition $\{W_1, W_2\}$.

Over all vertices of $G$, choose $w$ to minimize $\delta(w)$. Without loss of the generality, we may assume that $w \in W_2$. Let $l = \lfloor(\delta(w) - 1)/2\rfloor$. Note that $l \geq 0$. Since $\delta(w) \geq 2l + 1$ and $|W_1| = |W_2|$, there exists a set $S$ of $l$ vertices in $W_1$ that are not adjacent to $w$. Let $T$ be a set of $l+1$ neighbors of $w$ in $W_2$. Assign positive opinions $+1$ to $(W_1 - S) \cup (T \cup \{w\})$ and negative opinions $-1$ to $S \cup (W_2 - T - \{w\})$. This opinion function, say $f$, has weight 4. We claim that it is a strict majority function, with all of $W_1 \cup \{w\}$ voting aye, i.e., $f[x] \geq 1$ for all $x \in W_1 \cup \{w\}$. Since every vertex of $W_1$ has at least $2l + 1$ more neighbors in $W_1$ than in $W_2$, $f[x] \geq l$ for all $x \in W_1$. Because $\delta(w) \leq 2l + 2$, the switch of $l + 1$ opinions from negative to positive in $N[w] \cap W_2$ yields $f[w] \geq 1$. □

**Theorem 38 (Henning and Hind [35]).** *If $T$ is a tree, then $\gamma_{smaj}(T) \leq 2$.*

**Problem 8 (Broere et al. [5]).** Characterize those graphs $G$ for which $\gamma_s(G) = \gamma_{maj}(G)$.

**Problem 9 (Holm [37]).** What is the relationship between the majority domination number of a graph and that of its complement?

The concept of a signed total $k$-subdominating function in graphs was initially developed by Harris et al. [24]. A *signed total k-subdominating function* of a graph $G = (V, E)$ is a function $f : V \to \{-1, 1\}$ for which at least $k$ vertices $v$ of $G$ satisfy $f(N(v)) \geq 1$. The *signed total k-subdomination number* of G, denoted by $\gamma_{stk}(G)$, is the minimum weight of a signed total $k$-subdominating function of $G$. In the special case where $k = |V|$, the signed total $k$-subdomination number is the signed total domination number $\gamma_{st}$ which is discussed in Section 2.4. Similarly, we can define the *upper signed total k-subdomination number* $\Gamma_{stk}(G)$, the *signed majority total domination number* $\gamma^t_{maj}$ [74, 75], and the *strict signed majority total domination number* $\gamma^t_{smaj}$. Let $C_f(G) = \{v \in V : f(N(v)) \geq 1 :\}$. A signed total $k$-subdominating function of $G$ of weight $\gamma_{stk}(G)$ is called a $\gamma_{stk}(G)$-function. By the definition, we easily obtain the following proposition.

**Proposition 11.** *Let $f$ be a signed total k-subdominating function of G and $v \in C_f(G)$. If $d_G(v)$ is even, then $f(N(v)) \geq 2$ while if $d_G(v)$ is odd, then $f(N(v)) \geq 1$.*

Harris et al. [24] determined the values of $\gamma_{stk}$ for a cycle $C_n$ and a path $P_n$. They also established some lower bounds on $\gamma_{stk}$ of graphs and an upper bound on $\gamma_{stk}$ for trees.

**Theorem 39 (Harris et al. [24]).** *Let G be a graph of order n with minimum degree $\delta$ and maximum degree $\Delta$, $f$ a $\gamma_{stk}(G)$-function, and $l$ the number of vertices with even degree in $C_f(G)$. Then*

$$\gamma_{stk}(G) \geq \frac{2k(1+\Delta) + (\delta - 3\Delta)n + 2l}{\Delta + \delta}.$$

The next result gave a lower bound on the signed total $k$-subdominating function of a graph in terms of its degree sequence.

**Theorem 40 (Harris et al. [24]).** *Let G be a graph of order n with degrees $d_1 \leq d_2 \leq \ldots \leq d_n$, and let $f$ be a $\gamma_{stk}(G)$-function and $l$ the number of vertices with even degree in $C_f(G)$. Then*

$$\gamma_{stk}(G) \geq \frac{l + k + \sum_{i=1}^{k} d_i}{d_n} - n.$$

As an immediate consequence of Theorem 39 or Theorem 40, we have the following results.

**Corollary 4 (Harris et al. [24]).** *If $G$ is an $r$-regular graph of order $n$ ($r \geq 1$), then*

$$\gamma_{stk}(G) \geq \begin{cases} \left(\dfrac{r+1}{r}\right)k - n & \text{if } r \text{ is odd} \\ \left(\dfrac{r+2}{r}\right)k - n & \text{if } r \text{ is even.} \end{cases}$$

**Corollary 5 (Harris et al. [24]).** *If $G$ is a graph of order $n$ with maximum degree $\Delta$, then*

$$\gamma_{stk}(G) \geq k - 2n + \frac{k + 2m}{\Delta}.$$

Harris et al. [24] observed that $\gamma_{stk}(K_{1,n-1}) = 2 - n$ for $n \geq 3$, $1 \leq k < n$. This implies that the signed total $k$-subdomination number of a tree can be arbitrarily large negative if $k$ is less than the order of the tree.

Next they gave upper bounds on signed total $k$-subdomination number of a tree.

**Theorem 41 (Harris et al. [24]).** *For a tree $T$ of order $n \geq 2$,*

$$\gamma_{stk}(T) \leq \begin{cases} -1 & \text{if } k = (n+1)/2 \\ 2k - n & \text{otherwise,} \end{cases}$$

*and these bounds are sharp.*

When $T$ is nontrivial tree of even order $n$, and $k = n/2 + i$ for some integer $i$ with $0 \leq i \leq 3$, Theorem 41 can be improved slightly.

**Theorem 42 (Harris et al. [24]).** *For a tree $T$ of order $n \geq 2$, and any integer $k$ with $n/2 \leq k \leq n$ and $k = n/2 + i$, where $0 \leq i \leq 3$,*

$$\gamma_{stk}(T) \leq 2(k-1) - n,$$

*unless $T$ is a path, in which case $\gamma_{stk}(T) = 2k - n$.*

Trees of even order $n$ achieving the maximum possible total $k$-subdomination number (namely, $2k - n$) when $n/2 \leq k \leq n/2 + 3$ were characterized in [24]. In 2018, they [25] further characterized those trees of order $n$ achieving the maximum possible total $k$-subdomination number (namely, $2k - n$) when $n$ is odd and $(n + 3)/2 \leq k \leq (n + 5)/2$.

Xing et al. [74, 75] computed the exact value of $\gamma_{maj}^t$ for some special graphs, such as path $P_n$, cycle $C_n$, and complete bipartite $K_{m,n}$ and gave a lower bound on $\gamma_{maj}^t$ for a graph.

**Theorem 43 (Xing et al. [74]).** *If $G$ is a graph of order $n$ with minimum degree $\delta$ and maximum degree $\Delta$, then*

$$\gamma^t_{maj}(G) \geq \frac{\delta - 2\Delta + 1}{\delta + \Delta} n.$$

Motivated by the results in Theorems 37 and 38, we propose the following problem.

**Problem 10.** Determine the sharp upper bounds on $\gamma^t_{maj}$ and $\gamma^t_{smaj}$ for trees and graphs, respectively.

As far as we know there is only little literature available on the upper signed $k$-subdomination of graphs. In 2017, Muthuselvi and Arumugam [55] determined the upper signed $k$-subdomination number $\Gamma_{maj}$ of special graphs $K_n$, $K_{1,n-1}$, and $K_{m,n}$.

## 3 Minus Domination in Graphs

The minus domination problem is a proper generalization of the classical domination problem in a sense. A "sociological" motivation has been suggested by [15]. Let our graph be the model of a network of people. An edge means that the joined vertices are somehow closely related (acquaintances, neighbors, or the like). The label $-1, 0, +1$ indicates the opinion of every "vertex" about some controversial question (i.e., negative, undecided, or positive, respectively). Then it may be interesting to know for a given network whether, e.g., the negative opinions can abound, although every "vertex" observes a majority of positive opinions in his own neighborhood. Other motivations may come from facility location problems. In this section we will survey main results on minus domination, minus total domination, and minus $k$-subdomination of a graph.

### *3.1 Minus Domination in Graphs*

A *minus dominating function* of a graph $G = (V, E)$ is a function of the form $f : V \to \{-1, 0, 1\}$ such that $f[v] \geq 1$ for all $v \in V$. The *minus domination number* for a graph $G$ is $\gamma^-(G) = \min\{w(f) : f$ is a minus dominating function on $G\}$. Likewise, the *upper minus domination number* for a graph $G$ is $\Gamma^-(G) = \max\{w(f) : f$ is a minimal minus dominating function on $G\}$. Minus domination is similar in many ways to ordinary domination, but has different properties. In [15] various properties of the minus domination number are presented.

The problem to determine the value of $\gamma^-(G)$ is NP-complete, even when restricted to bipartite graphs, chordal graphs, planar graphs with maximum degree four [10, 13], chordal bipartite graphs, and bipartite planar graphs [47]. Damaschke proved that, unless $P = NP$, the value of $\gamma^-$ cannot be approximated in polynomial time within a factor $1+\varepsilon$ for some $\varepsilon > 0$, not even for graphs with maximum degree four [10]. Faria et al. [18] further showed that the minus domination problem is NP-complete for split graphs and they showed that there are polynomial-time algorithms for solving the minus domination problem in the classes of graphs including cographs, distance-hereditary graphs, and strongly chordal graphs. Moreover, there are linear-time algorithms for solving the minus domination problem on trees [13] and interval graphs [18].

By definition, we see that every signed dominating function is also a minus dominating function, so $\gamma^-(G) \leq \gamma_s(G)$. Let $D$ be a minimum dominating set in a graph $G = (V, E)$. Assigning to each vertex of $D$ the value 1 and to all other vertices the value 0, we produce a special minus dominating function of $G$ and so $\gamma^-(G) \leq \gamma(G)$. Does there exist a cubic graph for which $\gamma^- < \gamma$? Henning and Slater [36] constructed a cubic graph $G$ of order 52 satisfying $\gamma^-(G) \leq 14$ and $\gamma(G) = 15$. Chen et al. [8] constructed an infinite family of cubic graphs of order $n$ in which the difference $\gamma - \gamma^-$ can be arbitrarily large when $n \to \infty$. Let $G$ be a graph of order $n$. It is well-known that $\gamma(G) \leq n/2$. Reed [59] proved that $\gamma(G) \leq 3n/8$ if $\delta(G) \geq 3$ and conjectured that $\gamma(G) \leq \lceil n/3 \rceil$ if $G$ is cubic. For the difference $\gamma(G) - \gamma^-(G)$, it was known in [15] that $\gamma(G) - \gamma^-(G) \leq (n-4)/5$ if $G$ is a tree and the upper bound is sharp. Dunbar et al. [14] and Zelinka [78] independently showed that $\gamma^-(G) \geq n/4$ for a cubic graph of order $n$. This means that if $G$ is cubic, then by Reed's result, we have $\gamma(G) - \gamma^-(G) \leq n/8$. Furthermore, if Reed's conjecture is true, then $\gamma(G) - \gamma^-(G) \leq n/12$. Finally, they proposed the following problem.

**Problem 11 (Chen et al. [8]).** For a cubic graph $G$ of order $n$, what is the best possible upper bound for $\gamma(G) - \gamma^-(G)$?

Dunbar et al. [15] showed that $\alpha(G) \leq \Gamma^-(G)$ for every graph $G$, where $\alpha(G)$ is the independence number of $G$. Hence we have the following chain of inequalities.

**Proposition 12 (Henning and Slater [36]).** *For any graph G, we have*

$$\gamma^-(G) \leq \gamma(G) \leq i(G) \leq \alpha(G) \leq \Gamma^-(G).$$

In this special case when $G$ is a cubic graph, Henning and Slater obtained the following chain of inequalities.

**Proposition 13 (Henning and Slater [36]).** *For every cubic graph G, we have*

$$\gamma^-(G) \leq \gamma(G) \leq i(G) \leq \alpha(G) \leq \Gamma(G) \leq IR(G) \leq \gamma_s(G) \leq \Gamma_s(G),$$

*where $i(G)$ and $IR(G)$ are, respectively, the independent domination number and upper irredundance number of G.*

Up to now, some basic relationships among the above parameters are not very clear for general graphs, even for cubic graphs. For example, we do not know yet whether there exists a cubic $G$ such that $\Gamma^-(G) < \Gamma(G)$. Essentially no progress has been made on the following problem.

**Problem 12 (Henning and Slater [36]).** For a cubic graph $G$ of order $n$, how large is the ratio $(\Gamma(G) - \Gamma^-(G))/n$ or $(\Gamma^-(G) - \Gamma(G))/n$?

Kang and Cai [38] constructed an infinite family of 3-connected cubic graphs such that the difference $\Gamma^- - \gamma_s$ can be arbitrarily large, which disproved an open problem posed by Henning et al. [36].

We shall focus on lower bounds for the minus domination number of a graph. It is easily seen that a minus dominating function is minimal if and only if for every vertex $v \in V$ with $f(v) \geq 0$, there exists a vertex $u \in N[v]$ with $f[u] = 1$. This observation plays an important role in the aspect of establishing bounds on $\gamma^-$ or $\Gamma^-$.

**Theorem 44 (Damaschke [10]).** *Let $G$ be a graph with maximum degree $\Delta$. If $\Delta \leq 3$, then $\gamma^- \geq n/5$, while if $\Delta \leq 4$ and $d_2 = 0$, where $d_2$ is the number of 2-degree vertices in $G$, then $\gamma^- \geq n/7$, and these bounds are tight.*

Zelinka [81] showed that for complete bipartite graphs $K_{p,q}$ $(q \leq p)$, $\gamma^-(K_{p,q}) = 1$ if $q = 1$ and $\gamma^-(K_{p,q}) = 2$ otherwise. In [78] he established a sharp lower bound on $\gamma^-(G)$ for a cubic graph $G$. Dunbar et al. [14] generalized this result to $k$-regular graphs. They showed that $\gamma^-(G) \geq n/(k+1)$ for a $k$-regular graph $G$ of order $n$. Kang and Shan [43] further generalized this result to a general graph.

**Theorem 45 (Kang and Shan [43]).** *If $G$ is a graph of order $n$ and size $m$ with minimum degree $\delta$, then $\gamma^-(G) \geq n - 2m/(\delta + 1)$.*

**Theorem 46 (Kang and Shan [43]).** *For every tree $T$ of order $n$, $\gamma^-(T) \geq (n + 2 - s)/3$, where $s$ is the number of vertices of degree 1, and this bound is sharp.*

As observed by Füredi and Mubayi [21] a well-known probabilistic bound for the size of a transversal of a set system implied that $\gamma^-(G) = O((n/\delta)\log\delta)$ for a graph $G = (V, E)$ of order $n$ with minimum degree $\delta$. Indeed, consider the set system with $V$ as the ground set and with the $n$ closed neighborhoods $N[v]$, $v \in V$, as sets. By a simple and well-known probabilistic argument, it is possible to pick a transversal of size $O((n/r)\log r)$ for such a set system (see, e.g., [1] for arguments of this type). The vertices of the transversal are assigned $+1$'s and the other vertices get 0's, which defines a minus dominating function of $G$. Füredi and Mubayi [21] conjectured that this bound is asymptotically the best possible. Matoušek [53] partially confirmed their conjecture.

**Theorem 47 (Matoušek [53]).** *There are constants $C$ and $c > 0$ such that for all integers $r \geq C$ and for all $n$'s that are multiples of $4r$, there exists a bipartite $r$-regular multigraph $G$ of order $n$, in which each vertex has at least $r/2$ distinct neighbors and such that $\gamma^-(G) \geq c(n/r)\log r$.*

In 1999, Dunbar et al. [15] conjectured that $\gamma^-(G) \geq 4(\sqrt{n+1}-1)-n$ for a bipartite graph $G$ order $n$. Later, this conjecture was proved by several separate subsets of the authors [42, 51, 71, 76]. Kang et al. [41] further extended the result to $k$-partite graphs for $k \geq 2$.

**Theorem 48 (Kang et al. [41]).** *If $G=(V,E)$ is a k-partite graph of order n with $k \geq 2$, then*

$$\gamma^-(G) \geq \frac{2k}{k-1}\left(-1+\sqrt{1+\frac{2(k-1)}{k}n}\right)-n$$

*and this bound is sharp.*

Kang et al. also obtained another sharp lower bound on $\gamma^-$ for bipartite graphs.

**Theorem 49 (Kang et al. [41]).** *Let $G(X,Y)$ be a bipartite graph of order n and size m and let $\delta_X = min\{d(v) \mid v \in X\}$ and $\delta_Y = min\{d(v) \mid v \in Y\}$, then*

$$\gamma^-(G) \geq \lceil n-(m/\delta+m/(1+max(\delta_X,\delta_Y)))\rceil,$$

*and this bound is sharp.*

In 2008, by exploiting Turán's theorem, Shan et al. [62] continue to extend the result in Theorem 48 to $K_{k+1}$-free graphs and characterize the extremal graphs.

**Theorem 50 (Shan et al. [62]).** *For any integer $k \geq 2$, let $G=(V,E)$ be a $K_{k+1}$-free graph of order n, then*

$$\gamma^-(G) \geq \frac{2k}{k-1}\left(-1+\sqrt{1+\frac{2(k-1)}{k}n}\right)-n$$

*with equality if and only if $G \in \mathscr{H}(k,s)$, where $\mathscr{H}(k,s)$ is defined as in Theorem 7.*

*Proof.* Let $f : V \to \{+1, 0, -1\}$ be a minus dominating function on $G$ with $f(V(G)) = \gamma^-(G)$ and let $Q$ be the set of vertices in $V$ that are assigned the value 0. Further, let $G' = G - Q$ and $n' = |V(G')|$. Then $G'$ is still a graph without $(k+1)$-cliques and $n' \leq n$. Clearly, $f' = f|_{G'}$ is a signed dominating function on $G'$, so $\gamma_s(G') \leq f'(V(G')) = f(V(G))$. By Theorem 7, we have

$$\gamma^-(G) \geq \gamma_s(G') \geq \frac{2k}{k-1}\left(-1+\sqrt{1+\frac{2(k-1)}{k}n'}\right)-n'.$$

We define

$$h(x) = \frac{2k}{k-1}\left(-1+\sqrt{1+\frac{2(k-1)}{k}x}\right)-x.$$

It is easy to check that $h'(x) \leq 0$ when $x \geq 3$, so $h(x)$ is a strictly monotone decreasing function on the variable $x \geq 3$. This implies that

$$\gamma^-(G) \geq \gamma_s(G') \geq \frac{2k}{k-1}\left(-1+\sqrt{1+\frac{2(k-1)}{k}n}\right)-n$$

if $n' \geq 3$. If $n' \leq 2$, then each vertex in $G'$ is assigned the value $+1$, so no vertices in $G$ are assigned the value $-1$, and thus $\gamma^-(G) = f(V(G')) \geq 1$. Note that $1 \geq 2k/(k-1)\left(-1+\sqrt{1+2(k-1)n/k}\right) - n$ for $k \geq 2$, so the results follow.

We further characterize the extremal graphs attaining this bound. If the equality holds, i.e.,

$$\gamma^-(G) = 2k/(k-1)\left(-1+\sqrt{1+2(k-1)n/k}\right) - n,$$

then $h(n') = h(n)$. Observing the fact that $h(x)$ is a strictly monotone function on variable $x$ when $x \geq 3$, we have $n' = n$. Hence $Q = \emptyset$. Thus $f$ is also a minimum signed dominating function, i.e., $\gamma_s(G) = 2k/(k-1)\left(-1+\sqrt{1+2(k-1)n/k}\right) - n$. The characterization follows from Theorem 7. □

As an immediate consequence of Theorems 7 and 50, we obtain the following extremal result on the minus domination and signed domination of a graph containing no $(k+1)$-cliques.

**Theorem 51 (Shan et al. [62]).** *For any integer $k \geq 2$, let $G = (V, E)$ be a graph of order $n$ with no $(k+1)$-cliques, then the following statements are equivalent:*

(1) $\gamma_s(G) = \dfrac{2k}{k-1}\left(-1+\sqrt{1+\dfrac{2(k-1)}{k}n}\right) - n.$

(2) $\gamma^-(G) = \dfrac{2k}{k-1}\left(-1+\sqrt{1+\dfrac{2(k-1)}{k}n}\right) - n.$

(3) *$G \in \mathscr{H}(k, s)$, where $\mathscr{H}(k, s)$ is defined as in Theorem 7.*

Minus dominating functions can also be extended to minus $k$-dominating functions. For any integer $k \geq 1$, a *minus $k$-dominating function* of a graph $G = (V, E)$ is a function $f : V \to \{-1, 0, 1\}$ satisfying $f[v] \geq k$ for every $v \in V$. The minimum of the values of $f(V)$, taken over all minus $k$-dominating functions $f$, is the *minus $k$-domination number* of $G$, denoted by $\gamma_k^-(G)$. In 2016, Dehgardi [11] presents lower bounds on the minus $k$-domination number for regular graphs and $t$-partite graphs. For a general graph $G$, the author obtained the following lower bound.

**Theorem 52 (Dehgardi [11]).** *If $G$ is a connected graph of order $n$ and minimum degree $\delta \geq k-1$, then $\gamma_k^-(G) \geq \sqrt{1+4(k+1)n} - n - 1$.*

Whether these results in Theorems 48-51 are true for the minus $k$-domination number is an open question.

We now turn to upper bounds on the minus domination number of a graph. So far, no nontrivial upper bound on $\gamma^-$ for an arbitrary graph is known. In [36] Henning and Slater asked for the upper bounds on $\Gamma^-$ for a cubic graph. Kang and Cai [38] answered this problem by giving a sharp upper bound of $\Gamma^-(G) \leq 5n/8$ for cubic graphs $G$ of order $n$. In 2000, Kang and Cai [39] further generalized this result to $k$-regular graphs.

**Theorem 53 (Kang and Cai [39]).** *If $G$ is a $k$-regular graph of order $n$, $k \geq 1$, then*

$$\Gamma^-(G) \leq \begin{cases} \dfrac{k^2+1}{(k+1)^2}n & \text{for } k \text{ odd,} \\ \dfrac{k+1}{k+3}n & \text{for } k \text{ even} \end{cases}$$

*and these bounds are sharp.*

So far, we only obtain the bounds on $\gamma^-(G)$ and $\Gamma^-(G)$ for some special classes of graphs. It should be possible to find sharp bounds on the two parameters of general graphs in terms of maximum degree and minimum degree.

**Problem 13.** Find sharp lower bounds on $\gamma^-(G)$ and sharp upper bounds on $\Gamma^-(G)$ for general graphs $G$.

In general, we know that $\Gamma^-$ and $\Gamma_s$ are not comparable. However, when every vertex in a graph has even degree, every minimal signed dominating function is also a minimal minus dominating function, we have the following:

**Theorem 54 (Kang and Cai [39]).** *For every Eulerian graph $G$, $\Gamma_s(G) \leq \Gamma^-(G)$.*

Meanwhile, they also proposed the following conjecture.

*Conjecture 2 (Kang and Cai [39]).* For any cubic graph $G$, $\Gamma^-(G) \leq \Gamma_s(G)$.

In 2006, Shang and Yuan [64] partly settled this conjecture.

**Theorem 55 (Shang and Yuan [64]).** *For any claw-free cubic graph $G$ on $n$ vertices, $\Gamma^-(G) \leq n/2$.*

We recall that for every $k$-regular graph $G$ of order $n$, $\gamma_s(G) \geq 2n/(k+1)$ if $k$ is odd. So $n/2 \leq \gamma_s(G) \leq \Gamma_s(G)$ for a cubic graph $G$ of order $n$. As an immediate consequence of Theorem 55, the following result follows.

**Theorem 56 (Shang and Yuan [64]).** *For claw-free cubic graphs $G$, $\Gamma^-(G) \leq \Gamma_s(G)$.*

**Problem 14.** Let $k$ is an odd positive integer $k \geq 5$. For $k$-regular graphs or claw-free $k$-regular graphs $G$, are $\Gamma^-(G)$ and $\Gamma_s(G)$ comparable? If so, how do they compare?

Dunbar et al. [15] observed that there exist an outerplanar graph, a chordal graph, and a bipartite graph $G$ such that $\gamma^-(G) \leq -k$. They further posed the following question: For every positive integer $k$ and positive integer $m$, does there exist a graph $G$ with girth $m$ and $\gamma^-(G) \geq -k$? In 1999, Lee et al. [48] gave a positive answer to this open problem.

The Cartesian product of graphs is the most basic of graph products. As mentioned before, the study of signed domination of Cartesian product graphs was initiated by Haas and Wexler [22].

For signed and minus domination of graph products, we propose the following open problems.

**Problem 15.** What is the relationship between $\gamma_s(G\Box H)$ (resp. $\gamma^-(G\Box H)$) and each of $\min\{\gamma_s(G), \gamma_s(H)\}$ (resp. $\min\{\gamma^-(G), \gamma^-(H)\}$), $\max\{\gamma_s(G), \gamma_s(H)\}$ (resp. $\max\{\gamma^-(G), \gamma^-(H)\}$) and $\gamma_s(G) \times \gamma_s(H)$ (resp. $\gamma^-(G) \times \gamma^-(H)$)?

**Problem 16.** For a grid graph $P_r \Box P_s$ $(r, s \geq 1)$, what are sharp bounds on $\gamma^-(P_r \Box P_s)$? Further, determine the exact value of $\gamma^-(P_r \Box P_s)$.

**Problem 17.** For a $n$-cube $Q_n$, what is the precise value of $\gamma_s(Q_n)$ (resp. $\gamma^-(Q_n)$)?

## 3.2 *Minus Total Domination in Graphs*

Harris and Hattingh [23] developed an analogous theory for minus total domination that arises when we simply change "closed" neighborhood in the definition of minus domination to "open" neighborhood. The minus total domination number $\gamma_t^-$ and upper minus total domination number $\Gamma_t^-$ can be defined similarly. They showed that the decision problem for the minus total domination number of a graph is NP-complete, even when restricted to bipartite graphs or chordal graphs. However, there exists a linear-time algorithm for computing $\gamma_t^-(T)$ of a tree $T$.

It is easy to see that a minus total dominating function $f$ on a graph $G = (V, E)$ is minimal if and only if for every vertex $v \in V$ with $f(v) \geq 0$, there exists a vertex $u \in N(v)$ with $f(N(u)) = 1$.

In [46] the authors obtained a sharp lower bound on the minus total domination number of $k$-partite graphs.

**Theorem 57 (Kang et al. [46]).** *If $G = (V, E)$ is a $k$-partite graph of order $n$ with no isolated vertex, then*

$$\gamma_t^-(G) \geq 2\sqrt{\frac{k}{k-1}n} - n,$$

*and this bound is sharp.*

Note that $\gamma_t^-(G) \le \gamma_{st}(G)$ for any graph $G$. Hence

$$\gamma_{st}(G) \ge 2\sqrt{\frac{k}{k-1}n} - n$$

for a $k$-partite graph of order $n$ with no isolated vertex. Shan and Cheng [61] characterized the extremal graphs with equality by reproving the inequality.

**Theorem 58 (Shan and Cheng [61]).** *If $G = (V, E)$ is a $k$-partite graph of order $n$ with no isolated vertex, then the following statements are equivalent:*

(1) $\gamma_{st}(G) = 2\sqrt{\frac{k}{k-1}n} - n$.
(2) $\gamma_t^-(G) = 2\sqrt{\frac{k}{k-1}n} - n$.
(3) $G \in \mathscr{T}$*, where* $\mathscr{T}$ *is defined as in Theorem 24.*

In [61] the authors also obtained two other results involving in lower bounds for the minus total domination number $\gamma_t^-$.

**Theorem 59 (Shan and Cheng [61]).** *Let $G = (V, E)$ be a $k$-partite graph of order $n$ with $\delta(G) \ge 1$ and let $c = \lceil (\delta(G)+1)/2 \rceil$. Then*

$$\gamma_{st}(G) \ge \frac{k}{k-1}\left(-(c-1) + \sqrt{(c-1)^2 + 4\frac{k-1}{k}cn}\right) - n$$

*and this bound is sharp.*

The class of graphs $\mathscr{F} = \{G_{2,r} \mid r \ge 1\}$ is a subclass of $\mathscr{T}$. Clearly, each $G_{2,r}$ of $\mathscr{F}$ is a bipartite graph of order $n = 2r^2$ with vertex classes $X_1, X_2$ and $|X_i| = r^2$.

**Theorem 60 (Shan and Cheng [61]).** *If $G$ is a triangle-free graph of order $n$ with $\delta(G) \ge 1$, then*

$$\gamma_t^-(G) \ge 2\sqrt{2n} - n,$$

*where equality holds if and only if $G \in \mathscr{F}$.*

Combining the observation that $\gamma_s^t(G) \ge \gamma_t^-(G)$ for a graph $G$ with Theorem 61, we have the following equivalent formulations.

**Theorem 61 (Shan and Cheng [61]).** *If $G$ is a triangle-free graph of order $n$ with $\delta(G) \ge 1$, then the following statements are equivalent:*

(1) $\gamma_{st}(G) = 2\sqrt{2n} - n$.
(2) $\gamma_t^-(G) = 2\sqrt{2n} - n$.
(3) $G \in \mathscr{F}$.

*Remark 1.* By using Turán's theorem and a similar approach as described in the proof of Theorem 7, the above results can be extended to $K_{k+1}$-free graphs.

In [77] the authors established upper bounds on $\Gamma_t^-$ for small-degree regular graphs and characterized the extremal graphs by defining a family of cubic graphs $\mathscr{F}_1 = \{G_{k,l} | \; k \geq 1, l \geq 0\}$ and a family of 4-regular graphs $\mathscr{F}_2 = \{H_{k,l} \mid k \geq 1, 0 \leq l \leq k\}$.

The family of cubic graphs $\mathscr{F}_1 = \{G_{k,l} | \; k \geq 1, l \geq 0\}$ is defined as follows. For two integers $k \geq 1$, $l \geq 0$, let $G_{k,l}$ be a cubic graph with vertex set $\cup_{i=1}^{5} A_i$ with $|A_i| = a_i, i = 1, 2, \ldots, 5$ where all $a_i$s are integers satisfying $a_1 = 2k, a_2 = 2l, a_3 = 3a_1 = 6k, a_4 = 2a_2 = 4l$, and $a_5 = a_3 + 2a_4 = 6k + 8l$, and $A_1$ and $A_4$ are two independent sets. The edge set of $G_{k,l}$ is constructed as follows:

Add $3a_1$ edges between $A_1$ and $A_3$ so that each vertex in $A_1$ has degree 3, while each vertex in $A_3$ has degree 1. Add $3k$ edges joining vertices of $A_3$ so that $A_3$ induces a 1-regular graph. Add $l$ edges joining vertices of $A_2$ so that $A_2$ also induces a 1-regular graph. Add $2a_2$ edges between $A_2$ and $A_4$ so that each vertex in $A_2$ has degree 3, while each vertex in $A_4$ has degree 1. Add $a_5(= a_3 + 2a_4)$ edges between $A_3 \cup A_4$ and $A_5$ in such a way that each vertex of $A_5$ is adjacent to precisely a vertex of $A_3 \cup A_4$, and each vertex in $A_3$ is adjacent to precisely one vertex of $A_5$, while each vertex of $A_4$ is adjacent to precisely two vertices of $A_5$. Finally, add $a_5$ edges joining vertices of $A_5$ so that $A_5$ induces a 2-regular graph. By our construction, $G_{k,l}$ is a cubic graph of order $n = 14(k + l)$. The graph $G_{1,1}$ is exhibited in Figure 5.

**Theorem 62 (Yan et al. [77]).** *If $G$ is a cubic graph of order $n$, then*

$$\Gamma_t^-(G) \leq \frac{5}{7}n$$

*with equality if and only if $G \in \mathscr{F}_1$.*

Let $G_{k,l} \in \mathscr{F}_1$ with $k \geq 1, l \geq 0$. Clearly, the function $f$ that assigns $-1$ to each vertex of $A_1 \cup A_2$ and $+1$ to all other vertices is a minimal minus total dominating function on $G_{k,l}$ with $w(f) = 10(k + l) = 5n/7$. Therefore, $\Gamma_t^-(G_{k,l}) = 5n/7$.

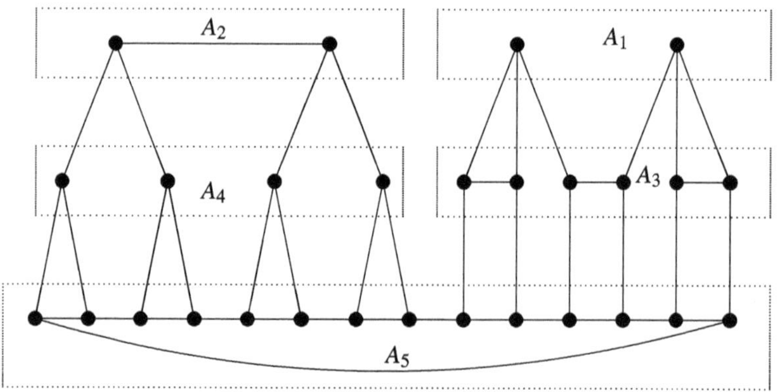


**Fig. 5** The graph $G_{1,1}$.

A family $\mathscr{F}_2$ of 4-regular graphs is defined as follows. For two integers $k \geq 1$, $0 \leq l \leq k$, let $H_{k,l}$ be a graph with vertex set $\cup_{i=1}^{8} A_i$ with $|A_i| = a_i$, for $1 \leq i \leq 8$, where all $a_i$s are integers satisfying $a_1 = a_3 = 2l$, $a_2 = a_4 = k$, $a_5 = a_6 = 4l$, $a_7 = 4(k-l)$, and $a_8 = 4(k+3l)$, and where $A_2$, $A_4$, $A_5$, and $A_6$ are independent sets. The edge set of $H_{k,l}$ is constructed as follows:

Add $l$ edges joining vertices of $A_1$ (resp. $A_3$) so that $A_1$ (resp. $A_3$) induces a 1-regular graph. Add $2(k-l)$ edges joining vertices of $A_7$ so that $A_7$ induces a 1-regular graph also. Add $6(k+3l)$ edges joining vertices of $A_8$ so that $A_8$ induces a 3-regular graph. Add $2l$ edges between $A_1$ and $A_3$ so that each vertex in $A_1$ is adjacent to precisely one vertex of $A_3$ and each vertex in $A_3$ is also adjacent to precisely one vertex of $A_1$, so each vertex of $A_1 \cup A_3$ has degree 2. Add $4l$ edges between $A_1$ (resp. $A_3$) and $A_5$ (resp. $A_6$) so that each vertex in $A_1$ (resp. $A_3$) is adjacent to two vertices of $A_5$ (resp. $A_6$) and each vertex of $A_5$ (resp. $A_6$) is adjacent to one vertex of $A_1$ (resp. $A_3$), so each vertex in $A_1 \cup A_3$ has degree 4, while each vertex in $A_5 \cup A_6$ has degree 1. Add $4k$ edges between $A_2$ (resp. $A_4$) and $A_6 \cup A_7$ (resp. $A_5 \cup A_7$) so that each vertex of $A_2 \cup A_4$ has degree 4, while each vertex of $A_6$ (resp. $A_5$) is adjacent to a vertex of $A_2$ (resp. $A_4$) and each vertex of $A_7$ is, respectively, adjacent to a vertex of $A_2$ and $A_4$. Then each vertex in $A_5 \cup A_6$ has degree 2, while each vertex in $A_7$ has degree 3. Finally, add $4(k+3l)$ edges between $A_5 \cup A_6 \cup A_7$ and $A_8$ in such a way that each vertex $A_5 \cup A_6$ is adjacent to precisely two vertices of $A_8$, and each vertex of $A_7$ is adjacent to precisely one vertex of $A_8$, while each vertex of $A_8$ is adjacent to precisely one vertex of $A_5 \cup A_6 \cup A_7$.

By the construction, each vertex in $H_{k,l}$ has degree 4. So $H_{k,l}$ is a 4-regular graph with order $n = \sum_{i=1}^{8} a_i = 10(k+2l)$. Figure 6 shows the graph $H_{1,1}$.

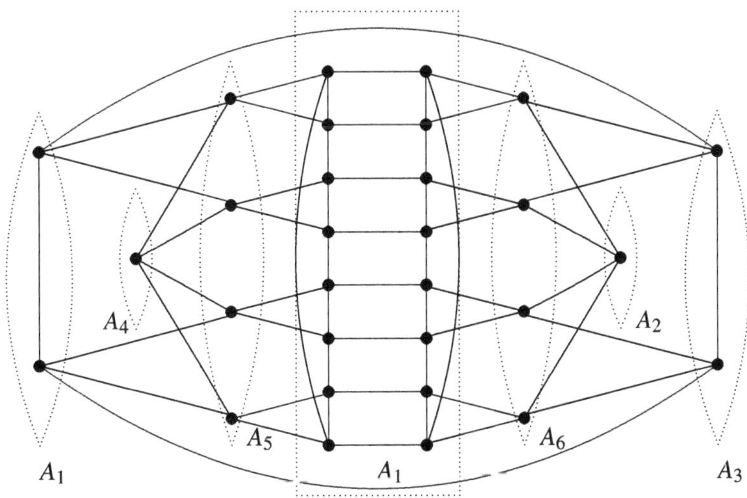


**Fig. 6** The graph $H_{1,1}$.

**Theorem 63 (Yan et al. [77]).** *If $G$ is a 4-regular graph of order $n$, then*

$$\Gamma_t^-(G) \leq \frac{7}{10}n$$

*with equality if and only if $G \in \mathscr{F}_2$.*

Let $H_{k,l} \in \mathscr{F}_2$ be a 4-regular graph of order $10(k+2l)$. Let $f$ be a function on $H_{k,l}$ which assigns to every vertex of $A_1 \cup A_2$ and $A_3 \cup A_4$ the value $-1$ and 0, respectively, and to all vertices of $\cup_{i=5}^{8} A_i$ the value $+1$. It is easy to see that $f$ is a minimal minus total dominating function with weight $w(f) = \sum_{i=5}^{8} a_i - (a_1 + a_2) = 8(k+2l) - (k+2l) = 7(k+l) = 7n/10$. Consequently, $\Gamma_t^-(H_{k,l}) = 7n/10$.

Wang and Shan [72] gave an upper bound on $\Gamma_t^-(G)$ for 5-regular graphs and characterized the extremal graphs by defining a class of 5-regular graphs $\mathscr{F}_3$. They also showed that the difference $\Gamma_t^- - \gamma_{st}$ can be made arbitrary large.

**Theorem 64 (Wang and Shan [72]).** *If $G$ is a 5-regular graph of order $n$, then*

$$\Gamma_t^-(G) \leq \frac{13}{17}n,$$

*with equality if and only if $G \in \mathscr{F}_3$.*

In general, we do not know whether the parameters $\Gamma_t^-$ and $\Gamma_{st}$ are comparable. However, note the fact that every minimal signed total dominating function is also a minimal minus total dominating function, we immediately have the following result.

**Theorem 65 (Yan et al. [77]).** *If $G$ is a graph of odd order, then $\Gamma_{st}(G) \leq \Gamma_t^-(G)$.*

**Theorem 66 (Yan et al. [77]).** *Let $G$ be a cubic graph of order $n$. Then the following statements are equivalent:*

(1) $\Gamma_{st}(G) = \frac{5}{7}n$.
(2) $\Gamma_t^-(G) = \frac{5}{7}n$.
(3) $G \in \mathscr{F}_1$.

**Theorem 67 (Wang and Shan [72]).** *Let $G$ be a 5-regular graph of order $n$. Then the following statements are equivalent:*

(1) $\Gamma_{st}(G) = \frac{13}{17}n$.
(2) $\Gamma_t^-(G) = \frac{13}{17}n$.
(3) $G \in \mathscr{F}_3$.

Li and Lu [49] further gave a sharp bound on $\Gamma_t^-$ for 6-regular graphs. Wu et al. [73] presented a sharp bound on $\Gamma_t^-$ for $k$-regular graphs, when $k$ is odd. In 2017, Li and Lu [50] finally gave a sharp upper bound on $\Gamma_t^-$ for $k$-regular graphs, when $k$ is even.

**Theorem 68 (Wu et al. [73], Li and Lu [50]).** *For a k-regular graph G of order n,*

$$\Gamma_t^-(G) \leq \begin{cases} \dfrac{k^2+1}{k^2+2k-1}n & \text{for } k \text{ odd,} \\ \dfrac{k^2-k+2}{k^2+k}n & \text{for } k \text{ even} \end{cases}$$

*and these bounds are sharp.*

However, the authors did not characterize the regular graphs with equalities. We propose the following open problems.

**Problem 18.** Characterize the graphs attaining the bounds in Theorem 68. Furthermore, for extremal graphs $G$ with equalities, does the quality $\Gamma_t^-(G) = \Gamma_{st}(G)$ hold?

**Problem 19.** For a $2r$-regular graph, are $\Gamma_t^-$ and $\Gamma_{st}$ comparable? If so, how do they compare?

**Problem 20.** Find sharp bounds on $\gamma_t^-(G)$ and $\Gamma_t^-(G)$ for general graphs $G$.

## 3.3 Minus k-Subdomination in Graphs

Minus $k$-subdomination was first introduced and studied by Broere et al. [4]. A *minus k-subdominating function* for a graph $G = (V, E)$ is defined as a function $f : V \to \{-1, 0, 1\}$ such that $f[v] \geq 1$ for at least $k$ vertices of $G$. The *minus k-subdomination number* of $G$, denoted by $\gamma_k^-(G)$, is equal to $\min f(V)$: $f$ is a minus $k$-subdominating function of $G$. In this special case where $k = |V|$, $\gamma_k^-(G)$ is the minus domination number. Similarly, we can define the upper minus $k$-subdomination number $\Gamma_k^-(G)$ of $G$.

Alon (see, [5]) proved that $\gamma_{maj}(G) \leq 2$ for a connected graph $G$. Let $k$ be an integer such that $1 \leq k \leq \lceil |V|/2 \rceil$. Since every (weak) majority domination function is a signed $k$-subdominating function, it follows that $\gamma_{ks}(G) \leq \gamma_{maj}(G)$. This implies that if $G$ is connected, then $\gamma_k^-(G) \leq 2$.

Minimal minus $k$-subdominating functions have been characterized in [4]. Let $f$ be a minus $k$-subdominating function for the graph $G$. We use three sets for such an $f$: $B_f = \{v \in V \mid f(N[v]) = 1\}$, $P_f = \{v \in V \mid f(v) \geq 0\}$, and $C_f = \{v \in V \mid f(N[v]) \geq 1\}$. For $A, B \subseteq V$ we say that $A$ *dominates* $B$, denoted by $A \succ B$, if for each $b \in B$, $N[b] \cap A \neq \emptyset$.

**Theorem 69 (Broere et al. [4]).** *A minus k-subdominating function f on G is minimal if and only if for each k-subset K of* $C_f$, *we have* $B_f \cap K \succ P_f$.

*Proof.* If $B_f \cap K \succ P_f$ for each $k$-subset $K$ of $C_f$, then clearly $f$ is minimal. Conversely, suppose that $f$ is a minimal minus $k$-subdominating function on $G$. Let $K \subseteq C_f$ and $|K| = k$. For any $v \in P_f$, if $N[v] \cap (B_f \cap K) = \emptyset$, then we can replace $f(v)$ by $f(v) - 1$ as $f(v) \geq 0$, the resulting function is still a minus $k$-subdominating function on $G$, a contradiction. Hence $N[v] \cap (B_f \cap K) \neq \emptyset$, i.e., $B_f \cap K \succ P_f$. □

**Theorem 70 (Broere et al. [4]).** *If $P_n$ is a path on $n \geq 2$ vertices and $1 \leq k \leq n-1$, then $\gamma_k^-(P_n) = \lceil k/3 \rceil + k - n + 1$.*

We illustrate Theorem 70 with an example $P_6 = v_1 v_2 \ldots v_6$. If $k = 3$, then let $f(v_1) = f(v_2) = f(v_3) = -1$, $f(v_4) = f(v_5) = 1$, and $f(v_6) = 0$. It is easy to see that $f$ is a minimal minus $k$-subdominating function on $P_6$ with $w(f) = -1 = \lceil k/3 \rceil + k - n + 1$. If $k = 4$, then let $f(v_1) = f(v_2) = -1$, $f(v_3) = f(v_4) = f(v_6) = 1$, and $f(v_5) = 0$ and clearly $f$ is a minimal minus $k$-subdominating function on $P_6$ with $w(f) = 1 = \lceil k/3 \rceil + k - n + 1$.

**Theorem 71 (Hattingh and Ungerer [28]).** *If $T$ is a tree of order $n \geq 2$ and $1 \leq k \leq n-1$, then $\gamma_k^-(T) \geq k - n + 2$.*

In [27] the extremal graphs achieved this lower bound in Theorem 71 were characterized. In fact, Theorem 71 supplements the following earlier result due to [15].

**Theorem 72 (Dunbar et al. [15]).** *If $T$ is a tree, then $\gamma^-(T) \geq 1$ with equality holds if and only if $T$ is a star $K_{1,n}$.*

Computing the exact value of $\gamma_k^-$ may in general be difficult. The following result provides the value on $\gamma_k^-$ of a cycle $C_n$ and a comet $C_{s,t}$ (see Section 2.5).

**Theorem 73 (Hattingh and Ungerer [28]).** *If $C_n$ is a cycle on $n \geq 3$ vertices and $1 \leq k \leq n-1$, then*

$$\gamma_k^-(C_n) = \begin{cases} \lceil (n-2)/3 \rceil & \text{if } k = n-1 \text{ and } k \equiv 0 \text{ or } k \equiv 1 \ (mod 3), \\ 2\lfloor (2k+4)/3 \rfloor - n & \text{otherwise.} \end{cases}$$

**Theorem 74 (Hattingh and Ungerer [29]).** *Let $p, s,$ and $t$ be positive integers such that $p = s + t$ and let $k$ be an integer such that $1 \leq k \leq p-1$. If $s, t \geq 2$, then*

$$\gamma_k^-(C_{s,t}) = \begin{cases} k - p + 2 & \text{if } 1 \leq k \leq s, \\ \lceil (k-s+1)/3 \rceil + k - p + 1 & \text{if } s+1 \leq k \leq p. \end{cases}$$

By simply changing the "closed neighborhood" in the definition of the minus $k$-subdominating function to the "open neighborhood," we can define the minus total $k$-subdominating function. The *minus total $k$-subdomination number* $\gamma_{tk}^-(G)$ and the *upper minus total $k$-subdomination number* $\Gamma_{tk}^-(G)$ can be defined similarly.

The motivation for studying the total $k$-subdomination number is rich and varied from a modeling perspective (see, e.g., [25]).

Harris, Hattingh, and Henning [25] showed that the decision problem of the minus total $k$-subdomination number is NP-complete for bipartite graphs and also present cubic time algorithms to compute the minus total $k$-subdomination and minus $k$-subdomination numbers of a tree. However, there has been no progress on bounds of the minus total $k$-subdomination number of graphs.

**Acknowledgments** We thank Haichao Wang and Yacai Zhao for their helpful suggestions and carefully proofreading most of the text which led to improvements in the presentation. This work was supported in part by the National Nature Science Foundation of China (grant numbers 11971298 and 11871329).

## References

1. Alon N, Spencer J (1992) The Probabilistic Method. New York: Wiley
2. Bange DW, Barkauskas AE, Host LH, Slater PJ (1996) Generalized domination and efficient domination in graphs. Discrete Math 159:1–11
3. Beineke L, Henning MA (1997) Opinion function in trees. Discrete Math 167/168:127–139
4. Broere I, Dunbar JE, Hattingh JH (1998) Minus $k$-subdomination in graphs. Ars Combin 50:177–186
5. Broere I, Hattingh JH, Henning MA, McRae AA (1995) Majority domination in graphs. Discrete Math 138: 125–135
6. Chang GJ, Liaw SC, Yeh HG (2002) $k$-subdomination in graphs. Discrete Appl Math 120:55–60
7. Chen W, Song E (2008) Lower bounds on several versions of signed domination number. Discrete Math 308:1837–1846
8. Chen Y, Cheng TCE, Ng CT, Shan E (2005) A note on domination and minus domination numbers in cubic graphs. Appl Math Lett 18:1062–1067
9. Cockayne EJ, Mynhardt CM (1996) On a generalisation of signed dominating functions of graphs. Ars Combin. 43:235–245
10. Damaschke P (2001) Minus domination in small-degree graphs. Discrete Appl Math 108:53–64
11. Dehgardi N (2016) The minus $k$-domination numbers in graphs. Commun Comb Optim 1:15–28
12. Delić D, Wang C (2010) Upper signed $k$-domination in a general graph. Inform. Process Lett 110:662–665
13. Dunbar J, Goddard W, Hedetniemi ST, Henning MA, McRae A (1996) The algorithmic complexity of minus domination in graphs. Discrete Appl Math 68:73–84
14. Dunbar J, Hedetniemi ST, Henning MA, McRae A (1996) Minus domination in regular graphs. Discrete Math 149:311–312
15. Dunbar J, Hedetniemi ST, Henning MA, McRae A (1999) Minus domination in graphs. Discrete Math 199:35–47
16. Dunbar J, Hedetniemi ST, Henning MA, Slater PJ (1995) Signed domination in graphs. Eds. Alavi Y and Schwenk A, Graph theory, Combinatorics, and Applications, (Wiley, New York, 1995), pp. 311–322
17. Erdös P, Gallai T (1960) Graphs with given degree of vertices. Mat Lapok 11:264–274
18. Faria L, Hon WK, Kloks T, Liu HH, Wang TM, Wang YL (2016) On complexities of minus domination. Discrete Optimization 22:6–19

19. Favaron O (1996) Signed domination in regular graphs. Discrete Math 158:287–293
20. Fiedler M (1973) Algebraic connectivity of graphs. Czech. Math J 23:298–305
21. Füredi Z, Mubayi D, Signed domination in regular graphs and set-systems. J Combin Theory, Ser B 76:223–239
22. Haas R, Wexler TB (2002) Bounds on the signed domination number of a graph. Electron Notes Discrete Math 11:742–750
23. Harris L, Hattingh JH (2004) The algorithm complexity of certain functional variations of total domination in graphs. Australas J Combin 29:143–156
24. Harris L, Hattingh JH, Henning MA (2006) Total $k$-subdominating functions on graphs. Australas. J. Combin. 35:141–154
25. Harris L, Hattingh JH, Henning MA (2018) Further results on total $k$-subdominating functions in trees. J Combin Math Combin Comput 104:65–73
26. Hattingh JH, Henning MA, Slater PJ (1995) On the algorithmic complexity of signed domination in graphs. Australas J Combin 12:101–112
27. Hattingh JH, McRae AA, Ungerer E (1998) Minus $k$-subdomination in graphs. III. Australas J Combin 17: 69–76
28. Hattingh JH, Ungerer E (1997) Minus $k$-subdomination in graphs. II. Discrete Math 171:141–151
29. Hattingh JH, Ungerer E (1998) The signed and minus $k$-subdomination numbers of comets. Discrete Math 183:141–152
30. Hattingh JH, Ungerer E, Henning MA (1998) Partial signed domination in graphs. Ars Combin 48:33–42
31. Haynes TW, Hedetniemi ST, Slater PJ (1998) Fundamentals of Domination in Graphs. New York: Marcel Dekker
32. Henning MA (1996) Domination in regular graphs. Ars Combin 43:263–271
33. Henning MA (1998) Dominating functions in graphs. Domination in Graphs: Vol. II, Marcel-Dekker, New York, pp. 31–62
34. Henning MA (2004) Signed total domination in graphs. Discrete Math 278:109–125
35. Henning MA, Hind HR(1998) Strict majority functions on graphs. J Graph Theory 28:49–56
36. Henning MA, Slater PJ (1996) Inequalities relating domination parameters in cubic graphs. Discrete Math 158:87–98
37. Holm TS (2001) On majority domination in graphs. Discrete Math 239:1–12
38. Kang L, Cai MC (1998) Minus domination number in cubic graphs. Chinese Sci Bull 43:444–447
39. Kang L, Cai MC (2000) Upper minus domination in regular graphs. Discrete Math 219:135–144
40. Kang L, Dang C, Cai MC, Shan E (2002) Upper bounds for $k$-subdomination number of graphs. Discrete Math 247:229–234
41. Kang L, Kim HK, Sohn MY (2004) Minus domination number in $k$-partite graphs. Discrete Math 277:295–300
42. Kang L, Qiao H, Shan E, Du DZ (2003) Lower bounds on the minus domination and $k$-subdomination numbers. Theoret Comput Sci 296:89–98
43. Kang L, Shan E (2000) Lower bounds on domination function in graphs. Ars Combin 56:121–128
44. Kang L, Shan E (2006) Signed total domination in nearly regular graphs. J Shanghai Univ 10: 4–8
45. Kang L, Shan E (2007) Dominating functions with integer values in graphs—a survey. J. Shanghai University 11(5):437–448
46. Kang L, Shan E, Caccett L (2006) Total minus domination in $k$-partite graphs, Discrete Math 306 (2006) 1771-1775.
47. Lee CM, Chang MS (2008) Variations of $Y$-dominating functions on graphs. Discrete Math 308:4185–4204
48. Lee J, Sohn MY, Kim HK (1999) A note on graphs with large girth and small minus domination number. Discrete Appl Math 91:299–303

49. Li Z, Lu X (2012) Upper minus total domination number of 6-regular graph. International J Pure and Appl Math 75(3): 319–326
50. Li Z, Lu X (2017) Upper minus total domination number of regular graphs. Acta Math Appl Sin 33(1): 69–74
51. Liu H, Sun L (2004) On the minus domination number of graphs. Czechoslovak Math J 54:883–887
52. Matoušek J (2000) On the signed domination in graphs. Combinatorica 20:103–108
53. Matoušek J (2001) Lower bound on the minus-domination number. Discrete Math 233:361–370
54. Merris R (1994) Laplacian matrices of graphs: a survey. Linear Algebra Appl 197/198:143–176
55. Muthuselvi A, Arumugam S (2017) Upper majority domination number of a graph. Arumugam S et al. (Eds.): ICTCSDM 2016, LNCS 10398, pp. 197–202, Springer
56. Nikiforov V (2007) Bounds on graph eigenvalues I. Linear Algebra Appl 420:667–671
57. Ore O (1962) Theory of Graphs. Amer Math Soc Transl 38:206–212
58. Poghosyan A, Zverovich V (2010) Discrepancy and signed domination in graphs and hypergraphs. Discrete Math 310:2091–2099
59. Reed BA (1996) Paths, stars and the number three. Combin Probab Comput 5:277–295
60. Shan E, Cheng TCE (2009) Upper bounds on the upper signed total domination number of graphs. Discrete Appl Math 157:1098–1103
61. Shan E, Cheng TCE (2008) Remarks on the minus (signed) total domination in graphs. Discrete Math 308:3373–3380
62. Shan E, Cheng TCE, Kang L. (2008) An application of the Turán theorem to domination in graphs. Discrete Appl Math 156:2712–2718
63. Shan E, Kang L. (2020) Variations of signed and minus domination: A survey. submitted to J Combin Math Combin Comput.
64. Shang W, Yuan J (2006) Upper minus domination in a claw-free cubic graph. Discrete Math 306:2983–2988
65. Shi W, Kang L, Wu S, Bounds on Laplacian eigenvalues related to total and signed domination of graphs. Czechoslovak Math J 60:315–325
66. Tang H, Chen Y (2008) A note on the signed domination number of graphs. Discrete Math 308:3416–3419
67. Turán P (1941) On an extremal problem in graph theory. Math Fiz Lapok 48:436–452 (in Hungarian)
68. Volkmann L (2016) Lower bounds on the signed total $k$-domination number of graphs. Aequationes Math 90:271–279
69. Wang C (2012) The signed $k$-domination number in graphs. Ars Combin 106:205–211
70. Wang C, Mao J (2001) Some more remarks on domination in cubic graphs. Discrete Math 237:193–197
71. Wang C, Mao J (2002) A proof of a conjecture of minus domination in graphs. Discrete Math 256:519–521
72. Wang H, Shan E (2009) Upper minus total domination of a 5-regular graph. Ars Combin 91: 429–438
73. Wu J, Miao Z, Lu C (2008) Upper minus total domination of regular graph. Acta Math Appl Sin 31(5):861–870
74. Xing H, Sun L (2005) On signed majority total domination in graphs. Czechoslovak Math J 2005, 55(130):341–348
75. Xing H, Sun L, Chen X (2005) On a generalization of signed total dominating functions of graphs. Ars Combin 77:205–215
76. Xu B (2003) On signed domination and minus domination in graphs. J. Math. Res. Exposition 23(4):586–590
77. Yan H, Yang X, Shan E (2007) Upper minus total domination in small-degree regular graphs. Discrete Math 307: 2453–2463
78. Zelinka B (1996) Some remarks on domination in cubic graphs. Discrete Math 158:249–255

79. Zelinka B (2001) Signed total domination number of a graph. Czechoslovak Math J 51: 225–229
80. Zelinka B (2003) On a problem concerning $k$-subdomination numbers of graphs. Czechoslovak Math J 53:627–629
81. Zelinka B (2006) Signed and minus domination in bipartite graphs. Czechoslovak Math J 56 (2006) 587–590
82. Zhang Z, Xu B, Li Y, Liu L (1999) A note on the lower bounds of signed domination number of a graph. Discrete Math 195:295–298
83. Zheng Y, Wang J, Feng Q (2013) Kernelization and lower bounds of the signed domination problem. Fellows M, Tan X, Zhu B (Eds.): FAW-AAIM 2013, LNCS 7924, pp. 261–271

# Fractional Dominating Parameters

**Wayne Goddard and Michael A. Henning**

## 1 Introduction

For a graph $G = (V, E)$ with vertex set $V$ and edge set $E$ and for a real-valued function $f: V \to \mathbb{R}$, the *weight* of $f$ is $\mathrm{w}(f) = \sum_{v \in V} f(v)$. Further, for $S \subseteq V$ we define $f(S) = \sum_{v \in S} f(v)$; in particular, this means that $\mathrm{w}(f) = f(V)$. For a vertex $v$ in $V$, the open neighborhood $N(v)$ of $v$ consists of all neighbors of $v$; that is, $N(v) = \{u \in V \mid uv \in E\}$. The closed neighborhood $N[v]$ of $v$ consists of $v$ and all neighbors of $v$; that is, $N[v] = N(v) \cup \{v\}$. For notational convenience, we denote $f(N[v])$ by $f[v]$.

Let $f: V \to \{0, 1\}$ be a function which assigns to each vertex of a graph an element of the set $\{0, 1\}$. We say $f$ is a *dominating function* if for every $v \in V$, $f[v] \geq 1$. The function $f$ is a *minimal dominating function* if there does not exist a dominating function $g: V \to \{0, 1\}$, $f \neq g$, for which $g(v) \leq f(v)$ for every $v \in V$. This is equivalent to saying that a dominating function $f$ is minimal if for every

The research of the author Michael A. Henning was supported in part by the University of Johannesburg.

W. Goddard (✉)
School of Computing and School of Mathematical and Statistical Sciences, Clemson University, 29634 Clemson, SC, USA

Department of Mathematics and Applied Mathematics, University of Johannesburg, Auckland Park 2006, Johannesburg, South Africa
e-mail: goddard@clemson.edu

M. A. Henning
Department of Mathematics and Applied Mathematics, University of Johannesburg, Auckland Park 2006, Johannesburg, South Africa
e-mail: mahenning@uj.ac.za



T. W. Haynes et al. (eds.), *Topics in Domination in Graphs*, Developments in Mathematics 64, https://doi.org/10.1007/978-3-030-51117-3_10

vertex $v$ such that $f(v) > 0$, there exists a vertex $u \in N[v]$ for which $f[u] = 1$. The domination number and upper domination number of a graph $G$ can be defined as $\gamma(G) = \min\{\mathrm{w}(f) \mid f$ is a dominating function on $G\}$ and $\Gamma(G) = \max\{\mathrm{w}(f) \mid f$ is a minimal dominating function on $G\}$, respectively.

## 2 Fractional Domination

We consider here the generalization of domination where vertices have fractional weights in the range [0, 1]. A real-valued function $f: V \to [0, 1]$ is called a *fractional dominating function* of $G$ if $f[v] \geq 1$ for each $v \in V$. The minimum weight of a fractional dominating function of graph $G$ is the *fractional domination number* $\gamma_f(G)$ of $G$. Thus, $\gamma_f(G) = \min\{\mathrm{w}(f) \mid f$ is a fractional dominating function for $G\}$. While ideas about fractional domination are found in Farber [15] and others, the parameter was formally defined in 1987 by S. T. Hedetniemi reporting on results in [25] at the Eighteenth Southeastern Conference, and in 1988 by Domke, Hedetniemi, and Laskar [12] and in 1990 by Grinstead and Slater [22]. For general information about fractional graph parameters, see [32].

The fractional domination number is readily viewed as a linear program. Thus we can talk of minimum, rather than infimum. The dual of the fractional domination number is the *fractional packing number*. This is the linear programming relaxation of the packing number.

A real-valued function $f: V \to [0, 1]$ is called a *fractional packing function* of $G$ if $f[v] \leq 1$ for each $v \in V$. A fractional packing function $f$ is maximal if there does not exist a fractional packing function $g: V \to \{0, 1\}$, $f \neq g$, for which $g(v) \geq f(v)$ for every $v \in V$. This is equivalent to saying that a fractional packing function $f$ is maximal if for every vertex $v$ with $f(v) < 1$, there exists a vertex $u \in N[v]$ such that $f[u] = 1$. The maximum weight of a fractional packing function of graph $G$ is the *fractional packing number* $\rho_f(G)$ of $G$.

Define $N$ as the (closed) neighborhood matrix (that is, the adjacency matrix with the 0's on the diagonal replaced by 1's). We have:

| *Fractional domination* $\gamma_f$ | *Fractional packing* $\rho_f$ |
|---|---|
| minimize $\vec{1}^t\vec{x} = \sum_{i=1}^{n} x_i$ | maximize $\vec{1}^t\vec{y} = \sum_{i=1}^{n} y_i$ |
| subject to: $\begin{cases} N\vec{x} \geq \vec{1} \\ x_i \geq 0 \end{cases}$ | subject to: $\begin{cases} N\vec{y} \leq \vec{1} \\ y_i \text{ unrestricted} \end{cases}$ |

Thus, by the fundamental theorem of linear programming, it follows that:

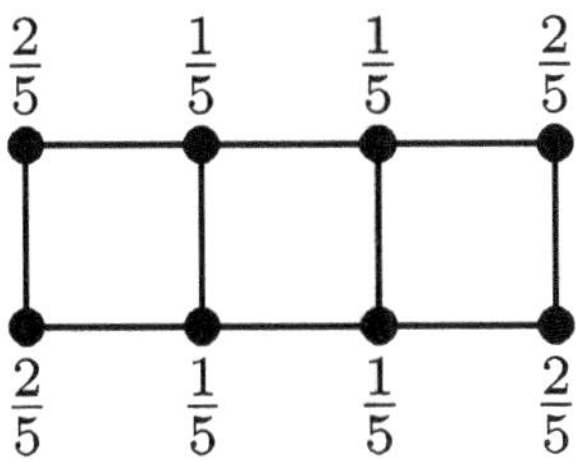


**Fig. 1** A fractional weighting in a prism $P_4 \,\square\, K_2$

**Theorem 1** *For any graph G, it holds that* $\gamma_f(G) = \rho_f(G)$.

For example, the function $f$ that assigns the weights to the vertices of the prism $G = P_4 \,\square\, K_2$ as illustrated in Figure 1 is a fractional dominating function and a fractional packing function of $G$ of weight $\mathrm{w}(f) = \frac{12}{5}$. Thus by Theorem 1, $\gamma_f(G) = \rho_f(G) = \frac{12}{5}$.

There are immediate bounds involving the minimum and maximum degrees. In particular, weight $1/(\Delta(G)+1)$ on every vertex is a packing function of $G$ and weight $1/(\delta(G)+1)$ on every vertex is a domination function of $G$. Thus for all graphs, we have the following result first observed by Grinstead and Slater [22] and Domke et al. [12].

**Observation 1 ([12, 22])** *If G is a graph of order n with minimum degree* $\delta(G) = \delta$ *and maximum degree* $\Delta(G) = \Delta$, *then*

$$\frac{n}{\Delta+1} \le \rho_f(G) = \gamma_f(G) \le \frac{n}{\delta+1}.$$

As a consequence of Observation 1 follows the value for a regular graph:

**Theorem 3 ([12, 22])** *If G is r-regular graph of order n, then* $\gamma_f(G) = \frac{n}{r+1}$.

Like the ordinary domination number, the parameter $\gamma_f(G)$ ranges between 1 and the order of the graph. Domke et al. [12] showed that $\gamma_f(G) = 1$ if and only if $G$ has a dominating vertex. It is trivial that $\gamma_f(G)$ equals the order if and only if the graph $G$ is empty. It is also trivial that $\gamma_f(G - e) \ge \gamma_f(G)$ where $e$ is any edge in the graph $G$.

Grinstead and Slater [22] observed an averaging argument: the mean of a collection of dominating functions is itself a dominating function. One can use this argument to "smooth" the function. Thus, for example, one may assume for each orbit of the automorphism group that the weight is the same for all vertices in the orbit. Results on complete bipartite (and complete multipartite graphs) given in [22] and in Domke and Laskar [14] follow:

**Theorem 4 ([14, 22])** *For the complete bipartite graph* $K_{r,s}$ *it holds that*

$$\gamma_f(K_{r,s}) = \frac{r(s-1)+s(r-1)}{rs-1}.$$

The above formula also follows by linear programming duality, since one can readily specify a fractional dominating function and a fractional packing function of the same weight.

For $k \geq 1$ an integer, a function $f : V \to \{0, 1, \ldots, k\}$ is called a *k-dominating function* of a graph $G = (V, E)$ if $f[v] \geq k$ for each $v \in V$. The *k-domination number*, denoted $\gamma_{\{k\}}(G)$, of $G$ is the minimum weight of a $\{k\}$-dominating function of $G$. We note that the characteristic function of a dominating set of $G$ is a $\{1\}$-dominating function, and so $\gamma_{\{1\}}(G) = \gamma(G)$. The lower bound with domination can be trivially generalized to $\{k\}$-domination, namely, $\gamma_f(G) \leq \gamma_{\{k\}}(G)/k$. Indeed, Domke et al. [13] showed that

**Theorem 5 ([13])** *For any graph G,*

$$\gamma_f(G) = \min_{k \in \mathbb{N}} \left\{ \frac{\gamma_{\{k\}}(G)}{k} \right\}.$$

Domke et al. [11] observed that for a tree, one has equality between the fractional and ordinary domination number. More generally, they proved that for a *block graph* (a graph where every block is complete) equality also holds. This result also follows from more general results proven by Domke et al. [13] involving the packing number.

**Theorem 6 ([11])** *If G is a block graph, then* $\gamma_f(G) = \gamma(G)$.

A graph is *chordal* if it contains no cycle of length greater than three as an induced subgraph. A *strongly chordal graph* is a chordal graph that also contains no induced trampoline, where a *trampoline* (or sun) consists of a $2n$-cycle $v_1 v_2 \ldots v_{2n} v_1$ in which the vertices $v_{2i}$ of even subscript form a complete graph on $n$ vertices. Iijima and Shibata [27] and Farber [15] showed that for strongly chordal graphs their fractional domination number equals their domination number.

**Theorem 7 ([15, 27])** *If G is a strongly chordal graph, then* $\gamma_f(G) = \gamma(G)$.

This result can be deduced from matrix properties. For, the neighborhood matrix $N$ of a graph is totally balanced if and only if the graph is strongly chordal, and linear programs with totally balanced matrices are guaranteed to have integral solutions (see for example [20]).

As was trivially observed, the ordinary domination number is at least the fractional domination number. This suggests the question of how large the domination number can be in terms of the fractional domination number. Chappell, Gimbel, and Hartman [5] showed that the ratio can be at most logarithmic in the order:

**Theorem 8 ([5])** *If G is a graph, then* $\gamma(G) \leq (1 + \ln(1 + \Delta(G)))\gamma_f(G)$.

This result is essentially best possible as there are $r$-regular graphs of order $n$ with domination number $\Theta(\frac{n \log r}{r})$; see Alon [1].

## 2.1 Efficient Domination

We say a function $f : V \to [0, 1]$ is an *efficient fractional dominating function* if for every vertex $v$ it holds that $f[v] = 1$. Equivalently, $N\vec{f} = \vec{1}$ where $\vec{1}$ denotes the all 1's vector in $\mathbb{R}^n$. For example, if $G$ is a regular graph of degree $r$, then the function $f$ that assigns to each vertex the value $1/(r+1)$ is an efficient fractional dominating function for $G$. If $G$ is a complete bipartite graph $K_{r,s}$ with partition $(R, S)$, then the function $f$ that assigns to each vertex of $R$ the value $(s-1)/(rs-1)$ and to each vertex of $S$ the value $(r-1)/(rs-1)$ is such a function.

Efficient fractional dominating functions were introduced and first studied in 1988 by Bange, Barkauskas, and Slater [3]. We call an efficient fractional dominating an *EFD-function* (standing for **E**fficient **F**ractional **D**ominating function). Grinstead and Slater [22] called a graph which has an EFD-function "fractionally efficiently dominatable." In 1996, Bange, Barkauskas, Host, and Slater [2] showed that:

**Theorem 9 ([2])** *If a graph G is fractionally efficiently dominatable, then every EFD-function of G has the same weight.*

We remark that testing for efficiency is a linear program calculation.

If a graph $G$ has an efficient dominating set $D$, then the characteristic function of $D$ represents a EFD-function. However, the converse is not true: having an efficient fractional dominating function does not imply an efficient dominating set. Consider, for example, any regular graph where $n$ is not a multiple of $r + 1$. Another is the graph formed from two 4-cycles by identifying a vertex of each (Figure 2).

The converse is true, however, in trees (see [21]), and indeed in block graphs (see [26]):

**Theorem 10 ([26])** *If B is a block graph, then B has a EFD-function if and only if it has an efficient dominating set.*

Further, note that even if $\gamma_f(G) = \gamma(G)$, it does not necessarily follow that there is an efficient dominating set of $G$. For example, the graph $G$ shown in Figure 3 satisfies $\gamma_f(G) = \gamma(G) = 3$ and has an EFD-function but does not possess an efficient dominating set.

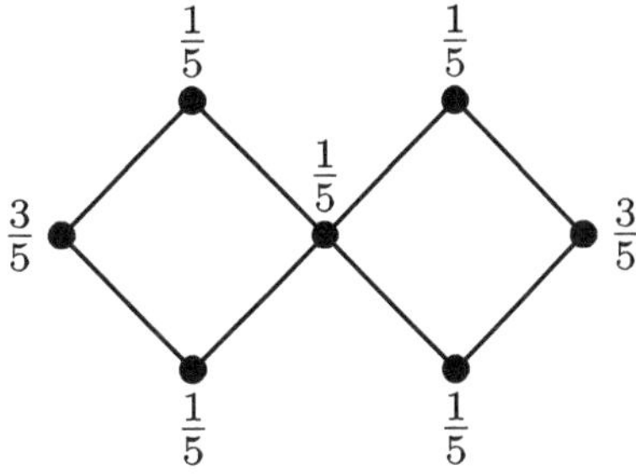


**Fig. 2** An efficient fractional dominating function in a graph

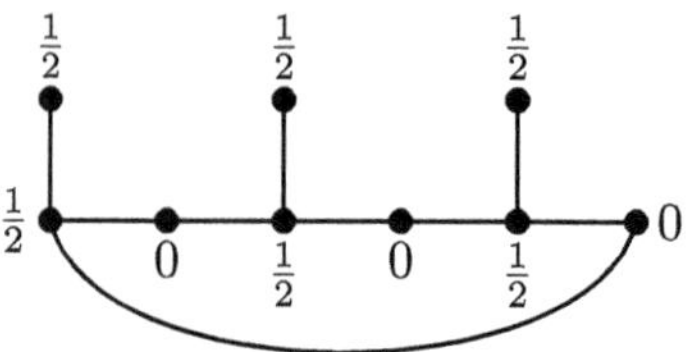


**Fig. 3** A graph with an EFD-function but no efficient dominating set

## 2.2 Graph Operations

The fractional domination number of the disjoint union of two graphs is just the sum of the individual fractional domination numbers. The result for the join was determined by Fisher [17]: if graphs $G$ and $H$ are not complete, then the optimal fractional dominating set is found by taking optimal fractional dominating sets of both $G$ and $H$ and then scaling all weights of $V(G)$ by some factor and all weights of $V(H)$ by some factor. Optimization yields the result for the join $G + H$ of $G$ and $H$, which consists of the disjoint union plus edges joining every vertex of $G$ to every vertex of $H$.

**Theorem 11 ([17])** *The fractional domination number of the join of noncomplete graphs $G$ and $H$ is given by*

$$\gamma_f(G + H) = 2 - \frac{\gamma_f(G) + \gamma_f(H) - 2}{\gamma_f(G)\gamma_f(H) - 1}.$$

Earlier, Fisher, Ryan, Domke, and Majumdar [18] determined the formula for the strong direct product $G \boxtimes H$ of $G$ and $H$, which has vertex set $V(G) \times V(H)$ and where distinct vertices $(u, u')$ and $(v, v')$ are adjacent in $G \boxtimes H$ if and only if $u = v$ and $u'v' \in E(H)$ or $u' = v'$ and $uv \in E(G)$ or $uv \in E(G)$ and $u'v' \in E(H)$. Their proof rests on the fact that the neighborhood matrix of $G \boxtimes H$ is the tensor product of the neighborhood matrices of $G$ and $H$.

**Theorem 12 ([18])** *For all graphs $G$ and $H$,*

$$\gamma_f(G \boxtimes H) = \gamma_f(G)\,\gamma_f(H).$$

From this they deduced the bounds for Cartesian product $G \,\square\, H$ of $G$ and $H$, which has vertex set $V(G) \times V(H)$, and edges between vertices $(g, h)$ and $(g', h')$ if either $gg' \in E(G)$ and $h = h'$, or $g = g'$ and $hh' \in E(H)$.

**Theorem 13 ([18])** *For all graphs $G$ and $H$,*

$$\gamma_f(G \,\square\, H) \geq \gamma_f(G)\,\gamma_f(H).$$

This established the fractional version of Vizing's conjecture. Related results are discussed in [16]. More on the strong product is given by John and Suen [28].

Hare [23] and Stewart and Hare [33] considered the fractional domination numbers of grids with fixed number of rows. Hare showed that the grid $P_2 \square P_m$ has an efficient dominating set whenever $m$ is odd, and calculated the fractional dominating number for the case that $m$ is even:

**Theorem 14 ([23])** *For $m \geq 1$,*

$$\gamma_f(P_2 \square P_m) = \begin{cases} \frac{1}{2}(m+1) & \text{if } m \text{ is odd} \\ \frac{m(m+2)}{2(m+1)} & \text{if } m \text{ is even.} \end{cases}$$

The case for $m = 4$ was illustrated in Figure 1. Also Rubalcaba and Walsh [31] noted that every minimal dominating function of $P_2 \square P_m$ is a packing function but not every maximal packing function is a dominating function. Values for the product $C_n \square P_m$ are given by Xu [34].

## 2.3 *Upper Fractional Domination*

A *minimal fractional dominating function* is a fractional dominating function where no value can be lowered. This is equivalent to saying that every vertex $u$ with positive weight has a vertex $v$ in its closed neighborhood such that $f[v] = 1$. The *upper fractional domination number* of $G$ is defined as $\Gamma_f(G) = \max\{\mathrm{w}(f) \mid f$ is a minimal fractional dominating function for $G\}$. The upper parameter $\Gamma_f(G)$ can be expressed as a maximum over a finite collection of linear programs. Hence we write the maximum rather than the supremum. But it is NP-complete in general; see [7]. Upper fractional domination in graphs was introduced and studied by Cheston et al. [7].

The domination and fractional domination parameters are related as follows.

**Observation 2** *For every graph $G$, we have $\gamma_f(G) \leq \gamma(G) \leq \Gamma(G) \leq \Gamma_f(G)$.*

Laskar et al. [30] provided the formula for the upper fractional domination number of the join of two graphs:

**Theorem 16 ([30])** *For all graphs $G$ and $H$,*

$$\Gamma_f(G + H) = \max\{\Gamma_f(G), \Gamma_f(H)\}.$$

Cheston and Fricke [6] established the following result.

**Theorem 17 ([6])** *Every (strongly) perfect graph $G$ has $\Gamma_f(G) - \Gamma(G) = \alpha(G)$.*

The convex combination of dominating functions is again dominating functions. Cockayne, Fricke, Hedetniemi, and Mynhardt [9] investigated when the convex combination of minimal dominating functions is again a minimal dominating function.

## 2.4 *Fractional Total Domination*

Fractional total domination is defined similarly to fractional domination. A real-valued function $f: V \to [0, 1]$ in a graph $G = (V, E)$ is called a *fractional total dominating function* of $G$ if the sum of the function values over any open neighborhood is at least 1. That is, for every $v \in V$, $f(N(v)) \geq 1$. The minimum weight of a fractional total dominating function of $G$ is the *fractional total domination number* $\gamma_{ft}(G)$ of $G$. Thus, $\gamma_{ft}(G) = \min\{\mathrm{w}(f) \mid f$ is a fractional total dominating function for $G\}$. The *upper fractional total domination number* of $G$ is defined as $\Gamma_{ft}(G) = \max\{\mathrm{w}(f) \mid f$ is a minimal fractional total dominating function for $G\}$, where a *minimal fractional total dominating function* is a fractional total dominating function where no value can be lowered.

The characteristic function of a total dominating set is trivially a fractional total dominating function; furthermore, the characteristic function of a minimal total dominating set is a minimal fractional total dominating function. The total domination and fractional total domination parameters are therefore related as follows, where $\Gamma_t(G)$ is the upper total domination number of $G$.

**Observation 3** *For every graph $G$, $\gamma_{ft}(G) \leq \gamma_t(G) \leq \Gamma_t(G) \leq \Gamma_{ft}(G)$.*

Fricke, Hare, Jacobs, and Majumdar [19] established some classes of graphs where the upper total domination and upper fractional total domination numbers are equal.

**Theorem 19 ([19])** *If $G$ is a bipartite graph with no induced cycle of length congruent to* 2 *modulo* 4, *then $\Gamma_{ft}(G) = \Gamma_t(G)$.*

As with the fractional domination number, the fractional total domination number is readily viewed as a linear program, where here the linear program uses the adjacency matrix $A$ rather than the neighborhood matrix $N$.

| **Fractional total domination** $\gamma_{ft}$ | | **Fractional open packing** $\rho_{fo}$ | |
|---|---|---|---|
| minimize | $\vec{1}^t\vec{x} = \sum_{i=1}^n x_i$ | maximize | $\vec{1}^t\vec{y} = \sum_{i=1}^n y_i$ |
| subject to: | $\begin{cases} A\vec{x} \geq \vec{1} \\ x_i \geq 0 \end{cases}$ | subject to: | $\begin{cases} A\vec{y} \leq \vec{1} \\ y_i \text{ unrestricted} \end{cases}$ |

Thus, by the fundamental theorem of linear programming, it follows that:

**Theorem 20** *For any graph $G$, it holds that $\gamma_{ft}(G) = \rho_{fo}(G)$.*

For example, if $G$ is an $r$-regular graph of order $n$, then the function $f$ that assigns to each vertex the value $1/r$ is a fractional total dominating function and a fractional open packing function of $G$ of weight $\mathrm{w}(f) = n/r$. Thus by Theorem 20, $\gamma_{ft}(G) = \rho_{fo}(G) = n/r$.

Fisher [17] showed that the fractional domination number of a graph and the fractional total domination number of its complement are related as follows.

**Theorem 21 ([17])** *For any graph G it holds that*

$$\frac{1}{\gamma_f(G)} + \frac{1}{\gamma_{ft}(\overline{G})} = 1.$$

**Proof.** If $\gamma_f(G) = 1$, then $G$ has a dominating vertex and so $\overline{G}$ has an isolated vertex. Since such a vertex can have arbitrarily large weight in an open packing, the result follows if we interpret such a graph to have $\gamma_{ft} = \infty$ and $1/\infty = 0$.

So assume $\gamma_f(G) > 1$. Consider a minimum fractional dominating function $g$ of $G$ and a maximum fractional packing $h$ of $G$. By above we know these both have weight $\gamma_f(G)$. Now for every vertex $v$, the total weight of $g$ outside $v$'s closed neighborhood $N[v]$ is at most $\gamma_f(G) - 1$. It follows that the function $g/(\gamma_f(G) - 1)$ is a fractional open packing of $\overline{G}$. That is,

$$\gamma_{ft}(\bar{(G)}) \geq \frac{\gamma_f(G)}{\gamma_f(G) - 1}.$$

Similarly, the total weight of $h$ outside $N[v]$ is at least $\gamma_f(G) - 1$, the function $h/(\gamma_f(G) - 1)$ is a fractional total dominating function of $\overline{G}$, and

$$\gamma_{ft}(\bar{(G)}) \leq \frac{\gamma_f(G)}{\gamma_f(G) - 1}.$$

The result follows. □

## 3 $\mathcal{P}$-Domination

In the spirit of fractional domination, Bange, Barkauskas, Host, and Slater [2] generalized domination to $\mathcal{P}$-domination for an arbitrary subset $\mathcal{P}$.

Given $\mathcal{P} \subseteq \mathbb{R}$, a function $f: V \to \mathcal{P}$ is a $\mathcal{P}$-dominating function of a graph $G = (V, E)$ if the sum of the function values over any closed neighborhood is at least 1. That is, for every $v \in V$, $f[v] \geq 1$. The $\mathcal{P}$-domination number, denoted $\gamma_{\mathcal{P}}(G)$, of $G$ is defined to be the infimum of $f(V)$ taken over all $\mathcal{P}$-dominating functions $f$.

This concept encompasses several common parameters. When $\mathcal{P} = \{0, 1\}$ we obtain the standard domination number; when $\mathcal{P} = [0, 1]$, we obtain the fractional domination number discussed above; when $\mathcal{P} = \{-1, 0, 1\}$ we obtain the minus domination number, and when $\mathcal{P} = \{-1, 1\}$ we obtain the signed domination number. We remark that discussion of the minus and signed domination numbers can be found in the chapter by Shan and Kang in this book, while discussion of algorithms and complexity of minus and signed domination can be found in the chapter by Hedetniemi, McRae, and Mohan in the companion book [**?** ].

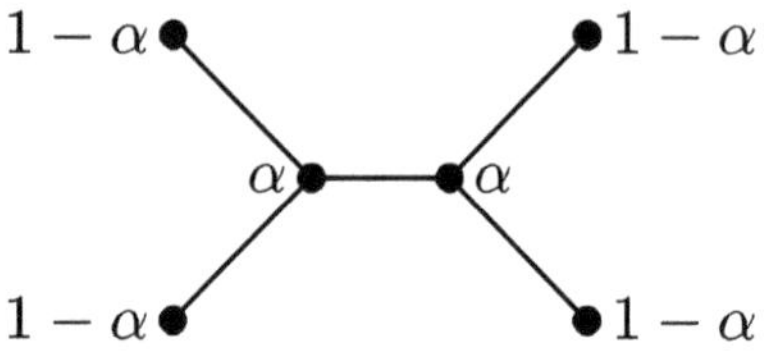


**Fig. 4** A double-star $T$ with $\gamma_{\mathbb{Z}}(T) = -\infty$

One important consequence of allowing negative weights is that for a graph $G$ its $\mathcal{P}$-domination number might be negative. For example, consider a double-star $T$ of order 6 with two vertices of degree 3. If $\mathcal{P} = \mathbb{Z}$ and $\alpha$ is a positive integer, then placing weight $\alpha$ on the two central vertices and weight $1 - \alpha$ on every leaf as illustrated in Figure 4 produced a $\mathcal{P}$-dominating function of total weight $2 - 2\alpha$. One can make $\alpha$ arbitrarily large, implying that $\gamma_{\mathbb{Z}}(T) = -\infty$.

For a subset $\mathcal{P} \subseteq \mathbb{R}$ and a graph $G = (V, E)$, a function $f : V \to \mathcal{P}$ is an *efficient $\mathcal{P}$-dominating function* if for every vertex $v$ it holds that $f[v] = 1$. Earlier, Bange et al. [2] when they established Theorem 9 had actually shown:

**Theorem 22** *If $\mathcal{P}$ be a subset of the reals $\mathbb{R}$ and if $f_1$ and $f_2$ are two arbitrary efficient $\mathcal{P}$-dominating functions for a graph $G$, then* $\mathrm{w}(f_1) = \mathrm{w}(f_2)$.

A function is *nonnegative* if all the function values are nonnegative. A function which is both nonnegative and efficient $\mathcal{P}$-dominating is called an *NE$\mathcal{P}$D-function* (standing for **N**onnegative **E**fficient $\mathcal{P}$-**D**ominating function). In particular, a NE **R**D-function (standing for **N**onnegative **E**fficient **R**eal **D**ominating function) in $G$ is a nonnegative efficient real dominating function in $G$. For example, if $\mathcal{P} = \mathbb{R}$ and $G$ is a regular graph of degree $k$, then the function $f$ that assigns to each vertex the value $1/(k+1)$ is a NE **R**D-function for $G$ noting that $f(v) > 0$ and $f[v] = 1$ for every vertex $v$ in the graph $G$.

In [21], we showed that the question about unbounded weights is directly connected to NE **R**D-functions.

**Theorem 23 ([21])** *For any graph $G$,*

$$\gamma_{\mathbb{R}}(G) = \begin{cases} \mathrm{w}(f) & \text{if } G \text{ has an NE } \mathbf{R}\text{D-function } f, \\ -\infty & \text{otherwise.} \end{cases}$$

The proof of Theorem 23 uses linear programming duality and complementary slackness. The dual of real domination is efficient packing. If the dual is feasible, that is, there is an efficient packing function, then that efficient packing function is also an efficient dominating function, and the solution is the size of the efficient dominating function. If the dual is not feasible, then the primal has an unbounded solution.

It follows from Theorem 23 that, for any subset $\mathcal{P}$ of $\mathbb{R}$, if a graph $G$ has an NE$\mathcal{P}$D-function $f$, then $\gamma_{\mathcal{P}}(G) = \gamma_{\mathbb{R}}(G) = \mathrm{w}(f)$, since $f$ is an NE **R**D-function of $G$, and so $\mathrm{w}(f) \geq \gamma_{\mathcal{P}}(G) \geq \gamma_{\mathbb{R}}(G) = \mathrm{w}(f)$.

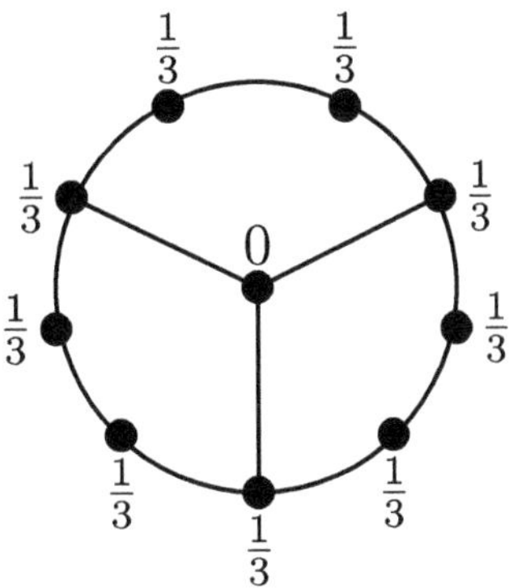


**Fig. 5** A graph $G$ with no efficient dominating set satisfying $\gamma_{\mathbb{R}}(G) = \gamma(G)$

Another consequence of Proposition 23 is that if $\{0, 1\} \subseteq \mathcal{P} \subseteq \mathbb{R}$, and graph $G$ has an efficient dominating set, then $\gamma_{\mathcal{P}}(G) = \gamma(G)$. This follows as the characteristic function of the efficient dominating set is an NE$\mathcal{P}$D-function for $G$. The converse, however, is not true. For example, the graph $G$ shown in Figure 5 has an NE$\mathcal{P}$D-function $f$ as illustrated, and so $\gamma_f(G) = \mathrm{w}(f) = 3$. Furthermore, it is evident that $\gamma(G) = 3$. However, the graph $G$ does not possess an efficient dominating set. Hence, $\gamma_{\mathbb{R}}(G) = \gamma(G)$ does not necessarily imply that $G$ has an efficient dominating set.

If one takes a $\mathbb{Q}$-dominating function $f$ and multiplies all the weights by the least common multiple of the weights' dominators, one obtains a $\mathbb{Z}$-dominating function. Therefore if there is a $\mathbb{Q}$-dominating function of arbitrarily negative weight, then there is a $\mathbb{Z}$-dominating function too of arbitrarily negative weight. Thus, if $\gamma_{\mathbb{Q}}(G) = -\infty$, then $\gamma_{\mathbb{Z}}(G) = -\infty$.

## 3.1 Graph Classes

By the results of Iijima and Shibata [27] and Farber [15] mentioned earlier, it follows from Theorem 23 that:

**Theorem 24 ([15, 21])** *For any strongly chordal graph G,*

$$\gamma_{\mathbb{R}}(G) = \begin{cases} \gamma(G) & \text{if } G \text{ has an NE } \mathbf{R}\text{D-function,} \\ -\infty & \text{otherwise.} \end{cases}$$

For real domination there is no result analogous to Vizing's conjecture. For example, the path $P_3$ has an efficient dominating set, but the grid $P_3 \square P_3$ does not have an NE **R**D-function. Hence, $\gamma_{\mathbb{R}}(P_3 \square P_3) = -\infty$ while $\gamma_{\mathbb{R}}(P_3) = 1$.

However, Theorem 12 on the strong direct product $G \boxtimes H$ of graphs $G$ and $H$ does generalize. The reason is that, if $g$ and $h$ are dominating functions, then the function $f\colon V(G) \times V(H) \to \mathbb{R}$ defined by $(a, b) \mapsto g(a)h(b)$ (and denoted by

$f = g \otimes h$) is a dominating function of $G \boxtimes H$, and if $g$ and $h$ are both efficient then so is $f$. Furthermore, $w(f) = \mathrm{w}(g) \times \mathrm{w}(h)$. That is,

**Theorem 25** *For all graph G and H, for the*

$$\gamma_{\mathbb{R}}(G \boxtimes H) = \begin{cases} \gamma_{\mathbb{R}}(G) \cdot \gamma_{\mathbb{R}}(H), & \text{if } \gamma_{\mathbb{R}}(G) \text{ and } \gamma_{\mathbb{R}}(H) \text{ both positive,} \\ -\infty, & \text{otherwise.} \end{cases}$$

Brešar, Henning, and Klavzar [4] studied the $\{k\}$-domination number in the Cartesian products of graphs, mostly related to Vizing's conjecture.

**Theorem 26 ([4])** *For any graphs G and H,*

$$\gamma_{\{k\}}(G)\gamma_{\{k\}}(H) \leq k(k+1)\gamma_{\{k\}}(G \,\square\, H).$$

Theorem 26 simplifies to $\gamma(G)\gamma(H) \leq 2\gamma(G \,\square\, H)$ in the special case when $k = 1$, which is the result of Clark and Suen [8].

### *3.2 Hardness Results*

If $\mathcal{P} = \mathbb{R}$ or $\mathbb{Q}$, then the determination of $\gamma_{\mathcal{P}}(G)$ can be formulated in terms of solving a linear programming problem, and so can be computed in polynomial-time (see [24] and [29]). On the other hand, the determination of the domination, signed domination and minus domination numbers has been shown to be NP-complete. We [21] showed that the problem is NP-hard provided $\mathcal{P}$ contains 0 and 1 and is bounded from above. On the other hand, there is a linear-time algorithm for finding a minimum $\mathcal{P}$-dominating function in a tree $T$ in many cases.

However, it remains an open problem to determine the complexity of $\mathbb{Z}$-domination. A graph-theoretic proof that shows that $\mathbb{Z}$-domination is a member of NP is not known. This is, however, a consequence of the general result that integer programming is in NP.

## 4 Other $\mathcal{P}$-Parameters

One can define $\mathcal{P}$-analogues of other graphical parameters. In some cases, such as independence number and total domination, one obtains the analogue of Theorem 23; in other cases, such as upper domination, there is an even simpler characterization of the real version. We discuss here only the independence and upper domination numbers.

Let $G = (V, E)$ be a graph where $|V| = n$ and $|E| = m$. Let $I$ denote the $n \times m$ incidence matrix of $G$. We say a function $f: V \to \mathcal{P}$ is a *$\mathcal{P}$-independence function* of $G$ if for every edge $e$ the sum of the values (weights) assigned under $f$ to the two ends of $e$ is at most 1. The *$\mathcal{P}$-independence number* $\alpha_{\mathcal{P}}(G)$ of $G$ is defined to be the supremum of $w(f)$ taken over all $\mathcal{P}$-independence functions $f$. If $\mathcal{P} = \{0, 1\}$, then one obtains the ordinary independence number.

An obvious lower bound on $\alpha_{\mathbb{R}}$ is $n/2$ attained by assigning to every vertex a weight of $1/2$. We say a function $g: E \to \mathcal{P}$ is an *efficient $\mathcal{P}$-matching function* of a graph $G = (V, E)$ if for every vertex $v$ the sum of the values assigned under $g$ to the edges incident with $v$ is 1. One can obtain a result similar to Proposition 23:

**Theorem 27** *For any graph G on n vertices,*

$$\alpha_{\mathbb{R}}(G) = \begin{cases} \frac{1}{2}n, & \text{if } G \text{ has a nonnegative efficient } \mathbb{R}\text{-matching function,} \\ +\infty, & \text{otherwise.} \end{cases}$$

The upper domination number is defined as the cardinality of the largest minimal dominating set of $G$ (see for example, [10]). We say a $\mathcal{P}$-dominating function $f$ is a *minimal $\mathcal{P}$-dominating function* if there does not exist a $\mathcal{P}$-dominating function $h$, $h \neq f$, such that $h(v) \leq f(v)$ for every $v \in V$. The *upper $\mathcal{P}$-domination number* for $G$ is $\Gamma_{\mathcal{P}}(G) = \sup\{ f(V) \mid f: V \to \mathcal{P}$ is a minimal $\mathcal{P}$-dominating function on $G\}$.

**Theorem 28** *Let $\mathcal{P} = \mathbb{Z}$, $\mathbb{Q}$ or $\mathbb{R}$. Then for any connected graph $G = (V, E)$,*

$$\Gamma_{\mathcal{P}}(G) = \begin{cases} 1, & \text{if } G \text{ is complete,} \\ +\infty, & \text{otherwise.} \end{cases}$$

## 5 Conclusion

In this chapter, we have surveyed some results concerning dominating functions in which negative weights are allowed. We have focused our attention on fractional, integer, or real dominating functions in graphs, and discussed the idea of $\mathcal{P}$-dominating functions in graphs, which provides numerous interesting theoretical and computational questions.

## References

1. N. Alon, Transversal numbers of uniform hypergraphs, Graphs Combin. 6 (1990) 1–4.
2. D. W. Bange, A. E. Barkauskas, L. H. Host, P. J. Slater, Generalized domination and efficient domination in graphs, Discrete Math. 159 (1996) 1–11.
3. D. W. Bange, A. E. Barkauskas, P. J. Slater, Efficient dominating sets in graphs, in: Applications of discrete mathematics (Clemson, SC, 1986), SIAM, Philadelphia, PA, 1988, pp. 189–199.
4. B. Brešar, M. A. Henning, S. Klavžar, On integer domination in graphs and Vizing-like problems, Taiwanese J. Math. 10 (5) (2006) 1317–1328.

5. G. Chappell, J. Gimbel, C. Hartman, Approximations of the domination number of a graph, J. Combin. Math. Combin. Comput. 104 (2018) 287–297.
6. G. A. Cheston, G. Fricke, Classes of graphs for which upper fractional domination equals independence, upper domination, and upper irredundance, Discrete Appl. Math. 55 (3) (1994) 241–258.
7. G. A. Cheston, G. Fricke, S. T. Hedetniemi, D. P. Jacobs, On the computational complexity of upper fractional domination, Discrete Appl. Math. 27 (3) (1990) 195–207.
8. W. E. Clark, S. Suen, An inequality related to Vizing's conjecture, Electron. J. Combin. 7 (2000) Note 4, 3.
9. E. J. Cockayne, G. Fricke, S. T. Hedetniemi, C. M. Mynhardt, Properties of minimal dominating functions of graphs, Ars Combin. 41 (1995) 107–115.
10. E. J. Cockayne, S. T. Hedetniemi, Towards a theory of domination in graphs, Networks 7 (1977) 247–261.
11. G. S. Domke, S. Hedetniemi, R. Laskar, R. Allan, Generalized packings and coverings of graphs, Congr. Numer. 62 (1988) 259–270.
12. G. S. Domke, S. T. Hedetniemi, R. C. Laskar, Fractional packings, coverings, and irredundance in graphs, Congr. Numer. 66 (1988) 227–238.
13. G. S. Domke, S. T. Hedetniemi, R. C. Laskar, G. Fricke, Relationships between integer and fractional parameters of graphs, in: Graph Theory, Combinatorics, and Applications, Vol. 1 (Kalamazoo, MI, 1988), Wiley, New York, 1991, pp. 371–387.
14. G. S. Domke, R. C. Laskar, The bondage and reinforcement numbers of $\gamma_f$ for some graphs, Discrete Math. 167/168 (1997) 249–259.
15. M. Farber, Domination, independent domination, and duality in strongly chordal graphs, Discrete Appl. Math. 7 (1984) 115–130.
16. D. C. Fisher, Domination, fractional domination, 2-packing, and graph products, SIAM J. Discrete Math. 7 (1994) 493–498.
17. D. C. Fisher, Fractional dominations and fractional total dominations of graph complements, Discrete Appl. Math. 122 (2002) 283–291.
18. D. C. Fisher, J. Ryan, G. Domke, A. Majumdar, Fractional domination of strong direct products, Discrete Appl. Math. 50 (1994) 89–91.
19. G. H. Fricke, E. O. Hare, D. P. Jacobs, A. Majumdar, On integral and fractional total domination, Congr. Numer. 77 (1990) 87–95.
20. D. R. Fulkerson, A. J. Hoffman, R. Oppenheim, On balanced matrices, Math. Programming Stud. (1) (1974) 120–132.
21. W. Goddard, M. A. Henning, Real and integer domination in graphs, Discrete Math. 199 (1999) 61–75.
22. D. L. Grinstead, P. J. Slater, Fractional domination and fractional packing in graphs, Congr. Numer. 71 (1990) 153–172.
23. E. O. Hare, $k$-weight domination and fractional domination of $P_m \times P_n$, Congr. Numer. 78 (1990) 71–80.
24. L. G. Hačijan, A polynomial algorithm in linear programming, Dokl. Akad. Nauk SSSR 244 (5) (1979) 1093–1096.
25. S. M. Hedetniemi, S. T. Hedetniemi, T. V. Wimer, Linear time resource allocation for trees, Technical Report URI-014, Dept. Mathematical Sciences, Clemson Univ. (1987).
26. M. A. Henning, G. Kubicki, Real domination in graphs, J. Combin. Math. Combin. Comput. 26 (1998) 147–160.
27. K. Iijima, Y. Shibata, A bipartite representation of a triangulated graph and its chordality, Research Report, Gunma University, Japan (1979).
28. N. John, S. Suen, Graph products and integer domination, Discrete Math. 313 (3) (2013) 217–224.
29. N. Karmarkar, A new polynomial-time algorithm for linear programming, Combinatorica 4 (4) (1984) 373–395.
30. R. Laskar, A. Majumdar, G. Domke, G. Fricke, A fractional view of graph theory, Sankhyā Ser. A 54 (1992) 265–279.

31. R. Rubalcaba, M. Walsh, Minimum fractional dominating functions and maximum fractional packing functions, Discrete Math. 309 (2009) 3280–3291.
32. E. R. Scheinerman, D. H. Ullman, Fractional graph theory, Dover, 2011.
33. L. S. Stewart, E. O. Hare, Fractional domination of $P_m \times P_n$, Congr. Numer. 91 (1992) 35–42.
34. B. Xu, Fractional domination of the Cartesian products in graphs, J. Math. Res. Appl. 35 (2015) 279–284.

# Roman Domination in Graphs

**Mustapha Chellali, Nader Jafari Rad, Seyed Mahmoud Sheikholeslami, and Lutz Volkmann**

This chapter is concerned with the concept Roman domination in graphs, which was introduced in 2004 by Cockayne, Dreyer, S.M. Hedetniemi and S.T. Hedetniemi based on the strategies for defending the Roman Empire presented by Stewart in [80] and ReVelle and Rosing [71]. Since then, more than 180 papers have been published on this topic, where several new variations were defined and that will be exposed in another chapter. In this chapter we will only survey the results on the Roman domination number as well as those when the structure of the graph is modified by the addition of edges/vertices or removing edges/vertices.

## 1 The Roman Domination Problem

For a graph $G = (V, E)$, let $f : V \to \{0, 1, 2\}$ be a function, and let $(V_0, V_1, V_2)$ be the ordered partition of $V$ induced by $f$, where $V_i = \{v \in V : f(v) = i\}$ for

M. Chellali
LAMDA-RO Laboratory, Department of Mathematics, University of Blida, B.P. 270, Blida, Algeria
e-mail: m_chellali@yahoo.com

N. Jafari Rad
Department of Mathematics, Shahed University, Tehran, Iran
e-mail: n.jafarirad@gmail.com

S. M. Sheikholeslami
Department of Mathematics, Azarbaijan Shahid Madani University, Tabriz, Iran
e-mail: s.m.sheikholeslami@azaruniv.ac.ir

L. Volkmann (✉)
Lehrstuhl II für Mathematik, RWTH Aachen University, Templergraben 55, D-52056 Aachen, Germany
e-mail: volkm@math2.rwth-aachen.de



T. W. Haynes et al. (eds.), *Topics in Domination in Graphs*, Developments in Mathematics 64, https://doi.org/10.1007/978-3-030-51117-3_11

$i = 0, 1, 2$. There is a $1-1$ correspondence between the functions $f : V \to \{0, 1, 2\}$ and the ordered partitions $(V_0, V_1, V_2)$ of $V$. So we will write $f = (V_0, V_1, V_2)$.

As defined in [25], a function $f : V \to \{0, 1, 2\}$ is a *Roman dominating function* (or just RDF) if every vertex $u$ for which $f(u) = 0$ is adjacent to at least one vertex $v$ for which $f(v) = 2$. The weight $w(f)$ of a Roman dominating function $f$ is the value $f(V) = \sum_{u \in V} f(u)$. The *Roman domination number* of a graph $G$, denoted by $\gamma_R(G)$, is the minimum weight of an RDF on $G$. A function $f = (V_0, V_1, V_2)$ is called a $\gamma_R$-function (or $\gamma_R(G)$-function when we want to refer $f$ to $G$), if it is a Roman dominating function and $f(V) = \gamma_R(G)$.

In the fourth century, the Roman Empire which dominated large areas of three continents (Europe, North Africa, and part of the Near East) began to lose its power and it was very difficult to secure all the conquered regions. Figure 1 provides a graph region corresponding to the Roman Empire with its eight provinces, where each vertex represents a region connected to specific vertices (neighboring regions). A region is *secured* by an army stationed there, and a region having no army can be protected by an army sent from a neighboring region.

With only four field armies (each consisting of six mobile legions), Emperor Constantine the Great chose to place two field armies in Rome and two others in Constantinople. Moreover, he decreed that a field army cannot be sent to a neighboring region if it leaves its original region unprotected. In other words, every region with no field army must be adjacent to a region that has at least two legions. It should be noted the deployment of armies adopted by Constantine the Great did

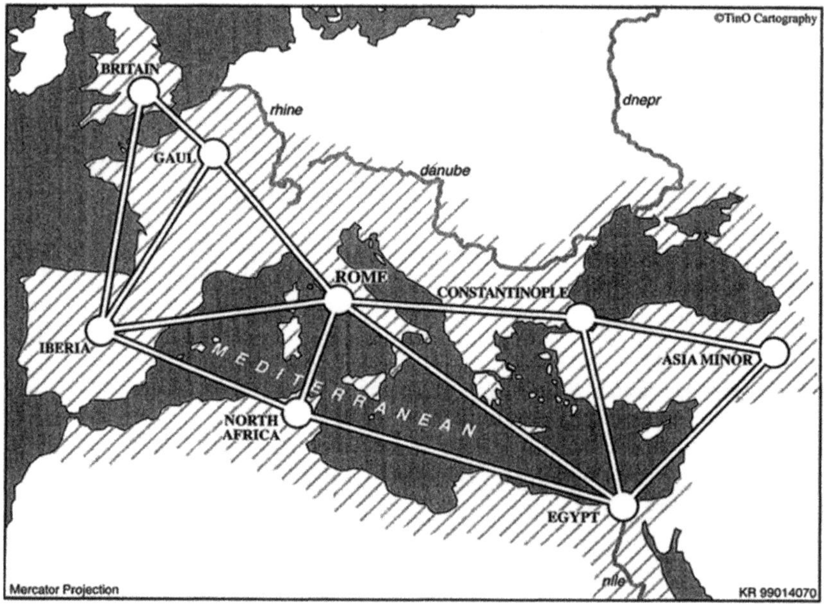


**Fig. 1** Map of the Roman Empire with its eight provinces ([71]).

not secure the region of Britain. However, placing two armies in Rome and an army in each of Britain and Asia Minor could have protected all of the regions of the Roman Empire. This defense strategy prompted the definition of Roman dominating function given in [25].

## 1.1 Properties of Roman Dominating Functions

We begin with the following results, presented in [25] which give basic properties of Roman dominating functions.

**Proposition 1.1 ([25])** *Let $f = (V_0, V_1, V_2)$ be any $\gamma_R(G)$-function of a graph $G$. Then*

**(a)** *The subgraph $G[V_1]$ induced by $V_1$ has maximum degree* 1.
**(b)** *No edge of $G$ joins $V_1$ and $V_2$.*
**(c)** *Each vertex of $V_0$ is adjacent to at most two vertices of $V_1$.*
**(d)** *$V_2$ is a minimum dominating set of $G[V_0 \cup V_2]$.*
**(e)** *Let $H = G[V_0 \cup V_2]$. Then each vertex $v \in V_2$ has at least two private neighbors relative to $V_2$ in $H$.*
**(f)** *If $v$ is isolated in $G[V_2]$ and has precisely one private neighbor, say $w \in V_0$, then $N(w) \cap V_1 = \emptyset$.*
**(g)** *Let $k_1$ equal the number of non-isolated vertices in $G[V_2]$, and let $C = \{v \in V_0 : |N(v) \cap V_2| \geq 2\}$. Then $|V_0| \geq |V_2| + k_1 + |C|$.*

**Proposition 1.2 ([25])** *Let $f = (V_0, V_1, V_2)$ be a $\gamma_R(G)$-function of an isolate-free graph $G$ such that $|V_1|$ is a minimum. Then*

**(a)** *$V_1$ is independent, and $V_0 \cup V_2$ is a vertex cover.*
**(b)** *$V_0$ dominates $V_1$.*
**(c)** *Each vertex of $V_0$ is adjacent to at most one vertex of $V_1$.*
**(d)** *Let $v \in V_2$ have exactly two private neighbors $w_1$ and $w_2 \in V_0$, relative to $V_2$. Then there do not exist vertices $y_1, y_2 \in V_1$ such that $(y_1, w_1, v, w_2, y_2)$ is the vertex sequence of a path $P_5$.*
**(e)** *$|V_0| \geq 3|V|/7$, and this bound is sharp even for trees.*

The two previous propositions allowed Cockayne et al. to establish the following result.

**Corollary 1.3** *For any nontrivial connected graph $G$,*

$$\gamma_R(G) = \min\{2\gamma(G - S) + |S| : S \textit{ is a packing}\}.$$

Favaron et al. [35] were interested in the minimum and maximum values of $|V_0|$, $|V_1|$, and $|V_2|$ for a $\gamma_R(G)$-function $f = (V_0, V_1, V_2)$. Let $\mathcal{R}$ be the family of graphs $G$ obtained from a connected graph $H$ such that each vertex of $H$ is identified with the central vertex of a $P_5$. Let $\mathcal{R}'$ be the family of graphs of $\mathcal{R}$ constructed from a graph $H$ having a vertex of degree $|V(H)| - 1$.

**Theorem 1.4 ([35])** *Let $f = (V_0, V_1, V_2)$ be a $\gamma_R(G)$-function of a connected graph $G$ of order $n \geq 3$. Then*

**(1)** $1 \leq |V_2| \leq \frac{2n}{5}$ *and a graph $G$ admits a $\gamma_R(G)$-function such that $|V_2| = \frac{2n}{5}$ if and only if $G$ belongs to $\mathcal{R} \cup \{C_5\}$.*
**(2)** $0 \leq |V_1| \leq \frac{4n}{5} - 2$ *and a graph $G$ admits a $\gamma_R(G)$-function such that $|V_1| = \frac{4n}{5} - 2$ if and only if $G$ belongs to $\mathcal{R}' \cup \{C_5\}$.*
**(3)** $\frac{n}{5} + 1 \leq |V_0| \leq n - 1$ *and a graph $G$ admits a $\gamma_R(G)$-function such that $|V_0| = \frac{n}{5} + 1$ if and only if $G$ belongs to $\mathcal{R}' \cup \{C_5\}$.*

## 1.2 Bounds on $\gamma_R$

Let $f = (V_0, V_1, V_2)$ be a $\gamma_R(G)$-function of a graph $G$ of order $n$. Since every vertex in $V_2$ has at least one private neighbor in $V_0$, we have $w(f) = |V_1| + 2|V_2| \leq |V_1| + |V_2| + |V_0| = n$, and so $\gamma_R(G) \leq n$. Clearly the equality is attained if and only if $\Delta(G) \leq 1$. This trivial upper bound has been improved by Chambers, Kinnersley, Prince, and West [17] as follows.

**Proposition 1.5 ([17])** *If $G$ is a graph of order $n$ with maximum degree $\Delta$, then $\gamma_R(G) \leq n - \Delta + 1$.*

It is noteworthy that although the upper bound of the Proposition 1.5 is easy to prove, the problem of characterizing the graphs achieving this bound seems to be quite difficult. Indeed, Bouchou, Blidia, and Chellali [15] have proved that the problem of deciding whether $\gamma_R(G) = n - \Delta + 1$ is co-NP-complete. However, they gave characterizations of trees, regular and semiregular graphs attaining the upper bound of Proposition 1.5, which we now present. Recall that a graph $G$ is called *semiregular* if $\Delta(G) - \delta(G) = 1$. We give first a necessary condition for connected graphs $G$ of order $n$ with $\gamma_R(G) = n - \Delta + 1$. For any vertex $v \in V(G)$, we write $\overline{N}(v) = V(G) - N[v]$.

**Proposition 1.6 ([15])** *Let $G$ be a graph of order $n$ with maximum degree $\Delta$. If $\gamma_R(G) = n - \Delta + 1$, then for every vertex $v$ of maximum degree we have:*

**(1)** *Every vertex of $N(v)$ is adjacent to at most two vertices in $\overline{N}(v)$.*
**(2)** *Each component of $G\left[\overline{N}(v)\right]$ is either $K_1$ or $K_2$.*

Let $T$ be a tree with a unique vertex of maximum degree $\Delta \geq 3$, say $x$, where each leaf of $T$ is at distance at most three from $x$ and such that all components in $T - x$ are paths of order at most 5 (for example, see Figure 2). Let $H_i$ be a component of $T - x$ of order $i$, where $i \in \{1, 2, \ldots, 5\}$. Note that since $T$ is a tree, $H_i$ contains exactly one vertex of $N(x)$. Let $n_1$ be the number of components $H_1$, $n_2$ the number of components $H_2$, $n_3$ the number of components $H_3$ having a leaf belonging to $N(x)$, $n_4$ the number of components $H_3$ whose center vertices belong to $N(x)$, $n_5$ the number of components $H_4$ having a support vertex belonging to

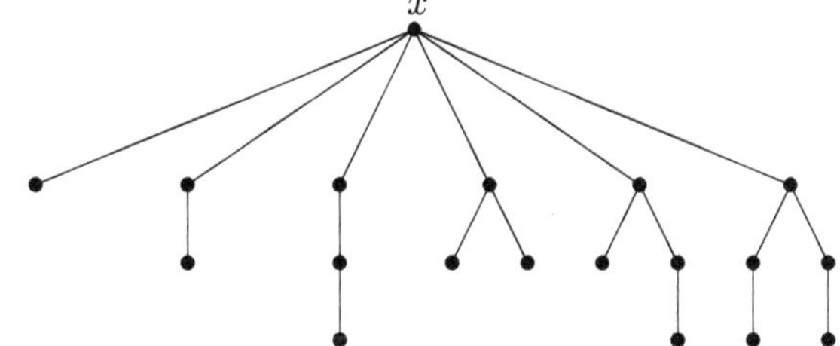


**Fig. 2** Example of a tree $T$, where each component of $T - x$ is a path of order at most 5.

$N(x)$, and $n_6$ be the number of components $H_5$ whose center vertices belong to $N(x)$. Hence $\sum_{i=1}^{6} n_i = \Delta(T)$.

Let $\mathcal{T}$ be the family of trees $T$ with only two adjacent vertices of maximum degree three such that every leaf of $T$ is at distance at most two from a vertex of maximum degree. Note that any tree $T \in \mathcal{T}$ has order $n \in \{6, 7, 8, 9, 10\}$.

**Theorem 1.7 ([15])** *Let $T$ be a nontrivial tree of order $n$ and maximum degree $\Delta$. Then $\gamma_R(T) = n - \Delta + 1$ if and only if $T \in \{P_2, P_3, P_4, P_5\}$ or $T \in \mathcal{T}$ or $T$ is one of the trees defined above such that either $n_4 + n_5 + n_6 = 0$ and ($n_1 \geq 1$ or $n_2 \geq 2$) or $n_4 + n_5 + n_6 \neq 0$ and $n_1 + n_2 \geq 2$.*

For regular graphs, we have:

**Theorem 1.8 ([15])** *Let $G$ be a $\Delta$-regular graph of order $n$ with $\Delta \geq 1$. Then $\gamma_R(G) + \Delta = n + 1$ if and only if $\Delta \in \{1, n-3, n-2, n-1\}$.*

For semiregular graphs, we have:

**Theorem 1.9 ([15])** *Let $G$ be a semiregular graph of order $n$ and maximum degree $\Delta$. Then $\gamma_R(G) + \Delta = n + 1$ if and only if $G$ fulfills one of the following:*

**(a)** $\Delta \geq n - 3$,
**(b)** *$G = pK_1 \cup qK_2$ with $p > 1$, $q \geq 1$ and $p + 2q = n$,*
**(c)** *$G = qK_2 \cup H$ with $2q + |V(H)| = n$, where $H \in \{P_3, P_4, P_5, C_3, C_4, C_5\}$ if $q \neq 0$ and $H \in \{P_3, P_4, P_5\}$ if $q = 0$,*
**(d)** *$G$ is isomorphic to one of the nine graphs in Figure 3.*

Chambers et al. [17] established an upper bound for the Roman domination number in terms of the order and characterized the extremal graphs attaining the bound.

**Theorem 1.10 ([17])** *If $G$ is a connected graph of order $n \geq 3$, then $\gamma_R(G) \leq \frac{4n}{5}$, with equality if and only if $G$ is $C_5$ or is obtained from $\frac{n}{5} P_5$ by adding a connected subgraph on the set of centers of the components of $\frac{n}{5} P_5$.*

Restricted to graphs with minimum degree at least two, Chambers et al. [17] improved the upper bound of Theorem 1.10. Let $\mathcal{B} = \{C_4, C_5, C_8, F_1, F_2\}$, where $F_1$ and $F_2$ are the graphs illustrated in Figure 3. Let also $F$ be the connected graph obtained from two disjoint cycles $C_5$, $(x_1, y_1, z_1, z_2, y_2, x_1)$ and $(x_2, y_3, z_3, z_4, y_4, x_2)$, by adding a new vertex $w$ and edges $wx_1$ and $wx_2$.

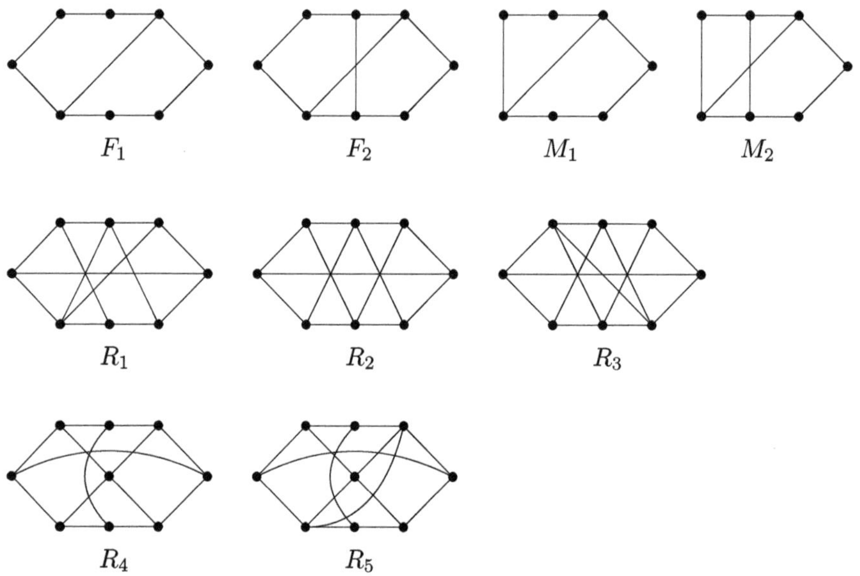


**Fig. 3** Connected semiregular extremal graphs of order $n \in \{7, 8, 9\}$

**Theorem 1.11 ([17])** *If $G$ is a connected graph of order $n$ with $\delta(G) \geq 2$ and $G \notin \mathcal{B}$, then*

$$\gamma_R(G) \leq 8n/11.$$

*Moreover, if $n \geq 9$, then $\gamma_R(G) = 8n/11$ if and only if*

**(1)** $n = 11$ *and $G$ is isomorphic to $F$ plus a subset of one of* $\{y_1y_3, y_1y_4, y_2y_3, y_2y_4\}$, $\{wz_1, y_1y_3, y_1y_4\}$, *or* $\{wz_1, wz_3, y_1y_3\}$ *added as edges, or*
**(2)** $n > 11$ *and $G$ consists of disjoint copies of the graphs $F$, $F + wz_1$, and $F + wz_1 + wz_3$ with additional edges connecting copies of $w$.*

In [12], Bermudo gave an upper bound on the Roman domination number for graphs $G$ containing none of the two following induced subgraphs. Let $B_1$ be the graph obtained from a path $P_7$ whose vertices are labeled in order $v_1, v_2, \ldots, v_7$ by adding the edge $v_3v_7$. Let $B_2$ be the graph obtained from $B_1$ by adding the edge $v_2v_6$.

**Theorem 1.12 ([12])** *Let $G$ be a graph of order $n \geq 15$, minimum degree $\delta \geq 2$, which does not contain any induced subgraph isomorphic to $B_1$ or $B_2$. Then* $\gamma_R(G) \leq \frac{12}{17}n$.

For 2-connected graphs, Liu and Chang [63] obtained the following upper bound.

**Theorem 1.13 ([63])** *For any 2-connected graph $G$ of order $n$,* $\gamma_R(G) \leq \max\{\lceil 2n/3 \rceil, 23n/34\}$.

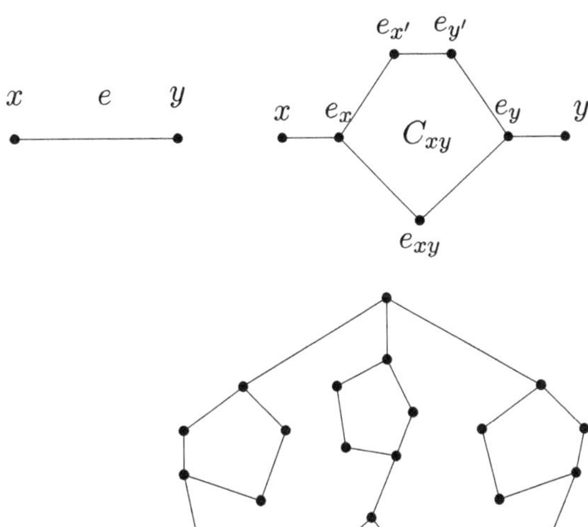


**Fig. 4** Replacement of an edge $e = xy$ by a 5-cycle $C_e$.

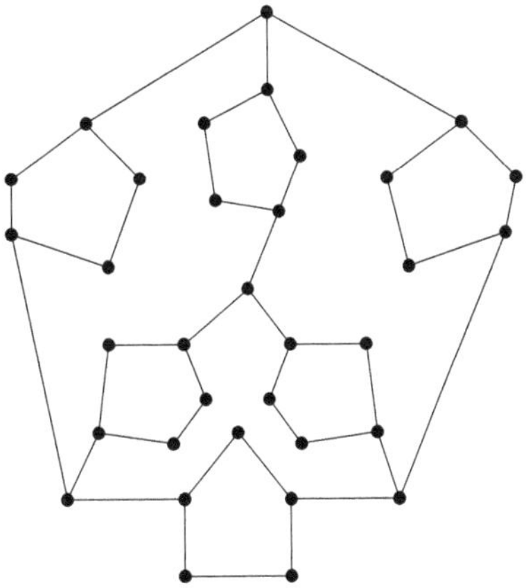

**Fig. 5** Explosion graph of $K_4$

Liu and Chang [63] provided a characterization of 2-connected graphs $G$ of order $n$ with Roman domination number $23n/34$ when $23n/34 > 2n/3$.

The explosion graph $G'$ of a multigraph $G$ is the graph obtained by replacing each edge $e = xy$ of the original graph by a 5-cycle $C_e = C_{xy}$ such that $x$ and $y$ are adjacent to two non-adjacent vertices in the 5-cycle, respectively (see Figures 4 and 5).

A graph $H'$ is an *almost-explosion graph* of a loopless multigraph $H$ if $H'$ can be obtained from the explosion graph of $H$ by adding some or none of the edges in $\{ye_x, xe_y\}$ for each $e = xy \in E(H)$. Now, consider $k$ graphs $G_1, G_2, \ldots, G_k$, each of them isomorphic to an almost-explosion graph of $K_4$, and let $\mathcal{F}_k$ be the family of graphs obtained from a disjoint union of these almost-explosion graphs $G_i$'s by adding suitable edges between vertices of the original $K_4$'s.

**Theorem 1.14 ([63])** *If $G$ is a 2-connected graph and $23n/34 > 2n/3$, then $\gamma_R(G) = 23n/34$ if and only if $G \in \mathcal{F}_k$ for some $k$.*

Restricted to graphs with minimum degree at least three, Liu and Chang [64] obtained the following sharp upper bound on the Roman domination number.

**Theorem 1.15 ([64])** *If $G$ is a graph with $n$ vertices and $\delta(G) \geq 3$, then $\gamma_R(G) \leq 2n/3$.*

The sharpness of the previous bound is shown for a large family of cubic 3-connected graphs described in [64] as follows. For any integer $t \geq 3$, we construct

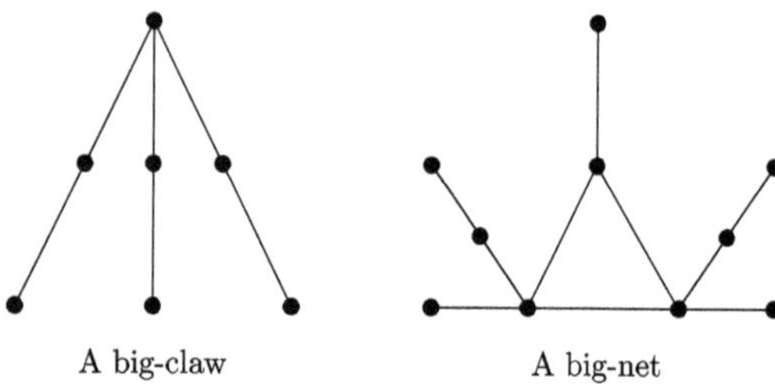


**Fig. 6** The big-claw and big-net graphs

the graph $G_t$ from the union of two disjoint $3t$-cycles $x_1, x_2, \ldots, x_{3t}, x_1$ and $y_1, y_2, \ldots, y_{3t}, y_1$ by adding edges $x_i y_{j_i}$ for $1 \le i \le 3t$, where $j_i = i$ if $i \equiv 1 \pmod 3$, $j_i = i + 1$ if $i \equiv 2 \pmod 3$, and $j_i = i - 1$ if $i \equiv 0 \pmod 3$.

The next result is stated for graphs with no induced subgraph isomorphic to neither a big-claw nor a big-net illustrated in Figure 6.

**Theorem 1.16 ([64])** *If $G$ is a connected big-claw-free and big-net-free graph with $n$ vertices, then $\gamma_R(G) \le \lceil 2n/3 \rceil$.*

Using a probabilistic method, Cockayne et al. [25] proved the following upper bound.

**Theorem 1.17 ([25])** *For a graph $G$ of order $n$ and minimum degree $\delta$,*

$$\gamma_R(G) \le \frac{2 + \ln(1+\delta)/2}{1+\delta} n.$$

Zverovich and Poghosyan [92] proved that the bound of Theorem 1.17 is asymptotically best possible. They also presented another probabilistic upper bound for the Roman domination number of a graph.

**Theorem 1.18 ([92])** *For any graph $G$ with $\delta = \delta(G) > 0$,*

$$\gamma_R(G) \le 2\left(1 - \frac{2^{\frac{1}{\delta}}\delta}{(1+\delta)^{1+\frac{1}{\delta}}}\right)n.$$

We conclude this subsection by the following lower bound on the Roman domination number in terms of the order and maximum degree that can be found in [26].

**Proposition 1.19 ([26])** *For any graph $G$ of order $n$ and maximum degree $\Delta \ge 1$,*

$$\gamma_R(G) \ge \frac{2n}{\Delta + 1}.$$

In [79], the authors present the bound $\gamma_R(G) \ge \lceil \frac{2n}{\Delta+1} \rceil + \epsilon$ for graphs $G$ of order $n$ and maximum degree $\Delta \ge 1$, with $\epsilon = 0$ when $n \equiv 0, 1 \pmod{(\Delta + 1)}$ and

$\epsilon = 1$ when $n \not\equiv 0, 1 \pmod{(\Delta+1)}$. The following example shows that this bound is not correct in general. Let $H$ be the graph obtained from a path $v_1v_2\ldots v_{12}$ by adding a pendant edge at the vertices $v_2$, $v_5$, and $v_8$. Then $\gamma_R(H) = 8$, but it follows $\gamma_R(H) \geq 9$ from the bound above.

## *1.3 Relationships with Some Domination Parameters*

In this subsection we present various relations involving $\gamma_R$ and some other parameters including $\gamma$ and $\gamma_t$. By Proposition 1.1, for every RDF $f = (V_0, V_1, V_2)$ of $G$, $V_1 \cup V_2$ is a dominating set in $G$. Moreover, assigning a 2 to every vertex in a minimum dominating set $D$ of $G$ and a 0 to every vertex not in $D$ yields a RDF of $G$. Hence we have the following inequality.

**Theorem 1.20 ([25])** *For every graph $G$, $\gamma(G) \leq \gamma_R(G) \leq 2\gamma(G)$. Moreover, the lower bound is sharp if and only if $G$ is the empty graph.*

Graphs $G$ such that $\gamma_R(G) = 2\gamma(G)$ are called *Roman graphs*. As mentioned in [25], the only Roman paths and cycles are $P_{3k}$, $P_{3k+2}$, $C_{3k}$, and $C_{3k+2}$. Henning [42] gave a characterization of Roman trees. Notice that such a characterization is called *constructive* since it applies recursively operations to build extremal trees. Cockayne et al. gave the following equivalent conditions for Roman graphs whose characterization remains an open problem.

**Proposition 1.21 ([25])** *Let $G$ be a graph. Then the following conditions are equivalent.*

**(i)** *$G$ is a Roman graph.*
**(ii)** *$G$ has a $\gamma_R(G)$-function $f = (V_0, V_1, V_2)$ such that $|V_1| = 0$.*
**(iii)** *$\gamma(G) \leq \gamma(G - S) + |S|/2$, for every packing set $S \subseteq V$.*

The graphs $G$ such that $\gamma_R(G) \in \{\gamma(G) + 1, \gamma(G) + 2\}$ have been characterized in [25] as follows.

**Theorem 1.22 ([25])** *Let $G$ be a connected graph of order $n$. Then*

**(i)** *$\gamma_R(G) = \gamma(G) + 1$ if and only if there is a vertex $v \in V$ of degree $n - \gamma(G)$.*
**(ii)** *$\gamma_R(G) = \gamma(G) + 2$ if and only if:*

*(a) $G$ does not have a vertex of degree $n - \gamma(G)$.*
*(b) either $G$ has a vertex of degree $n - \gamma(G) - 1$ or $G$ has two vertices $v$ and $w$ such that $|N[v] \cup N[w]| = n - \gamma(G) + 2$.*

The class of trees $T$ such that $\gamma_R(T) \in \{\gamma(T) + 1, \gamma(T) + 2\}$ has been characterized in [25]. For a positive integer $t$, a *wounded spider* is a star $K_{1,t}$ with at most $t - 1$ of its edges subdivided. Likewise, for an integer $t \geq 2$, *a healthy spider* is a star $K_{1,t}$ with all of its edges subdivided. In a wounded spider, a vertex of degree $t$ will be called the *head vertex*, and the vertices that are distance two from the head vertex will be the *foot vertices*. If the wounded spider is the path $P_2$, then both vertices of the path are head vertices, while in the case of $P_4$, both end vertices of $P_4$ are foot vertices and both interior vertices are head vertices.

**Proposition 1.23 ([25])** *If $T$ is a tree of order $n \geq 2$, then:*

**(i)** $\gamma_R(T) = \gamma(T) + 1$ *if and only if $T$ is a wounded spider.*
**(ii)** $\gamma_R(T) = \gamma(T) + 2$ *if and only if (i) $T$ is a healthy spider or (ii) $T$ is a pair of wounded spiders $T_1$ and $T_2$, with a single edge joining $v \in V(T_1)$ and $w \in V(T_2)$, subject to the following conditions:*

*(a) if either tree is a $P_2$, then neither vertex in $P_2$ are joined to the head vertex of the other tree.*
*(b) $v$ and $w$ are not both foot vertices.*

In [87], Xing, Chen, and Chen completed the characterization of graphs $G$ for which $\gamma_R(G) = \gamma(G) + k$ for any integer $k$ such that $2 \leq k \leq \gamma(G)$ in response to the open question posed in [25]. We note that the proof given in [87] contained a logical mistake that was corrected by Wu and Xing in [86].

**Theorem 1.24 ([87])** *Let $G$ be a connected graph of order $n$ and the domination number $\gamma(G) \geq 2$. If $k$ is an integer such that $2 \leq k \leq \gamma(G)$, then $\gamma_R(G) = \gamma(G) + k$ if and only if:*

**(i)** *for any integer $s$ with $1 \leq s \leq k-1$, $G$ does not have a set $U_t$ of $t$ $(1 \leq t \leq s)$ vertices such that $\left|\cup_{v \in U_t} N[v]\right| = n - \gamma(G) - s + 2t$;*
**(ii)** *there exists an integer $l$ with $1 \leq l \leq k$, and $G$ has a set $W_l$ of $l$ vertices such that $\left|\cup_{v \in W_l} N[v]\right| = n - \gamma(G) - k + 2l$.*

The lower bound $\gamma_R(G) \geq \gamma(G)$ has been improved by Chellali, Haynes, and Hedetniemi [19] by providing a bound in terms of the domination number and the maximum degree. Before presenting the new bound, let us recall that a dominating set $S$ for which $|N[v] \cap S| = 1$ for all $v \in V$ is an *efficient dominating set* (see Bange et al. [10]). Thus, a set $S$ is an efficient dominating set if $S$ is both a dominating set and a packing in $G$. We note that not every graph has an efficient dominating set, for example, the cycle $C_5$ does not. But paths $P_n$ and cycles $C_{3k}$ admit such sets. Also, one of the well-known bounds on the domination number is that due to Walikar et al. [83] who proved that for every graph $G$, $\gamma(G) \geq n/(1 + \Delta(G))$. If $\mathcal{G}$ denotes the family of connected graphs $G$ of order $n$ such that $\gamma(G) = n/(1 + \Delta(G))$, then clearly all graphs of $\mathcal{G}$ admit an efficient dominating set in which every vertex has maximum degree. Graphs other than those in $\mathcal{G}$ have efficient dominating sets. For instance, the path $P_4$ has a unique efficient dominating set but $P_4 \notin \mathcal{G}$. Let $\mathcal{F}$ be the family of graphs $G$ such that $G$ is the cycle $C_4$ or the corona of any connected graph in $\mathcal{G}$.

**Theorem 1.25 ([19])** *Let $G$ be a nontrivial, connected graph with maximum degree $\Delta$. Then $\gamma_R(G) \geq \frac{\Delta+1}{\Delta}\gamma(G)$, with equality if and only if $G \in \mathcal{F}$.*

In [35], Favaron et al. gave a relation involving $\gamma_R$ and $\gamma$ for any connected graph as follows. We recall that $\mathcal{R}$ is the family of graphs $G$ obtained from a connected graph $H$ such that each vertex of $H$ is identified with the central vertex of a $P_5$.

**Theorem 1.26 ([35])** *For any connected graph G of order $n \geq 3$, $\gamma_R(G) + \frac{1}{2}\gamma(G) \leq n$, with equality if and only if G is $C_4$, $C_5$, $C_4 \circ K_1$ or $G \in \mathcal{R}$.*

In [13], Bermudo, Fernau, and Sigarreta proved a very interesting relationship between Roman domination and the differential of a graph. For a set $S$, let $B(S)$ be the set of vertices in $V - S$ that have a neighbor in the set $S$. The *differential* of a set $S$, as defined in [67], is $\partial(S) = |B(S)| - |S|$, and the maximum value of $\partial(S)$ for any subset $S$ of $V$ is the *differential* of $G$, denoted $\partial(G)$. So, the following unexpected result was shown in [13].

**Theorem 1.27 ([13])** *If G is a graph of order n, then $\gamma_R(G) = n - \partial(G)$.*

**Proof.** For every RDF $f = (V_0, V_1, V_2)$ we can consider $f' = (V_0', V_1', V_2')$, where $V_2' = V_2$, $V_0' = B(V_2)$, and $V_1' = V - (V_2 \cup B(V_2)) = C(V_2)$, to obtain $f(V) \geq f'(V)$. Therefore

$$\gamma_R(G) = \min\{f(V) : f = (V_0, V_1, V_2) \text{ is an RDF and } V_0 = B(V_2)\}.$$

Finally, using that $|V_2| = |B(V_2)| - \partial(V_2)$, we have

$$\begin{aligned} \gamma_R(G) &= \min_{V_2 \subseteq V}\{2|V_2| + |C(V_2)|\} = \min_{V_2 \subseteq V}\{|B(V_2)| - \partial(V_2) + |V_2| + |C(V_2)|\} \\ &= \min_{V_2 \subseteq V}\{n - \partial(V_2)\} = n - \max_{V_2 \subseteq V}\{\partial(V_2)\} = n - \partial(G). \end{aligned}$$

■

Bermudo et al. derived various results on the Roman domination number of a graph $G$ from known bounds on $\partial(G)$. In particular they obtained the following.

**Theorem 1.28 ([13])** *Let G be a connected graph G of order n. Then:*

**(i)** *If $n \geq 3$ and $\delta(G) \geq 1$, then $\gamma_R(G) \leq \frac{2}{5}n + \gamma(G)$.*
**(ii)** *If $n \geq 9$ and $\delta(G) \geq 2$, then $\gamma_R(G) \leq \frac{4}{11}n + \gamma(G)$.*

In another paper, Bermudo [12] showed that if $G$ is a graph of order $n \geq 9$ with minimum degree at least two, then $\partial(G) \geq \frac{3}{4}\gamma(G)$. So the next result becomes immediate from Theorem 1.27, which is an improvement of Theorem 1.26.

**Theorem 1.29 ([12])** *If G is a graph of order $n \geq 9$ with minimum degree at least two, then $\gamma_R(G) + \frac{3}{4}\gamma(G) \leq n$.*

Moreover, Bermudo [12] conjectured that $\gamma_R(G) + \gamma(G) \leq n$ for all graphs $G$ of order $n$ and minimum degree at least three.

We now turn our attention to another domination parameter, namely the total domination number to which the Roman domination number is related. In [41], Hedetniemi et al. established a relation involving these two parameters.

**Theorem 1.30 ([41])** *If $G$ is a graph without isolated vertices, then $\gamma_R(G) \geq \gamma_t(G)$.*

The authors [41] concluded their paper with the open problem on characterizing the graphs $G$ having $\gamma_R(G) = \gamma_t(G)$. It is worth noting that from [25] and [41], we now have that for any graph $G$ without isolated vertices,

$$\gamma(G) \leq \gamma_t(G) \leq \gamma_R(G) \leq 2\gamma(G) \leq 2\gamma_t(G).$$

The problem of characterizing graphs $G$ with equal Roman domination and total domination numbers has been considered by Chellali, Haynes, and Hedetniemi [18], where their main result was the following.

**Theorem 1.31 ([18])** *For every connected graph $G$ of order $n \geq 2$, $\gamma_R(G) = \gamma_t(G)$ if and only if $\gamma_t(G) = 2\gamma(G)$.*

According to Theorem 1.31, if $G$ is a graph having $\gamma_R(G) = \gamma_t(G)$, then $\gamma_R(G) = 2\gamma(G)$ and thus $G$ is a Roman graph. However, not all Roman graphs satisfy $\gamma_R(G) = \gamma_t(G)$. To see this, consider the double star $T$, where each support vertex is adjacent to at least three leaves. Then $\gamma_R(T) = 2\gamma(T)$, but $4 = \gamma_R(T) > \gamma_t(T) = \gamma(T) = 2$.

The authors [18] also obtained the following result for graphs with diameter two.

**Proposition 1.32 ([18])** *For every graph $G$ of diameter two, the following statements are equivalent:*

**(a)** $\gamma_R(G) = \gamma_t(G)$.
**(b)** $\gamma_t(G) = 2\gamma(G)$.
**(c)** $\gamma(G) = 1$.

Finally, we present some results relating Roman domination and 2-rainbow domination in graphs. For a positive integer $k$, a *$k$-rainbow dominating function* ($k$RDF) of a graph $G$ is a function $f : V(G) \to \mathcal{P}(\{1, \ldots, k\})$ such that for any vertex $v \in V(G)$ with $f(v) = \emptyset$ the condition $\cup_{u \in N(v)} f(u) = \{1, \ldots, k\}$ is fulfilled. The weight of a $k$RDF $f$ is defined as $w(f) = \sum_{v \in V(G)} |f(v)|$, and the minimum weight of a $k$RDF is called the *$k$-rainbow domination number* of $G$, denoted by $\gamma_{rk}(G)$. The concept of $k$-rainbow domination was introduced by Brešar, Henning, and Rall in 2008 [16].

We start with a result that was obtained independently by Chellali and Jafari Rad in [21], and Wu and Xing in [86].

**Theorem 1.33 ([21, 86])** *For every graph $G$,*

$$\gamma_{r2}(G) \leq \gamma_R(G) \leq \frac{3}{2}\gamma_{r2}(G).$$

Fujita and Furuya [37] proved an upper bound for the sum of $\gamma_R(G)$ and $\gamma_{r2}(G)$ in terms of the order of $G$.

**Theorem 1.34 ([37])** *If $G$ is a connected graph of order $n \geq 3$, then*

$$\gamma_R(G) + \gamma_{r2}(G) \leq \frac{3}{2}n.$$

Moreover, Fujita and Furuya conjectured that if $G$ is a connected graph of minimum degree at least 2 that is distinct from $C_5$, then $\gamma_R(G)+\gamma_{r2}(G) \leq \frac{4}{3}n$. This conjecture was settled in 2016 by Alvarado, Dantas, and Rautenbach [4] who gave in addition a characterization of extremal graphs attaining the bound of Theorem 1.34.

For $k \in \mathbb{N}$, let $\mathcal{T}_k$ be the set of all trees $T$ that arise from $k$ disjoint copies $a_1b_1c_1d_1, \ldots, a_kb_kc_kd_k$ of the path $P_4$ by adding some edges between vertices the $b_i$'s so that the resulting graph is a tree. Let $\mathcal{T} = \bigcup_{k\in\mathbb{N}} \mathcal{T}_k$. Let the class $\mathcal{K}$ of connected graphs be such that a connected graph $G$ belongs to $\mathcal{K}$ if and only if $G$ arises

- either from the unique tree in $\mathcal{T}_k$ by adding the edge $c_1c_2$,
- or from some tree in $\mathcal{T}_k$ by arbitrarily adding edges between vertices in $\{b_1, \ldots, b_k\}$.

**Theorem 1.35 ([4])** *Let $G$ is a connected graph of order $n \geq 3$. Then $\gamma_R(G) + \gamma_{r2}(G) = \frac{3}{2}n$ if and only if $G \in \mathcal{K}$.*

## 1.4 Nordhaus–Gaddum Type Results

For a graph parameter $\mu$, bounds on $\mu(G) + \mu(\overline{G})$ and $\mu(G)\mu(\overline{G})$ in terms of the order of $G$ are called results of "*Nordhaus–Gaddum*" type, to honor Nordhaus and Gaddum [68] who gave bounds on the sum and product of the chromatic numbers of a graph and its complement. Chambers et al. [17] were the first to investigate Nordhaus–Gaddum type bounds for Roman domination.

**Theorem 1.36 ([17])** *If $G$ is a connected graph of order $n \geq 3$, then $\gamma_R(G) + \gamma_R(\overline{G}) \leq n + 3$, with equality if and only if $G$ or $\overline{G}$ is $C_5$ or $\frac{n}{2}K_2$.*

By Theorem 1.36, if neither $G$ nor $\overline{G}$ is $C_5$ or $\frac{n}{2}K_2$, then $\gamma_R(G)+\gamma_R(\overline{G}) \leq n+2$. Jafari Rad et al. [50] and Bouchou et al. [15] independently characterized graphs $G$ with $\gamma_R(G) + \gamma_R(\overline{G}) = n + 2$. Consider the following families of graphs.

- $\mathcal{H}_0 = \{C_6, C_7, C_8, C_i \cup C_j$, where $i, j \in \{3, 4, 5\}\}$,
- $\mathcal{H}_1 = \{pK_1 \cup qK_2 : p \geq 1, q \geq 1$ and $p + 2q = n\}$,
- $\mathcal{H}_2 = \{qK_2 \cup H$ with $2q + |V(H)| = n$, where $H \in \{P_3, P_4, P_5, C_3, C_4, C_5\}$ if $q \neq 0$ and $H \in \{P_3, P_4, P_5\}$ if $q = 0\}$,
- $\mathcal{H}_3 = \{F_1, F_2, M_1, M_2\}$, (shown in Figure 3).

**Theorem 1.37 ([15, 50])** *Let $G$ be a graph of order $n \geq 3$ such that $G \notin \{C_5, \frac{n}{2}K_2, \overline{\frac{n}{2}K_2}\}$. Then $\gamma_R(G) + \gamma_R(\overline{G}) \leq n + 2$, with equality if and only if $G$ or $\overline{G}$ belongs to $\{K_n\} \cup \mathcal{H}_0 \cup \mathcal{H}_1 \cup \mathcal{H}_2 \cup \mathcal{H}_3$.*

**Theorem 1.38 ([17])** *If $G$ is a graph of order $n \geq 160$, then $\gamma_R(G)\gamma_R(\overline{G}) \leq \frac{16n}{5}$, with equality if and only if $G$ or $\overline{G}$ is $\frac{n}{5}C_5$.*

Bouchou et al. [15] improved Theorem 1.36 for graphs of order $n \geq 160$.

**Theorem 1.39 ([15])** *If $G$ is a graph of order $n \geq 160$ such that every component of $G$ or $\overline{G}$ is of order at least 3, then $\gamma_R(G) + \gamma_R(\overline{G}) \leq \frac{4n}{5} + 4$, with equality only when $G$ or $\overline{G}$ is $\frac{n}{5}C_5$.*

## 1.5 Algorithmic and Complexity Results

Cockayne et al. [25] stated that the Roman domination problem on trees can be solved in linear time. However, this algorithm can be found in the 2000 PhD thesis by Dreyer Jr. [31]. They also stated that Alice McRae (2000) in a private communication has constructed proofs showing that the decision problem corresponding to the Roman domination number is NP-complete even when restricted to chordal graphs, bipartite graphs, split graphs, or planar graphs. Since these results have never been published, Schnupp [78] has shown the NP-completeness of this problem for planar, split, and bipartite graphs. Moreover, linear-time algorithms for the problem on block graphs and bounded treewidth graphs have been proposed in [44, 70]. The complexity of the Roman domination problem for interval graphs was mentioned as an open question in [25], and Liedloff et al. [62] answered this problem by showing that there are linear-time algorithms to compute the Roman domination number for interval graphs and cographs. In addition, they proved that the Roman domination problem can be expressed as a LinEMSOL($\tau_1$) optimization problem, which therefore implies that the Roman domination problem can be solved in linear time on graphs $G$ with a bounded clique-width $k$, provided that a $k$-expression of $G$ is also a part of the input. Liedloff et al. [62] have also shown that there are polynomial-time algorithms for computing the Roman domination numbers of AT-free graphs and graphs with a $d$-octopus.

Liu et al. [65] continued the study of algorithmic aspect of Roman domination by considering a slightly more general setting as follows. Given real numbers $b \geq a > 0$, an *$(a, b)$-Roman dominating function* of a graph $G = (V, E)$ is a function $f : V \rightarrow \{0, a, b\}$ such that every vertex $v$ with $f(v) = 0$ has a neighbor $u$ with $f(u) = b$. For $b = a = 1$, this is a dominating function; and for $b = 2$ and $a = 1$, this is a Roman dominating function. The *$(a, b)$-Roman domination number* $\gamma_R^{(a,b)}(G)$ is the minimum weight of an $(a, b)$-Roman domination function of $G$. An *independent* (respectively, *connected* or *total*) *$(a, b)$-Roman dominating function* is an $(a, b)$-Roman dominating function $f$ such that $\{v \in V : f(v) \neq 0\}$ induces a subgraph without edges (respectively, that is connected or without isolated vertices).

For a weight function $w$ on $V$, the weight of $f$ is $w(f) = \sum_{v \in V} w(v) f(v)$. The *weighted* $(a, b)$*-Roman domination number* $\gamma_R^{(a,b)}(G, w)$ is the minimum weight of an $(a, b)$-Roman dominating function of $G$. Similarly, the weighted *independent* (respectively, *connected* or *total*) $(a, b)$*-Roman domination number* $\gamma_{iR}^{(a,b)}(G, w)$ (respectively, $\gamma_{cR}^{(a,b)}(G, w)$ or $\gamma_{tR}^{(a,b)}(G, w)$) can be defined.

**Theorem 1.40 ([65])** *For any fixed* $(a, b)$ *the* $(a, b)$*-Roman domination problem (the total/connected/independent* $(a, b)$*-Roman domination problems) is NP-complete for bipartite graphs.*

**Theorem 1.41 ([65])** *For any fixed* $(a, b)$ *the* $(a, b)$*-Roman domination problem (the total/connected/weighted independent (a,b)-Roman domination problems) is NP-complete for chordal graphs.*

Liu et al. [65] also used linear programming method of Farber [34], and gave linear-time algorithms for the weighted $(a, b)$-Roman domination problem with $b \geq a > 0$, and the weighted independent $(a, b)$-Roman domination problem with $2a \geq b \geq a > 0$ on strongly chordal graphs with a strong elimination ordering provided.

## 1.6 Minimal Roman Dominating Functions

One of the problems posed by Dreyer Jr. [31] in his 2000 PhD thesis was about developing other interesting definitions of minimal Roman dominating functions. This problem was posed following some attempts to define the upper Roman domination number $\Gamma_R(G)$. The first attempt was due to E. J. Cockayne who suggested that a Roman dominating function was minimal if no value assigned to a vertex can be reduced and still result in a Roman dominating function. Clearly, by assigning a 1 to every vertex of a graph $G$ provides a minimal Roman dominating set and so $\Gamma_R(G) = |V(G)|$ . The second attempt was due by Laskar who suggested that in addition to the Cockayne requirement that no value can be reduced, the Roman dominating function must assign the value 2 to at least one vertex. It is also clear in this case that if $G$ is a graph with a leaf, then assigning a 2 to every leaf and a 0 to its support vertex provides a minimal Roman dominating function and again $\Gamma_R(G) = |V(G)|$ . Therefore both definitions seem to be uninteresting. It was not until twelve years after the introduction of Roman domination in 2004 to see an "interesting" definition of $\Gamma_R(G)$ given in [20] which allowed to obtain the Roman domination chain analogically to the well-known domination chain stated in 1978 by Cockayne, Hedetniemi, and Miller [27] relating the six parameters $ir(G)$, $\gamma(G)$, $i(G)$, $\alpha(G)$, $\Gamma(G)$, and $IR(G)$.

**Theorem 1.42 ([27])** *For any graph* $G$,

$$ir(G) \leq \gamma(G) \leq i(G) \leq \alpha(G) \leq \Gamma(G) \leq IR(G).$$

**Definition 1.43 ([20])** *A function $f = (V_0, V_1, V_2)$ is called irredundant if (i) $V_1$ is an independent set, (ii) no vertex in $V_1$ is adjacent to a vertex in $V_2$, and (iii) every vertex in $V_2$ has a private neighbor in $V_0$, with respect to the set $V_2$. Moreover, an irredundant function is said to be maximal if increasing the value assigned to any vertex results in a function that is no longer irredundant.*

Notice that every graph $G$ has an irredundant $\gamma_R(G)$-function, and thus the Roman domination number could equivalently be defined to equal the minimum weight of an irredundant Roman dominating function. So therefore, the *upper Roman domination number*, denoted by $\Gamma_R(G)$, has been defined as being equal to the maximum weight of an irredundant Roman dominating function on $G$.

Independent Roman dominating functions were defined in [25] (without being studied) as Roman dominating functions $f = (V_0, V_1, V_2)$ for which the set $V_1 \cup V_2$ is an independent set.

**Definition 1.44** *The independent Roman domination number $i_R(G)$ equals the minimum weight of an independent Roman dominating function on $G$, and the Roman independence number $\alpha_R(G)$ equals the maximum weight of an irredundant, independent Roman dominating function on $G$.*

We note that a subsection will be devoted thereafter to the independent Roman domination number $i_R(G)$ detailing the results obtained on it.

Chellali et al. [20] showed that every independent Roman dominating function on $G$ with weight $i_R(G)$ is irredundant. Thereby, $i_R(G)$ and $\alpha_R(G)$ have been defined in terms of irredundant, independent Roman dominating functions, while $\gamma_R(G)$ and $\Gamma_R(G)$ have been defined in terms of irredundant Roman dominating functions. Therefore $\gamma_R(G) \leq i_R(G) \leq \alpha_R(G) \leq \Gamma_R(G)$ for all $G$.

**Definition 1.45** *The Roman irredundance number $ir_R(G)$ equals the minimum weight of a maximal irredundant function on $G$, and the upper Roman irredundance number $IR_R(G)$ equals the maximum weight of an irredundant function on $G$.*

According to the definitions of $\Gamma_R(G)$ and $IR_R(G)$, it is obvious that $\Gamma_R(G) \leq IR_R(G)$ for all $G$. Moreover, Chellali et al. [20] showed that every graph $G$ has a $\gamma_R$-function that is maximal irredundant, and thus $ir_R(G) \leq \gamma_R(G)$. It should be noted that irredundant functions need not to be Roman dominating functions as can seen by the graph of Figure 7.

Let $h$ be a function on $G$ that assigns the value 0 to each $a_i$ vertex and to vertices $c$ and $d$; the value 2 to vertex $e$, and the value 1 to each $b_i$ vertex. Then $h$ is a maximal irredundant function but not a Roman dominating function, since no $a_i$ vertex, for $1 \leq i \leq 4$, has a neighbor assigned a 2 under $h$. Also, for this example $ir_R(G) = 6 < \gamma_R(G) = 7$.

Taking into account the above definitions, the following main result is obtained by Chellali et al. in [20].

**Theorem 1.46 ([20])** *For any graph $G$,*

$$ir_R(G) \leq \gamma_R(G) \leq i_R(G) \leq \alpha_R(G) \leq \Gamma_R(G) \leq IR_R(G).$$

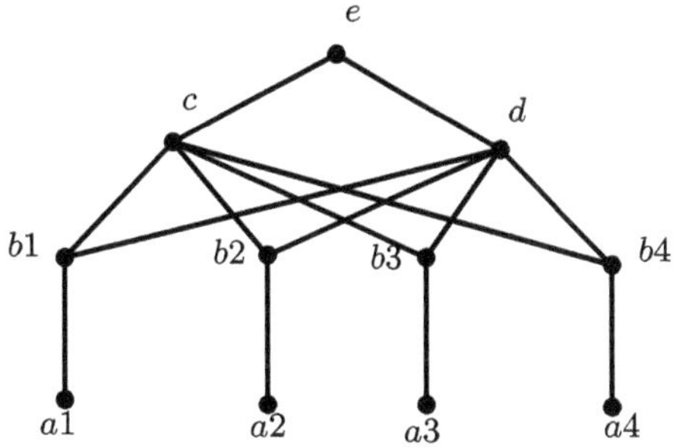


**Fig. 7** The graph $G$ with $ir_R(G) = 6$ and $\gamma_R(G) = 7$

Several examples of trees have been presented in [20] showing that the differences $\gamma_R - ir_R$, $\Gamma_R - \alpha_R$, and $IR_R - \Gamma_R$ can be arbitrarily large. Additional results obtained in [20] are gathered as follows.

**Theorem 1.47 ([20])** *Let G be a graph of order n. Then*

**(a)** $IR_R(G) \leq n$, *with equality if and only if* $\Gamma_R(G) = n$.
**(b)** *If* $\delta(G) \geq 1$, *then* $\Gamma_R(G) \geq 2\gamma(G)$.
**(c)** $\alpha(G) \leq \alpha_R(G) \leq 2\alpha(G)$.
**(d)** $ir(G) \leq ir_R(G)$.
**(e)** $\Gamma(G) \leq \Gamma_R(G)$.
**(f)** $IR(G) \leq IR_R(G)$.

It is worth noting that unlike $\gamma_R(G) \leq 2\gamma(G)$ for any graph $G$, it was shown in [20] that none of the parameters $ir_R$, $\Gamma_R$, and $IR_R$ can be bounded above by $2ir$, $2\Gamma$, and $2IR$, respectively. We close this subsection by a characterization of connected graphs $G$ of order $n$ such that $\alpha_R(G) = n$.

**Theorem 1.48 ([20])** *Let* $G = (V, E)$ *be a connected graph of order* $n \geq 2$. *Then* $\alpha_R(G) = n$ *if and only if its vertex set* $V$ *can be partitioned into three sets* $\{V_0, V_1, V_2\}$ *such that (i) the set* $V_1 \cup V_2$ *is an independent set, (ii)* $|V_0| = |V_2|$, *(iii)* $V_2$ *is a non-empty set of leaves, and (iv) no vertex in* $V_0$ *is adjacent to two or more vertices in* $V_2$.

## 1.7 Roman Domination in Special Graphs

This subsection will be devoted to presenting the exact values of the Roman domination number as well as some bounds on it for several classes of graphs. We begin by the following two results established in [25].

**Proposition 1.49** *For the classes of paths* $P_n$ *and cycles* $C_n$, $\gamma_R(P_n) = \gamma_R(C_n) = \lceil \frac{2n}{3} \rceil$.

**Proposition 1.50** *Let* $G = K_{m_1,\ldots,m_n}$ *be the complete n-partite graph with* $m_1 \leq m_2 \leq \ldots \leq m_n$.

**(a)** *If* $m_1 \geq 3$, *then* $\gamma_R(G) = 4$.
**(b)** *If* $m_1 = 2$, *then* $\gamma_R(G) = 3$.
**(c)** *If* $m_1 = 1$, *then* $\gamma_R(G) = 2$.

### 1.7.1 Cartesian Product of Graphs

The Cartesian products of two paths and two cycles are known as *grid graphs* and *torus graphs*, respectively, and the Cartesian product of any graph $G$ with a path $P_2$ is called *the prism graph* of $G$.

The Roman domination number of grid graphs was studied in [25, 28, 31, 69], where the following results were given.

**Proposition 1.51 ([25])** $\gamma_R(P_2 \square P_n) = n + 1$.

**Proposition 1.52 ([31])** *For* $k \geq 0, \gamma_R(P_3 \square P_{4k+i}) = 6k + i + 1$ *for* $i \in \{0, 1\}, \gamma_R(P_3 \square P_{4k+j}) = 6k + 2j$ *for* $j \in \{2, 3\}$.

*For all* $n \geq 1$, $\gamma_R(P_4 \square P_n) = 2n + 1$ *if* $n \in \{1, 2, 3, 5, 6\}$, *and* $\gamma_R(P_4 \square P_n) = 2n$ *otherwise.*

Dreyer Jr. has also proposed in [31] an algorithm for computing $\gamma_R(P_k \square P_n)$ for any fixed value of $k$ in $O(n)$-time. Using an algebraic approach Pavlič and Žerovnik in [69] presented an algorithm that runs in constant time for computing the Roman domination number of special classes of graphs including some prisms, grids, and torus.

**Proposition 1.53 ([69])** *For all* $n \geq 1$,

$$\gamma_R(P_5 \square P_n) = \begin{cases} 8 & \text{if } n = 3 \\ \lfloor \frac{12n}{5} \rfloor + 2 & \text{otherwise.} \end{cases}$$

$$\gamma_R(P_6 \square P_n) = \begin{cases} \lfloor \frac{14n}{5} \rfloor + 2 & \text{if } n < 5 \text{ or } n \in \{5k, 5k+3, 5k+4 \mid k \in \mathbb{N}\} \\ \lfloor \frac{14n}{5} \rfloor + 3 & \text{otherwise.} \end{cases}$$

$$\gamma_R(P_7 \square P_n) = \begin{cases} \lfloor \frac{16n}{5} \rfloor + 2 & \text{if } n \in \{1, 2, 4, 7, 5k \mid k \in \mathbb{N}\} \\ \lfloor \frac{16n}{5} \rfloor + 3 & \text{otherwise.} \end{cases}$$

$$\gamma_R(P_8 \square P_n) = \begin{cases} 9 & \text{if } n = 2 \\ 16 & \text{if } n = 4 \\ \lfloor \frac{18n}{5} \rfloor + 4 & \text{if } n \in \{5k+3 \mid k \in \mathbb{N}\} \\ \lfloor \frac{18n}{5} \rfloor + 3 & \text{otherwise.} \end{cases}$$

The authors [69] also gave the Roman graphs among $P_k \square P_n$ for $k \leq 8$. More precisely, those graphs for which $k = 1$ and $n \in \{3l + 2, 3l + 3 : l \in \mathbb{N}\}$; $k = 2$ and $n$ odd; $k = 3$ and $n \in \{4l + j : j = 1, 2, 3 \text{ and } l \in \mathbb{N}\}$; $k = 4$ and $n \notin \{1, 2, 3, 5, 6, 9\}$; $k = 5$ and $n \in \{1, 2, 3, 7, 5l, 5l + 1 : l \in \mathbb{N}\}$; $k = 6$ and $n \in \{1, 3, 5, 7, 8, 12, 15, 22\}$; $k = 7$ and $n \in \{2, 3, 4, 5, 6, 7, 8, 10, 11, 13, 16\}$; $k = 8$ and $n \in \{1, 4, 6, 7, 8\}$.

For the products $C_k \square P_n$ when $k \in \{3, 4, 5, 6, 7, 8\}$ and $C_k \square C_n$ when $k \in \{3, 4, 5, 6\}$, Pavlič and Žerovnik in [69] obtained the following.

**Proposition 1.54 ([69])** *For all $n \geq 1$,*
$\gamma_R(C_3\Box P_n) = \lfloor \frac{3n}{2} \rfloor + 1$, $\gamma_R(C_5\Box P_n) = 2n + 2$ *and* $\gamma_R(C_6\Box P_n) = \lfloor \frac{8n}{3} \rfloor + 2$.

$$\gamma_R(C_4\Box P_n) = \begin{cases} 3 & \text{if } n = 1 \\ 2n & \text{otherwise,} \end{cases}$$

$$\gamma_R(C_7\Box P_n) = \begin{cases} 3n+2 & \text{if } n \in \{1,2,4\} \\ 3n+3 & \text{otherwise,} \end{cases}$$

$$\gamma_R(C_8\Box P_n) = \begin{cases} 8 & \text{if } n = 2 \\ \lfloor \frac{7n}{2} \rfloor + 2 & \text{if } n \in \{3,4,8\} \\ \lfloor \frac{7n}{2} \rfloor + 3 & \text{otherwise.} \end{cases}$$

**Proposition 1.55 ([69])** *For all $n \geq 3$,*
$\gamma_R(C_3\Box C_n) = \lceil \frac{3n}{2} \rceil$ *and* $\gamma_R(C_4\Box C_n) = 2n$.

$$\gamma_R(C_5\Box C_n) = \begin{cases} 2n & \text{if } n \in \{5k \mid k \in \mathbb{N}\} \\ n+2 & \text{otherwise,} \end{cases}$$

$$\gamma_R(C_6\Box C_n) = \begin{cases} \lfloor \frac{8n}{3} \rfloor & \text{if } n \in \{6k \mid k \in \mathbb{N}\} \\ \lfloor \frac{8n}{3} \rfloor + 1 & \text{if } n \in \{6k+5, 18k+j \mid j = 3,8,13, k \in \mathbb{N}\} \\ \lfloor \frac{8n}{3} \rfloor + 2 & \text{otherwise.} \end{cases}$$

The Roman graphs among $C_k\Box P_n$ for $3 \leq k \leq 8$ are those for which: $k = 3$ and $n \in \{4l+1, 4l+2 : l \in \mathbb{N}\}$; $k = 4$ and $n \geq 2$; $k = 5$ and $n \in \{1,2,3\}$; $k = 6$ and $n \in \{1,3,4,6,6l+1,6l+3,6l+4,6l+6 : l \in \mathbb{N}\}$; $k = 7$ and $n \in \{2,4,2l+1 : l \in \mathbb{N}^*\}$; $k = 8$ and $1 \leq n \leq 6$. The Roman graphs among $C_k\Box C_n$ for $3 \leq k \leq 6$ are those for which: $k = 3$ and $n \in \{4l, 4l+1 : l \in \mathbb{N}\}$; $k = 4$ and $n \in \mathbb{N}$; $k = 5$ and $n \in \{3,4,5l,5l+1,5l+2,5l+4 : l \in \mathbb{N}\}$; $k = 6$ and $n \in \{6l, 6l+4, 18l+1, 18l+5, 18l+7 : l \in \mathbb{N}\}$.

In addition, Pavlič and Žerovnik [69] have also determined $\gamma_R(P_k\Box C_n)$ for $k \in \{2,3,4,5,6\}$. In [28], Currò established the following lower and upper bounds for the Roman domination number on grids.

**Theorem 1.56 ([28])** *For all $m, n \geq 3$, $\gamma_R(P_m\Box P_n) \geq \lfloor \frac{2mn+m+n-2}{5} \rfloor$.*

**Theorem 1.57 ([28])** *For all $m, n \geq 5$,*

$$\gamma_R(P_m\Box P_n) \leq \begin{cases} \lfloor \frac{2(mn+m+n)}{5} \rfloor - 1 & \text{if } m, n \equiv 0 \pmod 5 \\ \lfloor \frac{2(mn+m+n)}{5} \rfloor & \text{otherwise.} \end{cases}$$

Xueliang, Yuansheng, and Baoqi [89] have shown that $\gamma_R(C_{5n}\Box C_{5m}) = 10mn$. For the Cartesian product of any graphs $G$ and $H$, Wu [85] proved that $\gamma_R(G\Box H) \geq \gamma(G)\gamma(H)$. This result was improved by Yero and Rodríguez-Velázquez [90] when one of $G$ and $H$ admits an efficient dominating set.

**Theorem 1.58 ([90])** *Let $G$ and $H$ be two graphs. If $G$ has an efficient dominating set, then*

$$\gamma_R(G\Box H) \geq \gamma(G)\gamma_R(H).$$

Additional results on the Cartesian product of two graphs obtained by Yero and Rodríguez-Velázquez [90] are summarized as follows.

**Theorem 1.59 ([90])** *For any graphs $G$ and $H$,*

**(i)** $\gamma_R(G\Box H) \geq \dfrac{2\gamma(G)\gamma_R(H)}{3}$.

**(ii)** $\gamma_R(G\Box H) \geq \dfrac{\gamma(G)\gamma_R(H) + \gamma(G\Box H)}{2}$.

**(iii)** $\gamma_R(G\Box H) \leq \min\{|V(G)|\,\gamma_R(H), |V(H)|\,\gamma_R(G)\}$.

**(iv)** $\gamma_R(G\Box H) \leq 2\gamma(G)\gamma(H) + (|V(G)| - \gamma(G))(|V(H)| - \gamma(H))$.

**Theorem 1.60 ([90])** *For any graphs $G$ and $H$,*

*(i) If $G$ has at least one connected component of order greater than two, then*

$$\gamma_R(G\Box H) \leq (|V(G)| + 1)\gamma_R(H) - 2\gamma(H).$$

*(ii) If $G$ is a Roman graph, then*

$$\gamma_R(G\Box H) \leq 2\,|V(G)|\,(\gamma_R(H) - \gamma(H)) + 2\gamma(G)\,(2\gamma(H) - \gamma_R(H))\,.$$

Similar results are also obtained by Yero and Rodríguez-Velázquez [90] by considering the strong product of graphs.

### 1.7.2 Circulant Graphs

A *circulant graph* $C(n; S_c)$ is a graph with the vertex set $V(C(n; S_c)) = \{v_i \mid 0 \leq i \leq n-1\}$ and the edge set $E(C(n; S_c)) = \{v_i v_j \mid 0 \leq i, j \leq n-1, (i-j) \bmod n \in S_c\}$, $S_c \subseteq \{1, 2, \ldots, \lfloor n/2 \rfloor$, where subscripts are taken modulo $n\}$. Let $G$ be an $r$-regular graph with order $n$ $(r \geq 1)$, $m = \lfloor \frac{n}{r+1} \rfloor$, $t = n \pmod{r+1}$, then $n = (r+1)m + t$, where $0 \leq t \leq r$. Xueliang et al. [89] established the Roman domination number of the following circulant graph.

**Proposition 1.61 ([89])** *For $n \geq 7$,*

$$\gamma_R(C(n; \{1, 3\})) = \begin{cases} 2\lfloor \frac{n}{5} \rfloor & \text{if } n \equiv 0 \pmod 5 \\ 2\lfloor \frac{n}{5} \rfloor + 2 & \text{if } n \equiv 1, 2, 3 \pmod 5 \\ 2\lfloor \frac{n}{5} \rfloor + 3 & \text{if } n \equiv 4 \pmod 5). \end{cases}$$

**Proposition 1.62 ([89])** *For $n \geq 5$, $2 \leq k \leq \lfloor \frac{n}{2} \rfloor$, $n \neq 2k$*

$$\gamma_R(C(n; \{1, \ldots, k\})) = \begin{cases} 2\lfloor \frac{n}{2k+1} \rfloor & \text{if } n \equiv 0 \pmod{2k+1} \\ 2\lfloor \frac{n}{2k+1} \rfloor + 1 & \text{if } n \equiv 1 \pmod{2k+1} \\ 2\lfloor \frac{n}{2k+1} \rfloor + 2 & \text{if } n \equiv 2, \ldots, 2k \pmod{2k+1}). \end{cases}$$

### 1.7.3 Generalized Petersen Graphs

The *generalized Petersen graph* $P(n, k)$ is a graph of order $2n$ with $V(P(n,k)) = \{v_i, u_i : 0 \le i \le n-1\}$ and $E(P(n,k)) = \{v_i v_{i+1}, v_i u_i, u_i u_{i+k} : 0 \le i \le n-1,$ where subscripts are taken modulo $n\}$. We note that generalized Petersen graphs are 4-regular. The Roman domination number of generalized Petersen graphs has been addressed in several papers, where exact values have been obtained when $k \in \{1, 2, 3\}$.

**Proposition 1.63 ([89])** *For $n \ge 3$,*

$$\gamma_R(P(n,1)) = \begin{cases} n & \text{if } n \equiv 0 \pmod 4 \\ 4\lfloor \frac{n}{4} \rfloor + t + 1 & \text{if } n \equiv t \pmod 4 \text{ for } t \in \{1,2,3\}. \end{cases}$$

**Proposition 1.64 ([84])** *For $n \ge 5$, $\gamma_R(P(n,2)) = \lceil \frac{8n}{7} \rceil$.*

**Proposition 1.65 ([91])** *For $n \ge 5$,*

$$\gamma_R(P(n,3)) = \begin{cases} n & \text{if } n \equiv 0 \pmod 4 \\ n+1 & \text{if } n \equiv 1,3 \pmod 4 \\ n+2 & \text{if } n \equiv 2 \pmod 4. \end{cases}$$

For $k = 4$, the following upper bound on the Roman domination number of $P(n, 4)$ was given in [91] by using an integer programming formulation.

**Proposition 1.66 ([91])** *For $n \ge 13$,*

$$\gamma_R(P(n,4)) \le \begin{cases} \lceil \frac{14n}{13} \rceil + 1 & \text{if } n \equiv 6, 8, 12 \pmod{13} \\ \lceil \frac{14n}{13} \rceil & \text{otherwise.} \end{cases}$$

### 1.7.4 Lexicographic Product of Graphs

The *lexicographic product* of two graphs $G$ and $H$ is defined as the graph $G \cdot H$ with vertex set $V(G) \times V(H)$ and edge set $E(G \cdot H) = \{(u, v)(u', v') | uu' \in E(G)$ or $((u = u'$ and $vv' \in E(H))\}$. The Roman domination number of lexicographic product of graphs has been investigated by Šumenjak, Pavlič, and Tepeh in [81]. It was shown in [81] that if $G$ and $H$ are nontrivial connected graphs, then $\gamma_R(G \cdot H) \ge 2\gamma(G)$, with equality if and only if either $\gamma_R(H) = 2$ or $\gamma(G) = \gamma_t(G)$ and $\gamma_R(H) \ge 3$.

For disjoint subsets $A, B \subseteq V(G)$, an ordered couple $(A, B)$ is said to be a *dominating couple* of $G$ if for every vertex $x \in V(G) \setminus B$ there exists a vertex $w \in A \cup B$ such that $x \in N_G(w)$. The Roman domination number of the lexicographic product of graphs $G \cdot H$ is given as follows.

**Theorem 1.67 ([81])** *Let $G$ and $H$ be nontrivial connected graphs. Then*

$$\gamma_R(G \cdot H) = \begin{cases} 2\gamma(G) & \text{if } \gamma_R(H) = 2, \\ \zeta(G) & \text{if } \gamma_R(H) = 3, \\ 2\gamma_t(G) & \text{if } \gamma_R(H) = 4, \end{cases}$$

*where* $\zeta(G) = \min\{2|A| + 3|B| : (A, B)$ is a dominating couple of $G\}$.

In addition, Šumenjak, Pavlič, and Tepeh characterized Roman graphs among the lexicographic products of graphs.

**Theorem 1.68 ([81])** *Let $G$ and $H$ be nontrivial connected graphs. Then $G \cdot H$ is a Roman graph if and only if one of the following holds:*

1. $\gamma_R(H) = 2$ *or* $\gamma_R(H) \geq 4$
2. $\gamma_R(H) = 3$ *and there exists a minimum dominating couple* $(A, B)$, *such that* $B = \emptyset$.

### 1.7.5 Categorical Product of Graphs

The Roman domination number of categorical product of graphs has been investigated by Klobučar and Puljič [60, 61]. We summarize below some of their main results. We note that since $\Delta(G \times H) = \Delta(G) \times \Delta(H)$ and $\delta(G \times H) = \delta(G) \times \delta(H)$, any bound on the Roman domination number in terms of the order, maximum and minimum degrees yields a similar bound on $\gamma_R(G \times H)$.

**Proposition 1.69 ([61])** *For any graph $G$ without odd cycles,*

$$\gamma_R(P_2 \times G) = 2\gamma_R(G).$$

**Proposition 1.70 ([61])** *For $n \geq 2$, $\gamma_R(P_3 \times P_n) = \frac{3n}{2}$ if $n \equiv 0 \pmod 4$; $6\lfloor \frac{n}{4} \rfloor + 2$ if $n \equiv 1 \pmod 4$ and $6\lfloor \frac{n}{4} \rfloor + 2 + i$ if $n \equiv i \pmod 4$, where $i \in \{2, 3\}$.*

**Proposition 1.71 ([61])** *For $n \geq 2$, $\gamma_R(P_4 \times P_n) = 2n + 2$ if $n = 5$, and $\gamma_R(P_4 \times P_n) = 2n$ otherwise.*

**Proposition 1.72 ([60])** *For all $m, n \geq 2$,*
$\gamma_R(P_m \times P_n) \leq 4\left\lceil \frac{mn}{10} \right\rceil + 8\left\lceil \frac{m}{10} \right\rceil + 8\left\lceil \frac{n}{10} \right\rceil$, *and*
$\gamma_R(P_m \times P_n) > 4\left\lfloor \frac{mn-m-n+1}{10} \right\rfloor + 4\left\lfloor \frac{m-1}{10} \right\rfloor + 4\left\lceil \frac{n-1}{10} \right\rceil$.

Moreover, Klobučar and Puljič [60] concluded their paper by showing that $\lim_{m,n\to\infty} \frac{\gamma_R(P_m \times P_n)}{mn} = \frac{2}{5}$.

## 1.8 Roman Domatic Number

A set $\{f_1, f_2, \ldots, f_d\}$ of distinct Roman dominating functions on $G$ with the property that $\sum_{i=1}^{d} f_i(v) \leq 2$ for each $v \in V(G)$ is called a *Roman dominating family* (of functions) on $G$. The maximum number of functions in a Roman dominating family on $G$ is the *Roman domatic number* of $G$, denoted by $d_R(G)$. A graph with Roman domatic number three is illustrated in Figure 8. The Roman

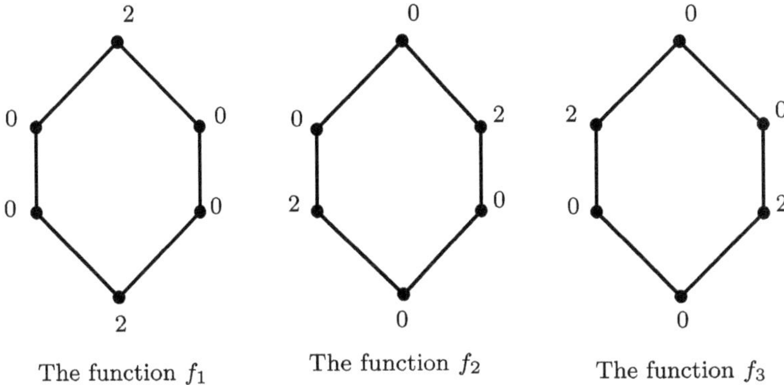


**Fig. 8** A graph with Roman domatic number 3

domatic number was introduced in 2010 by Sheikholeslami and Volkmann in [79] and has been studied in [82].

Upper bounds on the sum and product of $\gamma_R$ and $d_R$ in terms of the order of the graph have been established in [79].

**Theorem 1.73 ([79])** If $G$ is a graph of order $n$, then

$$\gamma_R(G) \cdot d_R(G) \leq 2n.$$

Combining the bound $\gamma_R(G) \geq \lceil \frac{2n}{\Delta+1} \rceil$ (see Proposition 1.19) with Theorem 1.73, we obtain the following corollary immediately.

**Corollary 1.74** If $G$ is a graph with maximum degree $\Delta \geq 1$, then $d_R(G) \leq \Delta+1$.

**Proposition 1.75 ([79])** If $G$ is a graph of order $n \geq 2$, then

$$\gamma_R(G) + d_R(G) \leq n + 2$$

with equality if and only if $\Delta(G) = 1$ or $G$ is a complete graph.

Since $\gamma_R(G) \geq 2$ for any graph $G$ of order $n \geq 2$, it follows from Theorem 1.73 that $d_R(G) \leq n$. It was shown in [79] that $d_R(G) = n$ if and only if $G$ is the complete graph on $n$ vertices. Moreover, a sharp upper bound on the Roman domatic number in terms of the minimum degree was also obtained in [79].

**Theorem 1.76 ([79])** For every graph $G$, $d_R(G) \leq \delta(G) + 2$.

The authors [79] gave the following example of graphs to illustrate the sharpness of Theorem 1.76. Let $G_i$ be a copy of $K_{k+3}$ with vertex set $V(G_i) = \{v_1^i, v_2^i, \ldots, v_{k+3}^i\}$ for $1 \leq i \leq k$ and let $G$ be the graph obtained from $\cup_{i=1}^k G_i$ by adding a new vertex $v$ attached to each $v_1^i$. Then $\delta(G) = k$ and $d_R(G) = \delta(G) + 2$.

The bound of Theorem 1.76 can be slightly improved for regular graphs as follows.

**Theorem 1.77** If $G$ is a $\delta$-regular graph of order $n$, then

$$d_R(G) \leq \delta + \epsilon$$

with $\epsilon = 1$ when $n \equiv 0, \frac{\delta+1}{2} (\bmod\ \delta + 1)$ and $\epsilon = 0$ when $n \not\equiv 0, \frac{\delta+1}{2} (\bmod\ \delta + 1)$.

**Proof.** If $\delta = 0$, then clearly $d_R(G) = 1$ and the result is immediate. Let now $\delta \geq 1, n = p(\delta+1)+r$ with integers $p \geq 1$ and $0 \leq r \leq \delta$, and let $\{f_1, f_2, \ldots, f_d\}$ be an RD family on $G$ such that $d = d_R(G)$. It follows that

$$\sum_{i=1}^{d} \omega(f_i) = \sum_{i=1}^{d} \sum_{v \in V} f_i(v) = \sum_{v \in V} \sum_{i=1}^{d} f_i(v) \leq \sum_{v \in V} 2 = 2n. \tag{1}$$

If $r = 0$, then we deduce from Proposition 1.19 that $\omega(f_i) \geq \gamma_R(G) \geq 2p$ for each $i \in \{1, 2, \ldots, d\}$. Suppose, to the contrary, that $d \geq \delta + 2$. Then we obtain

$$\sum_{i=1}^{d} \omega(f_i) \geq 2pd \geq 2p(\delta + 2) > 2n.$$

This is a contradiction to (1) and thus $d \leq \delta + 1$.

If $1 \leq r < \frac{\delta+1}{2}$, then Proposition 1.19 implies that $\omega(f_i) \geq \gamma_R(G) \geq 2p+1$ for each $i \in \{1, 2, \ldots, d\}$ and $\delta \geq 2$. Suppose, to the contrary, that $d \geq \delta + 1$. Then we obtain the contradiction

$$\sum_{i=1}^{d} \omega(f_i) \geq d(2p + 1) \geq (\delta + 1)(2p + 1) > 2n.$$

Therefore $d \leq \delta$, and the result is proved in this case.

If $r = \frac{\delta+1}{2}$, then as above $\omega(f_i) \geq \gamma_R(G) \geq 2p + 1$ for each $i \in \{1, 2, \ldots, d\}$. Suppose, to the contrary, that $d \geq \delta + 2$. Then we obtain the contradiction

$$\sum_{i=1}^{d} \omega(f_i) \geq d(2p + 1) \geq (\delta + 2)(2p + 1) > 2n.$$

Therefore $d \leq \delta + 1$.

Finally assume that $\frac{\delta+1}{2} < r \leq \delta$. Then Proposition 1.19 yields to $\omega(f_i) \geq \gamma_R(G) \geq 2p + 2$ for each $i \in \{1, 2, \ldots, d\}$. If we suppose to the contrary that $d \geq \delta + 1$, then we obtain

$$\sum_{i=1}^{d} \omega(f_i) \geq d(2p + 2) \geq (\delta + 1)(2p + 2) > 2n.$$

This contradiction to (1) implies that $d \leq \delta$ in this case, and the proof is complete. ∎

As an application of Theorems 1.76 and 1.77, the following Nordhaus–Gaddum type results have thus been established.

**Theorem 1.78 ([79])** For every graph $G$ of order $n$,

$$d_R(G) + d_R(\overline{G}) \leq n + 2. \tag{2}$$

**Theorem 1.79** If $G$ is a $k$-regular graph of order $n$, then

$$d_R(G) + d_R(\overline{G}) \leq n + 1, \tag{3}$$

and equality in (3) implies $n \equiv 0, \frac{k+1}{2} \pmod{(k+1)}$ and $n \equiv 0, \frac{n-k}{2} \pmod{(n-k)}$.

The complete graphs $K_n$ $(n \geq 2)$ are simplest examples of graphs showing that the bound of Theorem 1.79 is sharp. Exact values of the Roman domatic number have been determined for some classes of graphs including trees, cycles, fans, wheels, and complete bipartite graphs. In [79], it was shown that $d_R(T) = 2$ for any nontrivial tree $T$, while for cycles it was shown that $d_R(C_n) = 3$ if $n \equiv 0 \pmod 3$ and $d_R(C_n) = 2$ otherwise.

For $n \geq 3$, the *fan graph* $F_n$ is a graph obtained from a path $P_n$ by adding a new vertex attached to each vertex of $P_n$, and the *wheel graph* $W_n$ is a graph obtained from a cycle $C_n$ by adding a new vertex attached to each vertex of $C_n$. Tan et al. [82] have shown that for $n \geq 3$, $d_R(F_n) = 3$, $d_R(W_n) = 4$ if $n \equiv 0 \pmod 3$ and $d_R(W_n) = 3$ if $n \equiv 1, 2 \pmod 3$, while for the complete bipartite graphs, they proved that $d_R(K_{m,n}) = \max\{2, \min\{m, n\}\}$. Moreover, Tan et al. [82] have shown that the decision problem corresponding to the problem of computing $d_R(G)$ is NP-complete even when restricted to bipartite graphs. They also proved that there is a $(\ln n + O(\ln \ln n))$-approximation algorithm for Roman Domatic Number, where $n$ is the order of the input graph.

## 1.9 Summary and Open Problems

In this section, we have been interested in the problem of Roman domination by developing various aspects in order to better understand it. So we gave some properties on the Roman domination number as well as its relations with some domination parameters. The complexity and algorithmic aspects of the Roman domination problem have also been discussed. Many problems remain open, we give below a brief list of them.

1. Is it true that if $G$ is a graph of order $n$ with minimum degree at least three, then $\gamma_R(G) + \gamma(G) \leq n$? (It is a conjecture stated in [12]).
2. For which graphs $G$ is $\gamma_R(G) = 2\gamma(G)$?
3. Determine $\gamma_R(G_{m,n})$ for every $m$-by-$n$ grid graph $G_{m,n}$.
4. Characterize the trees $T$ with $\gamma_R(T) = \gamma_t(T) + 1$.

5. Characterize graphs $G$ (or at least trees) for which every $ir_R$-function is dominating?
6. For which graphs $G$ is $IR_R(G) = \Gamma_R(G)$?
7. For which graphs $G$ is $\Gamma_R(G) = \alpha_R(G)$?
8. Let $G$ be a graph of order $n$. Determine Nordhaus–Gaddum type results for $IR_R(G), \Gamma_R(G)$, and $\alpha_R(G)$.
9. Design an algorithm for computing the value of $\mu_R(T)$ for any tree $T$, where $\mu_R \in \{ir_R, \alpha_R, \Gamma_R, IR_R\}$.
10. Study the NP-completeness of the decision problems corresponding to the problem of computing $\mu_R(G)$, where $\mu_R \in \{ir_R, \alpha_R, \Gamma_R, IR_R\}$.

# 2 Changing and Unchanging with Respect to the Roman Domination

## 2.1 *Introduction*

It is of interest to know how the value of a graph parameter $\mu$ is affected when the graph is subjected to a change. The addition of a set of edges, the removal of a set of vertices/edges may increase or decrease $\mu$, or leave $\mu$ unchanged. Much has already been written when $\mu$ is the domination number or the total domination number (see, for example, [40, 43]). In this section, we consider the case where $\mu$ is the Roman domination number.

## 2.2 *Terminology*

Before going further, we need first to introduce some terminology that can be found in Jafari Rad et al. [49, 55], Chellali et al. [23], or Samodivkin [75]. According to the effects of vertex removal on the Roman domination number of a graph, we can partition the vertex set of a graph $G$ into three sets according to how their removal affects $\gamma_R(G)$. Let $V(G) = V_R^{=} \cup V_R^{+} \cup V_R^{-}$ such that:

$$V_R^{=} = \{v \in V(G) : \gamma_R(G - v) = \gamma_R(G)\},$$

$$V_R^{+} = \{v \in V(G) : \gamma_R(G - v) > \gamma_R(G)\},$$

$$V_R^{-} = \{v \in V(G) : \gamma_R(G - v) < \gamma_R(G)\}.$$

Accordingly, we have the following definitions. A graph $G$ of order at least two is *Roman domination vertex critical*, or just *$\gamma_R$-vertex critical*, if removing any vertex of $G$ decreases the Roman domination number. A graph $G$ of order at least two is *Roman domination vertex super-critical*, or just *$\gamma_R$-vertex super-critical*, if removing any vertex of $G$ increases the Roman domination number. We will later show that there is no $\gamma_R$-vertex super-critical graph. A graph $G$ of order at least two is *Roman domination vertex stable*, or just *$\gamma_R$-stable*, if removing any vertex of $G$ leaves the Roman domination number unchanged.

Since removing an edge from a graph cannot decrease the Roman domination number, we have $\gamma_R(G-e) \geq \gamma_R(G)$ for any edge $e \in E(G)$. Therefore the edge set of $G$ can be partitioned into $E_R^{=} = \{e \in E(G) : \gamma_R(G-e) = \gamma_R(G)\}$ and $E_R^{+} = \{e \in E(G) : \gamma_R(G-e) > \gamma_R(G)\}$. So, we say that a graph $G$ of size at least one is *Roman domination critical upon edge removal*, if for any $e \in E(G)$, $\gamma_R(G-e) > \gamma_R(G)$, and is *Roman domination stable upon edge removal*, if for any $e \in E(G)$, $\gamma_R(G-e) = \gamma_R(G)$. Likewise, $\gamma_R(G+e) \leq \gamma_R(G)$ for every edge $e \notin E(G)$. So, we say that a graph $G$ is *Roman domination edge critical*, or just *$\gamma_R$-edge critical*, if for every edge $e \in E(\overline{G})$, $\gamma_R(G+e) < \gamma_R(G)$, where $\overline{G}$ is the complement of $G$. A graph $G$ is *Roman domination stable upon edge addition*, if for any $e \in E(\overline{G})$, $\gamma_R(G+e) = \gamma_R(G)$.

Now, with respect to "changing" or "unchanging" of the Roman domination number upon the removal of a vertex or an edge or the addition of an edge, we have the following six families of graphs. We use acronyms to denote the following classes of graphs (C: changing; U: unchanging; V: vertex; E: edge; R: removal; A: addition). In our notation CVR refers to "changing vertex removal," CER refers to "changing edge removal," UVR refers to "unchanging vertex removal," and UER refers to "unchanging edge removal."

- $\mathcal{R}_{CVR}$ is the class of graphs $G$ such that $\gamma_R(G-v) \neq \gamma_R(G)$ for all $v \in V(G)$.
- $\mathcal{R}_{UVR}$ is the class of graphs $G$ such that $\gamma_R(G-v) = \gamma_R(G)$ for all $v \in V(G)$.
- $\mathcal{R}_{CER}$ is the class of graphs $G$ such that $\gamma_R(G-e) \neq \gamma_R(G)$ for all $e \in E(G)$.
- $\mathcal{R}_{UER}$ is the class of graphs $G$ such that $\gamma_R(G-e) = \gamma_R(G)$ for all $e \in E(G)$.
- $\mathcal{R}_{CEA}$ is the class of graphs $G$ such that $\gamma_R(G+e) \neq \gamma_R(G)$ for all $e \in E(\overline{G})$.
- $\mathcal{R}_{UEA}$ is the class of graphs $G$ such that $\gamma_R(G+e) = \gamma_R(G)$ for all $e \in E(\overline{G})$.

## 2.3 Changing

We will divide this subsection into three parts by focusing on "vertex removal," "edge removal," and "edge addition."

### 2.3.1 Vertex Removal

**Proposition 2.1 ([55])** *Let $v$ be a vertex in a graph $G$ such that $v \in V_R^{+}$, and let $f = (V_0, V_1, V_2)$ be any $\gamma_R(G)$-function. Then*

**(1)** $v \in V_2$,
**(2)** *$v$ has at least three private neighbors in $V_0$.*

As a consequence, no graph $G$ has $V_R^+ = V(G)$, and thus no graph is $\gamma_R$-vertex super-critical. The following is a keynote result in the study of $\gamma_R$-vertex critical graphs.

**Proposition 2.2 ([49])** *Let $v$ be a vertex in a graph $G$. Then $v \in V_R^-$ if and only if there is a $\gamma_R(G)$-function $f$ with $v \in V_1$. Moreover, if $v \in V_R^-$, then $\gamma_R(G - v) = \gamma_R(G) - 1$.*

Several families of $\gamma_R$-vertex critical graphs have been presented in [49] summarized as follows.

**Proposition 2.3 ([49])**

*(1) The ladder $P_2 \square P_n$ is $\gamma_R$-vertex critical if and only if $n \leq 2$.*
*(2) The graph $K_p \square K_q$ is $\gamma_R$-vertex critical if and only if $|p - q| \leq 1$.*
*(3) The circulant graph $C_n(\{1, 2, \ldots, r\})$ is $\gamma_R$-vertex critical if and only if $n \equiv 1 \text{ or } 2 \mod (2r + 1)$.*
*(4) The complete $r$-partite graph $K_{p_1, p_2, \ldots, p_r}$, with $r \geq 2$, is $\gamma_R$-vertex critical if and only if either $p_1 = p_2 = \cdots = p_r \in \{2, 3\}$, or $r = 2$ and $p_1 = p_2 = 1$, i.e., $K_{p_1, p_2, \ldots, p_r} \cong K_2$.*
*(5) A cycle $C_n$ is $\gamma_R$-vertex critical if and only if $n \equiv 1 \text{ or } 2 \pmod 3$.*

For graphs with minimum degree one, it is shown that any support vertex of a $\gamma_R$-critical graph is adjacent to exactly one leaf, and for block graphs, it was shown that $K_2$ is the only one that is $\gamma_R$-vertex critical ([49]). Recall that the corona cor($G$) of a graph $G$ is the graph obtained from $G$ by adding for each vertex $v \in V$ a new vertex $v'$ and the edge $vv'$. In [39], Hansberg et al. have shown that a connected unicyclic graph $G$ is $\gamma_R$-vertex critical if and only if $G =$cor($C_m$), where $m \equiv 1 \pmod 3$ or $G = C_n$, where $n \not\equiv 0 \pmod 3$.

For a given integer $k \geq 2$, a graph $G$ is called *$k$-$\gamma_R$-vertex critical* if $G$ is a $\gamma_R$-vertex critical graph and $\gamma_R(G) = k$. The $k$-$\gamma_R$-vertex critical graphs for some small values of $k$ have been studied in [55]. It can be seen that a graph $G$ is 2-$\gamma_R$-vertex critical if and only if $G = K_2$ or $\overline{K_2}$. It is proved in [55] that a graph $G$ of order $n \geq 4$ is 3-$\gamma_R$-vertex critical if and only if $n$ is even, and $G$ is an $(n - 2)$-regular graph. Note that $K_2 \cup K_1$ is a 3-$\gamma_R$-vertex critical graph which shows that the condition $n \geq 4$ is important in the previous result. The next result is a characterization of 4-$\gamma_R$-vertex critical graphs given in [55].

**Theorem 2.4 ([55])** *A graph $G$ of order $n \geq 5$ is 4-$\gamma_R$-vertex critical if and only if for any vertex $v$ there is a vertex $u$ of degree $n - 3$ which is not adjacent to $v$.*

The 4-$\gamma_R$-vertex critical graph consisting of 4 isolated vertices shows that the condition $n \geq 5$ is important in Theorem 2.4. The following provides an upper bound for the order of a $\gamma_R$-vertex critical graph in terms of the Roman domination number and maximum degree.

**Theorem 2.5 ([55])** *Let $G$ be a $\gamma_R$-vertex critical graph of order $n \geq 2$ with $\Delta(G) \geq 1$. If $\gamma_R(G)$ is odd, then*

$$n \leq \frac{\gamma_R(G) - 1}{2}(\Delta(G) + 1) + 1, \tag{4}$$

*and if $\gamma_R(G)$ is even, then*

$$n \leq \frac{\gamma_R(G) - 2}{2}(\Delta(G) + 1) + 2. \tag{5}$$

*Assume that equality holds in (4). If $\Delta(G) = 1$, then $G = sK_2 \cup tK_1$ with odd $t$, and if $\Delta(G) \geq 2$, then $G$ is regular.*

*Assume that equality holds in (5). If $\Delta(G) = 1$, then $G = sK_2 \cup tK_1$ with even $t$, and if $\Delta(G) \geq 2$, then there exists a $\gamma_R$-function $f = (V_0, V_1, V_2)$ such that $|V_1| = 2$.*

The complete $r$-partite graph $H = K_{p_1,p_2,\ldots,p_r}$ with $r \geq 2$ and $p_1 = p_2 = \ldots = p_r = 2$ is $\gamma_R$-vertex critical with $n(H) = 2r$, $\gamma_R(H) = 3$, $\Delta(H) = 2r - 2$ and equality in (4). Hence there exist $\gamma_R$-vertex critical graphs of arbitrary high maximum degree with equality in (4).

The complete $r$-partite graph $H = K_{p_1,p_2,\ldots,p_r}$ with $r \geq 2$ and $p_1 = p_2 = \ldots = p_r = 3$ is $\gamma_R$-vertex critical with $n(H) = 3r$, $\gamma_R(H) = 4$, $\Delta(H) = 3r - 3$ and equality in (5). Hence there exist $\gamma_R$-vertex critical graphs of arbitrary high maximum degree with equality in (5).

If $G$ is the disjoint union of a cycle of length 4 and an isolated vertex, then $G$ is $\gamma_R$-vertex critical with $\gamma_R(G) = 4$ with equality in (5). This example shows that equality in (5) does not imply regularity.

Further results for 4-$\gamma_R$-vertex critical graphs have been given by Martinez–Perez et al. [66]. The following provides an upper bound for the diameter of a $\gamma_R$-vertex critical graph.

**Theorem 2.6 ([55])** *For any $\gamma_R$-vertex critical graph $G$,*

$$\text{diam}(G) \leq \left\lceil \frac{3\gamma_R(G) - 5}{2} \right\rceil.$$

In addition, Jafari Rad and Volkmann [55] proposed the following conjecture. We recall that a graph $G$ is *$\gamma$-vertex critical* if the removal of each vertex decreases the domination number (see [40]).

**Conjecture 1 ([55])** *Any $\gamma$-vertex critical graph is $\gamma_R$-vertex critical.*

Note that a $\gamma_R$-vertex critical graph is not necessarily $\gamma$-vertex critical, as mentioned in [55] who showed that for any even $k \geq 4$ there is a $k$-$\gamma_R$-vertex critical graph which is not $\gamma$-vertex critical. However, if $G$ is a $\gamma_R$-vertex critical graph and $|V_1| = 1$ for any $\gamma_R(G)$-function $f = (V_0, V_1, V_2)$, then $G$ is $\gamma$-vertex critical.

### 2.3.2 Edge Removal

It has already been mentioned that the removal of an edge from $G$ cannot decrease $\gamma_R(G)$, however, it can increase it by at most one as shown in [55].

**Proposition 2.7** *If $e = xy$ is an edge of a graph $G$, then $\gamma_R(G - e) \leq \gamma_R(G) + 1$.*

According to Proposition 2.7, if $G$ is a Roman domination critical graph upon edge removal, then $\gamma_R(G - e) = \gamma_R(G) + 1$ for each edge $e \in E(G)$. Observe that if $G$ is Roman domination critical graph upon edge removal, then $\Delta(G) \geq 2$ (since for $\Delta(G) \leq 1$, no edge satisfies $\gamma_R(G - e) > \gamma_R(G)$). A characterization of Roman domination critical graphs upon edge removal was given in [55] as follows.

**Theorem 2.8 ([55])** *Let $G$ be a graph with $\Delta(G) \geq 2$. Then $G$ is Roman domination critical upon edge removal if and only if $G$ is a forest in which each component is an isolated vertex or a star of order at least* 3.

Since a star of order at least 3 is not Roman domination vertex critical, Theorem 2.8 implies the next result.

**Corollary 2.9** $\mathcal{R}_{CVR} \cap \mathcal{R}_{CER} = \emptyset$.

### 2.3.3 Edge Addition

It was shown in [49] that the addition of any edge to a graph $G$ can decrease $\gamma_R(G)$ by at most one, and therefore if $G$ is a $\gamma_R$-edge critical graph, then $\gamma_R(G + e) = \gamma_R(G) - 1$ for any edge $e \in E(\overline{G})$. A characterization of $\gamma_R$-edge critical graphs has been given in [49].

**Theorem 2.10 ([49])** *A graph $G$ is $\gamma_R$-edge critical if and only if for any two non-adjacent vertices $x, y$, there is a $\gamma_R(G)$-function $f = (V_0, V_1, V_2)$ such that* $\{f(x), f(y)\} = \{1, 2\}$.

For some classes of graphs, we have the following result which provides a characterization of those that are $\gamma_R$-edge critical graphs.

**Proposition 2.11 ([49])**

*(1) The ladder $P_2 \square P_n$ is $\gamma_R$-edge critical if and only if $n = 2$.*
*(2) The graph $K_p \square K_q$ is $\gamma_R$-edge critical if and only if $|p - q| \leq 1$.*
*(3) The circulant graph $C_n(\{1, 2, \ldots, r\})$ is $\gamma_R$-edge critical if and only if $r = 1$ and $n \in \{4, 5\}$, i.e., it is isomorphic to the cycle of length* 4 *or* 5.
*(4) The complete $r$-partite graph $K_{p_1, p_2, \ldots, p_r}$, where $r \geq 2$, is $\gamma_R$-edge critical if and only if $p_1 = p_2 = \ldots = p_r \in \{2, 3\}$.*

The $\gamma_R$-edge critical trees and unicyclic graphs have been characterized in [49] and [39], respectively. Let $T_1$ be a tree obtained from two copies of $P_5$ by adding an edge between the central vertices, and let $T_2$ be the tree obtained from $T_1$ by removing a leaf.

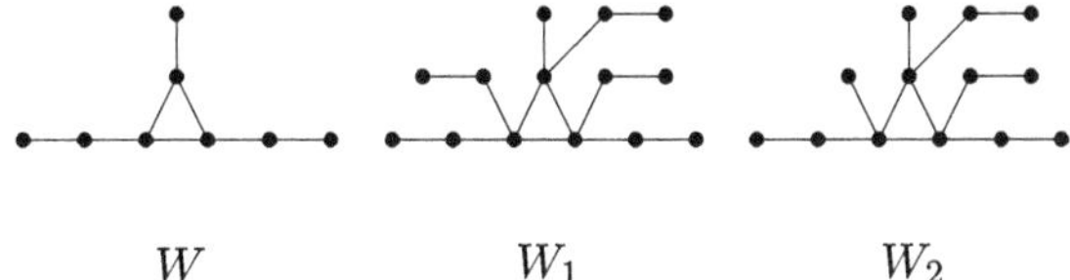


**Fig. 9** The graphs $W$, $W_1$, and $W_2$.

**Theorem 2.12 ([49])** *$T_1$ and $T_2$ are the only $\gamma_R$-edge critical trees.*

Let $W$ be the graph obtained from cor($C_3$) by subdividing two pendant edges. For any vertex $x$ of $W$ with $\deg_W(x) = 3$, we join $x$ to a leaf of a path $P_2$ to obtain a graph $W_1$. Note that $W_1$ has six leaves and six support vertices. Let $W_2$ be obtained from $W_1$ by removing one leaf. Note that $W$, $W_1$, and $W_2$ are $\gamma_R$-edge critical (Figure 9).

**Theorem 2.13 ([39])** *A connected unicyclic graph $G$ is $\gamma_R$-edge critical if and only if $G \in \{C_4, C_5, W, W_1, W_2\}$.*

Moreover, $\gamma_R$-edge critical graphs with precisely two cycles have been characterized in [48].

For an integer $k \geq 2$, we call a graph $G$, *$k$-$\gamma_R$-edge critical* if $G$ is $\gamma_R$-edge critical and $\gamma_R(G) = k$. Chellali et al. [24] obtained some properties of $\gamma_R$-edge critical graphs, and characterized those having a small Roman domination number. Clearly a graph $G$ of order $n \geq 2$ is 2-$\gamma_R$-edge critical if and only if $G$ is a complete graph. Let $\mathcal{H}$ be the class of all graphs $G$ of order at least $n \geq 3$ such that $\Delta(G) = n - 2$, and for any two non-adjacent vertices $x, y$ of $G$, $\Delta(G) \in \{\deg(x), \deg(y)\}$. As shown in [24], a graph $G$ is 3-$\gamma_R$-edge critical if and only if $G \in \mathcal{H}$. For graphs with large Roman domination number, the following was proved in [24].

**Theorem 2.14 ([24])** *A graph $G$ of order $n$ is $n$-$\gamma_R$-edge critical if and only if $n$ is even and $G = \frac{n}{2}K_2$ or $n$ is odd and $G = \frac{n-1}{2}K_2 \cup K_1$.*

**Theorem 2.15 ([24])** *A graph $G$ of order $n \geq 3$ is $(n-1)$-$\gamma_R$-edge critical if and only if $G = C_i \cup m_1K_1 \cup m_2K_2$, where $i \in \{3, 4, 5\}$, $m_1 \leq 1$, and $m_1 + 2m_2 = n - i$.*

A bound relating the diameter of $\gamma_R$-edge critical graphs to the Roman domination number was obtained by Chellali, et al. [24]. In addition, they have shown that for every even integer $n \geq 6$, there is an $n$-$\gamma_R$-edge critical graph $G$ with $\text{diam}(G) = 5$.

**Theorem 2.16 ([24])** *If $G$ is a $\gamma_R$-edge critical connected graph with $\gamma_R(G) > 3$, then* $\text{diam}(G) \leq 3\left\lceil (\gamma_R(G) - 3)/2 \right\rceil + 2$.

## 2.4 Unchanging

In this subsection we study the classes $\mathcal{R}_{UVR}$ and $\mathcal{R}_{UER}$ and $\mathcal{R}_{UEA}$.

### 2.4.1 Vertex Removal

The fact that $\gamma_R(P_n) = \gamma_R(C_n) = \left\lceil \frac{2n}{3} \right\rceil$ with a simple calculation shows that $P_n \in \mathcal{R}_{UVR}$ if and only if $n \equiv 0 \pmod 3$ and $C_n \in \mathcal{R}_{UVR}$ if and only if $n \equiv 0 \pmod 3$. A characterization of graphs belonging to $\mathcal{R}_{UVR}$ was given in [55].

**Theorem 2.17 ([55])** *For a graph $G$, $G \in \mathcal{R}_{UVR}$ if and only if $G$ has no isolated vertices, and for each vertex $v$, either*

*(1) for any $\gamma_R(G)$-function $f = (V_0, V_1, V_2)$, $v \in V_2$, and there is an RDF $g$ on $G - v$ with $w(g) = \gamma_R(G)$, or*
*(2) there is a $\gamma_R(G)$-function $f = (V_0, V_1, V_2)$ on $G$ such that $v \notin V_2$, and for any $\gamma_R(G)$-function $g = (V_0', V_1', V_2')$, $v \notin V_1'$.*

Chambers et al. [17] proved that if $G$ is a graph with $\delta(G) \geq 1$, then $\gamma_R(G) \leq 4n/5$. This bound was lowered for graphs in $\mathcal{R}_{UVR}$.

**Proposition 2.18 ([75])** *Let $G \in \mathcal{R}_{UVR}$ be a connected graph of order $n$. Then $\frac{2}{3}n \geq \gamma_R(G)$. If the equality holds, then for any $\gamma_R(G)$-function $f = (V_0, V_1, V_2)$, $V_2$ is an efficient dominating set of $G$ and each vertex of $V_2$ has degree 2. If $G$ has an efficient dominating set $D$ and each vertex of $D$ has degree 2, then $\frac{2}{3}n = \gamma_R(G)$.*

The bound in Proposition 2.18 is tight at least for all cycles $C_{3k}$, $k \geq 1$. Samodivkin [75] gave a constructive characterization of $\mathcal{R}_{UVR}$-trees who showed in addition that any tree in $\mathcal{R}_{UVR}$ has a unique Roman dominating function of minimum weight. It should also be noted that trees with unique Roman dominating functions have been characterized by Chellali and Jafari Rad [22].

A characterization of trees $T \in \mathcal{R}_{UVR}$ attaining equality in the bound of Proposition 2.18 was given by Samodivkin [75], while Hajian and Jafari Rad [38] characterized those graphs $G \in \mathcal{R}_{UVR}$ with minimum degree at least two achieving equality in Proposition 2.18.

**Theorem 2.19 ([38])** *Let $G \in \mathcal{R}_{UVR}$ be a connected graph of order $n$ with $\delta(G) \geq 2$. Then $\gamma_R(G) = \frac{2n}{3}$ if and only if $G$ is a cycle of order $3k$ for some integer $k$.*

**Corollary 2.20** *If $G \in \mathcal{R}_{UVR}$ is a connected graph of order $n$ with $2 \leq \delta(G) < \Delta(G)$, then $\gamma_R(G) \leq \frac{2n-2}{3}$. This bound is sharp.*

Additional upper bounds on the Roman domination number of graphs in $\mathcal{R}_{UVR}$ were obtained by Hajian and Jafari Rad [38].

**Theorem 2.21 ([38])** *If $G \in \mathcal{R}_{UVR}$ is a claw-free graph of order $n$ with $\delta(G) \geq 3$, then $\gamma_R(G) \leq \frac{4n-2}{7}$. This bound is sharp.*

**Theorem 2.22 ([38])** *If $G \in \mathcal{R}_{UVR}$ is a graph of order $n$ with $\delta(G) \geq 3$, then*

$$\gamma_R(G) \leq \frac{2n}{3}\left(\frac{1}{1 + \frac{\delta(G)-2)}{3\Delta(G)}}\right).$$

**Corollary 2.23** *If $G \in \mathcal{R}_{UVR}$ is a cubic graph of order n, then $\gamma_R(G) \le \frac{3n}{5}$.*

Theorem 2.22 has been improved for $C_5$-free graphs $G$ with $\delta(G) = 3$ and $\Delta(G) \ge 4$.

**Theorem 2.24 ([38])** *If $G \in \mathcal{R}_{UVR}$ is a $C_5$-free graph of order n with $\delta(G) = 3$ and $\Delta(G) \ge 4$, then*

$$\gamma_R(G) \le \frac{2n}{3}\left(\frac{\Delta(G) - 1/n}{\Delta(G) + 1/3}\right).$$

Theorem 2.22 has also been improved for graphs with minimum degree at least four as follows.

**Theorem 2.25 ([38])** *If $G \in \mathcal{R}_{UVR}$ is graph of order n with $\Delta(G) > 3\delta(G) - 6$ and $\delta(G) \ge 4$, then*

$$\gamma_R(G) \le \frac{2n}{3}\left(\frac{\Delta(G) - 1/n}{\Delta(G) + (\delta(G) - 2)/3}\right).$$

Since any planar graph has a vertex of degree at most five, we obtain the following.

**Corollary 2.26 ([38])** *If $G \in \mathcal{R}_{UVR}$ is a planar graph of order n with $\delta(G) \ge 4$ and $\Delta(G) \ge 10$, then $\gamma_R(G) \le \frac{2n}{3}(\frac{\Delta(G)-1/n}{\Delta(G)+(\delta(G)-2)/3})$.*

Samodivkin [77] studied Roman excellent graphs as those graphs $G$ in which every vertex is assigned a non-zero value under some $\gamma_R(G)$-function $f$. A characterization of Roman excellent trees was given by Samodivkin [77] who proved in addition that every tree in $\mathcal{R}_{UVR}$ is Roman domination excellent.

### 2.4.2 Edge Removal

Since $\gamma_R(P_n) = \gamma_R(C_n)$, we have $C_n \in \mathcal{R}_{UER}$. Furthermore, it is straightforward to verify that $P_n \in \mathcal{R}_{UER}$ if and only if $n \equiv 2 \pmod 3$. For graphs $G$ of order $n$ and maximum degree $\Delta(G) = n-1$, it is shown in [55] that $G \in \mathcal{R}_{UER}$ if and only if $G$ contains at least three vertices of degree $n-1$ or $G \cong K_2$. In particular, if $G$ is the complete $r$-partite graph $K_{p_1,p_2,\ldots,p_r}$ with $r \ge 2$ and $1 \le p_1 \le p_2 \le \ldots \le p_r$, then $G \in \mathcal{R}_{UER}$ if and only if $r = 2$ and $p_1 = p_2 = 1$ or $r \ge 3$ and $p_1 = p_2 = p_3 = 1$ or $p_1 \ge 2$.

Jafari Rad and Volkmann characterized the graphs $G$ belonging to $\mathcal{R}_{UER}$ as follows.

**Theorem 2.27 ([55])** *For a graph G, $G \in \mathcal{R}_{UER}$ if and only if for any edge $e = xy$ in G there is a $\gamma_R(G)$-function f which is also a $\gamma_R(G - e)$-function.*

### 2.4.3 Edge Addition

A graph $G$ is called $\gamma_R$*-EA-stable* if $G \in \mathcal{R}_{UEA}$. Chellali and Jafari Rad [23] have shown that there is no forbidden subgraph characterization of $\gamma_R$-EA-stable graphs, since for every graph $H$ they have provided a $\gamma_R$-EA-stable graph $G$ such that $H$ is an induced subgraph of $G$. However, a necessary and sufficient condition for a graph to be in $\mathcal{R}_{UEA}$ was given in [23].

**Theorem 2.28 ([23])** *A graph $G$ of order $n \geq 3$ is $\gamma_R$-EA-stable if and only if for every $\gamma_R(G)$-function $f = (V_0, V_1, V_2)$, $V_1 = \emptyset$.*

According to Theorem 2.28, if a graph $G$ is $\gamma_R$-EA-stable, then $\gamma_R(G) = 2|V_2|$ and thus no graph with an odd Roman domination number is $\gamma_R$-EA-stable. We next give a characterization of $\gamma_R$-EA-stable graphs with small even Roman domination number. For an integer $k$, a graph $G$ is called $k$-$\gamma_R$*-EA-stable* if $G$ is $\gamma_R$-EA-stable and $\gamma_R(G) = k$. Recall that $\gamma_R(G) \geq 2$ for every nontrivial graph $G$.

**Proposition 2.29 ([23])** *A connected graph $G$ of order $n \geq 2$ is 2-$\gamma_R$-EA-stable if and only if $G$ has a vertex $v$ with $\deg_G(v) = n - 1$.*

**Proposition 2.30 ([23])** *A connected graph $G$ of order $n$ is 4-$\gamma_R$-EA-stable if and only if $\Delta(G) \leq n-4$ and there are two vertices $x, y$ such that $N(x) \cup N(y) = V(G)$.*

For particular graphs, it was shown in [23] that the only $\gamma_R$-EA-stable paths are $P_2$ and $P_n$ with $n \equiv 0 \pmod 3$; the $\gamma_R$-EA-stable cycles are $C_n$ with $n \equiv 0 \pmod 3$. For $m \leq n$, $K_{m,n}$ is $\gamma_R$-EA-stable if and only if $m \neq 2$.

Chellali and Jafari Rad [23] have shown that any $\gamma_R$-EA-stable graph is Roman (graphs $G$ such that $\gamma_R(G) = 2\gamma(G)$). However, the converse is not true even for trees. To see this, consider a double star $T_{2,2}$ with two leaves attached at each support vertex. Then $T_{2,2}$ is a Roman tree but $\gamma_R(T_{2,2}+xy) = 3$, where $x$ is a support vertex and $y$ is a non-neighbor of $x$.

A graph $G$ is said to be a *strong Roman graph* if $V_1$ is empty for every $\gamma_R(G)$-function $f = (V_0, V_1, V_2)$. Therefore, Theorem 2.28 can also be stated as follows.

**Theorem 2.31 ([23])** *A graph $G$ is $\gamma_R$-EA-stable if and only if $G$ is a path $P_2$ or a strong Roman graph.*

It can be noted that a constructive characterization of strong Roman trees was also given in [23].

**Proposition 2.32 ([23])** *Let $G$ be a $\gamma_R$-EA-stable graph of order $n \geq 3$. Then*

**(i)** *For every vertex $v$, $\gamma_R(G - v) \geq \gamma_R(G)$.*
**(ii)** *$diam(G) \leq \frac{3\gamma_R(G)}{2} - 1$.*

**Theorem 2.33 ([23])** *If $G$ is a $\gamma_R$-EA-stable graph of order $n \geq 3$, then $n \leq \frac{\gamma_R(G)}{2}(1 + \Delta(G))$. The equality holds if and only if $\gamma_R(G) \equiv i_R(G)$ and each vertex of $V_2$ has maximum degree for every $\gamma_R$-function $(V_0, \emptyset, V_2)$.*

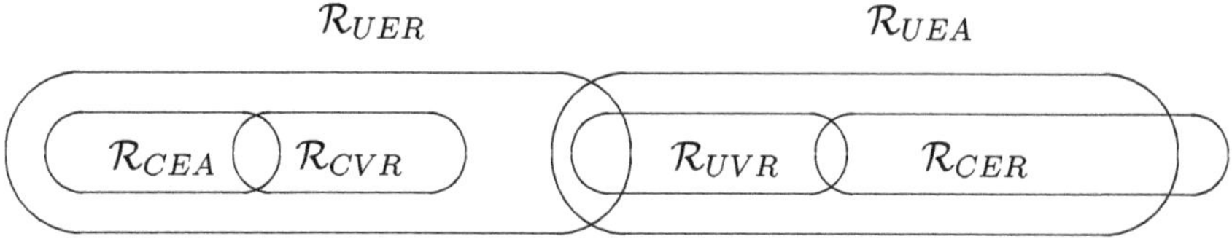


**Fig. 10** Classes of changing and unchanging graphs

**Theorem 2.34 ([23])** *If $G$ is a $\gamma_R$-EA-stable graph of order $n \geq 3$, then $n \geq \frac{3\gamma_R(G)}{2}$. Equality holds if and only if $\gamma_R(G) \equiv i_R(G)$, $V_2$ is a packing set for every $\gamma_R(G)$-function $(V_0, \emptyset, V_2)$ and each vertex of $V_2$ has exactly two private neighbors.*

## 2.5 Relations Between Families

Samodivkin [76] studied the relationship between the six classes $\mathcal{R}_{CVR}$, $\mathcal{R}_{UVR}$, $\mathcal{R}_{CER}$, $\mathcal{R}_{UER}$, $\mathcal{R}_{CEA}$, and $\mathcal{R}_{UEA}$ which can be seen in the Venn diagram of Figure 10 based on the results that follow.

**Theorem 2.35 ([76])** *Let a graph $G$ be in $\mathcal{R}_{CEA}$. Then all the following hold.*

*(i) A vertex $x \in V^{=}(G)$ if and only if there are $\gamma_R(G)$-functions $f_x$ and $g_x$ with $\{f_x(x), g_x(x)\} = \{0, 2\}$.*
*(ii) If $V^{=}(G)$ is not empty and $G[V^{=}(G)]$ is not a connected component of $G$, then each vertex in $V^{=}(G)$ has a neighbor in $V^{-}(G)$.*
*(iii) $G$ is in $\mathcal{R}_{UER}$.*

**Theorem 2.36 ([76])** *For an edge $e = uv$ of a graph $G$ is fulfilled $\gamma_R(G - e) = \gamma_R(G)$ if and only if there is a $\gamma_R(G)$-function $f_e$ such that at least one of the following holds:*

*(i) $f_e(u) = f_e(v)$,*
*(ii) at least one of $u$ and $v$ is in $V_1^{f_e}$,*
*(iii) $f_e(u) = 2$, $f_e(v) = 0$ and $v \notin pn[u, V_2^{f_e}]$,*
*(iv) $f_e(u) = 0$, $f_e(v) = 2$ and $u \notin pn[v, V_2^{f_e}]$.*

**Corollary 2.37** *Let $G$ be a graph with edges. Then for each edge $e$ incident to a vertex in $V^{-}(G)$, $\gamma_R(G - e) = \gamma_R(G)$. If $V^{-}(G)$ contains a vertex cover of $G$, then $G$ is in $\mathcal{R}_{UER}$. In particular, if $G$ is in $\mathcal{R}_{CVR}$, then $G$ is in $\mathcal{R}_{UER}$.*

It should be noted that for the establishment of the Venn diagram the cases that were vacuously true have not been considered. For example, (a) complete graphs are in both $\mathcal{R}_{CEA}$ and $\mathcal{R}_{UEA}$, and (b) edgeless graphs are in both $\mathcal{R}_{CER}$ and $\mathcal{R}_{UER}$. Therefore we exclude edgeless graphs and complete graphs.

**Theorem 2.38 ([76])** *Classes $\mathcal{R}_{CVR}$, $\mathcal{R}_{CEA}$, $\mathcal{R}_{CER}$, $\mathcal{R}_{UVR}$, $\mathcal{R}_{UER}$, and $\mathcal{R}_{UEA}$ are related as shown in the Venn diagram of Figure 10.*

Samodivkin [76] continued the study of the Venn diagram of Figure 10 by showing the following: a graph $G$ is in $\mathcal{R}_{CER} \cap \mathcal{R}_{UVR}$ if and only if $G = nK_{1,2}$, $n \geq 1$; a graph $G$ is in $(\mathcal{R}_{CER} \cap \mathcal{R}_{UEA}) - \mathcal{R}_{UVR}$ if and only if each component of $G$ is a star of order at least 4; and a graph $G$ is in $\mathcal{R}_{CER} - \mathcal{R}_{UEA}$ if and only if $\delta(G) = 0$ and each component of $G$ is an isolated vertex or a star of order at least 3. He also showed that for trees of order $n \geq 3$, $\mathcal{R}_{CER} - \mathcal{R}_{UEA} = \mathcal{R}_{UEA} \cap \mathcal{R}_{UER} = \mathcal{R}_{CVR} = \emptyset$.

## 2.6 Other Modifications

Changing and unchanging in Roman domination number according to other modifications have been studied. In [52, 56], Jafari Rad and Volkmann studied graphs for which contracting any edge decreases the Roman domination number. Jafari Rad studied in [47] Roman domination critical graphs upon edge subdivision and in [46] the graphs which he called *Roman domination bicritical graphs,* where the removal of any pair of vertices decreases the Roman domination number. Properties of Roman domination bicritical graphs are presented in [46] and Roman domination bicritical trees and unicyclic graphs are characterized. On the other hand, Amraee et al. [6] introduced the parameter *Roman domination stability number*, defined as the minimum number of vertices whose removal changes the Roman domination number. They presented various bounds and characterizations for the Roman domination stability number, and showed that the decision problem corresponding to the Roman domination stability number is NP-hard even when restricted to bipartite graphs. Bouchou and Blidia [14] associated indices, namely the *removal criticality index* and the *adding criticality index*, and computed these indices for paths and cycles. The *Roman domination subdivision number* $sd_{\gamma_R}(G)$ of a graph $G$ was introduced by Atapour et al. [8] defined as the minimum number of edges that must be subdivided (each edge in $G$ can be subdivided at most once) in order to increase the Roman domination number. Atapour et al. presented upper bounds on $sd_{\gamma_R}(G)$ for arbitrary graphs $G$ in terms of vertex degree. They also showed that the Roman domination subdivision number of a graph can be arbitrarily large, and presented several different conditions on $G$ which are sufficient to imply that $1 \leq sd_{\gamma_R}(G) \leq 3$. Further bounds on the Roman domination subdivision number are presented by Khodkar et al. [58, 59]. A constructive characterization of trees $T$ such that $sd_{\gamma_R}(T) = 2$ was given by Atapour et al. [7]. Finally, Amjadi et al. [5] introduced the Roman game domination subdivision number of a graph and presented sharp bounds on the Roman game domination subdivision number of a tree.

# 3 Roman Bondage and Roman Reinforcement

## 3.1 Roman Bondage

In 1983, Bauer, Harary, Nieminen, and Suffel [11] introduced the *bondage number* $b(G)$ of a non-empty graph $G$ defined as the minimum cardinality among all sets of edges $E' \subseteq E(G)$ for which $\gamma(G - E') > \gamma(G)$. For more details on the bondage number, the reader is referred to the surveys by Dunbar et al. [32] and Xu [88]. The *Roman bondage number* $b_R(G)$ of a graph $G$ has been introduced independently by Ebadi and PushpaLatha [33] and Jafari Rad and Volkmann [53], defined as the cardinality of a smallest set of edges $E' \subseteq E(G)$ for which $\gamma_R(G - E') > \gamma_R(G)$. Observe that if $\Delta(G) = 1$, then $\gamma_R(G) = |V(G)|$ and $\gamma_R(G - E') = \gamma_R(G)$ for any $E' \subseteq E(G)$. Therefore the Roman bondage number is only defined for a graph $G$ with maximum degree at least two. It has been shown by Bahremandpour et al. [9] that the problem corresponding to the Roman bondage number is NP-hard even for bipartite graphs. Hence it is interesting to look for bounds and exact values for the Roman bondage number in some classes of graphs. Here are some results obtained in [33, 53] giving the exact value of $b_R(G)$ for complete graphs, cycles, paths, complete $t$-partite graphs for $t \geq 2$, and Ladder graphs $P_2 \Box P_n$.

**Theorem 3.1 ([33, 53])** *If $G$ is a graph of order $n \geq 3$ with exactly $k \geq 1$ vertices of degree $n - 1$, then $b_R(G) = \lceil \frac{k}{2} \rceil$. In particular $b_R(K_n) = \lceil \frac{n}{2} \rceil$.*

**Theorem 3.2 ([33, 53])** *For all $n \geq 3$,*

**(i)** $b_R(P_n) = \begin{cases} 2 \text{ if } n \equiv 2 \pmod 3, \\ 1 \text{ otherwise.} \end{cases}$

**(ii)** $b_R(C_n) = \begin{cases} 3 \text{ if } n \equiv 2 \pmod 3, \\ 2 \text{ otherwise.} \end{cases}$

**Theorem 3.3 ([45])** *Let $G = K_{m_1, m_2, \ldots, m_t}$ be a complete $t$-partite graph with $m_1 = \cdots = m_i < m_{i+1} \leq \cdots \leq m_t$ and $n = \sum_{j=1}^{t} m_j$. Then*

$$b_R(G) = \begin{cases} \lceil \frac{i}{2} \rceil & \text{if } m_i = 1 \text{ (and } n \geq 3), \\ 2 & \text{if } m_i = 2 \text{ (and } i = 1), \\ i & \text{if } m_i = 2 \text{ (and } i \geq 2), \\ 4 & \text{if } m_i = 3 \text{ (and } i = t = 2), \\ n - 1 & \text{if } m_i = 3 \text{ (and } i = t \geq 3), \\ n - m_t & \text{if } m_i \geq 3 \text{ (and } m_t \geq 4). \end{cases}$$

**Theorem 3.4 ([9, 33])** *For all $n \geq 2$, $b_R(P_2 \Box P_n) = 2$.*

Independently, Ebadi et al. [33], and Jafari Rad and Volkmann [53], have shown that for every tree $T$ of order at least three, $b_R(T) \leq 3$. Moreover, Jafari Rad and Volkmann [53] showed that for unicyclic graphs $G$, $b_R(G) \leq 4$. Ebadi et al. [33] conjectured that $b_R(G) \leq n - 1$ for any graph of order $n \geq 3$. This conjecture was settled independently by Dehgardi et al. [30] and Akbari and Qajar [3]. Theorem 3.3 indicates that the equality $b_R(G) = n - 1$ is possible. Upper bounds on the Roman bondage number in terms of the degrees of the vertices are obtained in [53].

**Theorem 3.5 ([53])** *If $G$ is a graph, and $xyz$ a path of length 2 in $G$, then*

$$b_R(G) \leq \deg(x) + \deg(y) + \deg(z) - 3 - |N(x) \cap N(y)|. \tag{6}$$

*If $x$ and $z$ are adjacent, then*

$$b_R(G) \leq \deg(x) + \deg(y) + \deg(z) - 4 - |N(x) \cap N(y)|. \tag{7}$$

Applying Theorem 3.5 on a path of length 2 $xyz$ such that one of the vertices $x$, $y$ or $z$ has minimum degree, we get the following.

**Corollary 3.6 ([53])** *If $G$ is a connected graph of order $n \geq 3$, then*

$$b_R(G) \leq \delta(G) + 2\Delta(G) - 3.$$

Recall that for any connected graph $G$, the *average degree* $\deg_a(G)$ represents the value of the expression $\sum_{v \in V(G)} \deg(v)/|V(G)|$. Dehgardi et al. [29] obtained an upper bound on the Roman bondage number of graphs in terms of the average degree and maximum degree, as well as a lower bound on $|E(G)|$ in terms of the order $n$ of $G$, the maximum degree $\Delta$, and bondage number $b_R(G)$.

**Theorem 3.7 ([29])** *Let $G$ be a connected graph of order $n \geq 3$, average degree $\deg_a(G)$, and bondage number $b_R(G)$. Then*

$$b_R(G) \leq 2\deg_a(G) + \Delta(G) - 3 \text{ and } |E(G)| \geq (n/4)(b_R(G) - \Delta(G) + 3).$$

We observe that the two bounds are sharp for the cycle $C_n$ when $n \equiv 2 \pmod 3$.

The next upper bound involves the *edge-connectivity* $\lambda(G)$. Since $\lambda(G) \leq \delta(G)$, the next theorem is an improvement of Corollary 3.6.

**Theorem 3.8 ([53])** *If $G$ is a connected graph of order $n \geq 3$, then*

$$b_R(G) \leq \lambda(G) + 2\Delta(G) - 3.$$

It is well known that for planar graphs $G$, $\delta(G) \leq 5$, and so Corollary 3.6 leads to $b_R(G) \leq 2\Delta(G) + 2$ for planar graphs. This bound was improved in [54] as follows.

**Theorem 3.9 ([54])** *If $G$ is a connected planar graph of order $n \geq 3$, then $b_R(G) \leq \min\{2\Delta(G), \Delta(G) + 6\}$.*

Jafari Rad and Volkmann [54] then considered the girth of a connected planar graph and obtained the following bounds that improve Theorem 3.9.

**Theorem 3.10 ([54])** *Let $G$ be a connected planar graph of order $n \geq 3$.*

*(1) If $g(G) \geq r$ where $r \in \{4, 5, 6\}$, then $b_R(G) \leq \Delta(G) + 8 - r$.*
*(2) If $g(G) \geq 8$, then $b_R(G) \leq \Delta(G) + 1$.*

Since $b_R(C_n) = 3$ for a cycle $C_n$ of length $n \geq 8$ with $n \equiv 2 \pmod 3$, Theorem 3.10-(2) is best possible, at least for $\Delta = 2$. Moreover, Akbari et al. [2] have shown that $b_R(G) \leq 15$ for every planar graph $G$. Although finding a planar graph $G$ with $b_R(G) = 15$ remains open, Akbari et al. [2] constructed an infinite family of planar graphs $G$ with $b_R(G) = 7$. Restricted to planar graphs $G$ with minimum degree five, Samodivkin [74] showed that $b_R(G) \leq 14$. He has also studied the Roman domination number of graphs $G$ belonging to $\mathcal{R}_{UVR}$ (class of graphs $G$ such that $\gamma_R(G - v) = \gamma_R(G)$ for all $v \in V(G)$).

**Theorem 3.11 ([75])** *If a graph $G$ is in $\mathcal{R}_{UVR}$, then $b_R(G) \leq \delta(G)$.*

The bound stated in Theorem 3.11 is tight. For example, when (a) $G = C_{3k}$, $k \geq 1$, and (b) $\delta(G) = 1$. As an immediate consequence, if $T$ is a tree in $\mathcal{R}_{UVR}$, then $b_R(T) = 1$. Bounds involving the Roman bondage number and the bondage number of graphs under certain conditions were obtained by Bahremandpour et al. [9]. Note that $\beta(G)$ is the vertex covering number.

**Theorem 3.12 ([9])**

*1) For any connected graph $G$ of order $n \geq 3$ and $\gamma_R(G) = \gamma(G) + 1$, $b_R(G) \leq \min\{b(G), n_\Delta\}$, where $n_\Delta$ is the number of vertices with maximum degree in $G$.*
*2) For any Roman graph $G$, $b_R(G) \geq b(G)$, and this bound is sharp.*
*3) For any graph $G$, if $\gamma_R(G) = 2\beta(G)$, then $b_R(G) \geq \delta(G)$.*
*4) If $G \in \mathcal{R}_{CVR}$ and $\gamma_R(G) = 2\beta(G)$, then $b_R(G) \geq \delta(G) + 1$.*

If $\gamma_R(G) = 2$, then obviously $b_R(G) \leq \delta(G)$. In the case $\gamma_R(G) \geq 3$, Dehgardi et al. [30] proved that $b_R(G) \leq (\gamma_R(G) - 2)\Delta(G) + 1$. For graphs $G$ with $\gamma_R(G) \in \{3, 4\}$, Bahremandpour et al. [9] proved the following.

**Theorem 3.13 ([9])** *For any connected graph $G$ of order $n \geq 4$,*

$$b_R(G) \leq \begin{cases} \Delta(G) = n - 2 & \text{if } \gamma_R(G) = 3, \\ \Delta(G) + \delta(G) - 1 & \text{if } \gamma_R(G) = 4. \end{cases}$$

Akbari et al. [3] obtained an upper bound on the Roman bondage number of graphs in terms of the order and the Roman domination number.

**Theorem 3.14 ([3])** *For any connected graph $G$ of order $n \geq 3$, $b_R(G) \leq n - \gamma_R(G) + 5$.*

Roman bondage number of graphs on surfaces was studied by Samodivkin [72, 73], where he obtained upper bounds on $b_R(G)$ in terms of (a) the average degree and maximum degree, and (b) Euler characteristic, girth, and maximum degree. Samodivkin then showed that the Roman bondage number of every graph which admits a 2-cell embedding on a surface with non-negative Euler characteristic does not exceed 15. He also presented upper bounds for the Roman bondage number of a graph on topological surfaces in terms of maximum degree and orientable/non-orientable genus. Katagiri [57] presented a new upper bound for the Roman bondage number of a graph on a closed surface. We close this subsection by mentioning that Dehgardi et al. studied in [29] the fractional Roman bondage number defined as the optimal value of the linear programming relaxation of the integer programming formulation for Roman bondage. They determined the Roman fractional bondage number of some classes of graphs and presented different bounds on this parameter.

## 3.2 *Roman Reinforcement*

The concept of Roman reinforcement in graphs was introduced independently by Ebadi and PushpaLatha [33] and Jafari Rad and Sheikholeslami [51]. The *Roman reinforcement number* $r_R(G)$ of a graph $G$ is the minimum number of edges that must be added to $G$ in order to decrease the Roman domination number of $G$. Obviously, if $\gamma_R(G) \in \{1, 2\}$, then adding edges to $G$ will not decrease the Roman domination number, and thus it was defined $r_R(G) = 0$. Jafari Rad and Sheikholeslami gave a necessary and sufficient condition for graphs $G$ with $r_R(G) = 1$.

**Theorem 3.15 ([51])** Let $G$ be a connected graph of order $n \geq 3$. Then $r_R(G) = 1$ if and only if there is a $\gamma_R(G)$-function $f = (V_0, V_1, V_2)$ with $V_1 \neq \emptyset$.

According to Theorem 3.15, a connected graph $G$ with $\gamma_R(G) \geq 3$ satisfies $r_R(G) \geq 2$ if and only if for each $\gamma_R(G)$-function $f = (V_0, V_1, V_2)$, $V_1 = \emptyset$. Various bounds on the Roman reinforcement number of a graph have been obtained in [51], some of which are presented below.

**Theorem 3.16 ([51])** *For any graph $G$ on $n$ vertices, $r_R(G) \leq \lceil \frac{2n}{\gamma_R(G)} \rceil - 1$.*

**Theorem 3.17 ([51])** Let $G$ be a connected graph of order $n \geq 3$ and let $f = (V_0, V_1, V_2)$ be a $\gamma_R(G)$-function. Then

$$r_R(G) \leq \min\{\deg(u) \mid u \in V_2\}.$$

Moreover, if the subgraph induced by $V_2$ is isolate-free, then

$$r_R(G) \leq \min\{\deg(u) \mid u \in V_2\} - 1.$$

Using the Theorems 3.15, 3.16, and 3.17, the Roman reinforcement number for the following families of graphs can be obtained. Note that $r_R(P_n) = r_R(C_n) = 0$ for $n = 2, 3$.

**Theorem 3.18 ([33, 51])**

*1) For $n \geq 4$,*

$$r_R(P_n) = r_R(C_n) = \begin{cases} 2, & \text{if } n \equiv 0 \pmod 3 \\ 1, & \text{if } n \not\equiv 0 \pmod 3. \end{cases}$$

2) *For* $n \geq 2$, $r_R(P_2 \Box P_n) = 1$.
3) *Let* $G = K_{n_1,n_2,\ldots,n_m}$ *be the complete* $m$*-partite graph with* $2 \leq n_1 \leq n_2 \leq \cdots \leq n_m$. *Then*

$$r_R(G) = \begin{cases} 1 & \text{if } n_1 = 2 \\ n_1 - 2 & \text{if } n_1 \geq 3. \end{cases}$$

Some consequences of Theorem 3.17 are given in the following corollary.

**Corollary 3.19 ([51])**

1) *For any graph* $G$ *with no isolated vertex,* $r_R(G) \leq \Delta(G)$. *The bound is sharp for cycles* $C_{3k}$ $(k \geq 2)$.
2) *If a connected graph* $G$ *contains a path* $v_1v_2v_3v_4$ *or a cycle* $v_1v_2v_3v_4v_1$ *in which* $\deg(v_i) = 2$ *for* $i = 2, 3$ *and* $\deg(v_4) \leq 2$, *then* $r_R(G) \leq 2$. *This bound is sharp for paths* $P_{3k}$ *and cycles* $C_{3k}$ $(k \geq 2)$.
3) *For any tree* $T$ *of order* $n \geq 4$, $r_R(T) \leq \min\{\deg(v) : v \in S(T)\}$, *where* $S(T)$ *is the set of all support vertices of* $T$.

### *3.3 Summary and Open Problems*

In Sections 2 and 3, we were interested to know on changing and unchanging the Roman domination number according to some graph modifications such as the addition of a set of edges, the removal of a set of vertices/edges, contraction of edges or subdivisions of edges. Many properties, characterizations, and bounds have been given in these sections, but also many problems remain open, and which are given briefly below.

1. Is it true that any $\gamma$-vertex critical graph is $\gamma_R$-vertex critical? (This is a conjecture stated in [55]).
2. Give a characterization of every of the six classes of graphs shown in the Venn diagram of Figure 10.
3. Characterize all graphs $G \in \mathcal{R}_{UVR}$ of order $n$ with $\gamma_R(G) = \frac{2n}{3}$ and $\delta(G) = 1$.
4. Can the bounds of Theorems 2.24 and 2.25 be improved?
5. Characterize the trees $T$ with $b_R(T) = 1$, $b_R(T) = 2$, or $b_R(T) = 3$.
6. Is it true that the Roman bondage number of every planar graph is at most 7? (This is a conjecture stated in [2]).
7. Let $G$ be a connected graph of order $n \geq 4$ with $\gamma_R(G) \geq 3$. Is $b_R(G) \leq (\gamma_R(G) - 2)\Delta(G)$? It can be noted that by Corollary 3.6 this is obvious for graphs $G$ with $\gamma_R(G) \geq 5$.

## References


1. H. Abdollahzadeh Ahangar, J. Amjadi, M. Chellali, S. Nazari-Moghaddam and S.M. Sheikholeslami, Total Roman reinforcement numbers in graphs. *Discuss. Math. Graph Theory* 39 (2019) 787–803.
2. S. Akbari, M. Khatirinejad, and S. Qajar, A note on the roman bondage number of planar graphs. *Graphs Combin.* 29 (2013) 327–331.
3. S. Akbari and S. Qajar, A note on Roman bondage number of graphs. *Ars Combin.* 126 (2016) 87–92.
4. J.D. Alvarado, S. Dantas and D. Rautenbach, Averaging 2-rainbow domination and Roman domination. *Discrete Appl. Math.* 205 (2016) 202–207.
5. J. Amjadi, H. Karami, S.M. Sheikholeslami and L. Volkmann, Roman game domination subdivision number of a graph. *Trans. Combin.* 2 (2013) 1–12.
6. M. Amraee, N. Jafari Rad and M. Maghasedi, Roman domination stability in graphs. *Math. Reports* 21 (71) (2019) 193–204.
7. M. Atapour, A. Khodkar and S. M. Sheikholeslami, Trees whose Roman domination subdivision number is 2. *Util. Math.* 82 (2010) 227–240.
8. M. Atapour, S.M. Sheikholeslami and A. Khodkar, Roman domination subdivision number of a graph. *Aequationes Math.* 78 (2009) 237–245.
9. A. Bahremandpour, Fu-Tao Hu, S.M. Sheikholeslami and Jun-Ming Xu, On the Roman bondage number of a graph. *Discrete Math. Algorithms Appl.* 5 (2013) 1350001 (15 pages)
10. D. Bange, A. E. Barkauskas and P. J. Slater. Efficient dominating sets in graphs. In *Applications of Discrete Math.*, R. D. Ringeisen and F. S. Roberts, eds., SIAM, Philadelphia, (1988) 189–199.
11. D. Bauer, F. Harary, J. Nieminen and C. L. Suffel, Domination alteration sets in graphs. *Discrete Math.* 47 (1983) 153–161.
12. S. Bermudo, On the differential and Roman domination number of a graph with minimum degree two. *Discrete Appl. Math.* 232 (2017) 64–72.
13. S. Bermudo, H. Fernau, and J.M. Sigarreta, The differential and the Roman domination number of a graph. *Appl. Anal. Discrete Math.* 8 (2014) 155–171.
14. A. Bouchou and M. Blidia, Criticality indices of Roman domination of paths and cycles. *Australas. J. Combin.* 56 (2013), 103–112.
15. A. Bouchou, M. Blidia and M. Chellali, Extremal graphs for a bound on the Roman domination number. *Discuss. Math. Graph Theory* 40 (2020) 771–785.
16. B. Brešar, M.A. Henning and D.F. Rall, Rainbow domination in graphs. *Taiwanese J. Math.* 12 (2008) 201–213.
17. E.W. Chambers, B. Kinnersley, N. Prince and D.B. West, Extremal problems for Roman domination. *SIAM J. Discrete Math.* 23 (2009) 1575–1586.
18. M. Chellali, T.W. Haynes and S.T. Hedetniemi, Roman and total domination. *Quaestiones Math.* 38 (2015) 749–757.
19. M. Chellali, T.W. Haynes and S.T. Hedetniemi, Lower bounds on the Roman and independent Roman domination numbers. *Appl. Anal. Discrete Math.* 10 (2016) 65–72.
20. M. Chellali, T.W. Haynes and S.T. Hedetniemi, S.M. Hedetniemi and A. MacRae, A Roman domination chain. *Graphs Combin.* 32 (2016) 79–92.
21. M. Chellali and N. Jafari Rad, On 2-rainbow domination and Roman domination in graphs. *Australas. J. Combin.* 56 (2013) 85–93.
22. M. Chellali and N. Jafari Rad, Trees with unique Roman dominating functions of minimum weight. *Discrete Math. Algorithms Appl.* 6 (2014) 1450038 (10 pages).
23. M. Chellali and N. Jafari Rad, Roman domination stable graphs upon edge-addition. *Util. Math.* 96 (2015) 165–178.
24. M. Chellali, N. Jafari Rad and L. Volkmann, Some results on Roman domination edge critical graphs. *AKCE Int. J. Graphs Combin.* 9 (2012) 195–203.

25. E.J. Cockayne, P.A. Dreyer Jr., S.M. Hedetniemi and S.T. Hedetniemi, Roman domination in graphs. *Discrete Math.* 278 (2004) 11–22.
26. E.J. Cockayne, P.J.P. Grobler, W.R. Gründlingh, J. Munganga and J.H van Vuuren, Protection of a graph. *Util. Math.* 67 (2005) 19–32.
27. E.J. Cockayne, S.T. Hedetniemi and D.J. Miller, Properties of hereditary hypergraphs and middle graphs, *Canad. Math. Bull.* 21 (1978) 461–468
28. V. Currò, The Roman Domination Problem on Grid Graphs. PhD dissertation in Università degli Studi di Catania, 2014.
29. N. Dehgardi, B. Kheirfam, S.M. Sheikholeslami and O. Favaron, Roman fractional bondage number of a graph. *J. Combin. Math. Combin. Comput.* 87 (2013) 51–63.
30. N. Dehgardi, S.M. Sheikholeslami and L. Volkman, On the Roman $k$-bondage number of a graph. *AKCE Int. J. Graphs Combin.* 8 (2011) 169–180.
31. P.A. Dreyer, Jr., Applications and variations of domination in graphs. Ph.D. Thesis, Rutgers University, October 2000.
32. J.E. Dunbar, T.W. Haynes, U. Teschner, and L. Volkmann, *Bondage, insensitivity, and reinforcement*, in : T.W. Haynes, S.T. Hedetniemi, P.J. Slater (Eds.), Domination in Graphs: Advanced Topics, Marcel Dekker, New York, 1998, pp. 471–489.
33. K. Ebadi and L. PushpaLatha, Roman bondage and Roman reinforcement numbers of a graph. *Int. J. Contemp. Math.* 5 (2010) 1487–1497.
34. M. Farber, Domination, independent domination and duality in strongly chordal graphs. *Discrete Appl. Math.* 7 (1984) 115–130.
35. O. Favaron, H. Karami, R. Khoeilar and S.M. Sheikholeslami, On the Roman domination number of a graph. *Discrete Math.* 309 (2009) 3447–3451.
36. R. W. Frucht and F. Harary, On the corona of two graphs. *Aequationes Math.* 4 (1970) 322–325.
37. S. Fujita and M. Furuya, Difference between 2-rainbow domination and Roman domination in graphs. *Discrete Appl. Math.* 161 (2013) 806–812.
38. M. Hajian and N. Jafari Rad, On the Roman domination stable graphs. *Discuss. Math. Graph Theory* 37 (2017) 859–871.
39. A. Hansberg, N. Jafari Rad and L. Volkmann, Characterization of Roman domination critical unicyclic graphs. *Util. Math.* 86 (2011) 129–146.
40. T.W. Haynes, S.T. Hedetniemi, and P.J. Slater, *Fundamentals of Domination in Graphs*, Marcel Dekker, New York, 1998.
41. S.T. Hedetniemi, R.R. Rubalcaba, P.J. Slater and M. Walsh, Few Compare to the Great Roman Empire. *Congr. Numer.* 217 (2013) 129–136.
42. M.A. Henning, A characterization of Roman trees. *Discuss. Math. Graph Theory* 22 (2002) 325–334.
43. M.A. Henning and A. Yeo, *Total domination in graphs*. Springer Monographs in Mathematics (2013).
44. C.-H. Hsu, C.-S. Liu and S.-L. Peng, Roman domination on block graphs, in: Proceedings of the 22nd Workshop on Combinatorial Mathematics and Computation Theory, (2005) 188–191.
45. F.-T. Hu and J.-M. Xu, Roman bondage numbers of some graphs. *Australas. J. Combin.* 58 (2014) 106–118.
46. N. Jafari Rad, A generalization of Roman domination critical graphs. *J. Combin. Math. Combin. Comput.* 83 (2012) 33–49.
47. N. Jafari Rad, Roman domination critical graphs upon edge subdivision. *J. Combin. Math. Combin. Comput.* 93 (2015) 227–245.
48. N. Jafari Rad, Roman domination edge critical graphs having precisely two cycles. *Ars Combin.* 131 (2017) 355–372.
49. N. Jafari Rad, A. Hansberg and L. Volkmann, Vertex and edge critical Roman domination in graphs. *Util. Math.* 92 (2013) 73–88.
50. N. Jafari Rad and H. Rahbani, On a Nordhaus-Gaddum bound for Roman domination. *Discrete Math. Algorithms Appl.* (2018), under revision.
51. N. Jafari Rad and S.M. Sheikholeslami, Roman reinforcement in graphs. *Bull. Inst. Combin. Appl.* 61 (2011) 81–90.

52. N. Jafari Rad and L. Volkmann, Roman domination dot-critical trees. *AKCE Int. J. Graphs Combin.* 8 (2011) 75–83.
53. N. Jafari Rad and L. Volkmann, Roman bondage in graphs. *Discuss. Math. Graph Theory* 31 (2011) 763–773.
54. N. Jafari Rad and L. Volkmann, On the Roman bondage number of planar graphs. *Graphs Combin.* 27 (2011) 531–538.
55. N. Jafari Rad and L. Volkmann, Changing and unchanging the Roman domination number of a graph. *Util. Math.* 89 (2012) 79–95.
56. N. Jafari Rad and L. Volkmann, Roman domination dot-critical graphs. *Graphs Combin.* 29 (2013) 527–533.
57. M. Katagiri, Upper bounds for the Roman bondage number of graphs on closed surfaces. Annual reports of Graduate School of Humanities and Sciences (2017), http://nwudir.lib.nara-wu.ac.jp/dspace/handle/10935/4458.
58. A. Khodkar, B.P. Mobaraky and S.M. Sheikholeslami, Upper bounds for the Roman domination subdivision number of a graph. *AKCE Int. J. Graphs Comb.* 5 (2008) 7–14.
59. A. Khodkar, B.P. Mobaraky and S.M. Sheikholeslami, Roman domination subdivision number of a graph and its complement. *Ars Combin.* 111 (2013) 97–106.
60. Klobučar and Puljič, Some results for Roman domination number on cardinal product of paths and cycles. *Kragujevac J. Math.* 38 (2014) 83–94.
61. Klobučar and Puljič, Roman domination number on cardinal product of paths and cycles. *Croat. Oper. Res. Rev.* 6 (2015) 71–78.
62. M. Liedloff, T. Kloks, J. Liu and S.–L. Peng, Efficient algorithms for Roman domination on some classes of graphs. *Discrete Appl. Math.* 156 (2008) 3400–3415.
63. C.-H. Liu and G.J. Chang, Roman domination on 2-connected graphs. *SIAM J. Discrete Math.* 26 (2012) 193–205.
64. C.-H. Liu and G.J. Chang, Upper bounds on Roman domination numbers of graphs. *Discrete Math.* 312 (2012) 1386–1391.
65. C.-H. Liu and G.J. Chang, Roman domination on strongly chordal graphs. *J. Comb. Optim.* 26 (2013) 608–619.
66. A. Martinez-Perez and D. Oliveros, Critical properties on Roman domination graphs. arXiv:1311.4476v1 [math.CO] 18 Nov 2013.
67. J.L. Mashburn, T.W. Haynes, S.M. Hedetniemi, S.T. Hedetniemi and P.J. Slater, Differentials in graphs. *Util. Math.* 69 (2006) 43–54.
68. E.A. Nordhaus and J.W. Gaddum, On complementary graphs. *Amer. Math. Monthly* 63 (1956) 175–177.
69. P. Pavlič and J. Žerovnik, Roman domination number of the Cartesian products of paths and cycles. *Electronic J. Combin.* 19 (2012), Paper 19, 37pp.
70. S.-L. Peng and Y.-H. Tsai, Roman domination on graphs of bounded treewidth, in: Proceedings of the 24th Workshop on Combinatorial Mathematics and Computation Theory, 2007, 128–131.
71. C.S. ReVelle and K.E. Rosing, Defendens imperium romanum: a classical problem in military strategy. *Amer. Math. Monthly* 107 (2000) 585–594.
72. V. Samodivkin, Upper bounds for the domination subdivision and bondage numbers of graphs on topological surfaces. *Czechoslovak Math. J.* 63(138) (2013) 191–204.
73. V. Samodivkin, On the Roman bondage number of graphs on surfaces. *Int. J. Graph Theory Appl.* 1 (2015) 67–75.
74. V. Samodivkin, Upper bounds for domination related parameters in graphs on surfaces. *AKCE Int. J. Graphs Comb.* 13 (2016) 140–145.
75. V. Samodivkin, Roman domination in graphs: The class $R_{UVR}$. *Discrete Math. Algorithms and Appl.* 8 (2016) 1650049 (14 pp.).
76. V. Samodivkin, A note on Roman domination: changing and unchanging. *Australas. J. Combin.* 71 (2018) 303–311.
77. V. Samodivkin, Roman domination excellent graphs: trees. *Commun. Comb. Optim.* 3 (2018) 1–24.

78. M. Schnupp, Broadcast domination with flexible powers. Diplomarbeit, Univ. Jena, Germany, 2006.
79. S.M. Sheikholeslami and L. Volkmann, The Roman domatic number of a graph. *Appl. Math. Lett.* 23 (2010) 1295–1300.
80. I. Stewart, Defend the Roman Empire!. *Sci. Amer.* 281 (1999) 136–139.
81. T.K. Šumenjak, P. Pavlič and A. Tepeh, On the Roman domination in the lexicographic product of graphs. *Discrete Appl. Math.* 160 (2012) 2030–2036.
82. H. Tan, H. Liang, R. Wang and J. Zhou, Computing Roman domatic number of graphs. *Inform. Process. Lett.* 116 (2016) 554–559.
83. H.B. Walikar, B.D. Acharya, and E. Sampathkumar, Recent developments in the theory of domination in graphs, In *MRI Lecture Notes in Math., Mahta Research Instit., Allahabad,* Volume 1, 1979.
84. H. Wang, X. Xu, Y. Yang and C. Ji, Roman domination number of generalized Petersen Graphs $P(n, 2)$. *Ars Combin.* 112 (2013) 470–492.
85. Y. Wu, An improvement on Vizing's conjecture. *Inform. Process. Lett.* 113 (2013) 87–88.
86. Y. Wu and H. Xing, Note on 2-rainbow domination and Roman domination in graphs. *Appl. Math. Lett.* 23 (2010) 706–709.
87. H.-M. Xing, X. Chen and X.-G. Chen, A note on Roman domination in graphs. *Discrete Math.* 306 (2006) 3338–3340.
88. J.–M. Xu, On Bondage numbers of graphs: A Survey with Some Comments. *Int. J. Combin.* (2013), Article ID 595210, 34 pages.
89. F. Xueliang, Y. Yuansheng and J. Baoqi, Roman domination in regular graphs. *Discret Math.* 309 (2009) 1528–1537.
90. I.G. Yero and J.A. Rodríguez-Velázquez, Roman domination in Cartesian product graphs and strong product graphs. *Appl. Anal. Discrete Math.* 7 (2013) 262–274.
91. X. Zhang, Z. Shao and X. Xu, On the Roman domination numbers of generalized Petersen graphs. *J. Combin. Math. Combin. Comput.* 89 (2014) 311–320.
92. V. E. Zverovich and A. Poghosyan, On Roman, global and restrained domination in graphs. *Graphs Combin.* 27 (2011) 755–768.

# Rainbow Domination in Graphs

**Boštjan Brešar**

## 1 Introduction

Domination in graphs is often considered as a model for applications, where vertices in a dominating set provide a service or a product that has to be accessible in the neighborhood of every vertex of the network. In this sense, rainbow domination presents a more complex version of domination, where each color represents a different type of service, or product, and one seeks to distribute these service providers or products to nodes in the network in such a way that for those nodes that are not given any service, all types of services are available in their neighborhoods. While it is not clear how often this situation appears in practice, a stronger motivation for introducing rainbow domination lies in its immanent relation with domination in Cartesian products of graphs and potential consequences for the famous Vizing's conjecture. Rainbow domination was introduced in 2008 by Brešar, Henning, and Rall [13] and is formally defined as follows.

Let $G$ be a graph and let $f$ be a function that assigns to each vertex a set of colors chosen from the set $[k] = \{1, \ldots, k\}$; that is, $f\colon V(G) \to 2^{[k]}$. In this context, the elements of the set $[k]$ are called *colors* or *labels*. If for each vertex $v \in V(G)$ such that $f(v) = \emptyset$ we have

$$\bigcup_{x \in N(v)} f(x) = [k] \tag{1}$$

B. Brešar (✉)
Faculty of Natural Sciences and Mathematics, University of Maribor, Maribor, Slovenia

Institute of Mathematics, Physics and Mechanics, Ljubljana, Slovenia
e-mail: bostjan.bresar@um.si



T. W. Haynes et al. (eds.), *Topics in Domination in Graphs*, Developments in Mathematics 64, https://doi.org/10.1007/978-3-030-51117-3_12

then $f$ is a *k-rainbow dominating function* ($k$RDF) of $G$. The *weight*, $w(f)$, of a function $f$ is defined as $w(f) = \sum_{v \in V(G)} |f(v)|$. Given a graph $G$, the minimum weight of a $k$RDF is the *k-rainbow domination number of* $G$, and is denoted by $\gamma_{rk}(G)$. A $k$-rainbow dominating function of $G$ with weight $\gamma_{rk}(G)$ is a $\gamma_{rk}(G)$-*function.*

The *Cartesian product* $G \,\square\, H$ of graphs $G$ and $H$ is the graph whose vertex set is $V(G) \times V(H)$, and two vertices $(g_1, h_1)$ and $(g_2, h_2)$ are adjacent in $G \,\square\, H$ if either $g_1 = g_2$ and $h_1 h_2$ is an edge in $H$ or $h_1 = h_2$ and $g_1 g_2$ is an edge in $G$. There is a bijective correspondence between the set of all $k$-rainbow dominating functions of $G$ and the set of all dominating sets of $G \,\square\, K_k$, where $K_k$ is the complete graph on $k$ vertices. Indeed, given a $k$-rainbow dominating function $f$ of a graph $G$ (and letting $V(K_k) = \{h_1, \ldots, h_k\}$), the set

$$D_f = \bigcup_{g \in V(G)} \Big( \bigcup_{i \in f(g)} \{(g, h_i)\} \Big)$$

is a dominating set of $G \,\square\, K_k$. The reverse correspondence is also clear, and, in addition, $w(f) = |D_f|$. In particular, for every positive integer $k \geq 1$ and every graph $G$,

$$\gamma_{rk}(G) = \gamma(G \,\square\, K_k). \tag{2}$$

See Figure 1: the left figure shows a (minimum) dominating set of $C_4 \,\square\, K_3$, while the right figure presents the corresponding 3RDF in $C_4$.

Since rainbow domination is closely related to domination in Cartesian products of graphs, one of the motivations for its introduction was to make a progress in resolving the following famous conjecture, which was posed more than half a century ago.

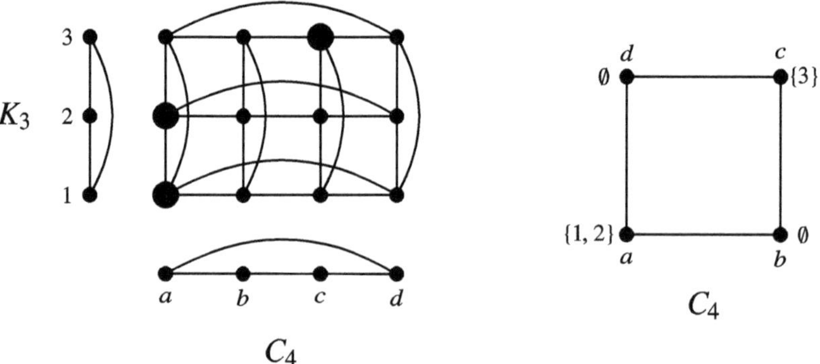


**Fig. 1** 3-Rainbow domination of $C_4$: the left figure presents $C_4 \,\square\, K_3$ in which vertices of a dominating set are enlarged; the right figure presents the corresponding 3RDF of $C_4$.

*Conjecture 1 (Vizing [73, 74])* For all graphs $G$ and $H$, $\gamma(G \square H) \geq \gamma(G)\gamma(H)$.

The equivalence between domination numbers of Cartesian products of graphs $G$ with complete graphs and rainbow domination numbers of $G$ has raised a number of questions that have led to various developments. In this survey, we present studies of rainbow domination, which reflect the current state-of-the-art and suggest problems of potential interest. To stay brief and focused, some of the topics are only referenced and the interested reader is invited to consult the corresponding papers.

In the next section we give some preliminary results and bounds on the rainbow domination numbers, and establish their values in some simple families of graphs. Section 3 is devoted to (standard and paired) domination in Cartesian products of graphs and connections with Vizing's conjecture, all in relation with rainbow domination. In Section 4, different bounds on the $k$-rainbow domination number of a graph are presented. The main focus is given to various upper bounds with respect to the order of a graph. In Section 5 this is further developed by an investigation of different functions, defined with respect to the rainbow domination number and some other parameters, such as minimum degree, order and radius of a graph, giving some insight into the asymptotic behavior of $\gamma_{rk}$. Section 6 presents an overview of the bounds on the rainbow domination numbers with respect to several other established domination invariants. In Section 7 we present computational complexity aspects of determining the rainbow domination numbers (NP-completeness results, efficient algorithms on some classes of graphs, and a discussion on possible approximation algorithms). In Section 8 we give a short list of additional topics that have been studied in relation with rainbow domination. This includes results in special classes of graphs, and some new invariants that are related to the rainbow domination numbers; we only mention the main concepts and give the corresponding references. Section 9 contains concluding remarks with some open problems.

## 2 Basic and Preliminary Observations

In this section, we present general bounds on the $k$-rainbow domination number in terms of the domination number, and establish $\gamma_{rk}(G)$ in some simple families of graphs $G$, which will be used also later.

Initial results on rainbow domination were established by Hartnell and Rall [46], who were using the language of domination in Cartesian products $G \square K_k$. One of their results, expressed in terms of rainbow domination, yields the following bounds for $\gamma_{rk}(G)$.

**Theorem 1 ([46])** *If $G$ is a graph of order $n$, then for any $k \geq 2$,*

$$\min\{n, \gamma(G) + k - 2\} \leq \gamma_{rk}(G) \leq k\gamma(G),$$

*and the bounds are sharp. In particular, $\gamma(G) \leq \gamma_{r2}(G) \leq 2\gamma(G)$.*

*Proof* The proof of the upper bound for the $k$-rainbow domination number of a graph $G$ is straightforward. Given a $\gamma(G)$-set $D$, the function $f\colon V(G) \to 2^{[k]}$ defined by $f(u) = [k]$ if $u \in D$, and $f(u) = \emptyset$ if $u \notin D$, is a $k$-rainbow dominating function of $G$ with $w(f) = k|D|$. This implies that $\gamma_{rk}(G) \le k\gamma(G)$.

For the proof of the lower bound, assume that on the contrary there exists a $k$RDF $f$ of $G$ with $w(f) < \min\{n, \gamma(G) + k - 2\}$. Since $w(f) < n$, there exists a vertex $x \in V(G)$ with $f(x) = \emptyset$. Hence, $\cup_{u\in N(x)} f(u) = [k]$, because $f$ is a $k$RDF. Let $\ell = \sum_{u\in N(x)} |f(u)|$ and let $B = \{u \in N(x) \mid f(u) \neq \emptyset\}$. Now, if $\ell - k + 1 \le |B|$, then let $C$ be an arbitrary subset of $B$ with $\ell - k + 1$ vertices, otherwise, we let $C = B$. Note that $\sum_{u\in B-C} |f(u)| \le k - 1$.

Since $w(f) < \gamma(G)+k-2$, we infer that the set $A = \{v \in V(G)-N[x] \mid f(v) \neq \emptyset\}$ has at most $\gamma(G) + k - \ell - 3$ vertices. This implies $|A \cup C| \le \gamma(G) - 2$, hence $A\cup C\cup\{x\}$ is not a dominating set. Let $z \in$ be a vertex of $G$, which is not dominated by any vertex of $A \cup C \cup \{x\}$. Clearly, $f(z) = \emptyset$, and note that if $z$ is adjacent to a vertex $v$ with $f(v) \neq \emptyset$, then $v \in B - C$. However, since $\sum_{u\in B-C} |f(u)| \le k - 1$, there exists a color $i \in [k]$ such that $i \notin f(v)$ for all vertices $v \in B - C$. This implies that $f$ is not a $k$RDF, a contradiction. Thus, $w(f) \ge \min\{n, \gamma(G) + k - 2\}$ for any $k$-rainbow dominating function $f$.

The sharpness of the lower bound can be demonstrated by the so-called Dutch windmill graphs $D_4^p$, $p \ge 1$, which are obtained from $p$ copies of the cycle $C_4$ all sharing a common vertex $u$. If $p > (k - 2)/2$, then $\gamma_{rk}(D_4^p) = k + p - 1$, where a $\gamma_{rk}(D_4^n)$-set $f$ is obtained by letting $f(u) = \{1, \dots, k - 1\}$, $f(x_i) = \{k\}$ for all vertices not adjacent to $u$, and $\emptyset$ to all other vertices. Since $\gamma(D_4^p) = p + 1$, we indeed have $\gamma_{rk}(D_4^p) = \gamma(D_4^p) + k - 2$.

For the upper bound consider the corona $G \circ K_t$ of an arbitrary graph $G$ and complete graph $K_t$. It is obtained from $G$ and $n(G)$ copies of $K_t$ by connecting each vertex of $G$ with all the vertices of its own copy of $K_t$. If $t \ge k$, then $\gamma_{rk}(G \circ K_t) = k \cdot n(G)$, and $\gamma(G \circ K_t) = n(G)$ implies the desired result. □

We remark that upper bounds similar to the one in Theorem 1 are known for several other domination invariants. In particular, $\gamma_{\{k\}}(G) \le k\gamma(G)$, where the parameter on the left side of the inequality is known as the $\{k\}$-domination number [32]. In addition, the Roman domination number, $\gamma_R(G)$, of a graph $G$, is bounded from above by $2\gamma(G)$ [27, 70].

The inequality $\gamma(G) \le \gamma_{r2}(G)$ in the above result can be generalized to the following inequality:

$$\gamma_{rk}(G) \le \gamma_{r(k+1)}(G), \tag{3}$$

which clearly holds for any graph $G$ and any $k \ge 1$.

Hartnell and Rall also characterized the graphs $G$, which attain the lower bound in the inequality $\gamma(G) \le \gamma_{r2}(G)$. (Clearly, by Theorem 1 there are no non-trivial graphs attaining $\gamma_{rk}(G) = \gamma(G)$ when $k \ge 3$.) These graphs are interesting also because they appear in several constructions, which attain the equality in the conjectured bound of Vizing's conjecture; see, e.g., Proposition 4.

**Theorem 2 ([46])** *For a connected graph $G$, $\gamma_{r2}(G) = \gamma(G)$ if and only if $G$ has a $\gamma(G)$-set $D$ that partitions into two non-empty subsets $D_1$ and $D_2$ such that $V(G) - N[D_1] = D_2$ and $V(G) - N[D_2] = D_1$.*

*Proof* Suppose that a connected graph $G$ has a $\gamma(G)$-set $D$ that partitions into two non-empty subsets $D_1$ and $D_2$ such that $V(G)-N[D_1] = D_2$ and $V(G)-N[D_2] = D_1$. Letting

$$f(u) = \begin{cases} \{1\} & \text{if } u \in D_1 \\ \{2\} & \text{if } u \in D_2. \\ \emptyset & \text{otherwise,} \end{cases}$$

it is easy to see that $f$ is a 2RDF of weight $\gamma(G)$, hence $\gamma_{r2}(G) = \gamma(G)$.

Conversely, let $\gamma_{r2}(G) = \gamma(G)$ for a connected graph $G$, and let $f$ be a $\gamma_{r2}(G)$-function. Since the set $D = \{v \in V(G) \mid f(v) \neq \emptyset\}$ is a dominating set of $G$, we infer that $|f(v)| = 1$ for every $v \in D$. Thus, $|D| = \gamma(G)$, and so $D$ is a $\gamma(G)$-set. Denote by $D_i = \{v \in V(G) \mid f(v) = \{i\}\}$ for $i \in \{1, 2\}$, and note that $\{D_1, D_2\}$ is a partition of $D$. Since $f$ is a 2RDF, every $x \notin D$ must have a neighbor in $D_1$ (and a neighbor in $D_2$), thus $D_1$ dominates $V(G) - D_2$. Clearly, no vertex of $D_1$ is adjacent to a vertex of $D_2$, because $D$ is a minimum dominating set, and so $D_2 = V(G) - N[D_1]$. In the same way we establish that $D_1 = V(G) - N[D_2]$. □

For the simplest example that demonstrates Theorem 2 consider $C_4$. Note that two non-adjacent vertices $v_1$ and $v_3$ form a $\gamma(C_4)$-set $D$, which can be partitioned into the sets $D_1 = \{v_1\}$ and $D_2 = \{v_3\}$.

We next present exact results for the 2-rainbow domination numbers in several families of graphs. The following lemma gives some intuition for dealing with rainbow domination, and will be used in the subsequent result. (Although similar auxiliary results have been known, they may not have been presented in this form, and so we also give a proof.)

**Lemma 1** *If $G$ is a graph with $\Delta(G) \leq k$, then there exists a $\gamma_{rk}(G)$-function $f$ such that $|f(v)| < k$ for every $v \in V(G)$.*

*Proof* Let $g$ be a $k$RDF of $G$ of minimum weight, that is, $g$ is a $\gamma_{rk}(G)$-function. We claim that there exists a $\gamma_{rk}(G)$-function $f$ with $|f(v)| < k$ for every $v \in V(G)$. Suppose that $|g(v)| = k$ for some $v \in V(G)$. Since $g$ is a $k$RDF with minimum weight, we infer that there exists a vertex $x_1 \in N(v)$ such that $g(x_1) = \emptyset$. Let $x_1, \dots, x_r$ be the neighbors of $v$ having $g(x_i) = \emptyset$. Note that $r \leq \Delta(G) \leq k$. Now, letting $f(x_i) = i$ for $i \in \{1, \dots, r-1\}$, $f(x_r) = \{r, r+1, \dots, k\}$, $f(v) = \emptyset$, and $f(u) = g(u)$ for the remaining vertices $u$ of $G$, we easily infer that $f$ is a $\gamma_{rk}(G)$-function. Since fewer vertices are mapped to the set of all $k$ colors by $f$ than there are such vertices with respect to $g$, induction completes the proof. □

**Proposition 1 ([14] and [13])**

*(i) For* $n \geq 1$, $\gamma_{r2}(P_n) = \lceil \frac{n+1}{2} \rceil$.

*(ii) For* $n \geq 3$, $\gamma_{r2}(C_n) = \begin{cases} \frac{n}{2} + 1 & \text{if } n \equiv 2 \,(\text{mod}\, 4) \\ \lceil \frac{n}{2} \rceil & \text{otherwise.} \end{cases}$

*Proof* Note that $\Delta(G) = 2$ when $G$ is a path or a cycle. Let $v_1, v_2, \ldots, v_n$ be the vertices of $G$ listed in the natural order. By Lemma 1, there exists a minimum 2-rainbow dominating function $f$ of $G$, in which $|f(v_i)| \leq 1$ for all $v_i \in V(G)$. If $G$ is the path $P_n$, this implies that $|f(v_1)| = 1$ for the leaf $v_1$; say $f(v_1) = 1$. Observe that we may assume without loss of generality that $f(v_2) = \emptyset$ and $f(v_3) = 2$. Continuing this way, we infer that $|f(v_k)| = 1$ for odd $k$, and $f(v_k) = \emptyset$ for even $k$. This yields (i).

The proof of (ii) is similar. Assuming without loss of generality that $f(v_1) = 1$, we may also assume that $f(v_2) = \emptyset$ and $|f(v_3)| = 1$. By the same reasoning, we infer that as long as $k < n$, we have $|f(v_k)| = 1$ for odd $k$, and $f(v_k) = \emptyset$ for even $k$. Now, if $n \equiv 0 \,(\text{mod}\, 4)$, $f(v_n) = \emptyset$, and so $w(f) = n/2$. If $n$ is odd, then $|f(v_n)| = 1$, and so $w(f) = (n+1)/2$. Finally, if $n \equiv 2 \,(\text{mod}\, 4)$, we get $f(v_1) = f(v_{n-1}) \neq \emptyset$, which implies that $|f(v_n)| = 1$, and so $w(f) = n/2 + 1$. □

Let $t \geq 2$. For $1 \leq r \leq t$, the graph $S_{t,r}$ is obtained from the star $K_{1,t}$ by subdividing $r$ of its edges by a single vertex. These graphs are sometimes called *spiders*. The *center* of the spider $S_{t,r}$ is the center of the star $K_{1,t}$ from which it is obtained. Note that only one of the vertices in $P_4 (= S_{2,1})$ may be called the center. The following result, which appeared implicitly in [77], establishes the 2-rainbow domination numbers of spiders. We will use this result in Section 4.

**Proposition 2 ([77])** *For* $t \geq 2$ *and* $1 \leq r \leq t$,

$$\gamma_{r2}(S_{t,r}) = \begin{cases} 3 & \text{if } t = 2 \\ 1 + r & \text{if } t \geq 3, r = t \\ 2 + r & \text{if } t \geq 3, r < t \end{cases}$$

*In particular, for every spider* $G$ *on* $n$ *vertices,* $\gamma_{r2}(G) \leq \frac{3}{4}n$, *where the equality holds if and only if* $G$ *is the path* $P_4$.

*Proof* The case $t = 2$ is covered by the graphs $P_4$ and $P_5$ and is clear.

Consider the graph $S_{t,t}$, for $t \geq 3$, with center vertex $c$. If $f$ is a 2RDF, then letting $f(x) = \emptyset$ for a leaf $x$ implies that its neighbor $y$ has $f(y) = \{1, 2\}$. Hence, if there exists a leaf $x$ with $f(x) = \emptyset$, then $w(f) \geq r + 1$. Otherwise, for all leaves $x$ we have $|f(x)| \geq 1$, and so $|f(c)| = 1$. This gives $\gamma_{r2}(S_{t,t}) \geq r + 1$. A function $f$ with $f(c) = \{1\}$ for the center $c$, $f(x) = \{2\}$ for the leaves $x$, and $f(y) = \emptyset$ for all other vertices $y$, is a 2RDF of weight $1 + r$, which is thus optimal.

Since in $S_{t,r}$, where $r < t$, center $c$ is adjacent to at least one leaf, we must either have $f(c) = \{1, 2\}$ or $|f(x)| = 1$ for all leaves $x$ adjacent to $c$. If there is more than one leaf adjacent to $c$, it is clear that an optimal 2RDF exists with $f(c) = \{1, 2\}$. If there is just one leaf $x$ adjacent to $c$, and $|f(c)| \leq 1$, then we must have $|f(x)| = 1$. However, if $f(c) = \emptyset$, then a non-leaf neighbor $y$ of $c$ must have $f(y) \neq \emptyset$. Hence also in this case one can quickly derive that there exists a $\gamma_{r2}(S_{t,r})$-function $f$ such that $f(c) = \{1, 2\}$. As in the previous paragraph we note that in an optimal 2RDF $f$, all leaves $x$ at distance 2 from $c$ should get $|f(x)| = 1$, which implies $\gamma_{r2}(S_{t,r}) \geq r+2$. Clearly, there exists a $\gamma_{r2}(S_{t,r})$-function with weight $r+2$.

Since $n = t + r + 1$, it is easy to verify that in each case, $\gamma_{r2}(S_{t,r}) \leq \frac{3}{4}n$, and that equality holds only when $r = 1$ and $t = 2$, since $\gamma_{r2}(S_{2,1}) = 3$. □

Like many other domination-type invariants, the rainbow domination number is hereditary on spanning supergraphs. Indeed, if $H$ is a spanning subgraph of $G$ with $V = V(H) = V(G)$, and $f: V \to 2^{[k]}$ is a $k$RDF of $H$, then $f$ is clearly also a $k$RDF of $G$. We state this basic observation in the following result.

**Proposition 3** *Let $k$ be a positive integer. If $G$ is a graph and $H$ a spanning subgraph of $G$, then $\gamma_{rk}(G) \leq \gamma_{rk}(H)$.*

## 3 Domination in Cartesian Products of Graphs

In this section we focus on results in Cartesian products of graphs that are related to rainbow domination. Some of the results consider domination and some others consider paired domination of Cartesian products of graphs.

The inability to resolve Vizing's conjecture motivated several authors to consider different kinds of domination concepts in Cartesian products and/or to focus on similar problems. A problem closely related to the conjecture is to characterize the graphs that achieve equality in the conjectured bound. Hartnell and Rall [46] used the graphs $G$ with $\gamma(G) = \gamma_{r2}(G)$ to construct several families of such graphs, one of which we describe next.

**Proposition 4 ([46])** *If $G$ is a graph with $\gamma(G) = \gamma_{r2}(G)$ and $H$ is the corona $H' \circ K_1$ over an arbitrary graph $H'$, then $\gamma(G \,\Box\, H) = \gamma(G)\gamma(H)$.*

*Proof* For a corona $H = H' \circ K_1$, let $\{V', L\}$ be the partition of $V(H)$, where $V'$ corresponds to the set $V(H')$ and $L$ is the set of leaves adjacent to vertices of $V'$. Note that $\gamma(H) = |V'| = |L|$. Let $G$ be a graph with $\gamma_{r2}(G) = \gamma(G)$, and let $f$ be a 2RDF of $G$. Let $V_i = \{u \in V(G) \mid f(u) = \{i\}\}$ for $i \in \{1, 2\}$, and let $V_0 = \{u \in V(G) \mid f(u) = \emptyset\}$. Note that $\{V_0, V_1, V_2\}$ is a partition of $V(G)$, and for every $x \in V_0$ there exists a vertex $y \in N_G(x) \cap V_1$ and a vertex $z \in N_G(x) \cap V_2$. We construct the set $D$ in $V(G \,\Box\, H)$ as follows:

$$D = (V_1 \times V') \cup (V_2 \times L),$$

and claim that $D$ is a dominating set of $G \Box H$.

Let $(g, h) \in V(G \Box H)$ be an arbitrary vertex, which is not in $D$. If $(g, h) \in V_1 \times L$, then there exists a vertex $g' \in V'$ such that $gg' \in E(G)$; hence $(g', h) \in D$ is a neighbor of $(g, h)$ that dominates $(g, h)$. In a similar way we check that $(g, h) \in V_2 \times V'$ is dominated by some $(g', h) \in V_1 \times L$. Finally, if $(g, h)$ is a vertex in $G \Box H$ such that $g \in V_0$, then let $y \in N_G(g) \cap V_1$ and $z \in N_G(g) \cap V_2$. Now, if $h \in V'$, then $(g, h)$ is dominated by $(y, h) \in D$, but if $h \in L$, then $(g, h)$ is dominated by $(z, h) \in D$. Since

$$|D| = |V_1 \times V'| + |V_2 \times L| = |V_1||V'| + |V_2||L| = (|V_1| + |V_2|)\gamma(H) = \gamma(G)\gamma(H),$$

we derive that $\gamma(G \Box H) \le \gamma(G)\gamma(H)$.

For the reversed inequality, let $h \in V'$ and let $\ell$ be its unique neighbor in $L$. Consider the subset $S$ of $V(G \Box H)$ consisting of the vertices that have the second coordinate in $\{h, \ell\}$, that is, $S = (V(G) \times \{h\}) \cup (V(G) \times \{\ell\})$. Note that for a vertex $g \in V(G)$ either $(g, h) \in D$, or $(g, \ell) \in D$, or there exists a vertex $g' \in N_G(g)$ such that $(g', \ell) \in D$. From this we easily derive that $|S \cap D| \ge \gamma(G)$, and so $|D| \ge |V'|\gamma(G) = \gamma(H)\gamma_{r2}(G)$. Thus, $\gamma(G \Box H) \ge \gamma(G)\gamma(H)$. □

There have been a number of different approaches to attack Vizing's conjecture, most of which have been described in the latest survey [12]. In particular, Clark and Suen [25] made a breakthrough by answering the question of Hartnell and Rall [45], whether there is a constant $c > 0$ such that the domination number of the Cartesian product of $G$ and $H$ is bounded as follows: $\gamma(G \Box H) \ge c\gamma(G)\gamma(H)$. Clearly, for $c < 1$ this inequality is weaker than the inequality in Vizing's conjecture. For $c = 1/2$, Clark and Suen provided a proof, which is based on a special partition of the Cartesian product of two graphs and double counting of the vertices in a minimum dominating set of $G \Box H$. (Rather than presenting an idea of the proof of Theorem 3 we will give a full proof of Theorem 4, which states a similar inequality related to rainbow domination, and uses the partition and double counting provided by Clark and Suen in their proof.)

**Theorem 3 ([25])** *For all graphs $G$ and $H$, $\gamma(G \Box H) \ge \frac{1}{2}\gamma(G)\gamma(H)$.*

Several small improvements of this bound have been proven by different authors, and several variations involving other domination invariants have been presented; see [12]. Brešar, Henning, and Rall asked [13], whether the bound of Clark and Suen can be improved by employing the 2-rainbow domination number as follows: is it true for all graphs $G$ and $H$ that

$$\gamma_{r2}(G \Box H) \ge \gamma(G)\,\gamma(H)\,? \tag{4}$$

To see that this would indeed be an improvement (if correct) of the bound of Clark and Suen, note that $\gamma_{r2}(G \Box H) \le 2\gamma(G \Box H)$; hence the truth of the inequality (4) implies $2\gamma(G \Box H) \ge \gamma_{r2}(G \Box H) \ge \gamma(G)\,\gamma(H)$, which is the Clark and Suen inequality. Clearly, the inequality (4) is weaker than Vizing's conjecture, because

$\gamma_{r2}(G \,\square\, H) \ge \gamma(G \,\square\, H)$. However, even a (much) weaker version of the proposed question in (4), where $\gamma_{r2}$ is replaced by $\gamma_{rk}$ for $k$ arbitrarily large, has not yet been resolved. It was asked in [12], whether there is (an arbitrarily large) integer $k$, for which the inequality

$$\gamma_{rk}(G \,\square\, H) \ge \gamma(G)\,\gamma(H)\,? \tag{5}$$

holds for all graphs $G$ and $H$. A weaker inequality similar to (5) was proven by Pilipczuk, Pilipczuk, and Škrekovski [56].

**Theorem 4 ([56])** *For all graphs $G$ and $H$ and every $k \ge 1$,*

$$\gamma_{rk}(G \,\square\, H) \ge \frac{k}{k+1}\gamma(G)\,\gamma(H).$$

*Proof* The framework of the proof is similar to the proof of Theorem 3 (cf. Clark and Suen [25]). Let $G$ be a graph and let $\{g_1, \dots, g_{\gamma(G)}\}$ be a minimum dominating set of $G$. Consider a partition $\{A_1, \dots, A_{\gamma(G)}\}$ of $V(G)$ chosen so that $g_i \in A_i$ and $A_i \subseteq N[g_i]$ for each $i$. Let $H_i = A_i \times V(H)$. For a vertex $h$ of $H$, the set of vertices $A_i \times \{h\}$ is called a *cell*, and is denoted by $C_i^h$.

Let $f$ be a $\gamma_{rk}(G \,\square\, H)$-function, and let $D_t = \{v \in V(G \,\square\, H) \mid t \in f(v)\}$ be the set of vertices $v \in V(G \,\square\, H)$ for which $f(v)$ contains the color $t \in \{1, \dots, k\}$. Let $D = \cup_{i=1}^k D_i$, and note that $D$ consists of the vertices $v \in V(G \,\square\, H)$ for which $f(v) \ne \emptyset$. Clearly, $D$ is a dominating set of $G \,\square\, H$, and $|D| \le \sum_{i-1}^k |D_i| = w(f)$.

Given a color $t \in [k]$, the cell $C_i^h$ is said to be *vertically t-undominated* by $f$ if for all $v \in A_i \times N_H[h]$, color $t$ is not in $f(v)$. Therefore, any vertex $(u, h)$ in such a cell $C_i^h$ either has $f(u, h) \ne \emptyset$ (that is, $(u, h) \in D$), or there exists a vertex $z$ in $N(u) \times \{h\}$ with $t \in f(z)$ (clearly, such $z$ does not belong to $C_i^h$, because the cell is vertically $t$-undominated).

Let $\ell_i^t$ be the number of vertically $t$-undominated cells $C_i^h$ in $H_i$. Projecting the vertices of $D_t \cap H_i$ onto $H$, note that they dominate all vertices of $H$ except for those that are projected from the vertically $t$-undominated cells $C_i^h$. We infer that $\gamma(H) \le |D_t \cap H_i| + \ell_i^t$, and summing up all $i \in [\gamma(G)]$, we get

$$\gamma(G)\gamma(H) \le |D_t| + \ell^t, \tag{6}$$

where $\ell^t$ is the number of all vertically $t$-undominated cells in $G \square H$ (that is, $\ell^t = \sum_{i=1}^{\gamma(G)} \ell_i^t$).

Now, for $h \in V(H)$ consider the subset $V(G) \times \{h\}$, and denote by $m_h^t$ the number of vertically $t$-undominated cells in $V(G) \times \{h\}$. Note that vertices of $D \cap (V(G) \times \{h\})$ dominate all vertically $t$-undominated cells in $D \cap (V(G) \times \{h\})$, and since every other cell in $V(G) \times \{h\}$ can be dominated by just one vertex, namely $(g_i, h)$, we infer

$$|D \cap (V(G) \times \{h\})| + (\gamma(G) - m_h^t) \ge \gamma(G),$$

which yields $|D \cap (V(G) \times \{h\})| \geq m_h^t$. Summing up all $h \in V(H)$ we get

$$m^t \leq |D|, \tag{7}$$

where $m^t = \sum_{h \in V(H)} m_h^t$, which is the number of all vertically $t$ undominated cells in $G \,\square\, H$. Using the fact that $\ell^t = m^t$, and combining the inequalities (6) and (7), we infer

$$\gamma(G)\gamma(H) \leq |D_t| + |D|\,. \tag{8}$$

Note that $t$ is still arbitrary, hence the above inequality holds for any chosen color in $[k]$. Since $\sum |D_t| = w(f)$, there exists a color $t$ such that $|D_t| \leq \frac{1}{k} w(f)$. Plugging such color $t$ to formula (8) and also using that $|D| \leq w(f)$, we get

$$\gamma(G)\gamma(H) \leq \frac{1}{k} w(f) + w(f),$$

which implies

$$\gamma_{rk}(G \square H) = w(f) \geq \frac{k}{k+1} \gamma(G)\,\gamma(H).$$

□

The following bounds on the domination number of the Cartesian product of graphs are easy to prove and are part of folklore:

$$\min\{n(G), n(H)\} \leq \gamma(G \,\square\, H) \leq \min\{\gamma(G)n(H), n(G)\gamma(H)\}. \tag{9}$$

The pairs of graphs that attain the upper bound clearly enjoy the inequality of Vizing's conjecture, hence there has been some interest in the class of graphs, for which the bound in the conjecture is trivially attained. Brešar and Rall [15] characterized the graphs that attain the lower bound in (9), and gave some partial results for the graphs attaining the upper bound. In particular, for any graph $G$ enjoying $\gamma_{rk}(G) = k\gamma(G)$ it follows that $\gamma(G \,\square\, K_k) = k\gamma(G)$, but more can be proved.

**Proposition 5 ([15])** *Let $G$ be a graph such that $\gamma_{rk}(G) = k\gamma(G)$, for some $k \geq 2$. If $H$ is any graph of order $k$, then $\gamma(G \,\square\, H) = k\gamma(G)$.*

The proof of Proposition 5 follows from the fact that $G \,\square\, H$ is a spanning subgraph of $G \,\square\, K_k$, and so $\gamma(G \,\square\, H) \geq \gamma(G \,\square\, K_k)$. Since $\gamma(G \,\square\, K_k) = k\gamma(G)$, this readily implies $\gamma(G \,\square\, H) = \gamma(G)n(H) = k\gamma(G)$.

To see that the converse of Proposition 5 does not hold, consider the graph $G$ in Figure 2. Let $f\colon V(G) \to 2^{[4]}$ be defined by $f(a_1) = \{1\}$, $f(a_2) = \{2\}$, $f(a_3) = \{3\}$, $f(b_1) = \{1\}$, $f(b_2) = \{1\}$, $f(b_3) = \{1\}$, $f(b) = \{4\}$, and $f(a) = \emptyset$. It is easy

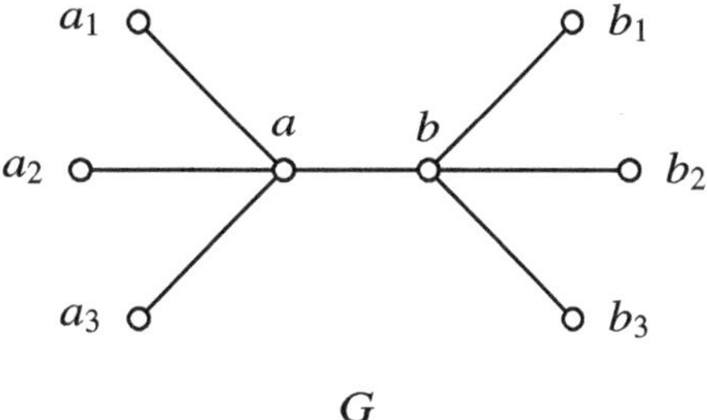


**Fig. 2** Graph $G$ with $\gamma_{r4}(G) < 4\gamma(G)$

to see that $f$ is a 4-rainbow dominating function, and thus $\gamma_{r4}(G) \le 7 < 4\gamma(G)$. Now, if $H = P_4$, one can verify that $\gamma(G \,\square\, H) = 8 = 4\gamma(G)$.

In the seminal paper [13], rainbow domination was used for bounding the paired domination number of the Cartesian product of graphs in which one of the factors is arbitrary and the vertex set of the other can be partitioned into $k$ sets, each of which is a minimum paired dominating set.

**Theorem 5 ([13])** *For an arbitrary graph $G$ and any graph $H$ whose vertex set can be partitioned into $k$ $\gamma_{\rm pr}(H)$-sets,*

$$\gamma_{\rm pr}(G \,\square\, H) \le \frac{1}{k}\,|V(H)|\,\gamma_{\rm rk}(G).$$

*Proof* Let $f$ be a $\gamma_{\rm rk}(G)$-function, and let $\pi = \{S_1, S_2, \ldots, S_k\}$ be a partition of $V(H)$ into $\gamma_{\rm pr}(H)$-sets. Note that $\gamma_{\rm pr}(H) = |S_i| = \frac{1}{k}\,|V(H)|$ for all $i \in \{1, \ldots, k\}$. We claim that the set

$$D = \bigcup_{x \in V(G)} \Big( \bigcup_{i \in f(x)} (\{x\} \times S_i) \Big)$$

is a paired dominating set of $G \,\square\, H$.

Clearly, the induced subgraph $(G \,\square\, H)[D]$ contains a perfect matching. This is because each vertex $(x, y) \in D$, which is in $\{x\} \times S_i$ for some $\gamma_{\rm pr}$-set $S_i$ in $H$, can be paired with the vertex $(x, z)$ in $\{x\} \times S_i$, where $y$ and $z$ are partners with respect to the perfect matching that exists in $H[S_i]$.

To see that $D$ is a dominating set of $G \,\square\, H$ we distinguish two cases. First, if $f(x) \ne \emptyset$ for $x \in V(G)$, then all vertices of $\{x\} \times V(H)$ are dominated by $\{x\} \times S_i$, where $i \in f(x)$. Second, consider the vertices in $\{x\} \times V(H)$, where $f(x) = \emptyset$. Since $f$ is a $k$RDF of $G$, we have $\bigcup_{u \in N[x]} f(u) = \{1, \ldots, k\}$. Now, for a vertex $(x, y) \in \{x\} \times V(H)$, there exists $i \in [k]$ and $S_i \in \pi$ such that $y \in S_i$; in addition, there exists $u \in N_G(x)$ such that $i \in f(u)$. Therefore, the vertex $(u, y) \in D$ dominates $(x, y)$. We infer that $D$ is indeed a paired dominating set of $G \,\square\, H$. □

Some interesting consequences of Theorem 5 can be derived for more specific families of graphs. For instance, the vertex set of the cycle $C_{4\ell}$ can clearly be partitioned into two $\gamma_{\rm pr}$-sets, which by Theorem 5 implies that $\gamma_{\rm pr}(G \,\square\, C_{4\ell}) \le$

$\frac{1}{2}|V(C_{4\ell})|\,\gamma_{r2}(G) = 2\ell\gamma_{r2}(G)$, where $G$ is an arbitrary graph and $\ell \geq 1$. When plugging $K_k \square K_2$ for $H$ in Theorem 5, and combining it with (2), the following result follows.

**Corollary 1 ([13])** *For an arbitrary graph $G$, $\gamma_{pr}(G \square (K_k \square K_2)) \leq 2\gamma_{rk}(G) = 2\gamma(G \square K_k)$. In particular, $\gamma_{pr}(G \square C_4) \leq 2\gamma_{r2}(G)$.*

Corollary 1 can be applied to derive the exact value of the paired domination number of $G \square C_4$ when $G$ is the cycle $C_n$ for $n \geq 3$:

$$\gamma_{pr}(C_n \square C_4) = 2\gamma_{r2}(C_n).$$

That is,

$$\gamma_{pr}(C_n \square C_4) = \begin{cases} n+2 & \text{if } n \equiv 2 \pmod 4 \\ 2\lceil \frac{n}{2} \rceil & \text{otherwise.} \end{cases}$$

The lower bound for $\gamma_{pr}(C_n \square C_4)$ follows from the general lower bound $\gamma_{pr}(G) \geq 2\lceil \frac{n(G)}{2\Delta(G)} \rceil$, which holds for any graph $G$ without isolated vertices. The upper bound is obtained by combining Corollary 1 with Proposition 1(ii). (The case $n \equiv 2 \pmod 4$ needs a separate verification of the inequality $\gamma_{pr}(C_n \square C_4) > n$.)

## 4 Bounds on the Rainbow Domination Number

In Section 2, lower and upper bounds on the rainbow domination number with respect to the domination number are presented. In Section 6, bounds on the rainbow domination number involving several other established domination invariants are given. This section is thus devoted to all other known types of bounds on the rainbow domination numbers.

One of the fundamental problems in studying a domination-type invariant is to find a general (sharp) upper bound for the invariant in terms of the order $n$ of a graph; more precisely, the desired bound is of the form $c \cdot n$, where $0 < c \leq 1$. Classical bounds of this type are $\gamma(G) \leq n/2$ due to Ore [55], and $\gamma_t(G) \leq 2n/3$ due to Cockayne et al. [26], where $G$ is a connected graph of order $n \geq 2$, respectively, $n \geq 3$. Imposing additional conditions on the minimum degree $\delta(G)$, better upper bounds are often obtained. For instance, McCuaig and Shepherd [53] proved that $\gamma(G) \leq 2n/5$ if $G$ is a connected graph of order $n \geq 8$ with $\delta(G) \geq 2$.

For the rainbow domination, the first result of this type is due to Wu and Rad [77], and is dealing with the 2-rainbow domination number.

**Theorem 6 ([77])** *For a connected graph $G$ of order $n \geq 3$,*

$$\gamma_{r2}(G) \leq \frac{3}{4}n.$$

*Proof* First note that it suffices to prove the inequality for trees. Indeed, if $G$ is a connected graph, and $T$ is a spanning tree of $G$, then, using Proposition 3, $\gamma_{r2}(G) \leq \gamma_{r2}(T)$. The proof is by induction on the order $n$ of a tree $T$. Base cases $n \in \{3, 4, 5\}$ can be easily checked. In addition, we check directly that the proof is correct for all trees $T$ with $\mathrm{diam}(T) \leq 3$. If $\mathrm{diam}(T) = 2$, then $T$ is a star $K_{1,n-1}$, and $\gamma_{r2}(T) = 2 < 3n/4$. If $\mathrm{diam}(T) = 3$, then $T$ has a dominating set with two adjacent vertices, hence $\gamma_{r2}(T) \leq 4 \leq 3n/4$ (as we are assuming $n \geq 6$).

Now, let $\mathrm{diam}(T) \geq 4$. Let $P$ be a longest path in $T$ chosen in such a way that the degree of a penultimate vertex $v$ of $P$ is maximized. Let $u$ be the non-leaf neighbor of $v$ on $T$. We distinguish several cases.

*Case 1.* $d_T(v) > 2$.
Let $T'$ be obtained from $T$ by deleting $v$ and its leaf neighbors, and let $n' = n(T')$. Since $\mathrm{diam}(T) \geq 4$, we infer that $n' \geq 3$, hence by the induction hypothesis, $\gamma_{r2}(T') \leq 3n'/4$. Let $f'$ be a $\gamma_{r2}(T')$-function. We define a function $f$ on $V(T)$ by letting $f(x) = f'(x)$ for $x \in V(T')$, $f(v) = \{1, 2\}$, and $f(x) = \emptyset$ for all leaf neighbors $x$ of $v$. Clearly, $f$ is a 2RDF of $T$, and $w(f) = w(f')+2 \leq 3n'/4+2 \leq 3(n-3)/4+2 < 3n/4$.

*Case 2.* $d_T(v) = 2 = d_T(u)$.
Let $T'$ be obtained from $T$ by deleting $u$, $v$ and the leaf neighbor $\ell$ of $v$. If $n' = n(T')$ equals 2, then $T$ is the path of order 5, and we are done by the initial step ($\gamma_{r2}(P_5) = 3 < 3n/4$). Hence, we may assume that $n' \geq 3$, and by the induction hypothesis, $\gamma_{r2}(T') \leq 3n'/4$. Let $f'$ be a $\gamma_{r2}(T')$-function. Let a function $f$ on $V(T)$ be defined by $f(x) = f'(x)$ for $x \in V(T')$, $f(v) = \{1, 2\}$, and $f(u) = f(\ell) = \emptyset$. Again it is obvious that $f$ is a 2RDF of $T$, and by the same computation as in Case 1 we get $w(f) < 3n/4$.

*Case 3.* $d_T(v) = 2$ and $d_T(u) > 2$.
By the choice of $P$, every neighbor of $u$, which is not on $P$ and is not a support vertex, has degree 2. First, suppose that every neighbor of $u$ is either a leaf or a support vertex. In this case $\mathrm{diam}(T) = 4$, and $T$ is a spider. By Proposition 2, $\gamma_{r2}(T) < 3n/4$ (the inequality is strict, since $T$ is not isomorphic to $P_4$). Finally, suppose that $u$ has a neighbor $t$, which is not a support vertex nor a leaf. In this case, $T - tu$ contains two components $T'$ and $T''$ such that $T''$ is a spider containing $u$. Now, $n(T') \geq 3$, hence by the induction hypothesis, $\gamma_{r2}(T') \leq 3n(T')/4$. By Proposition 2, we again infer $\gamma_{r2}(T'') \leq 3n(T'')/4$. Therefore, $\gamma_{r2}(T) \leq \gamma_{r2}(T') + \gamma_{r2}(T'') \leq 3n(T)/4$.

□

For a small example, which shows the sharpness of the inequality $\gamma_{r2}(G) \leq 3n(G)/4$, we can again take $P_4$. From the above proof it is clear that equality $\gamma_{r2}(T) = 3n(T)/4$ is possible only if in each inductive step, Case 3 is applied and the component $T''$ is isomorphic to $P_4$. This quickly yields the structure of the trees $T$ that achieve $\gamma_{r2}(T) = 3n(T)/4$. Moreover, Wu and Rad in [77] characterized all connected graphs $G$, which attain the equality in Theorem 6. Let $\mathcal{H}$ be the family of graphs that can be obtained from an arbitrary connected graph $H$ and $n(H)$ copies of $P_4$ by identifying each vertex of $H$ with a support vertex of its own copy of $P_4$.

**Theorem 7 ([77])** *For a connected graph G of order $n \geq 3$, the equality $\gamma_{r2}(G) = \frac{3}{4}n$ holds if and only if G is the corona $C_4 \circ K_1$ or G belongs to $\mathcal{H}$.*

Under the additional assumption for the minimum degree, $\delta(G) \geq 2$, Fujita and Furuya [35] improved the bound of Theorem 6 as follows.

**Theorem 8 ([35])** *If G is a connected graph of order n and $\delta(G) \geq 2$,*

$$\gamma_{r2}(G) \leq \frac{2}{3}n,$$

*and this bound is sharp.*

The proof of Theorem 8 is too involved to be presented here. For the family of graphs that attain the bound in the theorem, one can take the corona $G \circ K_2$, where $G$ is an arbitrary graph of order $n$. Indeed, $\gamma_{r2}(G \circ K_2) = 2n$, while $n(G \circ K_2) = 3n$.

Bounds of similar flavor have also been obtained for the 3-rainbow domination number. We start with a general bound for $\gamma_{r3}$ in connected graphs with no restriction on the minimum degree. It was proved by Fujita, Furuya, and Magnant [38].

**Theorem 9 ([38])** *If G is a connected graph of order $n \geq 4$, different from $P_4$, then*

$$\gamma_{r3}(G) \leq \frac{8}{9}n,$$

*and this bound is sharp.*

The bound in Theorem 9 is best possible in the strong sense, since the equality is attained for graphs of every feasible order $n$. That is, for $n = 9\ell$, where $\ell \geq 1$, we define the graphs $R_\ell$ as follows. First, let $Q$ be the graph obtained from the path $P_5$ and the path $P_4$, by adding an edge between a support vertex $x$ of $P_5$ and a support vertex of $P_4$. Note that $x$ is the unique central vertex of $Q$. Now, $R_\ell$ is obtained from the path $P_\ell$ and $\ell$ copies of the graph $Q$ by identifying each vertex of the path with the central vertex of its own copy of $Q$.

Furuya, Koyanagi, and Yokota [41] provided the bound for $\gamma_{r3}(G)$ when $\delta(G) \geq 2$.

**Theorem 10 ([41])** *If G is a connected graph of order $n \geq 8$ and $\delta(G) \geq 2$, then*

$$\gamma_{r3}(G) \leq \frac{5}{6}n.$$

The proof of Theorem 10 involves a large family of graphs $G$ that satisfy the equality $\gamma_{r3}(G) = \frac{5}{6}n$, and are described as follows [41]. Let $F$ be the graph $2K_3+e$, obtained from two disjoint triangles by adding an edge. Denote by $c(F)$ one of the central vertices of $F$, i.e., $c(F)$ is arbitrarily chosen among the two vertices with degree 3 in $F$. Now, given a graph $H$ construct a graph $F_H$ by taking $n(H)$ copies of $F$, and identifying each vertex of $H$ with the vertex $c(F)$ of its own copy of $F$. Let $\mathcal{F} = \{F_H \mid H \text{ is a tree}\}$. A graph $G$ is called *2-minimal* if $G$ is connected, $\delta(G) \geq 2$, and for every $e \in E(G)$, either $G - e$ is disconnected or $\delta(G - e) = 1$. Furuya, Koyanagi, and Yokota [41] proved the following result.

**Theorem 11 ([41])** *Let $G$ be a 2-minimal graph of order $n$. We have $\gamma_{r3}(G) \geq \frac{5}{6}n$ if and only if $G \in \{C_3, C_6, C_7\} \cup \mathcal{F}$.*

As proven in [41], every graph $G$ in $\mathcal{F}$ satisfies $\gamma_{r3}(G) = \frac{5}{6}n$. Note that every connected graph $G$ with $\delta(G) \geq 2$ contains a spanning subgraph $H$ which is 2-minimal. Therefore, by Theorem 11, we get $\gamma_{r3}(G) \leq \gamma_{r3}(H) \leq \frac{5}{6}n$ as soon as $n \geq 8$. This proves Theorem 10.

Next, we present some simple bounds on the rainbow domination numbers. Sheikholeslami and Volkmann [65] provided an upper and a lower bound with respect to the order and maximum degree of a graph.

**Proposition 6 ([65])** *If $G$ is a graph of order $n$ and maximum degree $\Delta$, then $\gamma_{rk}(G) \leq n - \Delta + k - 1$ and this bound is sharp.*

*Proof* Let $v$ be a vertex of $G$ with $d(v) = \Delta$. Define a function $f : V(G) \to 2^{[k]}$ by letting $f(v) = [k]$, $f(x) = \emptyset$ for all $x \in N(v)$, and $f(y) = \{1\}$ for every $y \in V(G) - N[x]$. Clearly, $f$ is a $k$RDF of $G$ and $w(f) = n - \Delta + k - 1$. The sharpness of the bound is ensured by any graph $G$ with $n > k$ and $\Delta(G) = n - 1$. □

**Proposition 7 ([65])** *If $G$ is a graph of order $n$ and maximum degree $\Delta$, then $\gamma_{r2}(G) \geq \left\lceil \frac{2n}{\Delta+2} \right\rceil$.*

Complete bipartite graphs $K_{2,s}$ attain the bound in Proposition 7; indeed, $\gamma_{r2}(K_{2,s}) = 2 = \frac{2(s+2)}{s+2} = \frac{2n}{\Delta+2}$, where $n = s + 2$ is the order of $K_{2,s}$, and $\Delta = s$ is the maximum degree of $K_{2,s}$. Wu and Rad [77] gave an upper bound for $\gamma_{r2}$ involving the diameter of a connected graph, which we slightly improve as follows.

**Proposition 8** *If $G$ is a connected graph of order $n$, then*

$$\gamma_{r2}(G) \leq n - \left\lfloor \frac{\mathrm{diam}(G)}{2} \right\rfloor .$$

*Proof* Let $d = \mathrm{diam}(G)$, and let $P$ be a path of length $d$ in $G$. By Proposition 1(i), there exists a 2RDF $f$ of $P$ with weight $\lceil \frac{d+2}{2} \rceil$. Extend $f$ from $V(P)$ to all vertices of $G$, by letting $f(v) = \{1\}$ if $v \notin V(P)$. Clearly, $f$ is a 2RDF of $G$ with $w(f) = n - (d + 1) + \lceil \frac{d+2}{2} \rceil = n - \lfloor \frac{d}{2} \rfloor$. □

We end this section with a Nordhaus–Gaddum type result due to Wu and Xing [79], which bounds the sum of the 2-rainbow domination numbers of a graph $G$ and its complement $\overline{G}$.

**Theorem 12 ([79])** *If $G$ is a graph with order $n \geq 3$, then*

$$5 \leq \gamma_{r2}(G) + \gamma_{r2}(\overline{G}) \leq n + 2,$$

*and the bounds are sharp.*

## 5 Asymptotic Behavior

Fujita, Furuya, and Magnant further investigated the upper bounds on $\gamma_{rk}(G)$ in terms of the order [38]. First, they considered the function

$$t(k) = \min\{t \mid \gamma_{rk}(G) \leq tn \text{ for a connected graph } G \text{ of sufficiently large order } n\},$$

which is described in the table:

| $k$ | 1 | 2 | 3 | $\geq 4$ |
|---|---|---|---|---|
| $t(k)$ | 1/2 | 3/4 | 8/9 | 1 |

The first three entries in the table are the theorem of Ore [55], Theorem 6, and Theorem 9, respectively. The last entry follows from $\gamma_{r4}(C_n) = n$ for any $n \geq 4$, and by using also (3).

Since $t(k)$ is fully described, another natural function was considered in [38]:

$$t'(k, d) = \min\{t' \mid \gamma_{rk}(G) \leq t'n \text{ for any graph } G \text{ of sufficiently large order } n \text{ and } \delta(G) \geq d\}.$$

The status of $t'(1, d)$ is well understood, because it is about domination number $\gamma$, which has been studied intensively; it is known that unless $d = 6$, we have $t'(1, d) = d/(3d - 1)$. For $k \geq 2$, results presented so far give $t'(2, 1) = 3/4$, $t'(2, 2) = 2/3$, $t'(3, 1) = 8/9$, $t'(3, 2) = 5/6$, and $t'(4, 1) = t'(4, 2) = 1$. Fujita, Furuya, and Magnant gave two general results about the behavior of $t'(k, d)$ [38].

**Theorem 13 ([38])** *Let $k$, $d$ and $n_0$ be positive integers such that $d \leq k - 4$. There exists a graph $G$ of order $n \geq n_0$ satisfying $\delta(G) = d$ and $\gamma_{rk}(G) = n$.*

Hence, $t'(d, k) = 1$ if $d \leq k - 4$. On the other hand, if $d$ is sufficiently larger than $k - 4$, then $t'(d, k) < 1$ as shown in [38]. First, it was shown, by using (2) and applying a well-known bound on the domination number by Caro and Roddity [17], that for any graph $G$ of order $n$ and $\delta(G) \geq d$,

$$\gamma_{\mathrm{r}k}(G) \leq \left(1 - (d + k - 1)\left(\frac{1}{d + k}\right)^{\frac{d+k}{d+k-1}}\right)kn.$$

Furthermore, if $d$ is an integer which is sufficiently large compared to $k$, then

$$\left(1 - (d + k - 1)\left(\frac{1}{d + k}\right)^{\frac{d+k}{d+k-1}}\right)k < 1.$$

These two observations yield the following result.

**Theorem 14 ([38])** *Given a positive integer $k$ there exists an integer $d$ (sufficiently large compared to $k$) and a number $t' \in (0, 1)$ such that*

$$\gamma_{\mathrm{r}k}(G) \leq t'n$$

*for every graph $G$ of order $n$ with $\delta(G) \geq d$.*

A natural question appears: given a positive integer $k$, how large must an integer $d$ be so that for every graph $G$ with $\delta(G) \geq d$ and sufficiently large order $n$ we have $\gamma_{\mathrm{r}k}(G) \leq t'n$, where $t' < 1$?

Fujita and Furuya gave a similar study of the asymptotic behavior of $\gamma_{\mathrm{r}k}(G)$ in which graphs $G$ with a given radius $\mathrm{rad}(G)$ are investigated [36]. Given positive integers $k, r, n$, they define $t^*(k, r; n)$ as the minimum value satisfying that $\gamma_{\mathrm{r}k}(G) \leq t^*(k, r; n) \cdot n$ for all connected graph $G$ of order $n$ and radius $r$; if no such graph exists, then $t^*(k, r; n) = \infty$. For $k \geq 1$ and $r \geq 1$, let

$$t^*(k, r) = \limsup_{n \mapsto \infty} t^*(k, r; n).$$

The case $r = 1$ can be easily analyzed. Note that graphs with $\mathrm{rad}(G) = 1$ have a universal vertex, hence for $n \geq k$, $\gamma_{\mathrm{r}k}(G) = k = (k/n) \cdot n$. This implies $t^*(k, 1; n) \leq k/n$, which gives $0 \leq t^*(k, 1) \leq \limsup_{n \mapsto \infty} t^*(k, 1; n) \leq \limsup_{n \mapsto \infty} k/n = 0$. Thus, $t^*(k, 1) = 0$.

It is clear that $t^*(k, r) \leq 1$ holds for any $r \geq 1$ and $k \geq 1$. Moreover, it was proved in [36] that

$$t^*(k, r) < 1 \tag{10}$$

always holds. On the other hand, for $r \geq 2$ and $k \geq 1$ the following lower bound was proved [36]:

$$t^*(k, r) \geq \frac{k}{k + 1}. \tag{11}$$

The exact value is also known for $t^*(k, 2)$; to give a flavor of this study we present the proof of this value from [36].

**Theorem 15 ([36])** *If $k \geq 2$ is a integer and $\epsilon > 0$ a real number, then there exists an integer $n_0$ such that $\gamma_{\rm rk}(G) < (\frac{k}{k+1} + \epsilon)n$ for every connected graph $G$ of order $n \geq n_0$ and* $\text{rad}(G) = 2$.

*Proof* Let $n_0$ be an integer such that $k < \frac{(k+1)\epsilon}{k} n_0$. Let $G$ be a connected graph of order $n \geq n_0$ with $\text{rad}(G) = 2$, and let $x$ be a central vertex of $G$. Assume to the contrary that $\gamma_{\rm rk}(G) \geq (\frac{k}{k+1} + \epsilon)n$.

Let $Y = N(N(x)) - N[x]$ (that is, $Y$ is the set of vertices at distance 2 from $x$), $X_1 = N(x) \cap N(Y)$, and $X_2 = N(x) - X_1$. Let $m_1 = |X_1|, m_1' = |X_2|$, and $m_2 = |Y|$. Let $f_i : V(G) \to 2^{[k]}$, $i \in [2]$, be defined as follows:

$$f_1(u) = \begin{cases} [k] & \text{if } u = x \\ \emptyset & \text{if } u \in N(x). \\ 1 & \text{otherwise.} \end{cases}$$

and

$$f_2(u) = \begin{cases} [k] & \text{if } u \in \{x\} \cup X_1 \\ \emptyset & \text{otherwise.} \end{cases}$$

It is clear that $f_i$ is a $k$RDF for both $i$. By definition of $f_1$ we have

$$(\frac{k}{k+1} + \epsilon)n \leq \gamma_{\rm rk}(G) \leq w(f_1) \leq k + m_2,$$

which gives

$$m_2 \geq (\frac{k}{k+1} + \epsilon)n - k. \tag{12}$$

By definition of $f_2$ we have

$$(\frac{k}{k+1} + \epsilon)n \leq \gamma_{\rm rk}(G) \leq w(f_2) \leq k(1 + m_1),$$

which gives

$$1 + m_1 \geq (\frac{1}{k+1} + \frac{\epsilon}{k})n. \tag{13}$$

Since $n = 1 + m_1 + m_1' + m_2$, by inequalities (12) and (13) we get

$$n - m_1' = 1 + m_1 + m_2 \geq (1 + \frac{k+1}{k}\epsilon)n - k.$$

This implies

$$k \geq m_1' + \frac{(k+1)\epsilon}{k}n \geq \frac{(k+1)\epsilon}{k}n_0\,,$$

a contradiction, by which the proof is complete. □

Combining (11) with Theorem 15 we infer the following result.

**Corollary 2** *For an integer $k \geq 1$, $t^*(k, 2) = \frac{k}{k+1}$.*

Another notion that falls under studies of asymptotic behavior of $\gamma_{rk}$ was presented by Chang, Wu, and Zhu [19] as an application of their algorithm for computing $\gamma_{rk}(T)$ of an arbitrary tree $T$ (the algorithm yields a $\gamma_{rk}(T)$-function; see Section 7). They define the following natural function on the set of all graphs $G$ (we denote this function by $m_r$, where index r refers to rainbow domination), and $n$ is the order of $G$:

$$m_r(G) = \min\{k \mid \gamma_{rk}(G) = n\}. \tag{14}$$

Since $\gamma_{rk}(G) = n$ for $k \geq n$, the function $m_r$ is well-defined. Next, for any vertex $x$ of an arbitrary graph $G$ and any non-empty set $S \subset N_G(x)$ let

$$d^*(x, S) = |S| + \min\{d(y) \mid y \in S\}.$$

Let $d^*(G)$ be the maximum of $d^*(x, S)$, where $x$ runs over all vertices of $G$ and $S$ runs over all non-empty subsets of $N_G(x)$. The lower bound $d^*(G) \geq \Delta(G) + 1$ was proved to hold in any non-trivial graph $G$ [19]. In addition, in any graph $G$, $d^*(G)$ can be determined in linear time; see [19].

In trees, the function $m_r$ can be squeezed between two values as follows.

**Theorem 16 ([19])** *For any tree $T$, $d^*(T) \leq m_r(T) \leq d^*(T) + 1$.*

The exact value of $m_r(T)$, where $T$ is a tree, can be obtained by first determining $\gamma_{rk}(T)$ for $k = d^*(T)$, using the algorithm from [19]. If the value is $n$, then $m_r(T) = d^*(T)$, otherwise $m_r(T) = d^*(T) + 1$. Since the algorithms to determine $d^*(T)$ and $\gamma_{rk}(T)$ are linear, so is determining $m_r(T)$ in trees.

## 6 Relations to Other Domination Invariants

A large majority of known results that relate rainbow domination with other domination concepts involve the 2-rainbow domination number and some other domination invariant(s) in which number 2 is an inherent part of the definition.

In the seminal paper by Brešar, Henning, and Rall [13], a new graph invariant, $\gamma_{w2}$, which can be viewed as the monochromatic version of $\gamma_{r2}$, was introduced. It appeared naturally in designing the algorithm to find an optimal 2RDF of a tree. (Unfortunately, the same invariant was introduced later by two different groups using two additional names, which is presented at the end of this section.) We give the definition of its generalized version, $\gamma_{wk}$, where $k$ is an arbitrary positive integer.

Let $G$ be a graph and let $f: V(G) \to [k]$. If for each vertex $v \in V(G)$ such that $f(v) = 0$ we have $f(N(v)) \geq k$, then $f$ is a *weak k-dominating function* of $G$. The *weight* of a function $f$ is defined as $f(V(G)) = \sum_{v \in V(G)} f(v)$, and the minimum weight of a weak $k$-dominating function of $G$ is the *weak k-domination number of G*, denoted by $\gamma_{wk}(G)$. (Note that in [13] the invariant was considered for $k = 2$ and was called the weak {2}-domination number.)

The following chain of inequalities follows from definitions of the involved invariants, and holds for any positive integer $k$ and any graph $G$:

$$\gamma(G) \leq \gamma_{wk}(G) \leq \gamma_{rk}(G) \leq k\gamma(G).$$

Almost all other known bounds involving $\gamma_{rk}$ are for $k = 2$, that is, they are about the 2-rainbow domination number.

Chellali, Haynes, and Hedetniemi conjectured [20] that in any graph $G$ with no isolated vertices, $\gamma_t(G) \leq \gamma_{r2}(G)$, and Furuya provided a proof of this conjecture [40]. Shao et al. [63] presented a structural characterization of the trees $T$ in which $\gamma_t(T) = \gamma_{r2}(T)$ by using 10 different operations for enlarging such trees. The inequality $\gamma_{r2}(G) \leq 2\gamma_t(G)$, which holds for any graph $G$ with no isolated vertices is a direct consequence of the bounds $\gamma_{r2}(G) \leq 2\gamma(G)$ and $\gamma(G) \leq \gamma_t(G)$.

Much attention was given to relationships between 2-rainbow domination and Roman domination. The latter is one of the most studied domination concepts; see [27, 70] for seminal papers on Roman domination. Given a graph $G$ a function $f : V \to \{0, 1, 2\}$ such that for all $v \in V(G)$, $f(v) = 0$ implies that there exists $w \in N(v)$ with $f(w) = 2$ is called a *Roman dominating function* of $G$. The *Roman domination number* of $G$, denoted by $\gamma_R(G)$, equals the minimum weight $f(V(G)) = \sum_{v \in V(G)} f(v)$ over all Roman dominating functions $f$ of $G$. It is easy to see that $\gamma_{r2}(G) \leq \gamma_R(G)$ for any graph $G$, cf. [79]. Indeed, if $f$ is a Roman dominating function of a graph $G$ with the minimum weight, then letting

$$g(u) = \begin{cases} \emptyset & \text{if } f(u) = 0 \\ \{1\} & \text{if } f(u) = 1. \\ \{1, 2\} & \text{if } f(u) = 2. \end{cases}$$

makes $g : V(G) \to 2^{\{1,2\}}$ a 2RDF with weight $f(V(G))$. Thus, $\gamma_{r2}(G) \leq f(V(G)) = \gamma_R(G)$. It was shown independently by Fujita and Furuya [35] and by Chellali and Rad [22] that $\gamma_R(G) \leq \frac{3}{2}\gamma_{r2}(G)$ for any graph $G$. In addition, it was shown in [22] that the bound can be improved in forests $T$ to read $\gamma_R(T) \leq \frac{4}{3}\gamma_{r2}(T)$, and it is sharp, since $\gamma_R(kP_5) = 4k$ and $\gamma_{r2}(kP_5) = 3k$.

Another six invariants that were compared to the 2-rainbow domination number in Bonomo et al. [11] can be intuitively described in the following table, which divides their defining properties into two criteria:

| | $f : V(G) \to \{0, 1, 2\}$ | $f : V(G) \to \{0, 1\}$ |
|---|---|---|
| Outer | Weak 2-domination ($\gamma_{w2}$) | 2-domination ($\gamma_2$) |
| Closed | $\{2\}$-domination ($\gamma_{\{2\}}$) | Double domination ($\gamma_{\times 2}$) |
| Open | Total $\{2\}$-domination ($\gamma_{t\{2\}}$) | Total double domination ($\gamma_{t\times 2}$) |

The first criterion is the range of the corresponding function $f$, which can be either $\{0, 1, 2\}$ or just $\{0, 1\}$. The second criterion distinguishes three possibilities with respect to which vertices need to be dominated and what kind of neighborhoods (closed or open) are considered for vertices to which a positive weight is assigned. There are three possibilities: (1) only vertices with zero weight need to be dominated ("outer domination"), (2) all vertices need to be dominated and vertices with positive weight dominate their closed neighborhoods ("closed domination"), and (3) all vertices need to be dominated and only open neighborhoods are dominated by vertices with positive weight ("open domination") [11]. Definitions of the six invariants are summarized in Table 1.

In each of the definitions, a function $f$ that assigns weights to vertices of a graph $G$ is involved. Its *weight* is $f(V(G)) = \sum_{v \in V(G)} f(v)$, and the invariant $\zeta(G)$ is defined as the minimum weight of a function that enjoys the particular condition for any $\zeta \in \{\gamma_{w2}, \gamma_{\{2\}}, \gamma_{t\{2\}}, \gamma_2, \gamma_{\times 2}, \gamma_{t\times 2}\}$. The concepts 2-domination, $\{2\}$-domination, and double domination are well known and were considered in a number of publications.

Table 2 summarizes the (lower and upper) bounds of 2-rainbow domination number in terms of nine other domination invariants—beside the six invariants from Table 1, there is also domination number, total domination number, and Roman domination number. Note that total domination number, total $\{2\}$-domination number, and double domination number are well-defined in graphs with no isolated vertices, while total double domination number is well-defined in graphs with minimum degree at least 2. Other invariants are well-defined in all graphs.

**Table 1** Summary of definitions of studied invariants.

| Name | Notion | Function | Condition | |
|---|---|---|---|---|
| Weak 2-domination | $\gamma_{w2}$ | $f : V \to \{0, 1, 2\}$ | $f(N(v)) \geq 2$ | if $f(v) = 0$ |
| $\{2\}$-domination | $\gamma_{\{2\}}$ | $f : V \to \{0, 1, 2\}$ | $f(N[v]) \geq 2$ | $\forall v$ |
| Total $\{2\}$-domination | $\gamma_{t\{2\}}$ | $f : V \to \{0, 1, 2\}$ | $f(N(v)) \geq 2$ | $\forall v$ |
| 2-domination | $\gamma_2$ | $f : V \to \{0, 1\}$ | $f(N(v)) \geq 2$ | if $f(v) = 0$ |
| Double domination | $\gamma_{\times 2}$ | $f : V \to \{0, 1\}$ | $f(N[v]) \geq 2$ | $\forall v$ |
| Total double domination | $\gamma_{t\times 2}$ | $f : V \to \{0, 1\}$ | $f(N(v)) \geq 2$ | $\forall v$ |

**Theorem 17** *The lower and upper bounds for $\gamma_{r2}(G)$ in terms of nine other invariants presented in Table 2 are correct and sharp; they hold for an arbitrary graph $G$ that is well-defined for the invariant involved in a particular inequality.*

Some of the bounds in Theorem 17 have already been presented in this chapter. In particular, L1 and U1 were presented in Theorem 1, and we mentioned above that L2 was proven by Furuya [40], while U2 and L3 are trivial. The bound U4 concerning Roman domination was first observed in [79], while L4 was shown in [22, 35]. All other bounds, namely U3, L5, L6, and U5-U9 were proven by Bonomo et al. [11]. In addition, it was proven in [11] that there does not exist a non-zero function $f$ such that $\zeta(G) \leq f(\gamma_{r2}(G))$ would hold for all graphs $G$, where $\zeta \in \{\gamma_2, \gamma_{\times 2}, \gamma_{t\times 2}\}$ (these results are presented as entries L7, L8, and L9 in Table 2).

To see that there is no non-zero function $f$ such that $\zeta(G) \leq f(\gamma_{r2}(G))$ for any $\zeta \in \{\gamma_2, \gamma_{\times 2}, \gamma_{t\times 2}\}$, consider the following example. Let $O_n$, $n \geq 3$, be the graph obtained from the multigraph $K_2^{(n)}$, which has two vertices that are connected with $n$ parallel edges, by subdividing each edge exactly twice (see Figure 3 with $O_4$ as the first graph from the left). Note that $\gamma_{r2}(O_n) = 4$, while $\gamma_2(O_n) = n+1$, $\gamma_{\times 2}(O_n) = n+3$, and $\gamma_{t\times 2}(O_n) = 2n+2$.

The sharpness of the bounds in Theorem 17 can be verified with the following examples (see also [11]). The bound L1 is attained by graphs $kC_4$, where $k \geq 1$, which can also be used for the bounds L2, L3, and L4. Indeed, $\gamma(kC_4) = \gamma_t(kC_4) = \gamma_{w2}(kC_4) = \gamma_{r2}(kC_4) = 2k$, while $\gamma_R(kC_4) = 3k$. In addition, Theorem 2 characterizes all connected graphs $G$ attaining $\gamma_{r2}(G) = \gamma(G)$. The bound U1 is attained by graphs $kK_2$ for any integer $k \geq 1$, which can be also used as examples attaining the bound U4, i.e., $\gamma(kK_2) = k$, and $\gamma_{r2}(kK_2) = \gamma_R(kK_2) = 2k$. Moreover, $\gamma_{t\{2\}}(kK_2) = 4k$, hence $kK_2$ can serve to prove the sharpness of the bound L6. For the bound U2, let $H$ be the tree on 6 vertices, where each of the two non-leaves is adjacent to two leaves; note that $\gamma_{r2}(kH) = 4k = 2\gamma_t(kH)$ (in Figure 3 graph $H$ is the second graph from the left). The bounds U3, U5, U6, U7, U8, and U9 are attained by graphs $K_n^{**}$, defined as follows. For $n \geq 2$, the graph $K_n^{**}$ is obtained from the complete graph of order $n$ by gluing two new triangles along each edge; that is, for each pair $x, y$ of vertices in the complete graph $K_n$ two vertices are added, each of which is adjacent only to $x$ and $y$. (The third graph from the left in Figure 3 is $K_3^{**}$.) Note that $\gamma_{r2}(K_n^{**}) = 2n-2$, while $\zeta(K_n^{**}) = n$

**Table 2** Sharp lower and upper bounds on $\gamma_{r2}(G)$ with respect to 9 other domination invariants. For an invariant $\zeta \in \{\gamma_2, \gamma_{\times 2}, \gamma_{t\times 2}\}$ there is no function $f$ such that $\zeta(G) \leq f(\gamma_{r2}(G))$ would hold for all graphs $G$. The entry with an asterisk, $\gamma_{r2}(G) \geq (\gamma_{\{2\}}(G)+1)/2$, holds only in graphs $G$ with at least one edge.

| | | 1 | 2 | 3 | 4 | 5 | 6 | 7 | 8 | 9 |
|---|---|---|---|---|---|---|---|---|---|---|
| L | $\gamma_{r2} \geq$ and $\gamma_{r2} \leq$ | $\gamma$ | $\gamma_t$ | $\gamma_{w2}$ | $\frac{2}{3}\gamma_R$ | $(\gamma_{\{2\}}+1)/2^*$ | $\gamma_{t\{2\}}/2$ | $\mathbf{0}\gamma_2$ | $\mathbf{0}\gamma_{\times 2}$ | $\mathbf{0}\gamma_{t\times 2}$ |
| U | $\gamma_{r2} \geq$ and $\gamma_{r2} \leq$ | $2\gamma_t$ | $2\gamma_{w2}-2$ | $\gamma_R$ | $2\gamma_{\{2\}}-2$ | $2\gamma_{t\{2\}}-2$ | $2\gamma_2-2$ | $2\gamma_{\times 2}-2$ | $2\gamma_{t\times 2}-2$ | |

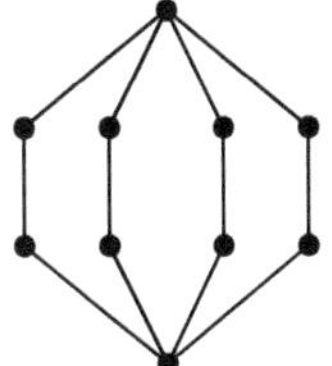
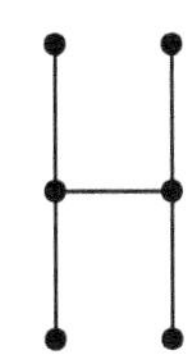
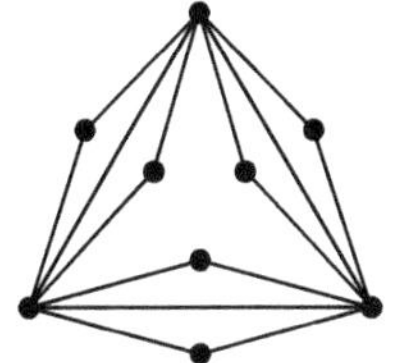
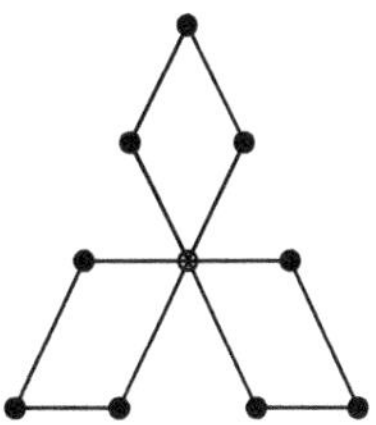

**Fig. 3** Graphs $O_4$, $H$, $K_3^{**}$, and $D_4^3$.

for any invariant $\zeta \in \{\gamma_{w2}, \gamma_{\{2\}}, \gamma_{t\{2\}}, \gamma_2, \gamma_{\times 2}, \gamma_{t\times 2}\}$, as soon as $n \geq 3$. Finally, to demonstrate the sharpness of the bound L5, which holds only for graphs with at least one edge, we use the family $D_4^n$, which are the graphs obtained from $n$ copies of $C_4$ by identifying $n$ vertices, one of each copy of $C_4$, to a single vertex; see Figure 3 again, where the graph on the right is $D_4^3$. One can prove that for $n \geq 3$, $\gamma_{r2}(D_4^n) = n + 1$ and $\gamma_{\{2\}}(D_4^n) = 2n + 1$.

Next, we present a couple of bounds that involve the 2-rainbow domination number of a graph $G$, the order, and an additional invariant of $G$.

**Theorem 18 ([79])** *If $G$ is a connected graph of order $n \geq 3$, then $\gamma_{r2}(G)+\frac{\gamma(G)}{2} \leq n$ with equality if and only if $G$ is the corona $C_4 \circ K_1$ or $G$ belongs to $\mathcal{H}$.*

Recall that the family $\mathcal{H}$ was defined in Section 4, and compare the above result with Theorem 7.

Fujita and Furuya [35] studied the interplay between 2-rainbow domination and Roman domination. They proved the following inequality, which was reproved by Alvarado, Dantas, and Rautenbach [2] who also provided a characterization of the extremal graphs.

**Theorem 19 ([2, 35])** *If $G$ is a connected graph of order $n \geq 3$, then $\gamma_{r2}(G) + \gamma_R(G) \leq \frac{3}{2}n$ with equality if and only if $G$ is the corona $C_4 \circ K_1$ or $G$ belongs to $\mathcal{H}$.*

It was conjectured in [35] that the bound in Theorem 19 can be improved to $\frac{4}{3}n(G)$ if a graph $G$ has minimum degree at least 2 and $G$ is not $C_5$. Alvarado, Dantas, and Rautenbach [2] proved this conjecture.

**Theorem 20 ([2])** *If $G$ is a connected graph of order $n \geq 3$, $\delta(G) \geq 2$ and $G$ is distinct from $C_5$, then $\gamma_{r2}(G) + \gamma_R(G) \leq \frac{4}{3}n$ and this bound is sharp.*

The sharpness of the bound in the above theorem is demonstrated by the graphs $G \circ K_2$, where $G$ is an arbitrary connected graph.

For an integer $k$, $0 \leq k \leq \gamma_{r2}(G)/2$, some kind of characterization of graphs $G$ with $\gamma_R(G) - \gamma_{r2}(G) = k$ was proven in [35]. On the other hand, it was shown in [3] that given a connected $K_4$-free graph $G$ and a positive integer $k$, it is NP-hard to decide whether $\gamma_R(G) - \gamma_{r2}(G) = k$. Alvarado et al. [3] provided also the following characterization of the graphs in which the equality of both invariants is hereditary.

**Theorem 21 ([3])** *A graph $G$ satisfies $\gamma_{\mathrm{r2}}(H) = \gamma_{\mathrm{R}}(H)$ for every induced subgraph of $G$ if and only if $G$ is $\{P_5, C_5, C_4\}$-free.*

The graphs $G$ in which $\gamma_{\mathrm{R}}(H) = \frac{3}{2}\gamma_{\mathrm{r2}}(H)$ for every induced subgraph $H$ of $G$ were also investigated in [3].

Chellali, Haynes, and Hedetniemi [20] considered relations between the 2-rainbow domination number and two other invariants—the weak Roman domination number and the secure domination number. We refer to [20] for definitions and results.

In a similar way as 2-rainbow domination can be viewed as the rainbow version of weak 2-domination, several other domination versions of known domination invariants were introduced in [11]. Let us present the idea of rainbow versions of domination invariants using the well-known example of 2-domination, which was introduced by Fink and Jacobson in the 1980s [34]. Given a graph $G$, a set $D \subseteq V(G)$ is a *2-dominating set* of $G$ if every $v \in V(G) - D$ has at least two neighbors in $D$. The *2-domination number*, $\gamma_2(G)$, of $G$ is the minimum cardinality of a 2-dominating set of $G$. Now, a *rainbow 2-dominating function* of $G$ is a function $f : V(G) \to \{\emptyset, \{1\}, \{2\}\}$ such that every $v \in V(G)$ with $f(v) = \emptyset$ has in its neighborhood vertices $u_1$ and $u_2$ such that $f(u_i) = \{i\}$ for $i \in \{1, 2\}$. The *rainbow 2-domination number* of G, $\widetilde{\gamma_2}(G)$, is the minimum weight $w(f) = |\{v \in V(G) \,|\, f(v) \neq \emptyset\}|$ over all rainbow 2-dominating functions $f$ of $G$. Bonomo et al. [11] introduced five rainbow invariants, three of which give new, interesting concepts. (Two of the rainbow invariants can be expressed through known invariants; notably, the rainbow version of the $\{2\}$-domination number equals $2\gamma(G)$, and the rainbow version of the total $\{2\}$-domination number equals $2\gamma_t(G)$.) These are the rainbow versions of 2-domination, double domination, and total double domination. A comprehensive study of these invariants, which produced sharp lower and upper bounds comparing 13 domination invariants to each other, was given in [11].

As mentioned in the beginning of this section, weak 2-domination was studied independently under two additional names. First, Chellali et al. [21] introduced and studied a graph invariant, which they called *Roman* $\{2\}$*-domination number*, and denoted it by $\gamma_{R2}$. The invariant coincides with the weak 2-domination number. In particular, in any graph $G$, $\gamma_{\mathrm{w2}}(G) = \gamma_{R2}(G)$. Second, the same invariant was studied in [47] under the name *Italian domination number*, and was denoted by $\gamma_I$. The main focus of investigation in [47] was on $\gamma_I(T)(=\gamma_{\mathrm{w2}}(T))$, where $T$ is a tree.

## 7 Computational Aspects

The decision version of the $k$-rainbow domination number can be stated as the following problem:

> $k$-RAINBOW DOMINATION PROBLEM
>
> *Instance:* A connected graph $G$ and positive integers $k$ and $t$.
> *Question:* Does $G$ have a $k$-rainbow dominating function of weight at most $t$?

The first result on this topic was given by Brešar and Kraner Šumenjak in [14], proving the NP-completeness of the 2-RAINBOW DOMINATION PROBLEM even when $G$ is a bipartite or a chordal graph. An extension of this result to an arbitrary positive integer $k$ was given by Chang, Wu, and Zhu [19]. The reduction of the problem is fairly simple and involves the DOMINATING SET PROBLEM. Let $k$ be fixed. Given a graph $G$ of order $n$, the graph $G'$ is obtained from $G$ by attaching to each vertex of $G$ a set of $k-1$ leaves. Formally, $V(G') = V(G) \cup \{v_2, v_3, \ldots, v_k \mid v \in V(G)\}$, and $E(G') = E(G) \cup \{vv_i \mid v \in V(G), 2 \leq i \leq k\}$. One can prove that $G$ has a dominating set of size at most $s$ if and only if $G'$ has a $k$RDF of weight at most $s + n(k-1)$. Later, Hon et al. [48] followed with NP-completeness of the $k$-RAINBOW DOMINATION PROBLEM in split graphs, which improves the same assertion about chordal graphs. We summarize these results in the next theorem.

**Theorem 22 ([19, 48])** *For a graph $G$ and a positive integer $k$, the $k$-*RAINBOW DOMINATION PROBLEM *is NP-complete even when $G$ is restricted to bipartite graphs or split graphs.*

A linear-time algorithm that gives a $\gamma_{r2}(T)$-function of an arbitrary tree $T$ was presented in the seminal paper [13]. In fact, the authors proved that $\gamma_{r2}(T) = \gamma_{w2}(T)$ for any tree $T$, and a linear-time algorithm to find a weak 2-dominating function of $T$ of minimum weight was given. Chang, Wu, and Zhu [19] extended this result by presenting a linear-time algorithm that computes a $\gamma_{rk}(T)$-function of an arbitrary tree $T$. A generalization of the rainbow dominating function was invented for this purpose in [19], and its monochromatic version was later used in [18] to establish $\gamma_{rk}(G)$ in an arbitrary block graph $G$. The latter function is defined as follows.

Let $G$ be a graph together with a *$k$-assignment*, which is a mapping $L$ that assigns to each vertex $v \in V(G)$ a label $L(v) = (a_v, b_v)$, where $a_v, b_v \in [k]_0$. A function $g\colon V(G) \to [k]_0$ is a *weak $\{k\}$-$L$-dominating function* of $G$ if for every $v \in V(G)$, $g(v) \geq a_v$, and $g(N(v)) \geq b_v$ whenever $g(v) = 0$. The *weak $\{k\}$-$L$-domination number* of $G$, which is denoted by $\gamma_{wkL}(G)$, is the minimum weight of a weak $\{k\}$-$L$-dominating function of $G$ (where the weight of $g$ is the sum of $g(v)$ over all vertices $v$ of $G$). Clearly, when $L(v) = (0, k)$ for every vertex $v \in V(G)$, the weak $k$-domination and weak $\{k\}$-$L$-domination coincide and we have $\gamma_{wkL}(G) = \gamma_{wk}(G)$. Chang, Li, and Wu [18] found a linear-time algorithm for determining a weak $\{k\}$-$L$-dominating function of weight $\gamma_{wkL}(G)$ for an arbitrary block graph, thus solving the problem of a weak $k$-domination number in block graphs. Furthermore, they proved that in any strongly chordal graph

$\gamma_{wk}(G) = \gamma_{rk}(G)$. Since the class of strongly chordal graphs contains the class of block graphs, the algorithm works also for establishing the $k$-rainbow domination number in block graphs; see [18] for more details.

Hon, Kloks, Liu, and Wang investigated algorithmic issues of the rainbow domination in several classes of perfect graphs [48]. Firstly, they observed that for a fixed $k$, the $k$-rainbow domination problem can be formulated in monadic second-order logic. By Courcelle's theorem this implies an existence of a linear-time algorithm for computing the invariant in graphs of bounded rankwidth (which includes the graphs with bounded treewidth). Secondly, they followed with some specific algorithms for determining $\gamma_{rk}(G)$, and $\gamma_{wkL}(G)$, in classes of cographs and trivially perfect graphs, respectively. In addition, they presented a polynomial-time algorithm to determine $\gamma_{r2}(G)$ in interval graphs $G$, which resolves the problem posed in [14].

We conclude this section with approximability issues concerning 2-rainbow domination numbers. The invariants $\gamma_{r2}$ and $\gamma_{w2}$ were studied along with a number of other domination parameters in [11]. In particular, the (in)approximability results for the domination number were applied to obtain the following results for the invariants of our interest.

**Theorem 23 ([11])** *There is a* $2(\ln(\Delta(G) + 2) + 1)$*-approximation algorithm for* $\gamma_{r2}$ *and there is an* $(\ln(\Delta(G) + 2) + 1)$*-approximation algorithm for* $\gamma_{w2}$.

**Theorem 24 ([11])** *For every* $\epsilon > 0$*, there is no polynomial-time algorithm approximating* $\gamma_{r2}$*, resp.* $\gamma_{w2}$*, for* $n$*-vertex split graphs without isolated vertices within a factor of* $(1/2 - \epsilon) \ln n$*, unless* $\mathrm{P} = \mathrm{NP}$.

The question remains whether one can fill the gap between the bounds in Theorems 23 and 24. More precisely, the following questions were posed in [11]:

1. Can the approximation ratios given by Theorem 23 be further improved (that is, by a constant factor)?
2. Can the factors $1/2 - \epsilon$ in the inapproximability bounds from Theorem 24 be improved to $1 - \epsilon$?

## 8 Additional Related Topics

This section gives a very brief presentation of additional topics that were studied in relation with rainbow domination. The main purpose is to give an idea in which directions the studies have been pursued. We first list some references that investigate rainbow domination in some specific classes of graphs, and then follow by a short presentation of some of the new invariants that arose from rainbow domination.

The class of generalized Petersen graphs has drawn a considerable attention when studying rainbow domination. Quite a few papers investigated the 2-rainbow domination number in this class of graphs [24, 61, 72, 75, 78, 80], while the 3-

rainbow domination number of generalized Petersen graphs was studied in [62]. The 2-rainbow domination number was considered also in Cartesian products of two cycles [67–69]. In particular, it was proved that $\gamma_{r2}(C_n \square C_5) = \gamma_{w2}(C_n \square C_5) = 2n$ [67]. An almost complete description of $\gamma_{r2}(G \circ H)$, where $G \circ H$ is the lexicographic product of graphs $G$ and $H$ was given in [49]. The 2-rainbow domination number of Sierpiński graphs and some of its variations was established in [52]. Recently, rainbow domination was studied in regular graphs [51]. In particular, exact values of $k$-rainbow domination numbers were established for all cubic Cayley graphs over abelian groups [51].

Several concepts that arise from $k$-rainbow dominating functions and the $k$-rainbow domination number of a graph have been introduced in the last decade. Most of them are generalizations of concepts that were studied with respect to dominating sets and the domination number of a graph. Generalizing the concept of domatic number as introduced in [28], Sheikholeslami and Volkmann introduced the *k-rainbow domatic number* of a graph [65]. For a given positive integer $k$, it is defined as the maximum number of $k$-rainbow dominating functions $f_1, \ldots, f_d$ such that $\sum_{i=1}^{d} |f_i(v)| \leq k$ for each $v \in V(G)$. The $k$-rainbow domatic number was studied further in [37, 54].

The *k-rainbow bondage number* of a graph $G$ with $\Delta(G) \geq 2$ was introduced in [29] as the minimum cardinality of a set $F \subseteq E(G)$ such that $\gamma_{rk}(G - F) = \gamma_{rk}(G) + 1$; see also [10, 58] for some further studies. In a sense dual to this concept is the following invariant introduced by Amjadi et al. [4]. The *k-rainbow reinforcement number* of a graph $G$ is the minimum number of edges that must be added to $G$ in order to decrease the $k$-rainbow domination number [4], see [57] for the complexity issues on reinforcement. Another related concept was introduced in [30] (see also [31, 33]), and is concerned with the minimum number of edges that must be subdivided (where each edge may be subdivided at most once) in order to increase the (2-)rainbow domination number of a graph.

Rainbow domination can also be defined in a natural way in digraphs, where the condition (1) from the definition of a $k$RDF is changed in such a way that the neighborhood $N(v)$ of a vertex $v$ is replaced by the in-neighborhood $N^-(v)$. The rainbow domination of a digraph was introduced by Amjadi et al. [5], and was studied also in [43, 44]. For additional concepts related to rainbow domination in digraphs cf. [9, 66].

Finally, we present a couple of variations of the rainbow domination number that restrict a $k$RDF in two different ways, in one case generalizing the total domination and in the other the independent domination, both to the context of rainbow domination. As we will see, the different authors generalized these invariants in (two) different ways.

Independent rainbow domination of graphs was first introduced by Chellali and Rad [23] (and then generalized by Shao et al. [64] from independent 2-rainbow domination to independent $k$-rainbow domination for arbitrary $k$) as follows. A $k$RDF $f$ is an *independent k-rainbow dominating function* if any two vertices in $G$ with non-empty value of $f$ are non-adjacent. The minimum weight of such a function is the *independent k-rainbow domination number* $i_{rk}(G)$ of $G$. Several papers considered the independent 2-rainbow domination [6–8, 23, 59], while bounds for the independent $k$-rainbow domination in bipartite and in general graphs were studied in [39]. A different variation of rainbow independent domination was introduced and studied by Kraner Šumenjak, Rall, and Tepeh [50]; see also [16] (note that the names of the invariants are also different with two words swapped). A $k$RDF function $f$ on a graph $G$ is a *k-rainbow independent dominating function* of $G$ if $|f(v)| \leq 1$ for every $v \in V(G)$ and from $f(u) = f(v)$, it follows that $uv \notin E(G)$. The minimum weight of such a function is the *k-rainbow independent domination number* $\gamma_{\mathrm{ri}k}(G)$ of $G$. In other words, rainbow independent domination in a graph $G$ coincides with independent domination in the Cartesian product of $G$ with $K_k$. In particular, $\gamma_{\mathrm{ri}k}(G) = i(G\Box K_k)$ for any graph $G$, which makes this invariant close to the original idea of rainbow domination (compare also with (2)).

Similarly, two different variations of rainbow domination, which generalize total domination, have been introduced. A *total k-rainbow dominating function* $f$ was introduced in [1] as a $k$RDF for which the subgraph of a graph $G$ induced by the set $\{v \in V(G) \mid f(v) \neq \emptyset\}$ has no isolated vertices. The corresponding invariant is called the *total k-rainbow domination number* of $G$. Alternatively, as introduced in [71], the *k-rainbow total domination number* $\gamma_{k\mathrm{rt}}(G)$ of a graph $G$ is defined as the minimum weight of a $k$RDF $f$ with the property that $f(v) = \{i\}$ for $v \in V(G)$ implies that there exists $u \in N_G(v)$ such that $i \in f(u)$. In other words, $k$-rainbow total domination of $G$ coincides with total domination of the Cartesian product of $G$ with $K_k$, and $\gamma_{k\mathrm{rt}}(G) = \gamma_t(G\Box K_k)$.

# 9 Concluding Remarks

A number of problems and directions for future research have been proposed in many papers concerning rainbow domination. Since the study of these concepts is still very vivid, one can expect some solutions and probably new questions in the following years. In this section, we restrict to a short list of ideas that deserve additional attention (the list is not exhaustive).

One the main motivations for introducing rainbow domination was to establish some new insights on domination number in Cartesian products of graphs, and, possibly, to approach Vizing's conjecture. Several results of this type were presented in Section 3. It would be very interesting to find new such connections. In particular, we repeat the following question: Is it true that for all graphs $G$ and $H$,

$$\gamma_{\mathrm{r}k}(G\,\Box\,H) \geq \gamma(G)\,\gamma(H)\,? \tag{15}$$

for an arbitrary fixed $k$. The case $k = 1$ is Vizing's conjecture, but also resolving the case $k = 2$ would be important, since it strengthens Theorem 3 due to Clark and Suen. We mention that a similar type of inequality was proved by Wu for the Roman domination number [76]. He proved that $\gamma_{\mathrm{R}}(G\,\Box\,H) \geq \gamma(G)\gamma(H)$ holds for all graphs $G$ and $H$. This is weaker than the inequality in Vizing's conjecture, yet it improves Theorem 3, since $\gamma(G\Box H) \leq \gamma_{R}(G\Box H) \leq 2\gamma(G\Box H)$. A number of other domination invariants could be investigated in this context, where one should determine whether the inequality (15) holds if $\gamma_{rk}$ on the left side of the inequality is replaced by a particular invariant. For instance, proving $\gamma_{\{2\}}(G\,\Box\,H) \geq \gamma(G)\,\gamma(H)$ would present an improvement of Theorem 3.

An interesting area of research is related to asymptotic behavior of $\gamma_{rk}$ in relation with various other graph parameters as presented in Section 5. We proposed a natural problem in Section 5 concerning the function $t'(k,d)$ defined in [38]. Let us rephrase this problem. Define the function $m_d$ on the set of all positive integers $k$ as follows:

$$m_d(k) = min\{d \mid t'(k,d) < 1\}.$$

By Theorem 14, the function $m_d$ is well-defined. In addition, by the results presented in Sections 4 and 5, it is known that $m_d(1) = m_d(2) = m_d(3) = 1$ and $m_d(4) > 2$. By Theorem 13 we also have $m_d(k) > k - 4$ for any $k \geq 5$. It would be interesting to better understand the function $m_d$ and thus the behavior of the function $t'(k,d)$.

The function $m_{\rm r}$, defined in (14), has a similar flavor as $m_d$, since it also considers some type of asymptotic behavior of $\gamma_{rk}$. In Section 5 we noted that $m_{\rm r}$ can be computed in linear time for trees, hence it is natural to ask what is the computational complexity for determining $m_{\rm r}(G)$ in other important classes of graphs $G$.

Concerning the function $t^*(k,r)$, its values for $r \in \{1,2\}$ were established and it was proven that $\frac{k}{k+1} \leq t^*(k,r) < 1$ for $r \geq 3$ and $k \geq 1$, cf. [36]. It would be interesting to establish exact values of $t^*(k,r)$ for more instances of $k$ and $r$. Other functions of similar flavor (involving other convenient parameters) might also be interesting.

The following relation between weak $k$-domination number and $k$-rainbow domination number is trivial: $\gamma_{rk}(G) \geq \gamma_{wk}(G)$. The question to determine known classes of graphs $G$ for which $\gamma_{rk}(G) = \gamma_{wk}(G)$ was initiated in [14]. A simple example where $\gamma_{r2}(G) > \gamma_{w2}(G)$ are cycles $C_{4k+2}$. Also, it is known that in dually chordal graphs [14] and cographs [48], $\gamma_{rk}(G)$ can be strictly larger than $\gamma_{wk}(G)$. An affirmative answer was presented in [18], where the equality was shown for strongly chordal graphs $G$. The equality may have positive consequences for finding efficient algorithms for establishing $\gamma_{rk}(G)$ in such classes of graphs $G$, as demonstrated in [14, 18, 48], but it is also of independent interest. In particular, it would be interesting to better understand the graphs in which $\gamma_{r2}(G) = \gamma_{w2}(G)$.

In [11], a dozen domination invariants in which number 2 is inherently involved in their definition was systematically considered, giving sharp lower and upper bounds between each pair of the invariants. In particular, 2-rainbow domination number was included, and also several other rainbow versions of known domination invariants were introduced. The definitions of the invariants involving number 2 can be extended in a natural way to the invariants involving an arbitrary positive integer $k$ (for instance, the rainbow 2-domination number of a graph $G$, as presented in Section 6, can be extended in a natural way to the rainbow $k$-domination number of $G$). A natural problem is to establish sharp bounds between $\gamma_{rk}(G)$ and other domination invariants in which a positive integer $k$ ($k \geq 2$) inherently appears, and, moreover, to generalize the results from [11].

We conclude the chapter with the problem of determining the 2-rainbow domination numbers of grid graphs, $P_m \square P_n$, where $m$ and $n$ are positive integers. In addition, the problem about the values in grid graphs is open for several other domination invariants mentioned in the previous paragraph and studied in [11]. The domination numbers, $\gamma(P_n \square P_n)$, of grid graphs were completely determined in [42]. Recently, the 2-domination numbers, $\gamma_2(P_m \square P_n)$, of grid graphs were also established [60], which could be useful for determining the 2-rainbow domination numbers of grid graphs. Since $\gamma_{r2}(P_m \square P_n) = \gamma(P_m \square P_n \square P_2)$, a solution to this problem would be the first step in the investigation of domination in 3-dimensional grid graphs.

**Acknowledgments** The author was in part supported by the Slovenian Research Agency (ARRS) under the grants P1-0297 and J1-9109.

## References

1. A. Abdollahzadeh, J. Amjadi, N.J. Rad, V. Samodivkin, Total $k$-rainbow domination numbers in graphs, Commun. Comb. Optim. 3 (2018) 37–50.
2. J.D. Alvarado, S. Dantas, D. Rautenbach, Averaging 2-rainbow domination and Roman domination, Discrete Appl. Math. 205 (2016) 202–207.
3. J.D. Alvarado, S. Dantas, D. Rautenbach, Relating 2-rainbow domination to Roman domination, Discuss. Math. Graph Theory 37 (2017) 953–961.
4. J. Amjadi, L. Asgharsharghi, N. Dehgardi, M. Furuya, S.M. Sheikholeslami, L. Volkmann, The $k$-rainbow reinforcement numbers in graphs, Discrete Appl. Math. 217 (2017) 394–404.
5. J. Amjadi, A. Bahremandpour, S.M. Sheikholeslami, L. Volkmann, The rainbow domination number of a digraph, Kragujevac J. Math. 37 (2013) 257–268.
6. J. Amjadi, M. Chellali, M. Falahat, S.M. Sheikholeslami, Unicyclic graphs with strong equality between the 2-rainbow domination and independent 2-rainbow domination numbers, Trans. Comb. 4 (2015) 1–11.
7. J. Amjadi, N. Dehgardi, N.M. Mohammadi, S.M. Sheikholeslami, L. Volkmann, Independent 2-rainbow domination in trees, Asian-Eur. J. Math. 8 (2015), 8 pp.
8. J. Amjadi, M. Falahat, S.M. Sheikholeslami, N.J. Rad, Strong equality between the 2-rainbow domination and independent 2-rainbow domination numbers in trees, Bull. Malays. Math. Sci. Soc. 39 (2016) 205–218.
9. J. Amjadi, N. Mohammadi, S.M. Sheikholeslami, L. Volkmann, The k-rainbow bondage number of a digraph, Discuss. Math. Graph Theory 35 (2015) 261–270.
10. J. Amjadi, A. Parnian, On the 2-rainbow bondage number of planar graphs, Ars Combin. 126 (2016) 395–405.
11. F. Bonomo, B. Brešar, L. Grippo, M. Milanič, M.D. Safe, Domination parameters with number 2: interrelations and algorithmic consequences, Discrete Appl. Math. 235 (2018) 23–50.
12. B. Brešar, P. Dorbec, W. Goddard, B. L. Hartnell, M. A. Henning, S. Klavžar and D. F. Rall, Vizing's conjecture: a survey and recent results, J. Graph Theory 69 (2012) 46–76.
13. B. Brešar, M.A Henning, D.F. Rall, Rainbow domination in graphs. Taiwanese J. Math. 12 (2008) 213–225.
14. B. Brešar, T. Kraner Šumenjak, On the 2-rainbow domination in graphs, Discrete Appl. Math. 155 (2007) 2394–2400.

15. B. Brešar, D.F. Rall, On Cartesian products having a minimum dominating set that is a box or a stairway, Graphs Combin. 31 (2015) 1263–1270.
16. S. Brezovnik, T. Kraner Šumenjak, Complexity of $k$-rainbow independent domination and some results on the lexicographic product of graphs, Appl. Math. Comput. 349 (2019) 214–220.
17. Y. Caro, Y. Roditty, On the vertex-independence number and star decomposition of graphs, Ars Combin. 20 (1985) 167–180.
18. G.J. Chang, B. Li, J. Wu, Rainbow domination and related problems on strongly chordal graphs, Discrete Appl. Math. 161 (2013) 1395–1401.
19. G.J. Chang, J. Wu, X. Zhu, Rainbow domination on trees, Discrete Appl. Math. 158 (2010) 8–12.
20. M. Chellali, T.W. Haynes, S.T. Hedetniemi, Bounds on weak roman and 2-rainbow domination numbers, Discrete Appl. Math. 178 (2014) 27–32.
21. M. Chellali, T.W. Haynes, S.T. Hedetniemi, A.A. McRae, Roman {2}-domination, Discrete Appl. Math. 204 (2016) 22–28.
22. M. Chellali, N.J. Rad, On 2-rainbow domination and Roman domination in graphs, Australas. J. Combin. 56 (2013), 85–93.
23. M. Chellali, N.J. Rad, Independent 2-rainbow domination in graphs, J. Combin. Math. Combin. Comput. 94 (2015) 133–148.
24. T. Chunling, L. Xiaohui, Y. Yuansheng, Z. Baosheng, Z. Xianchen, A lower bound for 2-rainbow domination number of generalized Petersen graphs $P(n, 3)$, Ars Combin. 102 (2011) 483–492.
25. W.E. Clark, S. Suen, An inequality related to Vizing's conjecture, Electron. J. Combin. 7 (2000) Note 4, 3 pp.
26. E.J. Cockayne, R.M. Dawes, S.T. Hedetniemi, Total domination in graphs, Networks 10 (1980) 211–219.
27. E.J. Cockayne, P.A. Dreyer Jr., S.T. Hedetniemi, S.M. Hedetniemi, Roman domination in graphs, Discrete Math. 278 (2004) 11–22.
28. E.J. Cockayne, S.T. Hedetniemi, Towards a theory of domination in graphs, Networks 7 (1977) 247–261.
29. N. Dehgardi, S.M. Sheikholeslami, L. Volkmann, The $k$-rainbow bondage number of a graph, Discrete Appl. Math. 174 (2014) 133–139.
30. N. Dehgardi, S.M. Sheikholeslami, L. Volkmann, The rainbow domination subdivision numbers of graphs, Mat. Vesnik 67 (2015) 102–114.
31. N. Dehgardi, M. Falahat, S.M. Sheikholeslami, A. Khodkar, On the rainbow domination subdivision numbers in graphs, Asian-Eur. J. Math. 9 (2016), 12 pp.
32. G. S. Domke, S. T. Hedetniemi, R. C. Laskar, and G. H. Fricke, Relationships between integer and fractional parameters of graphs. In Y. Alavi, G. Chartrand, O.R. Oellermann, and A.J. Schwenk, editors, *Graph Theory, Combinatorics, and Applications, Proc. Sixth Quad. Conf. on the Theory and Applications of Graphs* (Kalamazoo, MI 1988), Volume 2, pp. 371–387, 1991, Wiley.
33. M. Falahat, S.M. Sheikholeslami, L. Volkmann, New bounds on the rainbow domination subdivision number, Filomat 28 (2014) 615–622.
34. J. F. Fink and M. S. Jacobson, $n$-domination in graphs, in [*Graph theory with applications to algorithms and computer science* (Kalamazoo, Mich., 1984), Wiley-Intersci. Publ.], pp. 283–300, Wiley, New York, 1985.
35. S. Fujita, M. Furuya, Difference between 2-rainbow domination and Roman domination in graphs, Discrete Appl. Math. 161 (2013) 806–812.
36. S. Fujita, M. Furuya, Rainbow domination numbers on graphs with given radius, Discrete Appl. Math. 166 (2014) 115–122.
37. S. Fujita, M. Furuya, C. Magnant, $k$-rainbow domatic numbers, Discrete Appl. Math. 160 (2012) 1104–1113.
38. S. Fujita, M. Furuya, C. Magnant, General bounds on rainbow domination numbers, Graphs Combin. 31 (2015) 601–613.

39. S. Fujita, M. Furuya, C. Magnant, General upper bounds on independent $k$-rainbow domination, Discrete Appl. Math. 258 (2019) 105–113.
40. M. Furuya, A note on total domination and 2-rainbow domination in graphs, Discrete Appl. Math. 184 (2015) 229–230.
41. M. Furuya, M. Koyanagi, M. Yokota, Upper bound on 3-rainbow domination in graphs with minimum degree 2, Discrete Optim. 29 (2018) 45–76.
42. D. Gonçalves, A. Pinlou, M. Rao, S. Thomassé, The domination number of grids, SIAM J. Discrete Math. 25 (2011) 1443–1453.
43. G. Hao, J. Qian, On the rainbow domination number of digraphs, Graphs Combin. 32 (2016) 1903–1913.
44. G. Hao, D.A. Mojdeh, S. Wei, Z. Xie, Rainbow domination in the Cartesian product of directed paths, Australas. J. Combin. 70 (2018) 349–361.
45. B. Hartnell, D.F. Rall, Domination in Cartesian products: Vizing's conjecture, In [*Domination in graphs*, volume 209 of *Monogr. Textbooks Pure Appl. Math.*, Dekker, New York, 1998], 163–189.
46. B. Hartnell, D.F. Rall, On dominating the Cartesian product of a graph and $K_2$, Discuss. Math. Graph Theory 24 (2004) 389–402.
47. M.A. Henning, W. Klostermeyer, Italian domination in trees, Discrete Appl. Math. 217 (2017) 557–564.
48. W.-K. Hon, T. Kloks, H.-H. Liu, H.-L. Wang, Rainbow domination and related problems on some classes of perfect graphs, Topics in theoretical computer science, 121–134, Lecture Notes in Comput. Sci., 9541, Springer, [Cham], 2016.
49. T. Kraner Šumenjak, D.F. Rall, A. Tepeh, Rainbow domination in the lexicographic product of graphs, Discrete Appl. Math. 161 (2013) 2133–2141.
50. T. Kraner Šumenjak, D.F. Rall, A. Tepeh, On k-rainbow independent domination in graphs, Appl. Math. Comput. 333 (2018) 353–361.
51. B. Kuzman, On $k$-rainbow domination in regular graphs, Discrete Appl. Math. 284 (2020) 454–464.
52. J.-J. Liu, S.-C. Chang, C.-J. Lin, The 2-rainbow domination of Sierpiński graphs and extended Sierpiński graphs, Theory Comput. Syst. 61 (2017) 893–906.
53. W. McCuaig, B. Shepherd, Domination in graphs with minimum degree two, J. Graph Theory 13 (1989) 749–762.
54. D. Meierling, S.M. Sheikholeslami, L. Volkmann, Nordhaus-Gaddum bounds on the $k$-rainbow domatic number of a graph, Appl. Math. Lett. 24 (2011) 1758–1761.
55. O. Ore, *Theory of graphs*, American Mathematical Society Providence, R.I., 1962.
56. M. Pilipczuk, M. Pilipczuk, R. Škrekovski, Some results on Vizing's conjecture and related problems, Discrete Appl. Math. 160 (2012) 2484–2490.
57. N.J. Rad, On the complexity of reinforcement in graphs, Discuss. Math. Graph Theory 36 (2016) 877–887.
58. N.J. Rad, H. Kamarulhaili, On the complexity of some bondage problems in graphs, Australas. J. Combin. 68 (2017) 265–275.
59. A. Rahmouni, M. Chellali, Independent Roman {2}-domination in graphs, Discrete Appl. Math. 236 (2018) 408–414.
60. M. Rao, A. Talon, The 2-domination and Roman domination numbers of grid graphs, Discrete Math. Theor. Comput. Sci. 21 (2019) Paper No. 9, 14 pp.
61. Z. Shao, H. Jiang, P. Wu, S. Wang, J. Žerovnik, X. Zhang, J.-B. Liu, On 2-rainbow domination of generalized Petersen graphs, Discrete Appl. Math. 257 (2019) 370–384.
62. Z. Shao, M. Liang, C. Yin, X. Xu, P. Pavlič, J. Žerovnik, On rainbow domination numbers of graphs, Inform. Sci. 254 (2014) 225–234.
63. Z. Shao, S.M. Sheikholeslami, B. Wang, P. Wu, X. Zhang, Trees with equal total domination and 2-rainbow domination numbers, Filomat 32 (2018) 599–607.
64. Z. Shao, Z. Li, A. Peperko, J. Wan, J. Žerovnik, Independent rainbow domination of graphs, Bull. Malays. Math. Sci. Soc. 42 (2019) 417–435.

65. S.M. Sheikholeslami, L. Volkmann, The $k$-rainbow domatic number of a graph, Discuss. Math. Graph Theory 32 (2012) 129–140.
66. S.M. Sheikholeslami, L. Volkmann, The $k$-rainbow domination and domatic numbers of digraphs, Kyungpook Math. J. 56 (2016) 69–81.
67. Z. Stępień, A. Szymaszkiewicz, L. Szymaszkiewicz, M. Zwierzchowski, 2-rainbow domination number of $C_n \square C_5$, Discrete Appl. Math. 170 (2014) 113–116.
68. Z. Stępień, L. Szymaszkiewicz, M. Zwierzchowski, The Cartesian product of cycles with small 2-rainbow domination number, J. Comb. Optim. 30 (2015) 668–674.
69. Z. Stępień, M. Zwierzchowski, 2-rainbow domination number of Cartesian products: $C_n \square C_3$ and $C_n \square C_5$, J. Comb. Optim. 28 (2014) 748–755.
70. I. Stewart, Defend the Roman Empire!, Sci. Am. 281 (1999) 136–139.
71. A. Tepeh, Total domination in generalized prisms and a new domination invariant, Discuss. Math. Graph Theory (2020) doi: doi.org/10.7151/dmgt.2256
72. C. Tong, X. Lin, Y. Yang, M. Luo, 2-rainbow domination of generalized Petersen graphs $P(n, 2)$, Discrete Appl. Math. 157 (2009) 1932–1937.
73. V. G. Vizing, The Cartesian product of graphs, Vyčisl. Sistemy 9 (1963) 30–43.
74. V.G. Vizing, Some unsolved problems in graph theory, Uspehi Mat. Nauk 23 (1968) 117–134.
75. Y.-L. Wang, K.-H. Wu, A tight upper bound for 2-rainbow domination in generalized Petersen graphs, Discrete Appl. Math. 161 (2013) 2178–2188.
76. Y. Wu, An improvement on Vizing's conjecture, Inform. Process. Lett. 113 (2013) 87–88.
77. Y. Wu, N.J. Rad, Bounds on the 2-rainbow domination number of graphs, Graphs Combin. 29 (2013) 1125–1133.
78. K.-H. Wu, Y.-L. Wang, C.-C. Hsu, C.-C. Shih, On 2-rainbow domination in generalized Petersen graphs, Int. J. Comput. Math. Comput. Syst. Theory 2 (2017) 1–13.
79. Y. Wu, H. Xing, Note on 2-rainbow domination and Roman domination in graphs, Appl. Math. Lett. 23 (2010) 706–709.
80. G. Xu, 2-rainbow domination in generalized Petersen graphs $P(n, 3)$, Discrete Appl. Math. 157 (2009) 2570–2573.

# Eternal and Secure Domination in Graphs

**William F. Klostermeyer and C. M. Mynhardt**

## 1 Introduction

Dominating sets in graphs have long had a connection to facility-location problems in which a limited number of resources need to be placed on some of the vertices of a graph so as to be close to all other vertices of the graph, see, for example, [42, 43]. Viewing these limited resources as guards, we may think of a dominating set as being able to "protect" a graph against a single attack, such as a robbery, at any vertex, since each vertex either contains a guard or has a guard on a neighboring vertex. We may then define the more general problem of *graph protection* as the placement of mobile guards on the vertices of a graph to protect its vertices and edges against either single or long sequences of attacks. The modern study of models of graph protection was initiated in the late twentieth century by the appearance of four publications in quick succession that referred to the military strategy of Roman Emperor Constantine (Constantine The Great, 274–337 AD).

The seminal paper on the subject is Ian Stewart's "Defend the Roman Empire!" in *Scientific American*, December 1999 [82], which contains a reply to C. S. ReVelle's article "Can you protect the Roman Empire?", *Johns Hopkins Magazine*, April 1997 [79], and which is based on ReVelle and K. E. Rosing's "Defendens Imperium

Supported by the Natural Sciences and Engineering Research Council of Canada.

W. F. Klostermeyer
School of Computing, University of North Florida, Jacksonville, FL 32224-2669, USA
e-mail: wkloster@unf.edu

C. M. Mynhardt (✉)
Department of Mathematics and Statistics, University of Victoria, P.O. Box 3060 STN CSC, Victoria, BC, Canada V8W 3R4
e-mail: kieka@uvic.ca



T. W. Haynes et al. (eds.), *Topics in Domination in Graphs*, Developments in Mathematics 64, https://doi.org/10.1007/978-3-030-51117-3_13

Romanum: A Classical Problem in Military Strategy" in *American Mathematical Monthly*, August – September 2000 [80]. ReVelle's work [79] in turn is a response to the paper "'Graphing' an Optimal Grand Strategy" by J. Arquilla and H. Fredricksen [5], which appeared in *Military Operations Research* in 1995 and which is the oldest reference we could find that places the strategy of Emperor Constantine in a mathematical setting.

According to Roman mythology, Rome was founded by Romulus and Remus in 760 – 750 BC on the banks of the Tiber in central Italy. It started as a country town whose power gradually grew until it was the center of a large empire. In the third century AD Rome dominated not only Europe, but also North Africa and the Near East. The Roman army at that time was strong enough to use a *forward defense* strategy, deploying an adequate number of legions to secure on-site every region throughout the empire. However, the Roman Empire's power was greatly reduced over the following hundred years. By the fourth century AD, only twenty-five legions of the Roman army were available, which made a forward defense strategy no longer feasible.

According to E. N. Luttwak, *The Grand Strategy of the Roman Empire*, as cited in [80], to cope with the reducing power of the Empire, Constantine devised a new strategy called a *defense in depth* strategy, which used local troops to disrupt invasions. He deployed mobile Field Armies (FAs), units of forces consisting of roughly six legions powerful enough to secure any one of the regions of the Roman Empire, to stop the intruding enemy, or to suppress insurrection. By the fourth century AD there were only four FAs available for deployment, whereas there were eight regions to be defended (Britain, Gaul, Iberia, Rome, North Africa, Constantinople, Egypt, and Asia Minor) in the empire. See Figure 1 for an illustration showing this region as a graph.

An FA was considered capable of deploying to protect an adjacent region only if it moved from a region where there was at least one other FA to help launch it. The challenge that Constantine faced was to position four FAs in the eight regions of the empire. Consider a region to be *secured* if it has one or more FAs stationed in it already, and *securable* if an FA can reach it in one step. Constantine decided to place two FAs in Rome and another two FAs in Constantinople, making all regions either secured or securable – with the exception of Britain, which could only be secured after at least four movements of FAs.

It is mentioned in [5, 80, 82] that Constantine's "defense in depth" strategy was adopted during World War II by General Douglas MacArthur. When conducting military operations in the Pacific theater, he pursued a strategy of "island-hopping" – moving troops from one island to a nearby one, but only when he could leave behind a large enough garrison to keep the first island secure. The efficiency of Constantine's strategy under different criteria, and ways in which it can be improved, were also discussed in these three articles.

Constantine's strategy is now known in domination theory as **Roman domination**, a term that was coined by Cockayne, Dreyer, Hedetniemi and Hedetniemi [28]. Formally, a *Roman dominating function* on a graph $G = (V, E)$ is a function $f : V \to \{0, 1, 2\}$ satisfying the condition that every vertex $u$ with $f(u) = 0$

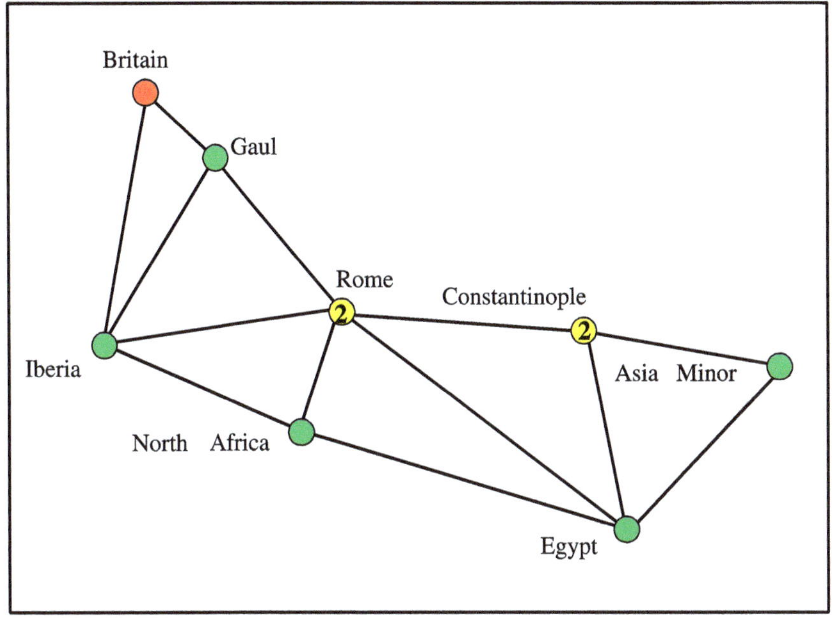


**Fig. 1** Map of the Roman Empire, fourth century AD

is adjacent to at least one vertex $v$ with $f(v) = 2$. **Weak Roman domination**, an alternative defense strategy that can be used if defense units can move without another unit being present, was introduced by Henning and Hedetniemi in [40]. A function $f : V \to \{0, 1, 2\}$ is a *weak Roman dominating function of* $G$ if each vertex $u$ with $f(u) = 0$ is adjacent to a vertex $v$ with $f(v) > 0$ such that the function $f' = (f - \{(v, f(v)), (u, 0)\}) \cup \{(v, f(v) - 1), (u, 1)\}$ also has the property that each vertex labeled 0 is adjacent to a vertex with positive label. **Secure domination**, introduced by Cockayne, Grobler, Gründlingh, Munganga and Van Vuuren in [30], is a defense strategy that can be used when it is not possible or desirable to station two defense units at the same location. A *secure dominating function* is a weak Roman dominating function $f$ such that $\{v \in V : f(v) = 2\} = \varnothing$. In this case the set $\{v \in V : f(v) = 1\}$ is a *secure dominating set* of $G$. Secure domination is discussed in the second part of this chapter.

In the first part of the chapter, we focus on defending the vertices of graphs against sequences of attacks, executed one at a time, by stationing guards at the vertices of the graph. At most one guard is stationed at each vertex, and guards that move in response to an attack do not return to their original positions before facing another attack. We refer to such models as *eternal* if the sequence of attacks is infinitely long, as they can be thought of as protecting a graph for eternity, and as *secure* in the case of single attacks (and the configuration of guards is a dominating set both before and after this single attack). These "dynamic" models of domination were first defined and studied by Burger, Cockayne, Gründlingh,

Mynhardt, Van Vuuren and Winterbach in an influential pair of papers from 2004 that defined the concept of eternal domination, then called infinite order domination, see [14, 15].

We begin by giving some definitions in the next section. Section 3 introduces the two main models of eternal domination. Results on eternal domination and m-eternal domination are presented in Sections 4 and 5, respectively. Secure domination is covered in Section 6. Open problems and conjectures are mentioned throughout the text where appropriate. We conclude by stating additional open problems in Section 7.

## 2 Definitions

Definitions pertaining to domination not given here can be found in the Glossary in this volume, while other definitions can be found in standard graph theory texts such as [25, 86]. For a graph $G = (V, E)$, a set $X \subseteq V$, and a vertex $v \in X$, the *private neighborhood* $\mathrm{pn}(v, X)$ *of* $v$ *with respect to* $X$ is the set of all vertices in $N[v]$ that are not contained in the closed neighborhood of any other vertex in $X$, i.e., $\mathrm{pn}(v, X) = N[v] - N[X - \{v\}]$. The elements of $\mathrm{pn}(v, X)$ are the *private neighbors of* $v$ *with respect to* $X$. It is possible that $v \in \mathrm{pn}(v, X)$; in this case we say that $v$ is a *self-private neighbor with respect to* $X$. No other private neighbor of $v$ with respect to $X$ belongs to $X$. The *external private neighborhood of* $v$ *with respect to* $X$ is the set $\mathrm{epn}(v, X) = \mathrm{pn}(v, X) - \{v\} = N(v) - N[X - \{v\}]$. The *internal private neighborhood of* $v$ *with respect to* $X$ is $\mathrm{ipn}(v, X) = \{x \in X : x$ is adjacent to $v$ but not to any vertex in $X - \{v\}\}$.

A *vertex cover* of $G$ is a set $C \subseteq V$ such that each edge of $G$ is incident with a vertex in $C$. The minimum cardinality of a vertex cover of $G$ is the *vertex covering number* $\tau(G)$ of $G$. It is clear that the independence number $\alpha(G)$ of $G$ equals the clique number $\omega(\overline{G})$ of the complement $\overline{G}$ of $G$. It is well known that $\alpha(G) + \tau(G) = n$ for all graphs $G$ of order $n$ (see, e.g., [25, p. 241]). The *matching number* (also called the *edge independence number*) $\alpha'(G)$ is the maximum cardinality of a matching of $G$. It is also well known that $\tau(G) \geq \alpha'(G)$ for all graphs, and that equality holds for bipartite graphs. The latter result is known as **König's theorem** (see, e.g., [25, Theorem 9.13]).

The *clique covering number* $\theta(G)$ is the minimum number $k$ of sets in a partition $V_1, ..., V_k$ of $V$ such that the subgraph $G[V_i]$ induced by each $V_i$ is a clique. Observe that $\theta(G)$ equals the chromatic number $\chi(\overline{G})$ of the complement $\overline{G}$ of $G$. Since $\chi(G) = \omega(G)$ if $G$ is perfect, and $G$ is perfect if and only if $\overline{G}$ is perfect [25, p. 203], it follows that $\alpha(G) = \theta(G)$ for all perfect graphs.

The *circulant graph* $C_n[a_1, \ldots, a_k]$, where $1 \leq a_1 \leq \cdots \leq a_k \leq \lfloor \frac{n}{2} \rfloor$, is the graph with vertex set $\{v_0, \ldots, v_{n-1}\}$ such that $v_i$ and $v_j$ are adjacent if and only if $i - j \equiv \pm a_\ell \pmod{n}$ for some $\ell \in \{1, \ldots, k\}$.

## 3 Eternal Domination Models

Let $\mathcal{D}_k$ be the collection of all dominating sets of $G$ of fixed cardinality $k$. For a set $D \in \mathcal{D}_k$, there is a single guard located on each vertex of $D$ and therefore we think of $D$ as a configuration of guards. A vertex $v$ is *occupied* if there is a guard on $v$, otherwise $v$ is *unoccupied*. We say that a (not necessarily dominating) set $X$ *protects* a vertex $v$, or $v$ is *protected* (by $X$), if $v$ or one of its neighbors is occupied.

Each eternal domination problem can be modeled as a two-player game, alternating between a *defender* and an *attacker*: the defender chooses $D_1 \in \mathcal{D}_k$ as well as each $D_i$, $i > 1$, while the attacker chooses the locations $r_1, r_2, \ldots$ of the attacks; we say the attacker *attacks* the vertices $r_i$. Thus, the game starts with the defender choosing $D_1$. For $i \geq 1$, the attacker attacks $r_i \in V - D_i$, and the defender *defends against* the attack by choosing $D_{i+1} \in \mathcal{D}_k$ subject to constraints (described below) that depend on the particular game. The defender wins the game if they can successfully defend the graph against any sequence of attacks, including sequences that are infinitely long, subject to the constraints of the game; the attacker wins otherwise. In other words, the attacker's goal is to force the defender into a configuration of guards that is not dominating.

For the **eternal domination problem**, $D_i \in \mathcal{D}_k$ for each $i \geq 1$, $r_i \in V - D_i$, and $D_{i+1} \in \mathcal{D}_k$ is obtained from $D_i$ by moving a guard to $r_i$ from an adjacent vertex $v \in D_i$. If the defender can win the game with the sets $\{D_i\} \subseteq \mathcal{D}_k$, then each such $D_i$ is an *eternal dominating set*. The smallest integer $k$ such that $\mathcal{D}_k$ contains eternal dominating sets is the *eternal domination number* $\gamma^{\infty}(G)$. This problem was first studied by Burger et al. in [15] in 2004 and will sometimes be referred to as the *one-guard moves* model. Shortly thereafter, Goddard, Hedetniemi and Hedetniemi published a second paper on the topic where they called it *eternal security* [36]. Subsequently, the problem has most often been called eternal domination. An illustrative example is $C_5$, the cycle of order five. Whereas $\gamma(C_5) = \alpha(C_5) = 2$, $\gamma^{\infty}(C_5) = 3$.

For the **m-eternal dominating set problem**, $D_i \in \mathcal{D}_k$ for each $i \geq 1$, $r_i \in V - D_i$, and $D_{i+1} \in \mathcal{D}_k$ is obtained from $D_i$ by moving guards to neighboring vertices. That is, each guard in $D_i$ may pass or move to an adjacent vertex, as long as one guard moves to $r_i$. Thus $r_i \in D_{i+1}$. The smallest integer $k$ such that $\mathcal{D}_k$ contains m-eternal dominating sets (defined similar to eternal dominating sets) is the m-eternal *domination number* $\gamma_{\mathrm{m}}^{\infty}(G)$. This "multiple guards move" version of the problem was introduced by Goddard et al. [36]. It is also called the "all-guards move" model and sometimes denoted $\gamma_{\mathrm{all}}^{\infty}$.

We say an attack on an unoccupied vertex $u$ is *defended* by a set $D \in \mathcal{D}_k$ or by (a guard on) a vertex $v \in D$ if both $D$ and $(D - \{v\}) \cup \{u\}$ belong to the collection $\{D_i\} \subseteq \mathcal{D}_k$ of eternal or m-eternal dominating sets.

It is clear that $\gamma(G) \leq \gamma_{\mathrm{m}}^{\infty}(G) \leq \gamma^{\infty}(G)$ for all graphs $G$. In contrast to the example above, observe that $\gamma_{\mathrm{m}}^{\infty}(C_5) = 2$ – with two guards on independent vertices, each attack on an unoccupied vertex can be defended by either a clockwise or a counterclockwise rotation of the guards.

Variations on the two above-mentioned graph protection models are obtained by imposing conditions on the induced subgraph $G[D_i]$. Thus one may define the *eternal total* (*connected*, respectively) *domination number* $\gamma_t^\infty(G)$ ($\gamma_c^\infty(G)$, respectively) and the m-eternal *total* (*connected*, respectively) *domination number* $\gamma_{\mathrm{mt}}^\infty(G)$ ($\gamma_{\mathrm{mc}}^\infty(G)$, respectively) in the obvious way. Eternal total domination and eternal connected domination were introduced by Klostermeyer and Mynhardt [54]. Discussions of many other different types of eternal graph protection problems can be found in the survey [59] by these authors. We briefly mention one such problem.

In the *eviction* model, each configuration $D_i$, $i \geq 1$, of guards is a dominating set. An attack occurs at a vertex $r_i \in D_i$ with at least one unoccupied neighbor. The next guard configuration $D_{i+1}$ is obtained from $D_i$ by moving the guard from $r_i$ to an unoccupied neighbor (i.e., this is a "one-guard moves" model). The size of a smallest eternal dominating set in the eviction model for $G$ is denoted $e^\infty(G)$. This model was introduced by Klostermeyer, Lawrence, and MacGillivray [47] and further studied in [53].

**Question 3.18** [47] *Is it true that $e^\infty(G) \leq \gamma^\infty(G)$ for all graphs $G$? (Note that strict inequality is possible – consider $C_5$, for example.)*

# 4 Eternal Domination

Consider an eternal dominating set $D$ of a graph $G$. A necessary condition for a guard on a vertex of $D$ to defend a neighboring vertex in a winning strategy for the defender is given below. A proof can be found in [59, Proposition 1].

**Proposition 4.1** *Let $D$ be an eternal dominating set of a graph $G$. For each $v \in D$, $G[\{v\} \cup \mathrm{epn}(v, D)]$ is a clique, and if $v \in D$ defends $u \in V(G) - D$, then $G[\{u, v\} \cup \mathrm{epn}(v, D)]$ is a clique.*

The converse of Proposition 4.1 is not true. Consider the graph $G$ in Figure 2. The set $D = \{x, y, z\}$ is an eternal dominating set of $G$ in which the guard on $x$ ($y$, $z$) defends $\{x, u, r\}$ ($\{y, v, s\}$, $\{z, w\}$). Also, $\mathrm{pn}(y, D) = \{y, v\}$ and $G[\{y, v, r\}]$ is a clique. Suppose, however, the guard on $y$ moves to $r$. If the next attack is at $s$, then only $z$ has a guard adjacent to $s$. But moving this guard to $s$ leaves $w$ unprotected. In the graph $H$ in Figure 2, $D = \{x, y\}$ is not an eternal dominating set, even though $\mathrm{pn}(x, D) \cup \{r\}$, $\mathrm{pn}(x, D) \cup \{w\}$, $\mathrm{pn}(y, D) \cup \{w\}$, $\mathrm{pn}(y, D) \cup \{s\}$ all induce cliques: first attack $r$; without loss of generality, $x$ defends $r$. Now attack $s$. If the guard on $y$ moves there, then $w$ is not protected; if the guard on $r$ moves there, then $u$ is not protected.

The example in Figure 2 raises an interesting question:

**Question 4.2** *If a dominating set $D$ is not an eternal dominating set, how long could it take, assuming each player plays optimally, to reach a non-dominating set by a sequence of attacks?*

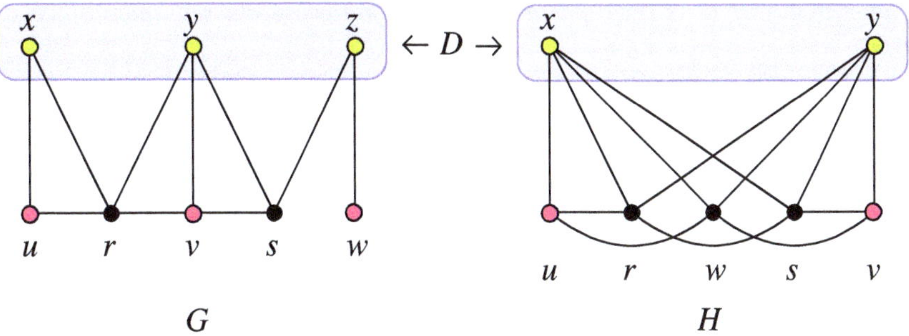


**Fig. 2** In $G$, $y$ does not defend $r$, and $D$ is not an eternal dominating set of $H$

## 4.1 Bounds for the Eternal Domination Number

As first observed by Burger et al. [15], it is straightforward to see that $\gamma^{\infty}$ lies between the independence and clique covering numbers.

**Fact 4.3** *For any graph G,* $\gamma(G) \leq \alpha(G) \leq \gamma^{\infty}(G) \leq \theta(G)$.

**Proof.** To see the lower bound, consider a sequence of consecutive attacks at the vertices of a maximum independent set. For the upper bound, observe that a single guard can defend all vertices of a clique. ■

As stated above, $C_5$, and more generally $C_{2n+1}$, $n \geq 2$, are examples of graphs with $\alpha(G) < \gamma^{\infty}(G)$.

The **Lovász theta function** [65] is another parameter that falls between $\alpha$ and $\theta$. An *orthonormal representation* of a graph $G = (V, E)$ is a family of unit vectors $\{u_i\}_{i \in V}$ such that $u_i \cdot u_j = 0$ whenever $i$ and $j$ are distinct and nonadjacent. The *Lovász theta function* $\vartheta(G)$ is defined by

$$\vartheta(G) = \min\{\theta : \text{there exists an orthonormal representation } \{u_i\}_{i \in V} \text{ of } G \text{ and a unit vector } c \text{ such that } (c \cdot u_i)^2 \geq \tfrac{1}{\theta} \text{ for every } i \in V\}.$$

Doug West [personal communication] asked whether $\gamma^{\infty}(G)$ is bounded from above by the Lovász theta function and this question remains open.

Since $\alpha(G) = \theta(G)$ for perfect graphs, the bounds in Fact 4.3 are tight for perfect graphs. A topic that has received much attention is finding classes of non-perfect graphs that satisfy one of the bounds in Fact 4.3. Before proceeding, we should point out that the independence number, eternal domination number, and clique covering number can vary widely, as shown by Klostermeyer and MacGillivray in [50].

**Theorem 4.4** [50] *For any positive integers c and d there exists a connected graph G such that* $\alpha(G) + c \leq \gamma^{\infty}(G)$ *and* $\gamma^{\infty}(G) + d \leq \theta(G)$.

Let $C_n^k$ denote the $k^{th}$ power (see [25, p. 105]) of the cycle of order $n$, where $2k + 1 < n$.

**Theorem 4.5** *If $G$ is a graph in one of the following classes, then $\gamma^{\infty}(G) = \theta(G)$.*

(a) [15] *Perfect graphs.*
(b) [15] *Any graph $G$ such that $\theta(G) \leq 3$.*
(c) [50] *$C_n^k$ and $\overline{C_n^k}$, for all $k \geq 1$, $n \geq 3$.*
(d) [78] *Circular-arc graphs (intersection graphs of a family of arcs of a circle).*
(e) [3] *$K_4$-minor-free graphs (a.k.a. series-parallel graphs, see, e.g., [86, p. 336] for definition).*
(f) [3] *$C_m \square C_n$; $P_m \square C_n$.*

### 4.1.1 Comparing $\gamma^{\infty}$, $\alpha$ and $\theta$

Goddard et al. [36] showed that

$$\text{if } \alpha(G) = 2, \text{ then } \gamma^{\infty}(G) \leq 3. \tag{1}$$

The Mycielski construction (see [25, p. 203]) yields triangle-free $k$-chromatic graphs for arbitrary $k$. The complements of these graphs have $\alpha = 2$, $\gamma^{\infty} = 3$, and $\theta = k$, and hence are examples of graphs with small eternal domination numbers and large clique covering numbers (see Theorem 4.8). The Grötzsch graph (which has order 11 and is obtained from $C_5$ via one application of the Mycielski construction) is the smallest 4-chromatic triangle-free graph, and its complement is the smallest known graph with $\gamma^{\infty} < \theta$ and indeed with $\alpha < \gamma^{\infty} < \theta$. Goddard et al. [36] also gave another example of a graph $G$ with $\alpha(G) < \gamma^{\infty}(G) < \theta(G)$: the circulant graph $C_{18}[1, 3, 8]$, which satisfies $\alpha = 6$, $\gamma^{\infty} = 8$ and $\theta = 9$.

Virgile [84, Section 2.3.1] also used Mycielski's construction (as part of a more elaborate construction) to disprove a conjecture by Klostermeyer and Mynhardt [57], namely that if $G$ is a graph such that $\gamma^{\infty}(G) = \theta(G)$, then $\gamma^{\infty}(G \square K_2) = \theta(G \square K_2)$ as well.

**Theorem 4.6** [84] *For each integer $k \geq 2$ there exists a graph $G_k$ such that $\alpha(G_k) = \gamma^{\infty}(G_k) = \theta(G_k) = \binom{k+1}{2} + 1$ and $\alpha(G_k \square K_2) = \gamma^{\infty}(G_k \square K_2) < \theta(G_k \square K_2)$.*

We next discuss the effect that joining a new vertex to all vertices of a graph that do not belong to a specific maximum independent set has on its eternal domination number. Let $G$ be any graph, let $I$ be any maximum independent set of $G$, and construct $G_{\alpha+1}$ by joining a new vertex $v$ to all vertices of $G - I$. Then $\alpha(G_{\alpha+1}) = \alpha(G) + 1$. Moreover, if $\gamma^{\infty}(G) = \alpha(G)$, then $\gamma^{\infty}(G_{\alpha+1}) = \alpha(G_{\alpha+1}) = \gamma^{\infty}(G) + 1$, because $G_{\alpha+1}$ can be defended with a new guard on $v$ which never needs to move, while the guards on vertices of $G$ defend attacks as in $G$. On the other hand, if $\gamma^{\infty}(G) > \alpha(G)$, then $\gamma^{\infty}(G_{\alpha+1}) = \gamma^{\infty}(G)$, as shown below. The proof that $\gamma^{\infty}(G_{\alpha+1}) \geq \gamma^{\infty}(G)$ rests on the observation in [49] that if a graph $H'$ is an induced subgraph of a graph $H$ and $\pi$ is any of the parameters $\alpha$, $\gamma^{\infty}$, $\theta$, then

$\pi(H') \leq \pi(H)$. This is trivially true for $\alpha$ and $\theta$. To see that it is true for $\gamma^\infty$, note that a sequence of attacks on $H$ but restricted to $H'$ requires $\gamma^\infty(H')$ guards, hence $\gamma^\infty(H) \geq \gamma^\infty(H')$.

**Proposition 4.7** *Let $G$ be a graph such that $\gamma^\infty(G) > \alpha(G)$, let $I$ be any maximum independent set, and let $G_{\alpha+1}$ be the graph obtained as described above. Then $\gamma^\infty(G_{\alpha+1}) = \gamma^\infty(G)$.*

**Proof.** As discussed above, $\gamma^\infty(G_{\alpha+1}) \geq \gamma^\infty(G)$. Let $v$ be the vertex added to $G$ to form $G_{\alpha+1}$. Place a guard on each vertex of $D'$, a minimum eternal dominating set of $G$. While vertices of $G$ are attacked, the guards duplicate their moves in $G$. Since $\gamma^\infty(G) \geq \alpha(G) + 1 > |I|$ and $v$ is adjacent to each vertex in $G - I$, $v$ is protected by any resulting configuration of guards. Suppose $v$ is attacked while the vertices of (say) the set $D$ are occupied. Let $D_I = D \cap I$ and $I' = \{u \in I - D : u \in \text{epn}(w, D) \text{ for some } w \in D\}$. Since $|D| > |I| \geq |D_I| + |I'|$, there exists a vertex $x \in D$ that does not belong to $I$ and also has no vertex in $I$ as external private neighbor. Therefore $v$ is adjacent to $x$ and to all vertices in $\text{epn}(x, D)$, hence $D'' = (D - \{x\}) \cup \{v\}$ is a dominating set of $G_{\alpha+1}$. Consider the next attack on a vertex $y$:

- If $y \in V(G) - I$, the guard on $v$ moves there and $(D'' - \{v\}) \cup \{y\}$ is an eternal dominating set of $G$ of the same type as $D$ and $D'$.
- If $y \in I$, then $I \nsubseteq D''$, hence (by counting) $D'' - I - \{v\} \neq \emptyset$. Since $D''$ is a dominating set of $G_{\alpha+1}$ (and neither $v$ nor any other vertex in $I$ dominates $y$), $y$ is dominated by vertices $z_1, \ldots, z_t \in D'' - I - \{v\}$. Since $v \in D''$, the only possible external private neighbors of the $z_i$ are vertices of $I$, and a counting argument similar to the case when $v$ was attacked shows that a guard on some $z_i$ can defend $y$. Thus $D''' = (D'' - \{z_i\}) \cup \{y\}$ is a dominating set of $G_{\alpha+1}$ of the same type as $D''$.

Consequently, $\gamma^\infty(G)$ guards suffice to defend any sequence of attacks on $G_{\alpha+1}$. ■

Klostermeyer and MacGillivray [49] proved the existence of graphs with $\gamma^\infty = \alpha$ and whose clique covering number is either equal to two (if $\alpha = 2$) or arbitrary otherwise. Proposition 4.7, the complements $G$ of Mycielski graphs and the construction $G \to G_{\alpha+1}$ can be used to prove the latter result. Note that all graphs thus constructed have triangles.

**Theorem 4.8** [49]

(a) *If $\alpha(G) = \gamma^\infty(G) = 2$, then $\theta(G) = 2$.*
(b) *For all integers $k \geq a \geq 3$ there exists a connected graph $G$ such that $\alpha(G) = \gamma^\infty(G) = a$ and $\theta(G) = k$.*

Goddard et al. [36] asked whether the eternal domination number can be bounded by a constant times the independence number. That this is impossible in general follows from the next two theorems. One of the main results on eternal domination is the following upper bound, due to Klostermeyer and MacGillivray [48]; also see [59, Theorem 6].

**Theorem 4.9** [48] *For any graph $G$,*

$$\gamma^{\infty}(G) \leq \binom{\alpha(G)+1}{2}.$$

Goldwasser and Klostermeyer [37] showed that this bound is tight for certain graphs. Specifically, let $G(n, k)$ be the graph with vertex set consisting of the set of all $k$-subsets of an $n$-set and where two vertices are adjacent if and only if their intersection is nonempty (thus $G(n, k)$ is the complement of a Kneser graph).

**Theorem 4.10** [37] *For each positive integer $t$, if $k$ is sufficiently large, then the graph $G(kt + k - 1, k)$ has eternal domination number $\binom{t+1}{2}$.*

Regan [78] found another graph for which the bound in Theorem 4.9 is tight: the circulant graph $C_{22}[1, 2, 4, 5, 9, 11]$. Theorems 4.9 and 4.10 show that it is impossible to find a constant $c$ such that $\gamma^{\infty}(G) \leq c\alpha(G)$ for all graphs $G$. As the graphs demonstrated by Theorem 4.10 are very large, it would be of interest to find other (smaller) graphs and families of graphs where the bound is tight.

Klostermeyer and MacGillivray [48] showed that the graph $G$ obtained by joining a new vertex to $m$ disjoint copies of $C_5$ satisfies $\alpha(G) = 2m$ and $\gamma^{\infty}(G) = 3m$, that is, $\gamma^{\infty}(G)/\alpha(G) = \frac{3}{2}$. This result and Theorem 4.8 can be placed in a more general setting, as explained by the same authors in [50].

A triple $(a, g, t)$ of positive integers is called *realizable* if there exists a connected graph $G$ with $\alpha(G) = a$, $\gamma^{\infty}(G) = g$ and $\theta(G) = t$. Theorem 4.9 shows that no triple with $g > \binom{a+1}{2}$ is realizable. The following theorem, stated in [50], provides a partial solution to the question of which triples are realizable.

**Theorem 4.11** *Let $(a, g, t)$ be a triple of positive integers such that $a \leq g \leq t$.*

(a) *The only realizable triple with $a = 1$ is $(1, 1, 1)$.*
(b) [15, 36, 49] *The only realizable triples with $a = 2$ are $(2, 2, 2)$ and $(2, 3, t)$, $t \geq 3$.*
(c) [15, 49, 50] *For all integers $a$, $g$ and $t$ with $3 \leq a \leq g \leq \frac{3}{2}a$ and $g \leq t$, the triple $(a, g, t)$ is realizable.*

The circulant $C_{21}[1, 3, 8]$, which satisfies $\gamma^{\infty}/\alpha = \frac{10}{6}$ (see [36]), shows that Theorem 4.11 does not characterize realizable triples, and was generalized to include all possible triples (up to the upper bound from Theorem 4.9) in [84, Section 2.2].

## 4.2 The Gamma-Theta Conjecture

The following problem is motivated by an error discovered in [50], where it is claimed that no such graph exists.

**Question 4.12** *Does there exist a graph $G$ with $\gamma(G) = \gamma^{\infty}(G)$ and $\gamma(G) < \theta(G)$?*

We rephrase this question as a conjecture, called the $\gamma - \theta$ *Conjecture*.

**Conjecture 4.13 (The $\gamma - \theta$ Conjecture)** *For every graph $G$, if $\gamma(G) = \gamma^{\infty}(G)$, then $\gamma(G) = \theta(G)$.*

Conjecture 4.13 is known to be true for perfect graphs, triangle-free graphs and graphs with maximum degree at most three [57], and for graphs without 4-cycles [46]. We prove this result in Theorem 4.16 to illustrate the basic proof technique used thus far for proving partial results on the $\gamma - \theta$ Conjecture. A definition and a lemma are needed first.

For a dominating set $D$, a vertex $x \in V - D$ is a *shared vertex* if $u, v \in D$ and $x$ is adjacent to both $u$ and $v$.

**Lemma 4.14** [46] *If $G$ is a graph without isolated vertices or 4-cycles, and $\gamma(G) = \gamma^{\infty}(G)$, then $G$ has a minimum eternal dominating set $D$ such that* $\text{epn}(v, D) \neq \emptyset$ *for each $v \in D$.*

**Proof.** Let $D$ be a minimum eternal dominating set of $G$ that maximizes the number of edges in $G[D]$. Suppose $\text{epn}(u, D) = \emptyset$ for some $u \in D$. Since $D$ is a minimum dominating set, $u$ is isolated in $G[D]$. Since $\deg u \geq 1$, $u$ is adjacent to a shared vertex $w$, where $w$ is also adjacent to $v \in D - \{u\}$. Since $G$ has no $C_4$, $N(u) \cap N(v) = \{w\}$ for each such vertex $v$. If $u$ defends $w$, then $D' = (D - \{u\}) \cup \{w\}$ is an eternal dominating set such that $G[D']$ has more edges than $G[D]$, a contradiction. Therefore $w$ is defended by some $x \in D$ such that $\text{epn}(x, D) \neq \emptyset$. By Proposition 4.1, $w$ is adjacent to each external private neighbor of $x$, and as shown above, $N(u) \cap N(x) = \{w\}$. Therefore $(D - \{u, x\}) \cup \{w\}$ is a smaller dominating set than $D$, a contradiction. ■

It is well known (see [9]) that any graph without isolated vertices has a minimum dominating set $D$ such that $\text{epn}(v, D) \neq \emptyset$ for each $v \in D$. Hence Lemma 4.14 will be redundant if the following question has a positive answer.

**Question 4.15** *Let $G$ be a graph such that $\gamma(G) = \gamma^{\infty}(G)$. Is it true that each $\gamma$-set is a $\gamma^{\infty}$-set?*

**Theorem 4.16** [46] *For any graph $G$ without 4-cycles, $\gamma(G) = \gamma^{\infty}(G)$ if and only if $\gamma(G) = \theta(G)$.*

**Proof.** Assume without loss of generality that $G$ has no isolated vertices. We only need to prove sufficiency. Suppose $\gamma(G) = \gamma^{\infty}(G) = k$. Let $D$ be a minimum eternal dominating set of $G$ such that $\text{epn}(v, D) \neq \emptyset$ for each $v \in D$; such an eternal dominating set exists by Lemma 4.14. If $\gamma(G) = 1$, then $G$ is complete and the statement holds. Hence we assume $\gamma(G) > 1$.

If each vertex of $G - D$ is an external private neighbor of a vertex in $D$, then, by Proposition 4.1, $\{\{x\} \cup \text{epn}(x, D) : x \in D\}$ is a clique cover of $G$ and the result follows. Hence assume $V - D$ has a shared vertex. For each $x \in D$, let $S_x$ denote the set of shared vertices **defended** by $x$ (from the initial configuration of guards on $D$). If $|S_x| \leq 1$ for each $x \in D$, then $R_x = \{x\} \cup S_x \cup \text{epn}(x, D)$ forms a clique (Proposition 4.1). Then $\{R_x : x \in D\}$ is a clique partition of $G$ into $\gamma(G)$ parts, provided that any vertex that lies in more than one $R_x$ is assigned to just one part in the partition.

Therefore, suppose that $w, w' \in S_u$ for some $u \in D$. Let $v$ be an external private neighbor of $u$, which we know exists. Then both $w, w'$ are adjacent to $v$ and thus $uwvw'u$ is a 4-cycle, a contradiction. Therefore, $|S_x| \leq 1$ for each $x \in D$, and the result follows. ■

It can also be shown that for any claw-free graph $G$ in which each vertex belongs to at most one triangle, $\gamma(G) = \gamma^\infty(G)$ if and only if $\gamma(G) = \theta(G)$. (Beginning with an independent minimum dominating set $D$, the claw-free condition ensures that each vertex in $V - D$ is adjacent to at most two elements of $D$. Then use the triangle condition to analyze the shared neighbors of any $u, v \in D$; there are at most two and each can be allocated to a (different) clique also containing $u$ or $v$.)

We conclude this section by addressing a question related to the $\gamma - \theta$ Conjecture. By Theorem 4.8(b), there exist connected graphs $G$ such that $\alpha(G) = \gamma^\infty(G) = a$ and $\theta(G) = k$ for any integers $k \geq a \geq 3$. All these graphs have triangles; hence the following question is of interest.

**Question 4.17** *Does there exist a triangle-free graph $G$ such that $\alpha(G) = \gamma^\infty(G) < \theta(G)$?*

## 4.3 The Fundamental Conjecture

Since eternal domination concerns the movement of guards, we may ask ourselves when a guard may actually move. This leads to the following conjecture, which was posed by Klostermeyer and MacGillivray [52], and which we term the *Fundamental Conjecture.*

**Conjecture 4.18 (The Fundamental Conjecture)** *Let $G$ be a graph with $\delta(G) \geq 1$ and minimum eternal dominating set $D$. For every vertex $v \in D$ with an unoccupied neighbor, there exists an eternal dominating set $D'$ such that $D' = (D - \{v\}) \cup \{u\}$, where $u \in N(v) - D$.*

In simple terms, we are asking whether any guard in an eternal dominating set is able to move (to some attacked neighbor) at any time, while still maintaining an eternal dominating set. That is, suppose a guard is on vertex $v$, where $v$ is an arbitrary vertex in a minimum eternal dominating set $D$. Is there a vertex $u$ adjacent to $v$ that can be attacked and then defended by the guard on $v$ so that the resulting configuration of guards is an eternal dominating set? The conjecture asserts that there always is such a vertex $u$.

Conjecture 4.18 is known to hold for isolate-free graphs $G$ with $\alpha(G) = 2$ or $\gamma^{\infty}(G) = \theta(G)$ [52]. The following weaker analog of Conjecture 4.18, which shows that any guard is able to move eventually, was proved in [52].

**Theorem 4.19** [52] *Let $G$ be a graph with no isolated vertices and $D$ a minimum eternal dominating set of $G$. For every vertex $u \in D$, there exists a minimum eternal dominating set $D'$ of $G$ such that (i) $u \notin D'$ and (ii) $D'$ reachable from $D$ by a sequence of guard moves.*

## 5 m-Eternal Domination

As mentioned in Section 3, m-eternal dominating sets are defined similarly to eternal dominating sets, except that when an attack occurs, each guard is allowed to move to a neighboring vertex to either defend the attacked vertex or to better position themselves for the future. This model was introduced by Goddard et al. [36]. As stated above, we sometimes refer to this as the "all-guards move" model of eternal domination.

Goddard et al. [36] determined $\gamma_{\mathrm{m}}^{\infty}(G)$ exactly for complete graphs, paths, cycles, and complete bipartite graphs. They also obtained the following fundamental bound.

**Theorem 5.1** [36] *For all graphs $G$, $\gamma(G) \leq \gamma_{\mathrm{m}}^{\infty}(G) \leq \alpha(G)$.*

**Outline of proof.** The left inequality is obvious. The right inequality is proved by induction on the order of $G$, the result being easy to see for small graphs. If $G$ has a vertex $v$ that is not contained in any maximum independent set, then $v$ is adjacent to at least two vertices of each maximum independent set of $G$. Therefore $\alpha(G - N[v]) \leq \alpha(G) - 2$. Hence (by induction) $G - N[v]$ can be protected by $\alpha(G) - 2$ guards. Since $K_{1,\deg(v)}$ is a spanning subgraph of $G[N[v]]$, $G[N[v]]$ can be protected by two guards. It follows that $\gamma_{\mathrm{m}}^{\infty}(G) \leq \alpha(G)$.

If each vertex of $G$ is contained in a maximum independent set, place a guard on each vertex of a maximum independent set $M$. Defend an attack on $v \in V(G) - M$ by moving all guards to a maximum independent set $M_v$ containing $v$. This is possible since Hall's Marriage Theorem ensures that there is a perfect matching between the symmetric difference of $M_v$ and $M$. ■

Theorem 5.1 places $\gamma_{\mathrm{m}}^{\infty}$ nicely in the inequality chain

$$\gamma(G) \leq \gamma_{\mathrm{m}}^{\infty}(G) \leq \alpha(G) \leq \gamma^{\infty}(G) \leq \theta(G).$$

Goddard et al. also claimed that $\gamma_{\mathrm{m}}^{\infty}(G) = \gamma(G)$ for all Cayley graphs $G$. This claim, however, is false, as was shown in the paper [10] by Braga, de Souza, and Lee. By computing $\gamma(G)$ and $\gamma_{\mathrm{m}}^{\infty}(G)$ for 7871 Cayley graphs of non-abelian groups, they found 61 connected Cayley graphs $G$ such that $\gamma_{\mathrm{m}}^{\infty}(G) = \gamma(G) + 1$. For all other connected Cayley graphs they investigated, $\gamma_{\mathrm{m}}^{\infty}(G) = \gamma(G)$.

The upper bound in Theorem 5.1 is not particularly good in general. For example, $K_{1,m}$ has independence number $m$ and can be defended with just two guards in this model. But equality holds for many graphs, such as $K_n$, $C_n$, and $P_2 \Box P_3$, just to name a few. By a careful analysis of the clique structure, it was shown in [11] that $\gamma_{\mathrm{m}}^{\infty}(G) = \alpha(G)$ for all proper interval graphs (a subclass of perfect graphs). Characterizing graphs with m-eternal domination number equal to the bounds in Theorem 5.1 remains open, as mentioned in Section 7.2. However, trees for which equality holds in the upper bound $\alpha$ were characterized by Klostermeyer and MacGillivray [51].

Define a *neo-colonization* to be a partition $\{V_1, V_2, \ldots, V_t\}$ of $G$ such that each $V_i$ induces a connected graph. A part $V_i$ is assigned weight one if it induces a clique, and $1 + \gamma_c(G[V_i])$ otherwise, where $\gamma_c(G[V_i])$ is the connected domination number of the subgraph induced by $V_i$. Then the *neo-colonization number* $\theta_c(G)$ is the minimum weight of any neo-colonization of $G$.

Rinemberg and Soulignac [81] showed that all interval graphs satisfy $\gamma_{\mathrm{m}}^{\infty}(G) = \theta_c(G)$. Goddard et al. [36] proved that $\gamma_{\mathrm{m}}^{\infty}(G) \le \theta_c(G) \le \gamma_c(G) + 1$. Klostermeyer and MacGillivray [50] proved that equality holds in the first inequality for trees.

**Theorem 5.2** [50] *If $T$ is a tree, then $\gamma_{\mathrm{m}}^{\infty}(T) = \theta_c(T)$.*

Chambers, Kinnersly, and Prince [24] gave a different upper bound for $\gamma_{\mathrm{m}}^{\infty}$. A proof is given below. A *branch vertex* of a tree is a vertex of degree at least three.

**Theorem 5.3** [24] *If $G$ is a connected graph of order $n$, then $\gamma_{\mathrm{m}}^{\infty}(G) \le \lceil \frac{n}{2} \rceil$.*

**Proof.** The proof is by induction on $n$, the result being easy to see for paths and cycles. Let $T$ be a spanning tree of $G$ with $r \ge 1$ branch vertices.

If $T$ has no vertex of degree two, then the subgraph of $T$ induced by the branch vertices is connected and, by [25, Theorem 3.7], $T$ has at least $r + 2$ leaves. Hence $n \ge 2r + 2$. Place a guard on each branch vertex and on one leaf. Whenever an unoccupied leaf $u$ is attacked, guards move so that $u$ and all branch vertices have guards. Hence $\gamma_{\mathrm{m}}^{\infty}(T) \le r + 1 \le \lceil \frac{n}{2} \rceil$.

If $T$ has a vertex $v$ of degree two, and $N(v) = \{u_1, u_2\}$, then at least one of the graphs $T - \{vu_i\}, i = 1, 2$, has a component of even order. Let $T_1$ be this component and let $T_2$ be the other component. Say $T_i$ has order $n_i$. By the induction hypothesis, $\gamma_{\mathrm{m}}^{\infty}(T_1) \le \frac{n_1}{2}$ and $\gamma_{\mathrm{m}}^{\infty}(T_2) \le \lceil \frac{n_2}{2} \rceil$. It follows that $\gamma_{\mathrm{m}}^{\infty}(T) \le \lceil \frac{n}{2} \rceil$ and therefore $\gamma_{\mathrm{m}}^{\infty}(G) \le \gamma_{\mathrm{m}}^{\infty}(T) \le \lceil \frac{n}{2} \rceil$. ■

The bound in Theorem 5.3 is exact for the corona of any graph because they have domination and clique covering numbers equal to half their order. It is also equal for odd length paths. It is not known which trees attain this bound [51], though partial results are given in [41]. The following small improvement was obtained by Henning, Klostermeyer and MacGillivray [41].

**Theorem 5.4** [41] *If $G$ is a connected graph with $\delta(G) \geq 2$ of order $n \neq 4$, then $\gamma_{\mathrm{m}}^{\infty}(G) \leq \left\lfloor \frac{n-1}{2} \right\rfloor$, and this bound is tight (e.g., for graphs obtained by joining a new vertex to one vertex of each of $k \geq 2$ disjoint copies of $C_4$).*

It is not hard to see that for many all-guards move models, the associated parameter is bounded above by $2\gamma$.

**Proposition 5.5** *For any connected graph $G$, $\gamma_{\mathrm{m}}^{\infty}(G) \leq 2\gamma(G)$, and the bound is tight for all values of $\gamma(G)$.*

**Outline of proof.** The result is trivial for $K_1$, so assume $|V(G)| \geq 2$. As shown in [9], every graph without isolated vertices has a minimum dominating set in which each vertex has an external private neighbor. Let $D$ be such a minimum dominating set of $G$. For each $v \in D$, place a guard at $v$ and at a private neighbor of $v$. This configuration is an m-eternal dominating set.

To see that the bound is tight for $\gamma = 1$, consider any star with at least three vertices. For $\gamma = 2$, consider the graph $G$ formed by joining two vertices by four internally disjoint paths of length three. It is routine to show that $\gamma_{\mathrm{m}}^{\infty}(G) = 4$.

For $\gamma = k \geq 3$, consider the cycle $C_{3k} = u_0, u_1, ..., u_{3k-1}, u_0$ and the $\gamma$-set $\{u_0, u_3, ..., u_{3k-3}\}$ of $C_{3k}$. For each $i = 0, ..., k-1$, add a new $u_{3i} - u_{3(i+1)(\mathrm{mod}\ 3k)}$ path of length three to form $G$. Then $\gamma(G) = k$, and again it can be shown that $\gamma_{\mathrm{m}}^{\infty}(G) = 2k$. ■

Klostermeyer and MacGillivray [51] characterized trees for which equality holds in the following bounds: $\gamma_{\mathrm{m}}^{\infty}(T) \leq \gamma_c(T) + 1$, $\gamma(T) \leq \gamma_{\mathrm{m}}^{\infty}(T)$, $\gamma_{\mathrm{m}}^{\infty}(T) \leq 2\gamma(T)$, and $\gamma_{\mathrm{m}}^{\infty}(T) \leq \alpha(T)$.

We now compare the m-eternal domination number and the vertex covering number. This may seem like an unusual pair of parameters to compare, but the comparison turns out to be interesting.

**Theorem 5.6**

(a) [55] *If $G$ is connected, then $\gamma_{\mathrm{m}}^{\infty}(G) \leq 2\tau(G)$.*
(b) [55] *If, in addition, $\delta(G) \geq 2$, then $\gamma_{\mathrm{m}}^{\infty}(G) \leq \tau(G)$.*
(c) [56] *If, in addition to (a) and (b), $G$ has girth seven or at least nine, then $\gamma_{\mathrm{m}}^{\infty}(G) < \tau(G)$.*
(d) [56] *For any nontrivial tree $T$, $\tau(T) \leq \gamma_{\mathrm{m}}^{\infty}(T) \leq 2\tau(T)$.*

It is not possible to relax the girth condition in Theorem 5.6(c) to girth less than five. Examples of graphs with girth less than five for which $\gamma_{\mathrm{m}}^{\infty}(G) = \tau(G)$ are given in [56]. The problem remains open for girths five, six, and eight, though it is believed that $\gamma_{\mathrm{m}}^{\infty}(G) < \tau(G)$ for such graphs. The trees where the bounds in Theorem 5.6(d) are tight are characterized in [56].

Braga, Reis, De Souza and Lee [12] proposed two heuristic methods suitable for practical applications of the m-eternal dominating set problem. Their methods use integer and constraint programming techniques to compute bounds on $\gamma_{\rm m}^{\infty}(G)$ for an input graph $G$. Let $Z = \{V_1, ..., V_k\}$ be a neo-colonization of $G$. Consider an ordered pair $\langle Z, Y\rangle$, where $Y$ is a set consisting of a connected dominating set $Y_i$ of $G[V_i]$ for each $V_i \in Z$ that does not induce a clique. The pair $\langle Z, Y\rangle$ is called a *neo-colonization setup* of $G$. The weight $w(\langle Z, Y\rangle)$ of $\langle Z, Y\rangle$ is given by the sum of 1 for every clique part of $Z$ and of $\min\{|Y_i| + 1, |V_i|\}$ for each non-clique part $V_i$ of $Z$. Notice that $w(\langle Z, Y\rangle)$ is bounded below by the weight of $Z$ and hence also by $\theta_c(G)$.

**Theorem 5.7** [12] *Let $\langle Z, Y\rangle$ be a neo-colonization setup of a graph $G$. There exists an efficient m-eternal defense strategy (i.e., a strategy with polynomial-time and polynomial-space requirements with respect to $|V(G)|$) of $G$ using $k = w(\langle Z, Y\rangle)$ guards.*

They showed that their programs $(i)$ run in a reasonable time, $(ii)$ produce a good quality upper bound on $\gamma_{\rm m}^{\infty}(G)$, and $(iii)$ output a structure from which one can derive an efficient strategy of defense of $G$ with $k$ guards, with the first two features being validated through experimentation on a total of 750 randomly generated graphs.

## 5.1 Multiple Guards on a Vertex

The results stated up until now in this chapter apply to the case when only one guard is allowed to occupy each vertex, and indeed, there is no advantage allowing multiple guards to occupy a single vertex in the "one guard moves" model [15]. A question stated in [36] is whether there is any advantage in allowing two guards to occupy the same vertex in the m-eternal domination problem. In response, Finbow, Gaspers, Messinger, and Ottoway [33] showed that there exist graphs for which it is an advantage in the all-guards move model to allow more than one guard on a vertex at a time. We sketch a proof of the simplest case of this result, using $\gamma_{\rm m}^{*\infty}(G)$ to denote the number of guards needed if more than one guard is allowed on a vertex at a time (and all guards are allowed to move in response to an attack). The graph $G_5$ in the proof is illustrated in Figure 3.

**Theorem 5.8** [33] *There exists a graph $G$ such that $\gamma_{\rm m}^{*\infty}(G) = 9$ and $\gamma_{\rm m}^{\infty}(G) = 10$.*

**Outline of proof.** Define a *gadget* to be the graph formed by taking two $K_4 - e$'s and combining one degree three vertex from each into a single vertex (so a gadget has seven vertices and two vertices of degree three). Form graph $G_5$ by taking five gadgets along with an additional vertex $x$; add an edge between $x$ and the vertices of degree three in each gadget.

To see that $\gamma_{\rm m}^{\infty}(G) = 10$, one can observe that there must be two guards in a gadget at some point in time: even though the domination number of a gadget is 1, because the degree two vertices in a gadget are independent and not adjacent to $x$

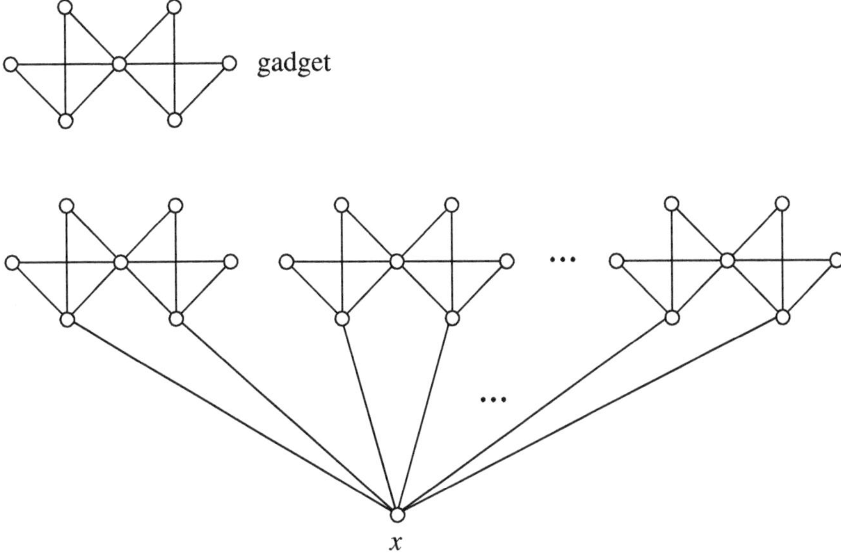


**Fig. 3** The graph $G_5$ in the proof of Theorem 5.8 is obtained by joining $x$ to the vertices of degree 3 in five gadgets.

and the dominating vertex in a gadget is not adjacent to $x$, it follows that we need at least two guards in each gadget at all times while there is no guard on $x$. If there is a guard on $x$, then at most one gadget can contain one guard.

On the other hand, by maintaining at least one guard in each gadget at all times and two guards on $x$, which move in and out of attacked gadgets, one can see that nine guards suffice to protect $G_5$. This is done by moving both guards from $x$ to the gadget where an attack occurs (i.e., that gadget contains three guards immediately after an attack) and moving two guards from the previously attacked gadget to $x$, while the remaining guard in the latter gadget moves to its degree six vertex. ■

The proof in [33] is more general than the sketch given above and shows there are graphs where $\gamma_{\mathrm{m}}^{*\infty}(G)$ and $\gamma_{\mathrm{m}}^{\infty}(G)$ can differ by any additive constant. The following question remains open.

**Question 5.9** *Does there exist a constant $c > 1$ such that $c\gamma_{\mathrm{m}}^{*\infty}(G) \geq \gamma_{\mathrm{m}}^{\infty}(G)$ for all graphs $G$?*

If any number of guards per vertex are allowed, then the bound in Theorem 5.3 can be improved to $\lceil \frac{n}{2} \rceil - 1$ when $\delta(G) \geq 2$ (with four small exceptions) [24]. It is not known whether their result holds if each vertex contains at most one guard. Under these conditions Nordhaus–Gaddum results were also shown in [24], for example, the following bound; they also characterize the graphs for which equality holds.

**Theorem 5.10** [24] $\gamma_{\mathrm{m}}^{\infty}(G) + \gamma_{m}^{\infty}(\overline{G}) \leq n + 1$.

## 5.2 Grids

Grid graphs, i.e., $P_m \Box P_n$, are a well-studied class of graphs in domination theory; see, for example, [42]. We sometimes refer to $P_m \Box P_n$ as the $m \times n$ *grid graph*. As shown by Goldwasser, Klostermeyer and Mynhardt [38], $\gamma_{\mathrm{m}}^{\infty}(P_2 \Box P_n) = \lceil \frac{2n}{3} \rceil$ for any $n \geq 2$, while $\gamma_{\mathrm{m}}^{\infty}(P_3 \Box P_n) = n$ for $2 \leq n \leq 8$. Based on these results, the next two theorems may seem surprising.

**Theorem 5.11** [38] *For* $n \geq 9$, $\gamma_{\mathrm{m}}^{\infty}(P_3 \Box P_n) \leq \lceil \frac{8n}{9} \rceil$.

In other words, the $3 \times 9$ grid graph can be defended by eight guards. It is then easy, as described in Theorem 5.11, to see that larger $3 \times n$ grids can be defended by at most $8n/9$ guards by partitioning such a grid into $3 \times 9$ grids (with perhaps one or two columns left over). Improving this bound has been the subject of a number of papers, e.g., Finbow, Messinger and Van Bommel [34], Finbow and Van Bommel [35], and Messinger and Delaney [69]. The first improved bounds were obtained in [34].

**Theorem 5.12** [34] *For* $n \geq 11$, $\left\lceil \frac{4n+6}{5} \right\rceil \leq \gamma_{\mathrm{m}}^{\infty}(P_3 \Box P_n) \leq \left\lceil \frac{6n+2}{7} \right\rceil$.

Theorem 5.12 shows that the bound in Theorem 5.11 is not tight in general, although it is tight for $n = 9, 10$, for example. Theorem 5.12 also disproved the conjecture in [38] that $\gamma_{\mathrm{m}}^{\infty}(P_3 \Box P_n) = \left\lceil \frac{4n+5}{5} \right\rceil$ for $n \geq 10$. Messinger and Delaney [69] developed a set of configurations for eternal dominating families which helped to reduce the upper bound.

**Theorem 5.13** [69] *For* $n \geq 12$,

$$\gamma_{\mathrm{m}}^{\infty}(P_3 \Box P_n) \leq \left\lceil \frac{4n+10}{5} \right\rceil + \begin{cases} 0 \text{ if } n \equiv 2, 4 \pmod 5 \\ 1 \text{ otherwise.} \end{cases}$$

Theorems 5.12 and 5.13 combined determined $\gamma_{\mathrm{m}}^{\infty}(P_3 \Box P_n)$ to within 3 of the exact value. Adapting configurations from [69], Finbow and Van Bommel [35] finally completed the determination of $\gamma_{\mathrm{m}}^{\infty}(P_3 \Box P_n)$. We summarize the results below.

**Theorem 5.14**

(a) [34, 38] *For* $n \leq 11$, $\gamma_{\mathrm{m}}^{\infty}(P_3 \Box P_n) = \left\lceil \frac{6n+2}{7} \right\rceil$.

(b) [35, 69] *For* $11 < n \leq 22$, $\gamma_{\mathrm{m}}^{\infty}(P_3 \Box P_n) = \left\lceil \frac{4n+6}{5} \right\rceil$.

(c) [35] *For* $n > 22$, $\gamma_{\mathrm{m}}^{\infty}(P_3 \Box P_n) = \left\lceil \frac{4n+7}{5} \right\rceil$.

Beaton, Finbow and MacDonald [7] continued the study of m-eternal domination in grid graphs and obtained the following results.

**Theorem 5.15** [7] *For any $n \in \mathbb{Z}^+$,*

(a) $\gamma_{\mathrm{m}}^{\infty}(P_4 \Box P_n) = 2\left\lceil \frac{n+1}{2} \right\rceil$, *with the exceptions* $\gamma_{\mathrm{m}}^{\infty}(P_4 \Box P_2) = 3$ *and* $\gamma_{\mathrm{m}}^{\infty}(P_4 \Box P_6) = 7$;
(b) $\left\lfloor \frac{10(n+1)}{7} \right\rfloor \leq \gamma_{\mathrm{m}}^{\infty}(P_6 \Box P_n) \leq \left\lceil \frac{8n}{5} \right\rceil + 8$.
(c) *In addition,* $\gamma_{\mathrm{m}}^{\infty}(P_5 \Box P_5) = 7$, $\gamma_{\mathrm{m}}^{\infty}(P_6 \Box P_6) = 10$, *and* $13 \leq \gamma_{\mathrm{m}}^{\infty}(P_7 \Box P_7) \leq 14$.

Van Bommel and Van Bommel [83] considered $5 \times n$ grids, determining exact values for $n \leq 12$ and proving the following bounds.

**Theorem 5.16** [83] $\lfloor \frac{6n+9}{5} \rfloor \leq \gamma_{\mathrm{m}}^{\infty}(P_5 \Box P_n) \leq \lfloor \frac{4n+3}{3} \rfloor$.

**Problem 5.17** *Determine the value of* $\gamma_{\mathrm{m}}^{\infty}(P_m \Box P_n)$. *In particular, is* $\gamma_{\mathrm{m}}^{\infty}(P_n \Box P_n) \leq \gamma(P_n \Box P_n) + c$, *for some constant c?*

The latter statement was conjectured to be true by Finbow and Klostermeyer [59]. A significant result on this conjecture was proven by Lamprou, Martin and Schewe [62]:

**Theorem 5.18** [62] *For any* $m, n \geq 16$, $\gamma_{\mathrm{m}}^{\infty}(P_m \Box P_n) \leq \gamma(P_m \Box P_n) + \mathcal{O}(m+n)$.

The proof of the preceding theorem utilizes an interesting guard movement strategy, which involves partitioning a large grid into smaller grids and moving (really rotating) the guards in each smaller grid upon each attack in a way that defends the attack and preserves a dominating set of the entire grid. This strategy requires $\gamma(P_m \Box P_n)$ plus additional guards initially placed around the perimeter of the grid.

**Conjecture 5.19** [38] *If* $\gamma_{\mathrm{m}}^{\infty}(P_3 \Box P_n) \leq r$, *then* $\gamma_{\mathrm{m}}^{\infty}(P_3 \Box P_{n+1}) \leq r + 1$.

The *strong grid* $P_m \boxtimes P_n$ is obtained from $P_m \Box P_n$ by joining vertices $(v_i, u_j)$ and $(v_k, u_\ell)$ if and only if $v_i v_k \in E(P_m)$ and $u_j u_\ell \in E(P_n)$. McInerney, Nisse and Perennes [67] considered the m-eternal domination number of strong grids. They showed that when $m \geq n$, $\gamma_{\mathrm{m}}^{\infty}(P_m \boxtimes P_n) = \lceil \frac{m}{3} \rceil \lceil \frac{n}{3} \rceil + O(m\sqrt{n})$, noting that $\gamma(P_m \boxtimes P_n) = \lceil \frac{m}{3} \rceil \lceil \frac{n}{3} \rceil$.

## 5.3 Cubic Graphs

The study of domination in cubic graphs has long history, with the best known upper bound currently being that of Kostochka and Stocker [60] which improved the famous result of Reed [77]:

**Theorem 5.20** [77] *If G is a graph of order n with* $\delta(G) \geq 3$, *then* $\gamma(G) \leq \frac{3}{8}n$.

**Theorem 5.21** [60] *If G is a connected cubic graph of order* $n \geq 10$, *then* $\gamma(G) \leq \frac{5}{14}n$.

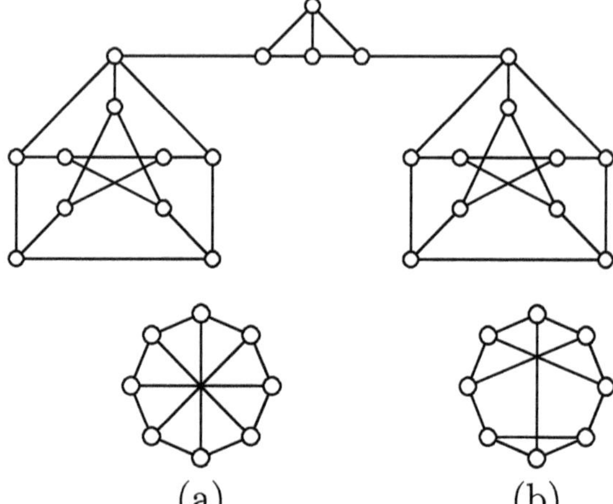


**Fig. 4** Graph with minimum degree three and $\gamma_{\mathrm{m}}^{\infty}(G) = 9$.

**Fig. 5** Two cubic graphs of order eight, each with $\gamma_{\mathrm{m}}^{\infty}(G) = 3$.

By Theorem 5.20, if $G$ is a cubic graph of order $n$, then $\gamma(G) \leq 3n/8$. However, there are cubic graphs $G$ with $\gamma_{\mathrm{m}}^{\infty}(G) > 3n/8$. For example, if $P$ is the Petersen graph (of order $n = 10$), then as first observed in [49], $\gamma_{\mathrm{m}}^{\infty}(P) = 4 = 2n/5$. It would be of great interest to find graphs, or families of graphs, with minimum degree three and $\gamma_{\mathrm{m}}^{\infty}(G) > 3n/8$. It does not appear to be easy to combine several Petersen graphs in some way to obtain such a graph with $\gamma_{\mathrm{m}}^{\infty}(G) > 3n/8$. We show an example in Figure 4 of one way of doing this to obtain a graph with $\gamma_{\mathrm{m}}^{\infty}(G) = 3n/8$. Two cubic graphs of order eight and $\gamma_{\mathrm{m}}^{\infty} = 3$ are exhibited in Figure 5. Except for the examples mentioned here we do not know any other cubic graphs with $\gamma_{\mathrm{m}}^{\infty}(G) \geq 3n/8$.

v Henning et al. [41] obtained an upper bound for cubic bipartite graphs.

**Theorem 5.22** [41] *If $G$ is a cubic bipartite graph of order $n$, then $\gamma_{\mathrm{m}}^{\infty}(G) \leq \frac{7}{16}n$.*

The complex proof of this theorem involves first removing a perfect matching $M$ from $G$ and considering the (even-length) cycles that remain. If all of these cycles are long enough, the theorem follows. If however, there are 4-cycles, then analyzing the edges from $M$ comes into play in an effort to reduce the number of guards needed, since $\gamma_{\mathrm{m}}^{\infty}(C_4) = 2$. It remains open to improve this result, as the next conjecture captures.

**Conjecture 5.23** [41] *If $G$ is a cubic bipartite graph of order $n$, then $\gamma_{\mathrm{m}}^{\infty}(G) \leq \frac{3}{8}n$.*

**Conjecture 5.24** [41] *If $G$ is a cubic graph of order $n$, then $\gamma_{\mathrm{m}}^{\infty}(G) \leq \frac{2}{5}n$.*

# 6 Secure Domination

The concept of secure domination was conceived by E. J. Cockayne and presented in a keynote lecture during a workshop at the University of South Africa in Pretoria, 2002. The papers [14, 15, 29, 30, 70] all resulted directly from this workshop but, due to the normal delays involved in the publication process, were not published in chronological order. As mentioned in the introduction, secure domination was first discussed by Cockayne et al. [30] and the other papers soon followed.

## 6.1 Terminology and Basic Results

A *secure dominating set* (*SDS*) of $G$ is a set $D \subseteq V$ with the property that for each $u \in V - D$, there exists $v \in D$ adjacent to $u$ such that $(D - \{v\}) \cup \{u\}$ is dominating. The minimum cardinality among all SDSs is the *secure domination number* $\gamma_s(G)$ of $G$. Naturally, $\gamma_s(G) \geq \gamma(G)$ for all graphs $G$. The obvious connection between secure domination and eternal domination can be seen in the pair of papers [14, 15].

We must point out an unfortunate instance of terminology confusion. The term *secure set* and the notation $\gamma_s(G)$ are also used for the problems discussed in, for example, [13, 44]. We shall not consider this different problem here; it is the secure domination version of the eviction problem mentioned in Section 3.

If $u \in V - D$ is defended by $v \in D$ and by no other vertex in $D$, then $v$ *uniquely defends* $u$; otherwise we also say that $v$ *jointly defends* $u$. If $v$ defends $u \in V - D$, it is easy to see that $\mathrm{pn}(v, D) \subseteq N[u]$. For $v \in D$ and $u \in \mathrm{epn}(v, D)$, if $v$ defends $u$, then $v$ uniquely defends $u$. This leads to the following result by Cockayne et al. [30]. Note the difference between Proposition 6.1 and Proposition 4.1 – the former gives a necessary and sufficient condition for $D$ to be an SDS, while, as mentioned in Section 4, the converse of the latter is false.

**Proposition 6.1** [30] *Let $D \subseteq V$. The vertex $v \in D$ defends $u \in V - D$ if and only if* $\mathrm{epn}(v, D) \cup \{v\} \subseteq N[u]$, *and $D$ is an SDS of $G$ if and only if for each $u \in V - D$, there exists $v \in D$ such that $G[\{u, v\} \cup \mathrm{epn}(v, D)]$ is complete.*

In contrast to the Fundamental Conjecture (Conjecture 4.18) for eternal domination, Burger and Van Vuuren showed that, for any graph $G$, any minimum secure dominating set $D$ of $G$, and any vertex $v \in D$ which has an unoccupied neighbor, $v$ is a defender.

**Theorem 6.2** [21] *Let $G$ be any graph.*

(a) *For every minimum secure dominating set $D$ of $G$, every vertex $v \in D$ such that $N(v) \not\subseteq D$ is a defender of at least one vertex in $V - D$.*
(b) *If $G$ has no isolated vertices, then there exists a minimum secure dominating set $D$ of $G$ in which every vertex is a defender of at least one vertex in $V - D$.*

Graphs with secure domination numbers 1, 2 or 3, respectively, were characterized in [21]. It is easily seen that $\gamma_s(G) = 1$ if and only if $G = K_n$. In order to characterize graphs with secure domination number 2, the following construction is required. Let $i, j$ be positive integers and let $k, \ell$ be nonnegative integers. Let $\Phi(i, j, k, \ell)$ denote the graph of order $i + j + k + \ell$ and size $\binom{i}{2} + \binom{j}{2} + k(i + 1) + \ell(j + 1)$ containing two vertex-disjoint cliques $K_i$ and $K_j$ of orders $i$ and $j$, respectively, together with two further disjoint independent sets $U_k$ and $W_\ell$ of vertices of cardinalities $k$ and $\ell$, respectively, to which the following edges are added: $(i)$ Each vertex in $K_i$ is joined to all vertices of $U_k$ (if $k > 0$). $(ii)$ Each vertex in $K_j$ is joined to all vertices of $W_\ell$ (if $\ell > 0$). $(iii)$ Some vertex $x \in V(K_i)$ is joined to all vertices in $W_\ell$, and some vertex $y \in V(K_j)$ is joined to all vertices in $U_k$. Note that $\Phi(1, 1, 0, 0) \cong \overline{K_2}$, $\Phi(1, 1, 1, 0) \cong K_{2,1}$, and, in general, $\Phi(1, 1, n - t, t) \cong K_{2,n}$ for $t \in \{0, ..., n\}$.

**Proposition 6.3** [21] *A graph has secure domination number 2 if and only if it is not complete and contains $\Phi(i, j, k, \ell)$ as spanning subgraph for some integers $i, j \geq 1$ and $k, \ell \geq 0$.*

A similar, but more complicated, construction that generalizes $\overline{K_3}$, $K_{3,1}$, and $K_{3,n}$ was used to characterize graphs with secure domination number 3.

## 6.2 Bounds on the Secure Domination Number

Many authors have obtained bounds, or improved bounds obtained by others, on the secure domination number for various graph classes. We first give bounds that hold for general graphs and then state better bounds that hold only for certain graph classes.

The *double domination number* $\gamma_{\times 2}(G)$ of a graph $G$ is the minimum cardinality of a set $S \subseteq V(G)$ such that every vertex of $V - S$ is adjacent to at least two vertices in $S$. Let $\alpha'$ denote the cardinality of a maximum matching of $G$.

### 6.2.1 Upper Bounds for General Graphs

It follows immediately from the definitions that $\gamma_s(G) \leq \gamma^{\infty}(G)$ for any graph $G$, hence also that $\gamma_s(G) \leq \theta(G)$. Since $\theta(G) \leq n - \omega(G) + 1$ for any $n$-vertex graph $G$, the bound $\gamma_s(G) \leq n - \omega(G) + 1$ also follows. We state a number of other upper bounds and give an outline of the proofs of the first two.

**Theorem 6.4** *Let $G$ be a connected graph of order $n$. Then*

(a) [29] $\gamma_s(G) \leq n - \alpha'$ *(the bound is tight, e.g., for $K_{1,n-1}$);*
(b) [58] $\gamma_s(G) \leq 2\alpha(G)$*;*
(c) [68] $\gamma_s(G) \leq \gamma(G) + \alpha(G) - 1$ *(the bound is trivially tight for $K_n$);*
(d) [26] $\gamma_s(G) \leq \gamma_{\times 2}(G)$ *(the bound is tight, e.g., for $C_5$);*

(e) [6] $\gamma_s(G) \leq n - \lfloor(4\,\mathrm{diam}(G) + 1)/7\rfloor$ *(the bound is tight, e.g., for $P_n$ and $K_{1,n-1}$).*

**Outline of proof.**

(a) Let $M = \{\{u_i v_i\} : 1 \leq i \leq \alpha'\}$ be a maximum matching of $G$. Observe that $D = V - \{u_1, \ldots u_{\alpha'}\}$ is a dominating set of $G$. Then for each $i \in \{1, \ldots, \alpha'\}$, $\mathrm{epn}(v_i, D) \subseteq \{u_i\}$ and $v_i$ can defend an attack at $u_i$.

(b) Let $A$ be a maximum independent set of $G$. For each $a \in A$, the subgraph induced by $\mathrm{pn}(a, A)$ is complete, otherwise $G$ has a larger independent set than $A$. If $\mathrm{epn}(a, A) \neq \emptyset$, choose arbitrary $x_a \in \mathrm{epn}(a, A)$. Define $D = A \cup \{x_a : a \in A \text{ and } \mathrm{epn}(a, A) \neq \emptyset\}$. Since $A$ is a dominating set, $\mathrm{epn}(v, D) = \emptyset$ for all $v \in D$, and $D$ is an SDS by Proposition 6.1. ■

A proof similar to that of Theorem 6.4(e) shows that $\gamma_s(G) \leq n - \lfloor 4c(G)/7 \rfloor$ for any graph $G$ of order $n$ and circumference $c(G) \geq 3$; this bound is trivially tight for cycles. Theorem 6.4(c) shows that the bound in part (b) is actually an inequality. Merouane and Chellali [68] improved the bound in (b) to $\frac{3}{2}\alpha(G)$ for triangle-free (and thus bipartite) graphs (see Theorem 6.7(a)), and ask whether $\gamma_s(G) \leq \frac{3}{2}\alpha(G)$ for every graph $G$. The same bound is obtained in [29] for claw-free graphs, as we state in the next subsection.

### 6.2.2 Bounds for Specific Graph Classes

The original paper [30] on secure domination also contains a lower bound on $\gamma_s(G)$ in terms of order and maximum degree in the case where $G$ is a triangle-free or $K_4$-free graph. Cockayne [27] showed that an improvement is possible for trees.

**Theorem 6.5** *Let $G$ be a graph of order $n$ and maximum degree $\Delta$.*

(a) [30] *If $G$ is triangle-free, then $\gamma_s(G) \geq n(2\Delta - 1)/(\Delta^2 + 2\Delta - 1)$.*

(b) [30] *If $G$ is $K_4$-free, then $\gamma_s(G) \geq n(2\Delta - 3)/(\Delta^2 + 2\Delta - 5)$. The bound is attained for infinitely many $n$.*

(c) [27] *If $G$ is a tree $T$ and $\Delta \geq 3$, then $\gamma_s(G) \geq (\Delta n + \Delta - 1)/(3\Delta - 1)$, and the bound is tight (e.g., for the generalized spider $S(3, \ldots, 3)$[1]).*

Part (a) of the next result follows from Theorem 6.4(a) and the fact that claw-free graphs of even order have perfect matchings and claw-free graphs of odd order have near-perfect matchings (see [66]).

**Theorem 6.6** [29] *Let $G$ be a connected claw-free graph of order $n$. Then*

(a) $\gamma_s(G) \leq \lceil \frac{n}{2} \rceil$;

(b) $\gamma_s(G) \leq \min\{2\gamma(G), 3\alpha(G)/2\}$.

*If, in addition, $G$ is $C_5$-free, then*

(c) $\gamma_s(G) \leq \alpha(G)$.

[1]For $\Delta \geq 3$, $S(3, \ldots, 3)$ is the tree obtained from $K_{1,\Delta}$ by subdividing each edge twice).

Thus, if $G$ is both claw-free and $C_5$-free, then $\gamma_s(G) \leq \alpha(G)$. A similar result holds if we simply require $G$ to have girth at least six (without adding the claw-free condition), as stated in Theorem 6.7(b); we include a proof. Theorem 6.7(c) gives a lower bound on the secure domination number of a tree in terms of its independence number.

**Theorem 6.7** [68]

(a) *For every triangle-free graph $G$, $\gamma_s(G) \leq \frac{3}{2}\alpha(G)$.*
(b) *For every graph $G$ with girth at least six, $\gamma_s(G) \leq \alpha(G)$.*
(c) *For every tree $T$, $\gamma_s(T) > \frac{\alpha(G}{2}$.*
(d) *For every tree $T$, $\gamma_s(T) \geq i(T)$ and this bound is tight (e.g., if $T$ is the corona of another tree).*

**Proof of (b).** We may assume that $G$ is connected. Let $D$ be a maximum independent set of $G$. Then $\text{pn}(v, D)$ induces a clique for each $v \in D$, otherwise $G$ has a larger independent set. Since the girth of $G$ is at least six, $\text{epn}(v, D)$ is independent and thus $|\text{epn}(v, D)| \leq 1$. We claim that $D$ is a secure dominating set of $G$. Suppose this is not the case. Let $u \in V - D$ be a vertex not defended by $D$ and define $D_u = N(u) \cap D$. If $u \in \text{epn}(y, D)$ for some $y \in D_u$, then $(D - \{y\}) \cup \{u\}$ is a dominating set of $G$, which is not the case. Hence we may assume that $|D_u| \geq 2$. If $\text{epn}(z, D) = \emptyset$ for some $z \in D_u$, then $(D - \{z\}) \cup \{u\}$ is a dominating set of $G$. Thus we may also assume that $|\text{epn}(v, D)| = 1$ for each $v \in D_u$. Since the girth of $G$ is at least six, $u$ is nonadjacent to the vertex in $\text{epn}(v, D)$ for every $v \in D_u$. Let $A = \bigcup_{v \in D_u} \text{epn}(v, D)$. Then $|A| = |D_u|$, and $A$ is independent, otherwise $G$ has a cycle of length five. Therefore $A \cup \{u\}$ is independent. But now $(D - D_u) \cup A \cup \{u\}$ is an independent set of $G$ larger than $D$, a contradiction. Hence $D$ is a secure dominating set of $G$, and it follows that $\gamma_s(G) \leq \alpha(G)$. ■

Trees for which equality holds in Theorem 6.7(d) were characterized in [64], while Mynhardt, Swart and Ungerer [70] gave a constructive characterization of trees with equal domination and secure domination numbers. The bound stated in Theorem 6.6(a) was improved in [22] for (not necessarily claw-free) graphs without endvertices.

**Theorem 6.8** [22] *If $G \not\cong C_5$ is a connected graph of order $n$ with $\delta(G) \geq 2$, then $\gamma_s(G) \leq n/2$ and this bound is tight.*

Equality in the bound given in Theorem 6.8 is satisfied by the class of connected graphs obtained from the disjoint union of 2-colored copies of $C_4$ and $C_6$ by joining some or all vertices of (say) color 1.

Arumugam, Ebadi, and Manrique [6] and Li, Shao, and Xu [63] also obtained bounds on $\gamma_s$ for trees; the lower bounds are identical while the upper bound given in the latter paper improves the bound given in the former. A characterization of trees with $\gamma_s(T) = \frac{n+2}{3}$ is given in [63].

**Theorem 6.9** *Let $T$ be a tree of order $n \geq 3$ with $\ell$ leaves and $t$ support vertices. Then*

(a) [6] $\left\lceil \frac{n+2}{3} \right\rceil \leq \gamma_s(T) \leq \left\lfloor \frac{n+k-1}{2} \right\rfloor$, *and the bounds are tight;*

(b) [63] $\left\lceil \frac{n+2}{3} \right\rceil \leq \gamma_s(T) \leq \left\lfloor \frac{2n+2\ell-t}{4} \right\rfloor$, *and the bounds are tight.*

Araki and Yumoto [2] considered secure domination in maximal outerplanar graphs. An inner face of a maximal outerplanar graph $G$ is an *internal triangle* if it is not adjacent to the outer face. A maximal outerplanar graph without internal triangles is called *stripped.*

**Theorem 6.10** [2] *Let $G$ be a maximal outerplanar graph of order $n \geq 3$. Then*

(a) $\gamma_s(G) \leq \left\lceil \frac{3n}{7} \right\rceil$ *and the bound is tight.*

(b) *If $G$ is stripped, then $\frac{n}{4} < \gamma_s(G) \leq \left\lceil \frac{n}{3} \right\rceil$ and the upper bound is tight.*

## 6.3 Secure Domination Critical and Stable Graphs

A graph $G$ is *secure domination critical with respect to edge removal*, abbreviated to *$\gamma_s$-ER-critical*, if $\gamma_s(G - e) > \gamma_s(G)$ for each edge $e$ of $G$ (in which case $\gamma_s(G - e) = \gamma_s(G) + 1$ for each edge $e$). Grobler and Mynhardt [39] gave constructive characterizations of $\gamma_s$-ER-critical graphs, bipartite $\gamma_s$-ER-critical graphs, and $\gamma_s$-ER-critical trees, respectively. Burger, De Villiers and Van Vuuren [17, 19] generalized $\gamma_s$-ER-critical graphs to $q$-ER-critical graphs: a graph $G$ is *$q$-ER-critical* if the smallest arbitrary subset of edges whose removal from $G$ necessarily increases the secure domination number has cardinality $q$.

In contrast, a graph $G$ is *$p$-ER-stable* if the largest arbitrary subset of edges whose removal from $G$ does not increase the secure domination number of the resulting graph has cardinality $p$. Burger, De Villiers and Van Vuuren [20] studied the problem of computing $p$-stable graphs for all admissible values of $p$ and determined the exact values of $p$ for which members of various infinite classes of graphs are $p$-stable. They also considered the problem of determining the largest value $\omega_n$ of $p$ for which a graph of order $n$ can be $p$-stable. In this regard, they formulated the following conjecture, which is true for all $n \in \{2, ..., 9\}$.

**Conjecture 6.11** [20] *For all $n \geq 2$, $\omega_n = n - 2$.*

It is true that $\omega_n \geq n - 2$ for all $n \geq 2$, because one needs to remove $n - 1$ edges from $K_{1,n-1}$ to increase $\gamma_s$. They also mention classes of graphs of order $n$ that are $n-2$-stable, namely $K_{1,n-1}$ for all $n$, $K_n - e$ if $n \equiv 0$ or $2 \pmod 3$, $K_{4,4}$ and $K_{3,3,3}$, and conjecture that this class is complete, i.e., consists of all $n - 2$-stable graphs.

## 6.4 Other Types of Secure Domination

By imposing additional conditions on secure dominating sets, one may obtain other types of secure domination. For example, if the subgraph induced by each secure dominating set has a perfect matching, we obtain several varieties of *secure paired domination* [45].

The set $S \subseteq V$ is a *restrained dominating set* if every vertex in $V - S$ is adjacent to a vertex in $S$ and to a vertex in $V - S$. By requiring a secure dominating set to be restrained, we create the concept of *restrained secure domination* [72, 73].

By considering the maximum cardinality of a minimal secure dominating set, we obtain the *upper secure domination number* [76]. A dominating set $S$ is a *perfect secure dominating set* if for each $v \in V - S$ there exists a **unique** $u \in S$ such that $v$ is adjacent to $u$ and $(S - \{u\}) \cup \{v\}$ is a dominating set [74]. If $D$ is a secure dominating set of $G$ and of $\overline{G}$, we get *global secure domination* [75]. We refer interested readers to the original publication for further information on these concepts, and only consider secure total domination in the next subsection.

### 6.4.1 Secure Total Domination

A *secure total dominating set* (*STDS*) of $G$ is a set $D \subseteq V$ with the properties that for each $u \in V - D$, there exists $v \in D$ adjacent to $u$ such that $D' = (D - \{v\}) \cup \{u\}$ is dominating, and such that the subgraphs of $G$ induced by $D$ and $D'$ have no isolated vertices. In this case we say that $v$ *totally defends* $u$. Evidently, $G$ has secure total dominating sets if and only if $G$ has no isolated vertices. The minimum cardinality among all STDSs is the *secure total domination number* $\gamma_{\text{st}}(G)$ of $G$. Secure total domination was introduced by Benecke, Cockayne and Mynhardt in [8], where an analogue of Proposition 6.1 appears. We rephrase their results to correct an error in the first part of the statement.

**Proposition 6.12** [8] *Let $D \subseteq V$ such that $G[D]$ has no isolated vertices. The vertex $v \in D$ totally defends $u \in V - D$ if and only if* $\text{ipn}(v, D) \cup \text{epn}(v, D) \cup \{v\} \subseteq N(u)$.

*The set $D$ is an STDS of $G$ if and only if* $\text{epn}(v, D) = \emptyset$ *for each $v \in D$, and for each $u \in V - D$ there exists $v \in D$ such that* $\text{ipn}(v, D) \cup \{v\} \subseteq N(u)$.

There exist graphs whose only STDS is the vertex set of the graph. These graphs were characterized by Benecke et al. [8]. Denote the set of leaves of a graph $G$ by $L(G)$, and the set of support vertices by $S(G)$.

**Proposition 6.13** [8] *For any graph $G$ of order $n$, $\gamma_{\text{st}}(G) = n$ if and only if $V - (L(G) \cup S(G))$ is independent.*

Secure total domination was also investigated by Klostermeyer and Mynhardt in [58], where the following bounds can be found. For $n \geq 1$, let $J_{2,n}$ be the graph obtained from $K_{2,n}$ by joining the two vertices of degree $n$ (or two nonadjacent vertices of $C_4$ if $n = 2$).

**Proposition 6.14** [58] *Let G be a graph without isolated vertices.*

(a) *If* $\delta(G) = 1$, *then* $\gamma_s(G) < \gamma_{st}(G)$.
(b) *If* $\delta(G) \geq 2$, $\gamma_s(G) \leq \gamma_{st}(G) \leq 2\gamma_s(G)$, *and both bounds are tight.*
(c) $\gamma_t(G) \leq \gamma_{st}(G)$. *If G is connected, then* $\gamma_{st}(G) = \gamma_t(G)$ *if only if* $\gamma_{st}(G) = 2$, *i.e., if and only if* $G = K_2$ *or* $J_{2,n}$ *is a spanning subgraph of G for some* $n \geq 1$.
(d) $\gamma_{st}(G) \leq 2\theta(G)$, *and the bound is tight (e.g., for coronas of paths).*
(e) $\gamma_{st}(G) < 3\alpha(G)$.

Duginov [32] improved the bound of Proposition 6.14(e).

**Theorem 6.15** [32] *For every graph G without isolated vertices,* $\gamma_{st}(G) \leq 2\alpha(G)$. *The bound is tight (trivially, for* $K_n$, $n \geq 2$*).*

Cabrera Martínez, Montejano and Rodríguez-Velázquez [23] in turn improved Duginov's bound as follows.

**Theorem 6.16** [23] *For every graph G without isolated vertices,* $\gamma_{st}(G) \leq \gamma(G) + \alpha(G)$. *The bound is tight, e.g., for graphs G obtained from an arbitrary graph H by joining each vertex of H to k new endvertices.*

In the light of Theorem 6.16, it is natural to ask whether the bound given in Proposition 6.14(b) can be improved to $\gamma_{st}(G) \leq \gamma(G) + \gamma_s(G)$. Cabrera Martínez et al. [23] showed that this is impossible in general and described classes of graphs for which it holds. Their paper appears in an open access journal and the classes and counterexamples are omitted here. They extended the second part of Theorem 6.14(c) to characterize graphs with $\gamma_{st}(G) = 3$. Let $\mathcal{G}$ be the family of graphs of order $n \geq 3$ obtained by joining a set $S$ of $n - 3$ vertices to the vertices of $P_3$ in such a way that $S$ is independent and each vertex in $S$ has degree 2.

**Theorem 6.17** [23] *A graph G satisfies* $\gamma_{st}(G) = 3$ *if and only if G has at most one universal vertex and contains a spanning subgraph* $H \in \mathcal{G}$.

Cabrera Martínez et al. [23] obtained a number of upper bounds for $\gamma_{st}$ that hold under different conditions. We mention three of them. Let $I_G$ denote the set of isolated vertices of the subgraph of $G$ induced by $V - (L(G) \cup S(G))$.

**Theorem 6.18** [23]

(a) *For any graph G of order n, diameter 2 and minimum degree* $\delta(G) \geq 3$, $\gamma_{st}(G) \leq \lfloor(5n - 2)/6\rfloor$. *The bound is tight, e.g., for the wheel* $W_5$ *of order 5 and for* $\overline{K_2} + P_3$.
(b) *For any graph G with minimum degree* $\delta(G) = 1$, $\gamma_{st}(G) \leq 2\alpha'(G) + |L(G)| - |S(G)| + |I_G|$. *An example of a graph for which his bound is tight is exhibited.*
(c) *For any graph G with minimum degree* $\delta(G) \geq 2$, $\gamma_{st}(G) \leq 2\alpha'(G) - \delta(G) + 2$. *The bound is tight for the graphs mentioned in (a).*

Finally, the lower bound $\gamma_{st}(T) \geq \gamma_s(T) \geq (\Delta n + \Delta - 1)/(3\Delta - 1)$ for a tree $T$ of order $n$ and maximum degree $\Delta$ of Theorem 6.5(b) combined with Proposition 6.14(b) was improved in [8]:

**Theorem 6.19** [8] *If $T$ is an $n$-vertex tree of maximum degree $\Delta \geq 3$, then $\gamma_{st}(T) \geq (4\Delta n + 4\Delta - 3n - 4)/(6\Delta - 5)$, and the bound is tight for all $\Delta \geq 3$.*

## *6.5 Complexity Results for Secure (Total) Domination*

### 6.5.1 Secure Domination

We conclude the section on secure domination with a summary of complexity results. Merouane and Chellali [68] showed that the decision version of the secure domination problem is NP-complete for bipartite graphs and split graphs (a subclass of chordal graphs). Burger, De Villiers and Van Vuuren [18] gave a linear algorithm for determining $\gamma_s(T)$ of an arbitrary tree. In [16] the same authors also presented two algorithms (a branch-and reduce algorithm and a branch-and-bound algorithm) for determining the secure domination number of a general graph $G$ of order $n$. The worst-case time complexities of both algorithms are $O(2^n)$.

Araki and Miyazaki [1] presented a linear-time algorithm for computing a minimum secure dominating set in a proper interval graph, while Pradhan and Jha [71] presented a linear-time algorithm to compute a minimum secure dominating set in block graphs, and showed that the secure domination problem is NP-complete for undirected path graphs[2] and chordal bipartite graphs.

Wang, Zhao and Deng [85] showed that the decision version of the secure domination problem is NP-complete for star convex bipartite graphs[3] and doubly chordal graphs[4]. They also proved that the secure domination problem (for general graphs and also for split graphs) cannot be approximated (by a polynomial time algorithm) within a factor of $(1 - \varepsilon) \ln |V|$ for any $\varepsilon > 0$ unless NP $\subseteq$ DTIME$(|V|^{O(\log\log|V|)})$, and that the secure domination problem is APX-complete for graphs with maximum degree 4. The secure domination problem in any graph with maximum degree $\Delta$ and minimum degree $\delta \geq 1$ can be approximated within an approximation ratio of $2(\ln(\Delta + 1) + 1)$.

[2] A graph is an *undirected path graph* if it is the intersection graph of a family of paths of a tree.

[3] A bipartite graph with partite sets $X$, $Y$ is *star convex bipartite* if there exists a vertex $x \in X$ (say) such that every vertex $y \in Y$ with $\deg(y) \geq 2$ is adjacent to $x$.

[4] A chordal graph is *doubly chordal* if its vertex set has a maximum neighborhood ordering (in addition to a perfect elimination ordering).

### 6.5.2 Secure Total Domination

Lad, Reddy and Kumar [61] showed that the decision problem of finding a minimum secure total dominating set is NP-complete for split graphs. In a paper published at about the same time as [61], Duginov [32] showed that the secure total dominating set problem is NP-complete for chordal bipartite graphs, planar bipartite graphs with girth at least $p$ (for any fixed $p \geq 4$) and maximum degree 3, split graphs, and graphs of separability[5] at most 2, and that the optimization version of this problem can be approximated in polynomial time within a factor of $c \ln |V|$ for some constant $c > 1$, but cannot be approximated in polynomial time within a factor of $c' \ln |V|$ for some constant $c' > 1$ (even for chordal graphs), unless P = NP.

# 7 Open Problems

We present a number of additional conjectures and open problems on some of the models discussed above.

## *7.1 Eternal Domination*

**Problem 7.1** *Study classes of graphs G such that (i)* $\gamma^{\infty}(G) = \alpha(G)$, *(ii)* $\gamma^{\infty}(G) = \theta(G)$.

As mentioned above, we know that $\gamma^{\infty}(G) = \theta(G)$ if $G$ is a series-parallel graph, so it makes sense to pose the following question.

**Problem 7.2** *Is it true that* $\gamma^{\infty}(G) = \theta(G)$ *if G is planar?*

A Vizing-like question was asked in [57].

**Problem 7.3** *Is it true for all graphs G and H that* $\gamma^{\infty}(G \Box H) \geq \gamma^{\infty}(G)\gamma^{\infty}(H)$?

It is easy to show that $\gamma^{\infty}(G \Box H) \geq \gamma(G)\gamma(H)$, because of the facts that $\gamma^{\infty}(G) \geq \alpha(G)$, $\gamma(G) \leq \alpha(G)$, and $\alpha(G \Box H) \geq \alpha(G)\alpha(H)$. It has also been shown that for all graphs $G$ and $H$ that $\gamma^{\infty}(G \boxtimes H) \geq \alpha(G)\gamma^{\infty}(H)$ [31], where $\boxtimes$ denotes the strong product. In that same paper, it was also shown that $\gamma^{\infty}(G \Box P_{2n+1}) > \gamma^{\infty}(G)\gamma^{\infty}(P_{2n+1})$ for all graphs $G$ without isolated vertices.

[5] A connected graph is a *graph of separability at most* 2 if it can be constructed from complete graphs and cycles by an iterative application of pasting along a vertex or an edge, where pasting along a vertex (respectively, an edge) is the graph operation that takes two vertex-disjoint graphs $G$ and $H$ with fixed vertices $u \in V(G)$, $v \in V(H)$ (fixed edges $\{u_1, u_2\} \in E(G)$ and $\{v_1, v_2\} \in E(H)$) and identifies the vertices $u$ and $v$ (respectively, identifies the vertex $u_1$ with $v_1$, and the vertex $u_2$ with $v_2$).

A special case of the last result is the following.

**Theorem 7.4** [31] *For any isolate-free graph $G$ of order $n$ such that $\gamma^{\infty}(G) \leq n/2$,*

$$\gamma^{\infty}(G\Box P_3) > \gamma^{\infty}(G)\gamma^{\infty}(P_3) = 2\gamma^{\infty}(G).$$

**Proof.** Arrange $G\Box P_3$ as three horizontal "layers" $G_1 \cong G_2 \cong G_3 \cong G$ (each vertex in $G_2$ is adjacent to its corresponding vertices of $G_1$ and $G_3$). Suppose $G\Box P_3$ can be defended by $2\gamma^{\infty}(G)$ guards. Attack vertices in $G\Box P_3$ so that all the guards are only on vertices of $G_1$ and $G_3$. If $\gamma^{\infty}(G) < n/2$, then $G_2$ contains an unprotected vertex, and if $\gamma^{\infty}(G) = n/2$ (and each vertex of $G_2$ is protected), then any attack of an unoccupied vertex in $G_1$ causes a vertex in $G_2$ to be unprotected. Both cases are impossible. ■

Interestingly, such a Vizing-like condition as described in Problem 7.3 was shown not to hold for all graphs $G$ in the all-guards-move model of eternal domination, see [57]. One can verify this by considering $P_3\Box P_3$.

The next question is most appropriately couched in the original model of eternal domination defined in [15], in which the complete attack sequence is constructed by the attacker in advance (rather than being constructed one attack at a time as the two players alternate turns, in which case the attacker can choose their next move based on the defender's previous move).

**Problem 7.5** *Find a function $f_G(n)$, whose growth rate is as small as possible, such that for any $n$-vertex graph $G$ and any given dominating set $D$, if $D$ can defend $G$ against all attack sequences of length at most $f(n)$, then $D$ is an eternal dominating set, i.e., $D$ can defend $G$ against all attack sequences.*

It seems that likely $f_G(n)$ can be bounded from above by an exponential function. That is, for example, after $2^n + 1$ attacks, some guard configuration must be repeated. Determining whether a better solution to this problem exists seems relevant to resolving the question of whether the decision problem "Is $D$ is an eternal dominating set for $G$?" lies in the complexity class PSPACE, as mentioned in [59]. This decision problem can be solved in exponential time, as one can construct and evaluate the auxiliary graph whose vertex set consists of all possible dominating sets of $G$ of cardinality $|D|$. This algorithm is described in, for example, [47], though it does not imply any solution to Problem 7.5. A related problem was considered in [4].

## 7.2 m-Eternal Domination

Recall the inequality chain $\gamma(G) \leq \gamma_{\mathrm{m}}^{\infty}(G) \leq \alpha(G) \leq \gamma^{\infty}(G) \leq \theta(G)$ from Section 5.

**Problem 7.6** *Describe classes of graphs having* $\gamma(G) = \gamma_{\mathrm{m}}^{\infty}(G)$, $\gamma^{\infty}(G) = \gamma_{\mathrm{m}}^{\infty}(G)$, $\gamma_{\mathrm{m}}^{\infty}(G) = \tau(G)$, *or* $\gamma_{\mathrm{m}}^{\infty}(G) = \alpha(G)$.

As shown by Braga et al. in [10], there exist connected Cayley graphs, necessarily of non-Abelian groups, whose m-eternal domination numbers exceed their domination numbers by one. This implies that there exist disconnected Cayley graphs $G$ such that $\gamma_{\mathrm{m}}^{\infty}(G) - \gamma(G)$ is an arbitrary positive integer. The picture for connected Cayley graphs is not so clear.

**Problem 7.7** *Does there exist a connected Cayley graph G such that* $\gamma_{\mathrm{m}}^{\infty}(G) > \gamma(G) + 1$?

**Problem 7.8** *Find conditions under which the bound* $\gamma_{\mathrm{m}}^{\infty}(G) \leq \lceil \frac{n}{2} \rceil$ *in Theorem 5.3 can be improved, and conditions under which equality holds.*

**Problem 7.9** *Find cubic graphs of order n such that* $\gamma_{\mathrm{m}}^{\infty}(G) \geq 3n/8$.

## 7.3 Secure Domination

**Problem 7.10** [68] *Is it true that* $\gamma_s(G) \leq \frac{3}{2}\alpha(G)$ *for every graph G?*

**Problem 7.11** [22] *Characterize the connected graphs G of order n with* $\delta(G) \geq 2$ *satisfying* $\gamma_s(G) = n/2$.

**Problem 7.12** *Can the lower bounds given in Theorem 6.5(a) and (b) be generalized to* $K_r$*-free graphs?*

It is shown in [30] that $\gamma_s(P_m \square P_n) \leq \frac{mn}{3} + 2$, leading to the following question.

**Problem 7.13** *Can a tighter bound be found for* $\gamma_s(P_m \square P_m)$?

**Problem 7.14** *Characterize the graphs G satisfying* $\gamma_{\mathrm{st}}(G) = 2\alpha(G)$.

## References

1. T. Araki, H. Miyazaki, Secure domination in proper interval graphs. *Discrete Appl. Math.* **247** (2018), 70–76.
2. T. Araki, I. Yumoto, On the secure domination numbers of maximal outerplanar graphs. *Discrete Appl. Math.* **236** (2018), 23–29.
3. M. Anderson, C. Barrientos, R. Brigham, J. Carrington, R. Vitray, J. Yellen, Maximum demand graphs for eternal security. *J. Combin. Math. Combin. Comput.* **61** (2007), 111–128.
4. M. Anderson, R. Brigham, J. Carrington, R. Dutton R. Vitray, J. Yellen, Mortal and eternal vertex covers. *J. Combin. Math. Combin. Comput.* **99** (2016), 3–21.
5. J. Arquilla, H. Fredricksen, "Graphing" an optimal grand strategy. *Military Operations Research* **1**(3) (1995), 3–17.
6. S. Arumugam, K. Ebadi, M. Manrique, Co-secure and secure domination in graphs. *Util. Math.* **94** (2014), 167–182.

7. I. Beaton, S. Finbow, J. A. MacDonald, Eternal domination of grids. *J. Combin. Math. Combin. Comput.* **85** (2013), 33–48.
8. S. Benecke, E. J. Cockayne, C. M. Mynhardt, Secure total domination in graphs. *Util. Math.* **74** (2007), 247–259.
9. B. Bollobas, E. J. Cockayne, Graph theoretic parameters concerning domination, independence, irredundance. *J. Graph Theory* **3** (1979), 241–250.
10. A. Braga, C. C. de Souza, O. Lee, A note on the paper "Eternal security in graphs" by Goddard, Hedetniemi, and Hedetniemi (2005). *J. Combin. Math. Combin. Comput.* **96** (2016), 13–22.
11. A. Braga, C. C. de Souza, O. Lee, The eternal dominating set problem for proper interval graphs, *Inform. Process. Lett.* **115** (2015), 582–587.
12. A. Braga, M. F. Reis, C. C. de Souza, O. Lee, Practical standpoint for m-eternal domination, manuscript (2019).
13. R. Brigham, R. Dutton, and S. T. Hedetniemi, Security in graphs. *Discrete Appl. Math.* **155** (2007), 1708–1714.
14. A. P. Burger, E. J. Cockayne, W. R. Gründlingh, C. M. Mynhardt, J. H. van Vuuren, W. Winterbach, Finite order domination in graphs. *J. Combin. Math. Combin. Comput.* **49** (2004), 159–175.
15. A. P. Burger, E. J. Cockayne, W. R. Gründlingh, C. M. Mynhardt, J. H. van Vuuren, W. Winterbach, Infinite order domination in graphs. *J. Combin. Math. Combin. Comput.* **50** (2004), 179–194.
16. A. P. Burger, A. P. de Villiers, J. H. van Vuuren, Two algorithms for secure graph domination. *J. Combin. Math. Combin. Comput.* **85** (2013), 321–339.
17. A. P. Burger, A. P. de Villiers, J. H. van Vuuren, The cost of edge failure with respect to secure graph domination. *Util. Math.* **95** (2014), 329–339.
18. A. P. Burger, A. P. de Villiers, J. H. van Vuuren, A linear algorithm for secure domination in trees. *Discrete Appl. Math.* **171** (2014), 15–27.
19. A. P. Burger, A. P. de Villiers, J. H. van Vuuren, Edge criticality in secure graph domination. *Discrete Optim.* **18** (2015), 111–122.
20. A. P. Burger, A. P. de Villiers, J. H. van Vuuren, Edge stability in secure graph domination. *Discrete Math. Theor. Comput. Sci.* **17** (2015), 103–122.
21. A. P. Burger, A. P. de Villiers, J. H. van Vuuren, On minimum secure dominating sets of graphs. *Quaestiones Math.* **39** (2016), 189–202.
22. A. P. Burger, M. A. Henning, J. H. van Vuuren, Vertex covers and secure domination in Graphs. *Quaestiones Math.* **31** (2008), 163–171.
23. A. Cabrera Martínez, L. P. Montejano, J. A. Rodríguez-Velázquez, On the secure total domination number of graphs. *Symmetry* **11** (2019), 12 pp.
24. E. Chambers, W. Kinnersly, N. Prince, Mobile eternal security in graphs. Manuscript (2006).
25. G. Chartrand, L. Lesniak, P. Zhang, *Graphs & Digraphs* (sixth edition), Chapman and Hall/CRC, Boca Raton, 2016.
26. M. Chellali, T. W. Haynes, S. T. Hedetniemi, Bounds on weak Roman and 2-rainbow domination numbers. *Discrete Appl. Math.* **178** (2014), 27–32.
27. E. J. Cockayne, Irredundance, secure domination and maximum degree in trees. *Discrete Math.* **307** (2007), 12–17.
28. E. J. Cockayne, P. A. Dreyer, S. M. Hedetniemi, S. T. Hedetniemi, Roman domination in graphs. *Discrete Math.* **278** (2004), 11–22.
29. E. J. Cockayne, O. Favaron and C.M. Mynhardt, Secure domination, weak Roman domination and forbidden subgraphs. *Bull. Inst. Combin. Appl.* **39** (2003), 87–100.
30. E. J. Cockayne, P. J. P. Grobler, W. R. Gründlingh, J. Munganga, J. H. van Vuuren, Protection of a graph. *Util. Math.* **67** (2005), 19–32.
31. K. Driscoll, W. Klostermeyer, E. Krop, P. Taylor, On eternal domination and Vizing-type inequalities, manuscript (2018).
32. O. Duginov, Secure total domination in graphs: bounds and complexity. *Discrete Appl. Math.* **222** (2017), 97–108.

33. S. Finbow, S. Gaspers, M.-E. Messinger, P. Ottoway, A note on the eternal dominating set problem, *Internat. J. Game Theory* **47** (2018), 543–555.
34. S. Finbow, M.-E. Messinger, M. van Bommel, Eternal domination in $3 \times n$ grids. *Australas. J. Combin.* **61** (2015), 156–174.
35. S. Finbow, M. van Bommel, The eternal domination number for $3 \times n$ grid graphs. *Australas. J. Combin.* **76** (2020), 1–23.
36. W. Goddard, S. M. Hedetniemi, S. T. Hedetniemi, Eternal security in graphs. *J. Combin. Math. Combin. Comput.* **52** (2005), 169–180.
37. J. Goldwasser, W. Klostermeyer, Tight bounds for eternal dominating sets in graphs. *Discrete Math.* **308** (2008), 2589–2593.
38. J. Goldwasser, W. Klostermeyer, C. M. Mynhardt, Eternal protection in grid graphs. *Utilitas Math.* **91** (2013), 47–64.
39. P. J. P. Grobler, C. M. Mynhardt, Secure domination critical graphs. *Discrete Math.* **309** (2009), 5820–5827.
40. M. A. Henning, S. T. Hedetniemi, Defending the Roman Empire – a new strategy. *Discrete Math.* **266** (2003), 239–251.
41. M. A. Henning, W. Klostermeyer, G. MacGillivray, Bounds for the m-eternal domination number of a graph. *Contribut. Discret. Math.* **12** (2017), 91–103.
42. T. W. Haynes, S. T. Hedetniemi, P. J. Slater, *Fundamentals of Domination in Graphs*. Marcel Dekker, New York, 1998.
43. J. Hooker, R. Garkinkel, C. Chen, Finite dominating sets for network location problems. *Operations Research* **39** (1991), 100–118.
44. P. Johnson, C. Jones, Secure-dominating sets in graphs. V.R. Kulli (Ed.), *Advances in Domination Theory II* (2013), 1–9.
45. J. Kang, C. M. Mynhardt, Secure paired domination in graphs. *AKCE Int. J. Graphs Comb.* **11** (2014), 177–197.
46. W. Klostermeyer, E. Krop, G. MacGillivray, On graphs with domination number equal to eternal domination number, manuscript, 2018.
47. W. Klostermeyer, M. Lawrence, G. MacGillivray, Dynamic dominating sets: the eviction model for eternal domination. *J. Combin. Math. Combin. Comput.* **97** (2016), 247–269.
48. W. Klostermeyer, G. MacGillivray, Eternal security in graphs of fixed independence number. *J. Combin. Math. Combin. Comput.* **63** (2007), 97–101.
49. W. Klostermeyer, G. MacGillivray, Eternally secure sets, independence sets, and cliques. *AKCE Int. J. Graphs Comb.* **2** (2005), 119–122.
50. W. Klostermeyer, G. MacGillivray, Eternal dominating sets in graphs. *J. Combin. Math. Combin. Comput.* **68** (2009), 97–111.
51. W. Klostermeyer, G. MacGillivray, Eternal domination in trees, to appear in *J. Combin. Math. Combin. Comput.* (2019).
52. W. Klostermeyer, G. MacGillivray, Eternal domination: criticality and reachability, *Discuss. Math. Graph Theory* **37** (2017), 63–77.
53. W. Klostermeyer, M.-E. Messinger, A. Ayello, An eternal domination problem in grids, *Theory and Applications of Graphs* **4** (2017).
54. W. Klostermeyer, C. M. Mynhardt, Eternal total domination in graphs. *Ars Combin.* **68** (2012), 473–492.
55. W. Klostermeyer, C. M. Mynhardt, Graphs with equal eternal vertex cover and eternal domination numbers. *Discrete Math.* **311** (2011), 1371–1379.
56. W. Klostermeyer, C. M. Mynhardt, Vertex covers and eternal dominating sets. *Discrete Appl. Math.* **160** (2012), 1183–1190.
57. W. Klostermeyer, C. M. Mynhardt, Domination, eternal domination, and clique covering. *Discuss. Math. Graph Theory* **35** (2015), 283–300.
58. W. Klostermeyer, C. M. Mynhardt, Secure domination and secure total domination in graphs. *Discuss. Math. Graph Theory* **28** (2008), 267–284.
59. W. Klostermeyer, C. M. Mynhardt, Protecting a graph with mobile guards, *Appl. Anal. Discr. Math.*, **10** (2016), 1–29.

60. A. V. Kostochka, C. Stocker, A new bound on the domination number of connected cubic graphs. *Sib. Elektron. Mat. Izv.* **6** (2009), 465–504.
61. D. Lad, P. V. S. Reddy, J. P. Kumar, Complexity issues of variants of secure domination in graphs. *Electron. Notes Discrete Math.* **63** (2017), 77–84.
62. I. Lamprou, R. Martin, S. Schewe, Eternally dominating large grids. *Theoret. Comput. Sci.*, **794** (2019), 27–46.
63. Z. Li, Z. Shao, J. Xu, On secure domination in trees. *Quaest. Math.* **40** (2017), 1–12.
64. Z. Li, J. Xu, A characterization of trees with equal independent domination and secure domination numbers. *Inform. Process. Lett.* **119** (2017), 14–18.
65. L. Lovász, On the Shannon capacity of a graph. *IEEE Trans. Inform. Theory* **25** (1979), 1–7.
66. L. Lovász, M. Plummer, *Matching Theory*, *North-Holland Mathematics Studies* (1986).
67. F. McInerney, N. Nisse, S. Perennes, Eternal domination in grids, manuscript (2018)
68. H. B. Merouane, M. Chellali, On secure domination in graphs, *Inform. Process. Lett.* **115** (2015), 786–790.
69. M. E. Messinger, A. Z. Delaney, Closing the gap: eternal domination on $3 \times n$ grids, *Contrib. Discrete Math.* **12** (2017), 47–61.
70. C. M. Mynhardt, H. C. Swart, E. Ungerer, Excellent trees and secure domination. *Util. Math.* **67** (2005), 255–267.
71. D. Pradhan, A. Jha, On computing a minimum secure dominating set in block graphs. *J. Comb. Optim.* **35** (2018), 613–631.
72. P. R. L. Pushpam, C. Suseendran, Secure restrained domination in graphs. *Math. Comput. Sci.* **9** (2015), 239–247.
73. P. R. L. Pushpam, C. Suseendran, Further results on secure restrained domination in graphs. *J. Discrete Math. Sci. Cryptogr.* **19** (2016), 277–291.
74. S. V. D. Rashmi, S. Arumugam, K. R. Bhutani, P. Gartland, Perfect secure domination in graphs. *Categ. Gen. Algebr. Struct. Appl.* **7** (2017), 125–140.
75. S. V. D. Rashmi, S. Arumugam, A. Somasundaram, Global secure domination in graphs. *Theoretical Computer Science and Discrete Mathematics*, 50–54, Lecture Notes in Comput. Sci., 10398, Springer, Cham, 2017.
76. S. V. D. Rashmi, A. Somasundaram, S. Arumugam, Upper secure domination number of a graph. *Electron. Notes Discrete Math.* **53** (2016), 297–306.
77. B. Reed, Paths, stars, and the number three. *Combin. Probab. Comput.* **5** (1996), 277–295.
78. F. Regan, *Dynamic variants of domination and independence in graphs*, graduate thesis, Rheinische Friedrich-Wilhelms University, Bonn, 2007.
79. C. S. ReVelle, Can you protect the Roman Empire? *Johns Hopkins Magazine* **50**(2), April 1997, 40.
80. C. S. ReVelle, K. E. Rosing, Defendens Imperium Romanum: A classical problem in military strategy. *Amer. Math. Monthly* **107**, Aug. – Sept. 2000, 585–594.
81. M. Rinemberg, F. Soulignac, The eternal dominating set problem for interval graphs, Manuscript (2018)
82. I. Stewart, Defend the Roman Empire! *Scientific American,* December 1999, 136–138.
83. C. M. van Bommel, M. F. van Bommel, Eternal domination numbers of $5 \times n$ grid graphs, *J. Combin. Math. Combin. Comput.* **97** (2016), 83–102.
84. V. Virgile, *Domination éternelle dans les graphes*, Master's Thesis, University of Montreal, 2018.
85. H. Wang, Y. Zhao, Y. Deng, The complexity of secure domination problem in graphs. *Discuss. Math. Graph Theory* **38** (2018), 385–396.
86. D. B. West, *Introduction to Graph Theory*, Prentice-Hall, 1996.

# Stratified Domination

**Gary Chartrand and Ping Zhang**

## 1 Introduction

A *vertex coloring* $c$ of a graph $G$ is an assignment of colors (elements of some set) to the vertices of $G$, one color to each vertex. That is, $c : V(G) \to S$ is a function, where $S$ is a set of colors. If the number of colors is large, then $S$ is typically chosen to be the set $[k] = \{1, 2, \ldots, k\}$ for some positive integer $k$. On the other hand, if the number of colors is relatively small, then $S$ is often chosen to be a set of actual colors, such as red, blue, green, etc. A vertex coloring of a graph $G$ then results in a partition $\{V_1, V_2, \ldots, V_k\}$ of the vertex set of $G$ into $k$ color classes, where for each $i \in [k]$, the set $V_i$ consists of the vertices of $G$ colored $i$. The best known and most studied vertex colorings are *proper colorings* where adjacent vertices are assigned different colors. In this case, each color class is an independent set of vertices, that is, the subgraph $G[V_i]$ induced by $V_i$ has no edges. The study of vertex colorings of graphs originated with the study of proper vertex colorings of planar graphs, which resulted from attempts to solve the famous Four Color Problem. Over the years, many other vertex colorings of graphs $G$ have been introduced where the resulting partition $\{V_1, V_2, \ldots, V_k\}$ of the vertex set of $G$ required each induced subgraph $G[V_i]$ to satisfy some prescribed property.

In 1992 Naveed Sherwani proposed studying vertex colorings of graphs $G$ resulting in a partition $\{V_1, V_2, \ldots, V_k\}$ of the vertex set of $G$ into $k \geq 2$ subsets in a way that did not specifically depend on a property of each induced subgraph $G[V_i]$, $i \in [k]$, but on some other requirement. A graph with such a $k$-coloring (or such a partition of its vertex set into $k$ subsets) was referred to as a *$k$-stratified graph*. This topic was initially studied in the doctoral dissertation of Rashidi [15].

G. Chartrand (✉) · P. Zhang
Department of Mathematics, Western Michigan University, Kalamazoo, MI 49008-5248, USA
e-mail: gary.chartrand@wmich.edu; ping.zhang@wmich.edu



T. W. Haynes et al. (eds.), *Topics in Domination in Graphs*, Developments in Mathematics 64, https://doi.org/10.1007/978-3-030-51117-3_14

Most of the interest in this subject has been centered around the case $k = 2$, that is, with 2-stratified graphs $F$, where then there is a partition $\{V_1, V_2\}$ of $V(F)$ into two subsets. In this case it is common to refer to the two colors as red and blue, where $V_1$ is the set of red vertices and $V_2$ is the set of blue vertices. Since $\{V_1, V_2\}$ is a partition of $V(F)$, there is always at least one red vertex and at least one blue vertex and so $F$ has order at least 2. When drawing a 2-stratified graph, the red vertices are typically represented by solid vertices and blue vertices by open vertices. Therefore, the graph $F$ of Figure 1 represents a 2-stratified graph of order 6 with four red vertices $u$, $w$, $x$, and $y$ and two blue vertices $v$ and $z$.

In 2003, Chartrand, Haynes, Henning, and Zhang [4] observed that stratified graphs could be used to look at domination from another point of view. In order to describe this, a *red-blue coloring* of a graph $G$ here refers to an assignment of the color red or blue to each vertex of $G$, where all vertices of $G$ may be assigned the same color. With each connected 2-stratified graph $F$, there are certain red-blue colorings of a graph $G$ that will be of special interest to us.

## 2 Domination Defined by Stratification

Let $F$ be a connected 2-stratified graph. Therefore, $F$ has at least one red vertex and at least one blue vertex. One blue vertex of $F$ is designated as the *root* of $F$ and is labeled $v$. Thus, $F$ is a 2-stratified graph *rooted at* a blue vertex $v$. Now let $G$ be a graph. By an *$F$-coloring* of $G$ is meant a red-blue coloring of $G$ having the property that every blue vertex $v$ of $G$ belongs to a copy of $F$ rooted at $v$. The *$F$-domination number* $\gamma_F(G)$ of $G$ is the minimum number of red vertices among all $F$-colorings of $G$. Since a red-blue coloring of $G$ in which every vertex of $G$ is colored red is vacuously an $F$-coloring of $G$, the $F$-domination number of $G$ is defined for every 2-stratified graph $F$, for every choice of a root of $F$, and for every graph $G$. The set of red vertices in an $F$-coloring of a graph is called an *$F$-dominating set*. An $F$-coloring of $G$ that results in $\gamma_F(G)$ red vertices is called a *$\gamma_F$-coloring*. Should it occur for a given 2-stratified graph $F$ that a graph $G$ of order $n$ contains a vertex belonging to no copy of $F$, then this vertex must be colored red in every $F$-coloring of $G$. In fact, if $G$ contains no subgraph isomorphic to $F$, then the only $F$-coloring of $G$ is the one in which every vertex of $G$ is colored red and so $\gamma_F(G) = n$.

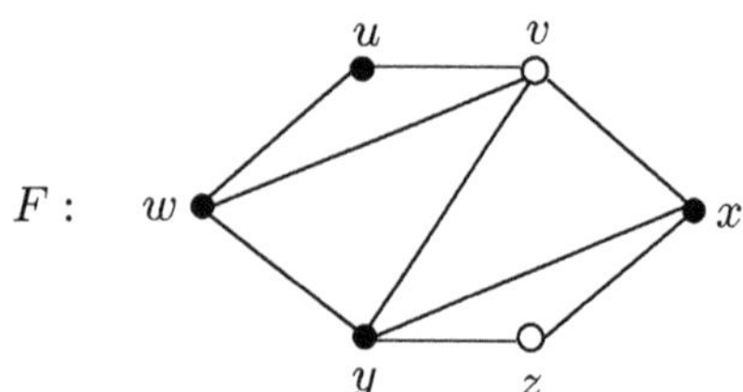


**Fig. 1** A 2-stratified graph

Since a 2-stratified graph must contain at least one red vertex and at least one blue vertex, there is only one connected 2-stratified graph of order 2, namely $F = K_2$, where one vertex of $F$ is colored red and the other vertex, necessarily the root, is blue. This 2-stratified graph $F$ is shown in Figure 2, along with a graph $G$ and two $F$-colorings of $G$.

In one $F$-coloring of the graph $G$ of Figure 2, there are three red vertices and in the other there are two red vertices. Therefore, $\gamma_F(G) \leq 2$. For every vertex $x$ of $G$, there are vertices of $G$ whose distance from $x$ is 2 or 3 and so if $x$ is the only vertex of $G$ colored red in a red-blue coloring of $G$, this coloring is not an $F$-coloring of $G$. That is, $\gamma_F(G) \neq 1$ and so $\gamma_F(G) = 2$.

One might notice for the 2-stratified graph $F = K_2$ of Figure 2 that an $F$-dominating set in an $F$-coloring of a graph $G$ is also a dominating set and so $\gamma(G) \leq \gamma_F(G)$. On the other hand, if we were to color the vertices in a minimum dominating set of $G$ red and all remaining vertices of $G$ blue, then the resulting red-blue coloring of $G$ has the property that every blue vertex of $G$ is adjacent to a red vertex of $G$; that is, this is an $F$-coloring of $G$ and so $\gamma_F(G) \leq \gamma(G)$. This results in the following observation.

**Observation 2.1 ([4])** *If $F$ is the 2-stratified graph $K_2$, then $\gamma_F(G) = \gamma(G)$ for every graph $G$.*

What we have seen illustrates the fact that with each connected 2-stratified graph $F$ rooted at some blue vertex, there is associated a certain type of domination and a corresponding domination parameter. While there is only one choice of domination and, thus, only one domination parameter when $F = K_2$, there are five choices when $F = P_3$, the path of order 3. These are all shown in Figure 3, where the five different 2-stratified graphs rooted at a blue vertex are denoted by $F_i$ for $1 \leq i \leq 5$.

The values of the five domination parameters $\gamma_{F_i}$ $(1 \leq i \leq 5)$ are shown in Figure 4 for the graph $G = P_4 \,\square\, K_2$.

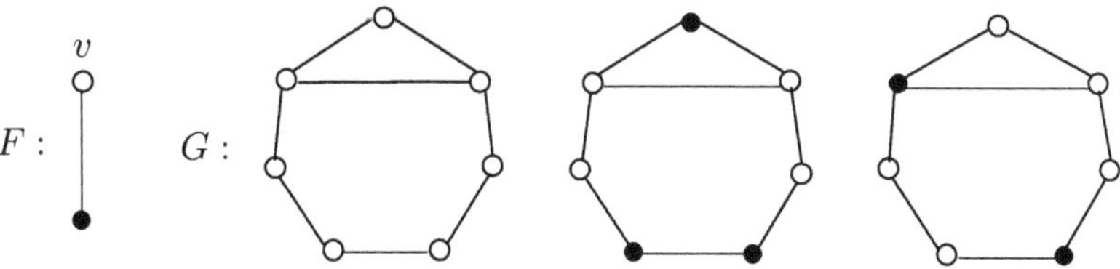


**Fig. 2** Two $F$-colorings of a graph

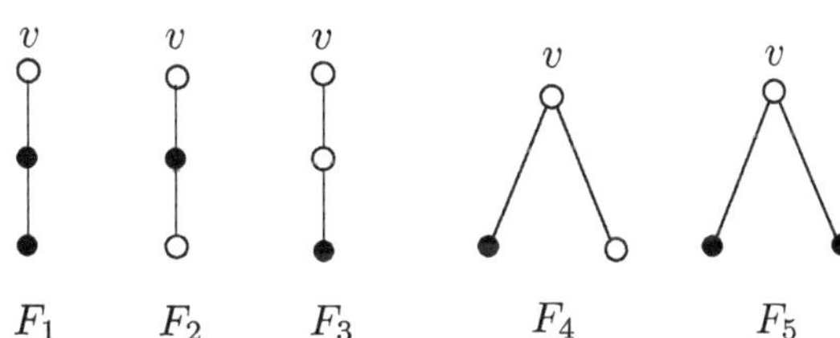


**Fig. 3** The five 2-stratified graphs obtained from $P_3$

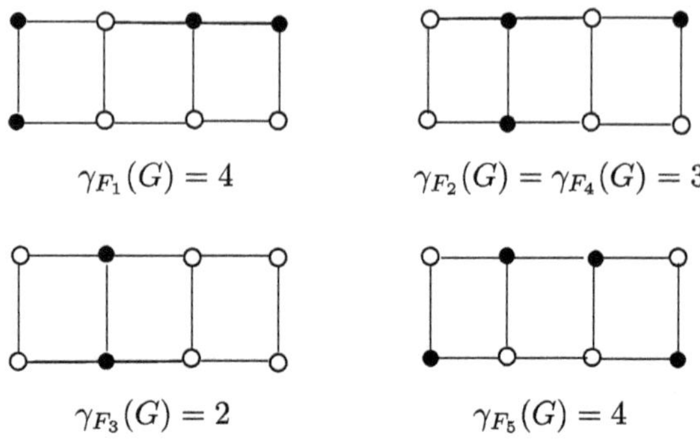


**Fig. 4** Domination parameters $\gamma_{F_i}$ $(1 \le i \le 5)$ for the graph $G = P_4 \,\square\, K_2$

For the 2-stratified graph $F_1$ (rooted at its only blue vertex), the parameter $\gamma_{F_1}$ is a familiar one. Recall that a set $S$ of vertices in a graph $G$ containing no isolated vertices is a *total dominating set* for $G$ if every vertex of $G$ is adjacent to some vertex of $S$. The minimum cardinality of a total dominating set for $G$ is the *total domination number* $\gamma_t(G)$ of $G$. A total dominating set of cardinality $\gamma_t(G)$ is called a $\gamma_t$*-set* for $G$. The domination number obtained from the 2-stratified graph $F_1$ is, in fact, the total domination number.

**Theorem 2.2 ([4])** *If $G$ is a graph without isolated vertices, then*

$$\gamma_{F_1}(G) = \gamma_t(G).$$

**Proof.** Let $S$ be a $\gamma_t$-set for $G$. By coloring each vertex of $S$ red and each vertex of $V(G) - S$ blue, an $F_1$-coloring of $G$ results. Thus, $\gamma_{F_1}(G) \le \gamma_t(G)$. We now show that $\gamma_t(G) \le \gamma_F(G)$ as well.

Among all $\gamma_{F_1}$-colorings of $G$, let $c$ be one for which the number of isolated vertices in the subgraph induced by the red vertices is minimum. Since each blue vertex in $G$ is adjacent to a red vertex, a minimum $F_1$-dominating set $S$ of red vertices is also a dominating set $S$ in $G$. Not only is every blue vertex adjacent to a red vertex but also every red vertex is adjacent to another red vertex, as we show next.

Assume, to the contrary, that there is a red vertex $u$ of $G$ adjacent only to blue vertices. That is, $u$ is an isolated vertex in the subgraph of $G$ induced by $S$. Let $v$ be a neighbor of $u$ in $G$. Since $v$ is a blue vertex and the red-blue coloring is a $\gamma_{F_1}$-coloring of $G$, it follows that $v$ belongs to a copy of $F_1$ rooted at $v$. So, $v$ must be adjacent to another red vertex $w$, which itself is adjacent to another red vertex, which in turn implies that $u \neq w$. If we were to interchange the colors of $u$ and $v$, a new $\gamma_{F_1}$-coloring $c'$ of $G$ is produced. However, the $\gamma_{F_1}$-coloring $c'$ has fewer isolated vertices in the subgraph induced by its red vertices, which contradicts the defining property of $c$. Therefore, every red vertex is adjacent to another red vertex, which implies that $S$ is a total dominating set of $G$ and so $\gamma_t(G) \le \gamma_{F_1}(G)$. Consequently, $\gamma_{F_1}(G) = \gamma_t(G)$. ∎

While the total domination number is only defined for graphs without isolated vertices, the $F_1$-domination number is defined for all graphs and, in this sense, is more general. Therefore, if $G$ is a graph without isolated vertices, then $\gamma_{F_1}(G) = \gamma_t(G)$; while if $G$ is a nonempty graph with $k \geq 1$ isolated vertices and $G = H + kK_1$, then $\gamma_{F_1}(G) = \gamma_t(H) + k$.

As is illustrated for the graph $G = P_4 \square K_2$ in Figure 4, every $F_i$-dominating set of $G$ is a dominating set of $G$ except when $i = 3$. For this graph $G$, the domination number $\gamma(G) = 3$, which is the same value as $\gamma_{F_2}(G)$ and $\gamma_{F_4}(G)$. That $\gamma_{F_2}(G) = \gamma(G)$ for $G = P_4 \square K_2$ is not a coincidence, as we show next.

**Theorem 2.3 ([4])** *If $G$ is a connected graph of order 3 or more, then*

$$\gamma_{F_2}(G) = \gamma(G).$$

**Proof.** Since the red vertices in any $F_2$-coloring of $G$ form a dominating set of $G$, it follows that $\gamma(G) \leq \gamma_{F_2}(G)$. Consequently, we need only show that $\gamma_{F_2}(G) \leq \gamma(G)$. Among all $\gamma$-sets of $G$, let $S$ be one so that the corresponding red-blue coloring has the maximum number of blue vertices $v$ that belong to a copy of $F$ rooted at $v$. We show that this red-blue coloring is, in fact, an $F_2$-coloring of $G$. Suppose that this is not the case. Then there is a blue vertex $v$ that does not belong to a copy of $F_2$ rooted at $v$.

Since $S$ is a dominating set of $G$, the blue vertex $v$ is adjacent to a (red) vertex $w \in S$. Because $v$ does not belong to a copy of $F_2$ rooted at $v$, it follows that $w$ is adjacent to no blue vertex other than $v$. Should it occur that $v$ is adjacent to a blue vertex $u$, then interchanging the colors of $v$ and $w$ produces a $\gamma$-set whose associated red-blue coloring contains more blue vertices $v'$ that belong to a copy of $F_2$ rooted at $v'$ than does the associated coloring of $S$, which is impossible. Hence, $v$ is not adjacent to any blue vertex in $G$. Suppose that $v$ is adjacent to a red vertex $x$ different from $w$. Necessarily, $x$ is not adjacent to any blue vertex different from $v$. However, in this case, $(S - \{w, x\}) \cup \{v\}$ is a dominating set of $G$ with $\gamma(G) - 1$ vertices, which is impossible. Therefore, $v$ is an end-vertex of $G$.

Since $G$ is a connected graph of order at least 3, it follows that $w$ must be adjacent to a vertex $y$ different from $v$ and, furthermore, $y$ must be a red vertex. If $y$ were an end-vertex of $G$, then $S - \{y\}$ would be a dominating set of $G$, which is impossible. If $y$ were adjacent only to red vertices, then here too $S - \{y\}$ would be a dominating set of $G$, again impossible. Therefore, $y$ must be adjacent to a blue vertex $z$. By interchanging the colors of $v$ and $w$, a $\gamma$-set is produced whose associated red-blue coloring contains more blue vertices $v'$ that belong to a copy of $F_2$ rooted at $v'$ than does the associated coloring of $S$, which cannot occur.

It therefore follows that every blue vertex $v$ of $G$ belongs to a copy of $F_2$ rooted at $v$, which implies that the red-blue coloring associated with $S$ is an $F_2$-coloring of $G$ and so $\gamma_{F_2}(G) \leq \gamma(G)$. Therefore, $\gamma_{F_2}(G) = \gamma(G)$. ■

We will omit discussing the 2-stratified graph $F_3$ for the present and turn to the 2-stratified graph $F_4$. First, we recall another well-known type of domination. A set $S$ of vertices in a graph $G$ is called a *restrained dominating set* if every vertex of $G$ not in $S$ is adjacent to both a vertex in $S$ and a vertex in $V(G) - S$. Since $S = V(G)$ is vacuously a restrained dominating set of $G$, every graph has a restrained dominating set. The *restrained domination number* $\gamma_r(G)$ of a graph $G$ is the minimum cardinality of a restrained dominating set for $G$. In an $F_4$-coloring of a graph, every blue vertex is adjacent to both a red vertex and a blue vertex. This leads us to the following observation.

**Observation 2.4** ([4]) *For every graph $G$, $\gamma_{F_4}(G) = \gamma_r(G)$.*

For a positive integer $k$, a set $S$ of vertices in a graph $G$ is a *$k$-dominating set* if every vertex not in $S$ is adjacent to at least $k$ vertices in $S$. The *$k$-domination number* $\gamma_k(G)$ is the minimum cardinality of a $k$-dominating set for $G$. For $k = 1$, we see that $\gamma_1(G) = \gamma(G)$; while for $k = 2$, we have the following observation.

**Observation 2.5** ([4]) *For every graph $G$, $\gamma_{F_5}(G) = \gamma_2(G)$.*

# 3 The Stratified Domination Number $\gamma_{F_3}$

The table below summarizes the four stratified domination parameters described in the preceding section and the well-known domination parameters that correspond to these.

| $i$ | 1 | 2 | 4 | 5 |
|---|---|---|---|---|
| $\gamma_{F_i}$ | $\gamma_t$ | $\gamma$ | $\gamma_r$ | $\gamma_2$ |

While the stratified domination number $\gamma_{F_3}$ does not correspond to any well-known domination parameter, this parameter has become the object of study in several papers. For example, Henning and Maritz [11] obtained the following three results, the first of which gives the value of this parameter for paths.

**Theorem 3.1 ([11])** *For each positive integer $n$,*

$$\gamma_{F_3}(P_n) = \left\lfloor \frac{n+7}{3} \right\rfloor + \left\lfloor \frac{n}{3} \right\rfloor - \left\lceil \frac{n}{3} \right\rceil.$$

**Proof.** We proceed by the Strong Form of Induction on the order $n$ of $P_n$. It is straightforward to see that the formula for $\gamma_{F_3}(P_n)$ holds for $n \in [5] = \{1, 2, 3, 4, 5\}$. This is the base of the induction. Assume for an integer $n \geq 6$ that $\gamma_{F_3}(P_i) = \left\lfloor \frac{i+7}{3} \right\rfloor + \left\lfloor \frac{i}{3} \right\rfloor - \left\lceil \frac{i}{3} \right\rceil$ for every integer $i$ with $i \in [n-1]$. We show that $\gamma_{F_3}(P_n) = \left\lfloor \frac{n+7}{3} \right\rfloor + \left\lfloor \frac{n}{3} \right\rfloor - \left\lceil \frac{n}{3} \right\rceil$. Let $P = P_n = (v_1, v_2, \ldots, v_n)$.

First, we claim that there exists a $\gamma_{F_3}$-coloring of $P$ in which $v_1$ and $v_4$ are colored red and $v_2$ and $v_3$ are colored blue. Let there be given a $\gamma_{F_3}$-coloring of $P$. Suppose, in this coloring, that $v_1$ is colored blue. Then $v_2$ must be blue as well and $v_3$ must be red. However, in this case, there is no copy of $F_3$ rooted at $v_2$, which is impossible. So, $v_1$ must be colored red. Consequently, $v_n$ must also be colored red. Therefore, in any $\gamma_{F_3}$-coloring of $P$, the two end-vertices of $P$ are red.

If $v_2$ is colored blue, then $v_3$ is blue and $v_4$ is red, which verifies the claim. On the other hand, suppose that $v_2$ is colored red in the given $\gamma_{F_3}$-coloring of $P$. If $v_3$ is blue, then $v_4$ is blue and $v_5$ is red. By interchanging the colors of $v_2$ and $v_4$, a new $\gamma_{F_3}$-coloring of $P$ is obtained in which $v_1$ and $v_4$ are red and $v_2$ and $v_3$ are colored blue, as desired. Suppose, though, that $v_3$ is red. Then $v_4$ must be blue, for otherwise $v_4$ is red and the vertices $v_2$ and $v_3$ could be recolored blue to produce an $F_3$-coloring of $P$ having two fewer red vertices, which is impossible. Thus, as claimed, $v_4$ is blue, which implies that $v_5$ is blue and $v_6$ is red. If we now recolor $v_2$ and $v_3$ blue and recolor $v_4$ and $v_5$ red, a new $\gamma_{F_3}$-coloring of $P$ is produced in which $v_1$ and $v_4$ are red and $v_2$ and $v_3$ are blue. This then verifies our claim.

Let there be given a $\gamma_{F_3}$-coloring of $P$ in which $v_1$ and $v_4$ are red and $v_2$ and $v_3$ are blue. Let $P' = (v_4, v_5, \ldots, v_n)$ be the subpath of $P$ of order $n-3$, where the colors of the vertices in $P'$ are those in $P$. This coloring of $P'$ is an $F_3$-coloring of $P'$ with $\gamma_{F_3}(P) - 1$ red vertices. Hence, $\gamma_{F_3}(P') \le \gamma_{F_3}(P) - 1$. On the other hand, any $\gamma_{F_3}$-coloring of $P'$ colors its end-vertices $v_4$ and $v_n$ red and can therefore be extended to an $F_3$-coloring of $P$ by coloring $v_1$ red and $v_2$ and $v_3$ blue. Therefore, $\gamma_{F_3}(P) \le \gamma_{F_3}(P') + 1$ and so $\gamma_{F_3}(P) = \gamma_{F_3}(P') + 1$. By the induction hypothesis,

$$\gamma_{F_3}(P') = \left\lfloor \frac{(n-3)+7}{3} \right\rfloor + \left\lfloor \frac{n-3}{3} \right\rfloor - \left\lceil \frac{n-3}{3} \right\rceil.$$

Thus,

$$\begin{aligned}\gamma_{F_3}(P_n) &= \left(\left\lfloor \frac{n+4}{3} \right\rfloor + 1\right) + \left(\left\lfloor \frac{n-3}{3} \right\rfloor + 1\right) - \left(\left\lceil \frac{n-3}{3} \right\rceil + 1\right) \\ &= \left\lfloor \frac{n+7}{3} \right\rfloor + \left\lfloor \frac{n}{3} \right\rfloor - \left\lceil \frac{n}{3} \right\rceil.\end{aligned}$$

Therefore, the formula for $\gamma_{F_3}(P_n)$ holds for each positive integer $n$. ■

Let $H_1 = P_6$ and for $k \ge 2$, let $H_k$ be the tree obtained from the disjoint union of the star $K_{1,k+1}$ and a subdivided star $S(K_{1,k})$ by joining a leaf of the star to the central vertex of the subdivided star. Figure 5 shows the trees $H_1$, $H_2$, and $H_3$. Let $\mathcal{H} = \{H_k : k \ge 1\}$.

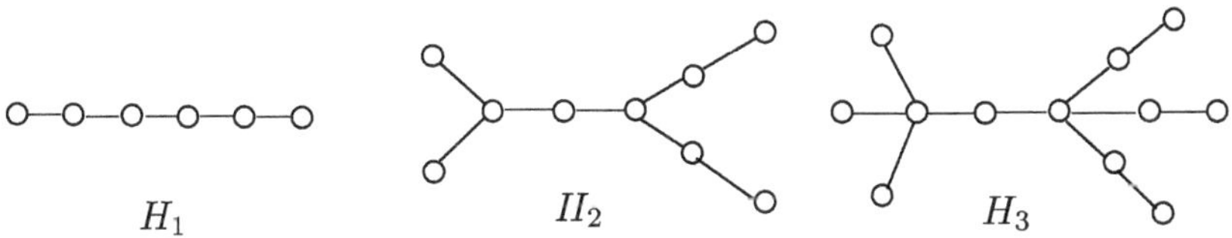


**Fig. 5** The trees $H_1$, $H_2$, and $H_3$

There is a sharp upper bound for the $F_3$-domination number of a tree that is not a star in terms of the order of the tree.

**Theorem 3.2 ([11])** *If $T$ is a tree of order $n$ with* $\text{diam}(T) \geq 3$, *then* $\gamma_{F_3}(T) \leq 2n/3$ *with equality if and only if* $T \in \mathcal{H}$.

For the tree $H_3$ of order 12 in $\mathcal{H}$, it follows by Theorem 3.2 that $\gamma_{F_3}(H_3) = 8 = 2n/3$. A $\gamma_{F_3}$-coloring of $H_3$ is shown in Figure 6.

When the diameter of a tree $T$ is at least 6, the upper bound for $\gamma_{F_3}(T)$ given in Theorem 3.2 is not sharp, but yet cannot be improved asymptotically.

**Corollary 3.3 ([11])** *If $T$ is a tree of order $n$ with* $\text{diam}(T) \geq 6$, *then* $\gamma_{F_3}(T) < 2n/3$ *and this bound is asymptotically best possible.*

**Proof.** By Theorem 3.2, $\gamma_{F_3}(T) < 2n/3$. It remains therefore to show that the upper bound $2n/3$ is asymptotically best possible. Let $\ell \geq 3$ be a fixed integer and let $k$ be a positive integer. Let $T$ be the tree obtained from $H_k$ by attaching a path of length $\ell$ at the central vertex $w$ of the subdivided star in $H_k$. (The tree $T$ is shown in Figure 7 for the case where $\ell = k = 3$.) Let $P$ be the resulting path of order $\ell + 1$ emanating from $w$. Then $\text{diam}(T) = \ell + 3 \geq 6$ and $n = |V(T)| = 3k + \ell + 3$.

If there is an $F_3$-coloring of $T$ in which $w$ is red, then all vertices of the subdivided star must be red as well. Furthermore, at least one end-vertex of $T$ in the star must be red and at least $\ell/3$ vertices of $P$ must be red (in addition to $w$). If there is an $F_3$-coloring of $T$ in which $w$ is blue, then all end-vertices of $T$ in both the star and the subdivided star must be red, as well as the center of the star must be red. In addition, at least $(\ell + 2)/3$ additional vertices (including at least one neighbor of $w$) are red. Consequently, there are at least $2k + (\ell + 5)/3$ red vertices in $T$. Therefore,

$$\lim_{k\to\infty} \frac{\gamma_{F_3}(G)}{n} = \lim_{k\to\infty} \frac{2k + (\ell + 5)/3}{3k + \ell + 3} = \lim_{k\to\infty} \frac{6k + \ell + 5}{9k + 3\ell + 9} = \frac{2}{3}$$

and so $\lim_{k\to\infty} \gamma_{F_3}(G) = \frac{2n}{3}$. ■

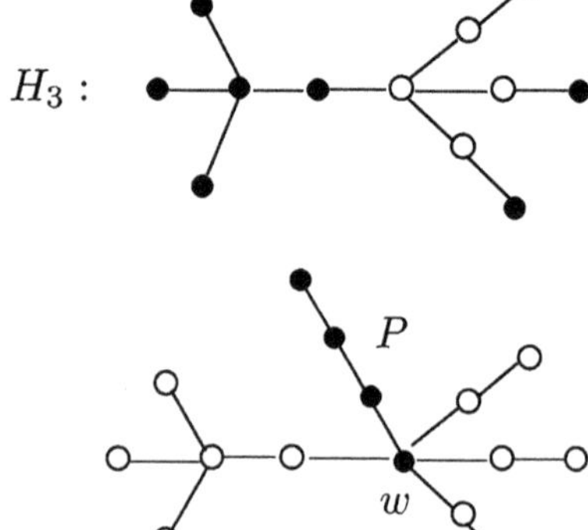


**Fig. 6** $\gamma_{F_3}(H_3) = 8 = 2n/3$

**Fig. 7** The tree $T$ in the proof of Corollary 3.3 when $\ell = k = 3$

The following realization result concerning Theorem 3.2 was presented in [10].

**Theorem 3.4 ([10])** *For every pair $k, n$ of integers with $4 \leq k \leq 2n/3$, there exists a tree $T_k$ of order $n$ with $\gamma_{F_3}(T_k) = k$.*

As stated in Theorem 3.2 and Corollary 3.3, if $T$ is a tree of order $n$ and diameter 3 or more, then $\gamma_{F_3}(T)$ is bounded above by $2n/3$. Of course, the minimum degree of every tree is 1. If $G$ is a connected graph of order $n$ with $\delta(G) \geq 2$, then Henning and Maritz [12] obtained an improved upper bound for $\gamma_{F_3}(G)$. To describe this result, we refer to the two graphs $H_1$ and $H_2$ shown in Figure 8 and let $S = \{C_4, C_5, C_8, H_1, H_2\}$.

**Theorem 3.5 ([12])** *If $G$ is a connected graph of order $n$ with $\delta(G) \geq 2$, then $\gamma_{F_3}(G) \leq (n-1)/2$ unless $G \in S$, in which case $\gamma_{F_3}(G) = \lceil n/2 \rceil$ for each $G \in S$.*

Henning and Maritz [12] also characterized those graphs $G$ of order $n \geq 9$ with $\delta(G) \geq 2$ for which $\gamma_{F_3}(G) = (n-1)/2$.

In [2], Chang, Kuo, Liaw, and Yan determined the $F_3$−domination numbers of Cartesian products of certain graphs, namely $P_m \square P_n$ and $C_m \square P_n$ for some specific values of $m$ and $n$. In [1], Chang, Chang, Kuo, and Poon showed that the $F_3$-domination problem is NP-complete for bipartite planar graphs and chordal graphs, They also described a linear-time algorithm for the $F_3$-domination problem for trees.

## 4 On 2-Stratified Triangle Domination

The five 2-stratified graphs $F_i$ $(1 \leq i \leq 5)$ are all based on the path $P_3$ of order 3. The only other connected graph of order 3 is the 3-cycle $C_3$. There are two 2-stratified graphs based on this graph, both shown in Figure 9, which are denoted by $F_6$ and $F_7$.

Necessarily, in any $F_6$-coloring or $F_7$-coloring of a graph, a vertex belonging to no triangle must be colored red. Since $F_6$ contains two red vertices while $F_7$

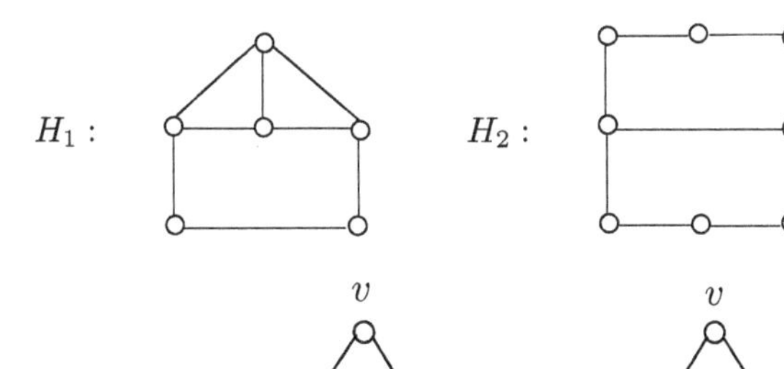


**Fig. 8** The graphs $H_1$ and $H_2$

**Fig. 9** The two 2-stratified $K_3$

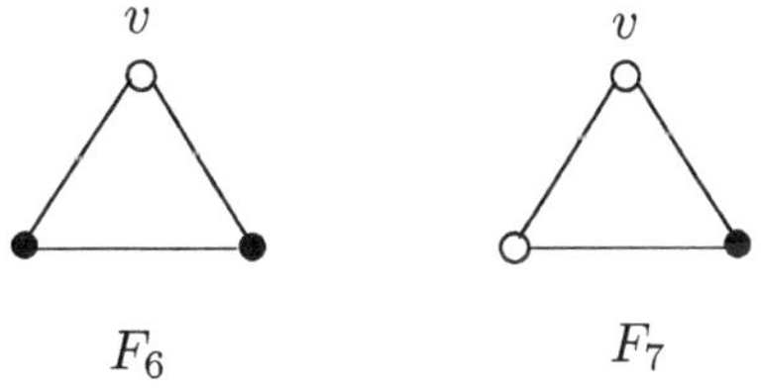

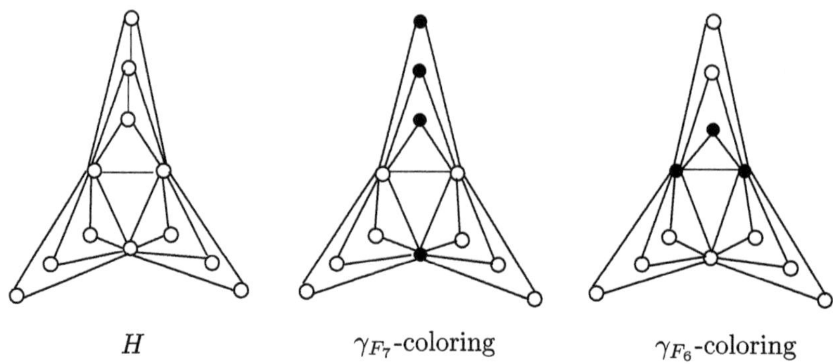


**Fig. 10** A graph $H$ with $\gamma_{F_7}(H) > \gamma_{F_6}(H)$.

contains one red vertex, this may suggest that $\gamma_{F_7}(G) \leq \gamma_{F_6}(G)$ for every graph $G$, but this is not the case. For example, for the graph $H$ of Figure 10, $\gamma_{F_7}(H) = 4$ and $\gamma_{F_6}(H) = 3$. Appropriate $\gamma_{F_7}$-colorings and $\gamma_{F_6}$-colorings of $H$ are given in Figure 10 as well.

In a $\gamma_{F_6}$-coloring of a graph in which every vertex lies on a triangle, every blue vertex belongs to a triangle where two-thirds of the vertices are red. In fact, in every $\gamma_{F_6}$-coloring of such a graph, at most two-thirds of the vertices of the graph are red. In the following result, it is useful to recall the following fact.

**Proposition 4.1 ([9])** *If $G$ is a graph of order $n$ without isolated vertices, then*

$$\gamma(G) \leq \frac{n}{2}.$$

**Theorem 4.2 ([4])** *If $G$ is a graph of order $n$ in which every vertex is in a triangle, then*

$$\gamma_{F_6}(G) \leq \frac{2n}{3}.$$

**Proof.** Suppose that the theorem is false. Then there exists a graph $G$ of order $n$ in which every vertex lies on a triangle but $\gamma_{F_6}(G) > 2n/3$. We may assume that every edge of $G$ also lies on a triangle for if $G$ contains edges belonging to no triangle, then the graph $G'$ obtained by deleting these edges from $G$ has the property that $\gamma_{F_6}(G') = \gamma_{F_6}(G)$.

Among all $\gamma_{F_6}$-colorings of $G$, we choose one that maximizes the number of red triangles (triangles all of whose vertices are red). For this $\gamma_{F_6}$-coloring, let $B = \{b_1, b_2, \ldots, b_k\}$ be the set of blue vertices and $R$ the set of red vertices. Therefore,

$$\gamma_{F_6}(G) = |R| = n - k > 2n/3$$

and so $n > 3k$. Thus, $|R| = n - k > 2k$.

The set $R$ is now partitioned into two sets $R_1$ and $R_2$ as follows. For each $i \in [k]$, there is a copy $T_i$ of $F_6$ rooted at the blue vertex $b_i$. Therefore, $T_i$ has two red vertices and one blue vertex $b_i$. The set $R_1$ is now defined as

$$R_1 = \left( \bigcup_{i=1}^{k} V(T_i) \right) - B;$$

that is, $R_1$ consists of all red vertices lying on at least one triangle $T_i$ $(1 \leq i \leq k)$. The set $R_2 = R - R_1$ consists of any remaining red vertices of $G$.

Since $| \cup_{i=1}^{k} V(T_i)| \leq 3k$ and $|B| = k$, it follows that $|R_1| = 2k - \ell$ for some integer $\ell$ with $0 \leq \ell \leq 2(k-1)$ as $R_1$ consists of at least two red vertices. Since $|R_1| + |R_2| = |R| > 2k$ and $|R_1| \leq 2k - \ell$, it follows that $|R_2| > \ell$ and so $|R_2| = \ell + r$ for some positive integer $r$.

We claim that every triangle of $G$ containing a red vertex of $R_2$ has two blue vertices. Suppose that this is not the case. Then there exists some vertex $x$ of $R_2$ that either belongs to a red triangle or to a triangle with exactly one blue vertex. If $x$ is in a red triangle, then $x$ may be recolored blue to produce an $F_6$-coloring of $G$ having fewer than $\gamma_{F_6}(G)$ red vertices, which is impossible. Suppose that $x$ belongs to a triangle $T$ with exactly one blue vertex, say $b_i$. Since $x \in R_2$, the triangle $T$ containing $x$ is not one of the triangles $T_j$ $(1 \leq j \leq k)$. On the other hand of course, $b_i$ belongs to the triangle $T_i$ whose other two vertices are red. If we interchange the colors of $x$ and $b_i$ (see Figure 11), a new $\gamma_{F_6}$-coloring of $G$ is produced that contains more red triangles than our original $\gamma_{F_6}$-coloring, which is also impossible. Therefore, as claimed, every triangle containing a (red) vertex of $R_2$ contains two blue vertices.

Next, we claim that $|B| \geq |R_2| + 1$, for if, instead, $|B| \leq |R_2|$, then the colors of the vertices in $B \cup R_2$ could be interchanged to produce an $F_6$-coloring of $G$ containing no more than $\gamma_{F_6}(G)$ red vertices but with more red triangles than in the original $\gamma_{F_6}$-coloring, which is not possible. Therefore, $|B| \geq |R_2| + 1$, as claimed.

For each $i \in [k]$, let $e_i$ be the edge in the triangle $T_i$ that is not incident with $b_i$ and let

$$E_B = \{e_i \, : \, i \in [k]\}.$$

That is, $E_B$ consists of those $k$ edges of $G$ that join two red vertices in the triangles $T_i$, where $i \in [k]$. Furthermore, let $H = G[E_B]$. Therefore, $V(H) = R_1$

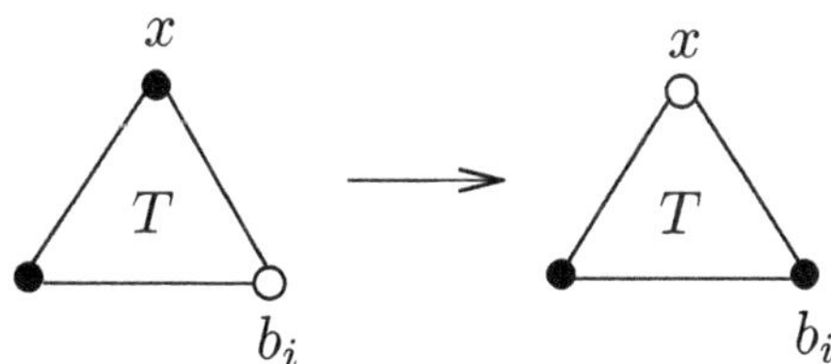


**Fig. 11** A step in the proof of Theorem 4.2

and $E(H) = E_B$. Let $R'_1$ be a $\gamma$-set of the graph $H$. Since $H$ has no isolated vertices, it follows by Proposition 4.1 that

$$|R'_1| \leq |V(H)|/2 = k - \ell/2.$$

We now interchange the colors of the vertices in $B \cup (R - R'_1)$. We claim that this new red-blue coloring of $G$ is also an $F_6$-coloring of $G$. Let $v$ be a blue vertex of $G$. Then $v \in R - R'_1$, which implies that either (1) $v \in R_2$ or (2) $v \in R_1 - R'_1$. If $v \in R_2$, then in the original red-blue coloring of $G$, $v$ belongs to a triangle with two blue vertices. Then, after the interchange of colors, $v$ belongs to a copy of $F_6$ rooted at $v$. Should it occur, however, that $v \in R_1 - R'_1$, then $v$ is adjacent to a vertex $u \in R'_1$. Thus, $uv = e_i$ for some $i \in [k]$. After the interchange of colors, the vertices $u$ and $b_i$ are red and so $v$ belongs to a copy of $F_6$ rooted at $v$. Therefore, as claimed, this new red-blue coloring of $G$ is an $F_6$-coloring of $G$.

Since

$$|R'_1| + |B| \leq 2k - \ell/2 < 2k + r = |R| = \gamma_{F_6}(G),$$

it follows that the number of red vertices in this $F_6$-coloring is less than $\gamma_{F_6}(G)$, which is impossible. Therefore, $\gamma_{F_6}(G) \leq 2n/3$. ■

The upper bound for $\gamma_{F_6}(G)$ given in Theorem 4.2 is sharp. For example, for the graph $G$ of order $n = 12$ shown in Figure 12, $\gamma_{F_6}(G) = 8 = 2n/3$.

We now turn our attention to the domination parameter $\gamma_{F_7}$. We saw in Theorem 4.2 that if $G$ is a graph of order $n$ in which every vertex is in a triangle, then $\gamma_{F_6}(G) \leq 2n/3$. The corresponding result for the 2-stratified graph $F_7$ is stated below.

**Theorem 4.3 ([4])** *If $G$ is a graph of order $n$ in which every vertex lies on a triangle, then*

$$\gamma_{F_7}(G) < \frac{n}{2}$$

*and this bound is asymptotically best possible.*

Henning and Maritz [11] stated that there may be an improved upper bound for $\gamma_{F_7}(G)$ for a graph $G$ in which every vertex lies on a triangle.

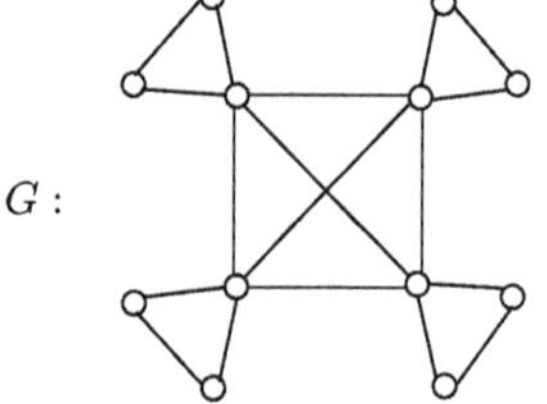


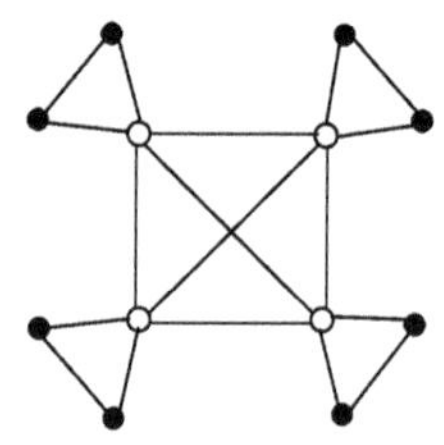

**Fig. 12** A graph $G$ of order $n$ with $\gamma_{F_6}(G) = 2n/3$

**Conjecture 4.4 ([11])** *If $G$ is a graph of order $n$ in which every vertex is in a triangle, then*

$$\gamma_{F_7}(G) \leq \frac{n}{2} - \frac{1}{8}\left(\sqrt{8n+1} - 1\right).$$

We now recall another well-known domination parameter. A dominating set of a graph $G$ that is independent is an *independent dominating set* for $G$. The minimum cardinality of an independent dominating set for $G$ is the *independent domination number* $i(G)$. It is known that for every graph, its independent domination number is at least as large as its domination number. Whatever the values of these two parameters may be, however, for a graph each of whose edges is on a triangle, the $F_7$-domination number lies between them.

**Theorem 4.5 ([4])** *If $G$ is a graph in which every edge lies on a triangle, then*

$$\gamma(G) \leq \gamma_{F_7}(G) \leq i(G).$$

**Proof.** Since, in every $F_7$-coloring of $G$, every blue vertex is adjacent to a red vertex, the set of red vertices in $G$ is a dominating set and so $\gamma(G) \leq \gamma_{F_7}(G)$. It therefore only remains to show that $\gamma_{F_7}(G) \leq i(G)$.

Let $S$ be a minimum independent dominating set of $G$. Thus, $|S| = i(G)$. Let each vertex of $S$ be colored red with all remaining vertices of $G$ colored blue. Therefore, every blue vertex is adjacent to a red vertex. By assumption, every edge of $G$ lies on a triangle. Also, $S$ is an independent set. Therefore, each blue vertex of $G$ is rooted in a copy of $F_7$. Consequently, this red-blue coloring of $G$ is an $F_7$-coloring of $G$, which implies that $\gamma_{F_7}(G) \leq i(G)$. ■

The two inequalities in the statements of Theorem 4.5 can be strict. To see this, we consider the graph $H$ of Figure 13, where seven vertices of $H$ are labeled as $u$, $v$, $w$, $w_1$, $w_2$, $w_3$, $x$. Since $\{u, w, x\}$ is a $\gamma$-set of $H$, it follows that $\gamma(H) = 3$. The red-blue coloring whose set of red vertices is $\{u, v, w, x\}$ is a $\gamma_{F_7}$-coloring of $H$. Hence, $\gamma_{F_7}(H) = 4$. Furthermore, $\{u, w_1, w_2, w_3, x\}$ is an $i$-set and so $i(H) = 5$. Therefore, $\gamma(H) < \gamma_{F_7}(H) < i(H)$ for the graph $H$. This is illustrated in Figure 13.

While Figure 13 shows a graph $H$ with $\gamma(H) = 3$, $\gamma_{F_7}(H) = 4$, and $i(H) = 5$, there are no restrictions on the possible values of these three parameters for a graph $G$ other than those given in Theorem 4.5 and that $\gamma(G) \geq 2$.

**Theorem 4.6 ([4])** *For every three integers $a, b, c$ with $2 \leq a \leq b \leq c$, there is a connected graph $G$ in which every edge lies on a triangle such that $\gamma(G) = a$, $\gamma_{F_7}(G) = b$, and $i(G) = c$.*

In the statement of Theorem 4.5, it is required that every edge of $G$ lies on a triangle of $G$. If, however, we require only that every vertex of $G$ lies on a triangle of $G$, then the conclusion does not follow. For example, for the graph $G$ of Figure 14, every vertex is on a triangle of $G$ but yet $\gamma(G) = i(G) = 3$ while $\gamma_{F_7}(G) = 4$.

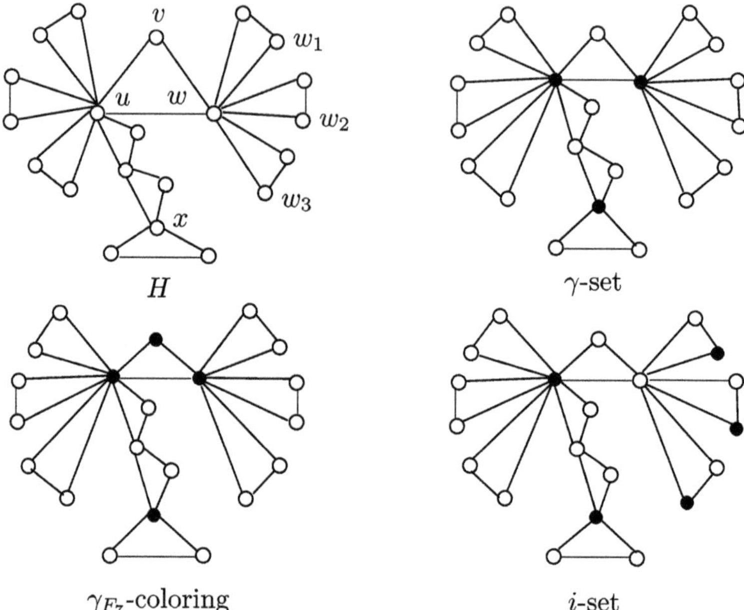


**Fig. 13** A graph $H$ with $\gamma(H) = 3$, $\gamma_{F_7}(H) = 4$, and $i(H) = 5$

**Fig. 14** A graph $G$ with $\gamma(G) = i(G) = 3$ and $\gamma_{F_7}(G) = 4$

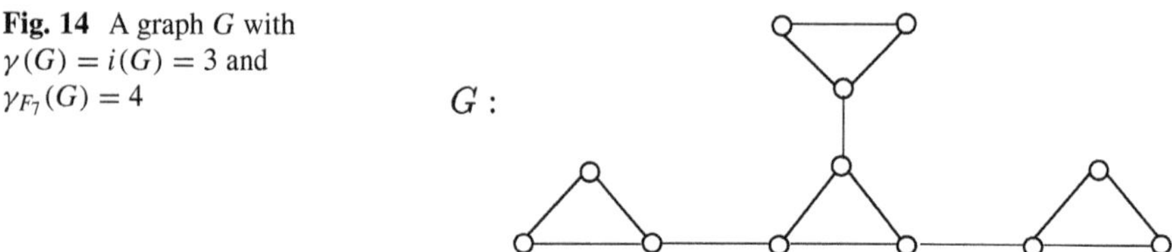


## 5 Stratified Domination for Further Study

Much of the research on stratified domination has been done when the 2-stratified graph $F$ is either $P_3$ or $C_3$ (also described in [5, 8, 10]). This is quite natural in the case of $P_3$ since, as we saw, this includes standard domination, total domination, restrained domination, and 2-domination. Also, these 2-stratified graphs are much easier to investigate. On the other hand, when $F = K_{1,3}$ (a claw) or $F = C_4$, stratified domination based on 2-stratifications of these two graphs has also been investigated, especially for prisms, in [3] and [13]. Prisms were the object of study as well in [6] and [7] when $F = C_5$ and $F = C_6$, respectively.

Another possibility deals with investigating alternative definitions of domination in terms of more than one 2-stratified graph. For an integer $k \geq 2$, let $H_1, H_2, \ldots, H_k$ be 2-stratified graphs, each rooted at a blue vertex and let $\mathcal{H} = \{H_1, H_2, \ldots, H_k\}$. By an *$\mathcal{H}$-coloring* of a graph $G$ is meant a red-blue coloring of the vertices of $G$ such that every blue vertex of $G$ is rooted at a copy of $H_i$ for

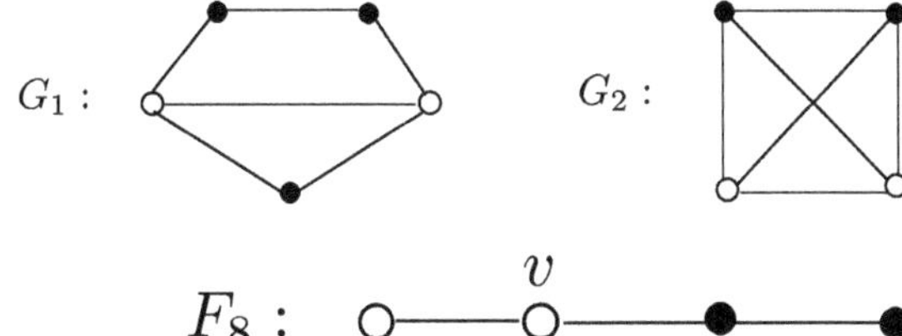


**Fig. 15** Examples of $\mathcal{H}$-colorings

**Fig. 16** The 2-stratified $P_4$

every integer $i$ with $1 \leq i \leq k$. The *$\mathcal{H}$-domination number* $\gamma_{\mathcal{H}}(G)$ is the minimum number of red vertices in an $\mathcal{H}$-coloring of $G$.

For example, let $F_1, F_2, \ldots, F_5$ be the five 2-stratified graphs of the path $P_3$ of order 3 shown in Figure 3 and let $F_6, F_7$ be the two 2-stratified graphs of the triangle $K_3$ shown in Figure 9. If $\mathcal{F} = \{F_1, F_2, \ldots, F_5\}$, then the red-blue coloring of $G_1$ in Figure 15 is an $\mathcal{F}$-coloring of $G_1$; while if $\mathcal{F} = \{F_1, F_2, \ldots, F_7\}$, then the red-blue coloring of $G_2$ in Figure 15 is an $\mathcal{F}$-coloring of $G_2$.

When $\mathcal{F} = \{F_1, F_4\}$, $\mathcal{F}$-colorings of graphs have been investigated in [14]. Thus, in such an $\mathcal{F}$-coloring of a graph $G$, every blue vertex is rooted at both a copy of $F_1$ and a copy of $F_4$. Therefore, an $\mathcal{F}$-coloring of a graph $G$ is the same as an $F_8$-coloring of $G$, where $F_8$ is the 2-stratified $P_4$ rooted at the vertex $v$, as shown in Figure 16.

In order to describe a result obtained on the $\mathcal{F}$-domination number of graphs for this set $\mathcal{F}$ of two 2-stratified graphs $P_3$, we define another additional term. A set $S$ of vertices in a graph $G$ is a *total restrained dominating set* of $G$ if $S$ is both a total dominating set and a restrained dominating set. The *total restrained domination number* of $G$ is the minimum cardinality of a total restrained dominating set for $G$ and is denoted by $\gamma_{tr}(G)$. The following results are due to Henning and Maritz [14].

**Theorem 5.1 ([14])** *Let $\mathcal{F} = \{F_1, F_4\}$. If $G$ is a graph without isolated vertices, then*

$$\max\{\gamma_t(G), \gamma_r(G)\} \leq \gamma_{\mathcal{F}}(G) \leq \gamma_{tr}(G).$$

**Proof.** By assigning the color red to those vertices belonging to a minimum total restrained dominating set of a graph $G$ and blue to the remaining vertices of $G$, an $\mathcal{F}$-coloring of $G$ is achieved, which implies that $\gamma_{\mathcal{F}}(G) \leq \gamma_{tr}(G)$. To verify the lower bound for $\gamma_{\mathcal{F}}(G)$, we first observe that the set of red vertices in an $\mathcal{F}$-coloring of a graph $G$ is a restrained dominating set of $G$ and so $\gamma_r(G) \leq \gamma_{\mathcal{F}}(G)$. Also, observe that the set of red vertices in an $\mathcal{F}$-coloring of a graph $G$ without isolated vertices need not be a total dominating set of $G$ as there could be isolated vertices in the subgraph of $G$ induced by the red vertices. However, every $\mathcal{F}$-coloring of $G$ is also an $F_1$-coloring of $G$. Therefore, there exists an $F_1$-coloring of $G$ with $\gamma_{\mathcal{F}}(G)$ red vertices.

Among all $F_1$-colorings of $G$ with $\gamma_{\mathcal{F}}(G)$ red vertices, we select one having the minimum number of isolated vertices in the subgraph of $G$ induced by its red vertices. Then, as shown in the proof of Theorem 2.2, every red vertex in such an $F_1$-coloring is adjacent to another red vertex. Thus, the red vertices form a total dominating set of $G$, which implies that $\gamma_t(G) \leq \gamma_{\mathcal{F}}(G)$. Therefore, $\max\{\gamma_t(G), \gamma_r(G)\} \leq \gamma_{\mathcal{F}}(G)$. ■

For $\mathcal{F} = \{F_1, F_4\}$, the $\mathcal{F}$-domination number has been determined for all cycles.

**Theorem 5.2 ([14])** *Let $\mathcal{F} = \{F_1, F_4\}$. For integers $n \geq 3$ and $i = 0, 1, 2, 3$,*

$$\gamma_{\mathcal{F}}(C_n) = \frac{n+i}{2}$$

*when $n \equiv i \pmod 4$.*

An upper bound for $\gamma_{\mathcal{F}}(G)$ has also been determined in terms of the order and maximum degree of a graph $G$.

**Theorem 5.3 ([14])** *Let $\mathcal{F} = \{F_1, F_4\}$. If $G$ is a connected graph of order $n$ where $2 \leq \delta(G) \leq \Delta(G) \leq n-2$, then*

$$\gamma_{\mathcal{F}}(G) \leq n - \Delta(G) + 1.$$

*Furthermore, this bound is sharp.*

**Theorem 5.4 ([14])** *Let $\mathcal{F} = \{F_1, F_4\}$. If $G$ is a connected graph of order $n \geq 4$ with $\delta(G) \geq 2$ where $C \neq C_7$, then*

$$\gamma_{\mathcal{F}}(G) \leq \frac{2n}{3}.$$

For an integer $k \geq 2$, let $H_1, H_2, \ldots, H_k$ be 2-stratified graphs, each rooted at a blue vertex and let $\mathcal{H} = \{H_1, H_2, \ldots, H_k\}$. By an *$\tilde{\mathcal{H}}$-coloring* of a graph $G$ is meant a red-blue coloring of the vertices of $G$ such that every blue vertex of $G$ is rooted at a copy of *exactly* one $H_i$, $1 \leq i \leq k$, and for each 2-stratified graph $H_i$, there is at least one blue vertex $v$ of $G$ such that there is a copy of $H_i$ rooted at $v$. The *$\tilde{\mathcal{H}}$-domination number* $\gamma_{\tilde{\mathcal{H}}}(G)$ is the minimum number of red vertices in an $\tilde{\mathcal{H}}$-coloring of $G$.

For example, let $F_1, F_2, \ldots, F_5$ be the five 2-stratified graphs of the path $P_3$ of order 3 shown in Figure 3. In Figure 17, an $\tilde{\mathcal{F}}$-coloring of a graph is shown for each of the ten different choices of $\mathcal{F} = \{F_i, F_j\}$ where $1 \leq i < j \leq 5$ as well as an $\tilde{\mathcal{F}}$-coloring of a graph for $\mathcal{F} = \{F_1, F_2, F_3\}$. In each red-blue coloring of a graph in Figure 17, we label a blue vertex $v$ by an integer $i \in [5]$ to indicate that $v$ is rooted at a copy of $F_i$.

1. The coloring of $G_1$ is an $\tilde{\mathcal{F}}$-coloring where $\mathcal{F} = \{F_1, F_2\}$.
2. The coloring of $G_2$ is an $\tilde{\mathcal{F}}$-coloring where $\mathcal{F}$ is $\{F_1, F_3\}$ or $\{F_1, F_4\}$.
3. The coloring of $G_3$ is an $\tilde{\mathcal{F}}$-coloring where $\mathcal{F} = \{F_1, F_5\}$.
4. The coloring of $G_4$ is an $\tilde{\mathcal{F}}$-coloring where $\mathcal{F} = \{F_2, F_3\}$.
5. The coloring of $G_5$ is an $\tilde{\mathcal{F}}$-coloring where $\mathcal{F} = \{F_2, F_4\}$.

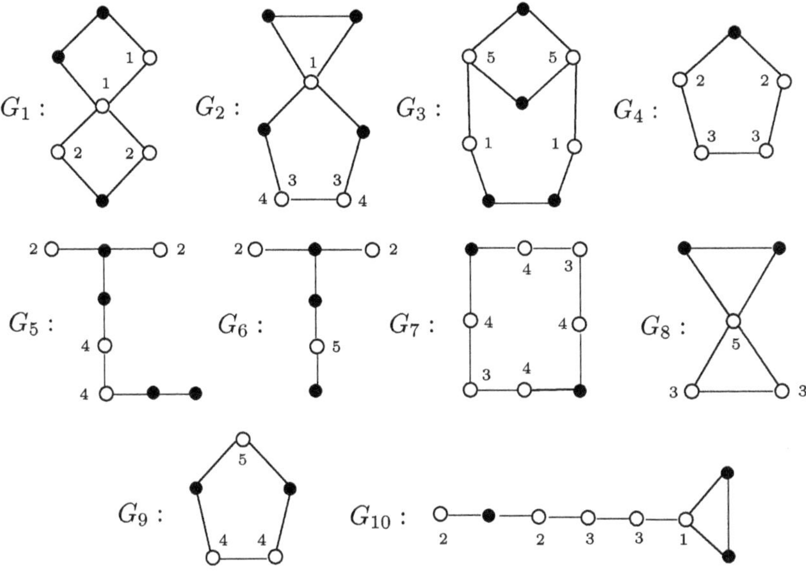


**Fig. 17** Examples of $\tilde{\mathcal{F}}$-colorings

6. The coloring of $G_6$ is an $\tilde{\mathcal{F}}$-coloring where $\mathcal{F} = \{F_2, F_5\}$.
7. The coloring of $G_7$ is an $\tilde{\mathcal{F}}$-coloring where $\mathcal{F} = \{F_3, F_4\}$.
8. The coloring of $G_8$ is an $\tilde{\mathcal{F}}$-coloring where $\mathcal{F} = \{F_3, F_5\}$.
9. The coloring of $G_9$ is an $\tilde{\mathcal{F}}$-coloring where $\mathcal{F} = \{F_4, F_5\}$.
10. The coloring of $G_{10}$ is an $\tilde{\mathcal{F}}$-coloring where $\mathcal{F} = \{F_1, F_2, F_3\}$.

Among the possible topics for investigation in this area are the following.

1. Which graphs have an $\tilde{\mathcal{H}}$-coloring for various sets $\mathcal{H}$?
2. For a given set $\mathcal{H}$, what is the minimum order of a graph possessing an $\tilde{\mathcal{H}}$-coloring?
3. For a given set $\mathcal{H}$, what is the minimum size of a graph of a fixed order possessing an $\tilde{\mathcal{H}}$-coloring?
4. Determine $\gamma_{\tilde{\mathcal{H}}}(G)$ for graphs $G$ belonging to some well-known classes of graphs.

## References


1. G. J. Chang, C. W. Chang, D. Kuo, and S. H. Poon, Algorithmic aspect of stratified domination in graphs. *Inform. Process. Lett.* **113** (2013), 861–865.
2. C. W. Chang, D. Kuo, S. C. Liaw, and J. H Yan, $F_3$−domination problem of graphs. *J. Comb. Optim.* **28** (2014), 400–413.
3. G. Chartrand, T. W. Haynes, M. A. Henning, and P. Zhang, Stratified claw domination in prisms. *J. Combin. Math. Combin. Comput.* **33** (2000), 81–96.

4. G. Chartrand, T. W. Haynes, M. A. Henning, and P. Zhang, Stratification and domination in graphs. *Discrete Math.* **272** (2003), 171–185.
5. G. Chartrand and P. Zhang, *Chromatic Graph Theory*, Second Edition. Chapman & Hall/CRC Press, Boca Raton (2020).
6. S. Ediz, On stratified domination in prisms. *Int. Math. Forum* **4** (2009), 2355–2362.
7. S. Ediz, On stratification and domination in prisms. *Appl. Math. Sci.* **5** (2011), 203–211.
8. R. Gera, *Stratification and Domination in Graphs and Digraphs*. Ph.D. Dissertation. Western Michigan University (2004).
9. T. W. Haynes, S. T. Hedetniemi, and P. J. Slater, *Fundamentals of Domination in Graphs*. Marcel Dekker, New York (1998).
10. T. W. Haynes, M. A. Henning, and P. Zhang, A survey of stratification and domination in graphs. *Discrete Math.* **309** (2009), 5806–5819.
11. M. A. Henning and J. E. Maritz, Stratification and domination in graphs II. *Discrete Math.* **286** (2004), 203–211.
12. M. A. Henning and J. E. Maritz, Stratification and domination in graphs with minimum degree two. *Discrete Math.* **301** (2005), 175–194.
13. M. A. Henning and J. E. Maritz, Stratification and domination in prisms. *Ars Combin.* **81** (2006), 343–358.
14. M. A. Henning and J. E. Maritz, Simultaneous stratification and domination in graphs with minimum degree two. *Quaestiones Mathematicae* **29**, (2006), 1–6.
15. R. Rashidi, *The Theory and Applications of Stratified Graphs*. Ph.D. Dissertation, Western Michigan University (1994).

# Global Domination

**Robert C. Brigham, Julie R. Carrington, and Ronald D. Dutton**

## 1 Introduction

The notion of a dominating set of a graph has been extended in a natural way to a collection of vertices that simultaneously dominates two or more graphs having the same vertex set. This concept was introduced independently by Sampathkumar (1989) [41] (who coined the term global domination used here) and Brigham and Dutton (1990) [8] (under the name factor domination). The following defines global dominating sets and related concepts.

**Definition 1** *Let $H = (V, E)$ be a graph having spanning subgraphs $F_i = (V, E_i)$, $1 \leq i \leq k$, where $E_1, E_2, \ldots, E_k$ partition $E$. The $F_i$ are called factors of $H$. Then $D_g \subseteq V$ is a global dominating set (GDS) if $D_g$ is a dominating set for each $F_i$, $1 \leq i \leq k$. The cardinality of a smallest such set, designated by $\gamma_g(F_1, F_2, \ldots, F_k)$, is the global domination number of the factoring. A dominating set $D_i$ for a factor $F_i$ is a local dominating set (LDS) and a minimum such set has cardinality designated by $\gamma_i$, called the local domination number of $F_i$.*

If the context makes the factors clear, the notation $\gamma_g(F_1, F_2, \ldots, F_k)$ is reduced to $\gamma_g$ or $\gamma_g(H)$. An example of a factoring is shown in Figure 1. It is easy to see that $\gamma(H) = 1$, $\gamma(F_1) = 2$, and $\gamma(F_2) = \gamma(F_3) = 3$. Since vertex $v_5$ is isolated in $F_2$,

R. C. Brigham (✉)
Department of Mathematics, University of Central Florida, 32816 Orlando, FL, USA
e-mail: robert.brigham@ucf.edu

J. R. Carrington
Department of Mathematical Sciences, Rollins College, 32789 Winter Park, FL, USA
e-mail: jcarrington@rollins.edu

R. D. Dutton
Department of Computer Science, University of Central Florida, 32816 Orlando, FL, USA
e-mail: dutton@cs.ucf.edu



T. W. Haynes et al. (eds.), *Topics in Domination in Graphs*, Developments in Mathematics 64, https://doi.org/10.1007/978-3-030-51117-3_15

it must be in any global dominating set. No set containing $v_5$ and two other vertices dominates every factor, but $\{v_2, v_3, v_4, v_5\}$ does. Thus $\gamma_g(H) = \gamma_g(F_1, F_2, F_3) = 4$.

Since both a global dominating set and $\cup_{i=1}^{k} D_i$ dominate all factors, we have

**Observation 2** *For any factoring of graph H,*

$$max\{\gamma_1, \gamma_2, \ldots, \gamma_k\} \leq \gamma_g \leq \gamma_1 + \gamma_2 + \cdots + \gamma_k.$$

The lower bound is achieved by $K_4$ with factors $F_1 = C_4$ and $F_2 = 2P_2$ for which $\gamma(F_1) = \gamma(F_2) = \gamma_g(K_4) = 2$. For an example of the sharpness of the upper bound, consider the graph $H$ of Figure 2 with factors $F_1$ and $F_2$ partitioning its edges. It can be seen that $\gamma(F_1) = \gamma(F_2) = 2$ and $\gamma_g(H) = 4$.

Both Sampathkumar [41] and Brigham and Dutton [8] concentrated on the important case of when there are two factors of $K_n$: a graph $G$ and its complement $\overline{G}$. In this special situation, we write $\gamma_g(G, \overline{G})$ as $\gamma_g(G)$, so then $\gamma_g(G) = \gamma_g(\overline{G})$. The value of $\gamma_g(G)$ has been computed for several common families of graphs.

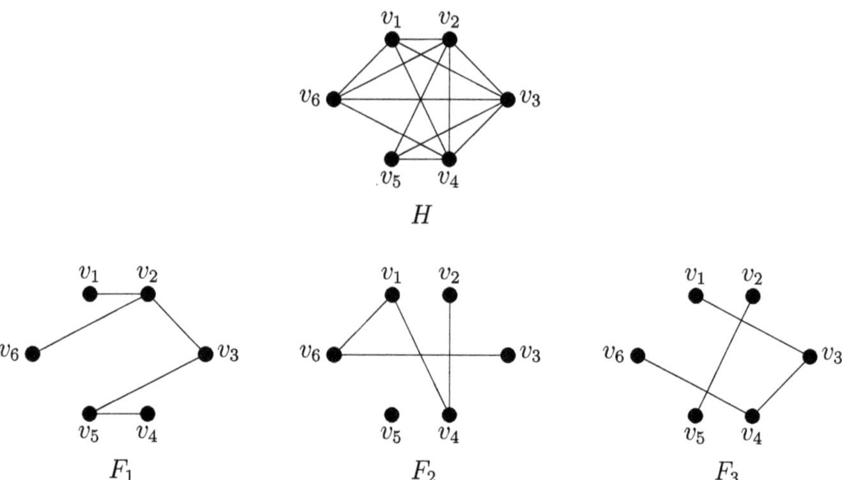


**Fig. 1** Example of factoring of a graph

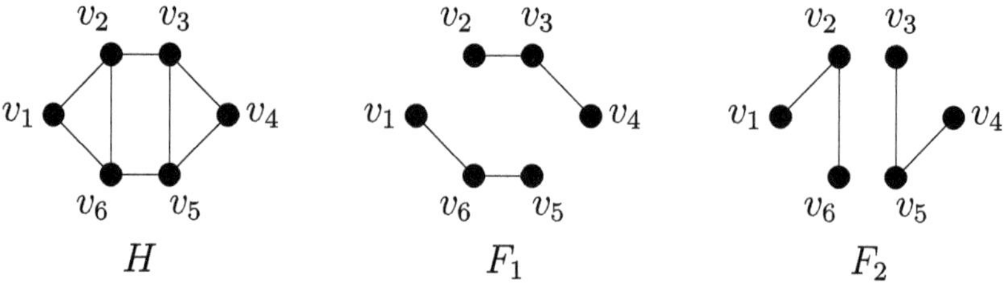


**Fig. 2** Example where $\gamma_g(H) = \gamma_g(F_1, F_2) = \gamma(F_1) + \gamma(F_2)$

**Theorem 3** *For the path $P_n$ on $n \geq 2$ vertices we have*

$$\gamma_g(P_n) = \begin{cases} 2 & \text{if } n = 2, 3 \\ \lceil n/3 \rceil & \text{if } n \geq 4 \end{cases}.$$

**Proof:** Let $P_n$ have vertices $v_1, v_2, \ldots, v_n$ in order. When $n = 2$, both $v_1$ and $v_2$ are isolated in $\overline{P_2}$ and thus both must be in a global dominating set. For $n = 3$, $v_2$ is isolated in $\overline{P_3}$ and hence must be in any global dominating set. Furthermore $v_1$ and $v_2$ dominate both $P_3$ and $\overline{P_3}$. For $n \geq 4$, it is well known $\gamma(P_n) = \lceil n/3 \rceil$ (see [10] page 85). If $n = 4$, $v_1$ and $v_4$ dominate both $P_4$ and $\overline{P_4}$ and $\lceil 4/3 \rceil = 2$. For $n \geq 5$, a minimum dominating set can always include $v_2$, and $v_5$ which also dominate $\overline{P_n}$. □

Values are given next for $K_n$, $C_n$, and $W_n$ which are the complete graph, cycle, and wheel, respectively, on $n$ vertices and for the complete $r$-partite graph $K_{n_1,n_2,\ldots,n_r}$.

$$\gamma_g(K_n) = n$$

$$\gamma_g(C_n) = \begin{cases} 3 & \text{if } n = 3, 5 \\ \lceil n/3 \rceil & \text{otherwise} \end{cases}$$

$$\gamma_g(W_n) = \begin{cases} 4 \text{ if } n = 4 \\ 3 \text{ otherwise} \end{cases}$$

$$\gamma_g(K_{n_1,n_2,\ldots,n_r}) = r$$

Definition 1 makes no assumption that the union of the factors must be the complete graph. Carrington (1992) [10] studied the general case extensively.

A modified definition of global domination, called *simultaneous domination*, has been considered by Dankelmann, Henning, Goddard, and Laskar (2006) [15] and Caro and Henning (2014) [9]. Here the definition is the same except the factors need not be edge disjoint. The smallest size of a simultaneous domination set of graph H with factors $F_1, F_2, \ldots, F_k$ is indicated by $\gamma_{sd}(F_1, F_2, \ldots, F_k)$ in all quoted results from these papers.

A possible application is a communication network, modeled by $H$, with $k$ edge disjoint subnetworks, represented by the factors $F_i$. The subnetworks might be required for reasons of security, redundancy, or limitation of recipients for different classes of messages. The number $\gamma_g(H)$ then represents the minimum number of "master" stations required so that a message issued simultaneously from all masters reaches all desired recipients after traveling over only one communication link, no matter which subnetworks are active.

As another example, suppose there is a collection $V$ of entities (people, countries, etc.) where the two entities for each pair are either friends or foes. Let G be the graph with vertex set $V$ and edge $v_iv_j$, $v_i, v_j \in V$, if and only if $v_i$ and $v_j$ are friends. Then the edges of the complement $\overline{G}$ correspond to pairs of vertices which are foes.

Then $\gamma_g(G)$ is the smallest set $D_g$ of entities such that any entity in $V - D_g$ is neither friends with all the entities of $D_g$ nor foes with all of them. Such a set might form an acceptable mediation panel for all parties.

Graph partitioning is important in the implementation of parallel algorithms. In this light, communication networks and their underlying graphs are often partitioned, either to impose a particular structure or to reflect routes taken by messages as they proceed through a network. See Berry (1990) [6], Bouloutas and Gopar (1989) [7], and Theimer and Lantz (1988) [45]. Carrington [10] discusses two global domination applications in detail and supplies references where such a partitioning is important. One of the applications involves binary constraint satisfaction problems and the other multicast messages in a communication network.

Section 2 presents general results with no restrictions on the definition and Section 3 deals with the special case of complementary factors. As with domination itself, some global domination research has considered restriction to special classes of graphs or branched to other forms of domination. Section 4 discusses several of these variations. Two of these, global total domination and global Roman domination, have received more interest than most and are covered in Sections 5 and 6, respectively. Finally Section 7 gives possible areas for further research that are mentioned in the literature.

Notation, whenever possible, agrees with that listed in the chapter "Glossary of Common Terms". A consequence is that some results presented in this chapter may not appear as presented in the original papers. This is certainly true with regard to independence numbers, covering numbers, packing numbers, and matching numbers whose symbols receive special attention in the chapter "Glossary of Common Terms". For example, Section 3 gives the bound $\gamma_g(G) \leq \min\{\alpha'(G)+1, \alpha'(\overline{G})+1\}$ when neither $G$ nor $\overline{G}$ has isolated vertices. Here $\alpha'(G)$ refers to the maximum cardinality of a maximal matching in $G$. The paper [8] from which this is taken actually presents the inequality as $\gamma_f \leq \min\{\beta_1, \overline{\beta_1}\} + 1$ in which $\beta_1 = \beta_1(G)$ means the same thing. This example also illustrates other points to keep in mind when turning to the original references. In general an invariant is shown here with an argument illustrating the graph to which it applies, as in $\alpha'(G)$, whereas the original paper may not. Furthermore, note that $\gamma_f$ has been changed to $\gamma_g(G)$, again to maintain a uniformity of notation. One exception to the above is the use of $n$ throughout to indicate the number of vertices of a graph. If the graph under consideration is clear, no argument is associated with $n$.

## 2 General Results

In this section we report results based on the definitions given in the introduction. Most unattributed ones can be found in [41] or [8].

## 2.1 Fundamental Facts

The global domination problem can be transformed into one of standard domination by constructing a graph $\hat{H}$ beginning with disjoint copies of $H, F_1, F_2, \ldots, F_k$ and joining vertex $u$ of the copy of $H$ to vertex $u$ of the copy of $F_i$ and to all vertices of this $F_i$ which are adjacent to $u$ in $F_i$, for all $u \in V(H)$ and $i = 1, 2, \ldots, k$. Thus $u$ in $\hat{H}$ dominates itself and every vertex that is adjacent to $u$ in every factor. It follows that $\gamma(\hat{H}) = \gamma_g(F_1, F_2, \ldots, F_k)$. The ideas behind this result can be seen in the simple example shown in Figure 3. A minimum global dominating set of $H$ with factors $F_1$ and $F_2$ is $\{v_1, v_2, v_4\}$ which also is a minimum dominating set of $\hat{H}$.

The following decision problem clearly is in NP.

k-FACTOR DOMINATING SET
INSTANCE: Graph $H = (V, E)$; factors $F_1, F_2, \ldots, F_k$ with $k \geq 2$; integer $M' \leq |V|$.
QUESTION: Is there a factor dominating set of size $M'$ or less?

K-FACTOR DOMINATING SET can be shown to be NP-Complete by transforming any instance of graph $H$ and integer $M$ of the NP-Complete problem DOMINATING SET (Garey and Johnson, 1979 [27]) to it by the polynomial transformation producing the join $H + (k-1)K_1$, where the vertices not in $H$ are $v_1, v_2, \ldots, v_{k-1}$, factor $F_i$ has only the edges incident to $v_i$ for $1 \leq i \leq k-1$, factor $F_k = H \cup (k-1)K_1$, and positive integer $M + (k-1)$. Then the instance of DOMINATING SET is a yes if and only if the corresponding instance of k-FACTOR DOMINATING SET is a yes. Carrington [10] and Carrington and Brigham (1992) [12] have shown that the problem remains NP-complete even when the factors are greatly simplified, namely, when each of the factors is a path; when there are four factors and each of them is a caterpillar; or when there are four factors and each of them is a forest all the components of which are $K_1$ or $K_2$.

A natural question following from Observation 2 is if, given any positive integers $a, a_1, a_2, \ldots, a_k$ satisfying $\max\{a_1, a_2, \ldots, a_k\} \leq a \leq a_1 + a_2 + \cdots + a_k$, there is a graph $G$ with factors $F_1, F_2, \ldots, F_k$ such that $\gamma(F_i) = a_i$ for $1 \leq i \leq k$ and $\gamma_g(F_1, F_2, \ldots, F_k) = a$. This is discussed in Carrington and Brigham (1991) [11] and in [10].

**Theorem 4 (Carrington [10])** *Consider a set of $k+1$ integers $\gamma_g, \gamma_1, \gamma_2, \ldots, \gamma_k$, such that $2 \leq \gamma_1 \leq \gamma_2 \leq \ldots \leq \gamma_k \leq \gamma_g \leq \gamma_1 + \gamma_2 + \ldots + \gamma_k$. Then there is a graph*

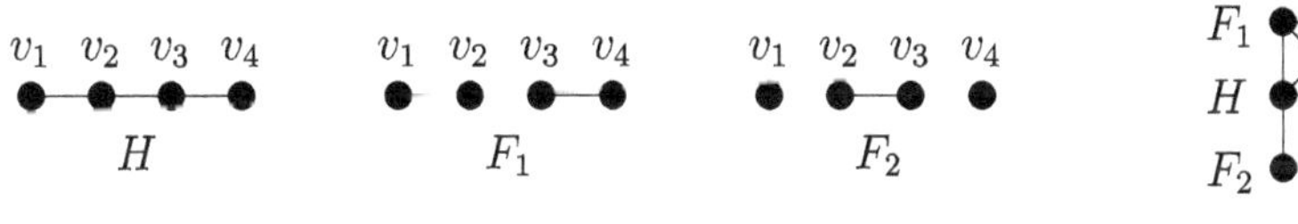


**Fig. 3** Example showing $\gamma(\hat{H}) = \gamma_g(F_1, F_2)$

*$G$ which can be factored into $k$ factors such that some $\gamma_g$ vertices form a minimum $GDS$ for the factoring and, for $1 \le i \le k$, some $\gamma_i$ vertices form a minimum $LDS$ for $F_i$ if*

1. *$k \in \{2, 3\}$ or*
2. *$k \in \{4, 5, 6\}$ in which cases the result holds if and only if $\sum_{i=1}^{k}[(k-1) - (\gamma_i - 1)(\gamma_g - k + 1)] \le \frac{(k-1)k}{2}$.*

The result is more complicated when $k \ge 7$.

Given a graph $H$ with $k$ factors, let $\beta_g^{''}(H)$ be the minimum number of edges such that all vertices are incident to at least one of the edges in all factors and $\alpha_g^{''}(H)$ be the maximum number of edges such that no two are adjacent in any factor. Carrington [10] develops the following Gallai type relation.

**Theorem 5 (Carrington [10])** *Let $H$ be a graph with factors $F_1, F_2, \ldots, F_k$ with no factor having an isolated vertex. Then $\alpha_g^{''}(H) + \beta_g^{''}(H) = kn$.*

**Proof:** It is shown in [10] that $\beta_g^{''}(H) = \sum_{i=1}^{k} \beta^{'}(F_i)$ and $\alpha_g^{''}(H) = \sum_{i=1}^{k} \alpha^{'}(F_i)$. Since no factor has an isolated vertex, Gallai's theorem gives $\beta^{'}(F_i) + \alpha^{'}(F_i) = n$ for $1 \le i \le n$ and the result follows. □

## 2.2 Bounds

Observation 2 provides basic lower and upper bounds on $\gamma_g$. Several others have been developed and a representative sample is given here. Let $H$ be the graph containing the factors.

Some early results include the following. When $k > \Delta(H)$, $\gamma_g(H) = n$ since every vertex will be isolated in at least one factor. Otherwise, $\gamma_g(H) \ge \gamma(H) + k - 2$. Let $Iso(H)$ be the collection of vertices isolated in at least one factor. Then $\gamma_g(H) \le \beta(H) + |Iso(H)|$ and $\gamma_g(H) \le n - \gamma(H) + |Iso(H)|$. Two results involving standard domination generalize in a straightforward way. The relation $\gamma \ge n/(\Delta + 1)$ generalizes to $\gamma_g(H) \ge nk/(\Delta(H) + k)$. Another familiar result is $\gamma + \varepsilon = n$, where $\varepsilon$ is the maximum number of end edges in a spanning forest of $G$ (Nieminen (1974) [35]). The parameter $\varepsilon$ can be generalized to $\varepsilon_g$, the cardinality of the largest set of vertices $X$ such that, in each $F_i$, there is a spanning forest in which $X$ is independent and each vertex of $X$ has degree one. Then the generalization is as follows.

**Theorem 6** *For any graph $H$ with factors $F_1, F_2, \ldots, F_k$, $\gamma_g(H) + \varepsilon_g(H) = n$.*

**Proof:** We may assume $k \le \Delta(H)$ and $\gamma_g(H) < n$. Let $D$ be a minimum global dominating set of $H$ with the given factors. For each vertex of $V(F_i) - D$, $1 \le i \le k$, select one edge between it and $D$. The subgraph of $F_i$ thus formed is a union of stars centered on the vertices of D and is a spanning forest of $F_i$. Furthermore, the vertices of $V(F_i) - D$ are independent and have degree one so it follows that $\varepsilon_g(H) \geq n - \gamma_g(H)$. Now suppose $X$ is a set of $\varepsilon_g(H)$ vertices satisfying the conditions given above. Then the vertices of $V(H) - X$ form a global dominating set and $\gamma_g(H) \leq n - \varepsilon_g(H)$. □

Dankelmann and Laskar (2003) [16] investigated the effect of minimum degree. Let $F_1, F_2, \ldots, F_k$, $k \geq 2$, be factors of the complete graph $K_n$. If $\delta(F_i) \geq 1$ for $i = 1, 2, \ldots, k$, then

$$\gamma_g(F_1, F_2, \ldots, F_k) \leq \begin{cases} (1 - 3^{-k/2})n & \text{if } k \text{ is even} \\ (1 - \frac{1}{2}3^{-(k-1)/2})n & \text{if } k \text{ is odd.} \end{cases}$$

When $k = 2$ this reduces to $\gamma_g(F_1, F_2) \leq \frac{2}{3}n$. The graph and factoring of Figure 2 show the bound is sharp.

Related results have been improved by Dankelmann et al. [15] (using simultaneous domination) to $\gamma_{sd}(F_1, F_2, \ldots, F_k) \leq (2k - 3)n/(2k - 2)$ when $k \geq 3$ and all factors are isolate-free, a bound that is sharp. Let $\delta' = \min\{\delta(F_i) | i = 1, 2, \ldots, k\}$. If $\delta' \geq 2$ and $k \leq e^{\delta'+1}/(\delta' + 1)$, then

$$\gamma_{sd}(F_1, F_2, \ldots, F_k) \leq \frac{\ln(\delta' + 1) + \ln(k) + 1}{\delta' + 1} n.$$

If $H$ is a graph with factors $F_1, F_2, \ldots, F_k$; $\delta(F_i) \geq \delta(H) \geq 1$ for $i = 1, 2, \ldots, k$; and $\overline{d}(H)$ is the average degree of $H$, then Caro and Henning [9] show

$$\gamma_{sd}(F_1, F_2, \ldots, F_k) \leq \left( \frac{\lceil \overline{d}(H) \rceil}{\lceil \overline{d}(H) \rceil + \delta} \right) n.$$

Furthermore, if each factor has minimum degree at least $\delta$ and $k \geq 2$, then

$$\gamma_{sd}(F_1, F_2, \ldots, F_k) \leq \left( 1 - \left( \frac{\delta}{\delta + 1} \right) \left( \frac{1}{k(\delta + 1)} \right)^{\frac{1}{\delta}} \right) n$$

and, if all factors are regular of the same positive degree, then

$$\gamma_{sd}(F_1, F_2, \ldots, F_k) \leq \left( \frac{k}{k + 1} \right) n.$$

Caro and Henning [9] also present several results when the factors are restricted to simple forms, illustrative examples of which are given here. When $r$ is a positive integer at most $n$, $n \equiv 0 \pmod{r}$, $k \geq 2$, and every factor consists of $n/r$ vertex disjoint copies of $K_r$ then

$$\gamma_{sd}(F_1, F_2, \ldots, F_k) \leq \begin{cases} \left(\frac{k}{2r}\right) n & \text{if } k \text{ is even} \\ \left(\frac{r(k+1)-2}{2r^2}\right) n & \text{if } k \text{ is odd.} \end{cases}$$

If $k \geq 2$, $n$ is even, and every factor is a 1-factor of $H$, that is, $n/2$ disjoint copies of $K_2$ and hence a perfect matching of $G$, then

$$\gamma_{sd}(F_1, F_2, \ldots, F_k) \leq \begin{cases} \left(\frac{k-1}{k}\right) n & \text{if } k \text{ is even} \\ \left(\frac{k}{k+1}\right) n & \text{if } k \text{ is odd.} \end{cases}$$

If $k \geq 2$, $n \equiv 0 \pmod 6$, and each factor is a $C_n$, then

$$\gamma_{sd}(F_1, F_2, \ldots, F_k) \leq \left(1 - \frac{1}{2}\left(\frac{2}{3}\right)^{k-2}\right) n.$$

If $k \geq 3$, $n \equiv 0 \pmod 5$ and each factor is $n/5$ vertex disjoint copies of $C_5$, then

$$\gamma_{sd}(F_1, F_2, \ldots, F_k) \leq \left(\frac{3}{5} + \frac{2}{5}\left(1 - \left(\frac{3}{5}\right)^{k-2}\right)\right) n.$$

## 3 The Special Case of Factors $G$ and $\overline{G}$

A wealth of bounds on the global domination number of complementary factors exists, that is, when there are two factors $G$ and $\overline{G}$ of $K_n$.

From Observation 2, we have $\gamma_g(G) \leq \gamma(G) + \gamma(\overline{G})$. Thus any Nordhaus–Gaddum bound on this sum is also an upper bound on $\gamma_g(G)$. Several such bounds are given by Dunbar, Haynes, and Hedetniemi (2005) [20]. One is $\gamma_g(G) \leq \gamma(G) + \gamma(\overline{G}) \leq \lfloor \frac{2n}{5} \rfloor + 3$ if each of $G$ and $\overline{G}$ is connected with minimum degree at least two. Other simple bounds are

1. $\gamma_g(G) < n$ if $G$ is not complete or empty,
2. $\gamma_g(G) \leq \gamma(G) + 1$ for any graph with a pendant vertex,
3. if $G$ or $\overline{G}$ is disconnected, then $\gamma_g(G) = \max\{\gamma(G), \gamma(\overline{G})\}$, and
4. $\gamma(G) \leq \gamma_g(G) \leq \gamma(G) + 1$ if $G$ is triangle-free.

Some that depend on the minimum and maximum degrees are

1. $\gamma_g(G) \leq \max\{\Delta(G) + 1, \Delta(\overline{G}) + 1\} = \max\{n - \delta(\overline{G}), n - \delta(G)\}$ or this bound minus one if $G$ is not complete, empty, or an odd cycle,
2. if $\gamma_g(G) > \max\{\gamma(G), \gamma(\overline{G})\}$, then $\gamma_g(G) \leq \min\{\Delta(G) + 1, \Delta(\overline{G}) + 1\}$, and

3. $\gamma_g(G) \leq \begin{cases} \delta(G)+2 & \text{if } \delta(G) = \delta(\overline{G}) \leq 2 \\ \max\{\delta(G)+1, \delta(\overline{G})+1\} & \text{otherwise.} \end{cases}$

Desormeaux, Gibson, and Haynes (2015) [18] showed that, if $\gamma(G) \geq 2$, then

$$\gamma_g(G) \leq \gamma(G) + 1 + \left\lceil \frac{\Delta(G) - \gamma(G)}{\gamma(G) - 1} \right\rceil.$$

They also proved that if $n \geq [\Delta(G)]^2 + 2$, then $\gamma_g(G) = \gamma(G)$.

Zverovich and Poghosyan (2011) [47] showed that for $\delta' = \min\{\delta(G), \delta(\overline{G})\} > 0$ that

$$\gamma_g(G) \leq \left(1 - \frac{\delta'}{2^{1/\delta'}(1+\delta')^{1+1/\delta'}}\right) n \text{ and } \gamma_g(G) \leq \frac{\ln(\delta'+1) + \ln 2 + 1)}{\delta'+1} n.$$

Bounds in [8] involving the diameter when $G$ and $\overline{G}$ are both connected are improved in [18]:

1. if $diam(G) \geq 3$, then $\gamma_g(G) \leq \gamma(G) + 2$,
2. if $diam(G) = 4$, then $\gamma_g(G) \leq \max\{4, \gamma(G)+1\}$, and
3. if $diam(G) \geq 5$, then $\gamma_g(G) = \gamma(G)$.

The preceding upper bound is sharp when $diam(G) = 3$. Furthermore, when $diam(G) = 3$ or $diam(G) = 4$, there is no forbidden subgraph characterization for graphs achieving the upper bound. When $G$ and $\overline{G}$ both have diameter 2, [18] shows that $\gamma_g(G) < 1 + \sqrt{n} + \sqrt{n \ln n}$.

Dutton (2011) [24] developed several bounds involving the *packing number* $\rho_2(G)$ of graph $G$, that is, the maximum number of vertices in a subset of the vertices of $G$ such that any two vertices in the subset have distance at least three. Bounds include

1. $\gamma_g(G) \leq \min\{\max\{\gamma(G)+1, \gamma(\overline{G})+1\}, \lfloor n/2 \rfloor\}$ if $rad(G) = rad(\overline{G}) = 2$, at least one of $G$ or $\overline{G}$ is triangle-free, and $\rho_2(G) \neq \rho_2(\overline{G})$,
2. $\gamma_g(G) \leq \min\{\max\{\gamma(G)+2, \gamma(\overline{G})+2\}, \lfloor n/2 \rfloor\}$ if $rad(G) = rad(\overline{G}) = 2$, neither $G$ nor $\overline{G}$ is triangle-free, and $\rho_2(G) \neq \rho_2(\overline{G})$,
3. $\gamma_g(G) \leq \min\{4, \lfloor n/2 \rfloor\}$ if $\rho_2(G) = \rho_2(\overline{G}) = 2$ and $G$ is not equal to a triangle with a pendant edge on two of its vertices, and
4. $\gamma_g(G) \leq \min\{\delta(G)+1, \delta(\overline{G})+1, \lfloor n/2 \rfloor\}$ if $\rho_2(G) = \rho_2(\overline{G}) = 1$ and $G$ is not $K_1$ or $C_5$.

Bounds have been developed dependent on several other invariants. For example,

1. $[2m - n(n-3)]/2 \leq \gamma_g(G) \leq n - \alpha(G) + 1$ if $G$ has no isolated vertices,
2. if both $G$ and $\overline{G}$ have no isolates, $\gamma_g(G) \leq \min\{\alpha'(G)+1, \alpha'(\overline{G})+1\}$,
3. $\gamma_g(G) \leq \max\{\chi(G), \chi(\overline{G})\}$,

4. $\gamma_g(G) \leq \min\{\omega(G) + \gamma(G) - 1, \omega(\overline{G}) + \gamma(\overline{G}) - 1\}$, where $\omega(G)$ is the size of a largest clique in $G$, and
5. $\gamma_g(G) \leq \max\{n - \kappa(G) - 1, n - \kappa(\overline{G}) - 1\}$.

# 4 Variations

As with standard domination, investigations of global concepts have considered special cases depending on the type of graph as well as alternate types of domination. Section 4.1 discusses some of the investigations into special families of graphs while Section 4.2 considers alternate forms of domination. Unless otherwise stated, the factoring is of graph $G$ and its complement.

## 4.1 Restrictions on Graphs

**(Bipartite and Unicyclic Graphs)** Arumugam and Kala (2009) [4] characterized bipartite graphs G for which $\gamma_g(G) = \gamma(G) + 1$ in the following theorem:

**Theorem 7** *Let G be a connected bipartite graph with bipartition $X, Y$ and $|X| \leq |Y|$. Then $\gamma_g(G) = \gamma(G) + 1$ if and only if either G is $K_2$ or every vertex in $X$ is adjacent to at least two pendant vertices and there exists a vertex in $Y$ which is adjacent to all vertices in $X$.*

The expression $\gamma_g(G) = \gamma(G) + 1$ can be replaced by $\gamma_g(G) = \beta(G) + 1$ in the above theorem. The characterization of unicyclic graphs for which $\gamma_g(G) = \beta(G) + 1$ is also derived.

**(Trees)** Trees are, of course, a special class of bipartite graphs. Both Rall (1991) [38] and Brigham and Dutton [8] have characterized those trees which achieve the upper value of $\gamma_g(G) = \gamma(G) + 1$.

**Theorem 8** *Let $T$ be a tree. Then $\gamma_g(T) = \gamma(T) + 1$ if and only if $T$ is a star with at least two vertices or $T$ has radius two and contains a vertex of degree at least two, all of whose neighbors have degree at least three.*

In a difficult multi-case argument based on the diameter, Mojdeh, Alishahi, and Chellali (2016) [34] characterize those trees $T$ for which $\gamma_g(T) = \gamma_g(T^2)$, where $T^2$ is the square of $T$. The square of a graph $G$ is the graph on the same vertex set as $G$ with two vertices adjacent in $G^2$ if and only if their distance in $G$ is at most 2.

**(Planar Graphs)** Enciso and Dutton (2008) [26] found several bounds on the global domination number of planar graphs $G$, including $\gamma_g(G) \leq \max\{\gamma(G) + 1, 4\}$.

**(Interval Graphs)** Maheswari, Lakshmi Naidu, Nagamuni Reddy, and Sudhakaraiah (2011) [33] have developed a construction that determines $\gamma_g(G)$ when $G$ is a connected interval graph based on the intervals $I_i = [a_i, b_i]$ for $1 \le i \le r$, $r \ge 2$, indexed in increasing order of their right endpoint. Letting interval $I_i$ be represented by its index $i$, $nbd[i]$ is the set of intervals intersecting $I_i$, $\max(i)$ is the largest interval in $nbd[i]$, $NI(i) = j$, where $j$ is the smallest index for which $b_i < a_j$ or null if there is no such $j$, and $Next(i) = \max\{nbd[NI(i)]\backslash nbd[\max(i)]\}$. The set of intervals is augmented by dummy intervals $I_0 = [a_0, b_0]$ and $I_{r+1} = [a_{r+1}, b_{r+1}]$ such that $b_0 < \min_{1\le i\le r}\{a_i\}$ and $a_{r+1} > b_r$. A directed graph $\hat{D} = (N, L)$ is constructed from the intervals, including $I_0$ and $I_{r+1}$, where $N$ corresponds to intervals that are not properly contained in other intervals. The directed edges are formed as follows: from $I_0$ to $I_j$ if and only if there is no interval $I_h$ such that $b_0 < a_h < b_h < a_j$, from $I_j$ to $I_{r+1}$ if and only if there is no interval $I_h$ such that $b_j < a_h < b_h < a_{r+1}$, and from $I_i$ to $I_j$ if and only if $j = Next(i)$. The main result shows that the vertices in a shortest directed path in $\hat{D}$ between $I_0$ and $I_{n+1}$ form a minimum global dominating set of $G$.

**(Changing and Unchanging)** Harary (1982) [29] defined changing and unchanging of an invariant of a graph $G$ as the study of how the value of the invariant changes when an edge is removed from $G$, an edge is added to $G$, or a vertex is removed from $G$. Interest has concentrated on when a result is true for every possible edge or every vertex of $G$. Let $e$ refer to an arbitrary edge of a graph.

Dutton and Brigham (2009) [25] showed

1. $\gamma_g(G) - 1 \le \gamma_g(G - e) \le \gamma_g(G) + 1$,
2. if $\gamma_g(G - e) = \gamma_g(G) - 1$ for some edge $e$ of $G$, then $\gamma_g(G - e) \le \gamma_g(G)$ for every edge of $G$,
3. if $\gamma_g(G - e) = \gamma_g(G) + 1$ for some edge $e$ of $G$, then $\gamma_g(G - e) \ge \gamma_g(G)$ for every edge of $G$, and
4. $\gamma_g(G - e) = \gamma_g(G) + 1$ for every edge of $G$ if and only if $G$ is a collection of at least two stars.

Other changing and unchanging problems appear to be more difficult. A characterization is given in [25] of graphs $G$ for which $\overline{G}$ is disconnected and $\gamma_g(G-e) = \gamma_g(G) - 1$ for every edge of $G$. Desormeaux, Haynes, and van der Merwe (2017) [19] studied graphs $G$ for which removing an arbitrary edge from $G$ and adding it to $\overline{G}$ decreases the global domination number. They show for non-empty graphs $G$ such that $\overline{G}$ is also non-empty and $\gamma(G) = 3$, that $\gamma_g(G-e) = \gamma_g(G) - 1$ for every edge of $G$ if and only if $G$ is one of the self-complementary graphs of order 5, that is, $G$ is either $C_5$ or a triangle with a pendant edge on two of its vertices.

Still and Haynes (2013) [44] characterized trees of domination number 2 and 3 whose global domination number remains the same when one of the changes occurs. A caterpillar is a tree which becomes a path (the spine) when all leaves are removed. A caterpillar with spine $(v_1, v_2, \ldots, v_r)$ can be coded by an $r$-tuple $(x_1, x_2, \ldots, x_r)$, where $x_i$ is the number of leaves adjacent to $v_i$ for $1 \le i \le r$. In the following let $T$ be a tree.

1. If $\gamma_g(T) = 2$, then $\gamma_g(T - e) = \gamma_g(T)$ for every edge $e$ of $T$ if and only if $T$ is $P_4$ or a non-trivial star,
2. if $\gamma_g(T) = 3$, then $\gamma_g(T - e) = \gamma_g(T)$ for every edge $e$ of $T$ if and only if $T$ is $P_7$ or the caterpillar $(1, 1, 1)$,
3. if $\gamma_g(T) = 2$, then $\gamma_g(T - v) = \gamma_g(T)$ for every vertex $v$ of $T$ if and only if $T$ is $P_3$, $P_4$, or $P_5$,
4. if $\gamma_g(T) = 3$, then $\gamma_g(T - v) = \gamma_g(T)$ for every vertex $v$ of $T$ if and only if $T$ is $P_8$, the caterpillar $(1, 0, 1, 0, 1)$, or $K_{1,3}$ with each edge subdivided exactly once,
5. if $\gamma_g(T) = 2$, then $\gamma_g(T + e) = \gamma_g(T)$ for every edge $e$ of $\overline{G}$ if and only if $T$ is $P_4$ or a non-trivial star, and
6. if $\gamma_g(T) = 3$, then $\gamma_g(T + e) = \gamma_g(T)$ for every edge $e$ of $\overline{G}$ if and only if $T$ is in one of seven infinite classes of caterpillars or $T$ is the caterpillar $(1, 1, 1)$ with spine $x$, $y$, $z$ adjacent to leaves $x'$, $y'$, $z'$, respectively, with new leaves adjacent to $x'$, $y'$, and $z'$ with at least one adjacent to $y'$.

## 4.2 Other Types of Global Domination

**(Global Connected Domination)** Kulli, Janakiram, and Soner (2009) [32] and Delić and Wang (2014) [17] have studied a global version of connected domination. A subset $D$ of the vertex set of a connected graph $G$ is a *global connected dominating set* of $G$ if it is both a global dominating set and a connected subgraph of $G$. However, the definitions differ in the following way. In [32] $\overline{G}$ must be connected and the vertices of $D$ must induce a connected subgraph in it, while those are not requirements in [17]. Let $\gamma_{gc}(G)$ represent the minimum size of a global connected dominating set that does not have to be connected in $\overline{G}$ and $\gamma_{gcb}(G)$ be the corresponding value when $D$ induces connected subgraphs in both $G$ and $\overline{G}$. Not surprisingly, the two definitions can lead to different results. Consider Figure 4 showing $P_4$ with vertices $v_1, v_2, v_3, v_4$ in order. A minimum connected dominating set is $\{v_2, v_3\}$ and this also is a dominating set of $\overline{P_4}$. Thus $\gamma_{gc}(P_4) = 2$. However, $\{v_2, v_3\}$ is not a connected dominating set of $\overline{P_4}$. In fact, the only connected dominating set of $P_4$ that is a connected dominating set of $\overline{P_4}$ is $\{v_1, v_2, v_3, v_4\}$, implying $\gamma_{gcb}(P_4) = 4$.

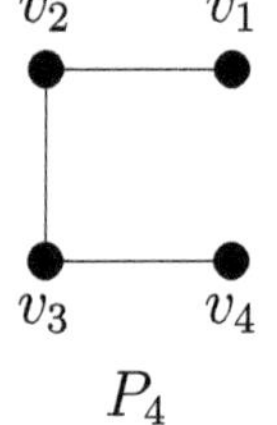


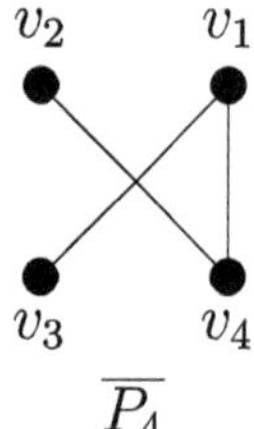


**Fig. 4** A graph with distinct values for $\gamma_{gc}(G)$ and $\gamma_{gcb}(G)$

Emphasizing the nonsymmetric aspect of the definition of $\gamma_{gc}$, [17] constructs connected graphs $G$ of order $n \geq 5$ such that $\overline{G}$ is connected and $\gamma_{gc}(G) \neq \gamma_{gc}(\overline{G})$. It also shows that, for $2 \leq k \leq n$, there is a connected graph $G$ such that $\gamma_{gc}(G) = k$ and a characterization of such graphs of order $n \geq 5$ is given when $3 \leq k \leq n-2$. Udayakumar and Sasireka (2010) [46] present results for $\gamma_{gc}(G)$ when $G$ is a Cartesian product.

**(Upper Global Domination)** The *upper global domination number* $\Gamma_g(G)$ is the maximum number of vertices in a minimal global dominating set of $G$. Rall (1991) [38] showed that $\Gamma_g(G) = \Gamma(G)$ when $G$ is a tree of diameter three or four or when $G$ has diameter at least five. He also points out there are trees of diameter three or four for which there are minimal global dominating sets that are not minimal dominating sets.

**(Double Global Domination)** Soner, Chaluvaraju, and Janakiram (2003) [43] discuss double global domination. A set $D \subseteq V(G)$ is a *double dominating set* if each vertex in $G$ is dominated by at least two vertices of $D$, and the minimum cardinality of such a set is denoted $dd(G)$. Such a $D$ is a *double global dominating set* if it is also a double dominating set of $\overline{G}$, and the minimum cardinality of such a set is designated $gdd(G)$. Several inequalities are given including

1. $\gamma_g(G) + 1 \leq gdd(G)$ with equality if $G = C_5$,
2. $\frac{2m(G)-n(n-5)}{3} \leq gdd(G)$,
3. $gdd(G) \leq 2\alpha'(G)$ if both $G$ and $\overline{G}$ have minimum degree at least 2,
4. $gdd(G) \leq 2\beta(G)$ if $G$ has minimum degree at least 2, and
5. $gdd(G) = dd(G)$ if $G$ has diameter at least 6.

**(Global Set Domination)** For connected graphs $G$, a subset $D$ of vertex set $V$ is a *set dominating set* if, for every subset $R$ of $V - D$, there is a non-empty subset $S$ of $D$ such that the subgraph induced by $S \cup R$ is connected. D is a *point-set dominating set* if $S$ is a single vertex and a *2 point-set dominating set* if $|S| \leq 2$. Global versions are possible if $G$ is *co-connected*, that is, both $G$ and $\overline{G}$ are connected. They have been studied, respectively, by Sampathkumar and Pushpa Latha (1994) [42], Pushpa Latha (1997) [36], and Gupta and Jain (2016) [28]. The associated domination and global domination numbers of $G$ are designated $\gamma_{set}(G)$ and $\gamma_{gset}(G)$, $\gamma_{1set}(G)$ and $\gamma_{g1set}(G)$, and $\gamma_{2set}(G)$ and $\gamma_{g2set}(G)$, respectively. A representative sample of results is given next, always assuming the graphs are co-connected.

1. For a tree $T$ with $n$ vertices and $\varepsilon$ leaves that is not a star, $\gamma_{gset}(T) = n - \varepsilon$,
2. if $G$ has a cut vertex, then $\gamma_{gset}(G) \leq \gamma_{set} + 1$, with a corresponding statement for global point-set domination,
3. if $G$ has diameter at least five, then $D$ is a minimal set dominating set if and only if $D$ is a minimal global set dominating set and $\gamma_{gset}(G) = \gamma_{set}(G)$,
4. if $diam(G) \geq 4$, $\gamma_{g1set}(G) = \gamma_{1set}(G)$ and $\gamma_{g2set}(G) = \gamma_{2set}(G)$, and
5. if $T$ is a tree, $\gamma_{g2set}(T) = \gamma_{2set}(T)+1$ if $diam(G) = 2$ and $\gamma_{g2set}(T) = \gamma_{2set}(T)$ otherwise.

**(Global Irredundance)** A vertex $x$ in a subset $S$ of the vertices of a graph $G$ is *irredundant* if $N[x] - N[S - \{x\}] \neq \emptyset$ and $S$ is *irredundant* if each $x \in S$ is. Dunbar and Laskar (1992) [21] and Dunbar, Laskar, and Monroe (1991) [22] discuss global versions of this concept. Set $S \subseteq V$ is a *global irredundant set* if each $x \in S$ is irredundant in either $G$ or $\overline{G}$, and is a *universal irredundant set* if each $x \in S$ is irredundant in both $G$ and $\overline{G}$. The symbols $ir_g(G)$ ($ir_u(G)$) and $IR_g(G)$ ($IR_u(G)$) represent the smallest and largest orders, respectively, of a maximal global (universal) irredundant set. The inequality $ir_g(G) \leq \gamma_g(G) \leq \Gamma_g(G) \leq IR_g(G)$ is a fundamental one relating four of the global invariants and parallels a corresponding result for the nonglobal versions (see Cockayne and Hedetniemi (1977) [13]). We also have

1. $IR_u(G) \leq IR(G) \leq IR_g(G)$, where $IR(G)$ is the maximum size of a maximal irredundant set of $G$,
2. $IR_g(G) \leq \Delta(G) + 1$ if $IR(G) < IR_g(G)$, and
3. $IR_u(G) \leq \Delta(G) + 1$, with equality in the latter for connected graphs with more than two vertices if and only if $G$ is $K_{r,r}$ minus a 1-factor.

As an example of the first of the above inequalities, consider $C_5$ with vertices labeled in order as $v_1, v_2, v_3, v_4$, and $v_5$. Then $IR_u(C_5) = IR(C_5) = 2$ where $\{v_1, v_3\}$ is both a maximum maximal irredundant set and a maximum maximal universal irredundant set. Furthermore, $\{v_1, v_2, v_3\}$ is a maximum global irredundant set so $IR_g(C_5) = 3$. Determining whether $IR_u(G) \geq M$, where $M$ is a positive integer, is an NP-complete problem. Nevertheless, the value of $IR_u(G)$ is known for many graph classes including paths and complete bipartite graphs. Furthermore, $IR_u(G) = 2$ for all trees except $K_{1,s}$ for which it is 1. On the other hand, $IR_u(G) \geq 2$ for any graph with diameter at least three. Graphs with $IR_g(G) \geq 3$ are equivalent to those containing an induced subgraph isomorphic to $K_3$, $\overline{K_3}$, $C_6$, $\overline{C_6}$, or $C_5$.

**(Global Domatic Number)** Sampathkumar (1989) [41] studied the *global domatic number* $d_g(G)$, that is, the maximum order of a partition of $V(G)$ such that each member of the partition is a global dominating set. For graph $G$, results include

1. $d_g(G) \leq \min\{\delta(G) + 1, \delta(\overline{G}) + 1\}$,
2. $d_g(G) \leq (n + 1)/2$,
3. $\gamma_g(G) + d_g(G) \leq n + 1$ with equality if and only if $G$ is complete or empty, and
4. if $G$ is a tree $T$, $d_g(T) \leq 2$.

**(Global Dominating-$\chi$-Coloring)** Sahul Hamid and Rajeswari (2014, 2017) [37, 40] introduced the *global dominating-$\chi$-coloring number* $gd_\chi(G)$ of a graph $G$, that is, the maximum number of color classes in a $\chi(G)$-coloring of $G$ that are global dominating sets of $G$. They derive its value for paths, cycles, wheels, complete graphs, complete multipartite graphs, and the Petersen graph. As an example consider $C_7$ illustrated in Figure 5 with a specific $\chi(G) = 3$ coloring. The set of vertices colored 1 and the set colored 2 both are global dominating sets of $C_7$. Thus $gd_\chi(G) \geq 2$. Any 3-coloring of $C_7$ has one color class with fewer than three

**Fig. 5** $gd_\chi(C_7) = 2$

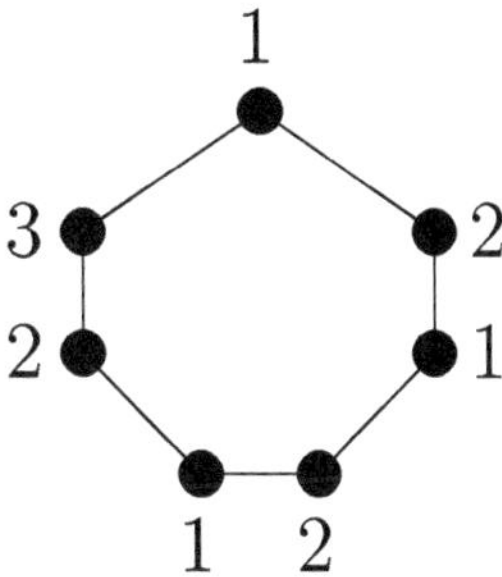


vertices. Since $\gamma_g(C_7) = 3$, at least one color class in any 3-coloring will not be a global dominating set and $gd_\chi(G) = 2$.

Further results, if $G$ is unicyclic with single cycle $C$, are

1. $gd_\chi(G) = 2$ if $C$ has even length at least 6,
2. $gd_\chi(G) = 1$ if $C$ has odd length and each of its vertices is a support vertex, and
3. $gd_\chi(G) = 2$ if $C$ is of odd length at least 7 and not all its vertices are support vertices.

For a tree $T$

1. $gd_\chi(T) = 0$ if and only if $T$ is a star or a double star and
2. $gd_\chi(T) = 1$ if and only if $T$ is obtained from a star by adding at least one pendant edge to at least two leaves of the star.

Bounds include $gd_\chi(G) \leq \delta(G) + 1$ and $gd_\chi(G) \leq \frac{n-\chi(G)s(G)}{\gamma_g(G)-s(G)}$, where $s(G)$ is the minimum cardinality of any color class in any $\chi$-coloring of $G$, and the latter bound is sharp. For integers $k > 1$ and $l$ with $0 \leq l \leq k$, there is a uniquely $k$-colorable connected graph $G$ with $\chi(G) = k$ and $gd_\chi(G) = l$.

## 5 Global Total Domination

A total dominating set $D$ of a graph $H$ with factors $F_1, F_2, \ldots, F_k$ is a *global total dominating set* if it is a total dominating set of every factor. Throughout this section we assume no factor has an isolated vertex. The size of a smallest global total dominating set is indicated by $\gamma_{gt}(H) = \gamma_{gt}(F_1, F_2, \ldots, F_k)$. Most of the research, as with other forms of global domination, has concentrated on the case when $H$ is complete with two factors $G$ and $\overline{G}$ and, as before, we write $\gamma_{gt}(G, \overline{G}) = \gamma_{gt}(G) = \gamma_{gt}(\overline{G})$.

Dunbar, Laskar, and Wallis (1992) [23] show that a minimal global total dominating set is not necessarily a minimal total dominating set and specify two graphs that illustrate this, one of which is shown in Figure 6. Here $\gamma_t(G) = 2$, $\gamma_{gt}(G) = 5$, and $\{v_1, v_2, v_3, v_4, v_6\}$ is a minimal (minimum in this case) global total

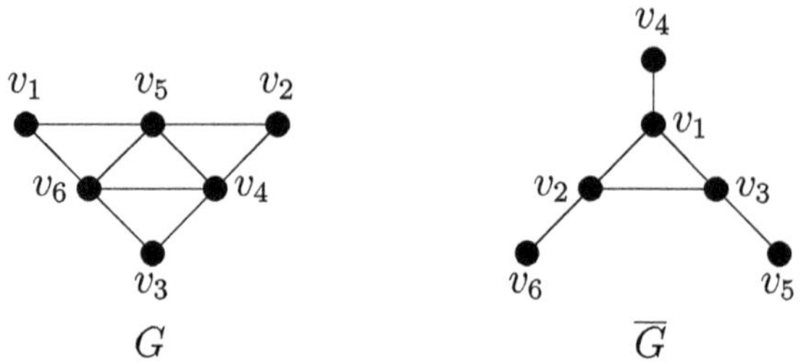

**Fig. 6** Example showing a graph whose minimum global total dominating set is not a minimum total dominating set

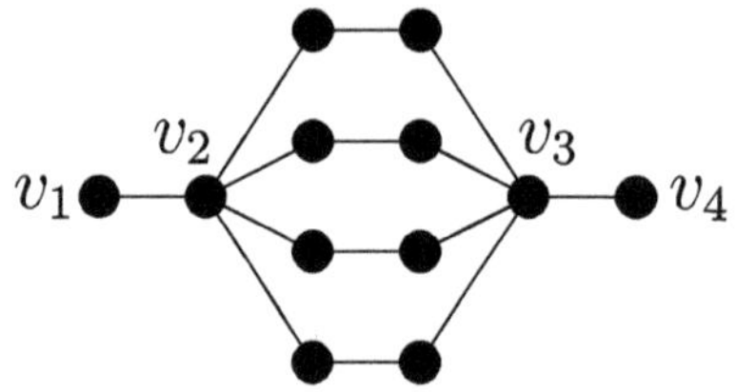

**Fig. 7** Example showing a graph $G$ with $\gamma_t(G) = \gamma_{gt}(G)$

dominating set. On the other hand, the graph of Figure 7 has $\gamma_t(G) = \gamma_{gt}(G) = 4$ and $\{v_1, v_2, v_3, v_4\}$ is a minimal (again minimum) global total dominating set.

Values of the global total domination number are given by Kulli and Janakiram (1996) [31] for paths, cycles, and complete bipartite graphs. The following theorem also is shown.

**Theorem 9** *Given graph G,*

1. *if* $diam(G) = 3$, *then* $\gamma_{gt}(G) \leq \gamma_t(G) + 2$,
2. *if* $diam(G) = 4$, *then* $\gamma_{gt}(G) \leq \gamma_t(G) + 1$, *and*
3. *if* $diam(G) \geq 5$, *then* $\gamma_{gt}(G) = \gamma_t(G)$.

**Proof:** Let $D$ be a minimum total dominating set of $G$. If $diam(G) = 3$, let $u$ and $v$ be vertices of distance 3 apart. No vertex of $G$ is adjacent to both $u$ and $v$, so $\{u, v\}$ is a total dominating set of $\overline{G}$ and $D \cup \{u, v\}$ is a global total dominating set of $G$. When $diam(G) = 4$, let $u$ and $v$ be distance 4 apart. Let $u_1 \in D \cap N[u]$. No vertex of $G$ is adjacent to both $u_1$ and $v$, so $\{u_1, v\}$ is a total dominating set of $\overline{G}$ and $D \cup \{v\}$ is a global total dominating set of $G$. If $diam(G) \geq 5$, then no vertex of $G$ is adjacent to every vertex of $D$, and so $D$ is also a total dominating set of $\overline{G}$. □

Using this result, Akhbari, Eslahchi, Jafari Rad, and Hasni (2015) [2] show for a tree $T$ that

1. if $diam(T) = 3$, then $\gamma_{gt}(T) = 4$,
2. if $diam(T) = 4$, then $\gamma_{gt}(T) = \gamma_t(T) + 1$, and
3. if $diam(T) \geq 5$, then $\gamma_{gt}(T) = \gamma_t(T)$.

They also note that the following problem is NP-complete.

GLOBAL TOTAL DOMINATING SET
INSTANCE: Graph $G$, positive integer $k$.

QUESTION: Does $G$ have a global total dominating set of cardinality at most $k$?

The proof employs a polynomial transformation from the known NP-complete problem TOTAL DOMINATING SET.

Dankelmann, Henning, Goddard, and Laskar (2006) [15] concentrate on the general case in their discussion of simultaneous domination. The symbol $\gamma_{sdt}$ is employed for these results to emphasize the domination is simultaneous. The following is shown.

**Theorem 10** *If there are at least two factors of $K_n$, all connected, then $\gamma_{sdt}(F_1, F_2, \ldots, F_k) \le n$ and the bound is sharp.*

**Proof:** Since the bound is obvious, only sharpness needs to be considered. Let $n$ be even and the vertices partitioned into $V_1$ and $V_2$ of equal size $n/2$. Let $F_1$ be a $K_{n/2}$ on $V_1$ along with a perfect matching between $V_1$ and $V_2$. Similarly $F_2$ is a $K_{n/2}$ on $V_2$ along with a different perfect matching between $V_1$ and $V_2$. Then every vertex of the graph is a leaf in one of the factors and thus it or a neighbor must be in any global total dominating set, implying $\gamma_{sdt}(F_1, F_2) = n$. □

In contrast, when the factors are $G$ and $\overline{G}$, Kulli and Janakiram (1996) [31] prove $\gamma_{gt}(G) = n$ if and only if $G$ is $P_4$, $mK_2$, or $\overline{mK_2}$, where $m \ge 2$.

Other bounds found in [15], assuming factors are of $K_n$, are

1. if $\delta(F_i) \ge 2$ for $1 \le i \le k$, then $\gamma_{sdt}(F_1, F_2, \ldots, F_k) \le \frac{2kn}{2k+1}$ and the bound is sharp,
2. $\gamma_{sdt}(F_1, F_2, \ldots, F_k) \le \frac{(\ln\delta + \ln k + 1)n}{\delta}$, where $\delta = \min\{\delta(F_i) | i = 1, \ldots, k\}$,
3. if $\delta(F_i) \ge 3$ for $i = 1, 2$, then $\gamma_{sdt}(F_1, F_2) \le n(1 - \sqrt{2/27})$ and the bound becomes $13n/21$ if $\delta(F_i) \ge 4$ for $i = 1, 2$, and
4. if $F_1$ and $F_2$ are $n$-cycles, $n \ge 3$ and $n \ne 5, 10$, then $\gamma_{sdt}(F_1, F_2) \le 3n/4$ and the bound is sharp for infinitely many $n$.

Jafari Rad and Sharifi (2015) [30] present additional bounds when the factors are $G$ and $\overline{G}$. They show a series of results depending on the value of $\delta' = \min\{\delta(G), \delta(\overline{G})\}$, including

1. if $\delta' = 1$, $\gamma_{gt}(G) \le \frac{2}{3}n + 1$,
2. if $\delta' = 2$, $\gamma_{gt}(G) \le \frac{22}{27}n$,
3. if $\delta' = 3$, $\gamma_{gt}(G) \le 0.683n$, and
4. If $\delta' > 3$ there are the two bounds $\gamma_{gt}(G) \le \left(1 - \frac{\delta'}{3^{\frac{1}{\delta'}}(1+\delta')^{1+\frac{1}{\delta'}}}\right) n$ and $\gamma_{gt}(G) \le \left(\frac{\ln(1+\delta') + \ln 3 + 1}{1+\delta'}\right) n$.

Other upper bounds appearing in [31] are $\gamma_{gt}(G) \le 2\beta(G)$ and, if $\min\{diam(G), diam(\overline{G})\} \ge 3$, then $\gamma_{gt}(G) \le \min\{n - \Delta(G) + 2 + \epsilon, \delta(G) + 4 - \epsilon\}$, where $\epsilon$ is 0 if $G$ is connected and 1 otherwise.

Lower bounds are not as common, but a few are found in [10, 31], including

1. $\gamma_{gt}(F_1, F_2, \ldots, F_k) \ge k + 1$ if $k$ is odd,

2. $\gamma_{gt}(F_1, F_2, \ldots, F_k) \geq k + 2$ if $k$ is even, and
3. $\gamma_{gt}(G) \geq 2m(G) - n(n - 3)$.

A few results concerning the structure of graphs with a given global total domination number appear in [2]. One shows, for any $r \geq 4$, there is a graph $G$ such that $diam(G) = diam(\overline{G}) = 2$ and $\gamma_{gt}(G) = r$. Furthermore, characterizations are given of graphs $G$ for which $\gamma_{gt}(G) = 4$ and $\gamma_{gt}(G) = n - 1$.

Whereas $\gamma_t(G)$ and $\gamma_{gt}(G)$ are the smallest order of a minimal total dominating set and minimal global dominating set of $G$, respectively, [23] defines $\Gamma_t(G)$ and $\Gamma_{gt}(G)$ as the largest order of a minimal total dominating set and minimal global total dominating set, respectively. Then, if $diam(G) > 4$, $\Gamma_{gt}(G) = \Gamma_t(G)$, paralleling a result for $\gamma_t(G)$ and $\gamma_{gt}(G)$.

## 6 Global Roman Domination

For graph $G = (V, E)$, a function $f : V \to \{0, 1, 2\}$ is a *Roman dominating function* if every vertex $u$ for which $f(u) = 0$ is adjacent to at least one vertex $v$ for which $f(v) = 2$. An introduction to Roman domination can be found in Cockayne, Dreyer Jr., Hedetniemi, and Hedetniemi (2004) [14]. A Roman dominating function is a *global Roman dominating function* if it also is a Roman dominating function for $\overline{G}$. The *weight* of such a function is $\sum_{u \in V} f(u)$ and the minimum weight of a Roman dominating function (global Roman dominating function) is denoted $\gamma_R(G)(\gamma_{gR}(G))$. Such minimum functions are $\gamma_R(\gamma_{gR})$-functions. The function partitions $V$ into $\{V_0, V_1, V_2\}$, where $V_i = \{v \in G : f(v) = i\}$ for $i = 0, 1, 2$.

Roushini Leely Pushpam and Padmapriea (2016) [39] present the following basic inequality relating the global Roman domination number to the standard global domination number.

**Theorem 11** *For any graph $G$, $\gamma_g(G) \leq \gamma_{gR}(G) \leq 2\gamma_g(G)$.*

**Proof:** For any $\gamma_{gR}$-function note that $V_1 \cup V_2$ dominates $G$ and $\overline{G}$ implying the left inequality. Placing 2 on the vertices of any $\gamma_g$-set and 0's elsewhere shows the right inequality. □

The values of $\gamma_{gR}(G)$ when $G$ is a path, cycle, complete graph, complete bipartite graph, or a wheel are given. A main result characterizes graphs for which $\gamma_{gR}(G) = n$. This is the maximum possible value since assigning 1 to every vertex yields a weight of $n$. The proof is a lengthy multi-case argument, but several intermediate results are of interest in themselves, including

1. $\gamma_{gR}(G) = n$ implies $diam(G) \leq 3$,
2. if $diam(G) = 3$, then $\gamma_{gR}(G) = n$ if and only if $G$ is $C_3$ with pendant edges incident to two or three of its vertices or $P_4$, and
3. if $\omega(G) = 2$, then $\gamma_{gR}(G) = n$ if and only if $G$ is $P_3$, $C_4$, $C_5$, or $K_{1,3}$.

Abdollahzadeh Ahangar (2016) [1] presents many additional characterizations including several on the relationship between $\gamma_g(G)$ and $\gamma_{gR}(G)$. In particular those graphs for which $\gamma_{gR}(G) = \gamma_g(G) + i$ for $0 \leq i \leq 3$ are characterized. The results for $i = 0, 1, 2$ are summarized next, where graph $G$ is assumed connected.

1. $\gamma_{gR}(G) = \gamma_g(G)$ if and only if $G = K_n$,
2. if $n \geq 5$, then $\gamma_{gR}(G) = \gamma_g(G) + 1$ if and only if $G = K_n - e$, and
3. if $n \geq 9$ and $G$ is not $K_n$ or $K_n - e$, then $\gamma_{gR}(G) = \gamma_g(G) + 2$ if and only if $G$ is one of $K_{n-3} + \overline{K_3}$, $K_{n-3} + (K_2 \cup K_1)$, $K_{n-4} + 2K_2$, $K_{n-4} + P_4$, $K_{n-4} + C_4$, or $G$ has two vertices $u$ and $v$ such that $|((N(u) \cup N(v)) \backslash (N(u) \cap N(v))) \backslash \{u, v\}| = n - \gamma_g(G)$.

The result for $i = 3$ is somewhat more complicated. When $G$ is unicyclic, [1] shows $\gamma_{gR}(G) = \gamma_g(G) + 2$ if and only if $G$ is a star of order at least five with an edge added between two leaves or is a $C_4$ with vertices $x_1, x_2, x_3, x_4$, in order, with at least two pendant edges on each of $x_1$ and $x_3$.

Atapour, Sheikholeslami, and Volkmann (2015) [5] characterized trees with specific relations between their global Roman and Roman domination numbers. As a step they developed the inequality $\gamma_{gR}(G) \leq \gamma_R(G) + \delta(G) + 1$ for a graph $G$ of order at least 4, a bound that is sharp for stars $K_{1,r}$ with $r \geq 3$. This shows, for tree $T$ with at least four vertices, that $\gamma_{gR}(T) \leq \gamma_R(T) + 2$. Since $\gamma_R(G) \leq \gamma_{gR}(G)$, it follows, for any tree, that $\gamma_{gR}(T) = \gamma_R(T) + i$, where $i$ is 0, 1, or 2. The characterization is given for trees when $i$ is 1 or 2 and involves a set of graphs derived from spiders. Recall a *spider* is a star whose edges have been subdivided any number of times. If the underlying spider is formed from a $K_{1,r}$, it is designated $S_r$. The *center* of a spider is the vertex that is the center of the star from which it is derived. The graphs required for the characterization theorems are defined as follows:

1. $\beta_1$ is the collection of spiders $S_r$, $r \geq 2$, (except stars and $P_5$),
2. $\beta_2$ is the collection of trees obtained from spiders $S_{r_1}, S_{r_2}, \ldots, S_{r_j}$ (except $P_4$ and $P_5$), $j \geq 2$ and $r_i \geq 2$ for $1 \leq i \leq j$, with centers $y_1, y_2, \ldots, y_j$, respectively, by adding a new vertex $x$ and edges $xy_i$ for $1 \leq i \leq j$, and
3. $\beta_3$ is the collection of trees obtained from a tree in $\beta_2$, say $T_1$, and adding a new vertex $z$ and edge $xz$, where $x$ is the vertex added to construct $T_1$.

The characterization for trees $T$ of order at least 4 is

1. $\gamma_{gR}(T) = \gamma_R(T) + 2$ if and only if $T$ is $K_{1,r}$ for $r \geq 3$ and
2. $\gamma_{gR}(T) = \gamma_R(T) + 1$ if and only if $T \in \beta_1 \cup \beta_2 \cup \beta_3$.

Some conditions are given for when $\gamma_{gR}(T) = \gamma_R(T)$, including if $diam(T) \geq 7$ (a result that also holds for any connected graph) and if $T$ has a $\gamma_R(T)$-function $f$ such that $f(u) = f(v) = 2$, where $u$ and $v$ are vertices for which $d(u, v) \neq 2$.

Both [1] and [5] have several upper bounds, many of which are similar to those developed for other types of global domination. Representative ones are

1. $\gamma_{gR}(G) \leq n - \left\lceil \frac{g(G)}{3} \right\rceil$ if $\delta(G) \geq 2$ and girth $g(G) \geq 6$,

2. $\gamma_{gR}(G) \leq n - \left\lceil \frac{diam(G)+1}{3} \right\rceil$ if $n \geq 5$,
3. $\gamma_{gR}(G) \leq n - deg(u) - deg(v) + 2|N(u) \cap N(v)| + 2$, where $n \geq 4$ and $u$ and $v$ are nonadjacent vertices, and
4. $\gamma_{gR}(G) \leq 2n \left(1 - \frac{\delta'}{(1+\delta')^{1+\frac{1}{\delta'}}}\right)$, where $\delta' = \min\{\delta(G), \delta(\overline{G}\} \geq 1$.

As in other types of global domination, lower bounds are rare, but the following is shown: if $G$ is connected, then $\gamma_{gR}(G) \geq \left\lceil \frac{diam(G)+2}{2} \right\rceil$.

Amjadi, Nazari-Moghaddam, and Sheikholeslami (2017) [3] investigate a total version of the global Roman domination number. Defined for graphs $G$ with no isolated vertices, a function $f$ is a *total Roman dominating function* if it is a Roman dominating function with the additional property that the subgraph of $G$ induced by the set of vertices of $G$ having positive weight under $f$ has no isolated vertex. Such an $f$ is a *global total Roman dominating function* if $f$ also is a total Roman dominating function of $\overline{G}$. The minimum weights for total Roman domination and global total Roman domination functions are $\gamma_{tR}$ and $\gamma_{gtR}$, respectively. The fundamental inequality $\gamma_{gt}(G) \leq \gamma_{gtR}(G) \leq 2\gamma_{gt}(G)$ is shown, as well as characterizations of graphs achieving the bounds, that is,

1. $\gamma_{gtR}(G) = \gamma_{gt}(G)$ if and only if $G$ is $P_4$, $mK_2$, or $\overline{mK_2}$, where $m \geq 2$ and
2. $\gamma_{gtR}(G) = 2\gamma_{gt}(G)$ if and only if there is a $\gamma_{gtR}(G)$-function $f = (V_0, V_1, V_2)$ such that $V_1 = \emptyset$.

Many results involving the global total Roman domination number of trees $T$ are shown including $\gamma_{gtR}(T) = \gamma_{tR}(T)$ if $diam(T) \geq 5$. A characterization is given of trees for which $\gamma_{gtR}(T) = \gamma_{tR}(T)$. Further results relate $\gamma_{gtR}(G)$ to $\gamma_{tR}(G)$ and $\gamma_{gR}(G)$ including

1. $\gamma_{gtR}(G) = \gamma_{tR}(G)$ if $rad(G) \geq 5$ and $G$ is connected,
2. $\gamma_{gtR}(G) \leq \gamma_{tR}(G) + 1$ if $diam(G) \geq 5$, and
3. $\gamma_{gtR}(G) \leq 2(\gamma_{gR}(G) - 1)$ if $n \geq 4$.

## 7 Open Problems

Some of the open problems presented in the literature are listed here.

1. [10] Given a graph $G$, determine ways to factor it to achieve a global dominating set having defined characteristics.
2. [10] Given the global domination number of a graph and the domination number of each factor, find a simpler characterization of when such a factoring is possible.
3. [10] Investigate additional complexity results. For example, consider the case of two factors, each a path on n vertices. Is determining whether their global domination number is at most $M$ an NP-complete problem?

4. [16] If $F_1$ and $F_2$ are edge-disjoint factors of $K_n$ (not necessarily including all the edges of $K_n$) and $\delta(F_i) \geq 2$ for $i = 1, 2$, is $\gamma_g(F_1, F_2) \leq \frac{3}{5}n$.
5. [9] For all $n \geq 4$, determine the exact value of $\gamma_{sd}(C_n, C_n)$ and $\gamma_{sd}(P_n, P_n)$. An upper bound of $n/2$ is shown in [9] for these numbers when $n \geq 4$ is even and of $(n+1)/2$ for $\gamma_{sd}(C_n, C_n)$ when $n \geq 5$ is odd.
6. [9] Characterize the connected factors $F_1$ and $F_2$ on $n$ vertices that have a 1-factor and satisfy $\gamma_{sd}(F_1, F_2) = n/2$.
7. [4] Characterize graphs $G$ for which $\gamma_g(G) = \beta(G) + 1$.
8. [25] Prove or disprove: If $\gamma_g(G - e) = \gamma_g(G) - 1$ for every edge $e$ of $G$ and $\gamma_g(\overline{G} - e) = \gamma_g(\overline{G}) - 1$ for every edge $e$ of $\overline{G}$, then $G$ is self-complementary. That this is true when $\gamma_g(G) = 3$ is shown in [19].
9. [25] Prove or disprove: If $\gamma_g(G - e) = \gamma_g(G) - 1$ for every edge $e$ of $G$ and $\gamma_g(\overline{G} - e) = \gamma_g(\overline{G}) - 1$ for every edge $e$ of $\overline{G}$, then $\gamma_g(G - v) = \gamma_g(G) - 1$ for every vertex $v$ of $G$. This is true when $\gamma_g(G) = 3$ as is shown in [19].
10. [44] Characterize graphs $G$ such that $\gamma_g(G - e) = \gamma_g(G)$ for every edge $e$ of $G$ (or trees T for $\gamma_g(T) \geq 4$).
11. [44] Characterize graphs $G$ such that $\gamma_g(G - v) = \gamma_g(G)$ for every vertex $v$ of $G$ (or trees T for $\gamma_g(T) \geq 4$).
12. [44] Characterize graphs $G$ such that $\gamma_g(G + e) = \gamma_g(G)$ for every edge $e$ of $\overline{G}$ (or trees T for $\gamma_g(T) \geq 4$).
13. [21] Determine the complexity of $ir_g$, $ir_u$, and $IR_g$.
14. [21] Determine graphs $G$ for which $IR_g(G) = IR_u(G)$.
15. [37, 40] Characterize uniquely colorable graphs $G$ with $gd_\chi(G) = 0$ for which $\gamma(G) = \chi(G)$.
16. [37, 40] Given integers $a$ and $b$ with $a \leq b$, does there exist a uniquely colorable graph $G$ with $gd_\chi(G) = 0$ for which $\gamma(G) = a$ and $\chi(G) = b$?
17. [37, 40] Given integer $k \geq 1$, does there exist a uniquely colorable graph $G$ with $\gamma(G) \leq \chi(G)$ for which $gd_\chi(G) = k$?
18. [37, 40] Characterize the graphs $G$ for which $gd_\chi(G) = 0$, $gd_\chi(G) = \chi(G)$, and $gd_\chi(G) = \delta(G) + 1$.
19. [2] Determine if every graph $G$ with $\gamma_{gt}(G) = r$ can be obtained from a graph $H$ with $\gamma_{gt}(H) = r - 1$ by adding a new vertex and joining it to at least one and at most $|V(H)| - 1$ vertices of $H$.

The following problems were posed by the referee.

1. Consider all k-factorings $F_1, F_2, \ldots, F_k$ of the edges of a graph. Which ones give the minimum of the global domination number and which the maximum?
2. Consider all k-factorings $F_1, F_2, \ldots, F_k$ of the edges of a graph. Determine a method for finding which ones minimize $\sum_{i=1}^{k} \gamma(F_i)$ and which ones maximize it.
3. Suppose graph $G$ has a k-factoring and $G$ has a leaf. This means most of the factors will have an isolated vertex which will tend to increase the global domination number. How does the problem change if it is not necessary to dominate isolated vertices?

## References


1. H. Abdollahzadeh Ahangar, On the global Roman domination number in graphs, *Iran. J. Sci. Technol. Trans. A Sci.* 40 (2016) 157–163.
2. M. H. Akhbari, C. Eslahchi, N. Jafari Rad, and R. Hasni, Some remarks on global total domination in graphs, *Appl. Math. E-Notes* 15 (2015) 22–28.
3. J. Amjadi, S. Nazari-Moghaddam, and S. M. Sheikholeslami, Global total Roman domination in graphs, *Discrete Math. Algorithms Appl.* 9 (2017) 1750050, 13 pp.
4. S. Arumugam and R. Kala, A note on global domination in graphs, *Ars Combin.* 93 (2009) 175–180.
5. M. Atapour, S. M. Sheikholeslami, and L. Volkmann, Global Roman domination in trees, *Graphs Combin.* 31 (2015) 813–825.
6. L. Berry, Graph theoretic models for multicast communications, *Comput. Networks and ISDN Systems* 20 (1990) 95–99.
7. A. Bouloutas and P. M. Gopar, Some graph partitioning problems and algorithms related to routing in large computer networks, *Proceedings of Ninth International Conference on Distributed Computing Systems* (June, 1989) 110–117.
8. R. C. Brigham and R. D. Dutton, Factor domination in graphs, *Discrete Math.* 86 (1990) 127–136.
9. Y. Caro and M. A. Henning, Simultaneous domination in graphs, *Graphs Combin.* 30 (2014) 1399–1416.
10. J. R. Carrington, *Global Domination of Factors of a Graph.* Ph.D. Dissertation, University of Central Florida (1992).
11. J. R. Carrington and R. C. Brigham, Factor domination, *Congr. Numer.* 83 (1991) 201–211.
12. J. R. Carrington and R. C. Brigham, Global domination of simple factors, *Congr. Numer.* 88 (1992) 161–167.
13. E. J. Cockayne and S. T. Hedetniemi, Towards a theory of domination in graphs, *Networks* 7 (1977) 247–261.
14. E. J. Cockayne, P. A. Dreyer Jr., S. M. Hedetniemi, and S. T. Hedetniemi, Roman domination in graphs, *Discrete Math.* 278 (2004) 11–22.
15. P. Dankelmann, M. A. Henning, W. Goddard and R. Laskar, Simultaneous graph parameters: factor domination and factor total domination, *Discrete Math.* 306 (2006) 2229–2233.
16. P. Dankelmann and R. C. Laskar, Factor domination and minimum degree, *Discrete Math.* 262 (2003) 113–119.
17. D. Delić and C. Wang, The global connected domination in graphs, *Ars Combin.* 114 (2014) 105–110.
18. W. J. Desormeaux, P. E. Gibson, and T. W. Haynes, Bounds on the global domination number, *Quaest. Math.* 38 (2015) 563–572.
19. W. J. Desormeaux, T. W. Haynes, and L. van der Merwe, Global domination edge critical graphs, *Util. Math.* 104 (2017) 151–160.
20. J. E. Dunbar, T. W. Haynes, and S. T. Hedetniemi, Nordhaus-Gaddum bounds for domination sums in graphs with specified minimum degree, *Util. Math.* 67 (2005) 97–105.
21. J. Dunbar and R. Laskar, Universal and global irredundancy in graphs, *J. Combin. Math. Combin. Comput.* 12 (1992) 179–185.
22. J. Dunbar, R. Laskar, and T. Monroe, Global irredundant sets in graphs, *Congr. Numer.* 85 (1991) 65–72.
23. J. Dunbar, R. Laskar, and C. Wallis, Some global parameters of graphs, *Congr. Numer.* 89 (1992) 187–191.
24. R. D. Dutton, Global domination and packing numbers, *Ars Combin.* 101 (2011) 489–501.
25. R. D. Dutton and R. C. Brigham, On global domination critical graphs, *Discrete Math.* 309 (2009) 5894–5897.
26. R. I. Enciso and R. D. Dutton, Global domination in planar graphs, *J. Combin. Math. Combin. Comput.* 66 (2008) 273–278.

27. M. R. Garey and D. S. Johnson, *Computers and Intractability: A Guide to the Theory of NP-Completeness*, W. H. Freeman and Company, New York 1979.
28. P. Gupta and D. Jain, Global 2-point set domination number of a graph, *International Conference on Graph Theory and its Applications (Electron. Notes Discrete Math.)* 53 (2016) 213–224.
29. F. Harary, Changing and unchanging invariants for graphs, *Bull. Malaysian Math. Soc.* 5 (1982) 73–78.
30. N. Jafari Rad and E. Sharifi, Bounds on global total domination in graphs, *Comput. Sci. J. Moldova* 23 (2015) 3–10.
31. V. R. Kulli and B. Janakiram, The total global domination number of a graph, *Indian J. Pure Appl. Math.* 27 (1996) 537–542.
32. V. R. Kulli, B. Janakiram, and N. D. Soner, Connected global domination in graphs, *Acta Ciencia Indica Math.* 35 (2009) 161–164.
33. B. Maheswari, Y. Lakshmi Naidu, L. Nagamuni Reddy, and A. Sudhakaraiah, Minimum global dominating set of an interval graph, *Int.J. Math. Sci. Eng. Appl.* 5 (2011) 265–269.
34. D. A. Mojdeh, M. Alishahi, and M. Chellali, Trees with the same global domination number as their square, *Australas. J. Combin.* 66 (2016) 288–309.
35. J. Nieminen, Two bounds for the domination number of a graph, *J. Inst. Math. Appl.* 14 (1974) 183–187.
36. L. Pushpa Latha, The global point-set domination number of a graph, *Indian J. Pure Appl. Math.* 28 (1997) 47–51.
37. M. Rajeswari and I. Sahul Hamid, On global dominating-$\chi$-coloring of graphs, *Tamkang J. Math.* 48 (2017) 149–157.
38. D. F. Rall, Dominating a graph and its complement, *Congr. Numer.* 80 (1991) 89–95.
39. P. Roushini Leely Pushpam and S. Padmapriea, Global Roman domination in graphs, *Discrete Appl. Math.* 200 (2016) 176–185.
40. I. Sahul Hamid and M. Rajeswari, Global dominating sets in minimum coloring, *Discrete Math. Algorithms Appl.* 6 (2014) 1450044, 13pp.
41. E. Sampathkumar, The global domination number of a graph, *J. Math. Phys. Sci.* 23 (1989) 377–385.
42. E. Sampathkumar and L. Pushpa Latha, The global set-domination number of a graph, *Indian J. Pure Appl. Math.* 25 (1994) 1053–1057.
43. N. D. Soner, B. Chaluvaraju, and B. Janakiram, The double global domination number of a graph, *J. Indian Math. Soc. (N.S.)* 70 (2003) 191–195.
44. E. M. Still and T. W. Haynes, Global domination stable trees, *J. Combin. Math. and Combin. Comput.* 85 (2013) 237–252.
45. M. M. Theimer and K. A. Lantz, Finding idle machines in a workstation-based distributed system, *Proceedings of Eighth International Conference on Distributed Computing Systems* (June, 1988) 112–122.
46. D. Udayakumar and A. Sasireka, The global connected domination in Cartesian graphs, *Bull. Pure Appl. Sci. Sect. E Math. Stat.* 29 (2010) 217–223.
47. V. Zverovich and A. Poghosyan, On Roman, global and restrained domination in graphs, *Graphs Combin.* 27 (2011) 755–768.

# Power Domination in Graphs

**Paul Dorbec**

## 1 Introduction

Power domination is a variation of the domination problem motivated by the physical rules for monitoring electrical networks. It was first introduced as a graph parameter by Haynes *et al.* in 2002 [20]. Before giving details on the initial motivation and the physical background in Section 1.1, we give a general definition. One of the key concepts of power domination is that of *monitoring* vertices, as defined below.

**Definition 1 (Power Dominating Set)** Given a graph $G = (V, E)$ and a set $S \subseteq V$ of vertices, we define the set of vertices *monitored* by the set $S$ as follows:

- **Domination:** All vertices in $S$ and all neighbors of vertices in $S$ are monitored,
- **Propagation:** Whenever a vertex $v$ is monitored and all but one of its neighbors, say $w$, are monitored, then vertex $w$ is also monitored. In this case we say that vertex $v$ propagates to vertex $w$.

An initial set of vertices that eventually monitors the whole graph is called a *power dominating set*. The *power domination number* is the minimum order of a power dominating set, denoted $\gamma_P(G)$.

The first step, the so-called domination step, exactly matches the definition of a dominating set. Thus a dominating set is also a power dominating set, and we observe that

$$\forall G, \quad \gamma_P(G) \leq \gamma(G)\,.$$

P. Dorbec (✉)
Normandie Université, UNICAEN, ENSICAEN, CNRS, GREYC, 14000 Caen, France
e-mail: paul.dorbec@unicaen.fr



T. W. Haynes et al. (eds.), *Topics in Domination in Graphs*, Developments in Mathematics 64, https://doi.org/10.1007/978-3-030-51117-3_16

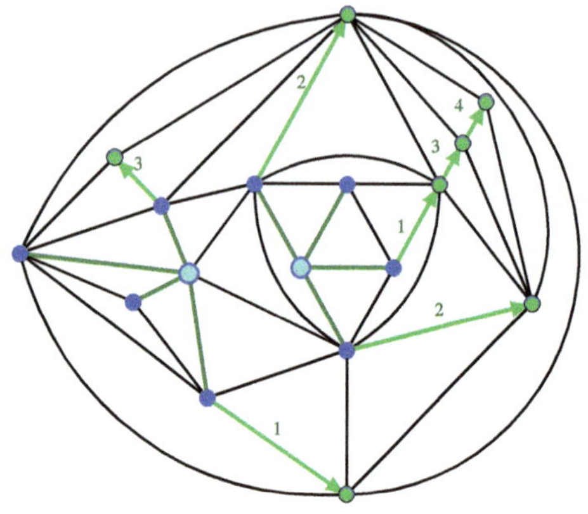


**Fig. 1** A graph with a power dominating set (in Cyan). The arrows show the propagation occurring, and their labels give the ordering of these propagation.

An example of a graph with a power dominating set is shown in Figure 1. The difference between the usual domination and power domination is best illustrated by the case of the path. Indeed, the power domination number of a path is equal to 1, and any vertex is a minimum power dominating set of the path. This comes from the fact that after the initial step of domination in the path, every monitored vertex has at most one unmonitored neighbor, and thus can propagate. So the set of monitored vertices increases until it covers the whole path.

Before continuing, we present a more formal definition that sometimes proves convenient. We define the sets $(\mathcal{P}^i)_{i\geq 0}$ of monitored vertices at stage $i$, following the definition introduced by Aazami in [1].

**Definition 2 (Monitored Vertices)** Let $G$ be a graph, and $S \subseteq V(G)$ a subset of vertices. We define the sets $(\mathcal{P}^i(S))_{i\geq 0}$ of vertices *monitored* by $S$ at step $i$ by the following rules.

- **Domination:** $\mathcal{P}^0(S) = N[S]$,
- **Propagation:** for $i \geq 0$,

$$\mathcal{P}^{i+1}(S) = \bigcup \left\{ N[v] \mid v \in \mathcal{P}^i(S), \left|N[v] \setminus \mathcal{P}^i(S)\right| \leq 1 \right\}.$$

Observe that the sequence of sets $(\mathcal{P}^i)_{i\geq 0}$ is a non-decreasing sequence. Moreover if at some stage $\mathcal{P}^i(S) = \mathcal{P}^{i+1}(S)$, then the sequence reaches a fixed point, which we denote $\mathcal{P}^\infty(S)$. Observe that the fixed point is necessarily reached for some $i < n$, so for a graph on $n$ vertices, $\mathcal{P}^\infty(S) = \mathcal{P}^n(S)$. Matching the earlier definition, we can now state that $S$ is a power dominating set of a graph $G$ if and only if $\mathcal{P}^\infty(S) = V(G)$.

With this notation, for a vertex $u$ in a path, $\mathcal{P}^i(\{u\}) = \{v \in V \mid dist(u, v) \leq i + 1\}$. In the example from Figure 1, the label $i$ on an arc shows that a vertex is first added to $\mathcal{P}^i(S)$.

## 1.1 Physical Motivation

In this section, we recall the history of the introduction of the parameter, with the successive definitions.

Power domination was first introduced by Baldwin et al. in [3], then described as a graph theoretical problem by Haynes et al. in [20]. The problem is motivated by the requirement for constant monitoring of power systems by placing a minimum number of phasor measurement units (PMU) in the network. A PMU placed at a bus measures the voltage of the bus plus the current phasors at that bus. Using Ohm and Kirchhoff current laws, it is then possible to infer from initial knowledge of the status of some part of the network the status of new branches or buses.

In Baldwin et al. [3], the following definitions are proposed:

> A *measurement-assigned subgraph*, called for short a measurement subgraph, is a subgraph which has a current measurement assigned to each of its branches. These are either actual measurement or calculated pseudo-measurement deduced from Kirchhoff's and Ohm's laws. [...] The *coverage* of a placement set of PMU's is the maximal spanning measurement subgraph that can be formed by this set, that is, the maximal observable sub-network that can be built from them.

They introduced the following formal definition of the spanning measurement subgraph:

**Definition 3 (Baldwin et al. [3])** A *spanning measurement subgraph* is constructed throughout the network on the grounds of the following rules:

Rule 1: Assign a current phasor measurement to each branch incident to a bus provided with a PMU;
Rule 2: Assign a pseudo-current measurement to each branch connecting two buses with known voltage;
Rule 3: Assign a pseudo-current measurement to a branch whose current can be inferred by using Kirchhoff's current law.

In terms of graphs, where buses are vertices and connecting branches are edges, we can describe the observability rules of a network with the following definition:

**Definition 4 (Haynes et al. [20])** Initially, set as monitored any vertex with a PMU and all edges incident to it. Then, expand iteratively the set of monitored edges and vertices with the following rules :

1. set as monitored any vertex incident to a monitored edge whose other end is monitored;
2. set as monitored any edge joining two monitored vertices;
3. if a vertex has all its incident edges monitored except one, set this one edge as monitored.

It was noticed by Dorfling and Henning in [15] that the power domination problem can be studied considering only vertices following the above definition. The coverage of a placement set $S$ of PMU is then simply the induced subgraph on the final set of monitored vertices. From this observation, we reach the definition used here, presented in Definition 1.

## 1.2 Relation with Zero Forcing Sets

It should be noted that there is a close relationship between power domination in graphs and *zero forcing sets*. Zero forcing sets were introduced in [2], together with the corresponding parameter $Z(G)$, which stands for the minimum size of a zero forcing set. Using the earlier definition, we can define a zero forcing set as a set of vertices $S$ such that applying only the propagation rule, the whole graph eventually gets monitored. In other words, it would follow Definition 2 where we define $\mathcal{P}^0(S)$ to be equal to $S$.

The motivation for the introduction of this parameter was that it is an upper bound for another parameter, called *maximum nullity* of a graph. For a graph $G$ of order $n$, the maximum nullity of $G$ corresponds to the maximum nullity (or corank) of a matrix in the set of symmetric $n \times n$ matrices having nonzero coefficients precisely where the adjacency matrix of the graph $G$ has nonzero values.

From the above definition of zero forcing sets, we easily infer that $Z(G) \leq \gamma_P(G)\Delta(G)$. Indeed, taking the vertices of a power dominating set plus all but one neighbor of each of them, one gets a zero forcing set of size at most $\gamma_P(G)\Delta(G)$. This was explicitly stated by Dean et al. in [9], where the first link between the two parameters was probably made.

Various later studies considered or just mentioned the link between zero forcing sets and power domination [5, 7, 16], or even between $k$-forcing sets and $k$-power domination that we define in Section 5 [17]. However, there are not many results for power domination that come from known results on zero forcing sets.

## 1.3 Algorithmic Aspects

We will not detail the algorithmic aspects here, since it is quite similar to the domination algorithms, and it is surveyed in the chapter dedicated to algorithms. It should just be remarked that power domination is NP-complete, with many possible reductions from the dominating set problem. However, as for the domination problem, there are polynomial algorithms for bounded treewidth graphs, using dynamic programming. The algorithms are slightly more involved than for domination, but use similar strategies.

## 2 Behavior by Small Graph Changes

As we explore in this section, one of the difficulties of power domination is that the power domination number has no monotonicity for any of the classical graph operations. For each of the usual unitary graph operations (vertex removal, edge removal, edge contraction), we give some examples and arguments illustrating the possible behavior of the power domination number. Detailed proofs of these results were given by Dorbec, Varghese, and Vijayakumar in [14].

### 2.1 Vertex Removal

In a graph, the removal of a vertex can have a similar effect on the power domination number than it can have on the domination number, that is, that it can slightly reduce the power domination number, or it can much increase it.

**Theorem 1 ([14])** *For the graph $G - v$ obtained by removing a vertex $v$ from a graph $G$, there is no upper bound to $\gamma_P(G - v)$ in terms of $\gamma_P(G)$. On the other hand, we have $\gamma_P(G - v) \geq \gamma_P(G) - 1$.*

As in domination, adding $v$ to a power dominating set of $G - v$ produces a power dominating set of $G$. On the other hand, removing the central vertex of a star greatly increases the power domination number. Less trivial examples are given in [14].

### 2.2 Edge Removal

Interestingly, the situation for edge removal is not as similar to the situation for the dominating sets. While removing an edge in a graph can only increase its domination number, it may decrease its power domination number.

**Theorem 2 ([14])** *Removing an edge $e$ from a graph $G$ results in a graph $G - e$ whose power domination number is bounded by*

$$\gamma_P(G) - 1 \leq \gamma_P(G - e) \leq \gamma_P(G) + 1 .$$

That removing an edge may increase the power domination number is of no surprise, as it is for domination. For a removed edge $e = uv$ in a graph $G$ with a power dominating set $S$, if $v$ is monitored no sooner than $u$ in the graph $G$, then $S \cup \{v\}$ is a power dominating set of $G - e$.

However, that removing an edge may also decrease the power domination number is less expected. Actually, this phenomenon comes from the fact that the removal may allow some propagation that was not possible before, as in the

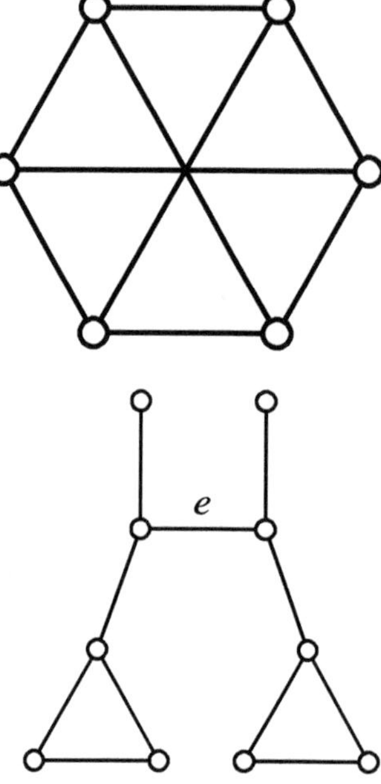


**Fig. 2** The graph $K_{3,3}$ has power domination number 2, removing any edge decrease this to 1.

**Fig. 3** A graph where contracting the edge $e$ makes the power domination number increase from 2 to 3.

example of the bipartite complete graph $K_{3,3}$ (see Figure 2), but also for many other examples.

## 2.3 Edge Contraction

Contracting an edge in a graph may easily result into the reduction of the domination number by one, the same is true for the power domination number. This happens, for example, when a minimum (power) dominating set contains both ends of the contracted edge, as in a double star. What happens in power domination but not in domination is that it may also increase the power domination number.

**Theorem 3 ([14])** *Let $G$ be a graph and $e$ be an edge in $G$. Then*

$$\gamma_P(G) - 1 \leq \gamma_P(G/e) \leq \gamma_P(G) + 1\,.$$

An (original) example where the power domination number of a graph increases when an edge is contracted is drawn in Figure 3. In this graph, the contraction of the edge $e$ merges two vertices with only one unmonitored neighbor, and thus prevents propagation.

To show that the power domination number increases by no more than one, just note that adding the newly formed vertex to a power dominating set of $G$ necessarily forms a power dominating set of $G/e$, preventing the previous phenomenon.

# 3 Bounds on General Families of Graphs

In this section, we present bounds on the power domination number of a graph under some general restrictions.

## 3.1 General Graphs

Let us first recall the initial general bound due to Haynes et al. [20]. They noted that the power domination number of a graph is always at least one, and that a dominating set of a graph is always also power dominating. We thus get

$$1 \leq \gamma_P(G) \leq \gamma(G)\,.$$

Haynes et al. proved that there is no forbidden subgraph characterization of the graphs reaching the upper bound. The proof is based on the following family of graphs. For any graph $G$, take the graph family $\mathcal{T}(G)$ of the graphs obtained by adding for each vertex $v \in V(G)$ two new vertices, namely $v_0$ and $v_1$, with the edges $vv_0$ and $vv_1$, and possibly or not the edge $v_0v_1$ (see Figure 4).

For all such graphs $G' \in \mathcal{T}(G)$, $\gamma_P(G') = \gamma(G') = |V(G)| = \frac{|V(G')|}{3}$. Now, observing that the initial graph $G$ may be any graph proves the above statement.

Actually, Zhao, Kang, and Chang proved in [32] that this construction plays a special role, while proving the following general bound. We denote by $\mathcal{T}$ the union of $\mathcal{T}(G)$ over all graphs $G$.

**Theorem 4 (Zhao, Kang, Chang [32])** *For any connected graph $G$ of order $n \geq 3$, $\gamma_P(G) \leq n/3$ with equality if and only if $G \in \mathcal{T} \cup \{K_{3,3}\}$.*

Note that this bound is an improvement of the same bound proved only for trees by Haynes et al. in [20].

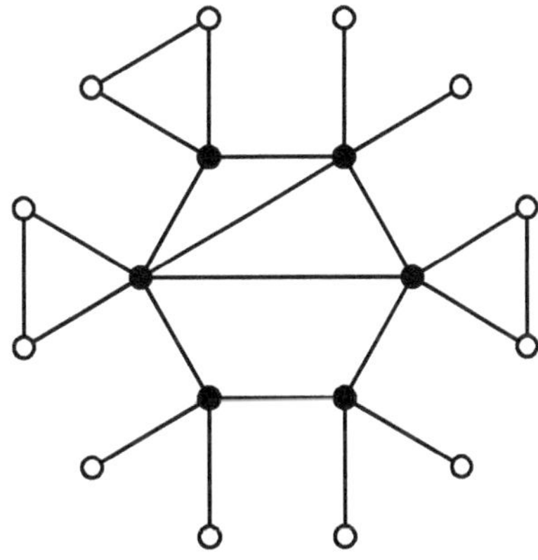

**Fig. 4** A graph from the family $\mathcal{T}$ built on a six vertices initial graph (on black vertices).

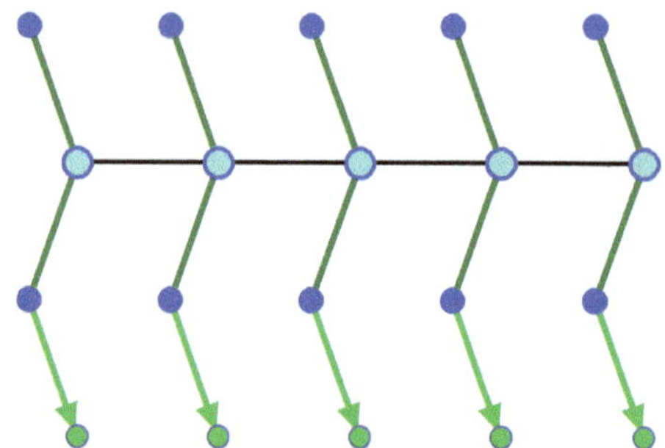

**Fig. 5** A graph attaining the bound of Conjecture 1. Black vertices form a minimum power dominating set.

To prove this result, they first recall (from [20]) that in any connected graph with maximum degree at least 3, there exists a minimum power dominating set containing only vertices of degree at least 3. Then, they show that the set $S$ can be chosen so that every vertex in $S$ has at least two private neighbors (by minimizing the number of edges in $G[S]$). The result follows.

This result can be seen as a generalization of the $\frac{n}{2}$ bound for domination, and the constructions look similar. Interestingly, the relationship between these two bounds is even more enlightened for generalized power domination, as observed in Section 5.

In [3], Baldwin et al. conjectured an upper bound on the size of a power dominating set. They considered the possibility of unknown power injections, which in graphs could be seen as a single leaf attached to the corresponding vertex/bus. That explains the unusual expression of the conjecture:

*Conjecture 1 (Baldwin et al. [3])* Let $G$ be a graph on $n$ vertices of which $k$ are of degree one. If no vertex of $G$ is adjacent to more than one leaf, then the power domination number of $G$ satisfies

$$\gamma_P(G) \leq \left\lceil \frac{2n-k}{6} \right\rceil.$$

To illustrate the conjecture and prove that this bound is tight if correct, they present the graph depicted in Figure 5. Note that though the above theorem proves it for graphs with no degree 1 vertex, the conjecture does not seem to have been considered on its own elsewhere.

## 3.2 *Regular Graphs*

For regular graphs, it seems that better bounds can be proved. Zhao, Kang, Chang [32] got the first results in that direction, using as an additional condition that the graph is claw-free. A few years later, Dorbec et al. proved in [11] the same

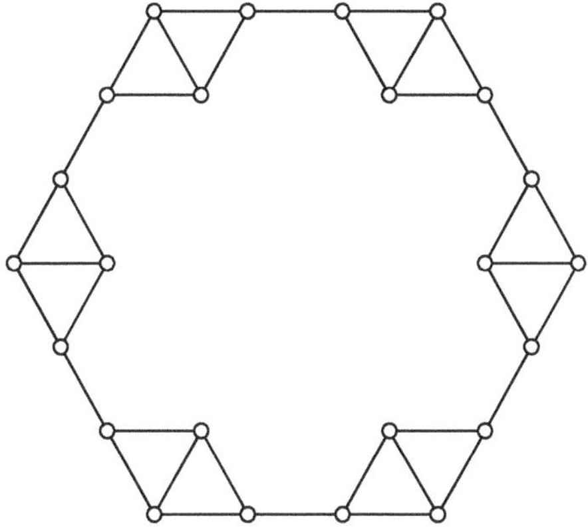

**Fig. 6** An example of a cubic graph reaching the bound of Theorem 5.

bound, dropping the claw-free condition, but still excluding the only known counter-example, which is $K_{3,3}$.

**Theorem 5 (Dorbec et al. [11])** *Let G be a connected cubic graph on n vertices. If G is not the complete bipartite graph $K_{3,3}$, then $\gamma_P(G) \leq \frac{n}{4}$.*

The bound is known to be tight thanks to the example of Zhao, Kang, Chang [32], consisting in a set of $K_4$ minus an edge, using the degree 2 vertices of $K_4 - e$ to join them into a cycle (see Figure 6). A minimum power dominating set is obtained by choosing one vertex of degree three from each of the six subgraphs $K_4 - e$.

The proof of this result is quite technical, but the idea behind the proof may be re-used. The strategy was, for a given initial set of vertices, to study what may happen at the boundary of the set of monitored vertices (the peripheral vertices). That the set does not continue to propagate gives quite some information (in particular in a cubic graph). Then one is likely to find a vertex whose addition to the set of selected vertices would greatly increase the number of monitored vertices, maintaining the expected bound. Structures that would prevent this are very special, and are retrospectively dealt with during the initial choice of the set $S$ of vertices.

Recently, Kang and Wormald [21] studied the power domination number on random cubic graphs. They proved that the power domination number of a random cubic graph of order $n$ is asymptotically almost surely between $0.033n$ and $0.068n$.

## *3.3 Maximal Planar Graphs*

Among the general bounds on the power domination number of a graph, there is a recent result on maximal planar graphs, by Dorbec, Gonzales, and Pennarun [10]. Applying a technique similar to the one for cubic graphs, Dorbec, Gonzales, and Pennarun [10] proved that maximal planar graphs satisfy the following inequality:

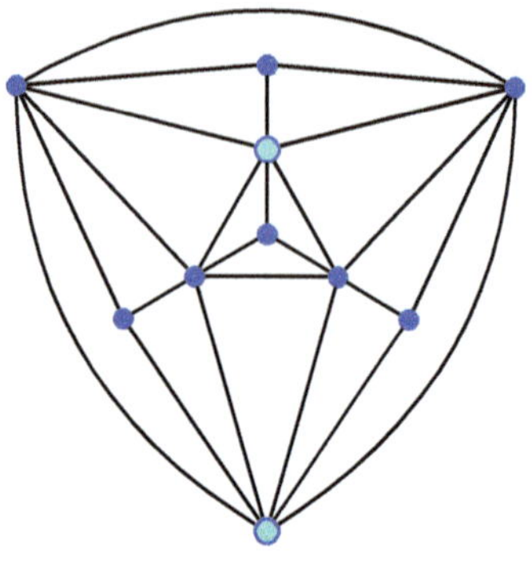

**Fig. 7** The triakis-tetrahedron, a maximal planar graph on 10 vertices with power domination number 2.

**Theorem 6 (Dorbec, Gonzales, Pennarun [10])** *Every maximal planar graph G on n vertices has a power domination number at most:*

$$\gamma_P(G) \leq \frac{n-2}{4}.$$

The known graphs for which this bound is tight have at most 10 vertices (see Figure 7 for the largest known example). This leads one to think that a better bound should exist. Until now, the maximal planar graphs having the largest known power domination number are obtained from a disjoint set of octahedra between which are added edges until reaching a maximal planar graph. A vertex specific to each octahedron is required to dominate such a graph, and thus this graph has power domination number $\frac{n}{6}$.

This is similar to the known bounds for the domination number of maximal planar graphs, the worst known family being maximal planar graphs obtained by adding edges between a set of disjoint $K_4$, which implies a domination number $\frac{n}{4}$, for an upper bound on the domination number of maximal planar graphs being at most $\frac{n}{3}$.

## 3.4 Bounded Diameter Graphs

Several attempts have been made to bound the power domination number in terms of the diameter of a graph. It was noted by Zhao and Kang that planar graphs with diameter at most 2 have a power domination number at most 2. However, they left as an open question whether the power domination number of a graph of diameter at most 2 could be large.

Soh and Koh proposed in [26] families of graphs with diameter at most 2 and unbounded power domination number. One family is simply the Cartesian product $K_n \square K_m$ of two complete graphs:

**Theorem 7 (Soh and Koh [26])** *For any $m \geq n \geq 2$, $\gamma_P(K_n \square K_m) = n - 1$.*

The other is an infinite graph, called the Rado graph, whose vertices are labelled with integers. In this graph, two vertices $x < y$ are adjacent if and only if the $x^{th}$ digit in the binary representation of $y$ is 1. Any $x$ and $y$ have as a common neighbor $(2^x + 2^y)$, which implies that the Rado graph has diameter 2. However, no propagation is possible in that graph since every vertex has infinitely many neighbors. Thus, the Rado graph has the same domination number and power domination number, and no finite subset of this graph is a dominating set. Note that the argument does not apply though to finite subgraphs of the Rado graph.

# 4 Recursively Defined Families

In this section, we consider the graph families for which explicit formulas are known.

Generally, when searching for a power dominating set of a graph, it is not too difficult to figure out what seems to be a good selection of vertices. This usually gives a pretty good upper bound to the power domination number of the graph. However, finding lower bounds turns out to be quite hard in general. In the following, we thus get into quite some details when original techniques are used to prove lower bounds.

To start with, a nice statement is that each vertex may be used for propagation only once, when it has exactly one unmonitored neighbor. Let a *peripheral vertex* be a monitored vertex with at least one unmonitored neighbor. What the earlier remark enables one to infer is that during the process of propagation, the number of peripheral vertices cannot increase. So we get as an invariant property that the number of peripheral vertices in the graph is no more than the sum of the degrees of the vertices in the initial set $S$. This was explicitly stated in [24], though it was implicitly used before. Though this seems a good handle to provide lower bounds, what makes this invariant not so easy to use is that at the end of the propagation steps, there are no peripheral vertices left. However, some of the later proofs show variants of this notion of peripheral vertices that are useful.

## 4.1 Products and Grids

In this section, we consider results on power domination in graph products, and in particular on the Cartesian product. We also consider other lattices, such as hexagonal and triangular grids.

It should be noted that a recent survey of results on graph products is given by Soh and Koh in [27], which is more detailed than what we present here, and should be referred to for having an exhaustive list of theorems on the topic. The same authors also surveyed earlier the results on the Cartesian product in [22].

### 4.1.1 Cartesian Product

Power domination in products of paths were among the first topics to be studied on power domination. Dorfling and Henning [15] studied the Cartesian product of two paths, i.e., grid graphs.

**Theorem 8 (Dorfling, Henning [15])** *The power domination number of the $n \times m$ grid $P_n \,\Box\, P_m$ for $m \geq n \geq 1$ is*

$$\gamma_P(P_n \,\Box\, P_m) = \begin{cases} \lceil \frac{n+1}{4} \rceil & \text{if } n \equiv 4 \pmod 8, \\ \lceil \frac{n}{4} \rceil & \text{otherwise.} \end{cases}$$

In their proof, they explicitly describe the shape of the set of monitored vertices by any initial subset of vertices in the grid. In addition, their proof also relies on a study of the cylinder, using the number of "columns" as an invariant, though, the use of an invariant is not explicit. The question on the cylinder (i.e., the product of a path and a cycle) was also studied later by Barrera and Ferraro [4] as well as the torus (product of two cycles).

The hypercube is also an interesting graph family for studying power domination. Actually, in the graph $G \,\Box\, K_2$, dominating one copy of $G$ is enough to power dominate the whole graph $G \,\Box\, K_2$. Therefore, we get that $\gamma_P(G \,\Box\, K_2) \leq \gamma(G)$ for any graph $G$. For the hypercube, this was observed by Dean et al. in [9]. They further conjectured that the domination number of $Q_n$ was equal to the power domination number of $Q_{n+1}$. But later on, Pai and Chiu [25] showed that $\gamma(Q_5) = 7$ while $\gamma_P(Q_6) = 6$, disproving their conjecture.

More results are proved by Varghese and Vijayakumar in [28] and by Soh and Koh in [27], in particular towards a characterization of graph products having power domination number equal to one. We refer the reader to the survey [27] for more details.

### 4.1.2 Strong Product

The study on the products of paths was continued by Dorbec, Mollard, Klavžar, and Špacapan in [13] with the three other classical products.

For the strong product, they prove the following lower bound:

**Theorem 9 (Dorbec et al. [13])** *Let $m \geq n \geq 2$. Then*

$$\gamma_P(P_m \boxtimes P_n) \geq \max\left\{\left\lceil \frac{m}{3} \right\rceil, \left\lceil \frac{m+n-2}{4} \right\rceil\right\}$$

Whenever $3n - m - 6 \not\equiv 4 \pmod 8$, one can construct an initial set $S$ which achieves this lower bound. For the remaining case, Soh and Koh claimed in [27] to have proved that the correct value is $\lceil \frac{m+n-2}{4} \rceil + 1$ though their proof is missing some details.

An interesting aspect of this lower bound is that it is proved by considering not just the number of peripheral vertices, but the number of *non-surrounded vertices*, that is of monitored vertices having at most seven monitored neighbors (recall that the maximal number of neighbors in the strong product of two paths is eight). This includes all peripheral vertices, but also the vertices from the border of the grid. The first step of the proof is then that the same invariant property can be proved for non-surrounded vertices, under some conditions (by proving that any propagation that would start on a vertex from the border could have been made from another vertex too). This invariant enables one to prove this lower bound, based on the number of vertices on the border of the grid.

Again, Soh and Koh in [27] claim the same results for the product of cycle, based on the same proof, but they do not explain how they adapt the invariant.

### 4.1.3 Direct Product

For the direct product of paths (which has two connected components), the first bound obtained in [13] can be stated as follows:

**Theorem 10 ([13])** *The power domination number of the direct product $P_n \times P_m$ for $m \geq n \geq 1$ is*

$$\gamma_P(P_n \times P_m) = \begin{cases} 2\lceil \frac{n}{4} \rceil & \text{if } n \text{ is even,} \\ 2\lceil \frac{m}{4} \rceil & \text{if } n \text{ is odd and } m \text{ even,} \end{cases}$$

*If both m and n are odd,*

$$\gamma_P(P_n \times P_m) \leq \max\left\{\left\lceil \frac{m}{4} \right\rceil + \left\lceil \frac{m-2}{4} \right\rceil, \left\lceil \frac{m+n}{6} \right\rceil + \left\lceil \frac{m+n-2}{6} \right\rceil\right\}$$

Actually, it is proved that one component has power domination number exactly $\max\left\{\lceil \frac{m}{4} \rceil, \lceil \frac{m+n}{6} \rceil\right\}$, whereas for the other component, the only lower bound proved is $\frac{n}{4}$. The proof is very technical, using a connection with a percolation process on a square grid as a reference.

The case of the product of two cycles was also considered in [27], who mentioned some earlier communications in a workshop on the topic. They also use the notion of peripheral vertices (called boundary vertices), with a nice trick. Their idea is to take a minimum power dominating set $S$ of the graph, remove one vertex $v$ from it, and then count the number of boundary vertices for $\mathcal{P}^\infty(S \setminus \{v\})$.

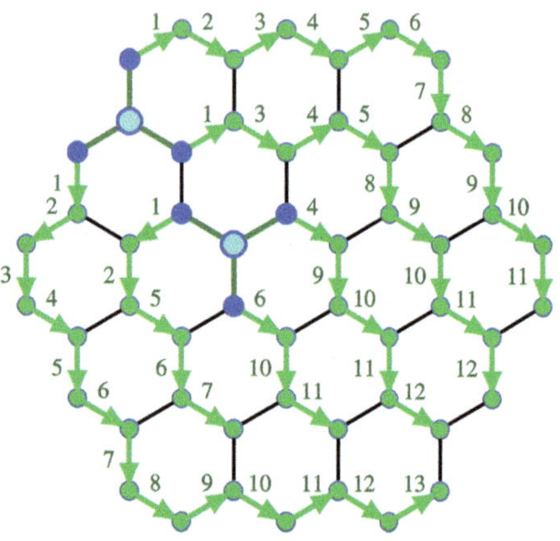

**Fig. 8** The hexagonal grid $HM(3)$.

#### 4.1.4 Lexicographic Product

The case of the lexicographic product is easier, since the role of the graph $G$ is much more important than the role of $H$ in $G \circ H$. Actually, unless the power domination number of $H$ is one, it is as good to totally dominate $G \circ H$ than to use propagation.

**Theorem 11 ([13])** *For any nontrivial graphs G and H, if G has no isolated vertices, then*

$$\gamma_P(G \circ H) = \begin{cases} \gamma(G);\ \gamma_P(H) = 1\,, \\ \gamma_t(G);\ \gamma_P(H) > 1\,. \end{cases}$$

#### 4.1.5 Hexagonal Grids and Triangular Grids

The first results on the hexagonal grid are given by Ferrero, Varghese[1], and Vijayakumar in [18]. They consider hexagonal grids with an hexagonal outer shape (see Figure 8) and give the exact power domination number of those grids. The method used is related to the method of Dorfling and Henning in [15] for the square grid.

**Theorem 12 (Ferrero, Varghese, Vijayakumar [18])** *Let $HM(n)$ be the hexagonal grid with an hexagonal outer shape whose side is made of n hexagons.*

$$\gamma_P(HM(n)) = \left\lceil \frac{2n}{3} \right\rceil .$$

[1]Seema Varghese who is cited here is actually the elder sister of Seethu Varghese who was cited elsewhere. Both of them were PhD students of Vijayakumar.

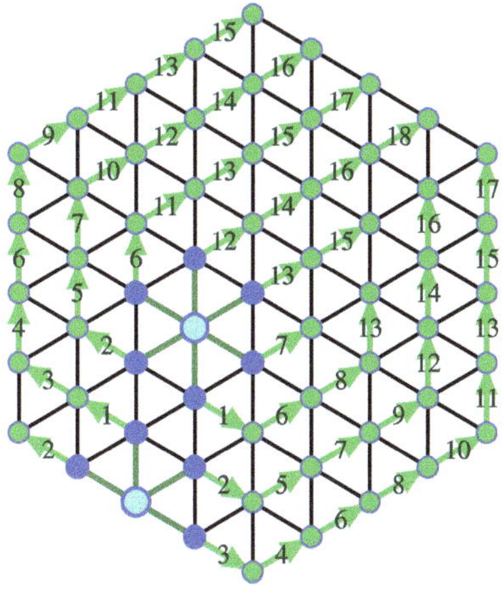

**Fig. 9** The triangular grid $T_5$.

When studying the triangular grid with hexagonal outer shape (see Figure 9), Bose, Pennarun, and Verdonschot [6] noticed a connection with that earlier result. They got a similar bound:

**Theorem 13 (Bose, Pennarun, Verdonschot [6])** *For $k \geq 1$, let $T_k$ be the triangular grid with hexagonal shape, whose side is made of k vertices. We have*

$$\gamma_P(T_k) = \left\lceil \frac{k}{3} \right\rceil .$$

However, to show the lower bound, they used a very different technique than Ferrero, Varghese, and Vijayakumar. Their nice idea is to consider a projection of the monitored set on one side, and then count the number of peripheral vertices of that projection when about half the vertices are monitored. They prove that this second number is a lower bound to the number of peripheral vertices before the projection. This results in a very original way of using the peripheral vertices invariant.

## 4.2 *Other Recursively Defined Families*

In this section, we survey the main recursively defined families for which the power domination number has been computed.

### 4.2.1 Generalized Petersen Graphs and Permutation Graphs

The case of generalized Petersen graphs was considered by both Barrera and Ferrero in [4] and by Xu and Kang in [31]. In [4], they suggest a more general study on Cayley graphs as a continuation of this study.

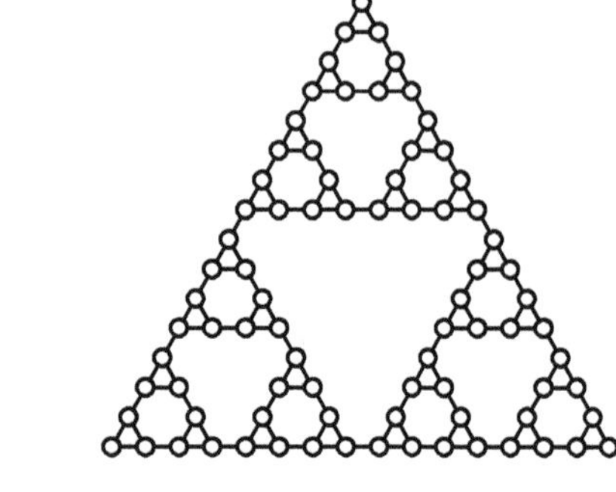

**Fig. 10** The Sierpiński graph $S_3^4$

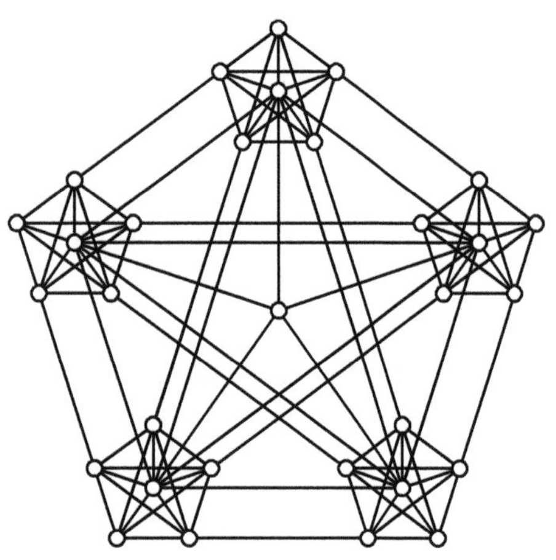

**Fig. 11** The WK-pyramidal network $WKP_{(5,2)}$.

Actually, the work was continued on *permutation graphs* which form a super-family of generalized Petersen graphs. Those were considered by Wilson in [30]. In his paper, Wilson mainly conjectured that in a permutation graph $G$ on $n$ vertices (that is based on two cycles on $\frac{n}{2}$ vertices), $\gamma_P(G) \leq \lceil \frac{n}{4} \rceil$. He proved that the bound in the conjecture is best possible. He also proposed a more detailed conjecture that holds all the open cases in his main conjecture.

### 4.2.2 Sierpiński Graphs

The case of the well-known Sierpiński graphs (see Figure 10) is dealt with by Dorbec and Klavžar in [12]. Exact values are given for all Sierpiński graphs, for power domination or generalized power domination as described in Section 5.

Another related family called WK-pyramidal networks was studied by Varghese and Vijayakumar in [28]. This family contains Sierpiński graphs as an induced subgraph, with the addition of (pyramidal) extra vertices in each clique (see Figure 11). Again, the generalized power domination number of almost all WK-pyramidal networks is explicitly given.

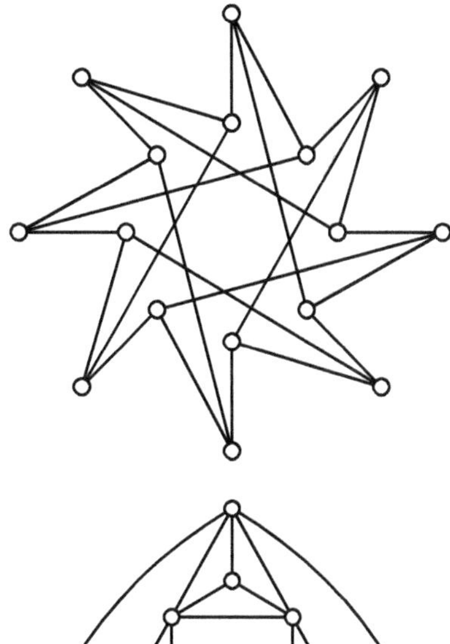

**Fig. 12** The Knödel graph $W_{3,16}$

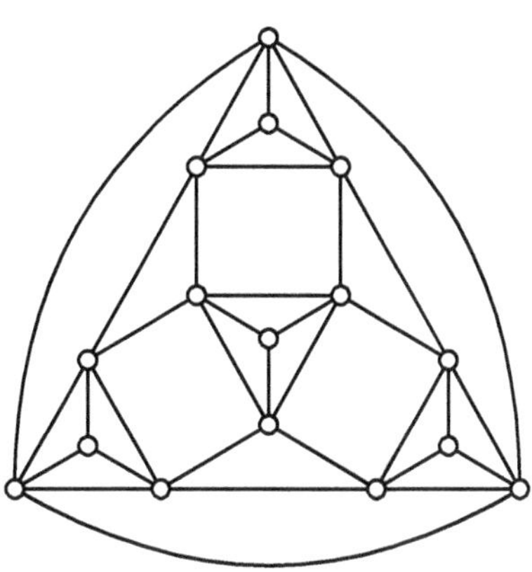

**Fig. 13** The Hanoi graph $HP_4^2$.

### 4.2.3 Other Families

We here present the other families which were considered, without much details. For De Bruijn graphs and Kautz graphs, upper bounds on the power domination number were first given by Kuo and Wu in [23], then the exact values were characterized by Grigorious, Kalinowski, and Stephen in [19].

The case of Knödel (see Figure 12) and Hanoi graphs (see Figure 13) was considered by Varghese, Vijayakumar, and Hinz in [29], who gave close formulas for the power domination number of subfamilies of these graphs.

## 5 Generalized Power Domination

A common generalization of domination and power domination was introduced by Chang *et al.* in [8], called $k$-power domination. In this section, we give its definition and state some of the main results on $k$-power domination.

## 5.1 Definition

The idea is to denote by $k$ the number of non-monitored neighbors of a vertex that a monitored vertex may propagate to. This gives the following definition:

**Definition 5 (Chang et al. [8])** Let $G$ be a graph, $S \subseteq V(G)$ and $k$ a non-negative integer. We define the sets $\left(\mathcal{P}_k^i(S)\right)_{i\geq 0}$ of vertices monitored by $S$ at step $i$ by the following rules.

- $\mathcal{P}_k^0(S) = N[S]$.
- $\mathcal{P}_k^{i+1}(S) = \bigcup N[v]$, $v \in \mathcal{P}_k^i(S)$ such that $\left|N[v] \setminus \mathcal{P}_k^i(S)\right| \leq k$.

Similarly as for power domination, a set $S$ of vertices is a $k$-power dominating set of a graph $G$ if $\mathcal{P}_k^\infty(S) = V(G)$.

Note that with this definition, $\gamma_{P,0}(G) = \gamma(G)$ and $\gamma_{P,1}(G) = \gamma_P(G)$, so we have a common generalization of domination and power domination. This may sound a bit artificial, though many results presented below tend to show that this makes sense.

## 5.2 First Results

One first thing to observe about $k$-power domination is that in any graph, a $k$-power dominating set is also by definition a $(k+1)$-power dominating set. We thus naturally have

$$\gamma(G) \geq \gamma_P(G) \geq \gamma_{P,2}(G) \geq \gamma_{P,3}(G) \geq \dots \tag{1}$$

It was noted in [8] that this inequality chain cannot be improved in a general setting:

*Remark 1 (Chang et al. [8])* For any finite non-increasing sequence of positive integers $(x_k)_{0\leq k\leq n}$, there exists a graph $G$ such that $\gamma_{P,k}(G) = x_k$ for $0 \leq k \leq n$.

The construction to prove this statement is a generalization of the construction of the family $\mathcal{T}$ that was used to show the $\frac{n}{3}$ bound for domination. Start initially with the corona of a complete graph $K_{x_0}$. This enforces $\gamma(G) = x_0$. Then to $x_1$ vertices of the complete graph, attach a second leaf, and possibly link it with the previous leaf. To $x_2$ of these vertices, attach a third leaf and add any number of links between the three leaves. Continue in such a way until $x_k = 1$ or the sequence is finished. See Figure 14 for an example.

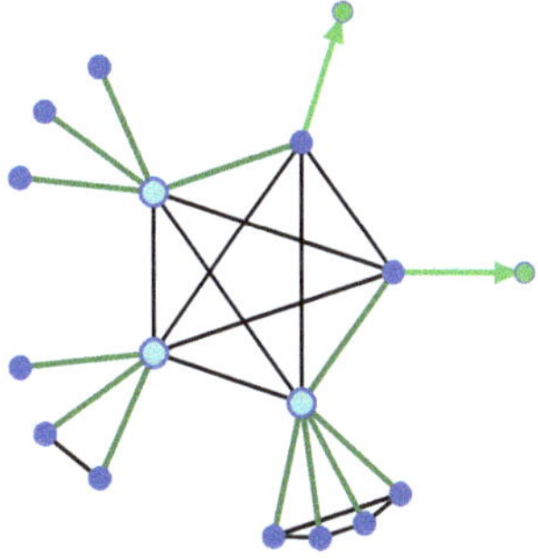

**Fig. 14** An example of a graph $G$ with $\gamma(G) = 5$, $\gamma_P(G) = 3 = \gamma_{P,2}(G)$, and $\gamma_{P,3}(G) = 1$. A minimum (1-)power dominating set is also drawn.

### 5.3 *Bounds for Generalized Power Domination*

Using this generalized setting, many results proved on power domination can be extended. We now summarize some of the main bounds found. Most of them were mentioned earlier, for the power domination number.

The first result is the following:

**Theorem 14 (Chang et al. [8])** *For any connected graph $G$ on $n \geq k+1$ vertices, we have $\gamma_{P,k}(G) \leq \frac{n}{k+2}$, and this bound is best possible.*

This result can be proved in the same way as the $\frac{n}{3}$ bound for (1-)power domination, and the examples reaching the bound are basically the same, obtained from any graph $G$, to which each vertex is attached $k+1$ leaves.

Interestingly enough, the (1-)power domination bound for regular graphs also extends well to generalized power domination, and this is actually how the result is proved in [11].

**Theorem 15 (Dorbec et al. [11])** *Let $k \geq 1$ and let $G$ be a connected $(k+2)$-regular graph of order $n$. If $G \neq K_{k+2,k+2}$, then $\gamma_{P,k}(G) \leq \frac{n}{k+3}$, and this bound is tight.*

It should be noted that the Sierpiński graphs in [12] and the WK-pyramidal networks in [28] are studied in the generalized setting.

## 6 Propagation Radius

Another parameter closely related to power domination was introduced to indicate the number of propagation steps required to monitor the whole graph from a minimum power dominating set.

It was introduced independently first by Dorbec and Klavžar in [12], where they called the parameter propagation radius (denoted $\mathrm{rad}_{P,k}(G)$ for $k$-power domination), and later by Ferrero et al. [16], who called it power propagation time

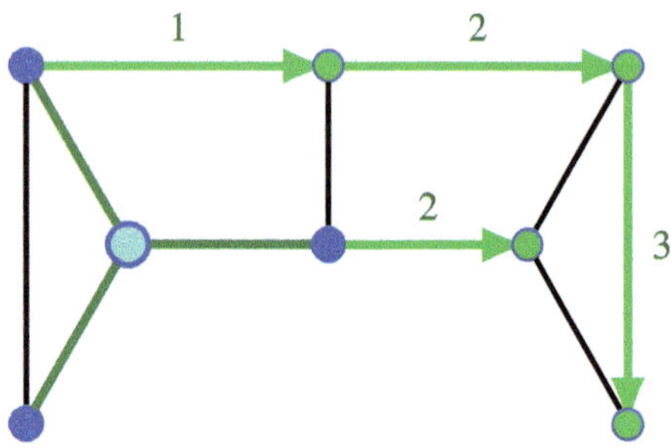

**Fig. 15** In the above graph with $\gamma_P(G) = 1$, $\text{rad}(G) = 2$ but $\text{rad}_{P,1}(G) = 4$

(denoted $ppt(G)$). We here stick to the earlier name and notation, namely the propagation radius.

**Definition 6 (Propagation Radius)** The radius of a $k$-power dominating set $S$ of a graph $G$ is defined by

$$\text{rad}_{P,k}(G, S) = 1 + \min\{i : \mathcal{P}^i_k(S) = V(G)\}.$$

The *$k$-propagation radius* of a graph $G$ can be expressed as

$$\text{rad}_{P,k}(G) = \min\{\text{rad}_{P,k}(G, S),\ S \text{ is a } k\text{-PDS of } G,\ |S| = \gamma_{P,k}(G)\}.$$

Note that the power propagation time, denoted $ppt(G)$, is defined so that it exactly matches the propagation radius, and we have $ppt(G) = \text{rad}_{P,1}(G)$. Note though that things are made a little confusing due to a slight difference between the notations of Aazami [1] and Chang et al. [8] for monitored sets at step $i$. In the first paper, $N[S]$ is the step 1 of monitoring (denoted $N[S] = S^{[1]}$ in [16]), while in the second, $N[S] = \mathcal{P}^0(S)$. Fortunately, the values coincide for the propagation radius.

The following early results are proved for the propagation radius:

**Theorem 16 (Dorbec, Klavžar [12])** *Let $G$ be a graph and $k$ a positive integer.*

- *We have $\gamma_P(G) = \gamma(G)$ if and only if* $\text{rad}_{P,1}(G) = 1$.
- *If $\Delta(G) \leq k + 1$, then $\gamma_{P,k}(G) = 1$ and* $\text{rad}_{P,k}(G) = \text{rad}(G)$.

It should be noted that $\gamma_{P,k}(G) = 1$ implies $\text{rad}_{P,k}(G) \geq \text{rad}(G)$ but not the equality, as is illustrated by the graph of Figure 15.

From the propagation radius, we can also infer a bound on the power domination number with the number of peripheral vertices. Since the number of peripheral vertices may not increase in 1-power domination, and these are the only vertices that can propagate, the peripheral vertices invariant translates to the following result:

**Theorem 17 (Liao [24])** *For any graph $G$ on $n$ vertices with maximum degree $\Delta$, we have*

$$\gamma_P(G) \geq \frac{n}{1 + \Delta\,\text{rad}_{P,1}(G)}$$

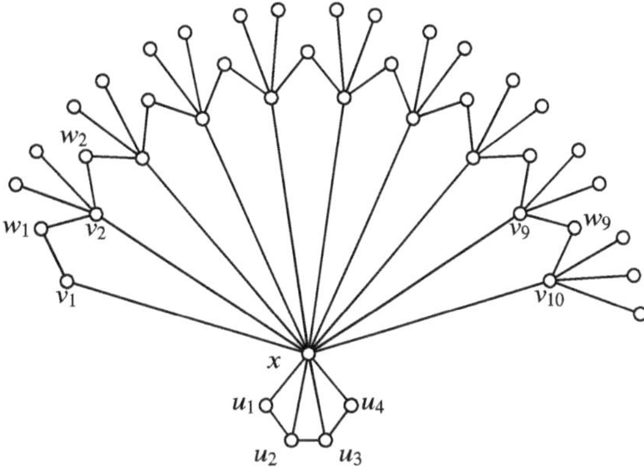


**Fig. 16** The peacock graph, with 3-power domination number 1, radius 2, but 3-propagation radius equal to 11.

This is easily verified by the fact that the number of peripheral vertices at all time is at most $\Delta\,\gamma_P(G)$, so the total number of vertices monitored is at most

$$\mathcal{P}^{\infty}(S) \leq \gamma_P(G) + \Delta\,\gamma_P(G)\,\mathrm{rad}_{\mathrm{P},1}(G)\,.$$

On its own, this result is difficult to use since there is no general relationship between the diameter or the radius of a graph and its power domination radius, as observed by Dorbec and Klavžar [12] and independently by Ferrero et al. [16]. Similar examples were given of graphs with small power domination number, small diameter, and very large propagation radius. One such example from [12] is depicted in Figure 16. It illustrates the statement for 3-power domination.

The number of articles where the propagation radius has been studied is not yet very large. In the initial paper [12], the propagation radius of Sierpiński graphs was computed in the same time as their $k$-power domination number. In [14], in the same time as the authors studied the changes on the generalized power domination number by canonical graph changes, they considered the possible evolution of the propagation radius (when the power domination number was modified). In [28], Varghese and Vijayakumar considered the propagation radius for WK-pyramidal networks, while in [29], Varghese, Vijayakumar, and Hinz studied the power domination number and propagation radius in Knödel graphs and Hanoi graphs.

## 7 Open Problems and Perspectives

In this final section, we present some research directions and questions that arise from the results surveyed here.

### 7.1 About the Relationship Between k and ℓ-Power Domination

One first question is about the link between the $k$ and $\ell$-power domination numbers of a graph for $k \neq \ell$. It was stated in Section 5 that when $k$ increases, the $k$-power domination number can only decrease. We have shown how these domination numbers can vary quite freely, but these examples use articulation points (*i.e.*, vertices whose removal disconnects the graph). The following questions come naturally, with a special interest in 2-connected graphs to avoid all the cases using articulation points.

**Question 1**

- Can we find a characterization of the 2-*connected* graphs such that $\gamma_{P,k}(G) = \gamma_{P,\ell}(G)$ for some $k < \ell$? And in particular for $\ell = k + 1$?
- Can we answer the same question in general?

### 7.2 About Regular Graphs

For regular graphs, a little is known already, and more could probably be proved. Everything that is known for power domination extends nicely to generalized power domination, and is summarized as follow. For a connected regular graph $G$ of order $n$ and degree $\Delta$, we have

- If $\Delta \leq k + 1$, $\gamma_{P,k}(G) = 1$.
- If $\Delta = k+2$, $\gamma_{P,k}(G) \leq \frac{n}{\Delta+1}$, with the single exception of the bipartite complete graph $K_{k+2,k+2}$.

A conjecture was made in [11] that the bound $\frac{n}{\Delta+1}$ could hold for larger $\Delta$, independently of $k$. Seemingly, some counter-examples were found to that conjecture at a workshop in Balatonalmádi in 2017, though there is no written evidence. The following question of an upper bound certainly is of interest, then.

**Question 2**

- What is the best possible upper bound on the power domination number of a connected regular graph with degree at least four?
- More generally, what is the best possible upper bound on the $k$-power domination number of a connected regular graph with degree at least $k + 3$?

### 7.3 Hypercubes and Products

Domination in graph products is a nice but difficult problem, as illustrated by Vizing's conjecture. However, power domination seems to behave quite differently on products, as we have seen in Section 4.1.

First, the question of the hypercube is still open. The intuition that $\gamma_P(Q_{i+1})$ would be equal to $\gamma(Q_i)$ for all $i$ was disproved, with the smallest counterexample being the pair $Q_5$-$Q_6$. The question remains open for generalized power domination, as mentioned below. We also present a general question for graph products.

**Question 3**

- Can more counterexamples be found to disprove that $\gamma_{P,k+1}(Q_{i+1}) = \gamma_{P,k}(Q_i)$ for $k \geq 1$? Can we predict the smallest counter-example for a given $k$?
- For some product $\otimes$, can we find some nontrivial way to relate $\gamma_{P,k}(G)$, $\gamma_{P,\ell}(H)$ and some $\gamma_{P,f(k,\ell)}(G \otimes H)$? Can such relations be completed with relations also on the propagation radii?

## 7.4 *Maximal Planar Graphs*

For planar graphs, we surveyed the known results in Section 3.3. The problems sound interesting and promising, but the initial problem remains open, with the following questions.

**Question 4**

- Is there an infinite maximal planar graph family that requires more than $\frac{n}{6}$ vertices to power dominate, as does the triakis-tetrahedron of Figure 7?
- More generally, what is the best $\alpha$ such that for all maximal planar graphs $G$, $\gamma_P(G) \leq \alpha|V| + O(1)$?
- Finally, what is the best possible upper bound on the size of a minimum $k$-power dominating set of a maximal planar graph?

Note that the second question is likely to be difficult, as the corresponding question for domination remains open.

## 7.5 *Propagation Radius*

To conclude, there are many other questions to explore concerning graph classes, in particular, the introduction of the propagation radius opens up many new questions.

**Question 5**

- For the main results proposed up to now on power domination, what can be said about the propagation radius?

## References

1. A. Aazami. Domination in graphs with bounded propagation: algorithms, formulations and hardness results. *J. Comb. Optim.*, 19(4):429–456, 2010.
2. AIM Minimum Rank Group. Zero forcing sets and the minimum rank of graphs. *Linear Algebra and its Applications*, 428(7):1628–1648, 2008.
3. T. L. Baldwin, L. Mili, M. B. Boisen, and R. Adapa. Power system observability with minimal phasor measurement placement. *IEEE Transactions on Power Systems*, 8(2):707–715, 1993.
4. R. Barrera and D. Ferrero. Power domination in cylinders, tori, and generalized Petersen graphs. *Networks*, 58(1):43–49, 2011.
5. K. F. Benson, D. Ferrero, M. Flagg, V. Furst, L. Hogben, V. Vasilevska, and B. Wissman. Zero forcing and power domination for graph products. *Australas. J. Combin.*, 70:221–235, 2018.
6. P. Bose, C. Pennarun, and S. Verdonschot. Power domination on triangular grids. *CoRR*, abs/1707.02760, 2017.
7. B. Brešar, C. Bujtás, T. Gologranc, S. Klavžar, G. Košmrlj, B. Patkós, Z. Tuza, and M. Vizer. Grundy dominating sequences and zero forcing sets. *Discrete Optim.*, 26:66–77, 2017.
8. G. J. Chang, P. Dorbec, M. Montassier, and A. Raspaud. Generalized power domination of graphs. *Discrete Appl. Math.*, 160(12):1691–1698, 2012.
9. N. Dean, A. Ilic, I. Ramírez, J. Shen, and K. Tian. On the power dominating sets of hypercubes. In *14th IEEE International Conference on Computational Science and Engineering, CSE 2011, Dalian, China, August 24–26, 2011*, pages 488–491, 2011.
10. P. Dorbec, A. González, and C. Pennarun. Power domination in maximal planar graphs. http://arxiv.org/abs/1706.10047, 2017.
11. P. Dorbec, M. A. Henning, C. Löwenstein, M. Montassier, and A. Raspaud. Generalized power domination in regular graphs. *SIAM J. Discrete Math.*, 27(3):1559–1574, 2013.
12. P. Dorbec and S. Klavžar. Generalized power domination: propagation radius and Sierpiński graphs. *Acta Appl. Math.*, 134:75–86, 2014.
13. P. Dorbec, M. Mollard, S. Klavžar, and S. Špacapan. Power domination in product graphs. *SIAM J. Discrete Math.*, 22(2):554–567, 2008.
14. P. Dorbec, S. Varghese, and A. Vijayakumar. Heredity for generalized power domination. *Discrete Math. Theor. Comput. Sci.*, 18(3):Paper No. 5, 11, Apr. 2016.
15. M. Dorfling and M. A. Henning. A note on power domination in grid graphs. *Discrete Appl. Math.*, 154(6):1023–1027, 2006.
16. D. Ferrero, L. Hogben, F. H. J. Kenter, and M. Young. Note on power propagation time and lower bounds for the power domination number. *J. Comb. Optim.*, 34(3):736–741, 2017.
17. D. Ferrero, L. Hogben, F. H. J. Kenter, and M. Young. The relationship between $k$-forcing and $k$-power domination. *Discrete Math.*, 341(6):1789–1797, 2018.
18. D. Ferrero, S. Varghese, and A. Vijayakumar. Power domination in honeycomb networks. *J. Discrete Math. Sci. Cryptogr.*, 14(6):521–529, 2011.
19. C. Grigorious, T. Kalinowski, and S. Stephen. On the power domination number of de Bruijn and Kautz digraphs. In *Combinatorial algorithms*, volume 10765 of *Lecture Notes in Comput. Sci.*, pages 264–272. Springer, Cham, 2018.
20. T. W. Haynes, S. M. Hedetniemi, S. T. Hedetniemi, and M. A. Henning. Domination in graphs applied to electric power networks. *SIAM J. Discrete Math.*, 15(4):519–529, 2002.
21. L. Kang and N. Wormald. Minimum power dominating sets of random cubic graphs. *J. Graph Theory*, 85(1):152–171, 2017.
22. K. M. Koh and K. W. Soh. Power domination of the Cartesian product of graphs. *AKCE Int. J. Graphs Comb.*, 13(1):22–30, 2016.
23. J. Kuo and W.-L. Wu. Power domination in generalized undirected de Bruijn graphs and Kautz graphs. *Discrete Math. Algorithms Appl.*, 7(1):1550003, 8, 2015.
24. C.-S. Liao. Power domination with bounded time constraints. *J. Comb. Optim.*, 31(2):725–742, 2016.

25. K.-J. Pai and W.-J. Chiu. A note on "on the power dominating sets of hypercubes". In *The 29th Workshop on Combinatorial Mathematics and Computation Theory*, 2012.
26. K. W. Soh and K. M. Koh. A note on power domination problem in diameter two graphs. *AKCE Int. J. Graphs Comb.*, 11(1):51–55, 2014.
27. K. W. Soh and K. M. Koh. Recent results on the power domination numbers of graph products. *New Zealand J. Math.*, 48:41–53, 2018.
28. S. Varghese and A. Vijayakumar. Generalized power domination in WK-pyramid networks. *Bull. Inst. Combin. Appl.*, 78:52–68, 2016.
29. S. Varghese, A. Vijayakumar, and A. M. Hinz. Power domination in Knödel graphs and Hanoi graphs. *Discuss. Math. Graph Theory*, 38(1):63–74, 2018.
30. S. Wilson. Power domination on permutation graphs. *Discrete Appl. Math.*, 262:169–178, 2019.
31. G. Xu and L. Kang. On the power domination number of the generalized Petersen graphs. *J. Comb. Optim.*, 22(2):282–291, 2011.
32. M. Zhao, L. Kang, and G. J. Chang. Power domination in graphs. *Discrete Math.*, 306(15):1812–1816, 2006.

www.ingramcontent.com/pod-product-compliance
Ingram Content Group UK Ltd.
Pitfield, Milton Keynes, MK11 3LW, UK
UKHW021832270726
14058UKWH00001B/109

* 9 7 8 3 0 3 0 5 1 1 1 9 7 *